Harmonic Analysis on Hypergroups

Approximation and Stochastic Sequences

SERIES ON MULTIVARIATE ANALYSIS

Editor: M M Rao ISSN: 1793-1169

Published

Vol. 1: Martingales and Stochastic Analysis
J. Yeh

Vol. 2: Multidimensional Second Order Stochastic Processes
Y. Kakihara

Vol. 3: Mathematical Methods in Sample Surveys
H. G. Tucker

Vol. 4: Abstract Methods in Information Theory
Y. Kakihara

Vol. 5: Topics in Circular Statistics
S. R. Jammalamadaka and A. SenGupta

Vol. 6: Linear Models: An Integrated Approach
D. Sengupta and S. R. Jammalamadaka

Vol. 7: Structural Aspects in the Theory of Probability:
A Primer in Probabilities on Algebraic-Topological Structures
H. Heyer

Vol. 8: Structural Aspects in the Theory of Probability (Second Edition)
H. Heyer

Vol. 9: Random and Vector Measures
M. M. Rao

Vol. 10: Abstract Methods in Information Theory (Second Edition)
Y. Kakihara

Vol. 11: Linear Models and Regression with R: An Integrated Approach
D. Sengupta and S. R. Jammalamadaka

Vol. 12: Stochastic Processes: Harmonizable Theory
M. M. Rao

Vol. 13: Hilbert and Banach Space-Valued Stochastic Processes
Y. Kakihara

Vol. 14: Harmonic Analysis on Hypergroups: Approximation and Stochastic Sequences
Rupert Lasser

Harmonic Analysis on Hypergroups

Approximation and Stochastic Sequences

Rupert Lasser
Technical University of Munich, Germany

NEW JERSEY • LONDON • SINGAPORE • BEIJING • SHANGHAI • HONG KONG • TAIPEI • CHENNAI • TOKYO

Published by

World Scientific Publishing Co. Pte. Ltd.
5 Toh Tuck Link, Singapore 596224
USA office: 27 Warren Street, Suite 401-402, Hackensack, NJ 07601
UK office: 57 Shelton Street, Covent Garden, London WC2H 9HE

Library of Congress Control Number: 2022053760

British Library Cataloguing-in-Publication Data
A catalogue record for this book is available from the British Library.

Series on Multivariate Analysis — Vol. 14
HARMONIC ANALYSIS ON HYPERGROUPS
Approximation and Stochastic Sequences

ISBN 978-981-126-619-5 (hardcover)
ISBN 978-981-126-620-1 (ebook for institutions)
ISBN 978-981-126-621-8 (ebook for individuals)

For any available supplementary material, please visit
https://www.worldscientific.com/worldscibooks/10.1142/13138#t=suppl

Preface

Harmonic analysis on hypergroups is the study of objects like sequences, functions, measures, functionals, operators, which are defined on hypergroups. Harmonic analysis on locally compact groups is based on the consideration of translates of such objects using the products $x \cdot y$ of two elements of the group. In the case of hypergroups the 'products' $\omega(x, y)$ are probability measures with compact support. Applying products $\omega(x, y)$, translates of such objects are defined and are in the focus of harmonic analysis on hypergroups. The other algebraic and topological properties of locally compact groups are adapted to hypergroups to achieve on the one hand a broad coverage of harmonic analysis topics from many mathematical fields and on the other hand to include harmonic analysis on locally compact groups.

We will combine (abstract) harmonic analysis on hypergroups with applied topics of spectral analysis and approximation by orthogonal expansions, see Chapter 4. Moreover, we investigate stochastic sequences satisfying stationarity based on hypergroups, see Chapter 6 and Chapter 7.

Charles Dunkl, Robert Jewett and René Spector developed three axiom schemes for locally compact hypergroups in the early 1970's. There are small technical differences in their definitions. In the meantime Jewett's paper [350] became the standard of the basic theory of hypergroups.

Structures related to hypergroups are already studied by Frobenius around 1900, see [234]. There are further papers from 1930 – 1940 dealing with algebraic objects or with shift operators related to hypergroups, e.g. see Wall [741] and Delsarte [174]. Furthermore, linearization of products of orthogonal polynomials related to polynomial hypergroups are already used around 1960 studying eigenvalue differential equations, e.g. see Hylleraas [343].

The **first chapter** starts with the prerequisites of topology and measure theory, which are basic for the definition of hypergroups. A hypergroup K is a locally compact Hausdorff space equipped with a convolution $\omega(x,y)$, an involution $\tilde{x}$ and a unit element e satisfying properties such as associativity or inversion. After the definition of hypergroups and the derivation of the extension of convolution and involution to the measure space $M(K)$, the translation operators on $C^b(K)$ and $C_0(K)$ are introduced. In the second part of Chapter 1 we describe important classes of hypergroups starting with polynomial hypergroups on $K = \mathbb{N}_0$ followed by hypergroups generated by families of multiplicative functions.

In the **second chapter** we begin with the discussion of the left-invariant Haar measures, and the translation operators and convolution operators on various Banach spaces are investigated. It is shown that $L^1(K)$ and $M(K)$ are Banach $*$-algebras. The next topic of this chapter is the study of representations of hypergroups. Equivalence results between bounded $*$-representations of K and bounded continuous positive definite functions on K are derived. Applying the Krein-Milman theorem it follows that a $*$-representation is irreducible exactly when the corresponding function is an extreme point of the set of all bounded continuous positive definite functions ψ with $\psi(e) = 1$.

In the **third chapter** we suppose that the hypergroup K is commutative. In 120 pages we present many results of harmonic analysis of such hypergroups, which may be essential for further investigations in this field. Three dual spaces of K ($\mathcal{X}^b(K)$, $\hat{K}$ and $\mathcal{S}(K)$) are in the center of research in sections $1-4$. Dual spaces of $L^1(K)$ and the reduced hypergroup C^*-algebra $C^*_\lambda(K)$ are identified with certain sets of multiplicative functions on K. Fourier transforms and Fourier-Stieltjes transforms are introduced, and the existence of a Plancherel measure π is shown. One should note that in general the dual spaces cannot be endowed with a dual hypergroup structure. Therefore all results concerning inverse Fourier transforms do not follow from results of the Fourier transformation of K (as in the group case). Conditions for equality of the three dual spaces are derived and the Pontryagin duality is studied. In section 5 regularity of $L^1(K)$ is investigated. Of course regularity of $L^1(K)$ depends on the choice of the dual space, on which the Fourier transforms are studied. Assuming that K is of $\frac{1}{2}$-subexponential growth we can use results about Beurling algebras in order to derive that $L^1(K)$ is $\mathcal{X}^b(K)$-regular. In the next section 6 results concerning (inverse) L^p-transforms, $1 \leq p \leq 2$, are shown. In particular, it is proved (for $1 < p < 2$) that the (inverse) L^p-transformations are surjective if and only if K is finite. In sections $7-8$ we prove Bochner-type theorems for continuous bounded functions on K. Furthermore we derive that for characters $\alpha \in \hat{K}$ holds $\alpha \in \mathcal{S}(K)$ if and only if the P_2-condition is satisfied in α. Considering continuous bounded functions on $\hat{K}$ we introduce a strong positive definiteness condition and derive a corresponding Bochner theorem. Extensions of this theorem are given for the case that K also is compact, which lead to a characterization of the Pontryagin duality for K. In section 9 we derive criteria characterizing Fourier-Stieltjes transforms generalizing results of Schoenberg or Eberlein from locally compact abelian groups to commutative hypergroups. In section 10 we study positive semicharacters γ on K and derive the modified hypergroup K_γ. Applying properties of K_γ we can derive that there exists exactly one positive character $\gamma \in \mathcal{S}(K)$. This character replaces the constant character 1 in the case that 1 is not an element of $\mathcal{S}(K)$. In section 11 harmonic analysis based on the dual space $\mathcal{S}(K)$ is the central topic. Without any assumption we construct a translation on $L^2(\mathcal{S}(K), \pi)$ and derive the following equivalence relations: K is discrete if and only if $\mathcal{S}(K)$ is compact, and K is compact if and only if $\mathcal{S}(K)$ is discrete. Assuming that property (F) is satisfied a translation of $C_0(\mathcal{S}(K))$ is constructed. Using this translation, a convolution on $M(\mathcal{S}(K))$ is defined such that $M(\mathcal{S}(K))$ is a commutative Banach $*$-algebra. Next we derive that the Fourier transform on $L^1(K)$ is surjective, or the inverse Fourier transform on $L^1(\mathcal{S}(K), \pi)$ is surjective, exactly when K is finite. For this result we have to suppose that $\mathcal{S}(K)$ is metrizable. Then we introduce a positive definiteness condition for continuous bounded functions on $\mathcal{S}(K) \cup \{1\}$ and derive a corresponding Bochner theorem. Section 12 is devoted to investigations of various types of invariant means on $L^\infty(K)$ and $UC(K)$. Furthermore given some $\alpha \in \hat{K}$, we study α-invariant functionals on $L^\infty(K)$ and $UC(K)$. In section 13 several versions of Reiter's P_1-condition are studied, several properties of spectral synthesis and equivalence relations are proved. For example it is shown that $P_1(\alpha, M)$ (with $\alpha \in \hat{K}$) is equivalent to the existence of α-invariant functionals on $L^\infty(K)$ or to the existence of bounded approximate identities in the maximal ideal $I(\alpha)$.

In **chapter 4** we suppose that $K = \mathbb{N}_0$ is a polynomial hypergroup generated by an orthogonal polynomial sequence $(R_n(x))_{n\in\mathbb{N}_0}$, i.e. the linearization coefficients $g(m,n;k)$ of the products $R_m(x)R_n(x)$ are positive or zero. In more than 200 pages we investigate problems of Fourier analysis, e.g. convergence of orthogonal expansions. Sections 1 and 2 contains the basic results concerning the three dual spaces which are subsets of $\mathbb{C}$, denoted by D, D_s and S. The Haar measure h on $\mathbb{N}_0$ is given by $h(n) = g(n,n;0)^{-1}$. The Plancherel measure π is the orthogonalization measure of $(R_n(x))_{n\in\mathbb{N}_0}$. First results concerning $A(S)$ and $A(D)$ are shown, e.g. a generalization of Wiener's theorem. Comparing the dual spaces a very interesting fact is the equivalence result: $1 \in S = \text{supp}\,\pi$ if and only if $\{\frac{H(n)}{a_n h(n)} : n \in \mathbb{N}_0\}$ is unbounded. We say that a polynomial hypergroup belongs to the Nevai class provided the recurrence coeffiecients of the generating orthogonal polynomial squence are convergent. In section 3 for polynomial hypergroups of the Nevai class the three dual spaces D, D_s and S are explicitly determined. In section 4 sufficient criteria are derived, which allow to show the nonnegativity of the linearization coefficients $g(m,n;k)$. Section 5 contains a large list of polynomial hypergroups with relevant informations concerning their dual spaces, their Haar weights and their Plancherel measure. The results of sections 3 and 4 are a very useful tool to derive the properties of the orthogonal polynomials studied in section 5. Section 6 has the goal to collect information about the size of the space $A(S)$ and the size of the sequence space $M(D_s)^\vee$ given a polynomial hypergroup generated by $(R_n(x))_{n\in\mathbb{N}_0}$. We will use known facts of corresponding spaces determined by another sequence $(P_n(x))_{n\in\mathbb{N}_0}$ of polynomials with degree $P_n = n$ and $P_n(1) = 1$, which are related to $(R_n(x))_{n\in\mathbb{N}_0}$ by properties which we denote by $R_n \overset{d}{\geq} P_n$ or (for the dual case) by $R_n \overset{c}{\geq} P_n$. Supposing these relations we find many examples of functions in $A(S)$ and sequences in $M(D_s)^\vee$. Finally we investigate condition (T), i.e. $T_n \overset{d}{\geq} P_n$, where $T_n(x)$ are the Chebyshev polynomials of the first kind. In section 7 we study summing sequences $(A_n)_{n\in\mathbb{N}_0}$, $A_n \subseteq \mathbb{N}_0$, or the Følner condition (F_p), $1 \leq \infty$. Such sequences play an important role in forming averages, for example constructing invariant means. A natural candidate for summing sequences is the canonical sequence $(S_n)_{n\in\mathbb{N}_0}$ where $S_n = \{0,1,...,n\}$. A useful tool to prove that $(S_n)_{n\in\mathbb{N}_0}$ is a summing sequence is property (H). In section 8 we study the existence of strongly invariant means on l^∞, and show that this is equivalent to $\{\frac{H(n)}{a_n h(n)} : n \in \mathbb{N}_0\}$ is unbounded. Section 9 is a continuation of section 13 in chapter 3. At first a general result is proved, which describes a sufficient condition such that $P_1(z,M)$ holds. Then conditions are derived, which imply the failure of $P_1(x,M)$. Finally we analyze the asymptotic behaviour of the structural function μ and define an important subset $\mathcal{T}$ of D_s. We can prove that $P_1(x,M)$ is satisfied provided $x \in \mathcal{T}$ and $\{\frac{H(n_m)}{a_{n_m} h(n_m)} : m \in \mathbb{N}\}$ is unbounded. In section 10 we study positive definite sequences. Of course we can use results of section 7 in chapter

3, e.g. Bochner's theorem. Several types of ergodic theorems are shown. The convergence of the Dirichlet-kernel $(D_n(x))_{n\in\mathbb{N}_0}$ in $L^2(D_s,a)$ to $\chi_{\{1\}}$ is shown, where $\varphi = \check{a} \in P^b(\mathbb{N}_0)$, provided $\frac{a_n h(n)}{H(n)} \to 0$ or $R_n(x) \to 0$ for all $x \in D_s\backslash\{1\}$ are valid. Then we fix some $y \in D_s$, $y \neq 1$, and consider the 'shifted' Dirichlet-kernel $(D_{n,y}(x))_{n\in\mathbb{N}_0}$ and derive similar convergence results. The limit is $\mu(y)\chi_{\{y\}}$, if $y \in \mathcal{T}$ and zero, when $y \notin \mathcal{T}$. Finally we study the mean value function Φ on D_s of $\varphi \in P^b(\mathbb{N}_0)$. In section 11 we investigate two topics: (1) Derive properties of a polynomial hypergroup in case that $\{h(n) : n \in \mathbb{N}_0\}$ is bounded. (2) Derive properties of a polynomial hypergroup depending on the Turan determinants of $(R_n(x))_{n\in\mathbb{N}_0}$. We mention that results of (2) yield a sufficient condition that the conjugate orthogonal polynomial sequence of $(R_n(x))_{n\in\mathbb{N}_0}$ generates a polynomial hypergroup. Section 12 contains a generalization of a theorem of Wiener, where the discrete part a_d of a complex Borel measure $a \in M(\mathbb{T})$ on the torus group is characterized. It is shown that, given $a \in M(D_s)$, we have $\mathcal{T} \cap \operatorname{supp} a_d = \emptyset$ if and only if $\lim\limits_{n\to\infty} \frac{1}{H(n)} \sum\limits_{k=0}^{n} |\check{a}(k)|^2 h(k) = 0$. In section 13 we introduce several homogeneous Banach spaces on S. Throughout section 13, it is assumed that S satisfies property (F). At first we collect results following from property (F) and the fact that S is compact. Obviously $L^p(S,\pi)$, $1 \leq p < \infty$, and $C(S)$ are homogeneous Banach spaces. We present and investigate the following homogeneous Banach spaces: The Wiener algebra $A(S)$, the $A^p(S)$-algebras, the Beurling algebra $A_*(S)$, the algebra $U_B(S)$, the Lipschitz algebra $\mathrm{lip}_B(\lambda)$. Finally we introduce a differential operator ∂_1 on a homogeneous Banach space $H_2^{(1)}(S)$. ($H_2^{(1)}(S)$ bears a resemblance to Sobolev spaces.) Section 14 consists of three parts: (i) Triangular schemes and approximate identities, (ii) Construction of concrete approximate identities, (iii) Selective approximate identities. Given a Banach space B of functions contained in $L^1(S,\pi)$, we search for triangular schemes $a_{n,k} \in \mathbb{C}$, $n \in \mathbb{N}_0$, $k = 0,...,n$, such that $\|A_n\varphi - \varphi\|_B \to 0$ for all $\varphi \in B$, where $A_n\varphi(x) = \sum\limits_{k=0}^{n} a_{n,k}\check{\varphi}(k)R_k(x)h(k)$. Then $(A_n)_{n\in\mathbb{N}_0}$ is called approximate identity. $(A_n)_{n\in\mathbb{N}_0}$ is an approximate identity with respect to B if and only if $\lim\limits_{n\to\infty} a_{n,m} = 1$ and $\|A_n\varphi\|_B \leq C\|\varphi\|_B$ for all $\varphi \in B$, $n \in \mathbb{N}_0$. The sequence $(A_n(x))_{n\in\mathbb{N}_0}$, $x \in S$, where $A_n(x) = \sum\limits_{k=0}^{n} a_{n,k}R_k(x)h(k)$, is called a kernel. A strong property of kernels is their summability. In part (i) we derive several conditions for kerrnels $(A_n(x))_{n\in\mathbb{N}_0}$, which are strongly connected with the property that $(A_n)_{n\in\mathbb{N}_0}$ is an approximate identity with respect to B. In part (ii) we present concrete examples: The Fejér-type kernel $(F_{2n}(x))_{n\in\mathbb{N}_0}$, the modified Fejér kernel $(\mathcal{F}_n(x))_{n\in\mathbb{N}_0}$, the de la Vallée-Poussin kernels and the Dirichlet kernel. In part (iii) selective approximate identities are studied, where the convergence is considered at points $y \in S$ for all functions $\varphi \in C(S)$. In addition to the point approximation a global approximation procedure on S is investigated. Examples of polynomial hypergroups such that S satisfies property (F) are collected in section 15. Furthermore, given an approximate identity $(A_n)_{n\in\mathbb{N}_0}$ with respect to $C(S)$ it is shown how special approximation kernels based on the triangular schemes $a_{n,k}$ can be used for several representation results, for example to derive (F) or $\overset{c}{\geq}$. In section 16 the Banach algebra $A(S) = (l^1(h))^\wedge|S$ and the Banach space $U(S)$ consisting of $\varphi \in C(S)$, such that $D_n\varphi$ converges uniformly on S towards φ, are investigated. It is shown that $A(S)$ is a proper subspace of $C(S)$. The Wiener-Tauberian property of $A(S)$ is derived provided $S = D_s = D$. Moreover it is shown how local information of a function $\varphi \in C(S)$ can be used to derive that $\varphi \in A(S)$. With regard to $U(S)$ it is proved that $A(S)$ is a proper subspace of $U(S)$. Using a result of Faber (from 1914) it follows that $U(S)$ is a proper subspace of $C(S)$ in case of $S = [a, b]$. An example with $U(S) = C(S)$ is presented in the case that the polynomial hypergroup is generated by the Little q-Legendre polynomials. Finally it is shown that $U(S)$ is character-invariant. Section 17 deals with almost convergent sequences with respect to polynomial hypergroups. For this, the invariant means play a decisive role as well as the convergence $(Lo1)$, i.e. $\lim\limits_{n\to\infty} \frac{1}{H(n)} \sum\limits_{k=0}^{n} L_m\alpha(k)h(k) = d(\alpha)$ uniformly in $m \in \mathbb{N}_0$, where $\alpha \in l^\infty$, $d(\alpha) \in \mathbb{C}$. We call $\alpha \in l^\infty$ $\mathcal{L}$-almost convergent to $d(\alpha)$, if $\mu(\alpha) = d(\alpha)$ for all invariant means μ on l^∞. Supposing that $\frac{a_n h(n)}{H(n)} \to 0$ we can show that α is $\mathcal{L}$-almost convergent to $d(\alpha)$ if and only if $(Lo1)$ is valid. Furthermore we derive several properties of $\mathcal{L}$-almost convergent sequences and consider a stronger form of almost convergence. Finally we analyze multipliers for almost convergent sequences for strongly amenable polynomial hypergroups.

In **chapter 5** we introduce weakly stationary random fields on commutative hypergroups K. That is a family $(X_a)_{a\in K}$ of square integrable complex-valued random variables on a probability space $(\Omega, \mathcal{F}, P)$, such that the covariance function $d(a, b) = E(X_a, \bar{X}_b) = \langle X_a, X_b \rangle$ is a continuous bounded function on $K \times K$ satisfying the stationarity condition $d(a, b) = \int\limits_K d(x, e)d\omega(a, \tilde{b})(x)$ for all $a, b \in K$. We write $d(a)$ instead of

$d(a,e)$. In section 1 it is shown that there exists a unique bounded regular Borel measure $\mu \in M(\hat{K})$ such that $d(a) = \int_{\hat{K}} \alpha(a)d\mu(\alpha)$ for every $a \in K$. (μ is called spectral measure of $(X_a)_{a\in K}$). Considering stochastic measures we have $X_a = \int_{\hat{K}} \alpha(a)dZ(a)$, where Z is a unique orthogonal stochastic measure on $\hat{K}$. It is shown that there is an isometric isomorphism Φ from $L^2(\hat{K},\mu)$ onto the Hilbert space $H(X_a, K)$, where $H(X_a, K)$ is the closure of span$\{X_a : a \in K\}$ in $L^2(\Omega, P)$. Moreover, translation operators $T_b : H(X_a, K) \to H(X_a, K)$ and corresponding multiplication operators $M_b : L^2(\hat{K},\mu) \to L^2(\hat{K},\mu)$ are introduced and it is shown that $T_b \circ \Phi = \Phi \circ M_b$. Finally it is proved that there exists a resolution E of the identity on $\hat{K}$ such that $T_a = \int_{\hat{K}} \alpha(a)dE(\alpha)$. These properties are very useful for investigations of weakly R_n-stationary random sequences. In section 2 a list of occasions is presented, where one meets weakly stationary random fields on commutative hypergroups.

In **chapter 6** weakly stationary random sequences on polynomial hypergroups are studied. We call them weakly R_n-stationary random sequences. Starting in section 1 with R_n-white noise $(Z_n)_{n\in\mathbb{N}_0}$ we introduce R_n-oscillations, R_n-moving average sequences and first-order R_n-autoregressive random sequences. The corresponding spectral measures are determined and it is shown that each first-order R_n-autoregressive random sequence with defining parameter α with $|\alpha| < 1$ has a representation as R_n-moving average sequence. Finally p-order R_n-autoregressive random sequences $(X_n)_{n\in\mathbb{N}_0}$ are introduced, and it is shown that such sequences have a representation as R_n-moving average sequence provided $P(x) = 1 - a_1R_1(x) - \ldots - a_pR_p(x) \neq 0$ for $x \in D$, where $1, a_1, \ldots, a_p$ $(a_p \neq 0)$ are the defining parameters of $(X_n)_{n\in\mathbb{N}_0}$. In section 2 it is assumed that the expected value of all X_n is constant, i.e. $E(X_n) = M$ for all $n \in \mathbb{N}_0$. Several linear unbiased estimators M_n of M are analyzed in order to find consistent estimators, that is, $E(|M_n - M|^2)$ tends to zero. It is shown that there exist consistent linear unbiased estimators if and only if $\mu(\{1\}) = 0$. Furthermore a description of the best linear unbiased estimators for M is given. In section 3 we investigate the problem of one-step prediction. Writing $H = H(X_n, \mathbb{N}_0)$ and $H_n = \text{span}\{X_0, \ldots, X_n\} \subseteq H$ we have to characterize that $\hat{X}_{n+1} \in H_n$ satisfying $\|\hat{X}_{n+1} - X_{n+1}\|_2 = \min\{\|Y - X_{n+1}\|_2 : Y \in H_n\}$. $(X_n)_{n\in\mathbb{N}_0}$ is called asymptotic R_n-deterministic if the prediction error $\delta_n = \|\hat{X}_{n+1} - X_{n+1}\|_2$ satisfies $\delta_n \to 0$ as $n \to \infty$. A first result describes $\hat{X}_{n+1}$ by a stochastic integral with the stochastic measure

Z on D_s. Then four different criteria are derived, where each one implies that $(X_n)_{n\in\mathbb{N}_0}$ is asymptotic R_n-deterministic. In section 4 the effect of filtering is studied. Given $(X_n)_{n\in\mathbb{N}_0}$, one can define $(Y_n)_{n\in\mathbb{N}_0}$ by $Y_n = a * X_n = \sum_{k=0}^{\infty} a_k T_k X_n h(k)$, where $a = (a_k)_{k\in\mathbb{N}_0} \in l^2(h)$. If $\mu = f\pi$, $f \in L^\infty(D_s, \pi)$, then $(Y_n)_{n\in\mathbb{N}_0}$ is a weakly R_n-stationary random sequence with spectral measure $\nu = |\mathcal{P}(a)|^2\mu$. The transition from $(X_n)_{n\in\mathbb{N}_0}$ to $(a * X_n)_{n\in\mathbb{N}_0}$ is called filtering. In section 5 we introduce R_n-ARMA sequences. Let $a_0, ..., a_q \in \mathbb{C}$, $a_q \neq 0$ and $b_0, ..., b_p \in \mathbb{C}$, $b_p \neq 0$. A weakly R_n-stationary random sequence $(X_n)_{n\in\mathbb{N}_0}$ is called R_n-ARMA sequence if $b_0X_n + b_1T_1X_n + ... + b_pT_pX_n = a_0Z_n + a_1T_1Z_nh(1) + ... + a_qT_qZ_nh(q)$ is satisfied. We show, given that $A(x)$, $B(x)$ are polynomials of degree q, respectively of degree p, and supposed that $B(x)$ has no zeros in $S = \operatorname{supp} \pi$, that the following result is true: If $(X_n)_{n\in\mathbb{N}_0}$ is a weakly R_n-stationary random sequence with spectral measure $d\mu(x) = |\frac{A(x)}{B(x)}|^2 d\pi(x)$, then there exists an R_n-white noise $(Z_n)_{n\in\mathbb{N}_0}$ such that $(X_n)_{n\in\mathbb{N}_0}$ is an R_n-ARMA sequence based on $(Z_n)_{n\in\mathbb{N}_0}$. In section 6 we generalize Wold's decomposition on the group $\mathbb{Z}$ to polynomial hypergroups on $\mathbb{N}_0$. Now denote by H_n the closure of span$\{X_k : k \geq n\}$ in $L^2(\Omega, P)$ and $S(X) = \bigcap_{n\in\mathbb{N}_0} H_n$. $(X_n)_{n\in\mathbb{N}_0}$ is called purely nondeterministic if $S(X) = \{0\}$ and deterministic if $S(X) = H = H(X_n, \mathbb{N}_0)$. Defining $X^s = T_n P_{S(X)} X_0$ and $X_n^r = T_n P_{S(X)} \perp X_0$, it is shown that $X_n = X_n^r + X_n^s$ for each $n \in \mathbb{N}_0$, where $(X_n^s)_{n\in\mathbb{N}_0}$ and $(X_n^r)_{n\in\mathbb{N}_0}$ are weakly R_n-stationary random sequences. $(X_n^s)_{n\in\mathbb{N}_0}$ is deterministic and $(X_n^r)_{n\in\mathbb{N}_0}$ is purely nondeterministic. Supposing that the spectral measure has the form $\mu = f\pi$, $f \in L^1(D_s, \pi)$, we derive criteria which imply that $(X_n)_{n\in\mathbb{N}_0}$ is purely nondeterministic, respectively, deterministic.

In **chapter 7** we discuss problems of difference equations of a special form, where R_n-stationarity of sequences in a Hilbert space H is a central topic. If $H = L^2(\Omega, P)$, there are strong connections to chapters 5 and 6. In section 1 we consider difference equations, where the inherent structure is induced by a polynomial hypergroup and the dynamics by a linear operator A on a vector space B. We define $a_nX(n+1) + b_nX(n) + c_nX(n-1) = A(X(n))$ for $n \in \mathbb{N}$, $X(0) \in B$ and $X(1) = A(X(0))$. The solutions are determined and three examples are analyzed. In section 2 the effect of R_n-stationarity is investigated. If we assume that an R_n-stationary solution $(X(n))_{n\in\mathbb{N}_0}$ is bounded, we obtain a spectral measure μ on D_s such that $\langle X(m), X(n)\rangle = \int_{D_s} R_m(x)R_n(x)d\mu(x)$. Using this representation several properties of $(X(n))_{n\in\mathbb{N}_0}$ are derived, e.g. ergodic properties. Section 3 contains several examples of bounded R_n-stationary sequences. In particular examples 4 and 5 can be used also in chapter 6. In section 4 the multipliers of bounded R_n-stationary sequences are characterized, and in section 5 the imaginary part of T_n-stationary sequences is determined.

In **chapter 8** we present further hypergroup classes. In section 1 we study the continuous actions of a compact group H on a locally compact Hausdorff space X. The results are essential for investigations of the H-orbit space X_H. Then we consider $[FIA]_B^-$-groups, where G is a locally compact group, B a subgroup of $\mathrm{Aut}(G)$, the group of topological automorphisms of G equipped with the Birkhoff topology. G is called $[FIA]_B^-$-group, if B is relatively compact. We study the H-orbit space G_H, where $H = \bar{B} \subseteq \mathrm{Aut}(G)$. Denote by $I(G)$ the group of inner automorphisms of G. If $I(G) \subseteq B$, then G_H is a commutative hypergroup and we can provide the following results: (1) $\mathcal{S}(G_H) = \widehat{G_H} = \mathcal{X}^b(G_H)$. (2) $\widehat{G_H}$ is a hypergroup under pointwise multiplication and (3) the Pontryagin duality holds for G_H. As examples we present $G = \mathbb{R}^n$ and $B = SO(n)$, where $(\mathbb{R}^n)_{SO(n)}$ can be identified with a Bessel-Kingman hypergroup, and the Dunkl-Ramirez hypergroup K based on p-adic integers, which is homeomorphic to $\mathbb{N}_0^* = \mathbb{N}_0 \cup \{\infty\}$, the one-point compactification of $\mathbb{N}_0$. In the third part of section 1 we study hypergroups which are based on a spherical projector $P : C_c(G) \to C_c(G)$ satisfying special properties, such that $H = \{\dot{x} = O_x : x \in G\}$ is a hypergroup, where O_x is the subset $O_x = \mathrm{supp}\,\mu_x$ of G and $\mu_x = P^*(\epsilon_x)$, P^* the adjoint operator of the extension of P to $C_0(G)$. As examples, the set of double cosets and the set of orbits are presented. A sufficient condition that the hypergroup H is commutative is proved and $\mathcal{X}^b(H)$ is identified with the set of P-spherical functions. In section 2 we present examples of hypergroups which are associated with special functions. We start with the Bessel-Kingman hypergroups $H_\alpha = [0,\infty[,\ \alpha > -\frac{1}{2}$, and determine the Haar measure m_α, the dual space $\widehat{H_\alpha}$ which is homeomorphic to H_α, and show that $\mathcal{S}(H_\alpha) = \widehat{H_\alpha} = \mathcal{X}^b(H_\alpha)$ and that the Pontryagin duality holds for H_α. The second example is the Naimark hypergroup $K = [0,\infty[$. $\mathcal{S}(K)$, $\hat{K}$ and $\mathcal{X}^b(K)$ are determined (these dual spaces are different), as well as the Haar measure m and the Plancherel measure π_K. Example 3 concerns the dual Jacobi hypergroups $H(\alpha,\beta) = [-1,1]$, which are discussed shortly. Example 4 contains the disc polynomial hypergroups $K_\alpha = \mathbb{N}_0 \times \mathbb{N}_0$ for $\alpha \geq 0$. The dual sppace $\widehat{K_\alpha}$ is homeomorphic to the disc D and is a hypergroup by pointwise multiplication, and the Pontryagin duality is valid for K_α. At the end we give a short survey of one-dimensional hypergroups defined on $[0, \infty[$ generated by eigenfunctions of a Sturm-Liouville problem, which is determined by a Sturm-Liouville fuction $A : [0, \infty[\to \mathbb{R}$. If there exist measures $\omega(x, y)$, such that the functions $\mu_f(x, y)$ defined by $\mu_f(x, y) = \int_0^\infty f(z) d\omega(x, y)(z)$, $f \in C^\infty([0, \infty[)$ are solutions of a Cauchy problem determined by A, then a convolution $\omega = \omega_A$ can be defined on $[0, \infty[$. If A satisfies two conditions (SL1) and (SL2), then $[0, \infty[$ with convolution ω_A (unit element 0 and involution equal to the identity mapping) is a Sturm-Liouville hypergroup.

Finally I want to thank Dr. Doris Jakubassa-Amundsen for typesetting the text. She was a great help to give this book a perfect form. Furthermore I want to thank my wife for her exceptional patience during the preparation of the manuscript.

Contents

Chapter 1

Hypergroups

1 Definition and basic properties

We start developing the theory of hypergroups following the concept of Jewett, see [350]. The approaches of Dunkl [192] and Spector [630] are very similar, but not exactly equivalent. We will rephrase the axioms of Jewett by starting with the convolution of points and then derive the extension to the space $M(K)$ of all regular complex Borel measures on K. In [350] or [96] it is required per axiom that there is a binary operation (convolution) on the vector space $M(K)$ under which $M(K)$ is an algebra. We will construct this convolution on $M(K)$ by extending the convolution of point measures. Since this procedure is very useful for the investigation of concrete examples, we will give a brief discussion of the necessary tools.

Let K be a locally compact Hausdorff space. We consider spaces $C^b(K), C_0(K)$ and $C_c(K)$ consisting of continuous complex-valued functions, those that are bounded, those that vanish at infinity and those with compact support, respectively.

A complex Radon measure Λ on K is a continuous linear functional on $C_c(K)$, i.e. for every compact $C \subseteq K$ there exists a constant α_C such that $|\Lambda(f)| \leq \alpha_C \|f\|_\infty$ for all $f \in C_c(K)$ with $\operatorname{supp} f \subseteq C$.

For complex Radon measures Λ on K write

$$\|\Lambda\| := \sup\{\, |\Lambda(f)| : f \in C_c(K),\ \|f\|_\infty \leq 1\}.$$

$\|\Lambda\|$ can be infinite. If $\|\Lambda\| < \infty$ the functional Λ can be uniquely extended to a continuous linear functional on $C_0(K)$ equipped with the $\|\cdot\|_\infty$-norm, i.e. Λ is an element of the dual space $C_0(K)^*$.

We have also to deal with measure and integration. We call a $[0,\infty]$-valued function μ defined on a σ-algebra $\mathcal{A}$, which is countably additive, a **positive** measure on $\mathcal{A}$. Frequently such a set-function is just called a measure. We add the adjective 'positive' to make clear that the range of μ is $[0, \infty]$, where the value ∞ is admissible.

In the context of hypergroups we have mainly to deal with regular complex Borel measures on K. Denote by $\mathcal{B}(K)$ the Borel σ-algebra on K. A set $\{A_k : k \in \mathbb{N}\}$ of $A_k \in \mathcal{B}(K)$ is called a partition of $A \in \mathcal{B}(K)$ whenever $A = \bigcup_{k=1}^{\infty} A_k$ and $A_k \cap A_j = \emptyset$ for $k \neq j$. A **complex Borel measure** μ on K is a complex-valued function on $\mathcal{B}(K)$, which satisfies $\mu(\emptyset) = 0$ and

$$\mu(A) = \sum_{k=1}^{\infty} \mu(A_k) \qquad \text{for all } A \in \mathcal{B}(K), \tag{1}$$

for every partition $\{A_k : k \in \mathbb{N}\}$ of A. One should observe that the convergence of the series in (1) is part of the assumption. One gets even absolute convergence of the series, since every rearrangement of the series also must converge to $\mu(A)$.

Given μ, define

$$|\mu|(A) := \sup\{\sum_{k=1}^{\infty} |\mu(A_k)| : \{A_k : k \in \mathbb{N}\} \text{ partition of } A\} \tag{2}$$

for $A \in \mathcal{B}(K)$. The set function $|\mu| : \mathcal{B}(K) \to [0, \infty]$ is called the **total variation** (measure) of μ. The total variation $|\mu|$ of a complex Borel measure μ has two important properties:

(i) $|\mu|$ is a positive Borel measure on K.
(ii) $|\mu(A)| \leq |\mu|(A) \leq |\mu|(K) < \infty$.

For the proof we refer to Theorem 6.2 and Theorem 6.4 of [610]. Note that the $[0, \infty]$-valued Borel measures on K do not form a subclass of the complex Borel measures on K. There exists a polar decomposition of μ, that is

(iii) $d\mu = h\, d|\mu|$, where h is a Borel-measurable function such that $|h(x)| = 1$ for every $x \in K$.

We refer again to [610,Theorem 6.12].

Finally we have to define the regularity of complex Borel measures. A complex Borel measure μ on K is called **regular**, if for each $A \in \mathcal{B}(K)$

$$\begin{aligned} |\mu|(A) &= \sup\{\, |\mu|(C) : C \subseteq A,\ C \text{ compact}\} \\ &= \inf\{\, |\mu|(V) : A \subseteq V,\ V \text{ open}\}. \end{aligned}$$

(Recall that $|\mu|$ is a bounded positive Borel measure.)

Define $M(K)$ to be the set of all regular complex Borel measures on K. (We use the notation $M(K)$ of [350].) It is clear that $M(K)$ is a normed linear space with norm $\|\mu\| = |\mu|(K)$, where addition and scalar multiplication are defined setwise. $M(K)$ is even a Banach space, see below.

If μ is a complex Borel measure, we define integration for bounded Borel-measurable functions f on K by

$$\int_K f(x)\, d\mu(x) = \int_K f(x)\, h(x)\, d|\mu|(x)$$

where h is the function of the polar decomposition of μ, see (iii). It is clear that for each complex Borel measure μ, the mapping $f \mapsto \int_K f(x)\, d\mu(x)$, $C^b(K) \to \mathbb{C}$ is a bounded linear functional on $C^b(K)$ with respect to the sup-norm. Considering the Banach space $C_0(K) \subseteq C^b(K)$ with sup-norm, Riesz' theorem (see [610,Theorem 6.19]) says that every bounded linear functional $\Lambda \in C_0(K)^*$ is represented by a unique $\mu \in M(K)$ such that

$$\Lambda(f) = \int_K f(x)\, d\mu(x) \qquad \text{for all } f \in C_0(K),$$

and $\|\Lambda\| := \sup\{\, |\Lambda(f)| : f \in C_0(K),\ \|f\|_\infty \leq 1\}$ is equal to $|\mu|(K)$.

In particular it follows that $M(K)$ and $C_0(K)^*$ are isometric isomorphic Banach spaces via the mapping $\mu \mapsto \Lambda_\mu$, $M(K) \to C_0(K)^*$, where $\Lambda_\mu(f) = \int_K f(x)\, d\mu(x)$ for all $f \in C_0(K)$.

Denote $M^+(K) = \{\mu \in M(K) : |\mu| = \mu\}$ and $M^1(K) = \{\mu \in M^+(K) : \mu(K) = 1\}$ the probability measures of $M(K)$. On $M^+(K)$ we consider the **cone topology** $\sigma(M^+(K), C_c(K) \cup \{1\})$ and the **Bernoulli topology** $\sigma(M^+(K), C^b(K))$. In [72,Proposition 1.4] it is shown that both topologies coincide on $M^+(K)$. Restricting to $M^1(K)$ it follows that the following three topologies on $M^1(K)$ coincide: $\sigma(M^1(K), C_c(K))$, $\sigma(M^1(K), C_0(K))$ and $\sigma(M^1(K), C^b(K))$.

The **support** $\operatorname{supp}(\mu)$ of $\mu \in M^+(K)$ is the complement of the largest open subset of K on which μ vanishes. It is apparent that $E = \operatorname{supp}(\mu)$ exactly when

(1) E is closed in K
(2) $\mu(E \cap U) > 0$ if $E \cap U \neq \emptyset$ and U is open in K
(3) $\mu(K \backslash E) = 0$.

We denote by $M_c^1(K) := \{\mu \in M^1(K) : \operatorname{supp}(\mu) \text{ is compact}\}$.

The collection of all nonvoid compact subsets of K is denoted by $\mathcal{C}(K)$. The set $\mathcal{C}(K)$ is given the **Michael topology**.

The Michael topology has a subbasis consisting of all compact sets of the form

$$\mathcal{C}_{U,V} = \{C \in \mathcal{C}(K) : C \cap U \neq \emptyset \text{ and } C \subseteq V\},$$

where U and V are open subsets of K, see [485]. In [485] this topology is called finite topology. In [388,Lemma 4.1] it is shown that in case of K being a metric space the Michael topology is equivalent to the Hausdorff topology. The Hausdorff topology on $\mathcal{C}(K)$ is determined by the metric D on $\mathcal{C}(K)$,

$$D(C_1, C_2) = \inf\{r > 0 : C_1 \subseteq V_r(C_2) \text{ and } C_2 \subseteq V_r(C_1)\}$$

for $C_1, C_2 \in \mathcal{C}(K)$, where

$$V_r(C) = \{x \in K : d(x, C) < r\}$$

and d is the metric on K.

Just two minor extension results have to be shown before we can formulate the definition of hypergroups.

Lemma 1.1.1 *Assume that there is a mapping $(x,y) \mapsto \omega(x,y)$, $K \times K \to M_c^1(K)$, which is continuous with respect to $\sigma(M_c^1(K), C_c(K))$ and such that $(x,y) \mapsto supp\,(\omega(x,y))$, $K \times K \to \mathcal{C}(K)$ is continuous with respect to the Michael topology. Then*

(a) for every continuous function $f : K \to \mathbb{C}$ the function $(x,y) \mapsto \omega(x,y)(f)$ is continuous.
(b) Let $a, b, c \in K$. The linear functionals on $C^b(K)$,

$$f \mapsto \int_K \omega(a,y)(f)\, d\omega(b,c)(y) \qquad \textit{and}$$

$$f \mapsto \int_K \omega(x,c)(f)\, d\omega(a,b)(x),$$

determine well-defined probability measures of $M(K)$, which we denote by $\omega(a, \omega(b,c))$ and $\omega(\omega(a,b),c)$, respectively.

Proof.

(a) Given $x_0, y_0 \in K$ and $\varepsilon > 0$ choose an open set U with compact closure $\overline{U}$ such that supp $(\omega(x_0,y_0)) \subseteq U$. Applying the continuity of $(x,y) \mapsto$ supp $(\omega(x,y))$ there exist neighbourhoods V_{x_0}, V_{y_0} of x_0 and y_0 with supp $(\omega(x,y)) \subseteq U$, for all $x \in V_{x_0}$, $y \in V_{y_0}$. The Lemma of Urysohn yields a function $g \in C_c(K)$ with $g|\overline{U} = f|\overline{U}$. According to the continuity of $(x,y) \mapsto \omega(x,y)$ there exist neighbourhoods U_{x_0}, U_{y_0} of x_0 and y_0 with $U_{x_0} \subseteq V_{x_0}$, $U_{y_0} \subseteq V_{y_0}$ such that

$$|\omega(x,y)(g) \; - \; \omega(x_0,y_0)(g)| < \varepsilon$$

for all $x \in U_{x_0}$, $y \in U_{y_0}$. Since for any $x \in U_{x_0}$, $y \in U_{y_0}$ we also have $\operatorname{supp} \omega(x,y) \subseteq U$, we obtain $\omega(x,y)(g) = \omega(x,y)(f)$, and the assertion follows.

(b) By (a) $y \mapsto \omega(a,y)(f)$ is a continuous map for each continuous function f on K. Since $\operatorname{supp} \omega(b,c)$ is compact, $\int_K \omega(a,y)(f)\, d\omega(b,c)(y)$ exists for each continuous function f on K. In particular, $f \mapsto \int_K \omega(a,y)(f)\, d\omega(b,c)(y)$ is a well-defined bounded linear functional on $C_0(K)$. By Riesz' theorem there exists a unique regular complex Borel measure on K, which we denote by $\omega(a,\omega(b,c))$. Obviously we have even $\omega(a,\omega(b,c)) \in M^1(K)$. In almost the same manner we obtain $\omega(\omega(a,b),c) \in M^1(K)$.

◇

Remark: Shortly we will show that $\omega(a,\omega(b,c))$ and $\omega(\omega(a,b),c)$ are elements of $M_c^1(K)$.

Lemma 1.1.2 *Let $x \mapsto \tilde{x}$ be a homeomorphism of K such that $\tilde{\tilde{x}} = x$ for all $x \in K$. For $\mu \in M_c^1(K)$ define $\tilde{\mu}(B) := \mu(\tilde{B})$ for every $B \in \mathcal{B}(K)$. Then $\tilde{\mu} \in M_c^1(K)$.*

Proof. It is apparent that $\tilde{\mu}$ is an element of $M^1(K)$. Moreover, $\operatorname{supp} \tilde{\mu} = (\operatorname{supp}(\mu))^\sim$ follows easily from the characterization of the support given above. ◇

Lemma 1.1.1 and Lemma 1.1.2 contain all the prerequisites for the definition of hypergroups.

Definition. Let K be a locally compact Hausdorff space. The quadruple $(K, \omega, \tilde{}, e)$ is called **hypergroup**, when

(H,ω) There is a mapping $\omega : K \times K \to M_c^1(K)$ which is continuous with respect to $\sigma(M_c^1(K), C_c(K))$ and such that $(x,y) \mapsto \operatorname{supp} \omega(x,y)$, $K \times K \to \mathcal{C}(K)$ is continuous with respect to the Michael topology. Furthermore ω satisfies the following associativity law $\omega(x,\omega(y,z)) = \omega(\omega(x,y),z)$ for all $x,y,z \in K$, where the probability measures $\omega(x,\omega(y,z))$ and $\omega(\omega(x,y),z)$ are defined in Lemma 1.1.1.

(H,$\tilde{}$) There is a homeomorphism $x \mapsto \tilde{x}$ on K such that $\tilde{\tilde{x}} = x$ and $\omega(x,y)^\sim = \omega(\tilde{y},\tilde{x})$ for all $x,y \in K$, where $\omega(x,y)^\sim$ is defined in Lemma 1.1.2.

(H,e) There exists a (necessarily unique) element $e \in K$ such that $\omega(e,x) = \epsilon_x = \omega(x,e)$ for all $x \in K$, and such that $e \in \operatorname{supp}(\omega(x,\tilde{y})$ if and only if $x = y$.

The mapping ω is called **convolution** and the mapping $\tilde{}$ is called **involution**. The element e is called **unit element**. If $\omega(x,y) = \omega(y,x)$ for all $x, y \in K$, we call K a **commutative** hypergroup. In the rest of this paragraph K will always be a hypergroup.

The hypergroup axioms are often numerated in various ways. We propose to use (H,ω), (H,$\tilde{}$) and (H,e) or the following numeration:

(H1) The mapping $\omega : K \times K \to M^1(K)$ is continuous with respect to $\sigma(M^1(K), C_c(K))$ such that associativity holds, i.e.

$$\omega(x, \omega(y,z)) = \omega(\omega(x,y), z) \qquad \text{for all } x, y, z \in K$$

(H2) $\operatorname{supp}(\omega(x,y))$ is compact for all $x, y \in K$.

(H3) The mapping $(x,y) \mapsto \operatorname{supp}(\omega(x,y))$, $K \times K \to \mathcal{C}(K)$ is continuous, where $\mathcal{C}(K)$ is given the Michael topology.

(H4) $\tilde{} : K \to K$ is a homeomorphism such that $\tilde{\tilde{x}} = x$ and $(\omega(x,y))\tilde{} = \omega(\tilde{y}, \tilde{x})$ for all $x, y \in K$.

(H5) There exists a unique element $e \in K$ such that

$$\omega(e,x) = \epsilon_x = \omega(x,e) \qquad \text{for all } x \in K.$$

(H6) We have $e \in \operatorname{supp}(\omega(x,\tilde{y}))$ if and only if $x = y$.

Obviously, (H1),...,(H6) are valid if and only if (H,ω), (H,$\tilde{}$) and (H,e) hold.

If the hypergroup K is **hermitian**, that is $\tilde{z} = z$ for every $z \in K$, then K is commutative. In fact, if K is hermitian, then $(\omega(x,y))\tilde{} = \omega(x,y)$, and hence $\omega(x,y) = (\omega(x,y))\tilde{} = \omega(\tilde{y},\tilde{x}) = \omega(y,x)$.

One should realise that in general a hypergroup K does not have an algebraic structure of its own. The properties of K are partly determined in the semigroup $M^1(K)$. Usually in the literature, the convolution $\omega(x,y)$ is written as $\varepsilon_x * \varepsilon_y$ or $\rho_x * \rho_y$. We prefer $\omega(x,y)$ to avoid confusions with convolutions induced by group structures. Often we just write K instead of $(K, \omega, \tilde{}, e)$.

Now we extend the convolution $\omega : K \times K \to M_c^1(K)$, which is continuous with respect to $\sigma(M_c^1(K), C_c(K)) = \sigma(M_c^1(K), C^b(K))$.
The point measures of $x \in K$ are denoted by ϵ_x. We will apply Radon product measures in order to construct a bilinear mapping $(\mu,\nu) \mapsto \mu * \nu$, $M(K) \times M(K) \to M(K)$, satisfying $\epsilon_x * \epsilon_y = \omega(x,y)$. For the construction we consider three Radon measures, which finally yield $\mu * \nu \in M(K)$.

One has to keep in mind that $\mathcal{B}(K)\otimes\mathcal{B}(K)$ - the smallest σ-algebra containing all subsets $A\times B\ \ (A,B\in\mathcal{B}(K))$ - can be a proper subset of $\mathcal{B}(K\times K)$. If K is second countable, then $\mathcal{B}(K)\otimes\mathcal{B}(K)=\mathcal{B}(K\times K)$, see [611,Lemma 7.6].

Theorem 1.1.3 *Let K be a locally compact Hausdorff space. Let $\omega : K\times K\to M_c^1(K)$ be a continuous map with respect to $\sigma(M_c^1(K),C_c(K))$ and such that $(x,y)\mapsto supp\,(\omega(x,y))$, $K\times K\to\mathcal{C}(K)$ is continuous with respect to the Michael topology. Let $\mu,\nu\in M(K)$. Then*

(i) The functional

$$\Lambda_{\mu,\nu}(F) := \int_K\left(\int_K F(x,y)\,d\mu(x)\right)d\nu(y) \qquad \text{for } F\in C_c(K\times K)$$

is a bounded complex Radon measure on $K\times K$ satisfying

$$\|\Lambda_{\mu,\nu}\| \ \le\ \|\mu\|\,\|\nu\|$$

$$\begin{aligned}\Lambda_{\mu,\nu}(F) &= \int_K\left(\int_K F(x,y)\,d\mu(x)\right)d\nu(y)\\ &= \int_K\left(\int_K F(x,y)\,d\nu(y)\right)d\mu(x) = \Lambda_{\nu,\mu}(F).\end{aligned}$$

(ii) Denote by $\mu\otimes\nu$ the unique regular complex Borel measure on $K\times K$ (with Borel σ-algebra $\mathcal{B}(K\times K)$) extending $\Lambda_{\mu,\nu}$. In particular, $\mu\otimes\nu\in M(K\times K)$ corresponds uniquely to $\Lambda_{\mu\otimes\nu}\in C_0(K\times K)^*$, so that $\|\mu\otimes\nu\|\ \le\ \|\mu\|\,\|\nu\|$ and

$$\Lambda_{\mu\otimes\nu}(F) \ =\ \int_{K\times K} F(x,y)\,d\mu\otimes\nu(x,y) \qquad \text{for } F\in C_0(K\times K),$$

and the mapping

$$F\ \mapsto\ \int_{K\times K} F(x,y)\,d\mu\otimes\nu(x,y), \qquad C^b(K\times K)\to\mathbb{C}$$

is a bounded linear functional on $C^b(K\times K)$.

(iii) The linear functional

$$\Lambda_{\mu*\nu}(f) \ =\ \int_{K\times K}\omega(x,y)(f)\,d\mu\otimes\nu(x,y) \qquad \text{for } f\in C^b(K)$$

satisfies $\|\Lambda_{\mu*\nu}\|\ \le\ \|\mu\otimes\nu\|\ \le\ \|\mu\|\,\|\nu\|$ and the restriction to $C_0(K)$ determines uniquely a regular complex Borel measure $\mu*\nu\in M(K)$ such that

$$\mu*\nu(f) \ =\ \int_{K\times K}\omega(x,y)(f)\,d\mu\otimes\nu(x,y)$$

for every $f\in C_0(K)$. Moreover $\|\mu*\nu\|\ \le\ \|\mu\|\,\|\nu\|$.

Proof.

(i) The first step is obvious. We have only to observe that Fubini's theorem can be applied, since $|\mu|$ and $|\nu|$ are bounded measures and $C_c(K \times K) \subseteq L^1(K \times K, |\mu| \times |\nu|)$, see [304,Theorem 13.8].
(ii) The existence of $\mu \otimes \nu \in M(K \times K)$ follows by Riesz' theorem.
(iii) By Lemma 1.1.1(a) the function $(x, y) \mapsto \omega(x, y)(f)$ is an element of $C^b(K \times K)$ for each $f \in C^b(K)$. Riesz' theorem again yields $\mu * \nu \in M(K)$.

◇

Remark: If $\mu = \epsilon_x,\ \nu = \epsilon_y$, clearly we have $\Lambda_{\epsilon_x * \epsilon_y}(f) = \omega(x, y)(f)$ for every $f \in C^b(K)$. We call $(\mu, \nu) \to \mu * \nu$ the **canonical extension** of $(\epsilon_x, \epsilon_y) \mapsto \omega(x, y)$.

Now we derive the positive continuity of the bilinear mappings $(\mu, \nu) \mapsto \mu \otimes \nu$, $M(K) \times M(K) \to M(K \times K)$ and $(\mu, \nu) \mapsto \mu * \nu$, $M(K) \times M(K) \to M(K)$, that is: If $\mu \geq 0,\ \nu \geq 0$, then $\mu \otimes \nu \geq 0$ and $\mu * \nu \geq 0$ and both mappings restricted to $M^+(K) \times M^+(K)$ are continuous with respect to the Bernoulli topology. Compare with subsection 2.4 of [350].

Theorem 1.1.4 *Assume that $\omega : K \times K \to M^1_c(K)$ fulfils the assumptions of Theorem 1.1.3. Consider the Borel measures $\mu \otimes \nu \in M(K \times K)$ and $\mu * \nu \in M(K)$ of Theorem 1.1.3. Then*

(1) The mapping $(\mu, \nu) \mapsto \mu \otimes \nu$, $M(K) \times M(K) \to M(K \times K)$ is positive continuous.
(2) The mapping $(\mu, \nu) \mapsto \mu * \nu$, $M(K) \times M(K) \to M(K)$ is positive continuous, and extends the mapping $(\epsilon_x, \epsilon_y) \mapsto \omega(x, y)$ uniquely to a bilinear positive-continuous mapping.

Proof.

(1) Obviously $\mu \otimes \nu \in M^+(K \times K)$ whenever $\mu, \nu \in M^+(K)$. For the continuity statement we apply that the cone topology on $M^+(K \times K)$ coincides with the Bernoulli topology. Let $(\mu_i)_{i \in I}$ and $(\nu_j)_{j \in J}$ be nets in $M^+(K)$ such that $\mu_i \to \mu$ and $\nu_j \to \nu$, $\mu \in M^+(K)$ and $\nu \in M^+(K)$ with respect to the Bernoulli topology. Applying Fubini's theorem ([304,Theorem 13.8]) we obtain

$$\mu_i \otimes \nu_j(1) \;=\; \mu_i(K)\, \nu_j(K) \;\to\; \mu(K)\, \nu(K) \;=\; \mu \otimes \nu(1).$$

Set $M := \|\mu \otimes \nu\| + 1$. There are indices $i_0 \in I$ and $j_0 \in J$ such that $\|\mu_i \otimes \nu_j\| < M$ for $i \geq i_0,\ j \geq j_0$. Consider the self-adjoint subalgebra

A of $C_0(K \times K)$, where

$$A = \{(x,y) \mapsto \sum_{k=1}^{n} g_k(x)\, h_k(y) : \; g_k, h_k \in C_c(K),\; n \in \mathbb{N}\}.$$

Using the theorem of Stone-Weierstraß it follows that A is dense in $C_0(K \times K)$ with respect to $\|\cdot\|_\infty$. Obviously for each $\eta \in A$ we obtain $\mu_i \otimes \nu_j(\eta) \to \mu \otimes \nu(\eta)$. Now let $f \in C_c(K \times K)$ and $\varepsilon > 0$. There exists some $\eta \in A$ such that $\|f - \eta\|_\infty < \frac{\varepsilon}{M}$. It follows for $i \geq i_0,\; j \geq j_0$

$$|\mu_i \otimes \nu_j(f) \;-\; \mu \otimes \nu(f)| \;\leq\; |\mu_i \otimes \nu_j(f - \eta)|$$

$$+\; |\mu_i \otimes \nu_j(\eta) \;-\; \mu \otimes \nu(\eta)| \;+\; |\mu \otimes \nu(\eta - f)|$$

$$\leq\; \varepsilon \;+\; |\mu_i \otimes \nu_j(\eta) \;-\; \mu \otimes \nu(\eta)| \;+\; \varepsilon.$$

Since $\mu_i \otimes \nu_j(\eta) \to \mu \otimes \nu(\eta)$ for $\eta \in A$, it follows $\mu_i \otimes \nu_j(f) \to \mu \otimes \nu(f)$, and the continuity of $(\mu,\nu) \mapsto \mu \otimes \nu$, $M^+(K) \times M^+(K) \to M^+(K \times K)$ (with respect to the cone topology) is shown.

(2) The positive continuity of $(\mu,\nu) \mapsto \mu * \nu$, $M^+(K) \times M^+(K) \to M^+(K)$ follows directly from (1). The uniqueness statement follows from the density of

$$M_e^+(K) \;=\; \{\mu = \sum_{k=1}^{n} c_k \epsilon_{x_k} : \; x_k \in K,\; c_k \geq 0,\; n \in \mathbb{N}\}$$

in $M^+(K)$ with respect to the cone topology.

◇

Each $\mu \in M(K)$ has a unique decomposition (Jordan decomposition)

$$\mu \;=\; \mu_1^+ \;-\; \mu_1^- \;+\; i\,(\mu_2^+ \;-\; \mu_2^-)$$

such that $\mu_i^+, \mu_i^- \in M^+(K)$ for $i = 1,2$, and the pairs (μ_1^+, μ_i^-), $i = 1,2$ are mutually singular. Here $\mu_1^+ - \mu_1^-$ is the Hahn decomposition of $\mu_1 = \mathrm{Re}(\mu)$ and $\mu_2^+ - \mu_2^-$ that of $\mu_2 = \mathrm{Im}(\mu)$.

The canonical extension

$$\mu * \nu(f) \;=\; \int_{K \times K} \omega(x,y)(f)\; d\mu \otimes \nu(x,y)$$

defined in Theorem 1.1.3 and analyzed in Theorem 1.1.4 yields the existence of a positive continuous extension of the convolution ω. Theorem 1.1.4 also yields the uniqueness among all positive continuous extensions. Applying the Jordan decomposition we conclude.

Corollary 1.1.5 *Suppose that $\omega : K \times K \to M_c^1(K)$ fulfils the assumptions of Theorem 1.1.3. For $\mu, \nu \in M(K)$ define*

$$\mu * \nu(f) = \int_{K\times K} \omega(x,y)(f)\, d\mu \otimes \nu(x,y) \qquad \text{for } f \in C^b(K).$$

Then $(\mu,\nu) \mapsto \mu * \nu$, $M(K) \times M(K) \to M(K)$ is the unique bilinear extension of ω which is continuous with respect to the Bernoulli topology $\sigma(M(K), C^b(K))$.

Remark: Obviously, $\omega(a, \omega(b,c))$ as defined in Lemma 1.1.1 is equal to $\epsilon_a * \omega(b,c)$.

Now we extend the involution $x \mapsto \tilde{x}$ to $M(K)$. For $\mu \in M_c^1(K)$ we have already introduced $\tilde{\mu} \in M_c^1(K)$ in order to formulate axiom (H,˜).

Theorem 1.1.6 *Assume that axioms (H,ω) and (H,˜) are satisfied.*

(1) Define $\tilde{f}(x) = f(\tilde{x})$ for $f \in C^b(K)$ and $\tilde{\mu}(f) = \mu(\tilde{f})$ for $\mu \in M(K)$. Then $\mu \mapsto \tilde{\mu}$ is an isometric isomorphism from $M(K)$ onto $M(K)$.
(2) For $\mu, \nu \in M(K)$ we have $(\mu * \nu)^\sim = \tilde{\nu} * \tilde{\mu}$.

Proof.

(1) The first result is apparent. (Note that $\tilde{\mu}$ is the image measure of $\mu \in M^+(K)$ under the mapping $x \mapsto \tilde{x}$.)
(2) For $F \in C^b(K \times K)$ put $\tilde{F}(x,y) = F(\tilde{x}, \tilde{y})$. Applying the construction in Theorem 1.1.3 we have

$$\int_{K\times K} \tilde{F}(x,y)\, d\mu \otimes \nu(x,y) = \int_{K\times K} F(x,y)\, d\tilde{\mu} \otimes \tilde{\nu}(x,y)$$
$$= \int_{K\times K} F(y,x)\, d\tilde{\nu} \otimes \tilde{\mu}(y,x).$$

Hence

$$\mu * \nu(\tilde{f}) = \int_{K\times K} \omega(x,y)(\tilde{f})\, d\mu \otimes \nu(x,y) = \int_{K\times K} \omega(\tilde{y}, \tilde{x})(f)\, d\mu \otimes \nu(x,y)$$
$$= \int_{K\times K} \omega(y,x)(f)\, d\tilde{\nu} \otimes \tilde{\mu}(y,x) = \tilde{\nu} * \tilde{\mu}(f)$$

for each $f \in C_c(K)$. It follows $(\mu * \nu)^\sim = \tilde{\nu} * \tilde{\mu}$.

◇

Next we investigate the convolution of subsets of K. For $A, B \subseteq K$ define

$$A * B = \bigcup_{x \in A, y \in B} \operatorname{supp} \omega(x,y) \quad \text{and } \tilde{A} = \{\tilde{x} : x \in A\}.$$

Proposition 1.1.7 *Assume that the mapping $\omega : K \times K \to M_c^1(K)$ has the property that $(x,y) \mapsto supp\, \omega(x,y)$, $K \times K \to \mathcal{C}(K)$ is continuous with respect to the Michael topology. Then $A * B$ is a compact subset of K whenever A and B are compact subsets of K.*

Proof. The continuity of $(x,y) \mapsto \operatorname{supp}(\omega(x,y))$ implies that the collection $\{\operatorname{supp}(\omega(x,y)) : x \in A, y \in B\} \subseteq \mathcal{C}(K)$ is a compact subset in $\mathcal{C}(K)$, provided A and B are compact. Using the following Lemma we get that $A * B$ is a compact subset of K. ◇

Lemma 1.1.8 *Let K be a locally compact Hausdorff space and $\mathcal{C}(K) = \{C \subseteq K : C \text{ compact}, C \neq \emptyset\}$ equipped with the Michael topology. If $\mathfrak{A}$ is a compact subset of $\mathcal{C}(K)$, then $B = \bigcup_{A \in \mathfrak{A}} A$ is a compact subset of K.*

Proof. Let $\{U_i : i \in I\}$, $U_i \subseteq K$ open, such that $B \subseteq \bigcup_{i \in I} U_i$. Consider all finite unions $V = \bigcup_{k=1}^{n} U_{i_k}$, where $\{i_1, ..., i_n\} \subseteq I$ and index this set of V's by j's, $j \in J$. Consider $\{V_j : j \in J\}$. For each $A \in \mathfrak{A}$ there is some $j \in J$ such that $A \subseteq V_j$. The Michael topology is determined by the subbasis $\mathcal{C}_{U,V}$. Hence we conclude that $\{\mathcal{C}_{K,V_j} : j \in J\}$ is an open cover of $\mathfrak{A}$, and thus there exists a finite set $\{j_1, ..., j_n\} \subseteq J$ such that $\mathfrak{A} \subseteq \bigcup_{j=1}^{n} \mathcal{C}_{K,V_j}$. It follows $B \subseteq \bigcup_{l=1}^{n} V_{jl}$, where each V_{jl} is a finite union of certain U_{i_k}. Hence B is covered by finitely many U_i, $i \in I$. ◇

By Proposition 1.1.7 we obtain the following result concerning the support of $\nu * \mu$.

From now on, if not stated otherwise, K will be a hypergroup.

Proposition 1.1.9 *Let $\mu, \nu \in M^+(K)$. Then*

$$supp\ (\mu * \nu) \;=\; ((supp\ \mu) * (supp\ \nu))^- \qquad (^-\ \textit{denotes closure}).$$

If μ and ν have compact support, then $\mu * \nu$ also has compact support and

$$supp\ (\mu * \nu) \;=\; (supp\ \mu) * (supp\ \nu).$$

Proof. If $z \notin ((\operatorname{supp}\ \mu) * (\operatorname{supp}\ \nu))^-$ select a neighbourhood U of z such that $U \cap ((\operatorname{supp}\ \mu) * (\operatorname{supp}\ \nu))^- = \emptyset$. For every continuous function f with $\operatorname{supp} f \subseteq U$ we have $\omega(x,y)(f) = 0$ for all $x \in \operatorname{supp}\ \mu$, $y \in \operatorname{supp}\ \nu$. Hence $\mu * \nu(f) = 0$, i.e. $z \notin \operatorname{supp}\ (\mu * \nu)$.

In order to show $((\text{supp } \mu) * (\text{supp } \nu))^- \subseteq \text{supp } (\mu * \nu)$ let $z \in (\text{supp } \mu) * (\text{supp } \nu)$, i.e. $z \in \text{supp } \omega(x,y)$ for some $x \in \text{supp } \mu$, $y \in \text{supp } \nu$. Given a neighbourhood U of z there exists a continuous function $f \geq 0$ with $\text{supp } f \subseteq U$ and $\omega(x,y)(f) > 0$. Using Lemma 1.1.1(a) we get $\mu * \nu(f) > 0$, that means $z \in \text{supp } (\mu * \nu)$.
If supp μ and supp ν are compact subsets of K, we know by Proposition 1.1.7 that $(\text{supp } \mu) * (\text{supp } \nu)$ is compact, and hence closed. ◇

Lemma 1.1.10 *For $A, B, C \subseteq K$ we have*

(i) $(A * B) * C = A * (B * C)$.
(ii) $(A * B) \cap C = \emptyset$ *if and only if* $(\tilde{A} * C) \cap B = \emptyset$.

Proof.

(i) Let $a \in A$, $b \in B$, $c \in C$. Using Proposition 1.1.9 we get
$$(\{a\} * \{b\}) * \{c\} = \text{supp}\,(\omega(a,b)) * \text{supp}\,(\varepsilon_c) = \text{supp}\,(\omega(a,b) * \varepsilon_c),$$
and just in the same way
$$\{a\} * (\{b\} * \{c\}) = \text{supp}\,(\varepsilon_a * \omega(b,c)).$$
The associativity law of (H,ω) gives
$$(\{a\} * \{b\}) * \{c\} = \{a\} * (\{b\} * \{c\})$$
and now it is straightforward to derive (i).

(ii) Axiom (H,e) yields $e \in \tilde{A} * B$ if and only if there exists some $x \in A \cap B$, and Axiom (H,$\tilde{\ }$) gives $(A * B)^{\tilde{}} = \tilde{B} * \tilde{A}$.
Therefore we have the following equivalent conditions:
$$(A * B) \cap C \neq \emptyset \iff e \in \tilde{C} * (A * B) = (\tilde{C} * A) * B = (\tilde{A} * C)^{\tilde{}} * B$$
$$\iff (\tilde{A} * C) \cap B \neq \emptyset.$$

◇

Now we are able to prove the following result concerning the left translation L_x and right translation R_x.

Theorem 1.1.11 *Given $x, y \in K$ define for a continuous function f on K the left translation L_x and the right translation R_x*
$$L_x f(y) := \omega(x,y)(f) \qquad \text{respectively} \qquad R_x f(y) := \omega(y,x)(f).$$
If $f \in C_c(K)$, then $L_x f \in C_c(K)$ and $R_x f \in C_c(K)$. If $f \in C_0(K)$, then $L_x f \in C_0(K)$ and $R_x f \in C_0(K)$. Finally, if $f \in C^b(K)$ then $L_x f \in C^b(K)$ and $R_x f \in C^b(K)$. Moreover for $f \in C^b(K)$ we have $\|L_x f\|_\infty \leq \|f\|_\infty$ and $\|R_x f\|_\infty \leq \|f\|_\infty$.

Proof. By Lemma 1.1.1(a) we know that $L_x f$ and $R_x f$ are continuous functions. Further for $f \in C^b(K)$ obviously $\|L_x f\|_\infty \le \|f\|_\infty$ holds. Now let $f \in C_c(K)$. If $L_x f(y) \neq 0$ then $\operatorname{supp} f \cap \operatorname{supp}(\omega(x,y)) \neq \emptyset$, i. e. $\operatorname{supp} f \cap (\{x\} * \{y\}) \neq \emptyset$. By Lemma 1.1.10 (ii) this is equivalent to $y \in \{\tilde{x}\} * \operatorname{supp} f$. Therefore

$$\{y \in K : L_x f(y) \neq 0\} \subseteq \{\tilde{x}\} * \operatorname{supp} f,$$

and since $\operatorname{supp} f$ is compact, we see that $\operatorname{supp} L_x f$ is compact, i. e. $L_x f \in C_c(K)$. Finally by the denseness of $C_c(K)$ in $C_0(K)$, we also get that $L_x f \in C_0(K)$ for every $f \in C_0(K)$. In the same manner the assertions concerning the right translation are proven. ◇

Recall that we have a convolution on the Banach space $M(K)$, see Corollary 1.1.5. In fact we can state the following result:

Theorem 1.1.12 *With the convolution the measure space $M(K)$ is a Banach algebra with unit.*

Proof. We only check the associativity law. The associativity of condition (H,ω) is equivalent to

$$\omega(y,z)(L_x f) = \omega(x,y)(R_z f) \qquad (*)$$

for all $x, y, z \in K$ and $f \in C_0(K)$. Therefore we get for $\lambda, \mu, \nu \in M(K)$ and $f \in C_0(K)$

$$\begin{aligned}
\lambda * (\mu * \nu)(f) &= \int_K \int_K L_u f(v)\, d(\mu * \nu)(v)\, d\lambda(u) \\
&= \int_K \int_K \int_K \omega(y,z)(L_u f)\, d\mu(y)\, d\nu(z)\, d\lambda(u) \\
&= \int_K \int_K \int_K \omega(u,y)(R_z f)\, d\mu(y)\, d\nu(z)\, d\lambda(u) \\
&= \int_K \int_K R_z f(v)\, d(\lambda * \mu)(v)\, d\nu(z) \\
&= (\lambda * \mu) * \nu(f).
\end{aligned}$$

◇

Since the measures $\omega(x,y)$ have compact support, the associativity formula $(*)$ is valid also for $f \in C^b(K)$ and it can be written as $R_x L_y f(z) =$

$L_y R_x f(z)$ for all $x, y, z \in K$, $f \in C^b(K)$. Together with the obvious identity $R_z f(u) = L_u f(z)$ formula $(*)$ implies a decomposition formula for the composition of translation operators.

Proposition 1.1.13 *Given $x, y \in K$ and $f \in C^b(K)$ we have*

$$L_x \circ L_y f(z) = \int_K L_u f(z)\, d\omega(y,x)(u)$$

and

$$R_x \circ R_y f(z) = \int_K R_u f(z)\, d\omega(x,y)(u)$$

for all $z \in K$.

Proof. By $(*)$ we get

$$\begin{aligned} L_x \circ L_y f(z) &= \omega(x,z)(L_y f) = \omega(y,x)(R_z f) \\ &= \int_K R_z f(u)\, d\omega(y,x)(u) = \int_K L_u f(z)\, d\omega(y,x)(u). \end{aligned}$$

In the same way one shows the formula for the right translation. ◇

We have already used that

$$M_e^+(K) = \{\mu = \sum_{k=1}^{n} c_k \epsilon_{x_k} : x_k \in K,\ c_k \geq 0,\ n \in \mathbb{N}\}$$

is dense in $M^+(K)$ with respect to $\sigma(M^+(K), C_c(K) \cup \{1\}) = \sigma(M^+(K), C^b(K))$. For the space $M(K)$ of complex measures we can use the regularity of the measures to show that

$$M_c(K) := \{\mu \in M(K) : \text{supp}\ (|\mu|)\ \text{compact}\}$$

is dense in $M(K)$ with respect to the norm.

Lemma 1.1.14 *Let K be a locally compact Hausdorff space. Then $M_c(K)$ is dense in $M(K)$ with respect to $\|\,.\,\|$.*

Proof. Let $\mu \in M(K)$, $\varepsilon > 0$. Since μ is regular, there exists a compact subset $C \subseteq K$ such that $|\mu|(K) - |\mu|(C) < \varepsilon$. Setting $\mu_c(A) = \mu(A \cap C)$ for each $A \in \mathcal{B}(K)$ we get a measure $\mu_c \in M(K)$ with compact support, i.e. supp $(|\mu_c|)$ is compact. Obviously $|\mu|(K \backslash C) + |\mu|(C) = |\mu|(K)$, and we obtain for each $f \in C_0(K)$, $\|f\|_\infty \leq 1$,

$$\left| \int_K f(x)\, d(\mu - \mu_c)(x) \right| = \left| \int_{K\backslash C} f(x)\, d\mu(x) \right| \leq |\mu|\,(K\backslash C) < \varepsilon.$$

Hence $\|\mu - \mu_c\| < \varepsilon$. ◇

The translation by elements of the hypergroup K can be extended to a module operation of $M(K)$ on $C^b(K)$ and $C_0(K)$. Given $\mu \in M(K)$ and $f \in C^b(K)$ we define

$$\mu * f(x) := \int_K \omega(\tilde{y}, x)(f)\, d\mu(y) = \tilde{\mu}(R_x f)$$

and

$$f * \mu(x) := \int_K \omega(x, \tilde{y})(f)\, d\mu(y) = \tilde{\mu}(L_x f)$$

for $x \in K$. Note that $\epsilon_{\tilde{y}} * f = L_y f$ and $f * \epsilon_{\tilde{y}} = R_y f$.

Theorem 1.1.15 *Given $\mu \in M(K)$ define for $f \in C^b(K)$,*

$$L_\mu f(x) := \mu * f(x)$$

and

$$R_\mu f(x) := f * \mu(x).$$

L_μ and R_μ are bounded linear operators from $C^b(K)$ into $C^b(K)$ and from $C_0(K)$ into $C_0(K)$.

Moreover, $\|L_\mu f\|_\infty \leq \|\mu\| \|f\|_\infty$ and $\|R_\mu f\|_\infty \leq \|\mu\| \|f\|_\infty$ hold for every $f \in C^b(K)$.

If $\mu \in M(K)$ has compact support and $f \in C_c(K)$ then $L_\mu f \in C_c(K)$ and $R_\mu f \in C_c(K)$.

Proof. First we show that $L_\mu f \in C^b(K)$ for $f \in C^b(K)$ and $\mu \in M(K)$ with compact support. Let $x_0 \in K$, $\varepsilon > 0$. Applying the compactness of $\operatorname{supp} \mu$ we obtain by Lemma 1.1.1(a) a neighborhood U of x_0 such that

$$|\omega(\tilde{y}, x)(f) - \omega(\tilde{y}, x_0)(f)| < \varepsilon$$

for all $x \in U$ and $y \in \operatorname{supp} \mu$. Hence we get

$$|\mu * f(x) - \mu * f(x_0)| < \varepsilon \|\mu\|$$

for all $x \in U$, and $L_\mu f \in C^b(K)$ is shown. Moreover, since the measures $\mu \in M(K)$ with compact support are dense in $M(K)$, we obtain the continuity of $L_\mu f$, too. The inequality $\|L_\mu f\|_\infty \leq \|\mu\| \|f\|_\infty$ obviously holds.

Next we consider $f \in C_c(K)$ and $\mu \in M(K)$ with compact support. If $L_\mu f(x) \neq 0$ we have $\operatorname{supp} \tilde{\mu} \cap \operatorname{supp} R_x f \neq \emptyset$. Since $\operatorname{supp} R_x f \subseteq \operatorname{supp} f * \{\tilde{x}\}$,

we see
$\text{supp}\,\tilde{\mu} \cap (\text{supp}\, f * \{\tilde{x}\}) \neq \emptyset$. Appealing to Lemma 1.1.10 we get $\text{supp}\,\tilde{f} * \text{supp}\,\tilde{\mu} \cap \{\tilde{x}\} \neq \emptyset$, i. e. $x \in \text{supp}\,\mu * \text{supp}\, f$ by means of (H,˜). Therefore we have $L_\mu f \in C_c(K)$.

To deal with $f \in C_0(K)$, $f \neq 0$, and arbitrary $\mu \in M(K)$ select for given $\varepsilon > 0$ a measure $\nu \in M(K)$ with compact support, $\nu \neq 0$, such that $\|\mu - \nu\| < \varepsilon/\|f\|_\infty$ and then $g \in C_c(K)$ with $\|f - g\|_\infty < \varepsilon/\|\nu\|$. We get $\|L_\mu f - L_\mu g\|_\infty < 3\varepsilon$, and $L_\mu f \in C_0(K)$ follows.

In the same way one can prove the statements for R_μ. ◇

2 Some examples

After having collected the most basic facts of hypergroups we introduce a few examples. For applications in time series analysis the hypergroup structures defined on the discrete time axis are the most important. Therefore we start with the so-called polynomial hypergroups on $\mathbb{N}_0 = \mathbb{N} \cup \{0\}$. In this class there exist a lot of concrete examples, see chapter 4.5. Among them there are hypergroups which have properties very different from groups. Then we present an approach to hypergroups which are generated by product formulas for special functions. Chapter 8 contains many further examples.

2.1 *Polynomial hypergroups*

Consider an orthogonal polynomial sequence $(R_n(x))_{n\in\mathbb{N}_0}$ defined on the real axis with respect to a probability measure $\pi \in M^1(\mathbb{R})$, i. e. degree $R_n = n$ for each $n \in \mathbb{N}_0$ and

$$\int_{\mathbb{R}} R_n(x) R_m(x)\, d\pi(x) = \delta_{n,m} h_m^{-1} \qquad (h_m > 0).$$

We assume that $R_n(1) \neq 0$ and suppose therefore (eventually after renorming) that for each $n \in \mathbb{N}_0$

$$R_n(1) = 1.$$

We note that $(R_n(x))_{n\in\mathbb{N}_0}$ is not ortho**normal**. The degree-condition and the orthogonality imply a recurrence relation of the following form

$$R_1(x) R_n(x) = a_n R_{n+1}(x) + b_n R_n(x) + c_n R_{n-1}(x) \tag{1}$$

for $n \in \mathbb{N}$ and starting with

$$R_0(x) = 1, \qquad R_1(x) = \frac{1}{a_0}(x - b_0) \tag{2}$$

with $a_n > 0$ for all $n \in \mathbb{N}_0$, $c_n > 0$ for all $n \in \mathbb{N}$ and $b_n \in \mathbb{R}$ for all $n \in \mathbb{N}_0$. A well-known result, usually referred to as Favard's theorem but in fact due to Stone and Perron, states that a polynomial sequence $(R_n(x))_{n\in\mathbb{N}_0}$ recursively defined by equation (1) and (2) is indeed an orthogonal polynomial sequence with respect to a certain measure $\pi \in M^1(\mathbb{R})$, see [138]. Note that $R_n(1) = 1$ implies $a_n + b_n + c_n = 1$ for each $n \in \mathbb{N}$, and $a_0 + b_0 = 1$. We will now extend the recurrence relation (1) a little bit.

Lemma 1.2.1 *Let $(R_n(x))_{n\in\mathbb{N}_0}$ be an orthogonal polynomial sequence (with respect to $\pi \in M^1(\mathbb{R})$) such that $R_n(1) = 1$. Then the products $R_m(x)R_n(x)$ can be linearized by*

$$R_m(x)R_n(x) = \sum_{k=|n-m|}^{n+m} g(m,n;k)\, R_k(x) \qquad (L)$$

with $g(m,n;k) \in \mathbb{R}$ for $k = |n-m|, \ldots, n+m$. Moreover we have $g(m,n;|n-m|) \neq 0$ and $g(m,n;n+m) \neq 0$.

Proof. Obviously the polynomials $R_n(x)$ form a basis in the real vector space of all polynomials. With the degree condition we have a unique representation

$$R_m(x)R_n(x) = \sum_{k=0}^{n+m} g(m,n;k)R_k(x),$$

where $g(m,n;k) \in \mathbb{R}$ for $k = 0,\ldots,n+m$, and $g(m,n;n+m) \neq 0$. It remains to show that $g(m,n;k) = 0$ for $k = 0,\ldots,|n-m|-1$, and $g(m,n;|n-m|) \neq 0$. For the first statement we may assume $m < n$. Given $k < n-m$ we have degree $(R_mR_k) < n$. Hence

$$0 = \int_{\mathbb{R}} R_m(x)R_k(x)R_n(x)\,d\pi(x) = \sum_{j=0}^{n+m} g(m,n;j) \int_{\mathbb{R}} R_j(x)R_k(x)\,d\pi(x)$$
$$= g(m,n;k)h_k^{-1}.$$

Since $h_k > 0$ we have $g(m,n;k) = 0$ for $k = 0,\ldots,|n-m|-1$. Assuming $g(m,n;n-m) = 0$ we get

$$0 = \int_{\mathbb{R}} R_{n-m}(x)R_m(x)R_n(x)\,d\pi(x) = \sum_{k=|n-2m|}^{n} g(m,n-m;k)$$
$$\times \int_{\mathbb{R}} R_k(x)R_n(x)\,d\pi(x) = g(m,n-m;n)h_n^{-1},$$

which contradicts $h_n > 0$ and $g(m,n-m;n) \neq 0$. $\diamond$

Obviously the following identities hold for the linearization coefficients $g(m,n;k)$, $k = |n-m|,\ldots,n+m$:

(i) $g(m,n;k) = g(n,m;k)$ for all $n,m \in \mathbb{N}_0$.
(ii) $g(0,n;n) = g(n,0;n) = 1$ for all $n \in \mathbb{N}_0$.
(iii) $g(1,n;n+1) = a_n$, $g(1,n;n) = b_n$ and $g(1,n;n-1) = c_n$ for all $n \in \mathbb{N}$.
(iv) $\sum_{k=|n-m|}^{n+m} g(m,n;k) = 1$ for all $m,n \in \mathbb{N}_0$.

Furthermore we have

(v) $g(n,n;0) = h_n^{-1}$ for all $n \in \mathbb{N}_0$.

In fact we have

$$h_n^{-1} = \int_{\mathbb{R}} R_n^2(x)\, d\pi(x) = \sum_{k=0}^{2n} g(n,n;k) \int_{\mathbb{R}} R_k(x)\, d\pi(x) = g(n,n;0).$$

More generally we see

$$\begin{aligned}\int_{\mathbb{R}} R_m(x) R_n(x) R_k(x)\, d\pi(x) &= \sum_{j=|m-n|}^{n+m} g(m,n;j) \int_{\mathbb{R}} R_j(x) R_k(x)\, d\pi(x) \\ &= g(m,n;k) h_k^{-1}\end{aligned}$$

and linearizing $R_m(x)R_k(x)$, we obtain

$$\begin{aligned}\int_{\mathbb{R}} R_m(x) R_n(x) R_k(x)\, d\pi(x) &= \sum_{j=|m-k|}^{m+k} g(m,k;j) \int_{\mathbb{R}} R_j(x) R_n(x)\, d\pi(x) \\ &= g(m,k;n) h_n^{-1}.\end{aligned}$$

Thus we have shown

(vi) $g(m,n;k)h(n) = g(m,k;n)h(k)$ for all $n,m \in \mathbb{N}_0$.

For the edge-coefficients $g(m,n;|n-m|)$ and $g(m,n;n+m)$ we have a simple recursion formula:

(vii) $g(m,n;n+m) = g(m-1,n;n+m-1)\dfrac{a_{n+m-1}}{a_{m-1}}$,
$g(m,n;n-m) = g(m-1,n;n-m+1)\dfrac{c_{n-m+1}}{a_{m-1}}$
for all $2 \le m \le n$.

In fact first write

$$\begin{aligned}R_1(x)(R_{m-1}(x)R_n(x)) &= \sum_{k=n-m+1}^{n+m-1} g(m-1,n;k) R_1(x) R_k(x) \\ &= g(m-1,n;n+m-1) a_{n+m-1} R_{n+m}(x) + \dots,\end{aligned}$$

where the remaining summands following $R_{n+m}(x)$ have degree smaller than $n+m$. Writing secondly

$$\begin{aligned}(R_1(x)R_{m-1}(x))R_n(x) &= (a_{m-1}R_m(x) + b_{m-1}R_{m-1}(x) + c_{m-1}R_{m-2}(x))R_n(x) \\ &= a_{m-1} g(m,n;n+m) R_{n+m}(x) + \dots.\end{aligned}$$

we see $a_{m-1}g(m,n;n+m) = g(m-1,n;n+m-1)a_{n+m-1}$. In a similar way the second equation of (vii) is proved. Applying (vii) it is straightforward to derive

(viii) $$g(m,n;n+m) = \frac{a_n a_{n+1} a_{n+2} \cdot \ldots \cdot a_{n+m-1}}{a_1 a_2 \cdot \ldots \cdot a_{m-1}}$$
$$g(m,n;n-m) = \frac{c_n c_{n-1} c_{n-2} \cdot \ldots \cdot c_{n-m+1}}{a_1 a_2 \cdot \ldots \cdot a_{m-1}}$$
for all $2 \leq m \leq n$.

In [405] a set of recursion formulas for all coefficients $g(m,n;k)$, $k = |n-m|,\ldots,n+m$ is given. In fact a direct but somehow tedious calculation shows:

(ix) If $2 \leq m \leq n$ then
$$g(m,n;n+m-1) = \frac{g(m-1,n;n+m-2)a_{n+m-2}}{a_{m-1}} + \frac{g(m-1,n;n+m-1)[b_{n+m-1}-b_{m-1}]}{a_{m-1}},$$
and for $k = 2,\ldots,2m-2$
$$g(m,n;n+m-k) = \frac{g(m-1,n;n+m-k-1)a_{n+m-k-1}}{a_{m-1}} + \frac{g(m-1,n;n+m-k+1)c_{n+m-k+1}}{a_{m-1}} + \frac{g(m-1,n;n+m-k)[b_{n+m-k}-b_{m-1}]}{a_{m-1}} - \frac{g(m-2,n;n+m-k)c_{m-1}}{a_{m-1}},$$
$$g(m,n;n-m+1) = \frac{g(m-1,n;n-m+1)[b_{n-m+1}-b_{m-1}]}{a_{m-1}} + \frac{g(m-1,n;n-m+2)c_{n-m+2}}{a_{m-1}}.$$

In our context those orthogonal polynomial sequences $(R_n(x))_{n\in\mathbb{N}_0}$ with nonnegative coefficients $g(m,n;k)$ are most important, since each one of these polynomial sequences generates a hypergroup structure on $\mathbb{N}_0$.

Theorem 1.2.2 *Let $(R_n(x))_{n\in\mathbb{N}_0}$ be an orthogonal polynomial sequence (with respect to $\pi \in M^1(\mathbb{R})$) such that $R_n(1) = 1$. Further assume that the coefficients $g(m,n;k)$ of (L) satisfy*
$$g(m,n;k) \geq 0 \qquad \text{for } k = |m-n|,\ldots,m+n.$$

Defining a convolution on $\mathbb{N}_0$ (with the discrete topology) by

$$\omega(m,n) = \sum_{k=|n-m|}^{n+m} g(n,m;k)\epsilon_k \qquad \text{for each } n,m \in \mathbb{N}_0$$

and taking the identity-mapping as involution, i. e. $\tilde{n} := n$, *the triplet* $(\mathbb{N}_0, \omega, \tilde{\ })$ *is a hypergroup with unit element* 0.

Proof. Note that $\omega(m,n)$ are indeed probability measures on $\mathbb{N}_0$ with compact support. Furthermore by the preceding remarks we have $0 \in \operatorname{supp}(\omega(m,n))$ if, and only if $m = n$. It remains to check the associativity (H,ω). Let $e_k(j) := \delta_{k,j}$. Then $\omega(m,n)(e_k) = g(m,n;k)$ and thus

$$\begin{aligned}
\epsilon_\ell * \omega(m,n)(e_k) &= \sum_{j=0}^{\infty} \omega(\ell,j)(e_k)\, d\omega(m,n)(j) = \sum_{j=|n-m|}^{n+m} g(m,n;j)g(j,\ell;k) \\
&= h_k \sum_{j=|n-m|}^{n+m} g(m,n;j) \int_{\mathbb{R}} R_\ell(x)R_j(x)R_k(x)\, d\pi(x) \\
&= h_k \int_{\mathbb{R}} R_\ell(x)(R_m(x)R_n(x))R_k(x)\, d\pi(x) \\
&= h_k \int_{\mathbb{R}} (R_\ell(x)R_m(x))R_n(x)R_k(x)\, d\pi(x) \\
&= \sum_{j=|\ell-m|}^{\ell+m} g(\ell,m;j)g(j,n;k) = \omega(\ell,m) * \epsilon_n(e_k).
\end{aligned}$$

Finally considering linear combination of e_k's we see

$$\epsilon_\ell * \omega(m,n) = \omega(\ell,m) * \epsilon_n.$$

◇

We call such hypergroups **polynomial hypergroups** on $\mathbb{N}_0$ generated by $(R_n(x))_{n\in\mathbb{N}_0}$. In order to emphasize the dependence on the orthogonal polynomial sequence
$(R_n(x))_{n\in\mathbb{N}_0}$ we sometimes write $\omega = \omega_R$. Obviously $(\mathbb{N}_0, \omega_R, \tilde{\ }, 0)$ are commutative hypergroups. Due to Favards's theorem every commutative hypergroup on $\mathbb{N}_0$ with identity involution and zero as unit element and with

$$\{n-1, n+1\} \subseteq \operatorname{supp}(\omega(1,n)) \subseteq \{n-1, n, n+1\}$$

for every $n \in \mathbb{N}$ is a polynomial hypergroup generated by a certain orthogonal polynomial sequence $(R_n(x))_{n\in\mathbb{N}_0}$. But one should note that the corresponding orthogonal polynomials depend on the choice of the starting polynomial $R_1(x)$, i.e. on the choice of $a_0, b_0 \in \mathbb{R}$. There is an abundance of orthogonal polynomial sequences $(R_n(x))_{n\in\mathbb{N}_0}$ satisfying the condition $R_n(1) = 1$ for all $n \in \mathbb{N}_0$ and the crucial nonnegativity condition $g(m,n;k) \geq 0$ for $k = |n-m|, \ldots, n+m$ and all $n, m \in \mathbb{N}_0$. But generally it is not easy to show the nonnegativity of the $g(m,n;k)$ for concrete $(R_n(x))_{n\in\mathbb{N}_0}$. A striking example is generated by the Chebyshev polynomials of the first kind $(T_n(x))_{n\in\mathbb{N}_0}$. They are defined by

$$T_n(x) = \cos(nt) \qquad \text{for } x \in [-1,1],$$

where $x = \cos t$, $t \in [0,\pi]$. It is well known that $(T_n(x))_{n\in\mathbb{N}_0}$ is orthogonal with respect to $\frac{1}{\pi}\chi_{[-1,1]}(1-x^2)^{-1/2}\,dx$. Obviously $T_n(1) = 1$ and the addition theorem for the cosine-function gives

$$T_m(x)T_n(x) = \frac{1}{2}T_{|n-m|}(x) + \frac{1}{2}T_{n+m}(x)$$

for all $n, m \in \mathbb{N}_0$, i.e. $g(m,n;|n-m|) = \frac{1}{2} = g(m,n;n+m)$ and $g(m,n;k) = 0$ otherwise. The generated hypergroup is called **Chebyshev hypergroup of first kind** on $\mathbb{N}_0$. The convolution is given by

$$\omega(m,n) = \frac{1}{2}\epsilon_{|n-m|} + \frac{1}{2}\epsilon_{n+m}.$$

Another example of a polynomial hypergroup on $\mathbb{N}_0$ with the property

$$\operatorname{supp}(\omega(m,n)) = \{|n-m|, n+m\} \qquad \text{for } m,n \in \mathbb{N}$$

is a **cosh-hypergroup** on $\mathbb{N}_0$. Here the convolution is defined by

$$\omega(m,n) = \frac{\cosh(n-m)}{2\cosh(n)\cosh(m)}\epsilon_{|n-m|} + \frac{\cosh(n+m)}{2\cosh(n)\cosh(m)}\epsilon_{n+m}.$$

There is a whole class of cosh-hypergroups on $\mathbb{N}_0$ depending on parameters $a > 0$, see subsection 4.5.7.3. The corresponding orthogonal polynomials with $a \neq 1$ will also be studied in subsection 4.5.7.3.

Here we have to note that those orthogonal polynomials $R_n(x)$ defined by equation (1) with

$$a_n = \frac{\cosh(n+1)}{2\cosh(n)\cosh(1)}, \quad b_n = 0, \quad c_n = \frac{\cosh(n-1)}{2\cosh(n)\cosh(1)}$$

and $R_0(x) = 1$, $R_1(x) = x$, in fact satisfy

$$R_m(x)R_n(x) = g(m,n;n-m)R_{n-m}(x) + g(m,n;n+m)R_{n+m}(x)$$

for $1 \leq m \leq n$, where

$$g(m,n;n-m) = \frac{\cosh(n-m)}{2\cosh(n)\cosh(m)}, \quad g(m,n;n+m) = \frac{\cosh(n+m)}{2\cosh(n)\cosh(m)}.$$

Formula (viii) immediately gives the form of both edge coefficients. That all other coefficients $g(m,n;k)$, $k = |n-m|+1,\dots,n+m-1$ are zero can be directly computed from their recursion formula in (ix) by induction on m.

Usually for the orthogonal polynomials $R_n(x)$ we only know the recurrence relation (1), whereas the linearization coefficients $g(m,n;k)$ are not explicitly known. But often we can derive from some general results the nonnegativity of the $g(m,n;k)$-coefficients. These results are very important for deriving polynomial hypergroups. So we have enclosed a separate paragraph (subsection 4.4) where the interested reader can find the relevant results. Here we only give a further class, the **ultraspherical polynomials** $R_n^{(\alpha,\alpha)}$, where $\alpha \geq -1/2$ is a fixed parameter. They are orthogonal with respect to $c_\alpha \chi_{[-1,1]}(1-x^2)^\alpha\,dx$, where $c_\alpha = \dfrac{\Gamma(2\alpha+2)}{2^{2\alpha+1}(\Gamma(\alpha+1))^2}$. With the normalization $R_n^{(\alpha,\alpha)}(1) = 1$ the recurrence relation (1) writes

$$R_1^{(\alpha,\alpha)}(x)R_n^{(\alpha,\alpha)}(x) = \frac{n+2\alpha+1}{2n+2\alpha+1}R_{n+1}^{(\alpha,\alpha)}(x) + \frac{n}{2n+2\alpha+1}R_{n-1}^{(\alpha,\alpha)}(x) \qquad \text{for } n \in \mathbb{N},$$

with $R_0^{(\alpha,\alpha)}(x) = 1, \quad R_1^{(\alpha,\alpha)}(x) = x$. That is

$$a_n = \frac{n+2\alpha+1}{2n+2\alpha+1}, \quad b_n = 0, \quad c_n = \frac{n}{2n+2\alpha+1}$$

for $n \in \mathbb{N}$, whereas $a_0 = 1$, $b_0 = 0$. The weights $h_n = g(n,n,0)^{-1}$ can be easily calculated from (viii). In fact we have for $n \in \mathbb{N}$

$$h_n = \frac{(2n+2\alpha+1)(2\alpha+1)_n}{(2\alpha+1)\,n!} = \frac{(2n+2\alpha+1)\Gamma(2\alpha+1+n)}{(2\alpha+1)\,\Gamma(1+n)\,\Gamma(2\alpha+1)},$$

where $(a)_n = a\cdot(a+1)\cdot\dots\cdot(a+n-1)$, $(a)_0 = 1$ denotes the Pochhammer symbol. One can even compute the corresponding $g(m,n;k)$ explicitly. For $m \leq n$ we have

$$g(m,n;n+m-j) = 0 \qquad \text{for } j \in \{1,3,\dots,2m-1\}$$

and

$$g(m,n;n+m-2j) = \frac{(m+n+\alpha+\frac{1}{2}-2j)n!m!(\alpha+\frac{1}{2})_j(\alpha+\frac{1}{2})_{m-j}(\alpha+\frac{1}{2})_{n-j}(2\alpha+1)_{m+n-j}}{(m+n+\alpha+\frac{1}{2}-j)j!(m-j)!(n-j)!(\alpha+\frac{1}{2})_{m+n-j}(2\alpha+1)_n(2\alpha+1)_m}$$

for $j = 0,1,\dots,m$, see [405] or [18,(5.7)]. Within the ultraspherical polynomials we find well–known examples. For $\alpha = -\frac{1}{2}$ we get the Chebyshev polynomials $T_n(x)$, which we have already introduced before. For

$\alpha = 0$ we obtain the Legendre polynomials, and for $\alpha = \frac{1}{2}$ we get the Chebyshev polynomials of the second kind.

In the latter case of the Chebyshev polynomials of the second kind the linearization coefficients reduce to the simple form

$$g(m, n; n + m - 2j) = \frac{m + n + 1 - 2j}{(n + 1)(m + 1)}$$

for $1 \leq m \leq n$ and $j = 0, \ldots, m$.

2.2 *Hypergroups generated by families of multiplicative functions*

The essential tool to construct polynomial hypergroups in the previous section is the linearization of the product of orthogonal polynomials, see Lemma 1.2.1. We extend this concept to derive hypergroups on other spaces such as $\mathbb{R}_0^+ = [0, \infty[$ or $[-1, 1]$. A product formula for a family of special functions will be the basis for our construction.

Theorem 1.2.3 *Let S be a set and K a locally compact Hausdorff space. Suppose there is a family $(\varphi_x)_{x \in K}$ of complex-valued functions $\varphi_x : S \to \mathbb{C}$ such that the following hold:*

(F1) There exist $e \in K$ and $s_0 \in S$ such that $\varphi_e(s) = 1$ for all $s \in S$ and $\varphi_x(s_0) = 1$ for all $x \in K$.

(F2) There exists a homeomorphism $\tilde{}$ on K such that $\varphi_{\tilde{x}}(s) = \overline{\varphi_x(s)}$ for all $s \in S$.

(F3) The functions $x \mapsto \varphi_x(s)$ are bounded and continuous for every $s \in S$.

(F4) If for $\mu \in M(K)$ we have $\int_K \varphi_x(s)\, d\mu(x) = 0$ for all $s \in S$, then $\mu = 0$.

(F5) For each $x, y \in K$ there exists a regular positive Borel measure $\omega(x, y)$ on K with compact support such that

$$\varphi_x(s)\, \varphi_y(s) = \int_K \varphi_z(s)\, d\omega(x, y)(z) \qquad (P)$$

for every $s \in S$.

(F6) We have $e \in supp\,(\omega(x, \tilde{y}))$ if and only if $x = y$.

(F7) The mapping $(x, y) \mapsto supp\,(\omega(x, y))$, $K \times K \to \mathcal{C}(K)$ is continuous.

Then K furnished with ω as convolution, $\tilde{}$ as involution and e as unit element, is a commutative hypergroup.

Proof. Since

$$1 = \varphi_x(s_0)\,\varphi_y(s_0) = \int_K \varphi_z(s_0)\,d\omega(x,y)(z) = \omega(x,y)(K),$$

we have $\omega(x,y) \in M^1(K)$. To prove (H,ω) we show that given any compact subset $A \subseteq K$, the linear span $T(A)$ of the functions $\alpha_s(x) = \varphi_x(s),\ x \in A,\ s \in S,$ is dense in $C(A)$.

In fact assume $f \in C(A)\backslash\overline{T(A)}$. Then there exist some $\nu \in M(A)$ such that $\nu(f) \neq 0$ and $\nu(\alpha_s) = 0$ for all $s \in S$. Let μ be the trivial extension of the complex measure ν on A to a complex measure on K and denote by $g \in C^b(K)$ a continuous extension of f. Then $\mu(g) = \nu(f) \neq 0$ and for all $s \in S$

$$\int_K \varphi_x(s)\,d\mu(x) = \int_A \alpha_s(x)\,d\nu(x) = 0,$$

which contradicts (F4).

Now let $x_0, y_0 \in K$ and fix compact neighbourhoods U_0 of x_0 and V_0 of y_0, and denote by

$$A := \bigcup_{x\in U_0, y\in V_0} \operatorname{supp}\, \omega(x,y).$$

Applying (F7) the subset $A \subseteq K$ is compact. Given $f \in C_0(K)$ and $\varepsilon > 0$ choose $g \in T(A)$ such that $\sup\limits_{x\in A} |g(x) - f(x)| < \varepsilon.$

Then for every $x \in U_0,\ y \in V_0$ we obtain

$$\left|\int_K f(z)\,d\omega(x,y)(z) - \int_K f(z)\,d\omega(x_0,y_0)(z)\right|$$

$$\leq 2\varepsilon + \left|\int_K g(z)\,d\omega(x,y)(z) - \int_K g(z)\,d\omega(x_0,y_0)(z)\right|$$

Since g has the form

$$g(z) = \sum_{i=1}^n a_i\,\varphi_z(s_i), \qquad z \in A,$$

with $s_1, ..., s_n \in S,\ a_1, ..., a_n \in \mathbb{C},$ we get for all $x \in U_0,\ y \in V_0$

$$\int_K g(z)\,d\omega(x,y)(z) = \sum_{i=1}^n a_i\,\varphi_x(s_i)\,\varphi_y(s_i).$$

Therefore we can find neighbourhoods $U \subseteq U_0$ of x_0 and $V \subseteq V_0$ of y_0 such that

$$\left|\int_K f(z)\,d\omega(x,y)(z) - \int_K f(z)\,d\omega(x_0,y_0)(z)\right| \leq 3\varepsilon$$

for $x \in U, \quad y \in V$, proving that $(x,y) \mapsto \omega(x,y)$ is continuous with respect to $\sigma(M_c^1(K), C_0(K))$. Note that now the canonical extension of ω is well-defined. To check the associativity law we observe that

$$(\varphi_x(s)\,\varphi_y(s))\,\varphi_z(s) = \int_K \varphi_u(s)\,\varphi_z(s)\,d\omega(x,y)(u)$$

$$= \int_K \int_K \varphi_w(s)\,d\omega(u,z)(w)\,d\omega(x,y)(u)$$

and

$$\varphi_x(s)\,(\varphi_y(s)\,\varphi_z(s)) = \int_K \varphi_x(s)\,\varphi_v(s)\,d\omega(y,z)(v)$$

$$= \int_K \int_K \varphi_w(s)\,d\omega(x,v)(w)\,d\omega(y,z)(v).$$

Condition (F4) implies $\epsilon_x * \omega(y,z) = \omega(x,y) * \epsilon_z$, and axiom (H,$\omega$) is shown.

From $\varphi_{\tilde{\tilde{x}}}(s) = \varphi_x(s)$ and $\overline{\varphi_x(s)\,\varphi_y(s)} = \varphi_{\tilde{x}}(s)\,\varphi_{\tilde{y}}(s)$ axiom (H, $\tilde{}$) follows. The other conditions of hypergroups hold obviously. ◇

A simple example is the **Chebyshev hypergroup** on $\mathbb{R}_0^+$. Let $S = K = \mathbb{R}_0^+$ and $\varphi_x(s) = \cos(xs)$. Then formula (P) is a well-known trigonometric identity. In fact putting

$$\omega(x,y) := \frac{1}{2}\,\epsilon_{|x-y|} + \frac{1}{2}\,\epsilon_{x+y}$$

we have for all $s \in \mathbb{R}_0^+$

$$\cos(xs)\,\cos(ys) = \frac{1}{2}\,\cos(|x-y|s) + \frac{1}{2}\,\cos((x+y)s)$$

$$= \int_{\mathbb{R}_0^+} \cos(zs)\,d\omega(x,y)(z).$$

By Theorem 1.2.3 we get a commutative hypergroup on $\mathbb{R}_0^+$ with the identity as involution and zero as unit element. Condition (F4) is satisfied by means of the uniqueness theorem for the Fourier-Stieltjes transform on $\mathbb{R}$. In fact given $\mu \in M(\mathbb{R}_0^+)$ let $\nu \in M(\mathbb{R})$ such that $\mu = \nu|\mathbb{R}_0^+ + \nu_1|\,]0,\infty[$, where ν_1 is the image measure of ν under the mapping $x \mapsto -x$. Then

$$0 = \int_{\mathbb{R}_0^+} \cos(xs)\,d\mu(x) = \int_{\mathbb{R}} e^{ixs}\,d\nu(x)$$

for all $s \in \mathbb{R}_0^+$ implies $\int_{\mathbb{R}} e^{ixs} d\nu(x) = 0$ for all $s \in \mathbb{R}$.

Now the uniqueness theorem yields $\nu = 0$, which means $\mu = 0$.

Another example is a **cosh-hypergroup** on $\mathbb{R}_0^+$. Let $K = S = \mathbb{R}_0^+$ and $\varphi_x(s) = \cos(xs)/\cosh(x)$. Again formula (P) reduces to the trigonometric identity above. Put now

$$\omega(x,y) := \frac{\cosh(x-y)}{2\cosh(x)\cosh(y)}\,\epsilon_{|x-y|} + \frac{\cosh(x+y)}{2\cosh(x)\cosh(y)}\,\epsilon_{x+y}\,.$$

Then for all $s \in \mathbb{R}_0^+$

$$\frac{\cos(xs)}{\cosh(x)}\,\frac{\cos(ys)}{\cosh(y)} = \frac{1}{2\cosh(x)\cosh(y)}\,(\cos(|x-y|s) + \cos((x+y)s))$$

$$= \int_{\mathbb{R}_0^+} \frac{\cos(zs)}{\cosh(z)}\,d\omega(x,y)(z).$$

Condition (F4) follows again by the uniqueness theorem on $\mathbb{R}$.
If $\mu \in M(\mathbb{R}_0^+)$ satisfies for each $s \in \mathbb{R}_0^+$

$$\int_{\mathbb{R}_0^+} \frac{\cos(xs)}{\cosh(x)}\,d\mu(x) = 0$$

we obtain that $\frac{1}{\cosh}\,\mu$ is the zero measure, and hence $\mu = 0$. Therefore Theorem 1.2.3 yields the cosh-hypergroup on $\mathbb{R}_0^+$ with identity as involution and zero as unit element.

A very interesting class of hypergroups on $\mathbb{R}_0^+$ is generated by Bessel functions. Let $S = K = \mathbb{R}_0^+$. Given $\alpha > -\frac{1}{2}$ consider

$$\varphi_x(s) = j_\alpha(xs)$$

where

$$j_\alpha(z) = \sum_{k=0}^{\infty} \frac{(-1)^k\,\Gamma(\alpha+1)}{2^{2k}k!\,\Gamma(k+\alpha+1)}\,z^{2k}, \qquad z \in \mathbb{C},$$

is the normalized Bessel function of order α.

Formula (P) follows from Sonine's product formula for Bessel functions, see [743,11.41]. In fact this identity gives

$$\varphi_x(s)\,\varphi_y(s) = c_\alpha \int_0^\pi \varphi_{(x^2+y^2-2xy\cos\varphi)^{1/2}}(s)\,\sin^{2\alpha}\varphi\,d\varphi$$

with $c_\alpha = \dfrac{\Gamma(\alpha+1)}{\Gamma(\alpha+\frac{1}{2})\,\Gamma(\frac{1}{2})}$. We obtain (P) by substituting $z = (x^2+y^2-2xy\cos\varphi)^{1/2}$, which results in

$$\varphi_x(s)\,\varphi_y(s) = \int_{\mathbb{R}_0^+} \varphi_z(s)\,d\omega(x,y)(z),$$

where

$$d\omega(x,y)(z) = c_\alpha\, 2z\, \frac{[(x+y+z)(x+y-z)(x-y+z)(-x+y+z)]^{\alpha-\frac{1}{2}}}{(2xy)^{2\alpha}}$$
$$\times\, \chi_{[|x-y|,x+y]}(z)$$

Condition (F4) is valid, as is shown in [617,Lemma 1.1]. Taking the identity map as involution and zero as unit element we get a commutative hypergroup on $\mathbb{R}_0^+$ for each $\alpha > -\frac{1}{2}$. This hypergroup is called **Bessel-Kingman hypergroup** of order α. For a specified discussion of Bessel-Kingman hypergroups we refer to section 8.2, Example 1. The limiting case $\alpha \to -\frac{1}{2}$ leads to the Chebyshev hypergroup on $\mathbb{R}_0^+$ from above. For special values of α the normalized Bessel functions of order α are represented as combinations of algebraic and trigonometric polynomials:

$$j_{1/2}(z) \;=\; \frac{\sin z}{z}, \qquad j_{3/2}(z) \;=\; \frac{3}{z^3}\,(\sin z - z\cos z).$$

An important example of a hypergroup on $K = [-1,1]$ is obtained by considering the dual space of the hypergroup generated by the ultraspherical polynomials $R_n^{(\alpha,\alpha)}(x)$, where $\alpha \geq -\frac{1}{2}$, see the examples of polynomial hypergroups in section 1.2.1.

Now the variable x will be an element of the hypergroup $K = [-1,1]$ and the degree n will be a parameter. In fact, the hypergroup $K = [-1,1]$ is the dual hypergroup of the discrete hypergroup $\mathbb{N}_0$ generated by $R_n^{(\alpha,\alpha)}(x)$, see subsection 4.5. Briefly we give the details.

Let $\alpha > -\frac{1}{2}$ be fixed. For $x,y \in\,]-1,1[$ Gegenbauer's addition formula [207,p.177], [254] says

$$R_n^{(\alpha,\alpha)}(x)\, R_n^{(\alpha,\alpha)}(y) \;=\; \int_{-1}^{1} R_n^{(\alpha,\alpha)}(z)\, K_{x,y}(z)\, (1-z^2)^\alpha\, dz,$$

for all $n \in \mathbb{N}_0$, where

$$K_{x,y}(z) \;=\; c_\alpha\, \frac{(1-x^2-y^2-z^2+2xyz)^{\alpha-\frac{1}{2}}}{[(1-x^2)(1-y^2)(1-z^2)]^\alpha}\, \chi_{I_{x,y}}(z),$$

and

$$I_{x,y} \;=\; \left[xy - \sqrt{(1-x^2)(1-y^2)}\,,\, xy + \sqrt{(1-x^2)(1-y^2)}\,\right],$$

and

$$c_\alpha \;=\; \frac{\Gamma(\alpha+1)}{\Gamma(\alpha+\frac{1}{2})\,\Gamma(\frac{1}{2})}\,.$$

Now let $K = [-1,1]$, $S = \mathbb{N}_0$, and consider $\varphi_x(n) = R_n^{(\alpha,\alpha)}(x)$ for $x \in K$, $n \in S$. With $d\omega(x,y) = K_{x,y}(z)\,(1-z^2)^\alpha\, dz$ for $x,y \in\,]-1,1[$

and $\omega(-1,x) = \omega(x,-1) = \epsilon_{-x}$, $\omega(1,x) = \omega(x,1) = \epsilon_x$ for $x \in [-1,1]$, we get a commutative hypergroup on $[-1,1]$. The unit element is $x = 1$, the involution is the identity map. Condition (F4) follows by Weierstraß's density theorem. We call this hypergroup the dual ultraspherical hypergroup.

The case $\alpha = -\frac{1}{2}$ leads to the dual Chebyshev hypergroup of first kind. Here we have $T_n(x) = \cos nt$, where $x = \cos t$, $x \in [-1,1]$ and $t \in [0,\pi]$. The trigonometric identity

$$\cos(n\vartheta)\,\cos(n\sigma) \;=\; \frac{1}{2}\left[\cos(n(\vartheta-\sigma)) + \cos(n(\vartheta+\sigma))\right]$$

can be written as

$$T_n(x)\,T_n(y) \;=\; \int_{-1}^{1} T_n(z)\,d\omega(x,y)(z)$$

for every $n \in \mathbb{N}_0$, where $\omega(x,y)$ is the discrete measure

$$\omega(x,y) \;=\; \frac{1}{2}\,\epsilon_{xy-\sqrt{(1-x^2)(1-y^2)}} + \frac{1}{2}\,\epsilon_{xy+\sqrt{(1-x^2)(1-y^2)}}\,.$$

Obviously we get a hypergroup on $K = [-1,1]$ with $x = 1$ as unit element and the identity map as involution.

Chapter 2

Basics of harmonic analysis on hypergroups

After having learned some outstanding examples of hypergroups we introduce basic tools of harmonic analysis which will be essential later on.

1 Haar measures

As before K will be a hypergroup. The regular Borel measures on K with values in $[0,\infty]$ are denoted by $M^\infty(K)$. We recall Riesz's representation theorem stating that for every positive linear functional Λ on $C_c(K)$ there exists a unique regular positive Borel measure $\mu \in M^\infty(K)$, such that

$$\Lambda(f) = \int_K f(x)\, d\mu(x)$$

for all $f \in C_c(K)$.

Definition. A regular positive Borel measure $m \in M^\infty(K)$, $m \neq 0$ is called **left-invariant**, if

$$\int_K f(x)\, dm(x) = \int_K L_y f(x)\, dm(x)$$

for all $f \in C_c(K)$, $y \in K$. There is of course an analogous definition of **right-invariant**.

Soon we will see that the left-invariance (respectively right-invariance) determines $m \in M^\infty(K)$ uniquely up to a multiplicative positive constant. We will call a left-invariant measure $m \in M^\infty(K), m \neq 0$, a **Haar measure** on K. Recently it was shown that on each locally compact hypergroup there exists a Haar measure. This result due to Chapovsky is not yet published in a mathematical journal. But it is available in the ArXiv. For K discrete or compact or commutative the existence of a Haar measure on K is known. For the discrete case the proof of existence is rather simple.

Theorem 2.1.1 *Let K be a discrete hypergroup. There exists a Haar measure m on K. If we assume $m(\{e\}) = 1$ we have*

$$m(\{x\}) = (\omega(\tilde{x}, x)(\{e\}))^{-1}$$

for all $x \in K$.

Proof. Using the associativity we obtain for $x, y, z \in K$

$$\sum_{t \in K} \omega(x,t)(\{e\})\,\omega(y,z)(\{t\}) = \epsilon_x * \omega(y,z)(\{e\})$$

$$= \omega(x,y) * \epsilon_z(\{e\}) = \sum_{t \in K} \omega(x,y)(\{t\})\,\omega(t,z)(\{e\})$$

By axiom (H,e) we have $\omega(x,y)(\{e\}) > 0$ if, and only if $x = \tilde{y}$. Hence the summation above reduces to

$$\omega(x,\tilde{x})(\{e\})\,\omega(y,z)(\{\tilde{x}\}) = \omega(x,y)(\{\tilde{z}\})\,\omega(\tilde{z},z)(\{e\}).$$

Now define

$$h(x) := (\omega(\tilde{x},x)(\{e\}))^{-1} \qquad \text{for } x \in K$$

and define the measure m by

$$m(E) = \sum_{x \in E} h(x) \qquad \text{for } E \subseteq K,\ E \text{ countable, and } m(E) = \infty \text{ else.}$$

The equation above now writes as

$$h(z)\omega(y,z)(\{\tilde{x}\}) = h(\tilde{x})\omega(x,y)(\{\tilde{z}\}),$$

and we get

$$\sum_{z \in K} h(z)\,\omega(y,z)(\{\tilde{x}\}) = \sum_{z \in K} h(\tilde{x})\,\omega(x,y)(\{\tilde{z}\})$$

$$= h(\tilde{x}) \sum_{z \in K} \omega(x,y)(\{\tilde{z}\}) = h(\tilde{x}) = m(\{\tilde{x}\}).$$

That is

$$\sum_{z \in K} L_y\, \chi_{\{\tilde{x}\}}(z)\, h(z) \;=\; \sum_{z \in K} \chi_{\{\tilde{x}\}}(z)\, h(z).$$

◇

In the case of polynomial hypergroups $K = \mathbb{N}_0$ the Haar measure m is the counting measure with weights $h(n) = g(n,n;0)^{-1}$, i. e. $m(E) = \sum_{n \in E} g(n,n;0)^{-1}$ for each $E \subseteq \mathbb{N}_0$. Note that we assume throughout that $h(0) = 1$, i. e. the orthogonalization measure π is a probability measure.

Now we consider the commutative compact case.

Proposition 2.1.2 *Let K be a compact hypergroup, suppose $f \in C(K)$ and define K_f to be the closure of the convex hull of $\{L_x f : x \in K\}$. Then*

(i) $x \mapsto L_x f$ *is a continuous mapping from K into $C(K)$,*
(ii) $L_x \circ L_y f = \int_K L_u f \, d\omega(y,x)(u)$ *for all $x, y \in K$ (the integral is a $C(K)$-valued Riemann-integral),*
(iii) K_f *is a convex compact subset of $C(K)$,*
(iv) $L_x(K_f) \subseteq K_f$ *for every $x \in K$.*

Proof.

(i) Applying Lemma 1.1.1(a) the compactness of K yields the continuity of $x \mapsto L_x f$.
(ii) By Proposition 1.1.13 follows that the $C(K)$-valued integral coincides with $L_x \circ L_y f$.
(iii) By (i) the set $\{L_x f : x \in K\}$ is compact. Hence the convex hull $\mathrm{co}\{L_x f : x \in K\}$ is totally bounded, see [609], Theorem 3.24, and therefore K_f is compact.
(iv) To prove $L_x(K_f) \subseteq K_f$ it is sufficient to show $L_x(L_y f) \in K_f$. Approximating in norm the $C(K)$-valued integral $\int_K L_u f \, d\omega(y,x)(u)$ by Riemann sums $\sum_i \omega(y,x)(E_i) L_{u_i} f$, where $\{E_i\}$ is a partition of K, statement (iv) follows by (ii).

◇

Now we apply the fixed point theorem of Markov-Kakutani [189,p.456] to derive a Haar measure for every commutative and compact hypergroup K.

Theorem 2.1.3 *Let K be a commutative and compact hypergroup. Then there exists a Haar measure m on K.*

Proof. For $f \in C(K)$ consider K_f as in Proposition 2.1.2. The family $\{L_x : x \in K\}$ of contracting linear operators $L_x : C(K) \to C(K)$ maps K_f into K_f and commutes. Hence by the fixed point theorem of Markov-Kakutani each K_f contains a function φ such that $L_x \varphi = \varphi$ for all $x \in K$. In particular, $\varphi(x) = \varphi(e)$, so that φ is constant. So far we have shown that to each $f \in C(K)$ corresponds at least one constant c which can be uniformly approximated on K by convex combinations of translates of f. Assume c' is another constant with the same property. We claim that $c = c'$. Pick $\varepsilon > 0$. There exist numbers $\lambda_1, \ldots, \lambda_n > 0$ and $\mu_1, \ldots, \mu_m > 0$ with

$\sum_{i=1}^n \lambda_i = 1 = \sum_{j=1}^m \mu_j$ and elements $x_1, \ldots, x_n \in K$ and $y_1, \ldots, y_m \in K$ such that

$$|c - \sum_{i=1}^{n} \lambda_i L_{x_i} f(y)| < \varepsilon \qquad \text{for all } y \in K.$$

and

$$|c' - \sum_{j=1}^{m} \mu_j L_{y_j} f(x)| < \varepsilon \qquad \text{for all } x \in K.$$

Put in the first inequality $y = y_j$ multiply by μ_j and add with respect to j. The result is

$$|c - \sum_{j=1}^{m} \sum_{i=1}^{n} \lambda_i \mu_j \omega(x_i, y_j)(f)| < \varepsilon.$$

Putting in the second inequality $x = x_i$ multiplying by λ_i and adding with respect to i gives

$$|c' - \sum_{i=1}^{n} \sum_{j=1}^{m} \lambda_i \mu_j \omega(y_j, x_i)(f)| < \varepsilon.$$

The commutativity of K implies $c = c'$.

It follows that to every $f \in C(K)$ corresponds a unique number $M(f) \in \mathbb{C}$ which can be uniformly approximated on K by convex combinations of translates of f. The following properties of M are obvious:

$$M(f) \geq 0 \quad \text{for } f \geq 0, \qquad\qquad M(1) = 1.$$

Further $M(\alpha f) = \alpha M(f)$ since $\alpha M(f) \in K_{\alpha f}$ for each $\alpha \in \mathbb{C}$ and $M(L_x f) = M(f)$ since $K_{L_x f} \subseteq K_f$.

Also one can prove that $M(f) + M(g)$ is contained in K_{f+g}. Hence $M(f) + M(g) = M(f + g)$. Now Riesz's representation theorem yields a regular positive probability measure $m \in M^\infty(K)$ which is left-invariant and satisfies

$$M(f) = \int_K f(x)\, dm(x)$$

for all $f \in C(K)$. $\diamond$

The proof of the existence of a Haar measure in the commutative (noncompact) case is rather involved and is given in [633]. An outline can also be found in [96], Theorem 1.3.15. The compact (noncommutative) case is dealt with in [96], Theorem 1.3.28, too.

So far we have studied the existence of a (left-invariant) Haar measure on K. As in the locally compact group case the Haar measure is unique up to a positive constant. But first we derive some basic properties of left-invariant Haar measures.

Proposition 2.1.4 *Let m be a left-invariant Haar measure on a hypergroup K. Then*

(i) $m(C) \leq m(\{x\} * C)$ *for every compact subset C of K,*
(ii) $m(V) > 0$ *for every open subset V of K ($V \neq \emptyset$), i. e. supp $m = K$.*

Proof.

(i) We already know that $\{x\} * C$ is compact, if C is compact. By the regularity of m, for $\varepsilon > 0$ there exists an open set $U \subseteq K$ such that $\{x\} * C \subseteq U$ and $m(U \setminus (\{x\} * C)) < \varepsilon$. Furthermore there exists $f \in C_c(K)$, $0 \leq f \leq 1$, $f \,|\, \{x\} * C = 1$, $f \,|\, K \setminus U = 0$. Hence

$$\int_K f(y)\, dm(y) - m(\{x\} * C) \leq m(U \setminus (\{x\} * C)) < \varepsilon.$$

For $y \in C$ follows $L_x f(y) = \omega(x, y)(f) = 1$, since $f|\mathrm{supp}\,(\omega(x, y)) = 1$. Thus we get by $L_x f \geq 0$

$$m(C) \leq \int_K L_x f(y)\, dm(y) = \int_K f(y)\, dm(y) \leq m(\{x\} * C) + \varepsilon.$$

As $\varepsilon > 0$ tends to zero, we have $m(C) \leq m(\{x\} * C)$.

(ii) Choose $x \in$ supp m. For $f \in C_c(K)$, $f \geq 0$ with $f(e) > 0$ follows $L_{\tilde{x}} f(x) = \omega(\tilde{x}, x)(f) > 0$ by (H,e). Since $L_{\tilde{x}} f$ is nonnegative and continuous we get

$$\int_K f(y)\, dm(y) = \int_K L_{\tilde{x}} f(y)\, dm(y) > 0.$$

This means $e \in$ supp m. Now consider an arbitrary $z \in K$ and $f \in C_c(K), f \geq 0$ with $f(z) > 0$. By $L_z f(e) = f(z) > 0$, we have

$$\int_K f(y)\, dm(y) = \int_K L_z f(y)\, dm(y) > 0,$$

where we have used $e \in$ supp m. Therefore $z \in$ supp m.

◇

Lemma 2.1.5 *Let $f \in C_c(K)$. Then for each $\varepsilon > 0$ there exists an open neighborhood W of e, such that $|f(x) - f(y)| < \varepsilon$ for all $x, y \in K$ with $\omega(\tilde{x}, y)(W) > 0$.*

Proof. Choose a compact neighborhood U of e with $\tilde{U} = U$, and consider $C = U * \operatorname{supp} f$. The set C is compact. Given $x \in C$ there exists a neighborhood U_x of x such that $|f(x) - f(y)| < \varepsilon/2$ for every $y \in U_x$. By (H3) there exists a neighborhood V_x of e such that $\{x\} * V_x * V_x \subseteq U_x$, and by the compactness of C follows $C \subseteq \bigcup_{i=1}^n \{x_i\} * V_{x_i}$. Put $W = U \cap (\bigcap_{i=1}^n V_{x_i})$. First notice that $\omega(\tilde{x}, y)(W) > 0$ is equivalent to $y \in \{x\} * W$ by Lemma 1.1.10. Now consider $x \notin C = U * \operatorname{supp} f$. For $y \in \{x\} * W$ we see that $y \notin \operatorname{supp} f$. In fact $y \in \operatorname{supp} f$ would imply $\{x\} * U \cap \operatorname{supp} f \neq \emptyset$ and then $x \in U * \operatorname{supp} f$ by Lemma 1.1.10. Therefore $f(x) = 0 = f(y)$, and the statement is shown for the case $x \notin C$. For $x \in C$ we have $x \in \{x_i\} * V_{x_i}$ for some i. Hence $\omega(\tilde{x}, y)(W) > 0$ implies $y \in \{x\} * W \subseteq \{x_i\} * V_{x_i} * W \subseteq \{x_i\} * V_{x_i} * V_{x_i}$. Thus $|f(x) - f(y)| \leq |f(x) - f(x_i)| + |f(x_i) - f(y)| < \varepsilon$. ◇

For a (left-invariant) Haar measure m and $f \in C_c(K)$ we get a bounded regular Borel measure $fm \in M(K)$, where $fm(E) := \int\limits_E f(x)\, dm(x)$, E Borel set in K.

Here and subsequently m is a left-invariant Haar measure on K.

Proposition 2.1.6 *Let $(k_i)_{i \in I}$ be a net of functions of $C_c(K)$ with $k_i \geq 0$, $\int\limits_K k_i(x)\, dm(x) = 1$ for every $i \in I$ and $\operatorname{supp} k_i \to \{e\}$, (i. e. for every neighborhood U of e there is an $i_0 \in I$ with $\operatorname{supp} k_i \subseteq U$ for all $i \geq i_0$). For every $f \in C_c(K)$ we have*

$$\lim_i \|(fm) * \widetilde{k_i} - f\|_\infty = 0,$$

where $\widetilde{k_i}(x) = k_i(\tilde{x})$.

Proof. Let $\varepsilon > 0$. Applying Lemma 2.1.5 there exists $i_0 \in I$ such that for all $i \geq i_0$ the following holds:

$$|f(x) - f(y)| < \varepsilon \qquad \text{for all } x, y \in K \text{ with } \omega(\tilde{x}, y)(k_i) > 0.$$

Hence for $i \geq i_0$ and each $x \in K$ we obtain

$$\begin{aligned} \left|(fm) * \widetilde{k_i}(x) - f(x)\right| &= \left|\int\limits_K \omega(\tilde{y}, x)(\widetilde{k_i}) f(y)\, dm(y) - f(x) \int\limits_K L_{\tilde{x}} k_i(y)\, dm(y)\right| \\ &\leq \int\limits_K \omega(\tilde{x}, y)(k_i) |f(y) - f(x)|\, dm(y) < \varepsilon \int\limits_K L_{\tilde{x}} k_i(y)\, dm(y) = \varepsilon. \end{aligned}$$

◇

The existence of nets $(k_i)_{i\in I}$ as in Proposition 2.1.6 (even with $\widetilde{k_i} = k_i$) can be derived by the Lemma of Urysohn. From Proposition 2.1.6 follows in particular:

Lemma 2.1.7 *Let $(k_i)_{i\in I}$ be a net of functions as in Proposition 2.1.6, suppose $\mu \in M(K)$. Then for every $h \in C_c(K)$*

$$\lim_i \int_K h(x)(\mu * k_i)(x)\, dm(x) = \int_K h(x)\, d\mu(x).$$

Proof. First we note an identity valid for $\mu, \nu \in M(K)$ and $f \in C_0(K)$:

$$\int_K L_\mu f(x)\, d\nu(x) = \int_K f(x)\, d\tilde{\mu} * \nu(x). \qquad (*)$$

In fact

$$\begin{aligned}\int_K L_\mu f(x)\, d\nu(x) &= \int_K \int_K \omega(\tilde{y}, x)(f)\, d\mu(y)\, d\nu(x)\\ &= \int_K \int_K \omega(y, x)(f)\, d\tilde{\mu}(y)\, d\nu(x) = \int_K f(x)\, d\tilde{\mu} * \nu(x).\end{aligned}$$

Hence we obtain by Proposition 2.1.6

$$\begin{aligned}\lim_i \int_K h(x)(\mu * k_i)(x)\, dm(x) &= \lim_i \int_K (\mu * k_i)(x)\, d(hm)(x)\\ = \lim_i \int_K k_i(x)\, d(\tilde{\mu} * (hm))(x) &= \lim_i \int_K \widetilde{k_i}(x)\, d\left(((hm)^{\sim} * \mu)\right)(x)\\ = \lim_i \int_K (hm) * \widetilde{k_i}(x)\, d\mu(x) &= \int_K h(x)\, d\mu(x).\end{aligned}$$

◇

Now we can show a very important property of the Haar measure.

Theorem 2.1.8 *Let $f, g \in C_c(K)$. Then for every $x \in K$*

$$\int_K L_x f(y) g(y)\, dm(y) = \int_K f(y) L_{\tilde{x}} g(y)\, dm(y).$$

Proof. Applying identity $(*)$ of Lemma 2.1.7 and bringing in a net $(k_i)_{i\in I}$ as in Proposition 2.1.6 with $\widetilde{k_i} = k_i$ we get

$$\int_K L_x f(y) g(y)\, dm(y) = \int_K \epsilon_{\tilde{x}} * f(y)\, d(gm)(y) = \int_K f(y)\, d(\epsilon_x * (gm))(y)$$

$$= \lim_i \int_K ((fm) * k_i)(y)\, d(\epsilon_x * (gm))(y)$$

$$= \lim_i \int_K (\epsilon_{\tilde{x}} * (fm) * k_i)(y) g(y)\, dm(y) = \int_K g(y)\, d(\epsilon_{\tilde{x}} * (fm))(y)$$

$$= \int_K (\epsilon_x * g)(y)\, d(fm)(y) = \int_K f(y) L_{\tilde{x}} g(y)\, dm(y).$$

$\diamond$

Remark: It is obvious that the statement of Theorem 2.1.8 is true for $f \in C_c(K)$ and $g \in C^b(K)$.

With Theorem 2.1.8 we are able to prove the following uniqueness theorem.

Theorem 2.1.9 *Let n be another left-invariant Haar measure on K. Then there exists a positive number c such that $n = cm$.*

Proof. In Lemma 2.1.5 we have shown that for $f \in C_c(K)$ and $\varepsilon > 0$ there exists a neighborhood W of e, such that $|f(y) - f(x)| < \varepsilon$ for all $x, y \in K$ with $y \in \{x\} * W$. Thus for every $x \in K$ and $z \in W$ follows

$$|R_z f(x) - f(x)| = \left| \int_K f(y)\, d\omega(x,z)(y) - f(x) \right|$$

$$\leq \int_K |f(y) - f(x)|\, d\omega(x,z)(y) < \varepsilon.$$

In particular by the theorem of dominated convergence we get

$$\lim_{z \to e} \int_K |R_z f(x) - f(x)|\, dn(x) = 0.$$

Now consider $f, g \in C_c(K)$, f and g nonzero and nonnegative, suppose $\varepsilon > 0$. Then there exists a neighborhood U of e with

$$\int_K |R_y f(x) - f(x)|\, dn(x) < \frac{\varepsilon}{2} \int_K f(x)\, dm(x)$$

and

$$\int_K |R_y g(x) - g(x)|\, dn(x) < \frac{\varepsilon}{2} \int_K g(x)\, dm(x)$$

for every $y \in U$. Further choose $h \in C_c(K)$ with $0 \le h \le 1,\ h = \tilde{h},\ h \,|\, K \setminus U = 0$ and $\int_K h(x)\, dm(x) > 0$. Hence by Theorem 2.1.8

$$\int_K h(x)\, dn(x) \int_K f(y)\, dm(y) = \int_K f(y) \int_K L_{\tilde{y}} h(x)\, dn(x)\, dm(y)$$
$$= \int_K \int_K f(y) \omega(\tilde{x}, y)(\tilde{h})\, dm(y)\, dn(x) = \int_K \int_K f(y) L_{\tilde{x}} h(y)\, dm(y)\, dn(x)$$
$$= \int_K L_x f(y) h(y)\, dm(y)\, dn(x) = \int_K h(y) \int_K R_y f(x)\, dn(x)\, dm(y).$$

Now $h(y) = 0$ for $y \in K \setminus U$. Thus

$$\left| \int_K h(y)\, dm(y) \int_K f(x)\, dn(x) - \int_K h(x)\, dn(x) \int_K f(y)\, dm(y) \right|$$
$$= \left| \int_K h(y) \int_K (f(x) - R_y f(x))\, dn(x)\, dm(y) \right|$$
$$< \frac{\varepsilon}{2} \int_K f(x)\, dm(x) \int_K h(y)\, dm(y).$$

That means

$$\left| \int_K f(x)\, dn(x) \Big/ \int_K f(x)\, dm(x) - \int_K h(x)\, dn(x) \Big/ \int_K h(x)\, dm(x) \right| < \frac{\varepsilon}{2}.$$

The same inequalitiy is valid for g and h. Combining both inequalities gives

$$\left| \int_K f(x)\, dn(x) \Big/ \int_K f(x)\, dm(x) - \int_K g(x)\, dn(x) \Big/ \int_K g(x)\, dm(x) \right| < \varepsilon.$$

This finally implies $n = cm$ for some $c > 0$. ◇

Finally we want to study how the left-invariant Haar measures fail to be right-invariant. Let m be a left-invariant Haar measure on K. Fix $x \in K$ and consider the positive linear functional Λ_x on $C_c(K)$ defined by

$$\Lambda_x(f) := \int_K R_{\tilde{x}} f(y)\, dm(y)$$

and let $m_x \in M^\infty(K)$ be the corresponding regular positive Borel measure on K. By the associative law (H1) we obtain

$$\int_K L_y f(z)\, dm_x(z) = \int_K R_{\tilde{x}} L_y f(z)\, dm(z) = \int_K L_y R_{\tilde{x}} f(z)\, dm(z)$$

$$= \int_K R_{\tilde{x}} f(z)\, dm(z) = \int_K f(z)\, dm_x(z)$$

for all $f \in C_c(K)$. Hence m_x is again a left-invariant Haar measure on K. By Theorem 2.1.9 there exists a number $\Delta(x) > 0$ such that $m_x = \Delta(x)\, m$. Note that $\Delta(x)$ is independent of the original choice of m. The function $\Delta : K \to]0, \infty[$ is called the **modular function** of K. By the definition of Δ and by approximating $f \in L^1(K)$ by functions of $C_c(K)$ we have

$$\int_K R_x f(y)\, dm(y) = \Delta(\tilde{x}) \int_K f(y)\, dm(y)$$

for every $f \in L^1(K)$, $x \in K$.

From Lemma 2.1.5 we obtain easily that $x \mapsto R_x f$ and $x \mapsto L_x f$ are uniformly continuous maps from K into $C_c(K)$ for each $f \in C_c(K)$. Hence $x \mapsto \int_K R_x f(y)\, dm(y)$ is continuous from K in $\mathbb{C}$, and so the modular function Δ is continuous. Furthermore, for $f \in C_c(K)$, $f \neq 0$, we have

$$\int_K R_x(R_y f)(z)\, dm(z) = \Delta(\tilde{x}) \int_K R_y f(z)\, dm(z) = \Delta(\tilde{x})\Delta(\tilde{y}) \int_K f(z)\, dm(z),$$

and by Fubini's theorem

$$\int_K \int_K R_u f(z)\, d\omega(x,y)(u)\, dm(z) = \int_K \int_K R_u f(z)\, dm(z)\, d\omega(x,y)(u)$$

$$= \int_K \Delta(\tilde{u})\, d\omega(x,y)(u) \int_K f(z)\, dm(z) = \omega(x,y)(\tilde{\Delta}) \int_K f(z)\, dm(z).$$

Hence $\Delta(x)\Delta(y) = \omega(x,y)(\Delta)$.

Let $f, g \in C_c(K)$, $f \neq 0$. We have $L_{\tilde{x}}\tilde{g}(y) = L_{\tilde{y}} g(x)$ and then by Theorem 2.1.8

$$\int_K f(x)\, dm(x) \int_K g(y)\, d\tilde{m}(y) = \int_K f(x)\, dm(x) \int_K \tilde{g}(y)\, dm(y)$$

$$= \int_K \int_K f(x)\, L_{\tilde{x}}\tilde{g}(y)\, dm(y)\, dm(x)$$

$$= \int_K \int_K f(x)\, L_{\tilde{y}} g(x)\, dm(x)\, dm(y) = \int_K \int_K L_y f(x)\, g(x)\, dm(x)\, dm(y)$$

$$= \int_K g(x) \int_K R_x f(y)\, dm(y)\, dm(x) = \int_K f(y)\, dm(y) \int_K \Delta(\tilde{x})\, g(x)\, dm(x).$$

Hence $\tilde{m} = \tilde{\Delta} m$. It follows $m = \tilde{\tilde{m}} = (\tilde{\Delta} m)^{\tilde{}} = \Delta \tilde{m} = \Delta\tilde{\Delta} m$, i.e. $\Delta\tilde{\Delta} = 1$. We collect the properties of Δ, shown above, in a theorem.

Theorem 2.1.10 *Let m be a left-invariant Haar measure on K. The modular function $\Delta : K \to\,]0,\infty[$ is uniquely determined by the equation*

$$\int_K R_x f(y)\, dm(y) \;=\; \Delta(\tilde{x}) \int_K f(y)\, dm(y) \qquad \textit{for every } f \in L^1(K).$$

The modular function is continuous and satisfies

(1) $\omega(x,y)(\Delta) \;=\; \Delta(x)\,\Delta(y) \;=\; \omega(y,x)(\Delta)$ *for each* $x,y \in K$,
(2) $m = \Delta\tilde{m}$ *and* $\Delta\tilde{\Delta} = 1$.

Moreover, we can derive the following fact.

Proposition 2.1.11 *For each pair $x,y \in K$ the modular function Δ is constant on supp $\omega(x,y)$ with constant value $\Delta(x)\Delta(y) = \omega(x,y)(\Delta)$.*

Proof. Let $x,y \in K$ and put $\nu = \omega(x,y) * \omega(\tilde{y},\tilde{x})$. Then $\nu \in M^1(K)$ and $\tilde{\nu} = \nu$. By Theorem 2.1.10 we obtain

$$\int_K \Delta(x)\, d\nu(x) \;=\; \int_K \int_K \omega(u,v)(\Delta)\, d\omega(x,y)(u)\, d\omega(\tilde{y},\tilde{x})(v)$$

$$= \int_K \int_K \Delta(u)\Delta(v)\, d\omega(x,y)(u)\, d\omega(\tilde{y},\tilde{x})(v) \;=\; \Delta(x)\Delta(y)\Delta(\tilde{x})\Delta(\tilde{y}) \;= 1,$$

and therefore

$$\int_K \left(\Delta(x) \,+\, \frac{1}{\Delta(x)}\right) d\nu(x) \;= 2.$$

Since $\nu(K) = 1$ this is only possible if Δ is constant on supp ν. Now, if $u,v \in$ supp $\omega(x,y)$ then supp $\omega(u,\tilde{v}) \subseteq$ supp ν. Hence $1 = \omega(u,\tilde{v})(\Delta) = \Delta(u)/\Delta(v)$. It follows $\Delta(u) = \Delta(v)$ for all $u,v \in$ supp $\omega(x,y)$, and the constant value is $\Delta(x)\Delta(y)$. ◇

Remark: If the modular function Δ is constant on K (with value 1), then K is called **unimodular**. Obviously every commutative hypergroup is unimodular.

Proposition 2.1.12 *If $\Delta(K)$ is a bounded subset of $]0,\infty[$, then K is unimodular.*

Proof. Let $\Delta(y) \le M$ for all $y \in K$. Suppose that there is some $x \in K$ such that $\Delta(x) \neq 1$. Since $\Delta\tilde{\Delta} = 1$ we can assume $\Delta(x) > 1$. Write ϵ_x^n for the n-fold convolution of ϵ_x. Each ϵ_x^n is a probability measure on K. Hence $\epsilon_x^n(\Delta) \le M$ for each $n \in \mathbb{N}$. However, by Theorem 2.1.10(1) we have $\epsilon_x^n(\Delta) = \Delta(x)^n \to \infty$ as n tends to infinity, a contradiction. ◇

Corollary 2.1.13 *Every compact hypergroup K is unimodular.*

Proof. Δ is continuous and so $\Delta(K)$ is bounded. ◇

For discrete hypergroups K the modular function can be easily calculated. We normalize the Haar measure m by $m(\{e\}) = 1$. For each $x \in K$ we obtain by Theorem 2.1.1

$$\Delta(x) = \Delta(x) \int_K \epsilon_e(y)\, dm(y) = \int_K R_{\tilde{x}} \epsilon_e(y)\, dm(y)$$
$$= \omega(x, \tilde{x})(\{e\}) \, \frac{1}{\omega(\tilde{x}, x)(\{e\})}.$$

Writing for the Haar weights $h(x) := \frac{1}{\omega(\tilde{x},x)(\{e\})}$, the following is true:

Proposition 2.1.14 *A discrete hypergroup is unimodular if and only if $h(x) = h(\tilde{x})$ for every $x \in K$.*

There exist non-unimodular discrete hypergroups, see [353].

2 Translation and convolution

So far we have investigated the convolution $\mu * \nu$ of measures $\mu, \nu \in M(K)$, the translation of $L_x f$ of functions $f \in C_c(K)$, $f \in C_0(K)$ and $f \in C^b(K)$, and the convolution $\mu * f$ of measures $\mu \in M(K)$ with functions $f \in C_c(K)$, $f \in C_0(K)$ and $f \in C^b(K)$ respectively. A basic tool was Lemma 1.1.1(a) stating that $(x, y) \mapsto \omega(x, y)(f)$ is a continuous mapping if f is continuous. We have to show a corresponding result for Borel functions.

Lemma 2.2.1 *Let $g : K \to [0, \infty]$ be a Borel measurable function. Then $(x, y) \mapsto \omega(x, y)(g)$; $K \times K \to [0, \infty]$ is Borel measurable.*

Proof. For convenience denote $\tilde{g}(x, y) = \omega(x, y)(g)$. By the theorem of monotone convergence it is sufficient to prove that $\widetilde{\chi_E}$ is Borel measurable if $E \in \mathcal{B}(K)$ is a Borel set. Let $\Sigma := \{E \in \mathcal{B}(K) : \widetilde{\chi_E}$ is Borel measurable$\}$. We claim that every open subset $V \subseteq K$ is an element of Σ. In fact the regularity of each measure $\omega(x, y)$ implies

$$\widetilde{\chi_V}(x, y) = \sup\{\tilde{g}(x, y) : g \in C_c(K),\ 0 \le g \le \chi_V\}.$$

We will show that $\Sigma = \mathcal{B}(K)$. Consider all subsystems of Σ, which contain all open subsets of K and are closed with respect to finite intersection. Let Σ_0 be a maximal subsystem of this type.

Pick some $A \in \Sigma_0$ and let

$$\Sigma_A := \{(B \cap A) \cup (C \setminus A) : B, C \in \Sigma_0\}.$$

It is easy to see that $\Sigma_0 \subseteq \Sigma_A$ and that Σ_A is closed with respect to finite intersection. Furthermore $\Sigma_A \subseteq \Sigma$, since given $B, C \in \Sigma_0$ we have

$$\chi_{(B\cap A)\cup(C\setminus A)} = \chi_{B\cap A} + \chi_C - \chi_{C\cap A}$$

and $B \cap A$, C, $C \cap A \in \Sigma_0 \subseteq \Sigma$. Thus Σ_A is a subsystem of Σ containing all open subsets of K and being closed with respect to finite intersection. The maximality of Σ_0 implies $\Sigma_0 = \Sigma_A$. By $(\emptyset \cap A) \cup (K \setminus A) = K \setminus A$ we thus have shown that $K \setminus A \in \Sigma_A = \Sigma_0$.

Next we deal with countable unions of $A_i \in \Sigma_0$. Let

$$\Sigma_1 := \left\{ \bigcup_{i=1}^{\infty} A_i : A_i \in \Sigma_0,\ A_i \subseteq A_{i+1} \right\}.$$

Obviously $\Sigma_0 \subseteq \Sigma_1$ and Σ_1 are closed with respect to finite intersection. Moreover $\Sigma_1 \subseteq \Sigma$. In fact if $A = \bigcup_{i=1}^{\infty} A_i \in \Sigma_1$, then $\chi_A = \lim_{i\to\infty} \chi_{A_i}$ and the theorem of monotone convergence yields the Borel measurability of $\tilde{\chi}_A$, i.e. $A \in \Sigma$. By the maximality of Σ_0 we obtain $\Sigma_1 = \Sigma_0$. Since Σ_1 is a σ-algebra containing all open subsets of K, we finally get

$\Sigma \subseteq \mathcal{B}(K) \subseteq \Sigma_0 \subseteq \Sigma.$ $\diamond$

The reader should notice that $\omega(x, y)(g) = \infty$ is possible, even when g is a Borel measurable function with finite values, i.e. $g : K \to [0, \infty[$.

We will study $L^p(K, m)$ for $1 \leq p < \infty$, where m is a left-invariant Haar measure on K. As usual

$$L^p(K, m) := \mathcal{L}^p(K, m)/\mathcal{N},$$

where

$$\mathcal{L}^p(K, m) := \left\{ f : K \to \mathbb{C} \text{ Borel measurable} : \int_K |f(x)|^p \, dm(x) < \infty \right\}$$

and $\mathcal{N} := \{f \in \mathcal{L}^p(K, m) : f = 0 \ m\text{-almost everywhere}\}$.

With $\|f\|_p := (\int_K |f(x)|^p \, dm(x))^{1/p}$ the linear space $L^p(K, m)$ becomes a Banach space.

$L^\infty(K, m)$ is the space of all m-essentially bounded Borel-functions on K, where now functions differing on locally m-zero sets are identified. The norm

$$\|f\|_\infty := \inf\{\alpha \geq 0 : \{x \in K : |f(x)| > \alpha\} \text{ is a locally } m\text{-zero set}\}$$

makes $L^\infty(K, m)$ a Banach space. We will write $L^p(K)$ or $\mathcal{L}^p(K)$ instead of $L^p(K, m)$ or $\mathcal{L}^p(K, m)$.

The next auxiliary (but very basic) result is the following companion of Lemma 2.2.1.

Lemma 2.2.2 *Let $g \in \mathcal{L}^1(K)$, $g : K \to [0, \infty]$ and suppose $x \in K$. Then $y \mapsto L_x g(y) = \omega(x,y)(g)$ is Borel measurable and*

$$\int_K g(z)\, dm(z) = \int_K \omega(x,y)(g)\, dm(y).$$

Proof. The statement on the Borel measurability is already proven in Lemma 2.2.1. Again it is sufficient to show the assertion for $g = \chi_E$, E a Borel set in K with $m(E) < \infty$.

Let $E \in \mathcal{B}(K)$ with $m(E) < \infty$. Choose some open subset U with $E \subseteq U \subseteq K$ and $m(U) < \infty$. Let

$$\Sigma = \left\{ F \in \mathcal{B}(K) : F \subseteq U \text{ and } m(F) = \int_K \chi_F(z)\, dm(z) = \int_K \omega(x,y)(\chi_F)\, dm(y) \right\}.$$

We claim that each open subset $V \subseteq U$ is an element of Σ. This follows as in Lemma 2.2.1 by

$$\widetilde{\chi_V}(x,y) = \sup\{\tilde{g}(x,y) : g \in C_c(K),\ 0 \le g \le \chi_V\}$$

and

$$\begin{aligned} m(V) &= \sup\left\{ \int_K g(z)\, dm(z) : g \in C_c(K),\ 0 \le g \le \chi_V \right\} \\ &= \sup\left\{ \int_K \omega(x,y)(g)\, dm(y) : g \in C_c(K),\ 0 \le g \le \chi_V \right\} \\ &= \int_K \omega(x,y)(\chi_V)\, dm(y). \end{aligned}$$

Applying the same procedure as in Lemma 2.2.1 we obtain that Σ is a σ–algebra and $\Sigma = \mathcal{B}(U) = \{B \cap U : B \in \mathcal{B}(K)\}$. In particular $E \in \Sigma$ and the proof is completed. ◇

Let us recall some facts that follow by Lemma 2.2.2. Given $f \in \mathcal{L}^1(K)$ we know

$$\int_K |f(z)|\, dm(z) = \int_K \omega(x,y)(|f|)\, dm(y),$$

i. e. $|\omega(x,y)(f)| \le \omega(x,y)(|f|) < \infty$ for m–almost every $y \in K$.

Hence the left translation

$$L_x f(y) = \omega(x,y)(f)$$

of $f \in \mathcal{L}^1(K)$ is defined for m–almost every $y \in K$. Since $L_x f \in \mathcal{N}$ if $f \in \mathcal{N}$, the left translation on $L^1(K)$ is well–defined.

The corresponding result is valid for $f \in \mathcal{L}^p(K),\ 1 \leq p < \infty$, since we obtain by Hölders inequality

$$|\omega(x,y)(f)|^p \leq \omega(x,y)(|f|^p).$$

Thus the left translation is well-defined on $L^p(K),\ 1 \leq p < \infty$.

Proposition 2.2.3 *For $f \in L^p(K),\ 1 \leq p < \infty$ and $x \in K$ we have*

$$\|L_x f\|_p \leq \|f\|_p.$$

Proof.

$$\begin{aligned}\|L_x f\|_p^p &= \int_K |L_x f(y)|^p \, dm(y) \leq \int_K L_x(|f|^p)(y)\, dm(y) \\ &= \int_K |f|^p(y)\, dm(y) = \|f\|_p^p.\end{aligned}$$

◇

Remark. In contrast to the group case the left translation is only a contracting operator and not isometric. Only for $f \in L^1(K),\ f \geq 0$ we have $\|L_x f\|_1 = \|f\|_1$ by Lemma 2.2.2.

Finally it remains to consider the case $p = \infty$.

Proposition 2.2.4 *For $f \in L^\infty(K)$ and $x \in K$ we have*

$$\|L_x f\|_\infty \leq \|f\|_\infty.$$

Proof. Write $f = g + h$, where $|h|$ is locally zero and $g(y) \leq \|f\|_\infty$ for every $y \in K$. When we have shown that $L_x|h|$ is also locally zero the proof is completed. Let A be a compact subset and let $B := \{x\} * A$. Obviously $L_x(|h|)(y) = L_x(\chi_B|h|)(y)$ for every $y \in A$, and since B is compact we obtain

$$\begin{aligned}\int_A L_x(|h|)(y)\, dm(y) &\leq \int_K L_x(\chi_B|h|)(y)\, dm(y) \\ &= \int_K (\chi_B|h|)(y)\, dm(y) = \int_B |h|(y)\, dm(y) = 0.\end{aligned}$$

◇

Now we extend Theorem 2.1.8 to $L^1(K)$. The statement of Theorem 2.1.8 may also be written as an identity in $M(K)$:

$$(L_x h)m = \epsilon_{\tilde{x}} * (hm)$$

for every $h \in C_c(K)$. Here $hm \in M(K)$ is the measure defined by $(hm)(g) = \int_K g(x)h(x)\,dm(x)$ for $g \in C_c(K)$. If h is an element of $L^1(K)$ this equation is also valid. In fact given $f \in L^1(K)$ there exists a sequence $(f_n)_{n\in\mathbb{N}}$ of functions $f_n \in C_c(K)$ such that $\|f - f_n\|_1 \to 0$ as $n \to \infty$. By Proposition 2.2.3 follows $\|L_x f - L_x f_n\|_1 \to 0$ as $n \to \infty$.

Now for $g \in C_c(K)$ we obtain from Theorem 2.1.8

$$\begin{aligned}\int_K g(y)\,d(\varepsilon_{\tilde{x}} * (fm))(y) &= \int_K L_{\tilde{x}} g(y) f(y)\,dm(y) = \lim_{n\to\infty} \int_K f_n(y) L_{\tilde{x}} g(y)\,dm(y) \\ &= \lim_{n\to\infty} \int_K L_x f_n(y) g(y)\,dm(y) = \int_K L_x f(y) g(y)\,dm(y) \\ &= \int_K g(y)\,d(L_x f)m(y).\end{aligned}$$

Thus we have shown:

Proposition 2.2.5 *For $f \in L^1(K)$ the following identity in $M(K)$ holds for every $x \in K$*

$$(L_x f)m = \varepsilon_{\tilde{x}} * (fm).$$

Now it is straightforward to derive the extension of Theorem 2.1.8 to $L^1(K)$.

Theorem 2.2.6 *For $f, g \in L^1(K)$ and $x \in K$ we have*

$$\int_K L_x f(y) g(y)\,dm(y) = \int_K f(y) L_{\tilde{x}} g(y)\,dm(y).$$

Proof. Let $g_n \in C_c(K)$ such that $\|g - g_n\|_1 \to 0$ as $n \to \infty$. Then

$$\begin{aligned}\int_K L_x f(y) g(y)\,dm(y) &= \lim_{n\to\infty} \int_K L_x f(y) g_n(y)\,dm(y) \\ &= \lim_{n\to\infty} \int_K g_n(y)\,d(\varepsilon_{\tilde{x}} * (fm))(y) \\ &= \lim_{n\to\infty} \int_K L_{\tilde{x}} g_n(y) f(y)\,dm(y) = \int_K f(y) L_{\tilde{x}} g(y)\,dm(y).\end{aligned}$$

◇

Remark: The statement of Theorem 2.2.6 is also valid if $f \in L^1(K)$ and $g \in C^b(K)$.

The next topic of this section is to introduce and study the convolution $\mu * f$, where $f \in L^p(K)$, $1 \leq p < \infty$, and $\mu \in M(K)$. One should notice that for $f \in L^1(K)$ the element $\mu * (fm) \in M(K)$ is already defined. Briefly we will see that $\mu * (fm)$ is absolutely continuous with respect to m.

Lemma 2.2.7 *Let $\mu \in M(K)$, $\mu \geq 0$ and suppose $f \in \mathcal{L}^p(K)$, $f \geq 0$, $1 \leq p < \infty$. For $x \in K$ we put*

$$\mu * f(x) := \int_K \omega(\tilde{y}, x)(f)\, d\mu(y).$$

Then $\mu * f(x)$ is finite for m–almost every $x \in K$, and $\mu * f \in \mathcal{L}^p(K)$ with $\|\mu * f\|_p \leq \|\mu\|\ \|f\|_p$.

Proof. We can assume $\mu(K) = 1$. By Hölders inequality follows

$$(\mu * f)^p(x) \leq \int_K (\omega(\tilde{y}, x)(f))^p\, d\mu(y) \leq \int_K \omega(\tilde{y}, x)(f^p)\, d\mu(y) = \mu * (f^p)(x).$$

By Fubini's theorem we get

$$\begin{aligned}\int_K (\mu * f)^p(x)\, dm(x) &\leq \int_K \int_K \omega(\tilde{y}, x)(f^p)\, d\mu(y)\, dm(x) \\ &= \int_K \int_K \omega(\tilde{y}, x)(f^p)\, dm(x)\, d\mu(y) = \int_K f^p(x)\, dm(x) < \infty,\end{aligned}$$

and hence $\mu * f(x)$ is finite for m-almost every $x \in K$, $\mu * f \in \mathcal{L}^p(K)$ and $\|\mu * f\|_p \leq \|\mu\| \|f\|_p$. ◇

We notice that $\mu * f \in \mathcal{N}$ if $f \in \mathcal{N}$. Furthermore for arbitrary $f \in \mathcal{L}^p(K)$ and $\mu \in M(K)$ we define first $|\mu| * f$, and then, applying the Jordan decomposition of f, we obtain $|\mu * f(x)| \leq |\mu| * |f|(x)$. Therefore the following result is valid.

Theorem 2.2.8 *Let $\mu \in M(K)$, $f \in L^p(K)$, $1 \leq p < \infty$. Then $\mu * f \in L^p(K)$ and $\|\mu * f\|_p \leq \|\mu\| \|f\|_p$.*

Proof. We already know that $\mu * f(x)$ is finite for m-almost every $x \in K$. Again we assume $\|\mu\| = 1$. As in the proof of Lemma 2.2.7 we obtain

$$|\mu * f(x)|^p \leq (|\mu| * |f|(x))^p \leq |\mu| * |f|^p(x),$$

and hence

$$\int_K |\mu * f(x)|^p \, dm(x) \le \int_K |\mu| * |f|^p(x) \, dm(x) = \int_K |f|^p(x) \, dm(x).$$

◇

Corollary 2.2.9 *For $f \in L^1(K)$, $\mu \in M(K)$ we have the following equality in $M(K)$*

$$\mu * (fm) = (\mu * f)m.$$

In particular $L^1(K)m$ is a left ideal in $M(K)$.

Proof. By Theorem 2.2.6 follows for $h \in C_c(K)$

$$\begin{aligned}
\mu * (fm)(h) &= \int_K \omega(x,y)(h) \, d\mu(x) \, d(fm)(y) \\
&= \int_K \int_K L_x h(y) f(y) \, dm(y) \, d\mu(x) = \int_K \int_K h(y) L_{\tilde{x}} f(x) \, dm(y) \, d\mu(x) \\
&= \int_K h(y) \int_K \omega(\tilde{x}, y)(f) \, d\mu(x) \, dm(y) = \int_K h(y) \mu * f(y) \, dm(y) \\
&= ((\mu * f)m)(h).
\end{aligned}$$

◇

The mapping $f \mapsto fm$, $L^1(K) \to M(K)$ is linear and isometric, i. e. $\|f\|_1 = \|fm\|$. It is clear how the convolution of $M(K)$ is transferred to $L^1(K)$. Given $f, g \in L^1(K)$ and $h \in C_c(K)$ we obtain from above

$$\begin{aligned}
(fm) * (gm)(h) &= \int_K h(x)(fm) * g(x) \, dm(x) \\
&= \int_K h(x) \int_K \omega(\tilde{y}, x)(g) \, d(fm)(y) \, dm(x) = \int_K h(x) f * g(x) \, dm(x),
\end{aligned}$$

where in the last equality we have used that

$$f * g(x) := \int_K f(y) L_{\tilde{y}} g(x) \, dm(y) = (fm) * g(x). \qquad (conv)$$

Theorem 2.2.10 *For $f, g \in L^1(K)$ define $f * g(x)$ as in (conv) above. Then $f * g(x)$ is finite for m-almost every $x \in K$, $f * g \in L^1(K)$ and $\|f * g\|_1 \le \|f\|_1 \|g\|_1$. With the convolution given in (conv) the Banach space $L^1(K)$ becomes a Banach algebra.*

Proof. We already know that $f * g(x) = (fm) * g(x)$ is finite for m-almost all $x \in K$. All the other assertions are already shown above. $\diamond$

We point out that $L^1(K)$ operates on $L^p(K)$, $1 < p < \infty$, by means of the embedding of $L^1(K)$ into $M(K)$. This operation is given by

$$f * g(x) = (fm) * g(x) = \int_K \omega(\tilde{y}, x)(g) f(y)\, dm(y).$$

The left translation is a continuous mapping from K into the $L^p(K)$-spaces for $1 \leq p < \infty$. In fact we have the following result.

Proposition 2.2.11 *Let $f \in L^p(K)$, $1 \leq p < \infty$. Then the mapping $x \mapsto L_x f$, $K \to L^p(K)$ is continuous.*

Proof. Suppose $x_0 \in K$, $\varepsilon > 0$. Choose some $h \in C_c(K)$ with $\|f - h\|_p < \varepsilon$, and choose some compact neighborhood U of x_0. By Lemma 1.1.1(a) there exists a neighborhood V of x_0 with $V \subseteq U$ and

$$|L_x h(y) - L_{x_0} h(y)|^p < \varepsilon / m(\tilde{U} * \text{supp } h)$$

for each $x \in V$ and all $y \in \tilde{U} * \text{supp}\, h$. Then

$$\|L_x h - L_{x_0} h\|_p^p = \int_{\tilde{U} * \text{supp}\, h} |L_x h(y) - L_{x_0} h(y)|^p\, dm(y) < \varepsilon$$

since for $y \notin \tilde{U} * \text{supp}\, h$ we obviously have $y \notin \{\tilde{x}\} * \text{supp}\, h$ for each $x \in V$. But that is equivalent to $\{x\} * \{y\} \cap \text{supp}\, h = \emptyset$ for every $x \in V$. Now we obtain by Proposition 2.2.3 for $x \in V$

$$\|L_x f - L_{x_0} f\|_p \leq \|L_x f - L_x h\|_p + \|L_x h - L_{x_0} h\|_p + \|L_{x_0} h - L_{x_0} f\|_p \leq 2\varepsilon + \varepsilon^{1/p}.$$

$\diamond$

Considering the sup-norm instead of the L_p-norm and following the lines of the proof above we can show:

Proposition 2.2.12 *Let $f \in C_0(K)$. Then the mapping $x \mapsto L_x f$, $K \to C_0(K)$ is continuous.*

Finally we prove a result concerning the convolution of elements of $L^p(K)$ and $L^q(K)$ where $\frac{1}{p} + \frac{1}{q} = 1$.

Proposition 2.2.13 *Let $1 < p < \infty$ and $1 < q < \infty$ with $\frac{1}{p} + \frac{1}{q} = 1$. For $f \in L^p(K)$ and $g \in L^q(K)$ define*

$$f * \tilde{g}(x) := \int_K f(y)\omega(\tilde{x}, y)(g)\, dm(y).$$

Then $f * \tilde{g} \in C_0(K)$ and $\|f * \tilde{g}\|_\infty \leq \|f\|_p \|g\|_q$.

Proof. By Hölders inequality and Proposition 2.2.3 we get

$$\left| \int_K f(y) L_{\tilde{x}} g(y)\, dm(y) \right| \leq \|f\|_p \|L_{\tilde{x}} g\|_q \leq \|f\|_p \|g\|_q.$$

Hence $f * \tilde{g}$ is defined for every $x \in K$ and $\|f * \tilde{g}\|_\infty \leq \|f\|_p \|g\|_q$. It remains to show that $f * \tilde{g} \in C_0(K)$. Using Hölders inequality again we see

$$|f * \tilde{g}(x) - f * \tilde{g}(x_0)| \leq \|f\|_p \|L_{\tilde{x}} g - L_{\tilde{x}_0} g\|_q,$$

and Proposition 2.2.11 gives the continuity of $f * \tilde{g}$. If f and g are elements of $C_c(K)$ we know $f * \tilde{g} \in C_c(K)$. For arbitrary $f \in L^p(K)$, $g \in L^q(K)$ choose $f_n, g_n \in C_c(K)$ with $\|f - f_n\|_p \to 0$ and $\|g - g_n\|_q \to 0$ as $n \to \infty$. Then

$$\begin{aligned} \|f * \tilde{g} - f_n * \tilde{g}_n\|_\infty &\leq \|f * (\tilde{g} - \tilde{g}_n)\|_\infty + \|(f - f_n) * \tilde{g}_n\|_\infty \\ &\leq \|f\|_p \|g - g_n\|_q + \|f - f_n\|_p \|g_n\|_q \to 0 \qquad \text{as } n \to \infty. \end{aligned}$$

Hence we get $f * \tilde{g} \in C_0(K)$. ◇

Remark. If $f \in L^1(K)$, $g \in L^\infty(K)$ then Proposition 2.2.11 applied to f yields $f * \tilde{g} \in C^b(K)$.

In Theorem 1.1.7 we have already stated that $M(K)$ is a Banach algebra with unit ϵ_e. $M(K)$ has also a canonical involution $\mu \mapsto \mu^*$ defined by

$$\mu^*(E) = \overline{\mu(\tilde{E})}$$

for each Borel set $E \subseteq K$.

Equivalently,

$$\int_K g(x)\, d\mu^*(x) = \int_K g(\tilde{x})\, d\overline{\mu}(x) = \overline{\int_K \overline{g(\tilde{x})}\, d\mu(x)}$$

for each $g \in C_0(K)$. $\mu \mapsto \mu^*$ is indeed an involution. We have only to check that $(\mu * \nu)^* = \nu^* * \mu^*$. For any $g \in C_0(K)$ we obtain

$$\begin{aligned} \int_K g(z)\, d(\mu * \nu)^*(z) &= \overline{\int_K \overline{g(\tilde{z})}\, d(\mu * \nu)(z)} \\ = \overline{\int_K \int_K \omega(x, y)(\overline{\tilde{g}})\, d\mu(x)\, d\nu(y)} &= \overline{\int_K \int_K \overline{\omega(\tilde{y}, \tilde{x})(g)}\, d\mu(x)\, d\nu(y)} \\ = \int_K \int_K \omega(y, x)(g)\, d\nu^*(y)\, d\mu^*(x) &= \int_K g(z)\, d(\nu^* * \mu^*)(z). \end{aligned}$$

Theorem 2.2.14 *$M(K)$ is a Banach $*$-algebra with unit ϵ_e.*

Having chosen a left-invariant Haar measure m on K we have seen that $L^1(K) = L^1(K,m)$ is a Banach algebra with convolution

$$f * g(x) := \int_K f(y)\, L_{\tilde{y}}g(x)\, dm(y) \;=\; \int_K f(y)\, R_x g(\tilde{y})\, dm(y)$$

for m-almost all $x \in K$, see Theorem 2.2.10.

It is rather easy to derive the following identities.

Proposition 2.2.15 *Let $f, g \in L^1(K)$. Then for m-almost all $x \in K$ we have:*

$$\text{(1)} \quad f * g(x) \;=\; \int_K L_x f(y)\, g(\tilde{y})\, dm(y)$$

$$\text{(2)} \quad f * g(x) \;=\; \int_K L_x f(\tilde{y})\, g(y)\, \Delta(\tilde{y})\, dm(y)$$

$$\text{(3)} \quad f * g(x) \;=\; \int_K f(\tilde{y})\, R_x g(y)\, \Delta(\tilde{y})\, dm(y).$$

Proof.

(1) Since $\omega(\tilde{y},x)(g) = \omega(\tilde{x},y)(\tilde{g})$, it follows by Theorem 2.2.6

$$\begin{aligned} f * g(x) &= \int_K f(y)\, R_x g(\tilde{y})\, dm(y) \;=\; \int_K f(y)\, L_{\tilde{x}}\tilde{g}(y)\, dm(y) \\ &= \int_K L_x f(y)\, g(\tilde{y})\, dm(y). \end{aligned}$$

(2) Since $m = \Delta\tilde{m}$, it follows by (1)

$$\begin{aligned} f * g(x) &= \int_K L_x f(y)\, g(\tilde{y})\, dm(y) \;=\; \int_K L_x f(y)\, g(\tilde{y})\, \Delta(y)\, d\tilde{m}(y) \\ &= \int_K L_x f(\tilde{y})\, g(y)\, \Delta(\tilde{y})\, dm(y). \end{aligned}$$

$$\text{(3)} \quad f * g(x) \;=\; \int_K f(y)\, R_x g(\tilde{y})\, dm(y) \;=\; \int_K f(\tilde{y})\, R_x g(y)\, \Delta(\tilde{y})\, dm(y).$$

◇

From now on we identify each function $f \in L^1(K)$ with the measure $fm \in M(K)$. The convolution of $f, g \in L^1(K)$ agrees with the convolution of fm and gm in $M(K)$. $L^1(K)$ is a left ideal of $M(K)$, see Corollary 2.2.9.

The involution of $f \in L^1(K)$ suiting to the involution on $M(K)$ is obtained in the following way. The Borel measure $\tilde{m}$ is a right-invariant measure. In fact, for $f \in C_c(K)$ we get

$$\int_K R_y f(x)\, d\tilde{m}(x) = \int_K R_y f(\tilde{x})\, dm(x) = \int_K L_{\tilde{y}} \tilde{f}(x)\, dm(x)$$
$$= \int_K \tilde{f}(x)\, dm(x) = \int_K f(x)\, d\tilde{m}(x).$$

We recall the relations $\tilde{m} = \tilde{\Delta}(m)$ and $m = \Delta \tilde{m}$.

If K is not unimodular, then the function Δ is unbounded. Hence the Banach spaces $L^1(K) = L^1(K,m)$ and $L^1(K,\tilde{m})$ are not the same. If $f \in L^1(K)$, then $\tilde{f} \in L^1(K,\tilde{m})$. Moreover, the mapping $f \mapsto \tilde{f}$ is an isometric isomorphism from the Banach space $L^1(K)$ onto the Banach space $L^1(K,\tilde{m})$. $L^1(K)$ can be mapped to $L^1(K,\tilde{m})$ in another way. For $f \in L^1(K)$ consider $M(f) = \Delta f$. Then $M(f) \in L^1(K,\tilde{m})$ (since $\int_K \Delta(x)|f(x)|d\tilde{m}(x) = \int_K |f(x)|dm(x)$), and M is an isometric isomorphism from $L^1(K)$ onto $L^1(K,\tilde{m})$. Composing the maps $f \mapsto \tilde{f}$ and M^{-1}, we get an isometric isomorphism of $L^1(K)$ onto itself. In particular, if $f \in L^1(K)$ the function

$$f^* := \tilde{\Delta}\, \tilde{\bar{f}}$$

is an element of $L^1(K)$.

Proposition 2.2.16 *If $f \in L^1(K)$ then $f^* = \tilde{\Delta}\tilde{\bar{f}} \in L^1(K)$, and the involution on $M(K)$ restricted to $L^1(K)$ is given by $f \mapsto f^*$, i.e. $(fm)^* = f^*m$.*

Proof. Let $f \in L^1(K)$. For $g \in C_0(K)$ we have

$$\int_K g(x)\, d(fm)^*(x) = \overline{\int_K \overline{g(\tilde{x})}\, f(x)\, dm(x)} = \int_K g(\tilde{x})\, \overline{f(x)}\, dm(x)$$
$$= \int_K g(\tilde{x})\, \overline{f(x)}\, \Delta(x)\, d\tilde{m}(x) = \int_K g(x)\, \overline{f(\tilde{x})}\, \Delta(\tilde{x})\, dm(x).$$

◇

Theorem 2.2.17 *$L^1(K)$ is a Banach $*$-subalgebra of $M(K)$.*

3 Representation of hypergroups

In this section we introduce and investigate representations of hypergroups K and representations of $M(K)$ and $L^1(K)$. In contrast to the group case we have to deal with representations which are in general not unitary.

Definition. Let H be a Hilbert space. A **representation** V of K by bounded operators on H is a mapping $x \mapsto V_x$, $K \to \mathcal{B}(H)$, such that

(i) $x \mapsto \langle V_x\xi, \eta\rangle$ is a continuous and bounded mapping on K for all $\xi, \eta \in H$.

(ii) $\int_K \langle V_z\xi, \eta\rangle \, d\omega(x,y)(z) = \langle V_x V_y \xi, \eta\rangle$
for all $x, y \in K$, $\xi, \eta \in H$.

(iii) $V_e = \mathrm{id}_H$.

The representation V is called **bounded**, if $\|V\| := \sup_{x\in K} \|V_x\| < \infty$.

V is said to be an $*$-representation if

(iv) $V_{\tilde{x}} = V_x^*$ for all $x \in K$.

If $\mathcal{A}$ is a subalgebra of $M(K)$ each algebra homomorphism from $\mathcal{A}$ into $\mathcal{B}(H)$, where H is a Hilbert space, is called a representation of $\mathcal{A}$ on H. If $\mathcal{A}$ is an $*$-subalgebra of $M(K)$ each $*$-algebra homomorphism from $\mathcal{A}$ into $\mathcal{B}(H)$ is called $*$-representation of $\mathcal{A}$ on H. If $T : \mathcal{A} \to \mathcal{B}(H)$ is a representation of $\mathcal{A}$ on H we usually write T_μ instead of $T(\mu)$. A representation T of $\mathcal{A}$ is said to be **nondegenerate** if there is no $\xi \in H$, $\xi \neq 0$, such that $T_\mu\xi = 0$ for each $\mu \in \mathcal{A}$.

The representation $T : \mathcal{A} \to \mathcal{B}(H)$ is called **bounded**, if

$$\|T\| := \sup_{\|\mu\|\leq 1} \|T_\mu\| < \infty.$$

Applying the theory of positive functionals on Banach $*$-algebras a result due to Gelfand-Naimark (see Theorem 21.22 in [304]) says, that every $*$-representation $T : \mathcal{A} \to \mathcal{B}(H)$, $\mathcal{A}$ a Banach $*$-subalgebra, is continuous, moreover $\|T_\mu\| \leq \|\mu\|$ for each $\mu \in \mathcal{A}$.

Proposition 2.3.1

(1) Let V be a bounded representation of K on a Hilbert space H, and let $\mathcal{A}$ be a subalgebra of $M(K)$. Then for each $\mu \in \mathcal{A}$ there exists a unique operator $T_\mu \in \mathcal{B}(H)$ such that

$$\langle T_\mu\xi, \eta\rangle = \int_K \langle V_x\xi, \eta\rangle \, d\mu(x) \qquad (*)$$

for all $\xi, \eta \in H$. The mapping $\mu \mapsto T_\mu$, $\mathcal{A} \mapsto \mathcal{B}(H)$ is a bounded representation of $\mathcal{A}$ on H, for which

$$\|T_\mu\| \leq \|V\| \, \|\mu\|.$$

(2) Let V be a bounded $*$-representation of K on H, and let $\mathcal{A}$ be an $*$-subalgebra of $M(K)$. Then the representation $\mu \mapsto T_\mu$ of (1) is an $*$-representation of $\mathcal{A}$ on H. If $L^1(K) \subseteq \mathcal{A}$, then T is nondegenerate.

Proof.

(1) Consider $\eta \mapsto \int_K \langle V_x\xi, \eta\rangle \, d\mu(x), \ H \to \mathbb{C}$. This mapping is conjugate-linear and

$$\left| \int_K \langle V_x\xi, \eta\rangle \, d\mu(x) \right| \ \leq \ \|V\| \, \|\xi\| \, \|\mu\| \, \|\eta\|. \qquad (**)$$

Hence there is a unique element of H, which we write as $T_\mu\xi$, such that

$$\langle T_\mu\xi, \eta\rangle \ = \ \int_K \langle V_x\xi, \eta\rangle \, d\mu(x).$$

The mapping $\xi \mapsto T_\mu\xi$ is linear for each fixed $\mu \in \mathcal{A}$ and by $(**)$ it follows $T_\mu \in \mathcal{B}(H)$ and $\|T_\mu\| \ \leq \ \|V\| \, \|\mu\|$ for each $\mu \in \mathcal{A}$. It is also clear that $\mu \mapsto T_\mu, \ \mathcal{A} \to \mathcal{B}(H)$ is linear.

Now we show $T_{\mu*\nu} = T_\mu T_\nu$ for $\mu, \nu \in \mathcal{A}$. Write $f(z) = \langle V_z\xi, \eta\rangle$. Since $\omega(x,y)(f) = \langle V_x V_y \xi, \eta\rangle$ we obtain by Theorem 1.1.3(iii)

$$\int_K \langle V_z\xi, \eta\rangle \, d\mu * \nu(z) \ = \ \int_K \int_K \omega(x,y)(f) \, d\mu(x) \, d\nu(y)$$

$$= \ \int_K \int_K \langle V_x V_y\xi, \eta\rangle \, d\mu(x) \, d\nu(y) \ = \ \int_K \langle T_\mu V_y\xi, \eta\rangle \, d\nu(y)$$

$$= \ \int_K \langle V_y\xi, T_\mu^*\eta\rangle \, d\nu(y) \ = \ \langle T_\nu\xi, T_\mu^*\eta\rangle \ = \ \langle T_\mu T_\nu\xi, \eta\rangle.$$

Hence $T_{\mu*\nu} = T_\mu T_\nu$.

(2) We have to prove $T_{\mu^*} = (T_\mu)^*$ if $V_{\tilde{x}} = V_x^*$. For all $\xi, \eta \in H$ we obtain

$$\langle T_{\mu^*}(\xi), \eta\rangle \ = \ \int_K \langle V_{\tilde{x}}\xi, \eta\rangle \, d\bar{\mu}(x) \ = \ \overline{\int_K \overline{\langle V_{\tilde{x}}\xi, \eta\rangle} \, d\mu(x)}$$

$$= \ \overline{\int_K \langle \eta, V_x^*\xi\rangle \, d\mu(x)} \ = \ \overline{\langle T_\mu\eta, \xi\rangle} \ = \ \langle \xi, T_\mu\eta\rangle \ = \ \langle (T_\mu)^*\xi, \eta\rangle.$$

To see that T is nondegenerate, suppose $\xi \in H, \ \xi \neq 0$. Select some $\eta \in H$ such that $\langle \xi, \eta\rangle = 1$. There exists a neighborhood U of $e \in K$ such that $|\langle V_x\xi, \eta\rangle \ - \ \langle \xi, \eta\rangle| \ < 1$. Set $f \ = \ \frac{1}{m(U)} \chi_U \ \in \ L^1(K) \subseteq \mathcal{A}$. Then

$$|\langle T_f\xi, \eta\rangle \ - \ \langle \xi, \eta\rangle| \ \leq \ \frac{1}{m(U)} \int_U |\langle V_x\xi, \eta\rangle \ - \ \langle \xi, \eta\rangle| \, dm(x) \ < 1.$$

Hence $\langle T_f\xi, \eta\rangle \neq 0$, in particular $T_f\xi \neq 0$.

◇

Remark:

(1) In contrast to [350] or [96] we define a representation of a hypergroup K first of all as a mapping from K into $\mathcal{B}(H)$ and extend it then to $M(K)$. The reader should note that (even for groups) there are representations of $M(K)$ which are not generated by $(*)$ as in the previous proposition, see [304,(23.28a)].

(2) We note that in general it cannot be assumed that the operators V_x are all unitary (as is the case for groups). In the case that the hypergroup is a locally compact group our definition of $*$-representations coincides with that of unitary representations on groups.

(3) Applying Theorem 21.22 in [304] we obtain that $\|T_\mu\| \leq \|\mu\|$ for all $\mu \in M(K)$, where T is generated by a bounded $*$-representation V of K as above. In particular, $\|V_x\| = \|T_{\epsilon_x}\| \leq \|\epsilon_x\| = 1$ for all $x \in K$. That means $\|V\| = \sup_{x\in K} \|V_x\| = 1$.

A representation V of K (respectively, T of $L^1(K)$) is called **cyclic**, if there exists a vector $\xi \in H$ such that the linear span of $\{V_x\xi : x \in K\}$ (respectively, $\{T_f\xi : f \in L^1(K)\}$) is dense in H. It is readily shown that each cyclic $*$-representation V of K (respectively, T of $L^1(K)$) is nondegenerate.

A closed subspace $H_0 \subseteq H$ is said to be invariant under V (respectively, under T) if $V_xH_0 \subseteq H_0$ for all $x \in K$ (respectively, $T_f(H_0) \subseteq H_0$ for all $f \in L^1(K)$).

For the special $*$-subalgebra $\mathcal{A} = L^1(K)$ of $M(K)$ we can show that every bounded cyclic representation T of $L^1(K)$ arises from a unique bounded representation V of K, and each nondegenerate $*$-representation T of $L^1(K)$ arises from a unique bounded $*$-representation V of K. Recall that $L^1(K)$ is a Banach $*$-algebra with involution $f^*(x) = \Delta(\tilde{x})\,\overline{f(\tilde{x})}$, see Theorem 2.2.17.

The following result about approximate identities will be used several times in connection with various applications.

Lemma 2.3.2 *For any neighborhood V of $e \in K$ with compact closure there exists $\varphi \in C_c(K)$, $\varphi \geq 0$, such that $\operatorname{supp}\varphi \subseteq V$ and $\int_K \varphi(x)\,dm(x) = 1$.*

Proof. Let U be a neighborhood of e such that $U * \tilde{U} \subseteq V$, and choose $f \in C_c(K)$, $f \geq 0$, $f \neq 0$ with $\operatorname{supp} f \subseteq U$. Put $\varphi = \frac{1}{c} f * \tilde{f}$, $c = \int_K f * \tilde{f}(y)\,dm(y) > 0$. Then φ is continuous, $\operatorname{supp}\varphi \subseteq U * \tilde{U} \subseteq V$, $\varphi \geq 0$ and $\int_K \varphi(y)\,dm(y) = 1$. ◇

Remark: The function $\varphi \in C_c(K)$ constructed in Lemma 2.3.2 is also positive definite. (For the definition of positive definiteness, see the next section 2.4.)

Proposition 2.3.3 *There exists a net of functions* $(\varphi_i)_{i\in I}$, $\varphi_i \in C_c(K)$, $\varphi_i \geq 0$, $\int_K \varphi_i(y)\, dm(y) = 1$ *for each* $i \in I$, *such that*

(1) $\lim\limits_i \varphi_i * f(x) = f(x)$ *pointwise, whenever* $f : K \to \mathbb{C}$ *is continuous.*

(2) $\lim\limits_i \|\varphi_i * f - f\|_\infty = 0$, *whenever* $f \in C_0(K)$.

(3) $\lim\limits_i \|\varphi_i * f - f\|_p = 0$, *whenever* $f \in L^p(K)$, $1 \leq p < \infty$.

Proof. Let $(V_i)_{i\in I}$ be a basis of neighborhoods of $e \in K$ with compact closure. Direct I by $i_1 \geq i_2$ if $V_{i_1} \subseteq V_{i_2}$. For each $i \in I$ choose $\varphi_i \in C_c(K)$ with $\operatorname{supp} \varphi_i \subseteq V_i$, $\varphi_i \geq 0$, $\|\varphi_i\|_1 = 1$ as in Lemma 2.3.2.

(1) For $x \in K$ we get

$$|\varphi_i * f(x) - f(x)| \leq \int_K \varphi_i(y)\, |L_{\tilde{y}} f(x) - f(x)|\, dm(y) \leq \sup_{y\in V_i} |L_{\tilde{y}} f(x) - f(x)|.$$

By Lemma 1.1.1 it follows $\lim_i \varphi_i * f(x) = f(x)$.

(2) Let $\varepsilon > 0$ and choose $g \in C_c(K)$ such that $\|g - f\|_\infty < \varepsilon$. The proof of Proposition 2.1.6 shows that $\lim_i \|\varphi_i * g - g\|_\infty = 0$. Since $\|\varphi_i * g - \varphi_i * f\|_\infty < \varepsilon$ for each $i \in I$, statement (2) follows.

(3) Using the $L^p(K)$-valued integral we have

$$\varphi_i * f - f = \int_K \varphi_i(y)\, (L_{\tilde{y}} f - f)\, dm(y)$$

and obtain

$$\|\varphi_i * f - f\|_p \leq \int_K \varphi_i(y)\, \|L_{\tilde{y}} f - f\|_p\, dm(y).$$

By Proposition 2.2.11 statement (3) follows.

◇

Remark: A net of functions satisfying (1) or (2) or (3) of Proposition 2.3.3 is called a **left approximate identity**.

We extend the statement of Proposition 1.1.13 to L^p-spaces.

Proposition 2.3.4 *Given* $x, y \in K$ *and* $f \in L^p(K)$, $1 \leq p < \infty$, *we have for the* $L^p(K)$*-valued integrals*

$$L_{\tilde{x}} \circ L_{\tilde{y}} f = \int_K L_z f\, d\omega(\tilde{y}, \tilde{x})(z) = \int_K L_{\tilde{z}} f\, d\omega(x, y)(z)$$

and

$$R_x \circ R_y f = \int_K R_z f\, d\omega(x, y)(z).$$

Proof. First note that $\operatorname{supp}\omega(x,y)$ is compact and $z \mapsto L_z f$ is a continuous mapping from K into $L^p(K)$. Hence the $L^p(K)$-valued integrals above are well-defined elements of $L^p(K)$. For $f \in C_c(K)$ Proposition 1.1.13 yields for each $u \in K$

$$L_{\tilde{x}} \circ L_{\tilde{y}} f(u) \;=\; \int_K L_z f(u)\, d\omega(\tilde{y},\tilde{x})(z) \;=\; \int_K L_{\tilde{z}} f(u)\, d\omega(x,y)(z).$$

That is

$$L_{\tilde{x}} \circ L_{\tilde{y}} f \;=\; \int_K L_z f\, d\omega(\tilde{y},\tilde{x})(z) \;=\; \int_K L_{\tilde{z}} f\, d\omega(x,y)(z)$$

and since translation and integration are bounded operators this equation is also valid for $f \in L^p(K)$. In the same way one shows the equation for the right translation. ◇

Theorem 2.3.5 *Let T be a bounded cyclic representation of $L^1(K)$ on a Hilbert space H. Then there exists a unique bounded representation V of K on H such that*

$$\langle T_f \xi, \eta\rangle \;=\; \int_K \langle V_x \xi, \eta\rangle\, f(x)\, dm(x) \qquad \textit{for every } \xi,\eta \in H,\ f \in L^1(K),$$

and $\|V\| = \|T\|$. Moreover, T and V have the same closed invariant subspaces.

Proof. By Proposition 2.3.3 there exists a net $(\varphi_i)_{i\in I}$ with $\varphi_i \in C_c(K)$, $\|\varphi_i\|_1 = 1$, $\varphi_i \geq 0$ such that $\lim_i \|\varphi_i * f - f\|_1 = 0$ for all $f \in L^1(K)$. Since $(L_{\tilde{x}}\varphi_i) * f = L_{\tilde{x}}(\varphi_i * f)$ we have

$$\lim_i \|(L_{\tilde{x}}\varphi_i) * f - L_{\tilde{x}} f\|_1 \;=\; 0 \qquad \text{for any } x \in K \text{ and } f \in L^1(K).$$

Hence

$$\lim_i \|T_{L_{\tilde{x}}\varphi_i}(T_f \eta) \;-\; T_{L_{\tilde{x}} f}(\eta)\| \;=\; 0 \qquad\qquad (*)$$

for every $\eta \in H$, $x \in K$, $f \in L^1(K)$.

Now select a cyclic vector $\xi \in H$ for T and put $H_0 = \{T_f \xi : f \in L^1(K)\}$. By $(*)$ we see that for each $x \in K$, $T(L_{\tilde{x}}\varphi_i)$ converge strongly on H_0 to an operator U_x on H_0, which satisfies

$$U_x(T_f \xi) \;=\; T_{L_{\tilde{x}} f}(\xi) \qquad \text{for every } f \in L^1(K).$$

The operator U_x is well-defined, because if $T_{f_1}\xi = T_{f_2}\xi$, then

$$T_{L_{\tilde{x}} f_1}(\xi) \;=\; \lim_i T_{L_{\tilde{x}}\varphi_i}(T_{f_1}\xi) \;=\; \lim_i T_{L_{\tilde{x}}\varphi_i}(T_{f_2}\xi) \;=\; T_{L_{\tilde{x}} f_2}(\xi).$$

Moreover, $\|T_{L_{\tilde{x}}\varphi_i}(T_f \xi)\| \;\leq\; \|T\|\, \|T_f \xi\|$ for each $i \in I$, $f \in L^1(K)$.

Therefore $\|U_x(T_f\xi)\| \leq \|T\| \, \|T_f\xi\|$, and so for each $x \in K$, U_x is a bounded operator on H_0, which can be extended uniquely to a bounded operator V_x on H, such that $\|V_x\| = \|U_x\|$.

We claim that V is a bounded representation of K on H. We know already that $\|V_x\| = \|U_x\| \leq \|T\|$ for all $x \in K$. Given $\eta \in H$ it follows for every $x, y \in K$ and some $f \in L^1(K)$

$$\|V_x\eta - V_y\eta\| \leq \|V_x\eta - V_x(T_f\xi)\| + \|U_x(T_f\xi) - U_y(T_f\xi)\|$$

$$+ \|V_y(T_f\xi) - V_y\eta\| \leq \|T\| \left(2\,\|\eta - T_f\xi\| + \|L_{\tilde{x}}f - L_{\tilde{y}}f\|_1 \, \|\xi\|\right).$$

Using Proposition 2.2.11 it follows that $x \mapsto \langle V_x\eta, \zeta\rangle$ is a continuous and bounded mapping on K for all $\eta, \zeta \in H$. Since $V_e(T_f\xi) = U_e(T_f\xi) = T_f\xi$ for each $f \in L^1(K)$ we have $V_e = \mathrm{id}_H$.

In order to show $\int\limits_K \langle V_z\eta, \zeta\rangle \, d\omega(x,y)(z) = \langle V_xV_y\eta, \zeta\rangle$ for all $\eta, \zeta \in H$ and $x, y \in K$ we apply Proposition 2.3.4 and obtain the following identity in H

$$T_{L_{\tilde{x}}L_{\tilde{y}}f}(\xi) = \int_K T_{L_{\tilde{z}}f}(\xi) \, d\omega(x,y)(z)$$

for all $f \in L^1(K)$. Note that $z \mapsto T_{L_{\tilde{z}}f}(\xi)$ is a bounded and continuous map from K into H. Hence

$$\langle V_xV_yT_f\xi, \zeta\rangle = \langle U_yT_f\xi, V_x^*\zeta\rangle = \langle T_{L_{\tilde{y}}f}(\xi), V_x^*\zeta\rangle$$

$$= \langle U_x(T_{L_{\tilde{y}}f}(\xi)), \zeta\rangle = \langle T_{L_{\tilde{x}}L_{\tilde{y}}f}(\xi), \zeta\rangle$$

$$= \left\langle \int_K T_{L_{\tilde{z}}f}(\xi) \, d\omega(x,y)(z), \zeta \right\rangle = \int_K \langle V_z(T_f\xi), \zeta\rangle \, d\omega(x,y)(z)$$

and it follows

$$\langle V_xV_y\eta, \zeta\rangle = \int_K \langle V_z\eta, \zeta\rangle \, d\omega(x,y)(z) \qquad \text{for all } \eta, \zeta \in H.$$

Till now we have shown that V is a bounded representation of K on H and $\|V\| \leq \|T\|$.

Given $f \in L^1(K)$ and $g \in C_c(K)$ we have

$$T_g \circ T_f = T_{g*f} = \int_K g(y) \, T_{L_{\tilde{y}}f} \, dm(y)$$

with an $L^1(K)$-valued integral on the right hand side. It follows for $\xi, \zeta \in H$

$$\langle T_g(T_f\xi), \zeta\rangle = \langle T_{g*f}\xi, \zeta\rangle = \left\langle \int_K g(y) \, T_{L_{\tilde{y}}f}(\xi) \, dm(y), \zeta \right\rangle$$

$$= \int_K g(y) \, \langle V_y(T_f\xi), \zeta\rangle \, dm(y)$$

and then

$$\langle T_g\eta, \zeta\rangle = \int_K g(y)\, \langle V_y\eta, \zeta\rangle\, dm(y) \qquad \text{for all } g \in L^1(K)$$

and $\eta, \zeta \in H$, where we used that $C_c(K)$ is dense in $L^1(K)$ and $\{T_f\xi : f \in L^1(K)\}$ is dense in H.

Moreover we have $\|T\| \leq \|V\|$, see Proposition 2.2.8. Finally, let W be another bounded representation of K on H, which generates T as in Proposition 2.3.1. Then

$$\int_K \langle V_x\eta, \zeta\rangle\, f(x)\, dm(x) = \langle T_f\eta, \zeta\rangle = \int_K \langle W_x\eta, \zeta\rangle\, f(x)\, dm(x)$$

for all $\eta, \zeta \in H$ and every $f \in L^1(K)$. Since $x \mapsto \langle (V_x - W_x)\eta, \zeta\rangle$ is bounded and continuous, it follows $\langle V_x\eta, \zeta\rangle = \langle W_x\eta, \zeta\rangle$ for all $\eta, \zeta \in H$ and $x \in K$. Hence $V_x = W_x$ for all $x \in K$.

Let $H_0 \subseteq H$ be invariant under V. Then $\langle T_f\eta, \zeta\rangle = 0$ for each $\eta \in H_0$, $\zeta \in H_0^\perp$ and every $f \in L^1(K)$. In particular, $T_f(H_0) \subseteq H_0^{\perp\perp} = H_0$. Now let H_0 be invariant under T. Assume there exist $x \in K$ and $\eta \in H_0$ such that $V_x\eta \notin H_0$. Choose some $\zeta \in H_0^\perp$ such that $\langle V_x\eta, \zeta\rangle = 1$. There exists some $f \in L^1(K)$ such that

$$0 \neq \int_K \langle V_x\eta, \zeta\rangle\, f(x)\, dm(x) = \langle T_f\eta, \zeta\rangle = 0,$$

a contradiction. ◇

Theorem 2.3.6 *Let T be a (bounded) nondegenerate $*$-representation of $L^1(K)$ on H. Then there exists a unique bounded $*$-representation V of K on H such that*

$$\langle T_f\xi, \eta\rangle = \int_K \langle V_x\xi, \eta\rangle\, f(x)\, dm(x) \qquad \textit{for all } f \in L^1(K),$$

and $\|V\| = 1$. Moreover, T and V have the same closed invariant subspaces.

Proof. (Recall that a cyclic $*$-representation of $L^1(K)$ is nondegenerate.)

Below we follow the lines of the proof of Theorem 2.3.5. Let H_0 be the linear span of $\{T_f\xi : f \in L^1(K), \xi \in H\}$. Since T is nondegenerate, H_0 is a dense subspace of H. In fact, if $\eta \in H_0^\perp$ then

$$0 = \langle \eta, T_f\xi\rangle = \langle T_{f^*}\eta, \xi\rangle \qquad \text{for all } \xi \in H,\ f \in L^1(K).$$

Since T is nondegenerate, η has to be zero. Therefore $\overline{H_0} = H$. Applying an approximate identity $(\varphi_i)_{i\in I}$, the calculation of Theorem 2.3.5 shows that $\sum_{j=1}^{n} T_{L_{\tilde{x}}\varphi_i}(T_{f_j}\xi_j)$ converges to $\sum_{j=1}^{n} T_{L_{\tilde{x}}f_j}(\xi_j)$ for $f_1, ..., f_n \in L^1(K)$; $\xi_1, ..., \xi_n \in H$ and $x \in K$. Given $x \in K$ define an operator $U_x : H_0 \to H_0$ by setting

$$U_x\left(\sum_{j=1}^{n} T_{f_j}\xi_j\right) := \sum_{j=1}^{n} T_{L_{\tilde{x}}f_j}(\xi_j) = \lim_i T_{L_{\tilde{x}}\varphi_i}\left(\sum_{j=1}^{n} T_{f_j}\xi_j\right).$$

U_x is well-defined because $\sum_{j=1}^{n} T_{f_j}\xi_j = 0$ implies

$$\sum_{j=1}^{n} T_{L_{\tilde{x}}f_j}(\xi_j) = \lim_i T_{L_{\tilde{x}}\varphi_i}\left(\sum_{j=1}^{n} T_{f_j}\xi_j\right) = 0.$$

Moreover, $\|T_{L_{\tilde{x}}\varphi_i}\| \leq \|L_{\tilde{x}}\varphi_i\| \leq 1$ for each $i \in I$, and hence U_x is a bounded operator on H_0, which can be extended uniquely to a bounded operator V_x on H such that $\|V_x\| = \|U_x\| \leq 1$.

In exactly the same manner as in Theorem 2.3.5 – by replacing $T_f\xi$ by finite sums $\sum_{j=1}^{n} T_{f_j}\xi_j$ – we can show that V is the unique bounded representation of K on H satisfying $\langle T_f\xi, \eta\rangle = \int_K \langle V_x\xi, \eta\rangle\, f(x)dm(x)$ for all $\xi, \eta \in H$, $f \in L^1(K)$. That T and V have the same closed invariant subspaces is already shown in Theorem 2.3.5. It remains to prove that $V_{\tilde{x}} = V_x^*$ for all $x \in K$. We note that

$$V_xT_f = T_{L_{\tilde{x}}f} = T_{\epsilon_x * f} \quad \text{and} \quad T_fV_x = T_{f*\epsilon_x},$$

which follows immediately from the definition of V_x. Since $T_{f^*} = (T_f)^*$, we obtain for $f, g \in L^1(K)$; $\xi, \eta \in H$,

$$\begin{aligned}\langle V_xT_f\xi, T_g\eta\rangle &= \langle T_{\epsilon_x*f}\xi, T_g\eta\rangle = \langle \xi, (T_{\epsilon_x*f})^*(T_g\eta)\rangle \\ &= \langle \xi, T_{(\epsilon_x*f)^*}(T_g\eta)\rangle = \langle \xi, T_{f^**\epsilon_{\tilde{x}}*g}(\eta)\rangle \\ &= \langle \xi, (T_f)^*V_{\tilde{x}}(T_g\eta)\rangle = \langle T_f\xi, V_{\tilde{x}}(T_g\eta)\rangle.\end{aligned}$$

Obviously this identity holds also for linear combinations

$$\sum_{j=1}^{n} T_{f_j}\xi_j \qquad \text{and} \qquad \sum_{k=1}^{m} T_{g_k}\eta_k,$$

and so $V_{\tilde{x}} = V_x^*$ follows. ◇

In the proof of Theorem 2.3.5 we have shown that the constructed bounded representation V on K is strongly continuous. A representation V of K is called **strongly continuous** if $x \mapsto \|V_x\xi\|,\ K \to [0,\infty[$ is continuous for each $\xi \in H$. This continuity condition corresponds to the strong operator topology on $\mathcal{B}(H)$, whereas we have chosen for the general definition of representations on K the weak operator topology, i.e. $x \mapsto \langle V_x\xi, \eta\rangle$ is continuous for all $\xi, \eta \in H$. Adapting some steps of the proof of Theorem 2.3.5 (also compare [304,Theorem 22.8]) we can show:

Proposition 2.3.7 *Let V be a bounded representation of K on a Hilbert space H. Then the mapping $x \mapsto \|V_x\xi\|$ is continuous for each $\xi \in H$.*

Proof. Let T be the representation on $\mathcal{A} = L^1(K)$ constructed from the representation V as in Proposition 2.3.1. In a first step we show that $\eta \in \{T_f\eta : f \in L^1(L)\}^-$ for each $\eta \in H$. Suppose on the contrary that there exists $\eta \in H$ such that $\eta \notin \{T_f\eta : f \in L^1(K)\}^-$. Then there is some $\xi \in H$ such that

$$\langle T_f\eta, \xi\rangle = 0 \qquad \text{for all } f \in L^1(K) \quad \text{and} \quad \langle \eta, \xi\rangle = 1.$$

By the continuity of $x \mapsto \langle V_x\eta, \xi\rangle$ in $x = e$ we get some neighborhood U of e with compact closure such that Re $\langle V_x\eta, \xi\rangle > \frac{1}{2}$ for all $x \in U$. Put $f = \chi_U$. Then

$$\text{Re}\,\langle T_f\eta, \xi\rangle = \int_K \text{Re}\,\langle V_x\eta, \xi\rangle\, f(x)\, dm(x) > \frac{1}{2}\, m(U) > 0,$$

a contradiction.

By the definition of T_f we get directly $V_xT_f = T_{\epsilon_x * f} = T_{L_{\tilde{x}}f}$. It follows for $x, x_0 \in K$ and $f \in L^1(K)$, $\eta, \xi \in H$,

$$|\langle V_xT_f\eta - V_{x_0}T_f\eta, \xi\rangle| \le \|V\|\,\|\eta\|\,\|\xi\|\,\|L_{\tilde{x}}f - L_{\tilde{x}_0}f\|_1.$$

We infer that

$$\|V_xT_f\eta - V_{x_0}T_f\eta\| \le \|V\|\,\|\eta\|\,\|L_{\tilde{x}}f - L_{\tilde{x}_0}f\|_1.$$

Let $\varepsilon > 0$. By Proposition 2.2.11 there is a neighborhood $W = W_f$ of x_0 such that $\|V_xT_f\eta - V_{x_0}T_f\eta\| < \frac{\varepsilon}{3}$ for every $x \in W$. Now choose $f \in L^1(K)$ such that $\|T_f\eta - \eta\| < \frac{\varepsilon}{3\,\|V\|}$. Then

$$\|V_x\eta - V_{x_0}\eta\| \le \|V_x\eta - V_xT_f\eta\| + \|V_xT_f\eta - V_{x_0}T_f\eta\|$$

$$+ \|V_{x_0}T_f\eta - V_{x_0}\eta\| < \varepsilon,$$

and the continuity of $x \mapsto \|V_x\eta\|$ in $x = x_0$ is shown. $\diamond$

Now we recall some standard terminology associated with representations on K. If V_1 and V_2 are representations of K on H_1 and H_2, respectively, an **intertwining** operator for V_1 and V_2 is a bounded linear operator $A \in \mathcal{B}(H_1, H_2)$ such that $AV_1(x) = V_2(x)A$ for all $x \in K$. The set of all intertwining operators for V_1 and V_2 is denoted by $\mathcal{C}(V_1, V_2)$. We write $\mathcal{C}(V)$ for $\mathcal{C}(V, V)$. $\mathcal{C}(V)$ is called **centralizer** of V. V_1 and V_2 are called **equivalent** if $\mathcal{C}(V_1, V_2)$ contains a bijective isometry. Note, if $A \in \mathcal{B}(H_1, H_2)$ is a bijective isometry, then $A^{-1} \in \mathcal{B}(H_2, H_1)$ is also a bijective isometry. Moreover, A is a bijective isometry if and only if A^*A an AA^* are the identity operators on H_1 and H_2, respectively. In particular, $A^* = A^{-1}$.

If $H_0 \subseteq H$ is a closed invariant subspace under V, then $V^{H_0}(x) = V_x|H_0$ defines a representation of K on H_0. V^{H_0} is said to be a **subrepresentation** of V on H_0. If V admits a closed invariant subspace H_0 which is nontrivial (i.e. $H_0 \neq \{0\}$, $H_0 \neq H$) and for which $H_0^\perp$ is also invariant, then V is called **reducible**. Otherwise V is **irreducible**.

Lemma 2.3.8 *Let V be a representation of K on a Hilbert space H, and let H_0 be a closed subspace of H. Both H_0 and $H_0^\perp$ are invariant under V if and only if H_0 is invariant under V and V^*, where $V^* = \{V_x^* : x \in K\}$.*

Proof. Assume H_0 and $H_0^\perp$ are invariant under V. Let $\xi \in H_0$, $\eta \in H_0^\perp$. Then $\langle V_x^*\xi, \eta\rangle = \langle \xi, V_x\eta\rangle = 0$. Hence H_0 is invariant under V^*. Conversely, if H_0 is invariant under V and V^*, consider $\xi \in H_0^\perp$, $\eta \in H_0$. Then $\langle V_x\xi, \eta\rangle = \langle \xi, V_x^*\eta\rangle = 0$. Hence $H_0^\perp$ is also invariant under V. $\diamond$

Remark: Let $K = \mathbb{N}_0$ be a polynomial hypergroup generated by $(R_n(x))_{n\in\mathbb{N}_0}$. By

$$V_n\begin{pmatrix} z_1 \\ z_2 \end{pmatrix} = \begin{pmatrix} 1 & R_n'(1) \\ 0 & 1 \end{pmatrix}\begin{pmatrix} z_1 \\ z_2 \end{pmatrix} = \begin{pmatrix} z_1 + R_n'(1)z_2 \\ z_2 \end{pmatrix}$$

a representation V of $K = \mathbb{N}_0$ on $H = \mathbb{C}^2$ is given. Note that differentiation of the linearization of the products $R_n(x)R_m(x)$ yields

$$R_n'(1) + R_m'(1) = \sum_{k=|n-m|}^{n+m} g(n,m;k)\, R_k'(1),$$

and property (ii) of representations follows immediately. The only nontrivial subspace H_0 invariant under V is $\mathbb{C}\binom{1}{0}$. In particular $H_0^\perp$ is not invariant under V.

If V is a $*$-representation of K on H, H_0 a closed invariant subspace of H, then $H_0^\perp$ is also invariant under V. This is obvious by Lemma 2.3.8. In particular we get the following result by Zorn's Lemma.

Proposition 2.3.9 *Every $*$-representation of K on H is a direct sum of cyclic representations.*

Proposition 2.3.10 *Let V be a representation of K on H. Let H_0 be a closed subspace of H, and P the orthogonal projection onto H_0. Then both H_0 and $H_0^\perp$ are invariant under V if and only if $P \in \mathcal{C}(V)$.*

Proof. Assume $P \in \mathcal{C}(V)$. For $\eta \in H_0$ we have $V_x\eta = V_xP\eta = PV_x\eta \in H_0$. Since $V_x^*P = (PV_x)^* = (V_xP)^* = PV_x^*$, H_0 is also invariant under V^*. By Lemma 2.3.8 both H_0 and $H_0^\perp$ are invariant under V. Conversely, assume that both H_0 and $H_0^\perp$ are invariant under V. For $\eta \in H_0$ it follows $V_xP\eta = V_x\eta \in H_0$, and therefore $PV_x\eta = V_x\eta = V_xP\eta$. If $\xi \in H_0^\perp$, then $V_xP\xi = 0 = PV_x\xi$. Hence $V_xP = PV_x$, i.e. $P \in \mathcal{C}(V)$. ◇

To apply Schur's Lemma for representations (not necessarily for $*$-representations) we have to deal with

$$\mathcal{C}_*(V) := \{T \in \mathcal{B}(H) : V_xT = TV_x,\ V_x^*T = TV_x^* \text{ for all } x \in K\}.$$

Theorem 2.3.11 (Schur's Lemma)
A representation V of K on H is irreducible if and only if $\mathcal{C}_*(V)$ contains only scalar multiples of the identity.

Proof. If V is reducible, there exists a nontrivial closed invariant subspace H_0 such that $H_0^\perp$ is also invariant. By Proposition 2.3.10 the orthogonal projection P onto H_0 is an element of $\mathcal{C}(V)$. Since $P = P^*$ we have even $P \in \mathcal{C}_*(V)$. Conversely, assume $T \in \mathcal{C}_*(V)$, $T \neq c\,\mathrm{id}$, $c \in \mathbb{C}$. Then $\mathrm{Re}\,T = \frac{1}{2}(T+T^*)$ and $\mathrm{Im}\,T = \frac{1}{2i}(T-T^*)$ are also elements of $\mathcal{C}_*(V)$, and at least one is different from a multiple of id. We may assume $\mathrm{Re}\,T$ is this operator. $\mathrm{Re}\,T$ is self-adjoint, and the spectral theorem tells us that each operator commuting with $\mathrm{Re}\,T$ commutes also with every orthogonal projection $P_E(\mathrm{Re}\,T)$, $E \subseteq \mathbb{R}$ a Borel set. Since all V_x and V_x^*, $x \in K$, commute with $\mathrm{Re}\,T$, all V_x and V_x^* commute with all orthogonal projections $P_E(\mathrm{Re}\,T)$. Hence $\mathcal{C}_*(V)$ contains nontrivial orthogonal projections. By Proposition 2.3.10 it follows that V is reducible. ◇

Corollary 2.3.12 *Assume that K is a commutative hypergroup.*

(1) Let V be an irreducible $*$-representation of K on a Hilbert space H. Then $\dim H = 1$.
(2) Let V be an irreducible representation of K on H consisting of normal operators V_x, $x \in K$. Then $\dim H = 1$.

Proof.

(1) The operators V_x all commute with one another. Hence every V_x belongs to $\mathcal{C}(V) = \mathcal{C}_*(V)$. Since V is irreducible, we have by Theorem 2.3.11 that $V_x = \alpha(x)\,\mathrm{id}$ with $\alpha(x) \in \mathbb{C}$ for each $x \in K$. In particular, every one-dimensional subspace of H is invariant under V.

(2) Since all V_x are normal, we get from the Theorem of Fuglede-Putnam-Rosenblum (see e.g. [609,Theorem 12.16]), that $\mathcal{C}(V) = \mathcal{C}_*(V)$, and $\dim H = 1$ follows as in (1).

◇

Remark: We want to point out that (even for commutative groups) there are $*$-representations which are not a direct sum of irreducible $*$-representations.

Corollary 2.3.13 *(1) Suppose V_1 and V_2 are irreducible $*$-representations of K on H_1 and H_2, respectively. If V_1 and V_2 are equivalent then* $\dim \mathcal{C}(V_1, V_2) = 1$. *Otherwise* $\mathcal{C}(V_1, V_2) = \{0\}$.

(2) Suppose V_1 and V_2 are irreducible representations of K on H_1 and H_2, respectively, consisting of normal operators $V_1(x), V_2(x)$, $x \in K$. If V_1 and V_2 are equivalent then $\dim \mathcal{C}(V_1, V_2) = 1$. *Otherwise* $\mathcal{C}(V_1, V_2) = \{0\}$.

Proof.

(1) Let $T \in \mathcal{C}(V_1, V_2)$. From $TV_1(x) = V_2(x)T$ follows $T^*V_2(\tilde{x}) = V_1(\tilde{x})T^*$ and vice versa. Hence $T^* \in \mathcal{C}(V_2, V_1)$, and we get

$$T^*TV_1(x) \;=\; T^*V_2(x)T \;=\; V_1(x)T^*T,$$

i.e. $T^*T \in \mathcal{C}(V_1) = \mathcal{C}_*(V_1)$, and

$$TT^*V_2(x) \;=\; TV_1(x)T^* \;=\; V_2(x)TT^*,$$

i.e. $TT^* \in \mathcal{C}(V_2) = \mathcal{C}_*(V_2)$. By Theorem 2.3.11 follows $T^*T = c_1\mathrm{id}$ and $TT^* = c_2\mathrm{id}$, $c_1, c_2 \in \mathbb{C}$. Hence $\|T\xi\|^2 = c_1\|\xi\|^2$ for all $\xi \in H_1$, and $\|T^*\eta\| = c_2\|\eta\|$ for all $\eta \in H_2$. It follows either $c_1 = c_2 = 0$ and $T = 0$, or $c_1 = c_2 > 0$ and $\frac{1}{\sqrt{c}}T$ is a bijective isometry. This shows $\mathcal{C}(V_1, V_2) = \{0\}$ exactly when V_1 and V_2 are not equivalent, and $\mathcal{C}(V_1, V_2)$ consists of scalar multiples of bijective isometries. If V_1 and V_2 are equivalent, let $S, T \in \mathcal{C}(V_1, V_2)$, $S \neq 0 \neq T$. Then $T^{-1}S = T^*S \in \mathcal{C}(V_1) = \mathcal{C}_*(V_1)$. By Theorem 2.3.11 holds $T^{-1}S = c\,\mathrm{id}$, $c \in \mathbb{C}$. Therefore $S = cT$, and so $\dim \mathcal{C}(V_1, V_2) = 1$.

(2) Let $T \in \mathcal{C}(V_1, V_2) \subseteq \mathcal{B}(H_1, H_2)$. From $TV_1(x) = V_2(x)T$ the normality of $V_1(x)$ and $V_2(x)$ yields $T(V_1(x))^* = (V_2(x))^*T$, see [390,Corollary 3.20], another version of the Fuglede-Putnam Theorem. Hence $T^*V_2(x) = V_1(x)T^*$, i.e. $T^* \in \mathcal{C}(V_2, V_1)$ Now we proceed as in (1) to show the statement.

◇

In Chapter 5 and 6 we will deal with many examples of representations in the context of weakly stationary random processes.

An important representation arises from the action of K on $L^2(K)$ by left (or right) translation. This is nothing but the representation $x \mapsto L_{\tilde{x}}$, $K \to \mathcal{B}(L^2(K))$ with $L_{\tilde{x}}f(y) = \epsilon_x * f(y) = \omega(\tilde{x}, y)(f)$, $f \in L^2(K)$. This representation is called **left regular representation** of K.

It is easily shown that $x \mapsto L_{\tilde{x}}$ is actually a $*$-representation of K. One has only to use Proposition 2.3.4 and Theorem 2.1.8.

The corresponding left regular representations of $M(K)$ or $L^1(K)$ are given by $L_\mu(g) = \mu * g$ for $\mu \in M(K)$, $g \in L^2(K)$ and $L_f(g) = f * g$ for $f \in L^1(K)$, $g \in L^2(K)$. The left regular representation is also **faithful**, i.e. $x \mapsto L_{\tilde{x}}$ is injective. In fact we have

Proposition 2.3.14 *The left regular representation is faithful.*

Proof. If $\mu \in M(K)$, $\mu \neq 0$, there exists some $g \in C_c(K)$ with $\int_K \tilde{g}(x)\, d\mu(x) \neq 0$, i.e. $L_\mu g(e) \neq 0$. By Theorem 1.1.15 the function $L_\mu(g) = \mu * g$ is continuous, and hence $L_\mu(g) \neq 0$ in $L^2(K)$. ◇

4 Positive definiteness and $*$-representations

There is a close link between $*$-representations of K and positive definite functions on K. Positive definiteness is also very important for investigating random fields indexed by K, and many other parts of harmonic analysis.

Definition. Let $\varphi : K \to \mathbb{C}$ be a Borel measurable, locally bounded function. We call φ positive definite if for all choices of $n \in \mathbb{N}$, $c_1, c_2, \ldots, c_n \in \mathbb{C}$ and $x_1, x_2, \ldots, x_n \in K$

$$\sum_{i=1}^{n}\sum_{j=1}^{n} c_i \overline{c_j}\, \omega(x_i, \tilde{x}_j)(\varphi) \geq 0.$$

In contrast to the group case a positive definite function need not to be bounded. Our main interest will be on bounded continuous positive definite functions. The set of these functions will be denoted by $P^b(K)$.

We can easily prove the following elementary properties of positive definite functions φ :

(i) $\bar{\varphi}$ and $\tilde{\varphi}$ are positive definite, too.
(ii) $\omega(x,\tilde{x})(\varphi) \geq 0$ for all $x \in K$, in particular $\varphi(e) \geq 0$.
(iii) $\varphi(\tilde{x}) = \overline{\varphi(x)}$ for all $x \in K$.
(iv) $|\varphi(x)|^2 \leq \varphi(e)\omega(x,\tilde{x})(\varphi)$ for all $x \in K$, in particular $|\varphi(x)|^2 \leq \varphi(e)\|\varphi\|_\infty$, if φ is bounded.
(v) Hence $\varphi(e) = \|\varphi\|_\infty$, if φ is bounded.
(vi) $2|\varphi(x)| \leq \varphi(e) + \omega(x,\tilde{x})(\varphi)$ for all $x \in K$.

Conditions (ii) to (vi) follow from

$$\varphi(e) + c\varphi(x) + \bar{c}\varphi(\tilde{x}) + |c|^2\omega(x,\tilde{x})(\varphi) \geq 0$$

for all $c \in \mathbb{C}$ and $x \in K$.

We add another inequality (Krein's inequality for hypergroups) for positive definite functions $\varphi \in P^b(K)$.

(vii) For $x, y \in K$ holds

$$\begin{aligned} |\varphi(x) - \varphi(y)|^2 &\leq \varphi(e)\,(\omega(x,\tilde{x})(\varphi) + \omega(y,\tilde{y})(\varphi) - 2\,\mathrm{Re}\,\omega(x,\tilde{y})(\varphi)) \\ &\leq 2\varphi(e)\,(\varphi(e) - \mathrm{Re}\,\omega(x,\tilde{y})(\varphi)). \end{aligned}$$

For the proof of (vii) consider the polynomial $P(\lambda)$, where

$$P(\lambda) = \sum_{i=1}^{3}\sum_{j=1}^{3} c_i\bar{c}_j\,\omega(x_i,\tilde{x}_j)(\varphi) \geq 0,$$

with

$$x_1 = e,\ x_2 = x,\ x_3 = y,\ c_1 = 1, \quad c_2 = \frac{\lambda|\varphi(x) - \varphi(y)|}{\varphi(x) - \varphi(y)}$$

and $c_3 = -c_2$ with $\lambda \in \mathbb{R}$. (If $\varphi(x) = \varphi(y)$, put $c_2 = \lambda$.)

A straightforward calculation shows that

$$\begin{aligned} P(\lambda) &= \lambda^2\,(\omega(x,\tilde{x})(\varphi) + \omega(y,\tilde{y})(\varphi) - \omega(x,\tilde{y})(\varphi) - \omega(y,\tilde{x})(\varphi)) \\ &\quad + 2\lambda\,|\varphi(x) - \varphi(y)| + \varphi(e). \end{aligned}$$

The discriminant of the quadratic polynomial $P(\lambda)$ cannot be positive. This is exactly the case when the inequality (vii) is valid. By (ii) and (v) we have $0 \leq \omega(z,\tilde{z}) \leq \varphi(e)$ for each $z \in K$, and the second inequality of (vii) follows.

We will derive other variants of Krein's inequality for hypergroups later on, see Corollary 2.4.7.

Let $\varphi : K \to \mathbb{C}$ be positive definite.

(viii) If $\varphi(e) = 1$ we have for every $x, y \in K$

$$|\omega(x,\tilde{y})(\varphi) - \varphi(x)\varphi(\tilde{y})|^2 \leq \left(\omega(x,\tilde{x})(\varphi) - |\varphi(x)|^2\right)\left(\omega(y,\tilde{y})(\varphi) - |\varphi(y)|^2\right)$$

In fact this inequality is equivalent to the nonnegativity of

$$\det \begin{pmatrix} 1 & \varphi(\tilde{x}) & \varphi(\tilde{y}) \\ \varphi(x) & \omega(x,\tilde{x})(\varphi) & \omega(x,\tilde{y})(\varphi) \\ \varphi(y) & \omega(\tilde{y},x)(\varphi) & \omega(y,\tilde{y})(\varphi) \end{pmatrix}.$$

Theorem 2.4.1 *A positive definite function φ continuous in e is continuous everywhere.*

Proof. We may assume that $\varphi(e) = 1$. In fact $\varphi(e) = 0$ implies $\varphi = 0$, see (iv). We apply inequality (viii). Let $x_0 \in K$, $\varepsilon > 0$, and choose a compact neighbourhood U of x_0. Since φ is locally bounded there is a constant $M > 0$ such that

$$\left(\omega(x,\tilde{x})(\varphi) - |\varphi(x)|^2\right)^{1/2} \leq M \quad \text{and} \quad |\varphi(x)| \leq M \qquad \text{for all } x \in U.$$

By the continuity of φ in e there is some compact neighbourhood $V = \widetilde{V}$ of e such that

$$\left(\omega(y,\tilde{y})(\varphi) - |\varphi(y)|^2\right)^{1/2} < \varepsilon/(2M) \quad \text{and} \quad |1 - \varphi(\tilde{y})| < \varepsilon/(2M) \quad \text{for all } y \in V.$$

Hence we obtain

$$\begin{aligned} &|\omega(\tilde{y},x)(\varphi) - \varphi(x)| \leq |\varphi(x) - \varphi(x)\varphi(\tilde{y})| + |\omega(\tilde{y},x)(\varphi) - \varphi(x)\varphi(\tilde{y})| \\ &\leq |\varphi(x)|\,|1 - \varphi(\tilde{y})| + \left(\omega(x,\tilde{x})(\varphi) - |\varphi(x)|^2\right)^{1/2}\left(\omega(y,\tilde{y})(\varphi) - |\varphi(y)|^2\right)^{1/2} \\ &< \varepsilon \qquad \text{for all } x \in U,\ y \in V. \end{aligned}$$

Now we can derive that φ is continuous in x_0. Put $k = \chi_V/m(V)$. For $x \in U$ we have

$$|k * \varphi(x) - \varphi(x)| \leq \int_K |\omega(\tilde{y},x)(\varphi) - \varphi(x)|k(y)\,dm(y) < \varepsilon$$

with the definition $k * \varphi(x) = \int_K \omega(\tilde{y},x)(\varphi)k(y)\,dm(y)$, where the integral exists, since φ is locally bounded. Now it is easy to check that for $x \in U$

$$k * \varphi(x) = k * (\chi_{U*V}\varphi)(x)$$

and k and $\chi_{U*V}\varphi$ are elements of $L^2(K)$. By Proposition 2.2.13 $k*(\chi_{U*V}\varphi)$ is a continuous function. Thus $k * \varphi$ equals on U a continuous function, and so φ is continuous in x_0. ◇

If φ_1 and φ_2 are positive definite functions on K, then $\lambda_1\varphi_1 + \lambda_2\varphi_2$ is positive definite for all $\lambda_1, \lambda_2 \geq 0$. This result is easily shown. One should note that the product of two positive definite functions on K is in general not positive definite. This fact is again in contrast to the group case.

If φ is positive definite, $\mu = \sum_{i=1}^{n} c_i \delta_{x_i}$ then $\mu * \mu^* = \sum_{i,j=1}^{n} c_i \bar{c}_j \omega(x_i, \tilde{x}_j)$ and so

$$\mu * \mu^* * \varphi(e) = \sum_{i,j=1}^{n} c_i \bar{c}_j \, \omega(x_i, \tilde{x}_j)(\varphi) \geq 0.$$

This result can be extended.

If $\mu \in M(K)$ has compact support, one can approximate μ by measures with finite support. Hence we have the following equivalent conditions.

Proposition 2.4.2 *Let $\varphi : K \to \mathbb{C}$ be a continuous function. The following statements are equivalent:*

(i) φ *is positive definite.*
(ii) $\int_K \varphi(x)\, d(\mu * \mu^*)(x) \geq 0$ *for all* $\mu \in M(K)$ *with compact support.*
(iii) $\int_K \varphi(x) f * f^*(x)\, dm(x) \geq 0$ *for all* $f \in C_c(K)$.

If in addition φ is bounded, these three conditions are equivalent to

(iv) $\int_K \varphi(x)\, d(\mu * \mu^*)(x) \geq 0$ *for all* $\mu \in M(K)$.
(v) $\int_K \varphi(x) f * f^*(x)\, dm(x) \geq 0$ *for all* $f \in L^1(K)$.

Important examples of $\varphi \in P^b(K)$ (i.e. φ is continuous, bounded and positive definite) are obtained by convolving $g \in L^2(K)$ with $\overline{\tilde{g}}$.

In fact by Proposition 2.2.13 we already know $\varphi = g * \overline{\tilde{g}} \in C_0(K)$. Furthermore it is easy to show for $c_1, \ldots, c_n \in \mathbb{C}$ and $x_1, \ldots, x_n \in K$ that

$$\sum_{i=1}^{n}\sum_{j=1}^{n} c_i \overline{c_j} \omega(x_i, \tilde{x}_j)(\varphi)$$
$$= \sum_{i=1}^{n}\sum_{j=1}^{n} c_i \overline{c_j} \int_K \int_K \overline{g(\tilde{z})} \omega(\tilde{z}, y)(g)\, dm(z)\, d\omega(x_i, \tilde{x}_j)(y)$$
$$= \sum_{i=1}^{n}\sum_{j=1}^{n} c_i \overline{c_j} \int_K \overline{g(\tilde{z})} \epsilon_{\tilde{z}} * \omega(x_i, \tilde{x}_j)(g)\, dm(z)$$

$$= \sum_{i=1}^{n}\sum_{j=1}^{n} c_i\overline{c_j} \int_K \overline{g(\tilde{z})} \epsilon_{\tilde{x}_j} * \omega(x_i, \tilde{z})(g)\, dm(z)$$

$$= \sum_{i=1}^{n}\sum_{j=1}^{n} c_i\overline{c_j} \int_K \overline{g(\tilde{z})} L_{\tilde{x}_j} (L_{x_i} g(\tilde{z}))\, dm(z)$$

$$= \sum_{i=1}^{n}\sum_{j=1}^{n} c_i\overline{c_j} \int_K \overline{L_{x_j} g(\tilde{z})} L_{x_i} g(\tilde{z})\, dm(z) = \int_K \left| \sum_{i=1}^{n} c_i L_{x_i} g(\tilde{z}) \right|^2 dm(z) \geq 0.$$

So we have shown

Proposition 2.4.3 *For every $g \in L^2(K)$ we have $\varphi = g * \overline{\tilde{g}}$ is an element of $P^b(K)$.*

In section 2.3 we have investigated the relation between nondegenerate $*$-representations of $L^1(K)$ and bounded $*$-representations V of K. We intend to include in this relation bounded continuous positive definite functions. First of all we show

Proposition 2.4.4 *If V is a bounded $*$-representation of K on a Hilbert space H and $\zeta \in H$, let $\varphi(x) = \phi_{\zeta,\zeta}(x) = \langle V_x\zeta, \zeta\rangle$. Then $\varphi \in P^b(K)$.*

Proof. In fact φ is continuous and $|\varphi(x)| \leq \|V\|\ \|\zeta\|^2 \leq \|\zeta\|^2$. Also, if $c_1, ..., c_n \in \mathbb{C}$; $x_1, ..., x_n \in K$,

$$\sum_{i=1}^{n}\sum_{j=1}^{n} c_i\bar{c}_j\ \omega(x_i, \tilde{x}_j)(\varphi) = \sum_{i=1}^{n}\sum_{j=1}^{n} c_i\bar{c}_j\ \langle V_{x_i} V_{\tilde{x}_j}\xi, \xi\rangle$$

$$= \sum_{i=1}^{n}\sum_{j=1}^{n} c_i\bar{c}_j\ \langle V_{\tilde{x}_j}\xi, V_{\tilde{x}_i}\xi\rangle = \|\sum_{i=1}^{n} c_i V_{\tilde{x}_i}\xi\|^2 \geq 0.$$

◇

Now we investigate the relations between continuous bounded positive definite functions on K, positive functionals on $L^1(K)$ and bounded $*$-representations on K.

A linear functional $p : L^1(K) \to \mathbb{C}$ is called positive, if $p(f * f^*) \geq 0$ for each $f \in L^1(K)$. Since $L^1(K)$ has a bounded approximate identity, see Proposition 2.3.3, Varopoulos Theorem (see [305,Theorem 32.27]) says that each positive linear functional on $L^1(K)$ is continuous.

Theorem 2.4.5 *Let $\varphi \in C^b(K)$. The following assertions are equivalent*

(1) $\varphi \in P^b(K)$.

(2) The linear functional $p_\varphi : L^1(K) \to \mathbb{C}$,

$$p_\varphi(f) = \int_K f(x)\,\varphi(x)\,dm(x)$$

is positive.

(3) There exists a bounded $*$-representation V of K on a Hilbert space H such that $\varphi(x) = \langle V_x\zeta, \zeta\rangle$ for some $\zeta \in H$.

Proof. (1) implies (2) by Proposition 2.4.2. Assume that p_φ is positive. We want to apply Theorem 21.24 of [304]. The assumptions of this Theorem are satisfied, since $L^1(K)$ has a bounded approximate identity, see [305,Corollary 32.28]. So we obtain that there exists a cyclic $*$-representation T of $L^1(K)$ on H with cyclic vector $\zeta \in H$ such that

$$p_\varphi(f) = \langle T_f\zeta, \zeta\rangle \qquad \text{for all } f \in L^1(K).$$

By Theorem 2.3.6 there exists a bounded $*$-representation V of K on H with

$$\langle T_f\xi, \eta\rangle = \int_K \langle V_x\xi, \eta\rangle\, f(x)\,dm(x) \qquad \text{for all } f \in L^1(K).$$

Write $\psi(x) = \langle V_x\zeta, \zeta\rangle$ for $x \in K$. We know already that $\psi \in P^b(K)$, see Proposition 2.4.4. Moreover,

$$\int_K \varphi(x)\,f(x)\,dm(x) = \langle T_f\zeta, \zeta\rangle = \int \psi(x)\,f(x)\,dm(x)$$

for every $f \in L^1(K)$. Since the functions φ and ψ are continuous, they are identical. So we have shown that (2) implies (3). Finally (3) $\Rightarrow$ (1) is exactly Proposition 2.4.4. ◇

The proof of the implication (2) $\Rightarrow$ (3) contains the following result.

Proposition 2.4.6 *If $\varphi \in L^\infty(K)$ such that the linear functional on $L^1(K)$, $p_\varphi(f) = \int_K f(x)\varphi(x)dm(x)$ is positive, then φ is locally m-almost everywhere equal to a function from $P^b(K)$. (Such functions $\varphi \in L^\infty(K)$ are also called* **of positive type***.)*

Corollary 2.4.7 *Let $\varphi \in P^b(K)$. Then*

(a) $|L_z\varphi(x) - L_z\varphi(y)|^2 \le 2\,\varphi(e)\,(\varphi(e) - \mathrm{Re}\, L_{\tilde{x}}\varphi(y))$ *for all $x, y, z \in K$.*

(b) $$\left|\sum_{i=1}^{n}\sum_{j=1}^{m} L_{y_j}\varphi(x_i)\, a_i\bar{b}_j\right|^2 \le \left(\sum_{i,j=1}^{n} L_{\tilde{x}_j}\varphi(x_i)\, a_i\bar{a}_j\right)\left(\sum_{i,j=1}^{m} L_{y_j}\varphi(\tilde{y}_i)\, b_i\bar{b}_j\right)$$
for all choices of $m,n \in \mathbb{N}$; $a_1,...,a_n \in \mathbb{C}$; $b_1,...,b_m \in \mathbb{C}$ *and* $x_1,...,x_n \in K$; $y_1,...,y_m \in K$.

(c) $$\left|\sum_{i=1}^{n} a_i\varphi(x_i)\right|^2 \le f(e) \sum_{i,j=1}^{n} L_{\tilde{x}_j}\varphi(x_i)\, a_i\bar{a}_j$$
for all choices of $n \in \mathbb{N}$; $a_1,...,a_n \in \mathbb{C}$ *and* $x_1,...,x_n \in K$.

Proof. We use that $\varphi(x) = \langle V_x\zeta, \zeta\rangle$ for all $x \in K$, V the $*$-representation of Theorem 2.4.5.

(a) We have

$$|L_z\varphi(x) - L_z\varphi(y)|^2 = |\langle V_zV_x\zeta,\zeta\rangle - \langle V_zV_y\zeta,\zeta\rangle|^2$$

$$= |\langle V_x\zeta - V_y\zeta, V_{\tilde{z}}\zeta\rangle|^2 \le \|V_x\zeta - V_y\zeta\|^2\, \|V_{\tilde{z}}\zeta\|^2.$$

Since $\|\varphi\|_\infty = \varphi(e)$, it follows $\|V_{\tilde{z}}\zeta\|^2 = \langle V_zV_{\tilde{z}}\zeta,\zeta\rangle = \omega(z,\tilde{z})(\varphi) \le \varphi(e)$, and since $\varphi(\tilde{x}) = \overline{\varphi(x)}$, we have

$$\|V_x\zeta - V_y\zeta\|^2 \le 2\,\varphi(e) - 2\,\mathrm{Re}\, L_{\tilde{x}}\varphi(y),$$

and (a) is shown.

(b) Let $\xi = \sum_{i=1}^{n} a_iV_{x_i}\zeta$, $\eta = \sum_{j=1}^{m} b_jV_{\tilde{y}_j}\zeta$. The Cauchy inequality

$$|\langle\xi,\eta\rangle|^2 \le \langle\xi,\xi\rangle\,\langle\eta,\eta\rangle$$

is exactly (b).

(c) Setting $m = 1$, $y_1 = e$, $b_1 = 1$, we get (c).

◇

We have to study the effect of the choice of the vector $\zeta \in H$ in Proposition 2.4.4 and Theorem 2.4.5 (3). First of all note that the bounded $*$-representation V in Proposition 2.4.4 and Theorem 2.4.5 (3) is not assumed cyclic. However, the mapping $x \mapsto \langle V_x\zeta,\zeta\rangle$ depends only on the cyclic subrepresentation of V on the closed invariant subspace H_0 generated by $\{V_x\zeta : x \in K\}$. Hence we can suppose in both statements that V is cyclic. Furthermore we have the following uniqueness theorem, which has a straightforward proof.

Theorem 2.4.8 *Assume that V_1 and V_2 are bounded cyclic $*$-representations of K with cyclic vector $\zeta_1 \in H_1$ and $\zeta_2 \in H_2$, respectively. If $\langle V_1(x)\zeta_1, \zeta_1\rangle = \langle V_2(x)\zeta_2, \zeta_2\rangle$ for all $x \in K$, then V_1 and V_2 are equivalent. More precisely, there exists a bijective isometry $A \in \mathcal{C}(V_1, V_2)$ such that $A\zeta_1 = \zeta_2$.*

Proof. For $x, y \in K$ we have

$$\langle V_1(x)\zeta_1, V_1(y)\zeta_1\rangle = \langle V_1(\tilde{y})V_1(x)\zeta_1, \zeta_1\rangle = \int_K \langle V_1(z)\zeta_1, \zeta_1\rangle \, d\omega(\tilde{y}, x)(z)$$

$$= \int_K \langle V_2(z)\zeta_2, \zeta_2\rangle \, d\omega(\tilde{y}, x)(z) = \langle V_2(x)\zeta_2, V_2(y)\zeta_2\rangle.$$

Set $A(V_1(x)\zeta_1) = V_2(x)\zeta_2$. The linear extension of A to the linear span of $\{V_1(x)\zeta_1 : x \in K\}$ yields an isometry of the linear span of $\{V_1(x)\zeta_1 : x \in K\}$ onto the linear span of $\{V_2(x)\zeta_2 : x \in K\}$. By continuity we get a bijective isometry $A \in \mathcal{B}(H_1, H_2)$ such that $A\zeta_1 = \zeta_2$. Given $\xi_2 \in H_2$ we obtain for $x, y \in K$,

$$\langle AV_1(y)V_1(x)\zeta_1, \xi_2\rangle = \langle V_1(y)V_1(x)\zeta_1, A^{-1}\xi_2\rangle$$

$$= \int_K \langle V_1(z)\zeta_1, A^{-1}\xi_2\rangle \, d\omega(y, x)(z) = \int_K \langle AV_1(z)\zeta_1, \xi_2\rangle \, d\omega(y, x)(z)$$

$$= \int_K \langle V_2(z)\zeta_2, \xi_2\rangle \, d\omega(y, x)(z) = \langle V_2(y)V_2(x)\zeta_2, \xi_2\rangle.$$

Hence we conclude that $AV_1(y)V_1(x)\zeta_1 = V_2(y)V_2(x)\zeta_2$, and therefore $V_2(y)AV_1(x)\zeta_1 = V_2(y)V_2(x)\zeta_2 = AV_1(y)V_1(x)\zeta_1$. It follows $V_2(y)A = AV_1(y)$, i.e. $A \in \mathcal{C}(V_1, V_2)$. ◇

Remark: We combine the results above: For each $\varphi \in P^b(K)$ there exists a bounded cyclic $*$-representation V of K on H with a cyclic vector $\zeta \in H$ such that $\varphi(x) = \langle V_x\zeta, \zeta\rangle$ for all $x \in K$. Those bounded cyclic $*$-representations that determine φ in this way, are equivalent. Therefore it makes sense to write V^φ for this representation.

Since supp $\omega(x, y)$ is compact for all $x, y \in K$, we have the following result.

Proposition 2.4.9 *$P^b(K)$ is a closed subset of $C^b(K)$ with respect to the topology of uniform convergence on compact subsets of K.*

We recall that the convolution of $f \in L^1(K)$ and $g \in L^\infty(K)$ is given by

$$f * g(x) = \int_K f(y)\, L_{\tilde{y}} g(x)\, dm(y),$$

see Lemma 2.2.7. As in Proposition 2.2.15 (1) we get $f * g(x) = \int_K L_x f(y) g(\tilde{y}) dm(y)$. Note that $f * g$ is an element of $C^b(K)$.

Proposition 2.4.10 *Let $\varphi \in P^b(K)$. Let V be a neighborhood of e with compact closure and put $f := \frac{1}{m(V)} \chi_V$. Then*

$$|\varphi(z) - f * \varphi(z)|^2 \leq 2\, \varphi(e)\, \frac{1}{m(V)} \int_V (\varphi(e) - Re\, \varphi(y))\, dm(y)$$

holds for all $z \in K$.

Proof. Since $\tilde{\varphi} = \bar{\varphi}$ we obtain for $z \in K$,

$$|\varphi(z) - f * \varphi(z)|^2 = \left| \int_K f(y) \varphi(z)\, dm(y) - \int_K f(y) L_{\tilde{y}} \varphi(z)\, dm(y) \right|^2$$

$$= \frac{1}{m(V)^2} \left| \int_K \chi_V(y)\, (\varphi(z) - L_{\tilde{y}} \varphi(z))\, dm(y) \right|^2$$

$$\leq \frac{1}{m(V)} \int_V |\varphi(z) - L_{\tilde{y}} \varphi(z)|^2\, dm(y) = \frac{1}{m(V)} \int_V |\varphi(\tilde{z}) - L_{\tilde{z}} \varphi(y)|^2\, dm(y).$$

Applying Corollary 2.4.7 (a) with $x = e$, we infer that

$$|\varphi(z) - f * \varphi(z)|^2 \leq 2\, \varphi(e)\, \frac{1}{m(V)} \int_V (\varphi(e) - \mathrm{Re}\, \varphi(y))\, dm(y).$$

◇

Proposition 2.4.11 *Let $f \in L^1(K)$ and let $(g_i)_{i \in I}$ be a net in $L^\infty(K)$ with $\|g_i\|_\infty \leq M$ for all $i \in I$. If this net converges in the weak $*$-topology to $g \in L^\infty(K)$, the net $(f * g_i)_{i \in I}$ converges to $f * g$ with respect to the uniform convergence on compact subsets of K.*

Proof. The linear functionals

$$\Phi_g(h) = \int_K h(y)\, g(\tilde{y})\, dm(y) = h * g(e)$$

and

$$\Phi_{g_i}(h) = \int_K h(y)\, g_i(\tilde{y})\, dm(y) = h * g_i(e), \qquad h \in L^1(K),$$

are continuous on $L^1(K)$, and since g_i converges in the weak $*$-topology to g we have $\lim_i \Phi_{g_i}(h) = \Phi_g(h)$. Furthermore,

$$|\Phi_{g_i}(h_1) - \Phi_{g_i}(h_2)| \leq \|g_i\|_\infty \|h_1 - h_2\|_1 \leq M \, \|h_1 - h_2\|_1.$$

Hence $\{\Phi_{g_i} : i \in I\}$ is an equicontinuous family of functions on $L^1(K)$. Ascoli's theorem implies that Φ_{g_i} converges to ϕ_g uniformly on compact subsets of $L^1(K)$. The mapping $x \mapsto L_x f$, $f \in L^1(K)$, is a continuous mapping from K into $L^1(K)$, see Proposition 2.2.11. Hence $\{L_x f : x \in C\}$, C a compact subset of K, is compact in $L^1(K)$. Since $f * g_i(x) = \Phi_{g_i}(L_x f)$ and $f * g(x) = \Phi_g(L_x f)$, it follows that $f * g_i$ converges uniformly on compact subsets of K towards $f * g$. ◇

Theorem 2.4.12 *Let $(\varphi_i)_{i \in I}$ be a net in $P^b(K)$ which converges in the weak $*$-topology to $\varphi \in P^b(K)$ and $\lim_i \varphi_i(e) = \varphi(e)$. Then $(\varphi_i)_{i\in I}$ converges to φ with respect to the uniform convergence on compact subsets of K.*

Proof. For any compact subset $C \subseteq K$ and $g \in C^b(K)$ we write $\|g\|_C = \sup\{|g(x)| : x \in C\}$. We consider $f \in L^1(K)$ of the form $f = \frac{1}{m(V)} \chi_V$, where V is a neighborhood of $e \in K$ with compact closure. We have

$$\|\varphi_i - \varphi\|_C \leq \|\varphi_i - f * \varphi_i\|_\infty + \|f * \varphi_i - f * \varphi\|_C + \|f * \varphi - \varphi\|_\infty.$$

Let $\varepsilon > 0$. Since φ is continuous, there exists a neighborhood V of e with compact closure such that

$$2\,\varphi(e)\,\frac{1}{m(V)} \int_V (\varphi(e) - \text{ Re } \varphi(y))\, dm(y) < \frac{\varepsilon^2}{9}.$$

By Proposition 2.4.10 we obtain

$$\|f * \varphi - \varphi\|_\infty < \frac{\varepsilon}{3}.$$

Moreover, since $(\varphi_i)_{i\in I}$ converges towards φ in the weak $*$-topology and since $\lim_i \varphi_i(e) = \varphi(e)$, there exists some $i_0 \in I$ such that

$$2\,\varphi_i(e)\,\frac{1}{m(V)} \int_V (\varphi_i(e) - \text{ Re } \varphi_i(y))\, dm(y) < \frac{\varepsilon^2}{9}$$

for all $i \geq i_0$. Again by Proposition 2.4.10 we infer that

$$\|\varphi_i - f * \varphi_i\|_\infty < \frac{\varepsilon}{3}$$

for $i \geq i_1$. Applying Proposition 2.4.11 there exists $i_1 \geq i_0$ such that

$$\|f * \varphi_i - f * \varphi\|_C < \frac{\varepsilon}{3}$$

for all $i \geq i_1$. Hence we get $\|\varphi_i - \varphi\|_C < \varepsilon$ for $i \geq i_1$. ◇

Corollary 2.4.13 *Let $(\varphi_n)_{n\in\mathbb{N}}$ be a sequence in $P^b(K)$ which converges pointwise on K to a $\varphi \in P^b(K)$. Then $(\varphi_n)_{n\in\mathbb{N}}$ converges to φ uniformly on compact subsets of K.*

Proof. Lebesgue's dominated convergence theorem implies that $(\varphi_n)_{n\in\mathbb{N}}$ converges to φ in the weak $*$-topology of $L^\infty(K)$. Now Theorem 2.4.12 yields the uniform convergence to φ on compact subsets of K. ◇

We have to investigate the implication of irreducibility of the bounded $*$-representation V in Theorem 2.4.5 on the corresponding positive definite function φ.

We denote by

$$P^1(K) = \{\varphi \in P^b(K) : \|\varphi\|_\infty = 1\} = \{\varphi \in P^b(K) : \varphi(e) = 1\}$$

and

$$P^0(K) = \{\varphi \in P^b(K) : \|\varphi\|_\infty \le 1\} = \{\varphi \in P^b(K) : \varphi(e) \le 1\}.$$

Proposition 2.4.14 *$P^0(K)$ is compact with respect to the weak $*$-topology of $L^\infty(K)$.*

Proof. $P^0(K)$ is obviously a bounded subset of $L^\infty(K)$. Let $\psi \in L^\infty(K)$ be a weak $*$-limit of a net $(\varphi_i)_{i\in I}$ in $P^0(K)$. The linear functionals $\Phi_{\varphi_i} \in L^\infty(K)$ determined by $\Phi_{\varphi_i}(h) = \int_K h(x)\varphi_i(x)dm(x),\ h \in L^1(K)$, are positive. Hence the linear functional Φ_ψ is also positive, and by Proposition 2.4.6 ψ is locally m-almost everywhere equal to a function from $L^\infty(K)$. Hence $P^0(K)$ is closed in the weak $*$-topology. By Alaoglu's theorem $P^0(K)$ is weak $*$-compact. ◇

Remark. $P^1(K)$ is in general not weak $*$-closed in $L^\infty(K)$, unless K is discrete.

We recall shortly the definition of extreme points. Let X be a vector space and $A \subseteq X$ a convex subset. A point $a \in A$ is called an **extreme point** of A if for all $x, y \in A$ and $\lambda \in]0,1[$ such that $a = \lambda x + (1-\lambda)y$, it follows that $x = y = a$. The set of extreme points of A is denoted by $\mathrm{ex}(A)$.

Considering the convex subsets $P^1(K)$ and $P^0(K)$ in $L^\infty(K)$, it is easy to derive for the constant functions $\varphi = 0$ and $\varphi = 1$ that $0 \in \mathrm{ex}(P^0(K))$ and $1 \in \mathrm{ex}(P^1(K))$. Moreover we have: If $\varphi \in P^0(K),\ 0 < \varphi(e) < 1,$ then $\varphi \notin \mathrm{ex}(P^0(K))$. In fact, put $\varphi_1 = \varphi/\varphi(e),\ \varphi_2 = 0$ and $\lambda = \varphi(e)$ to get $\varphi = \lambda\varphi_1 + (1-\lambda)\varphi_2$. If $\varphi \in P^1(K) \cap \mathrm{ex}(P^0(K))$, then obviously $\varphi \in \mathrm{ex}(P^1(K))$. Conversely, if $\varphi \in P^1(K)\backslash\mathrm{ex}(P^0(K))$, then $\varphi = \lambda\varphi_1 + (1-\lambda)\varphi_2,\ \varphi_1, \varphi_2 \in$

$P^0(K)$, $0 < \lambda < 1$. This implies $1 = \lambda\varphi_1(e) + (1-\lambda)\varphi_2(e)$, hence $\varphi_1(e) = \varphi_2(e) = 1$, and so $\varphi \notin \mathrm{ex}(P^1(K))$. Collecting those observations, we have

Proposition 2.4.15 *$0 \in ex(P^0(K))$, $1 \in ex(P^1(K))$ and*

$$ex(P^0(K)) = ex(P^1(K)) \cup \{0\}.$$

Now we can use the Krein-Milman theorem, see [609,3.21] to derive the following results.

Theorem 2.4.16 *(1) $P^0(K)$ is the weak $*$-closure of the convex hull of $ex(P^0(K))$.*
(2) The convex hull of $ex(P^1(K))$ is weak $*$-dense in $P^1(K)$.

Proof.

(1) $P^0(K) \subseteq L^\infty(K)$ is convex and weak $*$-compact, see Proposition 2.4.14. Hence the Krein-Milman theorem says that the convex hull of $\mathrm{ex}(P^0(K))$ is weak $*$-dense in $P^0(K)$.

(2) Let $\varphi \in P^1(K)$. By (1) and Proposition 2.4.15, φ is the weak $*$-limit of a net $(\varphi_i)_{i\in I}$, where the functions φ_i have the form

$$\varphi_i = c_1\psi_1 + \dots + c_n\psi_n + c_{n+1}\cdot 0, \quad c_1,\dots,c_{n+1} \geq 0,\ c_1+\dots+c_{n+1} = 1$$

and $\psi_1,\dots,\psi_n \in \mathrm{ex}(P^1(K))$.
Note that $\|\varphi\|_\infty = \varphi(e) = 1$ and $\|\varphi_i\|_\infty = \varphi_i(e) = c_1 + \dots + c_n \leq 1$. Since for each $\varepsilon > 0$ the set $\{g \in L^\infty(K) : \|g\|_\infty \leq 1-\varepsilon\}$ is weak $*$-closed, we infer that $\lim_i \varphi_i(e) = \lim_i \|\varphi_i\|_\infty = 1$.
Now set $\varrho_i = \varphi_i/\varphi_i(e)$ and $b_i = c_i/\varphi_i(e)$. Then $b_1,\dots,b_n \geq 0$, $b_1 + \dots + b_n = 1$ and $\varrho_i = b_1\psi_1 + \dots + b_n\psi_n$, i.e. ϱ_i is in the convex hull of $\mathrm{ex}(P^1(K))$, and $\lim_i \varrho_i = \varphi$ in the weak $*$-topology.

◇

Theorem 2.4.5 describes the relation between $\varphi \in P^b(K)$ and a cyclic bounded $*$-representation V of K via positive functionals on $L^1(K)$. There is another (more direct) way to construct a Hilbert space H and a bounded $*$-representation V of K on H, such that $\varphi(x) = \langle V_x\zeta, \zeta\rangle$, $\zeta \in H$.

Let $\varphi \in C^b(K)$, $\tilde{\varphi} = \bar{\varphi}$. We write $M_c(K) = \{\mu \in M(K) :$ supp $(|\mu|)$ compact$\}$. Denote by

$$F(\varphi) := \{\mu * \varphi : \mu \in M_c(K)\} \subseteq C^b(K).$$

If $\mu, \nu \in M_c(K)$ then $\mu * \nu \in M_c(K)$. Hence $F(\varphi)$ is a subspace of $C^b(K)$, which is invariant under convolution by $M_c(K)$ from the left hand side.

Write $g = \mu * \varphi$ and $h = \nu * \varphi$, whenever $\mu, \nu \in M_c(K)$, and define a sesquilinear form on $F(\varphi)$ by setting

$$\langle g, h\rangle_\varphi = \nu^* * \mu * \varphi(e).$$

Since $\tilde{\varphi} = \bar{\varphi}$, the definition of $\langle g, h\rangle_\varphi$ does not depend on the particular representation of g and h. In fact, by $\mu^* * \nu = (\nu^* * \mu)^*$ we get $\overline{\mu^* * \nu * \varphi(e)} = \nu^* * \mu * \varphi(e)$, and it follows that $\nu_1^* * \mu_1 * \varphi(e) = \nu_2^* * \mu_2 * \varphi(e)$, whenever $\mu_1 * \varphi = g = \mu_2 * \varphi$, $\nu_1 * \varphi = h = \nu_2 * \varphi$ for $\mu_1, \mu_2, \nu_1, \nu_2 \in M_c(K)$. Moreover, we have $\langle g, h\rangle_\varphi = \overline{\langle h, g\rangle_\varphi}$. If $\mu = \epsilon_x$ (i.e. $g = \epsilon_x * \varphi = L_{\tilde{x}}\varphi$) it follows

$$\langle L_{\tilde{x}}\varphi, h\rangle_\varphi = \nu^* * \epsilon_x * \varphi(e) = \overline{\epsilon_{\tilde{x}} * \nu * \varphi(e)} = \overline{\nu * \varphi(x)} = \overline{h(x)}.$$

Hence the **reproducing property** is valid:

$$h(x) = \langle h, L_{\tilde{x}}\varphi\rangle_\varphi \qquad \text{for each } h \in F(\varphi),\ x \in K.$$

Now we suppose in addition that $\varphi \in P^b(K)$. Note that $\varphi \in P^b(K)$ satisfies $\tilde{\varphi} = \bar{\varphi}$ and $\tilde{\varphi} \in P^b(K)$. If $g = \mu * \varphi$, $\mu \in M_c(K)$, we have

$$\langle g, g\rangle_\varphi = \mu^* * \mu * \varphi(e) = \int_K \tilde{\varphi}(z)\, d(\mu^* * \mu)(z) \geq 0.$$

Applying Cauchy's inequality it follows

$$|g(x)|^2 = |\langle g, L_{\tilde{x}}\varphi\rangle_\varphi|^2 \leq \langle g, g\rangle_\varphi\, \langle L_{\tilde{x}}\varphi, L_{\tilde{x}}\varphi\rangle_\varphi.$$

Hence $g = 0$ exactly when $\langle g, g\rangle_\varphi = 0$.

So far we have shown

Lemma 2.4.17 *Let $\varphi \in P^b(K)$ and $F(\varphi) := \{\mu * \varphi : \mu \in M_c(K)\}$. For $g = \mu * \varphi$, $h = \nu * \varphi$, if $\mu, \nu \in M_c(K)$, set*

$$\langle g, h\rangle_\varphi = \nu^* * \mu * \varphi(e).$$

Then $F(\varphi)$ is a pre-Hilbert space with scalar product $\langle ., .\rangle_\varphi$, and the reproducing property holds: $h(x) = \langle h, L_{\tilde{x}}\varphi\rangle_\varphi$ for all $h \in F(\varphi)$ and $x \in K$.

Denote by $H(\varphi)$ the completion of $F(\varphi)$. Since

$$\langle L_{\tilde{x}}\varphi, L_{\tilde{x}}\varphi\rangle_\varphi = \epsilon_x^* * \epsilon_x * \varphi(e) = \omega(\tilde{x}, x)(\varphi) \leq \varphi(e),$$

we have $|g(x)|^2 \leq \varphi(e)\, \langle g, g\rangle_\varphi$. Now, if $(g_n)_{n\in\mathbb{N}}$ is a Cauchy sequence in $F(\varphi)$ it follows that $(g_n)_{n\in\mathbb{N}}$ is also a Cauchy sequence in the Banach space $C^b(K)$ with respect to the sup-norm. Hence to each element $\eta \in H(\varphi)$ there corresponds a unique function $g \in C^b(K)$ with $g(x) = \lim_{n\to\infty} g_n(x)$ uniformly on K.

Now we define operators W_x, $x \in K$, on the pre-Hilbert space $F(\varphi)$, which finally lead to a bounded $*$-representation of K on $H(\varphi)$.

Lemma 2.4.18 *Let $\varphi \in P^b(K)$ and $\mu, \nu \in M_c(K)$, $x \in K$. Then*

(1) $x \mapsto \nu^* * \epsilon_x * \mu * \varphi(e)$ *is a continuous and bounded mapping on K.*
(2) $x \mapsto \mu^* * \epsilon_x * \mu * \varphi(e)$ *is positive definite.*

Proof.

(1) Write $g = \mu * \varphi$. Since

$$\nu^* * \epsilon_x * g(e) = \int_K L_{\tilde{x}} g(\tilde{y})\, d\nu^*(y)$$

we have to show that for each $\varepsilon > 0$, $x_0 \in K$, there exists a neighborhood V_{x_0} of x_0 such that

$$|\omega(x,y)(g) - \omega(x_0,y)(g)| < \varepsilon \qquad \text{for all } x \in V_{x_0},\ y \in C = \text{supp } \nu^*.$$

By Lemma 1.1.1(a) for each $y \in C$ there exist neighborhoods $U_{x_0}^y$ of x_0 and U_y of y such that

$$|\omega(x,z)(g) - \omega(x_0,y)(g)| < \varepsilon/2 \qquad \text{for each } z \in U_y,\ x \in U_{x_0}^y.$$

Let $C \subseteq \bigcup_{i=1}^n U_{y_i}$ and put $V_{x_0} = \bigcap_{i=1}^n U_{x_0}^{y_i}$, and the continuity of $x \mapsto \nu^* * \epsilon_x * g(e)$ in x_0 follows. This mapping is also bounded, since

$$|\nu^* * \epsilon_x * g(e)| \le \int_K |L_{\tilde{x}} g(\tilde{y})|\, d\nu^*(y) \le \|\varphi\|_\infty \|\mu\|\, \|\nu\|.$$

(2) Write $\psi(x) = \mu^* * \epsilon_x * \mu * \varphi(e)$. Then

$$\int_K \psi(x)\, d(\nu^* * \nu)(x) = \mu^* * \nu^* * \nu * \mu * \varphi(e)$$

$$= \int_K \tilde{\varphi}(z)\, d((\nu * \mu)^* * (\nu * \mu))(z) \ge 0.$$

◇

Theorem 2.4.19 *Let $\varphi \in P^b(K)$, $F(\varphi)$ the pre-Hilbert space determined in Lemma 2.4.17, and $H(\varphi)$ the completion of $F(\varphi)$.*

(1) For $g = \mu * \varphi$, $\varphi \in M_c(K)$, $x \in K$, define

$$W_x g = \epsilon_x * \mu * \varphi.$$

Then $W_x \in \mathcal{B}(F(\varphi), H(\varphi))$. If $h = \nu * \varphi$, $\nu \in M_c(K)$, then $x \mapsto \langle W_x g, h\rangle_\varphi$ is a continuous and bounded mapping on K.

(2) *Denote by* V_x *the unique extension of* W_x *to an operator* $V_x \in \mathcal{B}(H(\varphi))$ *with* $\|V_x\| = \|W_x\|$ *for all* $x \in K$. *Then* $x \mapsto V_x$ *is a bounded* $*$-*representation of* K *on* $H(\varphi)$ *and* $\varphi(x) = \langle V_x\varphi, \varphi\rangle_\varphi$.

Proof.

(1) We have

$$\langle W_x g, h\rangle_\varphi = \nu^* * \epsilon_x * \mu * \varphi(e),$$

and so $x \mapsto \langle W_x g, h\rangle_\varphi$ is continuous and bounded by Lemma 2.4.18 (1). Since

$$\langle W_x g, W_x g\rangle_\varphi = \mu^* * \epsilon_{\tilde{x}} * \epsilon_x * \mu * \varphi(e) = \omega(\tilde{x}, x)(\psi),$$

where $\psi(y) = \mu^* * \epsilon_y * \mu * \varphi(e)$, it follows by Lemma 2.4.18 (2) that

$$\langle W_x g, W_x g\rangle_\varphi \le \psi(e) = \int_K \varphi(\tilde{z})\, d(\mu^* * \mu)(z) = \langle g, g\rangle_\varphi.$$

Hence $W_x \in \mathcal{B}(F(\varphi), H(\varphi))$ and $\|W_x\| \le 1$.

(2) By (1), $x \mapsto \langle V_x\xi, \eta\rangle$ is a continuous and bounded mapping for all $\xi, \eta \in H(\varphi)$. For $g = \mu * \varphi$, $h = \nu * \varphi$ we obtain

$$\int_K \langle W_z g, h\rangle_\varphi \, d\omega(x,y)(z) = \int_K \nu^* * \epsilon_z * \mu * \varphi(e)\, d\omega(x,y)(z)$$

$$= \nu^* * \epsilon_x * \epsilon_y * \mu * \varphi(e) = \langle W_x W_y g, h\rangle_\varphi,$$

and it follows

$$\int_K \langle V_z\xi, \eta\rangle_\varphi \, d\omega(x,y)(z) = \langle V_x V_y \xi, \eta\rangle_\varphi$$

for all $\xi, \eta \in H(\varphi)$. Finally,

$$\langle g, W_{\tilde{x}} h\rangle_\varphi = (\epsilon_{\tilde{x}} * \nu)^* * \mu * \varphi(e) = \nu^* * \epsilon_x * \mu * \varphi(e) = \langle W_x g, h\rangle_\varphi$$

and therefore $V_{\tilde{x}} = V_x^*$ for all $x \in K$. $V_e = \mathrm{id}$ and $\|V\| = 1$ are obvious, as well as $\varphi(x) = \langle V_x\varphi, \varphi\rangle_\varphi$.

◇

Till now we have not applied that each vector $\xi \in H(\varphi)$ can be uniquely identified with a function $g = g_\xi \in C^b(K)$. In the following each element of $H(\varphi)$ will be interpreted as a function on K. In particular, for $g \in H(\varphi)$ the reproducing property is valid:

$$g(x) = \langle g, L_{\tilde{x}}\varphi\rangle_\varphi = \langle g, V_x\varphi\rangle_\varphi, \qquad x \in K.$$

Proposition 2.4.20 *Let $\varphi \in P^b(K)$ and let V be the $*$-representation of K on $H(\varphi)$ according to Theorem 2.4.19. Assume $H(\varphi) = H_0 \oplus H_0^\perp$, H_0 and $H_0^\perp$ nontrivial and invariant under V. Let P be the orthogonal projection onto H_0, and set $\varphi_1 := P\varphi$ and $\varphi_2 := \varphi - \varphi_1 = (id - P)\varphi$. If $\psi_1 \in H_0$ and $\psi_2 \in H_0^\perp$, then*

$$\psi_1(x) = \langle V_x\psi_1, \varphi_1\rangle_\varphi \qquad \text{and} \qquad \psi_2(x) = \langle V_x\psi_2, \varphi_2\rangle_\varphi.$$

Moreover, φ_1, $\varphi_2 \in P^b(K)$. In particular, if $\varphi \in ex(P^1(K))$ then V is irreducible.

Proof. We have

$$\psi_1(x) = \langle \psi_1, V_x\varphi\rangle_\varphi = \langle \psi_1, V_x(\varphi_1 + \varphi_2)\rangle_\varphi = \langle \psi_1, V_x\varphi_1\rangle_\varphi$$

and exactly in the same way $\psi_2(x) = \langle \psi_2, V_x\varphi_2\rangle_\varphi$ for every $x \in K$. For φ_1 and φ_2 it follows $\varphi_1(x) = \overline{\langle V_x\varphi_1, \varphi_1\rangle_\varphi}$, $\varphi_2 = \overline{\langle V_x\varphi_2, \varphi_2\rangle_\varphi}$, and so $\varphi_1, \varphi_2 \in P^b(K)$.

Finally let $\varphi \in \text{ex}(P^1(K))$. Assume V is reducible. That means we have $H(\varphi) = H_0 \oplus H_0^\perp$, H_0 as above. Then $\varphi_1(e) \neq 0 \neq \varphi_2(e)$ and $\varphi(x) = \varphi_1(x) + \varphi_2(x) = \varphi_1(e)\tilde{\varphi}_1(x) + \varphi_2(e)\tilde{\varphi}_2(x)$, where $\tilde{\varphi}_1(x) = \frac{1}{\varphi_1(e)}\varphi_1(x)$, $\tilde{\varphi}_2(x) = \frac{1}{\varphi_2(e)}\varphi_2(x)$, contradicting that φ is extremal. ◇

Proposition 2.4.21 *Let $\varphi = \varphi_1 + \varphi_2$, where $\varphi, \varphi_1, \varphi_2 \in P^b(K)$, and V the $*$-representation of K on $H(\varphi)$. Then there exists a positive operator $T \in \mathcal{B}(H(\varphi))$ with $\|T\| \leq 1$, $T \in \mathcal{C}(V)$ and*

$$\varphi_1(x) = T\varphi(x) = \langle T\varphi, V_x\varphi\rangle_\varphi,$$

and

$$\varphi_2(x) = (id - T)\varphi(x) = \langle (id - T)\varphi, V_x\varphi\rangle_\varphi.$$

In particular, φ_1 and φ_2 are elements of $H(\varphi)$.

Proof. For $g = \mu * \varphi$, $h = \nu * \varphi \in F(\varphi)$ set $\langle g, h\rangle_{\varphi_1} = \nu^* * \mu * \varphi_1(e)$. Since $\varphi_2 \in P^b(K)$ we have $\mu^* * \mu * \varphi_2(e) \geq 0$, and so $\mu^* * \mu * \varphi(e) \geq \mu^* * \mu * \varphi_1(e)$. It follows $0 \leq \langle g, g\rangle_{\varphi_1} \leq \langle g, g\rangle_\varphi$, and therefore

$$|\langle g, h\rangle_{\varphi_1}|^2 \leq \langle g, g\rangle_{\varphi_1} \langle h, h\rangle_{\varphi_1} \leq \langle g, g\rangle_\varphi \langle h, h\rangle_\varphi.$$

We infer that the sesquilinear form $\langle ., .\rangle_{\varphi_1}$ is continuous on $F(\varphi)$, and it can be uniquely extended to a continuous sesquilinear form on $H(\varphi)$. It follows that there exists an operator $T \in \mathcal{B}(H(\varphi))$ such that

$$\langle g, h\rangle_{\varphi_1} = \langle Tg, h\rangle_\varphi, \qquad g, h \in H(\varphi).$$

Obviously T is a positive operator on $H(\varphi)$, and $\langle Tg, g\rangle_\varphi \le \langle g, g\rangle_\varphi$ yields $\|T\| \le 1$. Moreover by the definition of $\langle .,. \rangle_{\varphi_1}$ we have $\langle V_x g, h\rangle_{\varphi_1} = \langle g, V_{\tilde{x}} h\rangle_{\varphi_1}$ and hence

$$\langle V_x Tg, h\rangle_\varphi = \langle Tg, V_{\tilde{x}} h\rangle_\varphi = \langle g, V_{\tilde{x}} h\rangle_{\varphi_1} = \langle V_x g, h\rangle_{\varphi_1}$$

$$= \langle TV_x g, h\rangle_\varphi$$

for each $g, h \in H(\varphi)$. It follows $V_x T = TV_x$ for $x \in K$. By the reproducing property we get

$$T\varphi(x) = \langle T\varphi, V_x\varphi\rangle_\varphi = \langle \varphi, V_x \varphi\rangle_{\varphi_1} = \epsilon_{\tilde{x}} * \varphi_1(e) = \varphi_1(x)$$

and

$$\varphi_2(x) = \varphi(x) - \varphi_1(x) = \langle (\mathrm{id} - T)\varphi, V_x\varphi\rangle_\varphi = (\mathrm{id} - T)\varphi(x).$$

◇

Corollary 2.4.22 *Let $\varphi \in P^1(K)$ and let V^φ be a cyclic bounded $*$-representation of K on H with a cyclic vector $\zeta \in H$ such that*

$$\varphi(x) = \langle V_x^\varphi \zeta, \zeta\rangle.$$

Then $\varphi \in ex(P^1(K))$ if and only if V^φ is irreducible.

Proof. By Theorem 2.4.8 we may choose for V^φ the $*$-representation constructed by Theorem 2.4.19. In Proposition 2.4.20 it is already shown that $\varphi \in \mathrm{ex}(P^1(K))$ implies that V is irreducible. If V is irreducible assume that $\varphi = \varphi_1 + \varphi_2$; $\varphi_1, \varphi_2 \in P^b(K)$. By Proposition 2.4.21 there exists $T \in \mathcal{C}(V)$, T positive, such that $T\varphi = \varphi_1$ and $(\mathrm{id} - T)\varphi = \varphi_2$. Since V is irreducible we get $T = \lambda\,\mathrm{id}$, $\lambda > 0$. So $\varphi_1 = \lambda\varphi$ and $\varphi_2 = (1-\lambda)\varphi$, and $\varphi \in \mathrm{ex}(P^1(K))$ follows. ◇

Lemma 2.4.23 *The linear span of $P^b(K) \cap C_c(K)$ is dense in $C_c(K)$ in the uniform norm.*

Proof. By Proposition 2.4.3 the set $P^b(K) \cap C_c(K)$ includes all functions of the form $g * \tilde{\overline{g}}$, $g \in C_c(K)$. By polarization the linear span of $P^b(K) \cap C_c(K)$ includes all functions of the form $f * \overline{\tilde{\overline{h}}}$ where $f, h \in C_c(K)$. Hence all functions $f * g$, $f, g \in C_c(K)$, are elements of the linear span of $P^b(K) \cap C_c(K)$. Given $f \in C_c(K)$ choose approximating functions of the form of Proposition 2.3.3. It follows that the linear span of $P^b(K) \cap C_c(K)$ is dense in $C_c(K)$ in the uniform norm. ◇

Theorem 2.4.24 *Let $x, y \in K$, $x \neq y$. Then there exists an irreducible $*$-representation V of K such that $V_x \neq V_y$.*

Proof. If $x \neq y$ there exists $f \in C_c(K)$ such that $f(x) \neq f(y)$. By Lemma 2.4.23 we can take f to be a linear combination of functions $\varphi_i \in P^b(K) \cap C_c(K)$. By Theorem 2.4.12 the weak $*$-topology restricted to $P^1(K)$ coincides with the topology of uniform convergence on compact subsets of K. Hence by Theorem 2.4.16 (2) each $\varphi \in P^1(K)$ can be approximated on $\{x, y\}$ closely enough by a convex combination of $\psi_j \in \text{ex}(P^1(K))$, and so there is a linear combination g of functions from $\text{ex}(P^1(K))$ which approximates on $\{x, y\}$ the function f, so that $g(x) \neq g(y)$. It follows that there must be a $\varphi \in \text{ex}(P^1(K))$ such that $\varphi(x) \neq \varphi(y)$. The corresponding $*$-representation $V = V^\varphi$ is irreducible by Corollary 2.4.22 and

$$\langle V_x \varphi, \varphi \rangle_\varphi \;=\; \varphi(x) \;\neq\; \varphi(y) \;=\; \langle V_y \varphi, \varphi \rangle_\varphi,$$

whence $V_x \neq V_y$. ◇

Remark: Theorem 2.4.24 is the extension of the Gelfand-Raikov Theorem from locally compact groups to hypergroups.

Chapter 3

Harmonic analysis on commutative hypergroups

Throughout this chapter we suppose K commutative. Clearly $\mu * \nu = \nu * \mu$ is valid for every $\mu, \nu \in M(K)$ as well as $f * g = g * f$ for all $f, g \in L^1(K)$. The Haar measure m satisfies $m = \tilde{m}$, where $\tilde{m} \in M(K)$ is defined by $\int_K f(x)\, d\tilde{m}(x) := \int_K f(\tilde{x})\, dm(x)$.
In fact it is easily seen that $\tilde{m}$ is a right-invariant Haar measure on K. The commutativity and the uniqueness of Haar measure imply that $m = c\tilde{m}$. Selecting some $f \in C_c(K)$ with $f(x) = f(\tilde{x})$ for each $x \in K$ and $\int_K f(x)\, dm(x) \neq 0$ we see that $c = 1$, and $m = \tilde{m}$ is shown.

1 Dual spaces

We shall thoroughly apply the Gelfand theory of commutative Banach $*$-algebras. The basic facts, which we will use, can be found in [356,2.2]. The involution in the Banach algebra $M(K)$ is given by $\mu^* := \overline{(\tilde{\mu})}$ and that in $L^1(K)$ is $f^* := \overline{(\tilde{f})}$. The reader should note that for general hypergroups the definition of f^* is $f^* = \overline{(\Delta \tilde{f})}$, see section 2.2. But in case of commutativity of K, the modular function Δ is 1.

We are now introducing two dual objects:

$$\mathcal{X}^b(K) := \{\alpha \in C^b(K) : \alpha \neq 0,\ \omega(x,y)(\alpha) = \alpha(x)\alpha(y) \text{ for all } x, y \in K\}$$

and

$$\hat{K} := \{\alpha \in \mathcal{X}^b(K) : \alpha(\tilde{x}) = \overline{\alpha(x)} \text{ for all } x \in K\}.$$

Clearly the constant function 1 is an element of $\hat{K}$, and $\alpha(e) = 1$ (since $\alpha(e)\alpha(x) = \alpha(x)$) and $\|\alpha\|_\infty = 1 = \alpha(e)$. The functions $\alpha \in \hat{K}$ are called **characters** of K.

For every $\alpha \in \mathcal{X}^b(K)$ put

$$h_\alpha(\mu) := \int_K \overline{\alpha(x)}\, d\mu(x), \qquad \mu \in M(K).$$

Clearly h_α is a continuous linear functional on $M(K)$ and on $L^1(K)$, too.

Furthermore

$$h_\alpha(\mu * \nu) = \int_K \overline{\alpha(z)}\, d\mu * \nu(z) = \int_K \int_K \overline{\omega(x,y)(\alpha)}\, d\mu(x)\, d\nu(y)$$

$$= \int_K \overline{\alpha(x)}\, d\mu(x) \int_K \overline{\alpha(y)}\, d\nu(y) = h_\alpha(\mu) h_\alpha(\nu)\,,$$

that means h_α is multiplicative, i.e. $h_\alpha \in \Delta(M(K))$ as well as $h_\alpha \in \Delta(L^1(K))$, (where $\Delta(A)$ is the structure space of the Banach algebra A, see [356,2.2].

For $\alpha \in \hat{K}$ we see

$$h_\alpha(\mu^*) = \int_K \overline{\alpha(x)}\, d\mu^*(x) = \overline{\int_K \alpha(x)\, d\mu(x)} = \overline{h_\alpha(\mu)}$$

and $h_\alpha(f^*) = \overline{h_\alpha(f)}$, i.e. $h_\alpha \in \Delta_s(M(K))$ and $h_\alpha \in \Delta_s(L^1(K))$ (where $\Delta_s(A)$ is the symmetric structure space of the Banach $*$-algebra A). As in the group case we can show that $\alpha \mapsto h_\alpha$ is a bijection from $\hat{K}$ onto $\Delta_s(L^1(K))$ respectively from $\mathcal{X}^b(K)$ onto $\Delta(L^1(K))$. But in contrast to commutative groups the two dual spaces $\hat{K}$ and $\mathcal{X}^b(K)$ can be different.

Theorem 3.1.1 *For the dual spaces $\hat{K}$ and $\mathcal{X}^b(K)$ of a commutative hypergroup is valid*

(i) For $\alpha \in \mathcal{X}^b(K)$ we have $h_\alpha \in \Delta(L^1(K))$,
and for $\alpha \in \hat{K}$ we have $h_\alpha \in \Delta_s(L^1(K))$.

(ii) If $h \in \Delta(L^1(K))$ then there exists $\alpha \in \mathcal{X}^b(K)$ such that $h_\alpha = h$.
If $h \in \Delta_s(L^1(K))$ then there exists $\alpha \in \hat{K}$ with $h_\alpha = h$.

(iii) If $\alpha, \beta \in \mathcal{X}^b(K)$, $\alpha \neq \beta$, then $h_\alpha \neq h_\beta$.

Proof.

(i) It remains to prove that $h_\alpha \neq 0$. Let $U = \{x \in K : |\alpha(x)|^2 > \frac{1}{2}\}$. The subset U is an open neighborhood of $e \in K$.
Choose some $f \in C_c(K)$, $0 \le f \le 1$, $f(e) = 1$, $f \,|\, K \setminus U = 0$, and put $g = \alpha f$. Then

$$h_\alpha(g) = \int_K \overline{\alpha(x)} g(x)\, dm(x) = \int_U |\alpha(x)|^2 f(x)\, dm(x) \ge \frac{1}{2} m(f) > 0.$$

(ii) Now consider $h \in \Delta(L^1(K))$. In particular $h \in L^1(K)^*, \|h\| = 1$. ($L^1(K)^*$ is the dual space of the Banach space $L^1(K)$.) Hence there exists $\varphi \in L^\infty(K)$ with $\|\varphi\|_\infty = 1$ and

$$h(f) = \int_K f(x)\overline{\varphi(x)}\,dm(x), \qquad \text{for all } f \in L^1(K).$$

Given $g \in L^1(K)$ and $f \in C_c(K)$ with $h(f) = 1$ (such a function f exists !), we obtain

$$\int_K g(y)\left(h(f)\overline{\varphi(y)}\right)\,dm(y) = h(f)h(g) = h(f * g) = \int_K \overline{\varphi(x)} f * g(x)\,dm(x)$$

$$= \int_K \int_K g(y)\omega(\tilde{y}, x)(f)\,dm(y)\overline{\varphi(x)}\,dm(x) = \int_K g(y)h(L_{\tilde{y}}f)\,dm(y).$$

Therefore $\overline{\varphi(y)} = h(f)\overline{\varphi(y)} = h(L_{\tilde{y}}f)$ for locally m-almost every $y \in K$. By Proposition 2.2.11 the mapping $y \mapsto L_{\tilde{y}}f$ is continuous. Moreover $|\varphi(y)| \leq \|h\|\|L_{\tilde{y}}f\|_1 \leq \|f\|_1$ and $\varphi(e) = h(f) = 1$. Finally we obtain for all $g \in L^1(K)$, $x \in K$ and for $f \in C_c(K)$ as above:

$$\int_K g(y)\overline{L_x\varphi(y)}\,dm(y) = \int_K L_{\tilde{x}}g(y)\overline{\varphi(y)}\,dm(y) = h(L_{\tilde{x}}g) = h(f * L_{\tilde{x}}g)$$

$$= h(L_{\tilde{x}}f * g) = h(L_{\tilde{x}}f)h(g) = \overline{\varphi(x)}h(g) = \int_K g(y)\overline{\varphi(x)\varphi(y)}\,dm(y).$$

Hence $L_x\varphi = \varphi(x)\varphi$ locally m-almost everywhere. Since $L_x\varphi$ and φ are continuous we even have $L_x\varphi(y) = \varphi(x)\varphi(y)$ for all $y \in K$, i.e. $\varphi \in \mathcal{X}^b(K)$.

If in addition $h \in \Delta_s(L^1(K))$ then it follows for $g \in L^1(K)$

$$\overline{\int_K g(x)\overline{\varphi(x)}\,dm(x)} = \overline{h(g)} = h(g^*)$$

$$= \int_K \overline{g(\tilde{x})\varphi(x)}\,dm(x) = \overline{\int_K g(x)\varphi(\tilde{x})\,dm(x)},$$

where we have used $m = \tilde{m}$. Hence $\varphi(\tilde{x}) = \overline{\varphi(x)}$, i.e. $\varphi \in \hat{K}$.

(iii) Suppose $h_\alpha(f) = h_\beta(f)$ for all $f \in L^1(K)$. Then

$$0 = \int_K f(x)\overline{(\alpha(x) - \beta(x))}\,dm(x),$$

and $\alpha = \beta$ follows. ◇

Appealing to the bijection between $\Delta_s(L^1(K))$ and $\hat{K}$ or between $\Delta(L^1(K))$ and $\mathcal{X}^b(K)$ we can transfer the weak $*$-topology of $\Delta_s(L^1(K)) \subseteq L^1(K)^*$ respectively $\Delta(L^1(K)) \subseteq L^1(K)^*$ onto $\hat{K}$ respectively $\mathcal{X}^b(K)$. With this topology (called Gelfand topology) $\hat{K}$ and $\mathcal{X}^b(K)$ become locally compact Hausdorff spaces. Soon we will see that this topology is identical with the compact-open topology on both $\hat{K}$ and $\mathcal{X}^b(K)$. Concerning the dual aspects we point out that in contrast to the group case we do not have $|\alpha(x)| = 1$. We only know that $|\alpha(x)| \leq 1$ for each $x \in K$. Another basic difference to groups is the fact that for $\alpha, \beta \in \hat{K}$ (or $\mathcal{X}^b(K)$) the product $\alpha\beta$ is in general not an element of $\hat{K}$ (or $\mathcal{X}^b(K)$). But in some situations $\alpha\beta$ can be decomposed by some probability measure $\omega(\alpha, \beta)$ on $\hat{K}$.

By a rather simple argumentation we can show a Proposition concerning the range of $\alpha \in \mathcal{X}^b(K)$.

Proposition 3.1.2 *Let $\alpha \in \mathcal{X}^b(K)$, $\alpha \neq 1$. Then*

$$\inf\{Re\,(\alpha(x)) : x \in K\} \leq 0 \qquad \text{and} \qquad \inf\{Im\,(\alpha(x)) : x \in K\} \leq 0.$$

Proof. Choose $x \in K$ with $\alpha(x) \neq 1$. For every $n \in \mathbb{N}$ we have

$$(\alpha(x))^n = \int_K \alpha(z)\, d\epsilon_x * \cdots * \epsilon_x(z).$$

($\epsilon_x * \cdots * \epsilon_x$ is the n-fold convolution of the point measure ϵ_x.) In particular

$$\mathrm{Re}\,((\alpha(x))^n) = \int_K \mathrm{Re}\,(\alpha(z))\, d\epsilon_x * \cdots * \epsilon_x(z)$$

and

$$\mathrm{Im}\,((\alpha(x))^n) = \int_K \mathrm{Im}\,(\alpha(z))\, d\epsilon_x * \cdots * \epsilon_x(z).$$

Let $K_1 := \inf\{\mathrm{Re}\,(\alpha(z)) : z \in K\}$ and $K_2 := \inf\{\mathrm{Im}\,(\alpha(z)) : z \in K\}$. Then $K_1 \leq \mathrm{Re}\,((\alpha(x))^n)$ and $K_2 \leq \mathrm{Im}\,((\alpha(x))^n)$ for all $n \in \mathbb{N}$. If $|\alpha(x)| = 1$, then obviously there exist $n \in \mathbb{N}$ and $m \in \mathbb{N}$ such that $\mathrm{Re}\,((\alpha(x))^n) < 0$ and $\mathrm{Im}\,((\alpha(x))^m) < 0$, (recall that $\alpha(x) \neq 1$). If $|\alpha(x)| < 1$ then $\lim_{n\to\infty}\mathrm{Re}\,((\alpha(x))^n) = 0$ and $\lim_{n\to\infty}\mathrm{Im}\,((\alpha(x))^n) = 0$. In any case we get $K_1 \leq 0$, $K_2 \leq 0$. ◇

Whereas for $\alpha \in \hat{K}$ we have $\tilde{\alpha} = \bar{\alpha}$, for $\alpha \in \mathcal{X}^b(K)$ we only see that $\tilde{\alpha} \in \mathcal{X}^b(K)$ and also $\bar{\alpha} \in \mathcal{X}^b(K)$. Since $\Delta_s(L^1(K))$ is a closed subset of

$\Delta(L^1(K))$ we see that $\hat{K}$ is also closed in $\mathcal{X}^b(K)$. The Fourier transform of $f \in L^1(K)$ is usually defined on $\hat{K}$. We will also include the transform, which is defined on the larger dual space $\mathcal{X}^b(K)$. Of course this transform will miss some nice properties.

For $\mu \in M(K)$ we define

$$\hat{\mu}(\alpha) := \int_K \overline{\alpha(x)}\, d\mu(x) \qquad \text{for } \alpha \in \hat{K}$$

and

$$\mathcal{F}\mu(\alpha) := \int_K \overline{\alpha(x)}\, d\mu(x) \qquad \text{for } \alpha \in \mathcal{X}^b(K).$$

For $f \in L^1(K)$ let

$$\hat{f}(\alpha) := \int_K \overline{\alpha(x)} f(x)\, dm(x) \qquad \text{for } \alpha \in \hat{K}$$

and

$$\mathcal{F}f(\alpha) := \int_K \overline{\alpha(x)} f(x)\, dm(x) \qquad \text{for } \alpha \in \mathcal{X}^b(K).$$

The functions $\hat{\mu}$ and $\mathcal{F}\mu$ are called **Fourier-Stieltjes transform** of μ, the functions $\hat{f}$ and $\mathcal{F}f$ are called **Fourier transform** of f.

An immediate consequence of the Gelfand theory is a Riemann-Lebesgue Lemma.

Theorem 3.1.3 *For the Fourier transform and Fourier-Stieltjes transform the following is true:*

(i) For $\mu \in M(K)$ we have $\hat{\mu} \in C^b(\hat{K})$ and $\mathcal{F}\mu \in C^b(\mathcal{X}^b(K))$.
(ii) For $f \in L^1(K)$ we have $\hat{f} \in C_0(\hat{K})$ and $\mathcal{F}f \in C_0(\mathcal{X}^b(K))$.

Proof. Identifying $\hat{K}$ and $\mathcal{X}^b(K)$ with $\Delta_s(L^1(K))$ and $\Delta(L^1(K))$, respectively, the Fourier transform becomes the Gelfand transform, yielding statement (ii). Considering $M(K)$ we get (i) again applying the Gelfand transform. ◇

The following theorem contains some properties for the Fourier-Stieltjes transformation. Restricting from $M(K)$ to $L^1(K)$ one gets the corresponding results for the Fourier transformation. The proof is easy and is omitted.

Theorem 3.1.4 *Let $\mu, \nu \in M(K)$, $\lambda \in \mathbb{C}$ and $x \in K$. Then*

(i) $\widehat{(\mu+\nu)} = \hat{\mu} + \hat{\nu}$, $\mathcal{F}(\mu+\nu) = \mathcal{F}\mu + \mathcal{F}\nu$.
(ii) $\widehat{(\lambda\mu)} = \lambda\hat{\mu}$, $\mathcal{F}(\lambda\mu) = \lambda\mathcal{F}\mu$.
(iii) $\widehat{(\mu*\nu)} = \hat{\mu}\hat{\nu}$, $\mathcal{F}(\mu*\nu) = (\mathcal{F}\mu)(\mathcal{F}\nu)$.
(iv) $\widehat{\mu^*} = \overline{\hat{\mu}}$, $\mathcal{F}(\overline{\mu})(\beta) = \overline{\mathcal{F}(\mu)(\overline{\beta})}$, $\mathcal{F}(\tilde{\mu})(\beta) = \mathcal{F}(\mu)(\tilde{\beta})$ *for all* $\beta \in \mathcal{X}^b(K)$.
(v) $\hat{\epsilon_x}(\alpha) = \overline{\alpha(x)}$ *for all* $\alpha \in \hat{K}$ *and* $\mathcal{F}(\epsilon_x)(\alpha) = \overline{\alpha(x)}$ *for all* $\alpha \in \mathcal{X}^b(K)$.
(vi) $\|\hat{\mu}\|_\infty \leq \|\mu\|$, $\|\mathcal{F}\mu\|_\infty \leq \|\mu\|$.

So far we have only studied the Gelfand topology on $\hat{K}$ and $\mathcal{X}^b(K)$. Now we show that this topology coincides with the compact-open topology.

Proposition 3.1.5 *Consider $\mathcal{X}^b(K)$ with the Gelfand topology. Then the mapping $(x, \alpha) \mapsto \alpha(x)$, $K \times \mathcal{X}^b(K) \to \mathbb{C}$ is continuous.*

Proof. Let $x_0 \in K$, $\alpha_0 \in \mathcal{X}^b(K)$. Choose some $f \in L^1(K)$ such that $\mathcal{F}f(\alpha_0) = 1$, and let $\varepsilon > 0$. By Proposition 2.2.11 there exists a neighborhood V_{x_0} of x_0 such that $\|L_{\tilde{x}}f - L_{\tilde{x}_0}f\|_1 < \varepsilon$. Then we obtain for $\alpha \in \mathcal{X}^b(K)$

$$\begin{aligned}|\alpha(x) - \alpha_0(x_0)| &= |\alpha(x) - \alpha_0(x_0)\mathcal{F}f(\alpha_0)| \\ &\leq |\alpha(x) - \alpha(x)\mathcal{F}f(\alpha)| + |\alpha(x)\mathcal{F}f(\alpha) - \alpha(x_0)\mathcal{F}f(\alpha)| \\ &\quad + |\alpha(x_0)\mathcal{F}f(\alpha) - \alpha_0(x_0)\mathcal{F}f(\alpha_0)| \\ &\leq |1 - \mathcal{F}f(\alpha)| + |\mathcal{F}(L_{\tilde{x}}f)(\alpha) - \mathcal{F}(L_{\tilde{x}_0}f)(\alpha)| \\ &\quad + |\mathcal{F}(L_{\tilde{x}_0}f)(\alpha) - \mathcal{F}(L_{\tilde{x}_0}f)(\alpha_0)|.\end{aligned}$$

Now by Theorem 3.1.4 (vi) follows

$$|\mathcal{F}(L_{\tilde{x}}f)(\alpha) - \mathcal{F}(L_{\tilde{x}_0}f)(\alpha)| \leq \|L_{\tilde{x}}f - L_{\tilde{x}_0}f\|_1 < \varepsilon$$

for $x \in V_{x_0}$. Consider a neighborhood of α_0 in the Gelfand topology of the form

$$U(\alpha_0, \varepsilon; f_1, \ldots, f_n) = \left\{\alpha \in \mathcal{X}^b(K) : |\mathcal{F}f_i(\alpha) - \mathcal{F}f_i(\alpha_0)| < \varepsilon \text{ for } i = 1, \ldots, n\right\},$$

where $\varepsilon > 0$, $n \in \mathbb{N}$ and $f_1, \ldots, f_n \in L^1(K)$. For $x_0 \in V_{x_0}$ and $\alpha \in U(\alpha_0, \varepsilon; f, L_{\tilde{x}_0}f)$, we see that $|\alpha(x) - \alpha_0(x_0)| < 3\varepsilon$. ◇

We recall that a neighborhood basis of $\alpha_0 \in \mathcal{X}^b(K)$ in the compact-open topology consists of the sets

$$V(\alpha_0, \varepsilon; C) = \left\{\alpha \in \mathcal{X}^b(K) : |\alpha(x) - \alpha_0(x)| < \varepsilon \text{ for all } x \in C\right\},$$

where $\varepsilon > 0$, $C \subseteq K$ is compact.

Theorem 3.1.6 *The Gelfand topology on $\mathcal{X}^b(K)$ and $\hat{K}$ coincides with the compact-open topology on $\mathcal{X}^b(K)$ and $\hat{K}$, respectively.*

Proof. Let $\alpha_0 \in \mathcal{X}^b(K)$. At first suppose $C \subseteq K$ compact and $\varepsilon > 0$. Choose some $f \in L^1(K)$ with $\mathcal{F}f(\alpha_0) = 1$. As shown in Proposition 3.1.5 there exists for each $x_0 \in K$ a neighborhood U_{x_0} of x_0 such that

$$|\alpha(x) - \alpha_0(x_0)| < 3\varepsilon$$

for every $\alpha \in U(\alpha_0, \varepsilon; f, L_{\tilde{x}_0} f)$. Since C is compact, there are $x_1, \ldots, x_n \in C$ such that $C \subseteq \bigcup_{i=1}^{n} U_{x_i}$. Hence for $x \in C$ and $\alpha \in U(\alpha_0, \varepsilon; f, L_{\tilde{x}_1} f, \ldots, L_{\tilde{x}_n} f)$, we have at first $x \in U_{x_i}$ for some i, and then

$$|\alpha(x) - \alpha_0(x)| \leq |\alpha(x) - \alpha_0(x_i)| + |\alpha_0(x_i) - \alpha_0(x)| < 6\varepsilon,$$

i.e. $U(\alpha_0, \varepsilon; f, L_{\tilde{x}_1} f, \ldots, L_{\tilde{x}_n} f) \subseteq V(\alpha_0, 6\varepsilon; C)$.

Conversely suppose $f_1, \ldots, f_n \in L^1(K)$ with $f_i \neq 0$ for $i = 1, \ldots, n$, and let $\delta > 0$. Then there exists some compact set $C \subseteq K$ such that

$$\int_{K \setminus C} |f_i(x)|\, dm(x) < \delta/4 \qquad \text{for } i = 1, \ldots, n.$$

Put $\varepsilon = \delta \min(1/\|f_1\|_1, \ldots, 1/\|f_n\|_1)/2$. For $\alpha \in V(\alpha_0, \varepsilon; C)$ it follows for $i = 1, \ldots, n$ that

$$\begin{aligned}|\mathcal{F}f_i(\alpha) - \mathcal{F}f_i(\alpha_0)| &\leq \int_C |f_i(x)||\alpha(x) - \alpha_0(x)|\, dm(x) + 2 \int_{K \setminus C} |f_i(x)|\, dm(x) \\ &< \varepsilon \|f_i\|_1 + \delta/2 \leq \delta.\end{aligned}$$

That means $V(\alpha_0, \varepsilon; C) \subseteq U(\alpha_0, \delta; f_1, \ldots, f_n)$. ◇

Remark: By Corollary 2.3.12 (1) each irreducible $*$-representation V of K (K commutative) is one-dimensional. Taking $H = \mathbb{C}$ we have $V_x(\xi) = \alpha(x)\xi$ for $x \in K$, $\xi \in \mathbb{C}$. The function α is a character of K, i.e. $\alpha \in \hat{K}$, and it is easy to check there is a bijective correspondence between $\hat{K}$ and the set of all (equivalent) irreducible $*$-representations of K. It is very informative to compare results of sections 3.1, 3.2 and 3.7 with those of sections 2.3 and 2.4.

Now we introduce

$$\mathcal{S}(K) := \left\{\alpha \in \hat{K} : |\hat{f}(\alpha)| \leq \|L_f\| \text{ for every } f \in L^1(K)\right\}$$

and

$$C^*_\lambda(K) := \{L_f : f \in L^1(K)\}^-,$$

the closure of $\{L_f : f \in L^1(K)\}$ in $\mathcal{B}(L^2(K))$. $C^*_\lambda(K)$ is called the **reduced hypergroup C^*-algebra** of K. (Recall that $L_f(g) = f*g,\ f \in L^1(K),\ g \in L^2(K)$, is the left regular representation.)

Proposition 3.1.7 *$\mathcal{S}(K)$ is a nonvoid closed subset of $\hat{K}$, and if $\alpha \in \mathcal{S}(K)$, then $\overline{\alpha} \in \mathcal{S}(K)$. For $f \in L^1(K)$ one has* $\sup\{|\hat{f}(\alpha)| : \alpha \in \mathcal{S}(K)\} = \|L_f\|$.

Proof. We shall apply Gelfand's theory for commutative C^*-algebras (see [356,2.4]). It is straightforward to prove that $C^*_\lambda(K)$ is a commutative C^*-algebra. Suppose $g \in L^1(K),\ g \neq 0$. Then $L_g \neq 0$ and the Gelfand transform $k \mapsto k(L_g);\ \Delta(C^*_\lambda(K))$
$\to \mathbb{C}$ is an element of $C_0(\Delta(C^*_\lambda(K)))$. Since functions of $C_0(\Delta(C^*_\lambda(K)))$ attain their supremum and since for commutative C^*-algebras the Gelfand transform is isometric, there exists some $k_g \in \Delta(C^*_\lambda(K))$ such that

$$|k_g(L_g)| = \sup\{|k(L_g)| : k \in \Delta(C^*_\lambda(K))\} = \|L_g\| > 0.$$

In particular the functions $f \mapsto k_g(L_f),\ L^1(K) \to \mathbb{C}$ is not zero, and using the properties of the left regular representation of $L^1(K)$ it is even an element of $\Delta_s(L^1(K))$. By Theorem 3.1.1 there exists $\alpha_g \in \hat{K}$ with

$$k_g(L_f) = h_{\alpha_g}(f) = \hat{f}(\alpha_g)$$

for every $f \in L^1(K)$. Since $|k_g(L_f)| \leq \|L_f\|$ for every $f \in L^1(K)$, we have $\alpha_g \in \mathcal{S}(K)$. In particular $\mathcal{S}(K)$ is nonvoid. Moreover we see that for each $g \in L^1(K),\ g \neq 0$ there is some $\alpha_g \in \mathcal{S}(K)$ such that

$$|\hat{g}(\alpha_g)| = \|L_g\|,$$

i.e. $\sup\{|\hat{g}(\alpha)| : \alpha \in \mathcal{S}(K)\} = \|L_g\|$. By the continuity of the Fourier transforms $\hat{f}$ it follows that $\mathcal{S}(K)$ is a closed subset of $\hat{K}$.

Finally, let $\alpha \in \mathcal{S}(K)$. To show that $\overline{\alpha} \in \mathcal{S}(K)$ notify that $\hat{f}(\overline{\alpha}) = \widehat{\tilde{f}}(\alpha)$, where $\tilde{f}(x) = f(\tilde{x})$. Moreover $L_{\tilde{f}}g(x) = L_f\tilde{g}(\tilde{x})$. For both identities we used $m = \tilde{m}$. Now it is obvious that $\overline{\alpha} \in \mathcal{S}(K)$. ◇

An immediate consequence of this result is a uniqueness theorem for the Fourier transform.

Theorem 3.1.8 *Let $f \in L^1(K),\ \mu \in M(K)$. If $\hat{f}(\alpha) = 0$ for all $\alpha \in \mathcal{S}(K)$, then $f = 0$. Likewise $\hat{\mu}\,|\,\mathcal{S}(K) = 0$ implies $\mu = 0$.*

Proof. If $\hat{f}\,|\,\mathcal{S}(K) = 0$ we get $L_f = 0$ and then $f = 0$ by the results shown before. Since $\widehat{(\mu * f)} = \hat{\mu}\hat{f}$ and $\mu * f \in L^1(K)$ we obtain from $\hat{\mu}\,|\,\mathcal{S}(K) = 0$

immediately $\mu * f = 0$ for all $f \in L^1(K)$. This implies $\mu = 0$, (compare the proof of Proposition 2.3.14). ◇

From Theorem 3.1.8 in particular follows that $\mathcal{S}(K)$ separates the points of K i.e. if $x, y \in K$, $x \neq y$ then there exists $\alpha \in \mathcal{S}(K)$ with $\alpha(x) \neq \alpha(y)$. In fact for $x \neq y$ we see that $\epsilon_x - \epsilon_y \in M(K)$ is not the zero-measure. Hence there exists $\alpha \in \mathcal{S}(K)$ with $\widehat{(\epsilon_x - \epsilon_y)}(\alpha) \neq 0$.

Two facts — straightforward to prove — should be noted at this place.

Theorem 3.1.9 *If K is discrete, then $\mathcal{X}^b(K)$, $\hat{K}$ and $\mathcal{S}(K)$ are compact. If K is compact, then $\mathcal{X}^b(K) = \hat{K} = \mathcal{S}(K)$ and $\hat{K}$ is discrete. Moreover, if K is compact, $\hat{K}$ is an orthogonal basis in $L^2(K)$.*

Proof. If K is discrete, the Banach $*$-algebra $L^1(K)$ has a unit element. Hence the structure spaces $\Delta(L^1(K))$ and $\Delta_s(L^1(K))$ are compact, i.e. $\mathcal{X}^b(K)$, $\hat{K}$ and $\mathcal{S}(K)$ are compact.

Now suppose that K is compact. Clearly $\mathcal{X}^b(K) \subseteq L^2(K)$. If α, $\beta \in \hat{K}$, $\alpha \neq \beta$, we get for each $x \in K$:

$$\alpha(x) \int_K \alpha(y)\overline{\beta(y)}\, dm(y) = \int_K L_x\alpha(y)\overline{\beta(y)}\, dm(y) = \int_K \alpha(y)\overline{L_{\tilde{x}}\beta(y)}\, dm(y)$$

$$= \overline{\beta(\tilde{x})} \int_K \alpha(y)\overline{\beta(y)}\, dm(y) = \beta(x) \int_K \alpha(y)\overline{\beta(y)}\, dm(y).$$

Choosing some $x \in K$ with $\alpha(x) \neq \beta(x)$, we see that $\int_K \alpha(y)\overline{\beta(y)}\, dm(y) = 0$, i.e. the elements of $\hat{K}$ are orthogonal. If $g \in L^2(K)$ and $0 = \int_K g(x)\overline{\alpha(x)}\, dm(x) = \hat{g}(\alpha)$ for every $\alpha \in \hat{K}$ the uniqueness of the Fourier transform implies $g = 0$. Therefore $\hat{K}$ is an orthogonal basis in $L^2(K)$. Since the Fourier transform of each $\alpha \in \hat{K} \subseteq L^1(K)$ has the form $\hat{\alpha}(\beta) = \delta_{\alpha,\beta}\|\alpha\|_2^2$, $\beta \in \hat{K}$, each α is isolated in $\hat{K}$ (recall that the Fourier transform is continuous). Hence $\hat{K}$ is discrete. It remains to prove $\mathcal{S}(K) = \hat{K} = \mathcal{X}^b(K)$. Suppose some $\beta \in \mathcal{X}^b(K) \setminus \mathcal{S}(K)$. By the uniqueness theorem there is some $\alpha \in \mathcal{S}(K)$ such that $\hat{\beta}(\alpha) \neq 0$. For every $x \in K$ we have

$$\beta(x)\hat{\beta}(\alpha) = \widehat{(L_x\beta)}(\alpha) = \alpha(x)\hat{\beta}(\alpha),$$

i.e. $\alpha = \beta$, which is a contradiction. ◇

In the next section we will show that $\mathcal{S}(K)$ is exactly the support of the Plancherel measure on $\hat{K}$. This paragraph we will close with a description of the Gelfand transform on the C^*-algebra $C^*_\lambda(K)$, which we already have studied in the proof of Proposition 3.1.7.

Theorem 3.1.10 *For $\mathcal{S}(K) \subseteq \hat{K}$ the following is valid:*

(i) For $\alpha \in \mathcal{S}(K)$ there exists $k_\alpha \in \Delta(C^*_\lambda(K))$ such that $k_\alpha(L_f) = \hat{f}(\alpha)$ for every $f \in L^1(K)$. Conversely for every $k \in \Delta(C^*_\lambda(K))$ there exists $\alpha \in \mathcal{S}(K)$ with $k(L_f) = \hat{f}(\alpha)$ for every $f \in L^1(K)$.
The mapping $\alpha \mapsto k_\alpha$, $\mathcal{S}(K) \to \Delta(C^*_\lambda(K))$ is a homeomorphism.

*(ii) The inverse Gelfand transformation $\psi \mapsto B_\psi$, $C_0(\mathcal{S}(K)) \to C^*_\lambda(K)$ with $\psi(\alpha) = k_\alpha(B_\psi)$ is an isometric *-isomorphism. In particular, given $\varphi \in C_0(\hat{K})$ there exists some $B_{\varphi|\mathcal{S}(K)}$ such that $\varphi(\alpha) = k_\alpha(B_{\varphi|\mathcal{S}(K)})$ and $\|\varphi\|_{\mathcal{S}(K)} = \|B_{\varphi|\mathcal{S}(K)}\|$. ($\|\varphi\|_{\mathcal{S}(K)} = \sup\{\,|\varphi(\alpha)| : \alpha \in \mathcal{S}(K)\}$.)*

Proof. Given $k \in \Delta(C^*_\lambda(K))$ we have already shown in the proof of Proposition 3.1.7 that there exists $\alpha \in \mathcal{S}(K)$ such that $k(L_f) = \hat{f}(\alpha)$ for every $f \in L^1(K)$. Conversely let $\alpha \in \mathcal{S}(K)$. For every operator $B \in C^*_\lambda(K)$ there is a sequence $(f_n)_{n\in\mathbb{N}}$, $f_n \in L^1(K)$ such that $\|B - L_{f_n}\| \to 0$ as $n \to \infty$. Since $\alpha \in \mathcal{S}(K)$ it follows that $(\hat{f}_n(\alpha))_{n\in\mathbb{N}}$ is a Cauchy sequence, and we define $k_\alpha(B) := \lim_{n\to\infty} \hat{f}_n(\alpha)$. Now it is routine to show that $k_\alpha(B)$ does not depend on the choice of the f_n, and that k_α satisfies $k_\alpha(B \circ C) = k_\alpha(B)k_\alpha(C)$ and $k_\alpha(B^*) = \overline{k_\alpha(B)}$ for $B, C \in C^*_\lambda(K)$. Hence $k_\alpha \in \Delta(C^*_\lambda(K))$ and $k_\alpha(L_f) = \hat{f}(\alpha)$ for every $f \in L^1(K)$. The injectivity of $\alpha \mapsto k_\alpha$, $\mathcal{S}(K) \to \Delta(C^*_\lambda(K))$ follows by Theorem 3.1.8. Obviously $\alpha \mapsto k_\alpha$ is an open mapping. Its continuity can be shown easily. The statements of (ii) are exactly the results of Gelfand's theorem. ◇

We have introduced $\mathcal{S}(K)$ based on the left regular representation on $L^1(K)$:

$$\mathcal{S}(K) = \{\alpha \in \hat{K} : |\hat{f}(\alpha)| \le \|L_f\| \text{ for all } f \in L^1(K)\}.$$

It is rather easy to derive the following result.

Proposition 3.1.11 *We have*

$$\begin{aligned}\mathcal{S}(K) &= \{\alpha \in \hat{K} : |\hat{\mu}(\alpha)| \le \|L_\mu\| \text{ for all } \mu \in M(K)\}\\ &= \{\alpha \in \hat{K} : |\hat{\mu}(\alpha)| \le \|L_\mu\| \text{ for all } \mu \in M_c(K)\}\\ &= \{\alpha \in \hat{K} : |\hat{f}(\alpha)| \le \|L_f\| \text{ for all } f \in C_c(K)\}.\end{aligned}$$

Proof. Obviously, $\{\alpha \in \hat{K} : |\hat{\mu}(\alpha)| \le \|L_\mu\| \text{ for all } \mu \in M(K)\} \subseteq \mathcal{S}(K)$. Let $\alpha \in \mathcal{S}(K)$ and $\mu \in M(K)$. Choose a net of functions $(\varphi_i)_{i\in I}$ as in Proposition 2.3.3. Then $\varphi_i * \mu \in L^1(K)$ and hence

$$|\widehat{\varphi_i * \mu}(\alpha)| \le \|L_{\varphi_i*\mu}\| = \sup\{\|\varphi_i * \mu * g\|_2 : g \in L^2(K),\ \|g\|_2 \le 1\}.$$

Since for each $g \in L^2(K)$ we have

$$\lim_i \|\varphi_i * \mu * g\|_2 = \|\mu * g\|_2 \quad \text{and} \quad \lim_i \widehat{\varphi_i * \mu}(\alpha) = \hat{\mu}(\alpha),$$

it follows $|\hat{\mu}(\alpha)| \leq \|L_\mu\|$. The second and third equality are valid, since M_c is dense in $M(K)$, and $C_c(K)$ is dense in $L^1(K)$. ◇

2 Plancherel's theorem

We continue investigating the dual space $\widehat{K}$ respectively $\mathcal{S}(K)$, coming up to the existence of the Plancherel measure on $\mathcal{S}(K)$.

Theorem 3.2.1 *The linear space $\{\widehat{f} : f \in C_c(K)\}$ is a dense self-adjoint subalgebra of $C_0(\widehat{K})$.*

Proof. Given $f, g \in C_c(K)$ we have $f * g \in C_c(K)$. Hence $A := \{\widehat{f} : f \in C_c(K)\}$ is a subalgebra of $C_0(\widehat{K})$. Moreover for $f \in C_c(K)$ the adjoint f^* is also an element of $C_c(K)$ and therefore A is self-adjoint. For $\alpha, \beta \in \widehat{K}$, $\alpha \neq \beta$ we find some $f \in L^1(K)$ such that $\widehat{f}(\alpha) \neq \widehat{f}(\beta)$, and given $\alpha \in \widehat{K}$ there is some $f \in L^1(K)$ with $\widehat{f}(\alpha) \neq 0$. Since $C_c(K)$ is dense in $L^1(K)$ these properties also hold for some $f \in C_c(K)$. The theorem of Stone-Weierstraß finally yields the statement. ◇

Lemma 3.2.2 *Let $\varphi \in C_c(\widehat{K})$. Then there exists exactly one continuous function $\check{\varphi} \in C_0(K) \cap L^2(K)$ such that*

$$B_{\varphi|\mathcal{S}(K)}(g) = \check{\varphi} * g \qquad \textit{for all } g \in L^2(K).$$

Proof. By Theorem 3.2.1 there exists a function $f \in C_c(K)$ such that $|\hat{f}(\alpha)| > 0$ for all $\alpha \in \operatorname{supp}\varphi$. Hence there exists $\psi \in C_c(\hat{K})$ with $\varphi = \psi(\hat{f}\,)^2$. According to Theorem 3.1.10 we have $B_{\varphi|\mathcal{S}(K)} = B_{\psi|\mathcal{S}(K)} \circ L_f \circ L_f$. Now $B_{\psi|\mathcal{S}(K)} f$ is an element of $L^2(K)$ and hence $\check{\varphi} := B_{\psi|\mathcal{S}(K)} f * f$ is an element of $C_0(K)$ by Proposition 2.2.13. Since $f \in C_c(K)$ we also have $\check{\varphi} \in L^2(K)$. For every $g \in L^2(K)$ we obtain

$$B_{\varphi|\mathcal{S}(K)}(g) = B_{\psi|\mathcal{S}(K)} \circ L_f \circ L_f(g) = \left(B_{\psi|\mathcal{S}(K)} f * f\right) * g = \check{\varphi} * g.$$

If for some $h \in L^2(K)$ one has $h * g = \check{\varphi} * g$ for all $g \in L^2(K)$, then $h = \check{\varphi}$ in $L^2(K)$, i. e. the uniqueness of $\check{\varphi}$. ◇

The mapping $\varphi \mapsto \check{\varphi}$, $C_c(\hat{K}) \to C_0(K)$ of Lemma 3.2.2 is linear, and we have $\check{\varphi} = 0$ if and only if $\|\varphi\|_{\mathcal{S}(K)} = \|B_{\varphi|\mathcal{S}(K)}\| = 0$.

Proposition 3.2.3 *Let $\varphi, \psi \in C_c(\hat{K})$ and $\mu \in M(K)$. Then*

(i) $(\overline{\varphi})^\vee = (\check{\varphi})^*$
(ii) $(\varphi\psi)^\vee = \check{\varphi} * \check{\psi}$
(iii) $(\hat{\mu}\varphi)^\vee = \mu * \check{\varphi}$.

Proof. First we note that for $h \in L^1(K)$ we have $B_{\hat{h}|\mathcal{S}} = L_h$. Using Theorem 3.1.4 we obtain $B_{\overline{\hat{h}}|\mathcal{S}(K)} = B_{\widehat{h^*}|\mathcal{S}(K)} = L_{h^*} = (L_h)^*$, hence $B_{\overline{\varphi}|\mathcal{S}(K)} = (B_{\varphi|\mathcal{S}(K)})^*$ for $\varphi \in C_0(\hat{K})$, and (i) follows. Applying the homomorphism property of the inverse Gelfand transform we get for $g \in L^2(K)$

$$(\varphi{\cdot}\psi)^\vee * g = B_{\varphi\psi|\mathcal{S}(K)}(g) = B_{\varphi|\mathcal{S}(K)}\left(B_{\psi|\mathcal{S}(K)}(g)\right) = \check{\varphi} * \left(\check{\psi} * g\right) = \left(\check{\varphi} * \check{\psi}\right) * g,$$

i.e. (ii) is valid. In the same way (iii) is shown. ◊

Proposition 3.2.4 *There exists a regular positive Borel measure π on $\hat{K}$ such that*

$$\check{\varphi}(e) = \int_{\hat{K}} \varphi(\alpha)\, d\pi(\alpha)$$

for all $\varphi \in C_c(\hat{K})$. Moreover $\mathcal{S}(K) = \operatorname{supp} \pi$.

Proof. Let $\varphi \in C_c(\hat{K})$, $\varphi \geq 0$. The square root ψ of φ satisfies $\check{\varphi} = \check{\psi} * (\check{\psi})^*$ and therefore

$$\check{\varphi}(e) = \int_K |\check{\psi}(y)|^2\, dm(y) \geq 0.$$

The mapping $\varphi \mapsto \check{\varphi}(e)$ is a positive linear functional on $C_c(\hat{K})$. Hence there exists a regular positive Borel measure $\pi \in M^\infty(\hat{K})$ on $\hat{K}$, such that

$$\check{\varphi}(e) = \int_{\hat{K}} \varphi(\alpha)\, d\pi(\alpha).$$

For $\varphi \in C_c(\hat{K})$, $\varphi \geq 0$, $\varphi(\alpha) > 0$, and $\alpha \in \mathcal{S}(K)$, we also have $\psi(\alpha) > 0$. Therefore $\check{\psi}$ is nonzero and $\check{\varphi}(e) > 0$ follows. That means $\mathcal{S}(K) \subseteq \operatorname{supp} \pi$. Conversely assuming $\alpha \in \operatorname{supp} \pi \setminus \mathcal{S}(K)$ we may choose some $\varphi \in C_c(\hat{K})$, $\varphi \geq 0$, $\varphi(\alpha) > 0$ and $\varphi|\mathcal{S}(K) = 0$. But then $\check{\varphi}(e) > 0$ is in contradiction to $\|\varphi\|_{\mathcal{S}(K)} = 0$. ◊

Remark: The proof above also shows that the Borel measure π is on $\mathcal{S}(K)$ uniquely determined.

Proposition 3.2.5 *Let $\varphi \in C_c(\hat{K})$. Then*

$$\int_{\hat{K}} |\varphi(\alpha)|^2 \, d\pi(\alpha) = \int_K |\check{\varphi}(x)|^2 \, dm(x).$$

The set $\{\check{\varphi} : \varphi \in C_c(\hat{K})\}$ is dense in $L^2(K)$.

Proof. For $\varphi \in C_c(\hat{K})$ we get

$$\int_{\hat{K}} |\varphi(\alpha)|^2 \, d\pi(\alpha) = (\varphi\overline{\varphi})^\vee(e) = \check{\varphi} * (\check{\varphi})^*(e) = \int_K |\check{\varphi}(x)|^2 \, dm(x).$$

In order to derive the second statement we show that given $h \in L^2(K)$ with

$$\int_K \check{\varphi}(x)\overline{h(x)} \, dm(x) = 0 \qquad \text{for all } \varphi \in C_c(\hat{K})$$

implies $h = 0$. Now for $\varphi \in C_c(\hat{K}),\ f \in C_c(K)$ we have $\hat{f}\varphi \in C_c(\hat{K})$, and by

$$0 = \int_K (\hat{f}\varphi)^\vee(x)\overline{h(x)} \, dm(x) = \int_K f * \check{\varphi}(x)\overline{h(x)} \, dm(x)$$

follows

$$0 = \int_K B_{\varphi|\mathcal{S}(K)}(f)(x)\overline{h(x)} \, dm(x) = \int_K f(x)\overline{(B_{\varphi|\mathcal{S}(K)})^* h(x)} \, dm(x).$$

Therefore $\int_K \check{\varphi}(x)\overline{h(x)} \, dm(x) = 0$ for all $\varphi \in C_c(\hat{K})$ implies $B_{\overline{\varphi}|\mathcal{S}(K)}(h) = (B_{\varphi|\mathcal{S}(K)})^*(h) = 0$ for all $\varphi \in C_c(\hat{K})$. Now consider $g \in L^1(K)$. Approximate $\hat{g} \in C_0(\hat{K})$ by $\varphi_n \in C_c(\hat{K})$ with respect to $\| \,.\, \|_\infty$. Therefore

$$\int_K \check{\varphi}(x)\overline{h(x)} \, dm(x) = 0 \qquad \text{for all } \varphi \in C_c(\hat{K})$$

implies $L_g h = g * h = 0$ for all $g \in L^1(K)$, i.e. $h = 0$. ◇

Theorem 3.2.6 (Plancherel-Levitan) *There exists a unique regular positive Borel measure $\pi \in M^\infty(\hat{K})$ on $\hat{K}$ with*

$$\int_K |f(x)|^2 \, dm(x) = \int_{\hat{K}} |\hat{f}(\alpha)|^2 \, d\pi(\alpha)$$

for all $f \in L^1(K) \cap L^2(K)$. The support of π is equal to $\mathcal{S}(K)$. The set $\{\hat{f} : f \in C_c(K)\}$ is dense in $L^2(\hat{K}, \pi)$. π is called **Plancherel measure**.

Proof. Consider π of Proposition 3.2.4. The mapping $\varphi \mapsto \check{\varphi}, C_c(\hat{K}) \to C_0(K) \cap L^2(K)$ has a unique extension to $L^2(\hat{K}, \pi)$. By Proposition 3.2.5 this extension is an isometric mapping from $L^2(\hat{K}, \pi)$ onto $L^2(K)$. (Note that the image of $L^2(\hat{K}, \pi)$ with respect to this isometry is complete and is dense in $L^2(K)$. Hence it is equal to $L^2(K)$.) It remains to prove the uniqueness of π. Assume

$$\int_{\hat{K}} |\hat{f}(\alpha)|^2 \, d\pi_0(\alpha) = \int_K |f(x)|^2 \, dm(x) = \int_{\hat{K}} |\hat{f}(\alpha)|^2 \, d\pi(\alpha)$$

for all $f \in C_c(K)$, where π_0 is another regular positive Borel measure on $\hat{K}$. Consider $\varphi \in C_c(\hat{K}),\ \varphi \geq 0$. Using Theorem 3.2.1 the set $\{\hat{f} : f \in C_c(K), \hat{f} \geq 0\}$ is sup-norm dense in $\{\varphi \in C_0(\hat{K}) : \varphi \geq 0\}$. Starting with some $f_0 \in C_c(K)$ with $\varphi \leq \hat{f}_0$ on $\operatorname{supp}\varphi$, we choose recursively $f_n \in C_c(K),\ n \in \mathbb{N}$, with $0 \leq \hat{f}_n \leq 1$ and

$$\varphi \leq \hat{f}_0 \hat{f}_1 \cdot \dots \cdot \hat{f}_n \leq \varphi + \frac{1}{n}.$$

Since $4\hat{f}\,\overline{\hat{g}} = |\hat{f} + \hat{g}|^2 - |\hat{f} - \hat{g}|^2 + i|\hat{f} + i\hat{g}|^2 - i|\hat{f} - i\hat{g}|^2$, we also have

$$\int_{\hat{K}} \hat{f}(\alpha)\overline{\hat{g}(\alpha)} \, d\pi_0(\alpha) = \int_{\hat{K}} \hat{f}(\alpha)\overline{\hat{g}(\alpha)} \, d\pi(\alpha)$$

for all $f, g \in C_c(K)$. Hence for each $n \in \mathbb{N}$

$$\int_{\hat{K}} \hat{f}_0(\alpha)\hat{f}_1(\alpha) \cdot \ldots \hat{f}_n(\alpha) \, d\pi_0(\alpha) = \int_{\hat{K}} \hat{f}_0(\alpha)\hat{f}_1(\alpha) \cdot \ldots \hat{f}_n(\alpha) \, d\pi(\alpha),$$

and the theorem of dominated convergence implies

$$\int_{\hat{K}} \varphi(\alpha) \, d\pi_0(\alpha) = \int_{\hat{K}} \varphi(\alpha) \, d\pi(\alpha).$$

◇

The mapping $\varphi \mapsto \check{\varphi},\ C_c(\hat{K}) \to C_0(K) \cap L^2(K)$ can be extended to an isometric isomorphism from $L^2(\hat{K}, \pi)$ onto $L^2(K)$, see the proof above. The inverse mapping $\mathcal{P} : L^2(K) \to L^2(\hat{K}, \pi)$ is called **Plancherel isomorphism**. $\mathcal{P}(f)$ will be called the **Plancherel transform** of $f \in L^2(K)$. For $f \in L^1(K) \cap L^2(K)$ the Fourier transform $\hat{f}$ and $\mathcal{P}(f)$ coincide as elements in $L^2(\hat{K},\pi)$. In order to prove this consider $f \in L^1(K) \cap L^2(K)$ and some $\varphi \in C_c(\hat{K})$. By Proposition 3.2.3 and Proposition 3.2.4 we obtain

$$\int_{\hat{K}} \hat{f}(\alpha)\overline{\varphi(\alpha)}\, d\pi(\alpha) = (\hat{f}\overline{\varphi})^\vee(e) = f * \check{\overline{\varphi}}(e) = \int_K f(x)\check{\overline{\varphi}}(\tilde{x})\, dm(x).$$

Furthermore

$$\int_{\hat{K}} \mathcal{P}(f)(\alpha)\overline{\varphi(\alpha)}\, d\pi(\alpha) = \int_K f(x)\overline{\mathcal{P}^{-1}\varphi(x)}\, dm(x) = \int_K f(x)\overline{\check{\varphi}(x)}\, dm(x),$$

since $\mathcal{P}^{-1}$ is isometric. Finally $\overline{\check{\varphi}(x)} = (\overline{\varphi})^*(\tilde{x}) = (\overline{\varphi})^\vee(\tilde{x})$ implies

$$\int_{\hat{K}} \hat{f}(\alpha)\overline{\varphi(\alpha)}\, d\pi(x) = \int_{\hat{K}} \mathcal{P}(f)(\alpha)\overline{\varphi(\alpha)}\, d\pi(\alpha)$$

for every $\varphi \in C_c(\hat{K})$. Therefore $\hat{f} = \mathcal{P}(f)$ in $L^2(\hat{K},\pi)$.

Moreover, the above equations can be extended to the **Parseval formula**

$$\int_K f(x)\, \overline{g(x)}\, dm(x) \quad = \quad \int_{\mathcal{S}(K)} \mathcal{P}(f)(\alpha)\, \overline{\mathcal{P}(g)(\alpha)}\, d\pi(\alpha).$$

Remark: We have already shown that $\alpha \in \mathcal{S}(K)$ implies $\bar{\alpha} \in \mathcal{S}(K)$, see the proof of Proposition 3.1.7. Since $\hat{f}(\bar{\alpha}) = \widehat{\tilde{f}}(\alpha)$, the uniqueness of π and $m = \tilde{m}$ yield that $\pi = \tilde{\pi}$.

If K is compact, then $\mathcal{X}^b(K) = \hat{K} = \mathcal{S}(K)$ and $\hat{K}$ is discrete (see Theorem 3.1.9). Thus π is determined by its weights on the points $\alpha \in \hat{K}$. The Fourier transform $\hat{\alpha}$ of $\alpha \in \hat{K}$ is given by

$$\hat{\alpha}(\beta) = \delta_{\alpha,\beta}\|\alpha\|_2^2,$$

compare the proof of Theorem 3.1.9. The Plancherel-Levitan theorem implies therefore

$$\|\alpha\|_2^2 = \int_K |\alpha(x)|^2\, dm(x) = \|\alpha\|_2^4 \pi(\{\alpha\}),$$

i.e. $\pi(\{\alpha\}) = 1/\|\alpha\|_2^2$.

For polynomial hypergroups on $\mathbb{N}_0$, see section 1.2.1, it is easy to check that the Plancherel measure π coincides with the orthogonalization measure. A general discussion of all facts of harmonic analysis in the context of polynomial hypergroups can be found in chapter 4.

3 Inversion theorem

In Proposition 3.1.5 we have shown that $(x,\alpha) \mapsto \alpha(x),\ K \times \hat{K} \to \mathbb{C}$ is a continuous mapping. Hence given $\varphi \in L^1(\hat{K},\pi)$ the function $\alpha \mapsto \varphi(\alpha)\alpha(x)$ is again an element of $L^1(\hat{K},\pi)$.

For $x \in K$ we define

$$\check{\varphi}(x) := \int\limits_{\hat{K}} \varphi(\alpha)\alpha(x)\, d\pi(\alpha).$$

The function $\check{\varphi}$ is called **inverse Fourier transform** of φ. Shortly we will see that for $\varphi \in C_c(\hat{K})$ the inverse Fourier transform coincides with the function $\check{\varphi}$, that we introduced in Lemma 3.2.2 and used basically throughout section 3.2

Lemma 3.3.1 *Let $\varphi \in C_c(\hat{K})$ and $x \in K$. Then*

$$(\widehat{\epsilon_x}\varphi)^{\vee} = L_{\tilde{x}}\check{\varphi},$$

where $^{\vee}$ is the mapping introduced in section 3.2.

Proof. Given $\varepsilon > 0$ there exists some $h \in C_c(K)$ such that

$$\|\varphi - \hat{h}\|_\infty < \varepsilon,$$

see Theorem 3.2.1. Hence $\|\widehat{\epsilon_x}\varphi - \widehat{\epsilon_x}\hat{h}\|_{\mathcal{S}(K)} < \varepsilon$, i. e. $\|B_{\widehat{\epsilon_x}\varphi|\mathcal{S}(K)} - L_{\epsilon_x * h}\| < \varepsilon$. Thus for all $g \in L^2(K)$, $\|g\|_2 \le 1$ we have $\|(\widehat{\epsilon_x}\varphi)^{\vee} * g - \epsilon_x * h * g\|_2 < \varepsilon$ and
$\|\check{\varphi} * (L_{\tilde{x}}g) - h * (L_{\tilde{x}}g)\|_2 < \varepsilon$, in particular

$$\|(\widehat{\epsilon_x}\varphi)^{\vee} * g - \check{\varphi} * (L_{\tilde{x}}g)\|_2 < 2\varepsilon.$$

Since we may choose $\varepsilon > 0$ arbitrarily small, we get

$$(\widehat{\epsilon_x}\varphi)^{\vee} * g = \check{\varphi} * L_{\tilde{x}}g = L_{\tilde{x}}\check{\varphi} * g,$$

and then $(\widehat{\epsilon_x}\varphi)^{\vee} = L_{\tilde{x}}\check{\varphi}$ in $L^2(K)$. Both functions being continuous we have even pointwise equality. ◇

Now it is easy to see that the two definitions of $\check{\varphi}$ coincide. Applying Proposition 3.2.4 and Lemma 3.3.1, we obtain for $\varphi \in C_c(\hat{K})$

$$\int\limits_{\hat{K}} \varphi(\alpha)\alpha(x)\, d\pi(\alpha) = (\widehat{\epsilon_{\tilde{x}}}\varphi)^{\vee}(e) = L_x\check{\varphi}(e) = \check{\varphi}(x).$$

We define an inverse transform for $a \in M(\hat{K})$, too. Put for $x \in K$

$$\check{a}(x) := \int\limits_{\hat{K}} \alpha(x)\, da(\alpha).$$

The function $\check{a}$ is called **inverse Fourier-Stieltjes transform** of a.

Proposition 3.3.2 *For the inverse transform the following is true:*

(i) *For* $a \in M(\hat{K})$ *we have* $\check{a} \in C^b(K)$ *and* $\|\check{a}\|_\infty \leq \|a\|$.
(ii) *For* $\varphi \in L^1(\hat{K}, \pi)$ *we have* $\check{\varphi} \in C_0(K)$.
(iii) $\{\check{\varphi} : \varphi \in C_c(\hat{K})\}$ *is sup-norm dense in* $C_0(K)$.

Proof. Whereas (i) is easily shown, for (ii) we already know that $\check{\varphi} \in C_0(K)$ if $\varphi \in C_c(\hat{K})$. For $\varphi \in L^1(\hat{K}, \pi)$ and $\varepsilon > 0$ select some $\psi \in C_c(\hat{K})$ such that $\|\varphi - \psi\|_1 < \varepsilon$. Hence $\|\check{\varphi} - \check{\psi}\|_\infty < \varepsilon$, and $\check{\varphi} \in C_0(K)$ follows. To prove (iii) we assume that $\{\check{\varphi} : \varphi \in C_c(\hat{K})\}$ is not dense in $C_0(K)$. Applying the Hahn-Banach theorem and Riesz' representation theorem there exists $\mu \in M(K)$, $\mu \neq 0$ such that $\int_K \check{\varphi}(x)\, d\mu(x) = 0$ for all $\varphi \in C_c(\hat{K})$. By Fubini's theorem we obtain

$$0 = \int_K \int_{\hat{K}} \varphi(\alpha)\alpha(x)\, d\pi(\alpha)\, d\mu(x) = \int_{\hat{K}} \varphi(\alpha)(\tilde{\mu})^\wedge(\alpha)\, d\pi(\alpha)$$

for all $\varphi \in C_c(\hat{K})$. Then $(\tilde{\mu})^\wedge = 0$ and from the uniqueness theorem (Theorem 3.1.8) follows a contradiction to $\mu \neq 0$. $\diamond$

The reader should note that in general $\hat{K}$ or $\mathcal{S}(K)$ are not endowed with a natural dual hypergroup structure. Therefore results concerning the inverse transforms do not follow from the corresponding results of the Fourier transform respectively Fourier-Stieltjes transform. A statement of this type is the following uniqueness theorem.

Theorem 3.3.3 *Let* $a \in M(\hat{K})$. *If* $\check{a} = 0$ *then* $a = 0$.

Proof. Assume that $a \neq 0$, but $\check{a} = 0$. Since $\{\hat{f} : f \in C_c(K)\}$ is dense in $C_0(\hat{K})$, see Theorem 3.2.1, we can find some $f \in C_c(K)$ with

$$\int_{\hat{K}} \hat{f}(\alpha)\, da(\alpha) \neq 0.$$

However we have also by Fubini's theorem

$$\int_{\hat{K}} \hat{f}(\alpha)\, da(\alpha) = \int_{\hat{K}} \int_K f(x)\overline{\alpha(x)}\, dm(x)\, da(\alpha) = \int_K \check{a}(\tilde{x}) f(x)\, dm(x) = 0.$$

$\diamond$

It is natural to pose the question whether Theorem 3.3.3 is valid if $a \in M(\mathcal{X}^b(K))$ is a measure on $\mathcal{X}^b(K)$, and the inverse Fourier-Stieltjes transform $(\text{in}\mathcal{F})(a)$ of $a \in M(\mathcal{X}^b(K))$ is defined by

$$(\text{in}\mathcal{F})(a)(x) \;=\; \int_{\mathcal{X}^b(K)} \alpha(x)\, da(\alpha), \qquad x \in K.$$

We can show that a uniqueness result for $(\text{in}\mathcal{F})$ holds exactly if a Stone-Weierstraß property is fulfilled, that means $\mathcal{F}(L^1(K))$ is norm-dense in $C_0(\mathcal{X}^b(K))$.

Theorem 3.3.4 *Equivalent are*

(1) $\mathcal{F}(L^1(K))$ *is dense in* $C_0(\mathcal{X}^b(K))$ *with respect to the sup-norm.*
(2) The uniqueness theorem is valid: If $(\text{in}\mathcal{F})(\bar{a}) = 0$ *for* $a \in M(\mathcal{X}^b(K))$, *then* $a = 0$.

Proof. (1) $\Rightarrow$ (2): Let $(\text{in}\mathcal{F})(\bar{a}) = 0$ and assume that $a \neq 0$. By assumption (1) there exists some $f \in L^1(K)$ with $\int_{\mathcal{X}^b(K)} \mathcal{F}f(\alpha)\, da(\alpha) \neq 0$, and as in the proof of Theorem 3.3.3 we have

$$\int_{\mathcal{X}^b(K)} \mathcal{F}f(\alpha)\, da(\alpha) \;=\; \int_{\mathcal{X}^b(K)} \int_K f(x)\, \overline{\alpha(x)}\, dm(x)\, da(\alpha)$$

$$=\; \int_K f(x)\, (\text{in}\mathcal{F})(\bar{a})(x)\, dm(x) \;=\; 0,$$

a contradiction.

(2) $\Rightarrow$ (1): Assume that the closure of $\mathcal{F}(L^1(K))$ is a proper subset of $C_0(\mathcal{X}^b(K))$. Then there exists some $a \in M(\mathcal{X}^b(K))$, $a \neq 0$, with

$$\int_{\mathcal{X}^b(K)} \mathcal{F}f(\alpha)\, da(\alpha) \;=\; 0 \qquad \text{for all } f \in L^1(K).$$

It follows

$$0 \;=\; \int_{\mathcal{X}^b(K)} \int_K f(x)\, \overline{\alpha(x)}\, dm(x)\, da(\alpha) \;=\; \int_K f(x)\, (\text{in}\mathcal{F})(\bar{a})(x)\, dm(x).$$

Hence $(\text{in}\mathcal{F})(\bar{a}) = 0$. But then by assumption (2) the measure a has to be zero, a contradiction. ◇

Remark: There are several commutative hypergroups K for which $\mathcal{F}(L^1(K))$ is not norm-dense in $C_0(\mathcal{X}^b(K))$. For example those polynomial hypergroups $K = \mathbb{N}_0$, for which $\mathcal{X}^b(K)$ contains an open circle $U_r(z_0) \subseteq \mathbb{C}$ around $z_0 \in \mathbb{C}$, see e.g. subsection 4.5.7.

In this case the Fourier transform $\mathcal{F}(f)|U_r(z_0)$ of each $f \in L^1(K)$ is a holomorphic function. Hence the continuous function $\varphi : \mathcal{X}^b(K) \to$

$\mathbb{C}$, $\varphi(z) = \bar{z}$ cannot be uniformly approximated on $\mathcal{X}^b(K)$ by Fourier transforms $\mathcal{F}(f)$, $f \in L^1(K)$.

To derive the inversion theorem we prove the following result.

Lemma 3.3.5 *Let $f, g \in L^1(K) \cap L^2(K)$ and $\mu \in M(K)$ as well as $a \in M(\hat{K})$. Putting $h = f * g$ we have*

(i) $h \in C_0(K) \cap L^1(K)$ *and* $\hat{h} \in C_0(\hat{K}) \cap L^1(\hat{K}, \pi)$,

(ii) $\int\limits_K h(x)\, d\tilde{\mu}(x) = \int\limits_{\hat{K}} \hat{h}(\alpha)\hat{\mu}(\alpha)\, d\pi(\alpha)$,

(iii) $\int\limits_{\hat{K}} \hat{h}(\alpha)\, da(\alpha) = \int\limits_K h(\tilde{x})\check{a}(x)\, dm(x)$.

Proof. The statements of (i) follow by Proposition 2.2.13 and Theorem 2.2.10 and from the fact that $\hat{h} = \hat{f}\hat{g}$, where $\hat{f}, \hat{g} \in L^2(\hat{K}, \pi)$. In order to show (ii) we apply the Plancherel-Levitan theorem (and $\tilde{m} = m$)

$$\begin{aligned}\int\limits_K h(x)\, d\tilde{\mu}(x) &= \int\limits_K \int\limits_K g(y)\omega(\tilde{y}, \tilde{x})(f)\, dm(y)\, d\mu(x) = \int\limits_K g(y)\mu * f(\tilde{y})\, dm(y) \\ &= \int\limits_K \mu * f(y)\overline{g^*(y)}\, dm(y) = \int\limits_{\hat{K}} \mathcal{P}(\mu * f)(\alpha)\overline{\mathcal{P}(g^*)(\alpha)}\, d\pi(\alpha) \\ &= \int\limits_{\hat{K}} \hat{\mu}(\alpha)\hat{f}(\alpha)\hat{g}(\alpha)\, d\pi(\alpha) = \int\limits_{\hat{K}} \hat{h}(\alpha)\hat{\mu}(\alpha)\, d\pi(\alpha).\end{aligned}$$

For (iii) note that

$$\int\limits_{\hat{K}} \hat{h}(\alpha)\, da(\alpha) = \int\limits_{\hat{K}} \int\limits_K h(x)\overline{\alpha(x)}\, dm(x)\, da(\alpha) = \int\limits_K h(\tilde{x})\check{a}(x)\, dm(x).$$

◇

Corollary 3.3.6 *Let $\mu \in M(K)$ and $a \in M(\hat{K})$. Then $\check{a} \in L^1(K)$ and $\mu = \check{a}m$ if and only if $\hat{\mu} \in L^1(\hat{K}, \pi)$ and $a = \hat{\mu}\pi$.*

Proof. First note that $\{f * g : f, g \in L^1(K) \cap L^2(K)\}$ is sup-norm dense in $C_0(K)$. In fact given some $g \in C_c(K)$ one can approximate g by $f_i * g$, $f_i \in C_c(K)$, with respect to $\|\,.\,\|_\infty$. (Compare Proposition 2.3.3.) Similarily $\{(f*g)^\wedge : f, g \in L^1(K) \cap L^2(K)\}$ is sup-norm dense in $C_0(\hat{K})$. For that apply Theorem 3.2.1 and approximate $g \in C_c(K)$ by $f_i * g$, $f_i \in C_c(K)$, with respect to $\|\,.\,\|_1$. The statement now follows directly from Lemma 3.3.5 (ii) and (iii). ◇

Theorem 3.3.7 (Inversion theorem) *Let $f \in L^1(K)$ such that $\hat{f} \in L^1(\hat{K}, \pi)$. Then $f = (\hat{f})^\vee$ in $L^1(K)$. If in addition f is continuous, then for all $x \in K$*

$$f(x) = \int_{\hat{K}} \hat{f}(\alpha)\alpha(x)\, d\pi(\alpha).$$

Proof. Put $\mu = fm$ and $a = \hat{f}\pi$, and apply Corollary 3.3.6. For continuous f one has pointwise equality, since $(\hat{f})^\vee$ is continuous, too. ◇

Of course we can derive from Corollary 3.3.6 also a dual inversion theorem.

Theorem 3.3.8 (Inversion theorem) *Let $g \in L^1(\hat{K}, \pi)$ such that $\check{g} \in L^1(K)$. Then $g = \widehat{\check{g}}$ in $L^1(\hat{K}, \pi)$. If in addition g is continuous, then for all $\alpha \in supp\, \pi = \mathcal{S}(K)$, we have*

$$g(\alpha) = \int_K \check{g}(x)\overline{\alpha(x)}\, dm(x).$$

Finally we present an inverse form of the Plancherel-Levitan theorem.

Theorem 3.3.9 *Let $\varphi \in L^1(\hat{K}, \pi) \cap L^2(\hat{K}, \pi)$. Then $\check{\varphi} \in L^2(K)$ and*

$$\int_K |\check{\varphi}(x)|^2\, dm(x) = \int_{\hat{K}} |\varphi(\alpha)|^2\, d\pi(\alpha).$$

Proof. In the notation of section 3.2 we only have to check that $\check{\varphi} = \mathcal{P}^{-1}(\varphi)$ in $L^2(K)$. In fact for $f \in C_c(K)$ we have

$$\begin{aligned}\int_K \check{\varphi}(x)\overline{f(x)}\, dm(x) &= \int_K \int_{\hat{K}} \varphi(\alpha)\alpha(x)\, d\pi(\alpha)\overline{f(x)}\, dm(x) \\ &= \int_{\hat{K}} \widehat{f^*}(\alpha)\varphi(\alpha)\, d\pi(\alpha) = \int_{\hat{K}} \varphi(\alpha)\overline{\hat{f}(\alpha)}\, d\pi(\alpha)\end{aligned}$$

on the one side and

$$\int_K \mathcal{P}^{-1}(\varphi)(x)\overline{f(x)}\, dm(x) = \int_{\hat{K}} \varphi(\alpha)\mathcal{P}(\bar{f})(\alpha)\, d\pi(\alpha) = \int_{\hat{K}} \varphi(\alpha)\overline{\hat{f}(\alpha)}\, d\pi(\alpha)$$

on the other side. Hence $\check{\varphi} = \mathcal{P}^{-1}(\varphi)$. ◇

Proposition 3.3.10 *Let $f, g \in L^2(K)$. Then $(\mathcal{P}(f)\mathcal{P}(g))^\vee = f * g$.*

Proof. Notice that $\mathcal{P}(f)\mathcal{P}(g) \in L^1(\hat{K}, \pi)$. From Lemma 3.3.1 we derive that $\mathcal{P}^{-1}(\mathcal{P}(g^*)\widehat{\epsilon_x}) = L_{\tilde{x}} g^*$. Hence

$$(\mathcal{P}(f)\mathcal{P}(g))^\vee(x) = \int_{\hat{K}} \mathcal{P}(f)(\alpha)\mathcal{P}(g)(\alpha)\,\alpha(x)\,d\pi(\alpha)$$

$$= \int_{\hat{K}} \mathcal{P}(f)(\alpha)\overline{\mathcal{P}(g^*)(\alpha)\overline{\alpha(x)}}\,d\pi(\alpha) = \int_K f(y) L_{\tilde{x}} g^*(y)\,dm(y) = f * g(x).$$

◇

4 Comparison of the dual spaces

In this section we investigate the dual objects $\mathcal{S}(K)$, $\hat{K}$ and $\mathcal{X}^b(K)$ of commutative hypergroups K in more detail.

Consider the Banach $*$-algebra $L^1(K) = L^1(K, m)$. For $f \in L^1(K)$ let

$$r(f) := \lim_{n\to\infty} \|f^n\|_1^{1/n},$$

where $f^n = f^{n-1} * f$ for $n \geq 2$ and $f^1 = f$. $r(f)$ is the **spectral radius** of $f \in L^1(K)$.

If K is not discrete, $L^1(K)$ does not have an identity. We adjoin the point measure ϵ_e to $L^1(K)$ within the measure algebra $M(K)$, and write $L^1(K)_e = L^1(K) \oplus \mathbb{C}\epsilon_e \subseteq M(K)$. $L^1(K)_e$ is a Banach $*$-algebra with identity. The norm of $f \in L^1(K)_e$ is denoted by $\|f\|$.

The **spectrum** of $f \in L^1(K)$ is the set

$$\sigma(f) = \{\lambda \in \mathbb{C} : \lambda\epsilon_e - f \text{ is not invertible in } L^1(K)_e\}.$$

By the Gelfand theory of commutative Banach algebras we have

$$r(f) = \max\{|\lambda| : \lambda \in \sigma(f)\} = \inf\{\|f^n\|_1^{1/n} : n \in \mathbb{N}\}$$

$$= \sup\{|\mathcal{F}(f)(\alpha)| : \alpha \in \mathcal{X}^b(K)\}.$$

Since $\sigma(f^*) = \overline{\sigma(f)}$, it follows $r(f) = r(f^*)$. By Proposition 3.1.7 it follows that

$$\|L_f\| = \sup\{|\hat{f}(\alpha)| : \alpha \in \mathcal{S}(K)\} \leq r(f) \leq \|f\|_1$$

for any $f \in L^1(K)$. Further, for $f \in L^1(K)$, put

$$s(f) := \sup\{|\hat{f}(\alpha)| : \alpha \in \hat{K}\}.$$

Then the four assignments $f \mapsto \|L_f\|$, $s(f), r(f)$ and $\|f\|_1$ are submultiplicative norms on $L^1(K)$ and

$$\|L_f\| \leq s(f) \leq r(f) \leq \|f\|_1.$$

Proposition 3.4.1 *Let K be a commutative hypergroup. Then $\mathcal{S}(K) = \hat{K}$ if and only if $\|L_f\| = s(f)$ for every $f \in L^1(K)$.*

Proof. We only have to show that $\|L_f\| = s(f)$ for all $f \in L^1(K)$ implies that $\mathcal{S}(K) = \hat{K}$. Assume that there is some $\alpha_0 \in \hat{K}\backslash\mathcal{S}(K)$. Then there exists $\varphi \in C_0(\hat{K})$ such that $\varphi(\alpha_0) = 1$ and $\varphi|\mathcal{S}(K) = 0$. By Theorem 3.2.1 we find some $f \in L^1(K)$ such that $\sup\{|\hat{f}(\beta)| : \beta \in \mathcal{S}(K)\} < |\hat{f}(\alpha_0)|$. Hence $\|L_f\| < s(f)$, a contradiction. $\diamond$

Lemma 3.4.2 *Let K be a commutative hypergroup, and let $\alpha \in \mathcal{X}^b(K)$. Then $\alpha \in \hat{K}$ if and only if $\mathcal{F}f(\alpha) \in \mathbb{R}$ for all $f \in L^1(K)$ such that $f = f^*$.*

Proof. Let $\alpha \in \hat{K}$ and $f \in L^1(K)$, $f = f^*$. Since $\hat{f}(\alpha) = \widehat{f^*}(\alpha) = \overline{\hat{f}(\alpha)}$, we have $\mathcal{F}f(\alpha) = \hat{f}(\alpha) \in \mathbb{R}$. Conversely, let $\alpha \in \mathcal{X}^b(K)$ such that $\mathcal{F}f(\alpha) \in \mathbb{R}$ for every $f = f^*$. Each $f \in L^1(K)$ can be written as $f = g + ih$ with $g = g^*$ and $h = h^*$. Then $f^* = g - ih$, and hence $\mathcal{F}f^*(\alpha) = \mathcal{F}g(\alpha) - i\mathcal{F}h(\alpha) = \overline{\mathcal{F}f(\alpha)}$. That is,

$$\int_K \overline{f(x)}\,\overline{\alpha(\tilde{x})}\,dm(x) = \mathcal{F}f^*(\alpha) = \overline{\mathcal{F}f(\alpha)} = \int_K \overline{f(x)}\,\alpha(x)\,dm(x)$$

for all $f \in L^1(K)$. It follows $\overline{\alpha(\tilde{x})} = \alpha(x)$ for all $x \in K$, i.e. $\alpha \in \hat{K}$. $\diamond$

Theorem 3.4.3 *Let K be a commutative hypergroup. Then the following statements are equivalent.*

(1) $\hat{K} = \mathcal{X}^b(K)$.
(2) $s(f) = r(f)$ *for all* $f \in L^1(K)$.
(3) $r(f * f^*) = r(f)^2$ *for all* $f \in L^1(K)$.
(4) $\mathcal{F}(f^2)(\alpha) \neq -1$ *for all* $f \in L^1(K)$, $f = f^*$, *and all* $\alpha \in \mathcal{X}^b(K)$.

Proof. (1) $\Rightarrow$ (2) is obvious.
If (2) holds, we obtain

$$r(f * f^*) = s(f * f^*) = \sup\{|\hat{f}(\alpha)|^2 : \alpha \in \hat{K}\} = s(f)^2 = r(f)^2$$

for each $f \in L^1(K)$, and this shows (3).
Now we prove that (3) $\Rightarrow$ (1). (The proof follows the one of [574,Lemma 4.2.1].) Assume that there exists some $\alpha \in \mathcal{X}^b(K)\backslash\hat{K}$. By Lemma 3.4.2 there is some $f \in L^1(K)$, $f = f^*$, such that $\mathcal{F}f(\alpha) = a + ib$, $a, b \in \mathbb{R}$, $b \neq 0$. Put $c = \frac{a}{b}$ and $g = \frac{1}{b}f$. Then $g = g^*$ and $\mathcal{F}g(\alpha) = c + i$. Denoting $\alpha^* = \overline{\tilde{\alpha}}$ we have $\mathcal{F}g(\alpha^*) = \overline{\mathcal{F}g(\alpha)}$, since $g = g^*$. Therefore $\mathcal{F}g(\alpha^*) = c - i$. Choose

some $h \in L^1(K)$ such that $\mathcal{F}h(\alpha) = 1$. Since $\mathcal{F}h^*(\alpha^*) = \overline{\mathcal{F}h(\alpha)}$, we also have $\mathcal{F}h^*(\alpha^*) = 1$. Temporarily, fix $m, n \in \mathbb{N}$ and set

$$\mu = (g + (ni - c)\epsilon_e)^m * h \in L^1(K).$$

Then $\mu^* = (g - (ni + c)\epsilon_e)^m * h^*$ and

$$\mu * \mu^* = ((g - c\epsilon_e)^2 + n^2\epsilon_e)^m * h * h^*.$$

The Fourier transforms of μ and μ^* are given by

$$\mathcal{F}\mu(\alpha) = (i(n+1))^m \quad \text{and} \quad \mathcal{F}\mu^*(\alpha^*) = (-i(n+1))^m.$$

It follows $(n+1)^m \leq r(\mu)$ and $(n+1)^m \leq r(\mu^*)$, and hence

$$\begin{aligned}(n+1)^{2m} &\leq r(\mu)r(\mu^*) = r(\mu * \mu^*) = r((g - c\epsilon_e)^2 + n^2\epsilon_e)^m * h * h^*) \\ &\leq ((r(g) + |c|)^2 + n^2)^m r(h * h^*).\end{aligned}$$

Therefore

$$(n+1)^2 \leq ((r(g) + |c|)^2 + n^2)(r(h * h^*))^{1/m}.$$

Now, letting $m \to \infty$ we obtain

$$(n+1)^2 \leq (r(g) + |c|)^2 + n^2,$$

and hence $1 + 2n \leq (r(g) + |c|)^2$. This contradiction shows (3) $\Rightarrow$ (1).

It remains to show (1) $\Leftrightarrow$ (4). If $\hat{K} = \mathcal{X}^b(K)$, we have for each $f = f^* \in L^1(K)$, and every $\alpha \in \mathcal{X}^b(K)$,

$$\mathcal{F}(f^2)(\alpha) = \widehat{f * f^*}(\alpha) = |f(\alpha)|^2 \geq 0.$$

Conversely assume that there is some $\alpha \in \mathcal{X}^b(K) \backslash \hat{K}$. By Lemma 3.4.2 there exists some $f \in L^1(K)$, $f = f^*$ such that $\mathcal{F}f(\alpha) = a + ib$, $a, b \in \mathbb{R}$, $b \neq 0$. Put

$$g = \frac{1}{b(a^2 + b^2)} (af^2 + (b^2 - a^2)f).$$

Then $g = g^*$ and $\mathcal{F}g(\alpha) = i$, and hence $\mathcal{F}(g^2)(\alpha) = -1$, which contradicts (4). ◇

Remark: From Theorem 3.4.3 we know that the spectral radius fulfils the C^*-condition $r(f * f^*) = r(f)^2$ for all $f \in L^1(K)$ if and only if $\hat{K} = \mathcal{X}^b(K)$. Since in that case $r(f) = \sup\{|\hat{f}(\alpha)| : \alpha \in \hat{K}\}$, r is a C^*-norm on the $*$-algebra $L^1(K)$. The completion of $L^1(K)$ with respect to r is a C^*-algebra which is isomorphic to $C_0(\hat{K})$. Note that $\{\hat{f} : f \in L^1(K)\}$ is a dense self-adjoint subalgebra of $C_0(\hat{K})$, see Theorem 3.2.1. This algebra is called **full hypergroup C^*-algebra** of K and is denoted by $C^*(K)$.

Corollary 3.4.4 *Let K be a commutative hypergroup. We have $\mathcal{S}(K) = \hat{K} = \mathcal{X}^b(K)$ if and only if $\|L_f\| = r(f)$ for every $f \in L^1(K)$.*

We are now going to show that a certain growth condition for the Haar measure ensures that $\mathcal{S}(K) = \hat{K} = \mathcal{X}^b(K)$.

Definition: A hypergroup K with a left-invariant Haar measure m is said to be of

(1) **polynomial growth** if for every compact subset $C \subseteq K$ there is some constant $\alpha = \alpha(C) \geq 0$ such that $m(C^n) = O(n^\alpha)$ as $n \to \infty$,
(2) **subexponential growth** if for every compact subset $C \subseteq K$ and $\delta > 0$ there exists a constant $M = M(C, \delta) > 0$ such that $m(C^n) \leq M(1+\delta)^n$ for all $n \in \mathbb{N}$.

Obviously polynomial growth implies subexponential growth. By Proposition 2.1.4 (i) we know that the sequence $(m(C^n))_{n\in\mathbb{N}}$ is increasing. The growth condition (1) and hence (2) are obviously satisfied if $m(C^n) \leq 1$ for all $n \in \mathbb{N}$. Hence we can restrict the study of the growth conditions to the case when $m(C) \geq 1$.

Proposition 3.4.5 *K is of subexponential growth if and only if $\lim\limits_{n\to\infty} (m(C^n))^{1/n} = 1$ for all compact subsets C of K with $m(C) \geq 1$.*

Proof. Let $\delta > 0$ and $M = M(C, \delta) > 0$ such that $m(C^n) \leq M(1+\delta)^n$ for all $n \in \mathbb{N}$, where $m(C) \geq 1$. Then $m(C^n)^{1/n} \leq M^{1/n}(1 + \delta)$ for each $n \in \mathbb{N}$, and we obtain $\limsup\limits_{n\to\infty}(m(C^n))^{1/n} \leq 1 + \delta$. Since this is true for each $\delta > 0$, $\limsup\limits_{n\to\infty}(m(C^n))^{1/n} \leq 1$. As $m(C^n) \geq 1$ for each $n \in \mathbb{N}$, we get $\lim\limits_{n\to\infty} (m(C^n))^{1/n} = 1$.

Conversely, let $\lim\limits_{n\to\infty} (m(C^n))^{1/n} = 1$ with $m(C) \geq 1$. Then for each $\delta > 0$ there exists some $N \in \mathbb{N}$ such that $1 \leq (m(C^n))^{1/n} \leq 1 + \delta$ for every $n \geq N$. Put $M := \max\{m(C), ..., m(C^{N-1})\} \geq 1$. Then $m(C^n) \leq M(1+\delta)^n$ for all $n \in \mathbb{N}$. ◇

Remark:

(1) If for every compact subset C of K there exists $\beta = \beta(C)$ with $0 \leq \beta < 1$ such that $m(C^n) = O(\exp(n^\beta))$, then K is of subexponential growth.
(2) It is easily shown that K is of subexponential growth exactly when for every compact subset C of K holds $m(C^n) = o(a^n)$ for each $a > 1$.

Theorem 3.4.6 *Let K be a commutative hypergroup of subexponential growth. Then $\mathcal{S}(K) = \hat{K} = \mathcal{X}^b(K)$.*

Proof. In view of Corollary 3.4.4 we have to show that $r(f) \leq \|L_f\|$ for every $f \in L^1(K)$. Assume that $f \neq 0$ and let $\varepsilon > 0$. Then there exist $g \in C_c(K)$, $g \neq 0$ and $d \in L^1(K)$ such that $\|d\|_1 < \varepsilon$ and $f = g + d$. Let $C = \text{supp}\, g$. We have $\text{supp}\,(g^n) \subseteq C^n$, which follows by induction from Proposition 1.1.9. (In fact, we use only the first step in the proof of Proposition 1.1.9.) Since

$$f^n = (g+d)^n = \sum_{k=0}^{n} \binom{n}{k} d^k\, g^{n-k},$$

we obtain

$$\|f^n\|_1 \leq \varepsilon^n + \sum_{k=0}^{n-1} \binom{n}{k} \varepsilon^k\, \|g^{n-k}\|_1$$

$$\leq \varepsilon^n + \sum_{k=0}^{n-1} \binom{n}{k} \varepsilon^k\, (m(C^{n-k}))^{1/2} \|g^{n-k}\|_2.$$

For $k = 0, ..., n-1$ we have

$$\|g^{n-k}\|_2 = \|L_{(g^{n-k-1})}(g)\|_2 \leq \|L_g\|^{n-k-1} \|g\|_2$$

$$= \frac{\|L_g\|^{n-k}}{\|L_g\|}\, \|g\|_2$$

and by Proposition 2.1.4 (i),

$$m(C^{n-k}) \leq m(C^n).$$

Therefore

$$\|f^n\|_1 \leq \varepsilon^n + (m(C^n))^{1/2} \frac{\|g\|_2}{\|L_g\|} \sum_{k=0}^{n-1} \varepsilon^k\, \|L_g\|^{n-k}.$$

Since

$$\|L_g\| \leq \|g\|_1 \leq m(C))^{1/2}\, \|g\|_2 \leq (m(C^n))^{1/2}\, \|g\|^2,$$

it follows

$$\|f^n\|_1 \leq (m(C^n))^{1/2} \frac{\|g\|_2}{\|L_g\|} (\varepsilon + \|L_g\|)^n.$$

Now for each $\delta > 0$ there exists $M = M(\delta, C) > 0$ such that $m(C^n) \leq M(1+\delta)^n$ for all $n \in \mathbb{N}$. Thus

$$\|f^n\|_1 \leq \frac{\|g\|_2}{\|L_g\|} (M(1+\delta)^n)^{1/2}(\varepsilon + \|L_g\|)^n,$$

and so

$$r(f) = \lim_{n\to\infty} \|f^n\|_1^{1/n} = (1+\delta)^{1/2}(\varepsilon + \|L_g\|).$$

Since $\delta > 0$ was arbitrary and

$$| \, \|L_f\| - \|L_g\| \, | \leq \|f - g\|_1 = \|d\|_1 < \varepsilon,$$

we obtain finally

$$r(f) \leq 2\varepsilon + \|L_f\|.$$

Since this is true for every $\varepsilon > 0$, we get $r(f) \leq \|L_f\|$. ◇

An interesting property of $\mathcal{S}(K)$ is the fact that the compact subsets of $\mathcal{S}(K)$ determine the topology of K.

Lemma 3.4.7 *Let U_{x_0} be a neighbourhood of $x_0 \in K$, and V a neighbourhood of $e \in K$ with compact closure such that $\tilde{V} = V$ and $\{x_0\} * V * V \subseteq U_{x_0}$. Let $g = \chi_V / \|L_{x_0}\chi_V\|_2 \in L^2(K)$. Then, for every $y \in K \backslash U_{x_0}$ we have $\|L_{x_0}g - L_y g\|_2 \geq 1$.*

Proof. Since $y \notin U_{x_0}$ we have $\{y\} \cap \{x_0\} * V * V = \emptyset$, and by Lemma 1.1.10(ii) follows $\{y\} * \tilde{V} \cap \{x_0\} * V = \emptyset$, which is equivalent to $\{\tilde{y}\} * V \cap \{\tilde{x}_0\} * V = \emptyset$. Since $\{z \in K : L_y g(z) \neq 0\} \subseteq \{\tilde{y}\} * V$, it follows

$$\|L_{x_0}g - L_y g\|_2^2 = \|L_{x_0}g\|_2^2 + \|L_y g\|_2^2 - 2 \operatorname{Re}\left(\int_K L_{x_0}g(z)\overline{L_y g(z)}\, dm(z)\right)$$

$$= \|L_{x_0}g\|_2^2 + \|L_y g\|_2^2 \geq 1.$$

◇

Lemma 3.4.8 *Given some neighbourhood U_{x_0} of $x_0 \in K$, there exist a compact subset $\Gamma \subseteq \mathcal{S}(K)$ and $\varepsilon > 0$ such that*

$$W(x_0, \Gamma, \varepsilon) := \{y \in K : |\alpha(y) - \alpha(x_0)| < \varepsilon \text{ for all } \alpha \in \Gamma\} \subseteq U_{x_0}.$$

Proof. Choose $g \in L^2(K)$ as in Lemma 3.4.7. We know that $\|L_{x_0}g - L_y g\|_2 < 1$ implies $y \in U_{x_0}$. Now there exists $f \in C_c(K)$ with $\|f\|_1 = 1$ and $\|f * g - g\|_2 < \frac{1}{3}$. Obviously, we have $\|L_x g - f * L_x g\|_2 =$

$\|L_x(f*g-g)\|_2 \leq \|f*g-g\|_2 < \frac{1}{3}$ for every $x \in K$, and using Proposition 3.1.7 we get

$$\|f * L_{x_0}g - f * L_y g\|_2 = \|(L_{x_0}f - L_y f) * g\|_2$$

$$\leq \sup\{|(L_{x_0}f - L_y f)^\wedge(\alpha)| : \alpha \in \mathcal{S}(K)\} \, \|g\|_2.$$

Since $\hat{f} \in C_0(\mathcal{S}(K))$, there exists a compact subset $\Gamma \subseteq \mathcal{S}(K)$ such that $|\hat{f}(\alpha)| < \frac{1}{6\|g\|_2}$ for all $\alpha \in \mathcal{S}(K)\backslash\Gamma$. For any $y \in W(x_0, \Gamma, \frac{1}{3\|g\|_2})$ and every $\alpha \in \mathcal{S}(K)$, we have

$$|(L_{x_0}f - L_y f)^\wedge(\alpha)| = |\hat{f}(\alpha)| \, |\alpha(x_0) - \alpha(y)| < \frac{1}{3\,\|g\|_2},$$

and thus

$$\|L_{x_0}g - L_y g\|_2 \leq \|L_{x_0}g - f * L_{x_0}g\|_2 + \|f * L_{x_0}g - f * L_y g\|_2 + \|f * L_y g - L_y g\|_2 < 1,$$

which shows $y \in U_{x_0}$. ◇

Theorem 3.4.9 *For every $x_0 \in K$ the family of subsets*

$$W(x_0, \Gamma, \delta) = \{y \in K : |\alpha(y) - \alpha(x_0)| < \delta \ \textit{for all } \alpha \in \Gamma\},$$

where $\Gamma \subseteq \mathcal{S}(K)$ compact, $\delta > 0$, is an open neighbourhood basis of x_0.

Proof. By Lemma 3.4.8 we only have to check that $W(x_0, \Gamma, \delta)$ is open. By Proposition 3.1.5 the mapping $(x, \alpha) \to \alpha(x)$, $K \times \mathcal{S}(K) \to \mathbb{C}$ is continuous. Applying the compactness of $\Gamma \subseteq \mathcal{S}(K)$ it is a routine exercise to derive that $W(x_0, \Gamma, \delta)$ is open. ◇

If the character space $\hat{K}$ is a hypergroup under pointwise multiplication, we can derive that $\mathcal{S}(K) = \hat{K}$.

$\hat{K}$ is a **hypergroup under pointwise multiplication**, if for all $\alpha, \beta \in \hat{K}$ there exists a probability measure $\omega(\alpha, \beta) \in M^1(\hat{K})$ with compact support such that

$$\alpha(x)\, \beta(x) = \int_{\hat{K}} \tau(x) \, d\omega(\alpha, \beta)(\tau) \qquad (P)$$

for all $x \in K$, satisfying the axioms (H1) – (H6), where the convolution is given by ω, the involution $\tilde{}$ is the complex conjugation, i.e. $\tilde{\alpha}(x) = \overline{\alpha(x)}$ for all $x \in K$, and the constant function 1 is the unit element.

By the uniqueness theorem, Theorem 3.3.3, the probability measure $\omega(\alpha, \beta)$ satisfying only (P) is uniquely determined by $\alpha, \beta \in \hat{K}$. It is obvious that (H4) and (H5) always hold.

Proposition 3.4.10 *Suppose that for every $\alpha, \beta \in \hat{K}$ there exists $\omega(\alpha,\beta) \in M^1(\hat{K})$ such that (P) is valid. Then the axioms (H1), (H4) and (H5) are satisfied.*

Proof. By the remark preceding the proposition we have only to verify (H1). Let $\varphi \in C_0(\hat{K})$ and $\varepsilon > 0$. By Theorem 3.2.1 there exists $f \in C_c(K)$ such that $\|\varphi - \hat{f}\|_\infty < \varepsilon$. Given $\alpha_0, \beta_0 \in \hat{K}$ we obtain

$$\left| \int_{\hat{K}} \varphi(\tau)\, d\omega(\alpha,\beta)(\tau) - \int_{\hat{K}} \varphi(\tau)\, d\omega(\alpha_0,\beta_0)(\tau) \right|$$

$$\leq 2\varepsilon + \left| \int_{\hat{K}} \hat{f}(\tau)\, d\omega(\alpha,\beta)(\tau) - \int_{\hat{K}} \hat{f}(\tau)\, d\omega(\alpha_0,\beta_0)(\tau) \right|$$

$$\leq 2\varepsilon + \int_K |f(x)|\, |\alpha(x)\,\beta(x) - \alpha_0(x)\,\beta_0(x)|\, dm(x).$$

Since $\hat{K}$ carries the topology of uniform convergence on compacta and since $\operatorname{supp} f$ is compact the weak $\star$-continuity of $(\alpha,\beta) \to \omega(\alpha,\beta)$ follows.

Applying Theorem 3.2.1 again, the associativity of the convolution product follows, as was to be shown. ◇

Remark: If property (P) holds, by Proposition 3.4.10 one only has to check (H2), (H3) and (H6) in order to assure that $\hat{K}$ is a hypergroup under pointwise multiplication.

Theorem 3.4.11 *Assume that $\hat{K}$ is a hypergroup under pointwise multiplication. Then the Plancherel measure π on $\hat{K}$ is translational invariant, i.e. π is a Haar measure on $\hat{K}$. In particular $\mathcal{S}(K) = \hat{K}$.*

Proof. Let $\varphi \in C_c(\hat{K})$. Applying Theorem 3.2.1 we can select $g_n \in L^1(K) \cap C_0(K)$ with $\widehat{g_n} \in L^1(\hat{K},\pi)$ such that $\|\varphi - \widehat{g_n}\|_\infty \leq \frac{1}{n}$, compare the proof of Corollary 3.3.6. Let $\alpha \in \hat{K}$. Then $\|L_\alpha\varphi - L_\alpha\widehat{g_n}\|_\infty \leq \frac{1}{n}$, and since $\operatorname{supp} L_\alpha\varphi$ is compact it follows

$$\lim_{n\to\infty} \int_{\hat{K}} L_\alpha\widehat{g_n}(\beta)\, d\pi(\beta) = \int_{\hat{K}} L_\alpha\varphi(\beta)\, d\pi(\beta).$$

For any $f \in L^1(K)$ we have

$$L_\alpha\hat{f}(\beta) = \widehat{\overline{\alpha} f}(\beta) \qquad \text{for all } \beta \in \hat{K}.$$

In fact,

$$L_\alpha\hat{f}(\beta) = \int_{\hat{K}} \hat{f}(\tau)\, d\omega(\alpha,\beta)(\tau) = \int_{\hat{K}}\int_K f(x)\,\overline{\tau(x)}\, dm(x)\, d\omega(\alpha,\beta)(\tau)$$

$$= \int_K f(x)\,\overline{\alpha(x)\beta(x)}\, dm(x) = \widehat{\overline{\alpha} f}(\beta).$$

Since the functions $\widehat{\overline{\alpha} g_n} = L_\alpha \widehat{g_n}$ are elements of $L^1(\hat{K}, \pi)$ the inversion theorem 3.3.7 implies

$$\int_{\hat{K}} L_\alpha \widehat{g_n}(\beta)\, d\pi(\beta) = \int_{\hat{K}} \widehat{\overline{\alpha} g_n}(\beta)\, d\pi(\beta) = \overline{\alpha(e)}\, g_n(e) = \int_{\hat{K}} \widehat{g_n}(\beta)\, d\pi(\beta).$$

Therefore we get

$$\int_{\hat{K}} L_\alpha \varphi(\beta)\, d\pi(\beta) = \int_{\hat{K}} \varphi(\beta)\, d\pi(\beta).$$

◇

We continue with some remarks on the second dual $\hat{\hat{K}}$ of K supposing that $\hat{K}$ is a hypergroup under pointwise multiplication. For $x \in K$ define $i(x) : \hat{K} \to \mathbb{C}$ by $i(x)(\alpha) = \overline{\alpha(x)}$. Then $i(x)$ is a continuous and bounded function fulfilling $i(x)(1) = 1$, $i(x)(\overline{\alpha}) = \overline{\overline{\alpha}(x)} = \alpha(x) = \overline{i(x)(\alpha)}$ and

$$\int_{\hat{K}} \overline{\tau(x)}\, d\omega(\alpha, \beta)(\tau) = \overline{\alpha(x)\beta(x)} = i(x)(\alpha)\, i(x)(\beta).$$

Thus $i(x) \in \hat{\hat{K}}$ for every $x \in K$. The mapping $i : K \to \hat{\hat{K}}$ is an isomorphic embedding of K into $\hat{\hat{K}}$. In fact $i(\tilde{x}) = \overline{i(x)}$ and $i(\omega(x,y))(\alpha) = \int_K i(z)(\alpha)\, d\omega(x,y)(z) = i(x)(\alpha)\, i(y)(\alpha)$ for every $\alpha \in \hat{K}$. (Note that we have extended $i : K \to C(\hat{K})$ to a mapping $i : M(K) \to C(\hat{K})$ by

$$i(\mu)(\alpha) = \int_K i(z)(\alpha)\, d\mu(z).)$$

Furthermore given $x, y \in K$, $x \neq y$, there exists $\alpha \in \hat{K}$ with $\alpha(x) \neq \alpha(y)$, see Theorem 3.1.8, i.e. $i(x) \neq i(y)$.

The mapping $i : K \to \hat{\hat{K}}$ is a homeomorphism by virtue of Theorem 3.4.9. In particular $i(K)$ is a closed subset of $\hat{\hat{K}}$. Summing up we have:

Theorem 3.4.12 *Assume that $\hat{K}$ is a hypergroup under pointwise multiplication. The mapping $i : K \to \hat{\hat{K}}$, $i(x)(\alpha) = \overline{\alpha(x)}$, is a homeomorphism from K onto a closed subset of $\hat{\hat{K}}$ fulfilling*

$$\int_K i(z)(\alpha)\, d\omega(x,y)(z) = i(x)(\alpha)\, i(x)(\alpha) \qquad \textit{for every } \alpha \in \hat{K}$$

and $i(\tilde{x}) = \overline{i(x)}$, $i(e) = 1$.
(We shall call $i : K \to \hat{\hat{K}}$ Pontryagin mapping, and say that the **Pontryagin duality** *is valid for K, if $i(K) = \hat{\hat{K}}$.*

If $K = G$ is a locally compact abelian group, the Pontryagin duality holds. As far as the author knows it is still an open question whether the Pontryagin duality holds when $\hat{K}$ is a hypergroup under pointwise multiplication.

However, we show next that this question has an affirmative answer at least when K is discrete.

Proposition 3.4.13 *Let K be a discrete hypergroup, such that $\hat{K}$ is a hypergroup under pointwise multiplication. Then the mapping $i : K \to \hat{\hat{K}}$ is surjective.*

Proof. By Theorem 3.4.11 we know $\mathcal{S}(K) = \hat{K}$ and π is a Haar measure on the compact hypergroup $\hat{K}$. Theorem 3.1.9 states that $\hat{\hat{K}} = \text{supp } \varrho$, where ϱ is the Plancherel measure on $\hat{\hat{K}}$. Assume that there exists some $\sigma_0 \in \hat{\hat{K}} \backslash i(K)$. We have $\sigma_0 \in C(\hat{K}) \subseteq L^1(\hat{K}, \pi)$ and the proof of Theorem 3.1.9 shows that $\hat{\sigma}_0(\sigma) = 0$ for each $\sigma \in \hat{\hat{K}}$, $\sigma \neq \sigma_0$. In particular, $\sigma_0^\vee(x) = \hat{\sigma}_0(i(x)) = 0$ for all $x \in K$. By Theorem 3.3.3 it follows that $\sigma_0 = 0$, a contradiction. ◇

Theorem 3.4.14 *Let K be a commutative hypergroup, such that $\hat{K}$ is a hypergroup under pointwise multiplication. Then $i(K) = \mathcal{S}(\hat{K})$ is valid.*

Proof. By Proposition 3.2.5 and Theorem 3.2.6 – applied to $\hat{K}$ – we obtain

$$\int_K |\check{\varphi}(x)|^2 \, dm(x) = \int_{\hat{K}} |\varphi(\alpha)|^2 \, d\pi(\alpha) = \int_{\hat{\hat{K}}} |\hat{\varphi}(a)|^2 \, d\varrho(a) \qquad (1)$$

for all $\varphi \in L^1(\hat{K}, \pi) \cap L^2(\hat{K}, \pi)$, where ϱ is the Plancherel measure on $\hat{\hat{K}}$. Furthermore, $\check{\varphi}(x) = \hat{\varphi}(i(x))$ for all $\varphi \in L^1(\hat{K}, \pi)$ and $x \in K$. By (1) it follows

$$\int_{\hat{\hat{K}}} |\hat{\varphi}(a)|^2 \, d(i(m))(a) = \int_{\hat{\hat{K}}} |\hat{\varphi}(a)|^2 \, d\varrho(a)$$

for all $\varphi \in L^1(\hat{K}, \pi) \cap L^2(\hat{K}, \pi)$, where $i(m)$ is the image measure of the Haar measure m on K and $i : K \to \hat{\hat{K}}$ the embedding of K into $\hat{\hat{K}}$. Since the Plancherel measure ϱ is uniquely determined by (1), we obtain that $\mathcal{S}(\hat{K}) = \text{supp } \varrho = \text{supp}\,(i(m)) = i(K)$. ◇

Corollary 3.4.15 *Let K be a commutative hypergroup, such that $\hat{K}$ is a hypergroup under pointwise multiplication of subexponential growth. Then the mapping $i : K \to \hat{\hat{K}}$ is surjective.*

Proof. Combine Theorem 3.4.6 and Theorem 3.4.14. $\diamond$

In the next subsection 3.5 we characterize the surjectivity of the Pontryagin map $i : K \to \hat{\hat{K}}$ in terms of a regularity condition.

Further results concerning the dual spaces of commutative hypergroups will be studied in subsection 3.11. There we shall derive a weak convolution product on $\mathcal{S}(K)$. The involution $\alpha \mapsto \bar{\alpha}$ on $\mathcal{S}(K)$ or $\hat{K}$ leads to an involution on the Banach spaces $M(\hat{K})$ and $L^1(\mathcal{S}(K), \pi)$ in the following way.

Let $a \in M(\hat{K})$. Define $a^* \in M(\hat{K})$ by setting

$$a^*(F) = \overline{a(F)} \qquad (i)$$

for each Borel set $F \subseteq \hat{K}$. Equivalently, a^* is defined by

$$\int_{\hat{K}} h(\alpha)\, da^*(\alpha) = \overline{\int_{\hat{K}} \overline{h(\bar{\alpha})}\, da(\alpha)}$$

for every $h \in C_0(\hat{K})$.

For $\varphi \in L^1(\mathcal{S}(K), \pi)$, considering $\varphi\pi \in M(\hat{K})$, we define

$$\varphi^*(\alpha) = \overline{\varphi(\bar{\alpha})} \qquad (ii)$$

for every $\alpha \in \mathcal{S}(K)$. Then

$$\int_{\mathcal{S}(K)} h(\alpha)\, \overline{\varphi(\bar{\alpha})}\, d\pi(\alpha) = \overline{\int_{\mathcal{S}(K)} \overline{h(\bar{\alpha})}\, \varphi(\alpha)\, d\pi(\alpha)} = \int_{\mathcal{S}(K)} h(\alpha)\, d(\varphi\pi)^*(\alpha)$$

for all $h \in C_0(\hat{K})$.

For the inverse Fourier transform, respectively for the inverse Fourier-Stieltjes transform one can easily prove

Proposition 3.4.16 *Let $a \in M(\hat{K})$ and $\varphi \in L^1(\mathcal{S}(K), \pi)$. For $a^* \in M(\hat{K})$ defined by (i) and $\varphi^* \in L^1(\mathcal{S}(K), \pi)$ defined by (ii), we have*

(1) $(a^*)^\vee(x) = \overline{\check{a}(x)}$ *for all* $x \in K$,

(2) $(\varphi^*)^\vee(x) = \overline{\check{\varphi}(x)}$ *for all* $x \in K$.

5 Regularity conditions and a functional calculus

Throughout this section K will again be a commutative hypergroup. We shall study regularity conditions of the algebra $L^1(K)$ and the vector space $L^1(\mathcal{S}(K), \pi)$.

A family $\mathcal{G}$ of complex valued functions on a T_1 topological space X is said to be regular if for every closed subset $E \subseteq X$ and $x \in X \backslash E$ there exists $f \in \mathcal{G}$ such that $f(x) \neq 0$ and $f|E = 0$.

Definition:

(1) $L^1(K)$ is called **$\mathcal{S}(K)$-regular** (**$\hat{K}$-regular** or **$\mathcal{X}^b(K)$-regular**), if $\mathcal{F}(L^1(K))|\mathcal{S}(K)$ is regular ($\mathcal{F}(L^1(K))|\hat{K}$ is regular or $\mathcal{F}(L^1(K))$ is regular), respectively.
(2) $L^1(\hat{K},\pi)$ is called **inverse regular**, if $L^1(\hat{K},\pi)^\vee$ is regular.

Remark: $\mathcal{X}^b(K)$-regularity of $L^1(K)$ is exactly the definition of regularity of the commutative Banach algebra $L^1(K)$ (see [356,Def.4.2.1]).

The following result is a direct consequence of the uniqueness theorem, Theorem 3.1.8.

Proposition 3.5.1 *(1) If $L^1(K)$ is $\mathcal{X}^b(K)$-regular, then $\mathcal{S}(K) = \hat{K} = \mathcal{X}^b(K)$.*
(2) If $L^1(K)$ is $\hat{K}$-regular, then $\mathcal{S}(K) = \hat{K}$.

Proof.

(1) Assume that there exists $\alpha \in \mathcal{X}^b(K)\backslash\mathcal{S}(K)$. If $L^1(K)$ is $\mathcal{X}^b(K)$-regular, there is $f \in L^1(K)$ such that $\hat{f}|\mathcal{S}(K) = 0$ and $\mathcal{F}f(\alpha) \neq 0$. By Theorem 3.1.8 it follows $f = 0$ in $L^1(K)$, and so $\mathcal{F}f(\alpha) = 0$, a contradiction. Hence $\mathcal{S}(K) = \hat{K} = \mathcal{X}^b(K)$.
(2) Supposing the existence of some $\alpha \in \hat{K}\backslash\mathcal{S}(K)$ the $\hat{K}$-regularity of $L^1(K)$ leads to a contradiction as in (1).

$\diamond$

At first we prove that $L^1(\hat{K},\pi)$ is inverse regular. Note that $L^1(\hat{K},\pi)^\vee$ is in general not an algebra with respect to pointwise multiplication.

Proposition 3.5.2 *Let E be a compact subset of K, and let $V \subseteq K$ be open, such that $V = \tilde{V}$ and $\tilde{V}$ is compact. Then there exists $\varphi \in L^1(\hat{K},\pi) \cap L^2(\hat{K},\pi)$ such that $0 \leq \check{\varphi} \leq 1$, $\check{\varphi}(x) = 1$ for all $x \in E$ and $\check{\varphi}(x) = 0$ for $x \notin E * V * V$. Also $\|\varphi\|_1^2 \leq m(E * V)/m(V)$.*

Proof. Consider χ_V and χ_{E*V}, the characteristic functions of V and $E * V$, respectively. Obviously, $\chi_V \in L^1(K) \cap L^2(K)$ and $\chi_{E*V} \in L^1(K) \cap L^2(K)$. By Proposition 3.3.10 we know

$$\varphi := \frac{1}{m(V)}\, \mathcal{P}(\chi_V)\, \mathcal{P}(\chi_{E*V}) \,\in L^1(\hat{K},\pi) \cap L^2(\hat{K},\pi)$$

and

$$\check{\varphi} = \frac{1}{m(V)}\, \chi_V * \chi_{E*V}.$$

For each $x \in K$ we have

$$\check{\varphi}(x) = \frac{1}{m(V)} \int_V L_x \chi_{E*V}(\tilde{y}) \, dm(y) = \frac{1}{m(V)} \int_V \omega(x, \tilde{y})(E * V) \, dm(y)$$

If $x \in E$ then supp $\omega(x, \tilde{y}) \subseteq E*V$ for all $y \in V$. Hence it follows that $\check{\varphi}(x) = 1$ for all $x \in E$. If $x \in K$ is such that $\check{\varphi}(x) \neq 0$, then supp $\omega(x, \tilde{y}) \cap E*V \neq \emptyset$ for some $y \in V$. By Lemma 1.1.10 (ii) it follows $x \in y * E * V \subseteq E * V * V$. Thus $\check{\varphi}(x) = 0$ whenever $x \notin E * V * V$. Obviously $0 \leq \check{\varphi}(x) \leq 1$ for all $x \in K$. Finally

$$\|\varphi\|_1^2 = \frac{1}{m(V)^2} \int_K \chi_V(x) \chi_{E*V}(x) \, dm(x) \leq \frac{1}{m(V)} m(E * V).$$

◇

The following theorem is an immediate consequence of Proposition 3.5.2.

Theorem 3.5.3 *$L^1(\hat{K}, \pi)$ is inverse regular.*

To deal with regularity conditions for $L^1(K)$ is much more challenging and delicate.

(1) If K is a compact commutative hypergroup, we know $\mathcal{S}(K) = \hat{K} = \mathcal{X}^b(K)$, $\mathcal{X}^b(K)$ is discrete and $\mathcal{X}^b(K)$ is an orthogonal basis in $L^2(K)$ (see Theorem 3.1.9). In particular, $L^1(K)$ is $\mathcal{X}^b(K)$-regular. In fact, for $\alpha \in \mathcal{X}^b(K)$ we have $\hat{\alpha}(\alpha) > 0$ and $\hat{\alpha}(\beta) = 0$ for all $\beta \in \mathcal{X}^b(K)$, $\beta \neq \alpha$.

(2) Let K be a discrete commutative hypergroup, such that $\mathcal{S}(K) = \hat{K} = \mathcal{X}^b(K)$, and $\mathcal{S}(K) = \{\alpha_i : i \in I\} \cup \{1\}$, where $(\alpha_i)_{i \in I}$ is a net of characters such that $\alpha_i \mapsto 1$. Moreover, assume that each character $\alpha \neq 1$ is an element of $L^1(K) \cap L^2(K)$. Such hypergroups exist. In fact, the little q-Legendre polynomials generate a polynomial hypergroup with these properties, see subsection 4.5.12.
If $\alpha \in \mathcal{X}^b(K)$, $\alpha \neq 1$, then $\hat{\alpha}(\alpha) > 0$ and $\hat{\alpha}(\beta) = 0$ for all $\beta \in \mathcal{X}^b(K)$, $\beta \neq \alpha$. Let $\alpha = 1$ and $E \subseteq \mathcal{X}^b(K)$ closed with $1 \notin E$. Then E is a finite set, say $E = \{\alpha_1, ..., \alpha_n\}$. Choose some $f \in L^1(K)$ such that $\hat{f}(1) = 1$. Denote $z_i := \hat{f}(\alpha_i)$, $i = 1, ..., n$, and select a polynomial $P(z) = \sum_{k=0}^{n} a_k z^k$, $z \in \mathbb{C}$, such that $P(1) = 1$ and $P(z_i) = 0$, $i = 1, ..., n$.
Define $g = \sum_{k=0}^{n} a_k f^k$, where f^k is the k-fold convolution of f, and $f^0 = \epsilon_e$. Then $\hat{g}(\alpha) = P(\hat{f}(\alpha))$ for each $\alpha \in \mathcal{X}^b(K)$. In particular, $\hat{g}(1) = P(1) = 1$ and $\hat{g}(\alpha_i) = P(z_i) = 0$, $i = 1, ..., n$, and we have shown that $L^1(K)$ is $\mathcal{X}^b(K)$-regular.

One way to derive regularity conditions for $L^1(K)$ is to use an appropriate functional calculus. We introduce **Beurling algebras** $L^1(\mathbb{R},\omega)$ and embed them via a functional calculus into $L^1(K)$. The weight function on $\mathbb{R}$ has to be non-quasianalytic. A good reference for the role played by non-quasianalytic weights for Beurling algebras is [356,section 4.7]. The regularity result for $L^1(G,\omega)$, G a locally compact abelian group, ω a non-quasianalytic weight, is due to Domar [181].

Let G be a locally compact abelian group with Haar measure dt. A Borel measurable function $\omega : G \to]0,\infty[$ is called a **weight function** if $\omega(s+t) \leq \omega(s)\omega(t)$. The Beurling algebra $L^1(G,\omega)$ is the set of all functions F on G such that

$$\|F\|_\omega := \int_G |F(t)|\,\omega(t)\,dt < \infty.$$

$L^1(G,\omega)$ is a Banach space, and with the convolution product, $L^1(G,\omega)$ is a Banach algebra.

If K is not discrete we adjoin the point measure ϵ_e to $L^1(K)$, and write $L^1(K)_e = L^1(K) \oplus \mathbb{C}\epsilon_e \subseteq M(K)$, see the beginning of section 3.4.

Now let $f \in L^1(K)$ such that $\mathcal{F}f$ is real-valued. The absolutely convergent series

$$\exp(itf) := \sum_{k=0}^{\infty} \frac{(it)^k f^k}{k!}, \qquad t \in \mathbb{R},$$

yields a family of elements in $L^1(K)_e$. Here, of course, $f^0 = \epsilon_e$ and f^k is the k-fold convolution product of f. We put

$$\omega_f(t) := \|\exp(-itf)\|, \qquad t \in \mathbb{R}.$$

Clearly, $\mathcal{F}(\exp(-itf))(\alpha) = \exp(-it\mathcal{F}f(\alpha))$ for each $\alpha \in \mathcal{X}^b(K)$. Hence $|\mathcal{F}(\exp(-itf))(\alpha)| = 1$ for all $\alpha \in \mathcal{X}^b(K)$. In particular

$$\omega_f(t) = \|\exp(-itf)\| \geq 1 \qquad \text{for all } t \in \mathbb{R} \text{ and } f \in L^1(K)$$

with $\mathcal{F}f(\mathcal{X}^b(K)) \subseteq \mathbb{R}$. It is straightforward to check that $\omega_f : \mathbb{R} \to]0,\infty[$ is also continuous. Finally we note that

$$\begin{aligned}\mathcal{F}(\exp(-i(s+t)f)(\alpha) &= \exp(-is\mathcal{F}f(\alpha)) \cdot \exp(-it\mathcal{F}f(\alpha)) \\ &= \mathcal{F}(\exp(-isf) * \exp(-itf))(\alpha) \qquad \text{for all } \alpha \in \mathcal{X}^b(K).\end{aligned}$$

Therefore,

$$\exp(-i(s+t)f) = \exp(-isf) * \exp(-itf)$$

and hence

$\omega_f(s+t) = \|\exp(-i(s+t)f)\| \leq \|\exp(-isf)\|\ \|\exp(-itf)\| = \omega_f(s)\omega_f(t).$

So far we have shown that for every $f \in L^1(K)$, such that $\mathcal{F}f$ is real-valued, ω_f is a weight-function on $G = \mathbb{R}$.

Before studying the Beurling algebras we will briefly investigate the subset $\{f \in L^1(K) : \mathcal{F}f(\mathcal{X}^b(K)) \subseteq \mathbb{R}\}$ of $L^1(K)$. In case of $\mathcal{X}^b(K) = \hat{K}$ it is straightforward to derive from the uniqueness theorem (Theorem 3.1.8) that $\hat{f}(\hat{K}) \subseteq \mathbb{R}$ if and only if $f = f^*$. One has only to use $\widehat{f^*}(\alpha) = \overline{\hat{f}(\alpha)}$ for all $\alpha \in \hat{K}$.

The situation is a little bit more involved when $\hat{K}$ is a proper subset of $\mathcal{X}^b(K)$. For a general $\alpha \in \mathcal{X}^b(K)$ we have $\mathcal{F}(f^*)(\alpha) = \overline{\mathcal{F}f(\alpha^*)}$, and $\alpha \neq \alpha^*$ whenever $\alpha \in \mathcal{X}^b(K)\backslash\hat{K}$. Recall that $\alpha^* = \bar{\tilde{\alpha}}$.

For $\alpha \in \mathcal{X}^b(K)$ let

$$S_\alpha(L^1(K)) := \{f \in L^1(K) : \mathcal{F}(f^*)(\alpha) = \overline{\mathcal{F}f(\alpha)}\}.$$

Then we have

(i) $S_\alpha(L^1(K)) = L^1(K)$ for all $\alpha \in \hat{K}$,
(ii) $S_\alpha(L^1(K)) \subsetneq L^1(K)$ for each $\alpha \in \mathcal{X}^b(K)\backslash\hat{K}$.

To check (ii) we have to show that there is some $f \in L^1(K)$ such that $\mathcal{F}f(\alpha) \neq \mathcal{F}f(\alpha^*)$. Since $\alpha \neq \alpha^*$ the Gelfand theory, see Theorem 3.1.1 (iii) yields the existence of such an f.

Proposition 3.5.4 *For $f \in L^1(K)$ the following statements are equivalent.*

(1) $\mathcal{F}f(\mathcal{X}^b(K)) \subseteq \mathbb{R}$,
(2) $f = f^*$ *and* $f \in S_\alpha(L^1(K))$ *for all* $\alpha \in \mathcal{X}^b(K)$.

Proof. (1) $\Rightarrow$ (2): We only have to show that $f \in S_\alpha(L^1(K))$ for all $\alpha \in \mathcal{X}^b(K)\backslash\hat{K}$. We already know that $f = f^*$, which implies

$$\mathcal{F}(f^*)(\alpha) = \mathcal{F}f(\alpha) = \overline{\mathcal{F}f(\alpha)} \qquad \text{for all } \alpha \in \mathcal{X}^b(K)\backslash\hat{K}.$$

(2) $\Rightarrow$ (1) : By assumption (2) we get $\overline{\mathcal{F}f(\alpha)} = \mathcal{F}(f^*)(\alpha) = \mathcal{F}f(\alpha)$ for each $\alpha \in \mathcal{X}^b(K)$, and hence $\mathcal{F}f(\mathcal{X}^b(K)) \subseteq \mathbb{R}$. ◇

Remark: It is easily checked that, for each $\alpha \in \mathcal{X}^b(K)$, $S_\alpha(L^1(K))$ is a closed $*$-subalgebra of $L^1(K)$. Hence

$$S(L^1(K)) := \bigcap_{\alpha \in \mathcal{X}^b(K)} S_\alpha(L^1(K))$$

is a closed $*$-subalgebra of $L^1(K)$. In general $S(L^1(K))$ is a very small algebra.

Proposition 3.5.5 *Let $f \in L^1(K)$, $\mathcal{F}f$ real-valued. Define the weight function on $\mathbb{R}$, $\omega_f(t) = \|\exp(-itf)\|$. Then $L^1(\mathbb{R},\omega_f)$ is a Beurling algebra. The structure space $\Delta(L^1(\mathbb{R},\omega_f))$ is homeomorphic to $\mathbb{R}$ in such a way that the Gelfand transform of any $F \in L^1(\mathbb{R},\omega_f)$ is given by*

$$\hat{F}(t) \;=\; \int_{\mathbb{R}} F(s)\, \exp(-ist)\, ds, \qquad t \in \mathbb{R}.$$

Proof. In order to apply [356, Lemma 2.8.6 and Proposition 2.8.7], we determine $R_+ := \inf\{\omega_f(t)^{1/t} : t > 0\}$ and $R_- := \sup\{\omega_f(-t)^{-1/t} : t > 0\}$. In the proof of Lemma 2.8.6 of [356] it is shown that $R_+ = \lim_{t\to\infty} \omega_f(t)^{1/t}$. Let $r(g)$ be the spectral radius of $g \in L^1(K)_e$. Then

$$r(g) \;=\; \lim_{n\to\infty} \|g^n\|^{1/n} \;=\; \inf_{n\in\mathbb{N}} \|g^n\|^{1/n}$$

$$=\; \sup\{|\mathcal{F}g(\alpha)| : \alpha \in \mathcal{X}^b(K)\}.$$

We claim that

$$R_+ \;=\; r(\exp(-if)) \;=1.$$

To see this, let $M = \sup\{\|\exp(-itf)\| : t \in [1,2]\}$. For each $\varepsilon > 0$ there exists $n_0 \in \mathbb{N}$ such that

$$\|\exp(-inf)\| \;\le\; (r(\exp(-if)) + \varepsilon)^n$$

for all $n \ge n_0$. For $t > n_0 + 1$ there exists $n \ge n_0$ such that $t - n \in [1,2]$. Hence, for $t > n_0 + 1$ and $n \ge n_0$,

$$\|\exp(-itf)\| \;\le\; \|\exp(-inf)\| \cdot \|\exp(-i(t-n)f)\|$$

$$\le\; M(r(\exp(-if)) + \varepsilon)^n.$$

It follows that

$$R_+ = \lim_{t\to\infty} \|\exp(-itf)\|^{1/t} \;\le\; r(\exp(-if))$$

$$=\; \sup\{|\mathcal{F}(\exp(-if))(\alpha)| : \alpha \in \mathcal{X}^b(K)\} \;=1.$$

Since $\|\exp(-itf)\| \ge r(\exp(-itf)) = 1$ for each $t \in \mathbb{R}$, $R_+ = 1$ follows.

For R_- we obtain in the same way,

$$R_- \;=\; \sup\{\|\exp(itf)\|^{-1/t} : t > 0\} \;=\; \frac{1}{\inf\{\|\exp(itf)\|^{1/t} : t > 0\}}$$

$$=\; r(\exp(if)) \;=1.$$

Since $R_+ = R_- = 1$, an application of [356,Proposition 2.8.7] yields that $\Delta(L^1(\mathbb{R}, \omega_f))$ can be identified with $\mathbb{R}$, and that the Gelfand transforms of $F \in L^1(\mathbb{R}, \omega_f)$ are then given as stated above. ◇

Another class of Beurling algebras which can be embedded into $L^1(K)$ can be defined on the group $G = \mathbb{Z}$, the integers.

Exactly as before we put, for $f \in L^1(K)$ with $\mathcal{F}f$ real-valued,

$$\omega_f(n) = \|\exp(-inf)\|, \qquad n \in \mathbb{Z}.$$

Each $\omega_f : \mathbb{Z} \to]0, \infty[$ is a weight function on $\mathbb{Z}$. The Beurling algebra $l^1(\mathbb{Z}, \omega_f)$ is the set of all functions F on $\mathbb{Z}$ such that

$$\|F\|_{\omega_f} := \sum_{n=-\infty}^{\infty} |F(n)|\, \omega_f(n) < \infty.$$

Equipped with the convolution product, $l^1(\mathbb{Z}, \omega_f)$ is a Banach algebra. Put $R_+ := \inf\{\omega_f(n)^{1/n} : n \in \mathbb{N}\}$, $R_- := \sup\{\omega_f(-m)^{-1/m} : m \in \mathbb{N}\}$. Then in the same way as in the proof of Proposition 3.5.5 we can show that $R_+ = R_- = 1$. By [356,Proposition 2.8.8] we get the following result.

Proposition 3.5.6 *Let $f \in L^1(K)$ such that $\mathcal{F}f$ is real-valued. Define the weight function on $\mathbb{Z}$, $\omega_f(n) = \|\exp(-inf)\|$. Then $l^1(\mathbb{Z}, \omega_f)$ is a Beurling algebra. The structure space $\Delta(l^1(\mathbb{Z}, \omega_f))$ is homeomorphic to the torus $\mathbb{T}$ in such a way that the Gelfand transform of any $F \in l^1(\mathbb{Z}, \omega_f)$ is given by*

$$\hat{F}(z) = \sum_{n=-\infty}^{\infty} F(n)\, z^{-n}, \qquad z \in \mathbb{T}.$$

Now we present an $L^1(\mathbb{R}, \omega_f)$-functional calculus, which is based on the Bochner integral. As reference for the Bochner integral we use [757, Ch.V.4 and Ch.V.5].

Theorem 3.5.7 *Let $f \in L^1(K)$ such that $\mathcal{F}f$ is real valued. For every $F \in L^1(\mathbb{R}, \omega_f)$ the Bochner integral*

$$\hat{F}(f) := \int_{\mathbb{R}} F(t)\, \exp(-itf)\, dt$$

is a well-defined element of $L^1(K)_e$. If $\int_{\mathbb{R}} F(t)\, dt = 0$, then $\hat{F}(f) \in L^1(K)$. Furthermore,

$$\mathcal{F}(\hat{F}(f))(\alpha) = \int_{\mathbb{R}} F(t)\, \exp(-it\mathcal{F}f(\alpha))\, dt = \hat{F}(\mathcal{F}f(\alpha))$$

for every $\alpha \in \mathcal{X}^b(K)$.

Proof. The map $t \mapsto \exp(-itf)$, $\mathbb{R} \to L^1(K)_e$, is continuous and thus separably-valued, see [757]. It follows that $t \mapsto F(t)\exp(-itf)$ is almost separably-valued. Furthermore $t \mapsto F(t)\exp(-itf)$ is weakly Borel-measurable, and hence strongly Borel measurable by Petti's theorem. Finally $F \in L^1(\mathbb{R},\omega_f)$ says exactly that $t \mapsto \|F(t)\exp(-itf)\|$ is integrable. Bochner's theorem yields that $\hat{F}(f) = \int_{\mathbb{R}} F(t)\exp(-itf)dt$ is a well-defined element of $L^1(K)_e$.

If $\int_{\mathbb{R}} F(t)dt = 0$ then $\int_{\mathbb{R}} F(t)\epsilon_e dt = \epsilon_e \int_{\mathbb{R}} F(t)dt = 0$, and hence

$$\hat{F}(f) = \int_{\mathbb{R}} F(t)\, \exp(-itf)\, dt = \int_{\mathbb{R}} F(t)\, (\exp(-itf) - \epsilon_e)\, dt.$$

Since the functions $\exp(-itf) - \epsilon_e$ are elements of $L^1(K)$, the Bochner integral $\hat{F}(f)$ is an element of $L^1(K)$, which is closed in $L^1(K)_e$.

Finally for each $F \in L^1(\mathbb{R},\omega_f)$ the Bochner integral $\hat{F}(f)$ is a limit of a sequence of functions χ_n, $n \in \mathbb{N}$, where each χ_n is an $L^1(K)_e$-valued function with only finite many values in $L^1(K)_e$. The Fourier transform of each $\int_{\mathbb{R}} \chi_n(t)dt \in L^1(K)_e$ is therefore a $\mathbb{C}$-valued integral of functions with finite values. Since the Fourier transform is norm-decreasing, we obtain

$$\mathcal{F}(\hat{F}(f))(\alpha) = \int_{\mathbb{R}} F(t)\, \exp(-it\mathcal{F}f(\alpha))\, dt$$

for each $\alpha \in \mathcal{X}^b(K)$. The second equality in the stated formula is exactly the formula in Proposition 3.5.5 with $s = \mathcal{F}f(\alpha)$. $\diamond$

In a similar way we can embed $l^1(\mathbb{Z},\omega_f)$ into $L^1(K)_e$. The proof of the following theorem is analogous to the one of Theorem 3.5.7.

Theorem 3.5.8 *Let $f \in L^1(K)$ such that $\mathcal{F}f$ is real-valued. For every $F \in l^1(\mathbb{Z},\omega_f)$ the series*

$$\hat{F}(f) = \sum_{n=-\infty}^{\infty} F(n)\exp(-inf)$$

is absolutely convergent and thus converges to an element of $L^1(K)_e$. If $\sum_{n=-\infty}^{\infty} F(n) = 0$, then $\hat{F}(f) \in L^1(K)$ and

$$\mathcal{F}(\hat{F}(f)(\alpha)) = \sum_{n=-\infty}^{\infty} F(n)\exp(-in\mathcal{F}f(\alpha)) = \hat{F}(\mathcal{F}f(\alpha)).$$

Now we can give a sufficient condition for regularity of $L^1(K)$. This condition is based on the non-quasianalyticity of the weight functions on $\mathbb{R}$ or $\mathbb{Z}$.

Definition: A weight function ω on $\mathbb{R}$ such that $\omega(t) \geq 1$ for all $t \in \mathbb{R}$ is called **non-quasianalytic** if

$$\sum_{n=-\infty}^{\infty} \frac{\ln(\omega(nt))}{1+n^2} < \infty \qquad \text{for all } t \in \mathbb{R}.$$

A weight function ω on $\mathbb{Z}$ such that $\omega(n) \geq 1$ for all $n \in \mathbb{Z}$ is called non-quasianalytic if

$$\sum_{n=-\infty}^{\infty} \frac{\ln \omega(n)}{1+n^2} < \infty.$$

Remark: In [356, Lemma 4.7.8 and Lemma 4.7.9] it is proved that $L^1(\mathbb{R},\omega)$ or $L^1(\mathbb{Z},\omega)$ is regular if ω is a non-quasianalytic weight function on $\mathbb{R}$ or on $\mathbb{Z}$, respectively.

Proposition 3.5.9 *Let E be a closed subset of $\mathcal{X}^b(K)$ and $\alpha \in \mathcal{X}^b(K)\backslash E$. Suppose there is some $f \in L^1(K)$ such that $\mathcal{F}f$ is real-valued, $\mathcal{F}f(E) \subseteq [-\eta,\eta]$ and $\mathcal{F}f(\alpha) > \eta$ for some $\eta > 0$ and*

$$\sum_{n=0}^{\infty} \frac{\ln \|\exp(-inf)\|}{1+n^2} < \infty.$$

Then there exists $g \in L^1(K)$ such that $\mathcal{F}g|E = 0$ and $\mathcal{F}g(\alpha) = 1$.

Proof. Since $\mathcal{F}f$ is real-valued we have

$$\begin{aligned}\mathcal{F}(\exp(-inf))(\beta) &= \exp(-in\mathcal{F}f(\beta)) = \overline{\exp(in\mathcal{F}f(\beta))} \\ &= \overline{\mathcal{F}(\exp(inf))(\beta)} = \mathcal{F}((\exp(inf))^*)(\beta)\end{aligned}$$

for all $\beta \in \hat{K}$. Therefore $\exp(-inf) = \exp(inf)^*$, and hence $\|\exp(-inf)\| = \|\exp(inf)\|$. In particular

$$\sum_{n=-\infty}^{\infty} \frac{\ln \|\exp(-inf)\|}{1+n^2} \leq 2\sum_{n=0}^{\infty} \frac{\ln \|\exp(inf)\|}{1+n^2} < \infty.$$

It is clear that $\|\exp(-itf)\| \leq \exp(\|f\|)_1)$ for $|t| \leq 1$. Hence

$$\begin{aligned}\|\exp(-itf)\| &\leq \|\exp(-i\lfloor t\rfloor f)\| \, \|\exp(-i(t-\lfloor t\rfloor)f\| \\ &\leq \|\exp(-i\lfloor t\rfloor f)\| \, \exp(\|f\|_1),\end{aligned}$$

and we obtain

$$\int_{\mathbb{R}} \frac{\ln \| \exp(-itf)\|}{1+t^2}\, dt \;\leq\; \exp(\|f\|_1) \sum_{n=-\infty}^{\infty} \frac{\ln \| \exp(-inf)\|}{1+n^2} < \infty.$$

By [356, Lemma 4.7.7 and Lemma 4.7.8] the Beurling algebra $L^1(\mathbb{R},\omega_f)$ with weight function $\omega_f(t) = \| \exp(-itf)\|$ is regular. Hence there exists $F \in L^1(\mathbb{R},\omega_f)$ such that $\hat{F}|[-\eta,\eta] = 0$ and $\hat{F}(\mathcal{F}f(\alpha)) = 1$. Now put $g = \hat{F}(f) \in L^1(K)_e$, the Bochner integral as constructed in Theorem 3.5.7. Since $\int_{\mathbb{R}} F(t)dt = \hat{F}(0) = 0$ it follows that $g \in L^1(K)$. Moreover, $\mathcal{F}g(\beta) = \hat{F}(\mathcal{F}f(\beta)) = 0$ for every $\beta \in E$ and $\mathcal{F}g(\alpha) = \hat{F}(\mathcal{F}f(\alpha)) = 1$. ◇

Now we will use the fact that $\widehat{L^1(K)}$ is dense in $C_0(\hat{K})$.

Theorem 3.5.10 *Suppose that $\mathcal{X}^b(K) = \hat{K}$, and suppose that*

$$\sum_{n=0}^{\infty} \frac{\ln \| \exp(-inf)\|}{1+n^2} < \infty \qquad (*)$$

for every $f \in C_c(K)$ such that $f = f^*$ and $\|f\|_1 \leq 1$. Then $L^1(K)$ is $\mathcal{X}^b(K)$-regular. In particular $\mathcal{S}(K) = \hat{K} = \mathcal{X}^b(K)$.

Proof. Let $E \subseteq \hat{K}$ be closed and let $\alpha \in \hat{K}\backslash E$. By Theorem 3.2.1 $\widehat{C_c(K)}$ is dense in $C_0(K)$. Hence there exists $g \in C_c(K)$ with $|\hat{g}(\beta)| \leq \frac{1}{2}$ for all $\beta \in E$ and $\hat{g}(\alpha) = 1$. In particular $\|g\|_1 \geq 1$, and setting $f = \frac{1}{2\|g\|_1}\,(g+g^*)$, we see that $\mathcal{F}f = \hat{f}$ is real-valued and $\|f\|_1 \leq 1$ and setting $\eta = \frac{1}{2\|g\|_1} > 0$ we have $\hat{f}(\beta) \in [-\eta,\eta]$ for each $f \in E$ and $\hat{f}(\alpha) = 2\eta$. It follows from Proposition 3.5.9 that $L^1(K)$ is $\mathcal{X}^b(K)$-regular. ◇

Of course, we have to find sufficient conditions which ensure that $(*)$ in Theorem 3.5.10 is satisfied. We begin by investigating the impact of growth conditions of $\omega_f(n) = \| \exp(-inf)\|$. In the proof of Proposition 3.5.5 we have shown that $\lim_{n\to\infty} \omega_f(n)^{1/n} = 1$, and applying the proof of Proposition 3.4.5 we obtain that for each $\delta > 0$ there exists a constant $M = M(f,\delta) > 0$ such that

$$\omega_f(n) \;\leq\; M\,(1+\delta)^n$$

for all $n \in \mathbb{N}$. (This is the subexponential growth condition for $(\omega_f(n))_{n\in\mathbb{N}}$, see subsection 3.4.) Strengthening this growth condition a little bit, we get the following result.

Corollary 3.5.11 *Suppose that $\mathcal{X}^b(K) = \hat{K}$ and suppose that for each $f \in C_c(K)$ with $f = f^*$ and $\|f\|_1 \leq 1$ there exists a constant $\beta = \beta(f) \in [0,1[$ such that $\| \exp(-inf)\| = O(\exp(n^\beta))$. Then $L^1(K)$ is $\mathcal{X}^b(K)$-regular.*

Proof. We only have to show that condition $(*)$ of Theorem 3.5.10 is satisfied. Since for some $M > 0$ we have

$$\ln(\| \exp(-inf)\|) \leq \ln(M \exp(n^\beta)) = \ln(M) + n^\beta,$$

with $\beta < 1$, condition $(*)$ is shown. ◇

Now we consider a growth condition for sets C^n, $n \in \mathbb{N}$, where C is a compact subset of K. If K is of polynomial growth, it is shown by [690] and [247], using a functional calculus of Dixmier [179] that $L^1(K)$ is $\mathcal{X}^b(K)$-regular. Note that we have $\mathcal{S}(K) = \hat{K} = \mathcal{X}^b(K)$, whenever K is of polynomial growth, see Theorem 3.4.6. We prove the $\mathcal{X}^b(K)$-regularity of $L^1(K)$ in case K satisfies the following condition:

For every compact $C \subseteq K$ there is some $\beta = \beta(C)$ with

$$0 \leq \beta < \frac{1}{2} \text{ such that } m(C^n) = O(\exp(n^\beta)). \qquad (**)$$

Condition $(**)$ is more general than polynomial growth, but not as general as subexponential growth.

If K satisfies $(**)$ we say that K is of $\frac{1}{2}$**-subexponential growth**.

Theorem 3.5.13 below is due to E.Perreiter [543,Theorem 5.6].

Lemma 3.5.12 *Let $f \in L^1(K)$ such that $\|f\|_1 \leq 1$. Then*

$$\left\| \sum_{k=n^2}^{\infty} \frac{(-i)^k n^k f^k}{k!} \right\|_1 \longrightarrow 0 \qquad \text{as } n \to \infty.$$

Proof. Using Stirling's formula it follows

$$\left\| \sum_{k=n^2}^{\infty} \frac{i^k n^k f^k}{k!} \right\|_1 \leq \sum_{k=n^2}^{\infty} \frac{n^k \|f\|_1^k}{k!} \leq \sum_{k=n^2}^{\infty} \frac{n^k}{k!}$$

$$\leq \frac{n^{n^2}}{(n^2)!} e^n \sim n^{n^2} e^n \frac{e^{n^2}}{\sqrt{2\pi n^2}(n^2)^{n^2}} = \frac{1}{\sqrt{2\pi}} e^{n^2+n} n^{-n^2-1} \longrightarrow 0$$

as $n \to \infty$. ◇

Theorem 3.5.13 *Suppose that K is of $\frac{1}{2}$-subexponential growth. Then $L^1(K)$ is $\mathcal{X}^b(K)$-regular.*

Proof. We know that $\mathcal{S}(K) = \hat{K} = \mathcal{X}^b(K)$ by Theorem 3.4.6. Let $f \in C_c(K)$ such that $\|f\|_1 \leq 1$ and $f = f^*$. We have $|\exp(-in\hat{f}(\alpha)) - 1| \leq n\,|\hat{f}(\alpha)|$. The Plancherel isomorphism yields $\exp(-inf) - \epsilon_e \in L^2(K)$ and

$\| \exp(-inf) - \epsilon_e \|_2 \leq n \, \|f\|_2$. Let $C = \text{supp}\, f$ and $\varepsilon > 0$. By Lemma 3.5.12 there exists $n_0 \in \mathbb{N}$ such that

$$\left\| \sum_{k=n^2}^{\infty} \frac{(-i)^k n^k f^k}{k!} \right\|_1 < \varepsilon$$

for all $n \geq n_0$. Set $C_n = C^{n^2-1}$. The growth condition implies that, for all $n \geq n_0$,

$$\int_{C_n} |\exp(-inf)(x) - \epsilon_e(x)| \, dm(x) \;\leq\; \| \exp(-inf) - \epsilon_e \|_2 \, (m(C_n))^{1/2}$$

$$\leq\; n \, \|f\|_2 \, (M \exp((n^2-1)^{\beta}))^{1/2},$$

where $M > 0$ and $0 \leq \beta = \beta(C) < \frac{1}{2}$. Hence we obtain

$$\| \exp(-inf) \| \;\leq\; 1 + \; \| \exp(-inf) - \epsilon_e \|_1$$

$$\leq\; 1 + \left\| \sum_{k=1}^{n^2-1} \frac{(-i)^k n^k f^k}{k!} \right\|_1 + \left\| \sum_{k=n^2}^{\infty} \frac{(-i)^k n^k f^k}{k!} \right\|_1$$

$$\leq\; 1 + \int_{C_n} |\exp(-inf)(x) - \delta_e(x)| \, dm(x) \; + \varepsilon$$

$$\leq\; 2n \, \|f\|_2 \, (M \exp((n^2-1)^{\beta}))^{1/2} \;\leq\; 2n \, \|f\|_2 \, M^{1/2} \, \exp(n^{2\beta})$$

for all $n \geq n_1$, where $n_1 \geq n_0$ is chosen large enough. Choosing some α with $0 \leq 2\beta < \alpha < 1$ we can refer to Corollary 3.5.11 and the statement follows. ◇

If $\hat{K}$ is a hypergroup under pointwise multiplication, we can accomplish regularity by carrying over methods for locally compact abelian groups to commutative hypergroups. Assume that $\hat{K}$ is a hypergroup under pointwise multiplication. We collect some facts, which are direct consequences of the hypergroup structure on $\hat{K}$.

In the proof of Theorem 3.4.11 we have already shown that

$$L_\alpha \hat{f}(\beta) \;=\; \widehat{\bar{\alpha} \cdot f}(\beta).$$

Since the translation operator is continuous and $L^1(K) \cap L^2(K)$ is dense in $L^2(K)$, we get for the Plancherel transform

$$L_\alpha(\mathcal{P}(f)) \;=\; \mathcal{P}(\bar{\alpha} \cdot f)$$

for all $f \in L^2(K)$. Now a supplementary result to Proposition 3.3.10 is easily shown.

Proposition 3.5.14 *Suppose that $\hat{K}$ is a hypergroup under pointwise multiplication. Let $f, g \in L^2(K)$. Then $\widehat{f \cdot g}(\alpha) = (\mathcal{P}(f) * \mathcal{P}(g))(\alpha)$ for all $\alpha \in \hat{K}$.*

Proof. Using $\mathcal{P}(f)(\alpha) = \mathcal{P}(\bar{f})(\bar{\alpha})$, the Parseval formula yields

$$\widehat{f \cdot g}(\alpha) = \int_K f(x)\, g(x)\, \overline{\alpha(x)}\, dm(x) = \int_{\hat{K}} \mathcal{P}(f)(\beta)\, \overline{\mathcal{P}(\alpha \cdot \bar{g})(\beta)}\, d\pi(\beta)$$

$$= \int_{\hat{K}} \mathcal{P}(f)(\beta)\, \mathcal{P}(\bar{\alpha} \cdot g)(\bar{\beta})\, d\pi(\beta) = \int_{\hat{K}} \mathcal{P}(f)(\beta)\, L_\alpha(\mathcal{P}(g))(\bar{\beta})\, d\pi(\beta)$$

$$= \mathcal{P}(f) * \mathcal{P}(g)(\alpha)$$

for each $\alpha \in \hat{K}$. ◇

Proposition 3.5.15 *Assume that $\hat{K}$ is a hypergroup under pointwise multiplication. Let $F \subseteq \hat{K}$ be compact, and let $V \subseteq \hat{K}$ be a Borel set such that V is symmetric (with respect to involution, which is the complex conjugation), $\pi(V) > 0$ and V has compact closure $\bar{V}$. Then there exists $f \in L^1(K)$ such that $\hat{f}(\alpha) = 1$ for all $\alpha \in F$ and $\hat{f}(\alpha) = 0$ for all $\hat{K} \backslash (F * V * V)$ and $0 \leq \hat{f} \leq 1$.*

Proof. The set $F * V$ is a Borel set contained in the compact set $F * \bar{V}$. Hence the characteristic functions χ_V and χ_{F*V} are elements of $L^2(\hat{K}, \pi)$. Let $g = \mathcal{P}^{-1}(\chi_V)$ and $h = \mathcal{P}^{-1}(\chi_{F*V})$. Then $f = \frac{1}{\pi(V)}\, g \cdot h$ is an element of $L^1(K)$ and by Proposition 3.5.14 it follows that for each $\alpha \in \hat{K}$,

$$\hat{f}(\alpha) = \frac{1}{\pi(V)}\, (\chi_V * \chi_{F*V})(\alpha) = \frac{1}{\pi(V)} \int_V L_\alpha(\chi_{F*V})(\bar{\beta})\, d\pi(\beta)$$

$$= \frac{1}{\pi(V)} \int_V \omega(\alpha, \bar{\beta})(F * V)\, d\pi(\beta).$$

In particular we have $0 \leq \hat{f}(\alpha) \leq 1$ for each $\alpha \in \hat{K}$, since the measures $\omega(\alpha, \bar{\beta})$ have norm 1. Now let $\alpha \in F$. Then for each $\beta \in V$, $\operatorname{supp} \omega(\alpha, \bar{\beta}) \subseteq F * V$, and hence $\hat{f}(\alpha) = 1$. Consider now $\alpha \in \hat{K}$ such that $\hat{f}(\alpha) > 0$, i.e. $\int_V \omega(\alpha, \bar{\beta})(F * V)\, d\pi(\beta) > 0$. Then $\omega(\alpha, \bar{\beta})(F * V) > 0$ for some $\beta \in V$, and it follows that

$$\operatorname{supp} \omega(\alpha, \bar{\beta}) \cap F * V \neq \emptyset.$$

Let $\gamma \in (\{\alpha\} * \{\bar{\beta}\}) \cap F * V$. By Axiom (H6) this is equivalent to

$$1 \in (\{\beta\} * \{\bar{\alpha}\}) * (F * V) = \{\bar{\alpha}\} * (\{\beta\} * F * V),$$

which implies $\alpha \in \{\beta\} * F * V \subseteq F * V * V$. Therefore, if $\alpha \in \hat{K} \backslash (F * V * V)$ then $\hat{f}(\alpha)$ has to be zero. ◇

Theorem 3.5.16 *Assume that $\hat{K}$ is a hypergroup under pointwise multiplication. Then $L^1(K)$ is $\hat{K}$-regular.*

Proof. Let $E \subseteq \hat{K}$ closed and $\alpha \in \hat{K}\backslash E$. Let U_α be a neighbourhood of α such that $U_\alpha \subseteq \hat{K}\backslash E$. There exists a symmetric and relatively compact neighbourhood V of $1 \in \hat{K}$ such that $\{\alpha\}*V*V \subseteq U_\alpha$. Then $E \subseteq \hat{K}\backslash(\{\alpha\}* V*V)$, and by Proposition 3.5.15 there exists $f \in L^1(K)$ such that $\hat{f}(\alpha) = 1$ and $\hat{f}(\beta) = 0$ for all $\beta \in E$. ◇

Remark:

(1) The $\frac{1}{2}$-subexponential growth of K is not necessary such that $L^1(K)$ is $\mathcal{X}^b(K)$-regular. The polynomial hypergroup $K = \mathbb{N}_0$ generated by the little q-Legendre polynomials is $\mathcal{X}^b(K)$-regular, see Example (2) at the beginning of this subsection. However, this hypergroup is of exponential growth.
(2) Several examples of polynomial hypergroups which are of $\frac{1}{2}$-subexponential growth are discussed in section 4.5.
(3) For hypergroups $K = G_H$ induced by $[FIA]_B^-$-groups G such that $B \supseteq I(G)$, we will show in subsection 8.1.2 that $L^1(K)$ is of subexponential growth and $\hat{K}$ is a hypergroup under pointwise multiplication. By Theorem 3.4.6 and Theorem 3.5.16 it follows that $L^1(K)$ is $\mathcal{X}^b(K)$-regular.

As announced at the end of subsection 3.4 we derive a characterization for the so-called Pontryagin duality. We use a regularity condition of $L^1(\hat{K},\pi)$.

Theorem 3.5.17 *Suppose that $\hat{K}$ is a hypergroup under pointwise multiplication. Then $i(K) = \hat{\hat{K}}$ if and only if $L^1(\hat{K},\pi)$ is $\hat{\hat{K}}$-regular.*

Proof. First suppose that $L^1(\hat{K},\pi)$ is $\hat{\hat{K}}$-regular. By Theorem 3.4.12 we know that $i(K)$ is a closed subset of $\hat{\hat{K}}$. Assume for a contradiction that $i(K)$ is a proper subset of $\hat{\hat{K}}$. Then there exists $\varphi \in L^1(\hat{K},\pi)$ with $\hat{\varphi}|i(K) = 0$ and $\hat{\varphi} \neq 0$. Then $\varphi \neq 0$ in $L^1(\hat{K},\pi)$, and so $\varphi\pi \in M(\hat{K})$ is a nonzero measure. Thus there exists some $\psi \in C_0(\hat{K})$ such that $\int\limits_{\hat{K}} \varphi(\alpha)\psi(\alpha)\, d\pi(\alpha) \neq 0$. By Theorem 3.2.1 there exists $f \in L^1(K)$ such that $\int\limits_{\hat{K}} \varphi(\alpha)\hat{f}(\alpha)\, d\pi(\alpha) \neq 0$. Using Fubini's theorem we obtain

$$0 \neq \int_{\hat{K}} \varphi(\alpha) \int_K f(x)\, \overline{\alpha(x)}\, dm(x)\, d\pi(\alpha) = \int_K f(x) \int_{\hat{K}} \varphi(\alpha)\, \overline{\alpha(x)}\, d\pi(\alpha)\, dm(x)$$

$$= \int_K f(\tilde{x}) \int_{\hat{K}} \varphi(\alpha)\, \alpha(x)\, d\pi(\alpha)\, dm(x)$$
$$= \int_K f(\tilde{x}) \int_{\hat{K}} \varphi(\alpha)\, \overline{i(x)(\alpha)}\, d\pi(\alpha)\, dm(x) = 0,$$

a contradiction.

Conversely, suppose that $i(k) = \hat{\hat{K}}$. Theorem 3.5.3 tells us that $L^1(\hat{K}, \pi)$ is $\hat{\hat{K}}$-regular. ◇

Remark: If $\hat{K}$ and $\hat{\hat{K}}$ are hypergroups under pointwise multiplication, then $L^1(\hat{K}, \pi)$ is $\hat{\hat{K}}$-regular by Theorem 3.5.16. The previous theorem tells us that $i(K) = \hat{\hat{K}}$.

6 L^p-transforms

We already know that the Fourier transform is a bounded linear operator from $L^1(K)$ into $L^\infty(\mathcal{S}(K), \pi)$ with operator norm ≤ 1, and that the Plancherel isomorphism maps $L^2(K)$ onto $L^2(\mathcal{S}(K), \pi)$. Moreover, both transformations coincide on $L^1(K) \cap L^2(K)$. Denote by $E(K)$ the space of all m-integrable simple functions on K. The Riesz-Thorin interpolation theorem, see [62], yields that

$$\|\hat{f}\|_q \leq \|f\|_p$$

for $\frac{1}{p} + \frac{1}{q} = 1$, $1 \leq p \leq 2$ and all $f \in E(K)$. Since each function$f \in C_c(K)$ can be uniformly approximated by functions of $E(K)$, this inequality is true for all $f \in C_c(K)$ and the mapping $f \mapsto \hat{f}$, $C_c(K) \to L^q(\mathcal{S}(K), \pi)$ can be extended uniquely to $L^p(K)$. This extension is called the **Hausdorff-Young transformation**, and the **Hausdorff-Young transform** of $f \in L^p(K)$, $1 \leq p \leq 2$, is also denoted by $\hat{f}$. So we have the following result. (Several results of this section are contained in [171].)

Theorem 3.6.1 (Hausdorff-Young) *Let $1 \leq p \leq 2$ and $\frac{1}{p} + \frac{1}{q} = 1$. The Hausdorff-Young transformation $f \mapsto \hat{f}$ is a linear mapping from $L^p(K)$ into $L^q(\mathcal{S}(K), \pi)$ such that $\|\hat{f}\|_q \leq \|f\|_p$.*

Applying the Riesz-Thorin interpolation in the same way as above we get for the inverse Fourier transformation $\varphi \mapsto \check{\varphi}$ the inequality $\|\check{\varphi}\|_q \leq \|\varphi\|_p$ for all $\varphi \in C_c(\mathcal{S}(K))$, where $1 \leq p \leq 2$, $\frac{1}{p} + \frac{1}{q} = 1$. Its extension $\varphi \mapsto \check{\varphi}$, $L^p(\mathcal{S}(K), \pi) \to L^q(K)$, is called the **inverse Hausdorff-Young transformation**.

Theorem 3.6.2 (Inverse Hausdorff-Young) *Let $1 \leq p \leq 2$, $\frac{1}{p} + \frac{1}{q} = 1$. The inverse Hausdorff-Young transformation $\varphi \mapsto \check{\varphi}$ is a linear mapping from $L^p(\mathcal{S}(K), \pi)$ into $L^q(K)$ with $\|\check{\varphi}\|_q \leq \|\varphi\|_p$.*

Remark: Proving the inverse Hausdorff-Young theorem 3.6.2 we have to take into account that $\mathcal{S}(K)$ or $\hat{K}$ is in general not a hypergroup.

In order to show the inversion theorem and the uniqueness theorem for the Hausdorff-Young transformation and the inverse Hausdorff-Young transformation we prove several properties of the transformation on $L^p(K)$ and $L^p(\mathcal{S}(K),\pi)$, $1 \leq p \leq 2$. To avoid errors we call the inverse Hausdorff-Young transform on $L^p(K)$ (on $L^p(\mathcal{S}(K),\pi)$) L^p-transform (or inverse L^p-transform).

Proposition 3.6.3 *Let* $1 \leq p_1, p_2 \leq 2$.

(i) Let $f \in L^{p_1}(K) \cap L^{p_2}(K)$. *Then the* L^{p_1}*-transform of* f *and the* L^{p_2}*-transform of* f *agree* π*-almost everywhere on* $\mathcal{S}(K)$.

(ii) Let $\varphi \in L^{p_1}(\mathcal{S}(K),\pi) \cap L^{p_2}(\mathcal{S}(K),\pi)$. *Then the inverse* L^{p_1}*-transform of* φ *and the inverse* L^{p_2}*-transform of* φ *agree* m*-almost everywhere on* K.

Proposition 3.6.3 is shown in the same way as for locally compact abelian groups, see [305,Theorem 31.26].

Lemma 3.6.4 *Let* $f, g \in L^2(K)$ *and* $h \in L^1(K)$. *Then*

$$\int_K f * g(y)\, h(\tilde{y})\, dm(y) = \int_K f(\tilde{y})\, g * h(y)\, dm(y).$$

Proof. We know that $f * g \in C_0(K)$, see Proposition 2.2.13, and $g * h \in L^2(K)$. Applying Fubini's theorem we obtain

$$\int_K f * g(y)\, h(\tilde{y})\, dm(y) = \int_K \int_K f(\tilde{x})\, L_y g(x)\, dm(x)\, h(\tilde{y})\, dm(y)$$

$$= \int_K f(\tilde{x}) \int_K L_y g(x)\, h(\tilde{y})\, dm(y)\, dm(x) = \int_K f(\tilde{x})\, g * h(x)\, dm(x).$$

◇

We have to extend the results of Proposition 3.2.3.

Proposition 3.6.5 *Let* $1 \leq p \leq 2$.

(i) Let $f \in L^p(K)$ *and* $\varphi \in L^p(\mathcal{S}(K),\pi)$. *Then* $(\hat{f} \cdot \varphi)^\vee(x) = f * \check{\varphi}(x)$ *for all* $x \in K$.

(ii) Let $\mu \in M(K)$ *and* $\varphi \in L^p(\mathcal{S}(K),\pi)$. *Then* $(\hat{\mu} \cdot \varphi)^\vee = \mu * \check{\varphi}$ m*-almost everywhere. In particular, for* $f \in L^1(K)$ *and* $\varphi \in L^p(\mathcal{S}(K),\pi)$ *we have* $(\hat{f} \cdot \varphi)^\vee = f * \check{\varphi}$ m*-almost everywhere.*

Proof.

(i) $\hat{f} \cdot \varphi$ is an element in $L^1(\mathcal{S}(K),\pi)$. Hence the inverse Fourier transform is well-defined. Let $f \in C_c(K)$, $\varphi \in C_c(\mathcal{S}(K))$. We have

$$
\begin{aligned}
(\hat{f}\varphi)^{\vee}(x) &= \int_{\mathcal{S}(K)} \hat{f}(\alpha)\varphi(\alpha)\alpha(x)\, d\pi(\alpha) \\
&= \int_{\mathcal{S}(K)} \int_K f(y)\, \overline{\alpha(y)}\, dm(y)\, \varphi(\alpha)\alpha(x)\, d\pi(\alpha)
\end{aligned}
$$

$$
= \int_K \int_{\mathcal{S}(K)} \alpha(x)\, \overline{\alpha(y)}\, \varphi(\alpha)\, d\pi(\alpha)\, f(y)\, dm(y) = \int_K L_x\check{\varphi}(\tilde{y})\, f(y)\, dm(y)
$$

$$
= f * \check{\varphi}(x) \qquad \text{for every } x \in K.
$$

For $\varphi \in L^p(\mathcal{S}(K), \pi)$ choose a sequence $(\varphi_n)_{n\in\mathbb{N}}$ in $C_c(\mathcal{S}(K))$ such that $\lim\limits_{n\to\infty} \|\varphi_n - \varphi\|_p = 0$. Then

$$
\|(\hat{f}\varphi)^{\vee} - f * \check{\varphi}\|_\infty \le \|\hat{f}(\varphi - \varphi_n)\|_1 + \|f * \check{\varphi}_n - f * \check{\varphi}\|_\infty
$$

$$
\le \|\hat{f}\|_q\, \|\varphi - \varphi_n\|_p + \|f\|_p\, \|\check{\varphi}_n - \check{\varphi}\|_q \le 2\, \|f\|_p\, \|\varphi - \varphi_n\|_p \longrightarrow 0
$$

as $n \to \infty$. Hence $(\hat{f}\varphi)^{\vee}(x) = f * \check{\varphi}(x)$ for every $x \in K$.
Finally, if $f \in L^p(K)$ choose a sequence $(f_n)_{n\in\mathbb{N}}$ in $C_c(K)$ such that $\lim\limits_{n\to\infty} \|f_n - f\|_p = 0$. Then we obtain as above that $(\hat{f}\cdot\varphi)^{\vee}(x) = f * \check{\varphi}(x)$ for all $x \in K$.

(ii) We already know that $(\hat{\mu}\cdot\varphi)^{\vee} = \mu * \check{\varphi}$ m-almost everywhere for $\mu \in M(K)$ and $\varphi \in C_c(K)$. For $\varphi \in L^p(\mathcal{S}(K), \pi)$ choose a sequence $(\varphi_n)_{n\in\mathbb{N}}$ in $C_c(\mathcal{S}(K))$ such that $\|\varphi - \varphi_n\|_p \to 0$ as $n \to \infty$. Then

$$
\begin{aligned}
\|(\hat{\mu}\varphi)^{\vee} - \mu * \check{\varphi}\|_q &\le \|\hat{\mu}(\varphi - \varphi_n)\|_p + \|\mu * (\varphi - \varphi_n)^{\vee}\|_q \\
&\le 2\, \|\mu\|\, \|\varphi - \varphi_n\|_p \longrightarrow 0
\end{aligned}
$$

as $n \to \infty$.

◇

Proposition 3.6.6 *Let $C \subseteq \mathcal{S}(K)$ be a compact set such that $\pi(C) > 0$. Then there exists a sequence $(f_n)_{n\in\mathbb{N}}$ in $C_c(K)$ such that*

$$
\|(f_n * f_n^*)^{\wedge} - \chi_C\|_1 \longrightarrow 0 \qquad \textit{as } n \to \infty.
$$

Proof. Let $0 < \varepsilon < \|\chi_C\|_2$. Using Theorem 3.2.6 we can choose some $f \in C_c(K)$ such that $\|\hat{f} - \chi_C\|_2 < \varepsilon/(3\|\chi_C\|_2)$. Then

$$
\|(f^* * f)^{\wedge} - \chi_C\|_1 = \|\, |\hat{f}|^2 - \chi_C\|_1
$$

$$
\le \int_{\mathcal{S}(K)} |\hat{f}(\beta) - \chi_C(\beta)| \cdot |\, |\hat{f}|(\beta) + \chi_C(\beta)|\, d\pi(\beta)
$$

$$
\le \|\hat{f} - \chi_C\|_2\, (\, \|\hat{f}\|_2 + \|\chi_C\|_2) < \varepsilon.
$$

◇

We have to extend the statement of Proposition 2.3.3 (3).

Proposition 3.6.7 *Let $1 \le p < \infty$. There exists a net $(\varphi_i)_{i\in I}$ of functions $\varphi_i \in C_c^+(K)$ such that $\varphi_i * f \to f$ in $L^p(K)$ for each $f \in L^p(K)$ (in norm) and such that $\widehat{\varphi_i}$ converges uniformly to 1 on compact subsets of $\mathcal{S}(K)$.*

Proof. Consider the neighbourhood basis of $e \in K$ consisting of sets

$$W(\Gamma,\varepsilon) = \{y \in K : |\alpha(y) - 1| < \varepsilon \text{ for all } \alpha \in \Gamma\},$$

where $\varepsilon > 0$ and Γ is a compact subset of $\mathcal{S}(K)$, see Lemma 3.4.8. For each $W = W(\Gamma,\varepsilon)$ choose a function $\varphi_W \in C_c^+(K)$ such that φ_W vanishes outside of W and $\int_K \varphi_W(x)dm(x) = 1$. Defining $(\varepsilon_1,\Gamma_1) \le (\varepsilon_2,\Gamma_2)$ whenever $\varepsilon_2 \le \varepsilon_1$, $\Gamma_1 \subseteq \Gamma_2$, we get a partial order, and the functions φ_W form a net.

Following the proof of Proposition 2.1.6 we obtain that $\|\varphi_W * f - f\|_p \to 0$ for each $f \in C_c(K)$. Since $C_c(K)$ is dense in $L^p(K)$ it follows $\|\varphi_W * f - f\|_p \to 0$ for every $f \in L^p(K)$. Furthermore

$$|\widehat{\varphi_W}(\alpha) - 1| \le \sup_{x\in W} |\alpha(x) - 1| < \varepsilon \qquad \text{for all } \alpha \in \Gamma,$$

whenever $W = W(\Gamma,\varepsilon)$, i.e. $\widehat{\varphi_W} \to 1$ uniformly on compact subsets of $\mathcal{S}(K)$. ◇

We will use the fact that if $f \in L^p \cap L^q$ for some $1 \le p < q \le \infty$, then $f \in L^r$ for every $p \le r \le q$. This is valid for any measure space (Ω,σ). Indeed, setting

$$E_1 = \{x \in \Omega : |f(x)| \le 1\}, \quad E_2 = \{x \in \Omega : |f(x)| > 1\},$$

one has for $q < \infty$

$$\|f\|_r^r = \int_{E_1} |f(x)|^r d\sigma(x) + \int_{E_2} |f(x)|^r d\sigma(x)$$

$$\le \int_{E_1} |f(x)|^p d\sigma(x) + \int_{E_2} |f(x)|^q d\sigma(x) \le \|f\|_p^p + \|f\|_q^q.$$

For $q = \infty$ an elementary estimate yields $f \in L^r$.

Proposition 3.6.8 *(1) Let $1 \le p \le 2$ and $f \in L^p(K)$. Suppose that the L^p-transform $\hat{f}$ of f belongs to $L^2(\mathcal{S}(K),\pi)$. Then $f \in L^2(K)$ and $f = \mathcal{P}^{-1}(\hat{f})$ m-almost everywhere.*

(2) Let $1 \le p \le 2$ and $\varphi \in L^p(\mathcal{S}(K),\pi)$. Suppose that the inverse L^p-transform $\check{\varphi}$ of φ belongs to $L^2(K)$. Then $\varphi \in L^2(\mathcal{S}(K),\pi)$ and $\varphi = \mathcal{P}(\check{\varphi})$ π-almost everywhere.

Proof.

(1) Take the net $(\varphi_i)_{i\in I}$ of Proposition 3.6.7. It follows
$$\varphi_i * f \in L^p(K)\cap C_0(K) \subseteq L^p(K)\cap L^\infty(K) \subseteq L^2(K).$$
Hence,
$$(\varphi_i * f)^\wedge = \mathcal{P}(\varphi_i * f) \in L^2(\mathcal{S}(K),\pi)\cap L^q(\mathcal{S}(K),\pi), \qquad \frac{1}{p}+\frac{1}{q}=1.$$
Furthermore, $(\varphi_i * f)^\wedge = \widehat{\varphi}_i\hat{f}$ π-almost everywhere. Indeed, choosing a sequence $(f_n)_{n\in\mathbb{N}}$ in $C_c(K)$ such that $\|f_n - f\|_p \to 0$ as $n\to\infty$ we have $(\varphi_i * f_n)^\wedge = \widehat{\varphi}_i\widehat{f_n}$ and hence
$$\|(\varphi_i*f)^\wedge - \widehat{\varphi}_i\hat{f}\|_q \le \|(\varphi_i*f)^\wedge - (\varphi_i*f_n)^\wedge\|_q + \|\widehat{\varphi}_i\widehat{f_n} - \widehat{\varphi}_i\hat{f}\|_q \longrightarrow 0$$
as $n\to\infty$. Thus $(\varphi_i * f)^\wedge = \widehat{\varphi}_i\hat{f}$ π-almost everywhere. Applying the Plancherel isomorphism we get
$$\|\varphi_i * f - \mathcal{P}^{-1}(\hat{f})\|_2 = \|(\varphi_i * f)^\wedge - \hat{f}\|_2 = \|\widehat{\varphi}_i\hat{f} - \hat{f}\|_2.$$
Given $\varepsilon > 0$ choose $\Gamma \subseteq \mathcal{S}(K)$ such that
$$\int_{\mathcal{S}(K)\setminus\Gamma} |(\widehat{\varphi}_i(\alpha)-1)\,\hat{f}(\alpha)|^2\,d\pi(\alpha) \le 4\int_{\mathcal{S}(K)\setminus\Gamma} |\hat{f}(\alpha)|^2\,d\pi(\alpha) < \frac{\varepsilon}{2}$$
for all $i\in I$. Then there exists $i_0\in I$ such that
$$\int_\Gamma |(\widehat{\varphi}_i(\alpha)-1)\,\hat{f}(\alpha)|^2\,d\pi(\alpha) < \frac{\varepsilon}{2}$$
for all $i\ge i_0$. Thus $\|\varphi_i * f - \mathcal{P}^{-1}(\hat{f})\|_2 \to 0$ and $\|\varphi_i * f - f\|_p \to 0$, and it follows $\mathcal{P}^{-1}(\hat{f}) = f$ m-almost everywhere.

(2) Applying Parseval's formula and Proposition 3.6.5 (ii) we obtain for $h\in L^1(K)$ and $\psi\in L^2(\mathcal{S}(K),\pi)$
$$\int_{\mathcal{S}(K)} \mathcal{P}(\check{\varphi})(\bar{\alpha})\,\hat{h}(\alpha)\,\psi(\alpha)\,d\pi(\alpha) = \int_K \check{\varphi}(x)\,\mathcal{P}^{-1}(\hat{h}\psi)(x)\,dm(x)$$
$$= \int_K \check{\varphi}(x)\,h*\check{\psi}(x)\,dm(x).$$
Write $\tilde{h}(x) := h(\tilde{x})$ and $\tilde{\psi}(\alpha) := \psi(\bar{\alpha})$. Since $\bar{\pi} = \pi$, we have $(\tilde{\psi})^\vee(y) = \check{\psi}(\tilde{y})$ and hence $h*(\tilde{\psi})^\vee(\tilde{x}) = \tilde{h}*\check{\psi}(x)$. Now assume that $\psi\in L^1(\mathcal{S}(K),\pi)\cap C_0(\mathcal{S}(K))$. Then $\psi\in L^r(\mathcal{S}(K),\pi)$ for $1<r<\infty$. Applying Proposition 3.6.5 (i) and Lemma 3.6.4 we obtain
$$\int_{\mathcal{S}(K)} \varphi(\bar{\alpha})\hat{h}(\alpha)\psi(\alpha)\,d\pi(\alpha)$$
$$= \int_K h(\tilde{x})(\varphi\tilde{\psi})^\vee(x)\,dm(x) = \int_K h(\tilde{x})\check{\varphi}*(\tilde{\psi})^\vee(x)\,dm(x)$$
$$= \int_K \check{\varphi}(\tilde{x})h*(\tilde{\psi})^\vee(x)\,dm(x) = \int_K \check{\varphi}(x)\tilde{h}*\check{\psi}(x)\,dm(x).$$

Assume that $h = \tilde{h}$. It follows

$$\int_{\mathcal{S}(K)} (\mathcal{P}(\breve{\varphi})(\bar{\alpha}) - \varphi(\bar{\alpha}))\, \hat{h}(\alpha)\psi(\alpha)\, d\pi(\alpha) \;=\; 0.$$

Choose $h_j \in C_c(K)$ with $\widetilde{h_j} = h_j$ and supp $h_j \to \{e\}$, such that $\widehat{h_j}$ converges uniformly to 1 on compact subsets of $\mathcal{S}(K)$, compare Proposition 3.6.7 and Proposition 2.1.6. It follows

$$\int_{\mathcal{S}(K)} (\mathcal{P}(\breve{\varphi})(\bar{\alpha}) - \varphi(\bar{\alpha}))\, \psi(\alpha)\, d\pi(\alpha) \;=\; 0. \qquad (*)$$

Now let $C \subseteq \mathcal{S}(K)$ be a compact subset, and choose a sequence $(f_n)_{n\in\mathbb{N}}$ in $C_c(K)$ such that $\|(f_n * f_n^*)^\wedge - \chi_C\|_1 \to 0$ as $n \to \infty$, see Proposition 3.6.6. Then there exists a subsequence $(f_{n_k} * f_{n_k}^*)_{n\in\mathbb{N}}$ such that $(f_{n_k} * f_{n_k}^*)^\wedge(\alpha) - \chi_C(\alpha) \to 0$ for π-almost all $\alpha \in \mathcal{S}(K)$. Inserting $\psi = (f_n * f_n^*)^\wedge$ in formula $(*)$, we infer that $\mathcal{P}(\breve{\varphi}) = \varphi$ π-almost everywhere on C. Since C was an arbitrary compact subset of $\mathcal{S}(K)$ we have $\mathcal{P}(\breve{\varphi}) = \varphi$ π-almost everywhere.

◇

Now we are able to prove an inversion theorem for the Hausdorff-Young transformations.

Theorem 3.6.9 *Let $1 \le p, r \le 2$. The L^p-transformations satisfy:*

(i) If $f \in L^p(K)$ such that $\hat{f} \in L^r(\mathcal{S}(K), \pi)$, then $f \in L^2(K)$ and $(\hat{f})^\vee = f = \mathcal{P}^{-1}(\hat{f})$ in $L^2(K)$. In particular, if $\hat{f} = 0$ then $f = 0$.

(ii) If $g \in L^p(\mathcal{S}(K), \pi)$ such that $\breve{g} \in L^r(K)$, then $g \in L^2(\mathcal{S}(K), \pi)$ and $(\breve{g})^\wedge = g = \mathcal{P}(\breve{g})$ in $L^2(\mathcal{S}(K), \pi)$. In particular, if $\breve{g} = 0$ then $g = 0$.

Proof.

(i) Let $f \in L^p(K)$ such that $\hat{f} \in L^r(\mathcal{S}(K), \pi)$. Then

$$\hat{f} \in L^q(\mathcal{S}(K), \pi) \cap L^r(\mathcal{S}(K), \pi) \subseteq L^2(\mathcal{S}(K), \pi), \qquad \text{where } \frac{1}{q} + \frac{1}{p} = 1.$$

Proposition 3.6.8 (1) yields $f \in L^2(K)$ and $\mathcal{P}^{-1}(\hat{f}) = f$. By Proposition 3.6.3 (ii) we have $\mathcal{P}^{-1}(\hat{f}) = (\hat{f})^\vee$.

(ii) is shown in a similar manner using Proposition 3.6.8 (2) and Proposition 3.6.3 (i).

◇

Remark: The case $r = 1$ in Theorem 3.6.9 is of particular interest. If $f \in L^p(K)$, $1 \le p \le 2$, and $\hat{f} \in L^1(\mathcal{S}(K), \pi)$, then

$$f(x) = \int_{\mathcal{S}(K)} \hat{f}(\alpha)\alpha(x)\, d\pi(\alpha) \qquad \text{for } m\text{-almost all } x \in K,$$

compare with Theorem 3.3.7. If $\varphi \in L^p(\mathcal{S}(K), \pi)$, $1 \le p \le 2$, and $\check{\varphi} \in L^1(K)$, then

$$\varphi(\alpha) = \int_K \check{\varphi}(x)\, \overline{\alpha(x)}\, dm(x) \qquad \text{for } \pi\text{-almost all } \alpha \in \mathcal{S}(K),$$

see Theorem 3.3.8.

We give further convolution results in the context of the Hausdorff-Young transformation.

Proposition 3.6.10 *Let $1 \le p \le 2$ and $\frac{1}{p} + \frac{1}{q} = 1$. For $f \in L^p(K)$ and $\mu \in M(K)$ we have*

$$(\mu * f)^\wedge = \hat{\mu}\hat{f} \qquad \pi\text{-almost everywhere on } \mathcal{S}(K).$$

Proof. By Theorem 2.2.8 we know that $\mu * f \in L^p(K)$. Let $(f_n)_{n\in\mathbb{N}}$ be a sequence in $L^1(K) \cap L^p(K)$ such that $\lim\limits_{n\to\infty} \|f_n - f\|_p = 0$. Then $(\mu * f_n)^\wedge = \hat{\mu}\widehat{f_n}$ by Theorem 3.1.4 (iii), and it follows $(\mu * f)^\wedge = \hat{\mu}\hat{f}$ π-almost everywhere on $\mathcal{S}(K)$. ◇

The Parseval formula, see subsection 3.2, can be generalized.

Proposition 3.6.11 *Let $1 \le p \le 2$ and $\frac{1}{p} + \frac{1}{q} = 1$. For $f \in L^p(K)$ and $\varphi \in L^p(\mathcal{S}(K), \pi)$ we have*

$$\int_K f(x)\, \overline{\check{\varphi}(x)}\, dm(x) = \int_{\mathcal{S}(K)} \hat{f}(\alpha)\, \overline{\varphi(\alpha)}\, d\pi(\alpha).$$

Proof. At first consider $f \in C_c(K) \subset L^2(K)$, $\varphi \in C_c(\mathcal{S}(K)) \subseteq L^2(\mathcal{S}(K), \pi)$. Parseval's formula yields $\langle f, \mathcal{P}^{-1}(\varphi)\rangle = \langle \mathcal{P}(f), \varphi\rangle$. For $f \in L^p(K)$ and $\varphi \in L^p(\mathcal{S}(K), \pi)$ choose $(f_n)_{n\in\mathbb{N}}$ in $C_c(K)$ such that $\lim\limits_{n\to\infty} \|f - f_n\|_p = 0$, and choose $(\varphi_n)_{n\in\mathbb{N}}$ in $C_c(\mathcal{S}(K))$ such that $\lim\limits_{n\to\infty} \|\varphi - \varphi_n\|_p = 0$. Then $\lim\limits_{n\to\infty} \|\check{\varphi} - \check{\varphi}_n\|_q = 0$ and $\lim\limits_{n\to\infty} \|\hat{f} - \widehat{f_n}\|_q = 0$. By Proposition 3.6.3 and Hölder's inequality the generalization of Parseval's identity follows. ◇

Now we can investigate the range of the Hausdorff-Young transformations. Let $1 < p < 2$, $\frac{1}{p} + \frac{1}{q} = 1$. We use the symbol T instead of $^\wedge$, and S instead of $^\vee$. We know that $T : L^p(K) \to L^q(\mathcal{S}(K), \pi)$ is a bounded injective operator from $L^p(K)$ into $L^q(\mathcal{S}(K), \pi)$, and $S : L^p(\mathcal{S}(K), \pi) \to L^q(K)$

is a bounded injective operator from $L^p(\mathcal{S}(K),\pi)$ into $L^q(K)$, see Theorem 3.6.9. The adjoint operator $T^* : L^P(\mathcal{S}(K),\pi) \to L^q(K)$ fulfils

$$T^*(\varphi)(f) = \int_{\mathcal{S}(K)} \hat{f}(\alpha)\,\overline{\varphi(\alpha)}\,d\pi(x) = \int_K f(x)\,\overline{\check{\varphi}(x)}\,dm(x)$$

$$= \int_K f(x)\,\overline{S(\varphi)(x)}\,dm(x)$$

for $\varphi \in L^p(\mathcal{S}(K),\pi)$ and each $f \in L^p(K)$, i.e. $T^*(\varphi) = S(\varphi)$ for all $\varphi \in L^p(\mathcal{S}(K),\pi)$. In the same manner it follows $S^*(f) = T(f)$ for all $f \in L^p(K)$.

Proposition 3.6.12 *Let $1 < p < 2$, $\frac{1}{p} + \frac{1}{q} = 1$. Then $L^p(K)^\wedge$ is a dense linear subspace of $L^q(\mathcal{S}(K),\pi)$, and $L^p(\mathcal{S}(K),\pi)^\vee$ is a dense linear subspece of $L^q(K)$.*

Proof. Since S is injective, T^* is injective and hence $T(L^p(K))$ is dense in $L^q(\mathcal{S}(K),\pi)$. Since T is injective, S^* is injective and $S(L^p(\mathcal{S}(K),\pi))$ is dense in $L^q(K)$. ◇

In case of $p = 2$ we know that the Plancherel transformation (and its inverse) is an isometric isomorphism. We show now that for $1 < p < 2$ the Hausdorff-Young transformations are onto if and only if K is finite. In order to prove this we apply two Lemmata which hold for general infinite measure spaces. The reader may find the proofs in [303]. We formulate the statements in the context of hypergroups as needed below.

Lemma 3.6.13 *Let K be an infinite hypergroup, and let $1 < p < \infty$. Then there exists a sequence $(f_n)_{n\in\mathbb{N}}$ of functions in $L^p(K)$ such that f_n converges weakly to zero in $L^p(K)$ and*

$$\|f_{n_1} + f_{n_2} + \ldots + f_{n_m}\|_p = m^{1/p}$$

for all subsets $\{f_{n_1}, f_{n_2}, \ldots, f_{n_m}\}$ of $(f_n)_{n\in\mathbb{N}}$ and $m \in \mathbb{N}$.

Lemma 3.6.14 *Let K be an infinite commutative hypergroup. Let $q \geq 2$ and $(\varphi_n)_{n\in\mathbb{N}}$ a sequence in $L^q(\mathcal{S}(K),\pi)$ which converges weakly to zero in $L^q(\mathcal{S}(K),\pi)$. Then there exists a subsequence $(\varphi_{n_k})_{k\in\mathbb{N}}$ of $(\varphi_n)_{n\in\mathbb{N}}$ and a positive constant A such that*

$$\|\varphi_{n_1} + \varphi_{n_2} + \ldots + \varphi_{n_m}\|_q \leq A\,m^{1/2}.$$

Theorem 3.6.15 *Let K be a commutative hypergroup, $1 < p < 2$. The following statements are equivalent*

(1) The Fourier transform $^\wedge : L^p(K) \to L^q(\mathcal{S}(K), \pi)$ *is surjective.*
(2) The inverse Fourier transform $^\vee : L^p(\mathcal{S}(K), \pi) \to L^q(K)$ *is surjective.*
(3) K *is finite.*

Proof. We have to prove (1) $\Rightarrow$ (3) and (1) $\Rightarrow$ (2). Assume that $^\wedge$ is surjective, and K is infinite. Then $^\wedge$ is a topological isomorphism from $L^p(K)$ onto $L^q(\mathcal{S}(K), \pi)$ and hence there exists $M > 0$ such that $\|f\|_p \leq M\ \|\hat{f}\|_q$ for all $f \in L^p(K)$. Let $(f_n)_{n\in\mathbb{N}}$ be a sequence of functions with the properties given in Lemma 3.6.13. In particular, $(\widehat{f_n})_{n\in\mathbb{N}}$ converges weakly to zero in $L^q(\mathcal{S}(K), \pi)$ by Parseval's formula in Proposition 3.6.11. Therefore by Lemma 3.6.14 there exists a subsequence $(\widehat{f_{n_k}})_{k\in\mathbb{N}}$ of $(\widehat{f_n})_{n\in\mathbb{N}}$ and a constant $A > 0$ such that $\|\widehat{f_{n_1}} + \widehat{f_{n_2}} + \ldots + \widehat{f_{n_m}}\|_q \leq Am^{1/2}$. It follows

$$\begin{aligned} m^{1/p} &= \|f_{n_1} + f_{n_2} + \ldots + f_{n_m}\|_p \\ &\leq M\ \|\widehat{f_{n_1}} + \widehat{f_{n_2}} + \ldots + \widehat{f_{n_m}}\|_q \leq M\,A\,m^{1/2} \end{aligned}$$

for all $m \in \mathbb{N}$. Since $\frac{1}{p} > \frac{1}{2}$ we have a contradiction. Thus K has to be finite.

Now assume that $S = {}^\vee$ is surjective. Then $T^* = S$ is a topological isomorphism, and hence $T = {}^\wedge$ is a topological isomorphism. Therefore K is finite, as just shown. $\diamond$

Remark: The case $p = 1$ is studied in subsection 3.11.2.

7 Bochner theorem

Positive definite functions on general hypergroups are already introduced in subsection 2.4. We suppose now that K is a commutative hypergroup.

Consider the inverse Fourier-Stieltjes transform $\check{a}$ of a positive measure $a \in M^+(\hat{K})$. We know $\check{a} \in C^b(K)$. For $n \in \mathbb{N}$, $c_1, \ldots, c_n \in \mathbb{C}$ and $x_1, \ldots, x_n \in K$ we have

$$\sum_{i=1}^n \sum_{j=1}^n c_i \overline{c_j} \omega(x_i, \tilde{x}_j)(\check{a}) = \sum_{i=1}^n \sum_{j=1}^n c_i \overline{c_j} \int_K \int_{\hat{K}} \alpha(z)\, da(\alpha)\, d\omega(x_i, \tilde{x}_j)(z)$$

$$= \sum_{i=1}^n \sum_{j=1}^n c_i \overline{c_j} \int_{\hat{K}} \alpha(x_i) \overline{\alpha(x_j)}\, da(\alpha) = \int_{\hat{K}} \left| \sum_{i=1}^n c_i \alpha(x_i) \right|^2 da(\alpha) \geq 0.$$

Therefore $\check{a} \in P^b(K)$.

A very important result states that the converse is valid, too. In fact the generalization of Bochner's theorem from locally compact abelian groups to commutative hypergroups says that for each $\varphi \in P^b(K)$ there exists a bounded positive measure $a \in M(\hat{K})$ with $\check{a} = \varphi$.

Theorem 3.7.1 (Bochner)
Let $\varphi \in P^b(K)$. Then there exists a bounded positive Borel measure $a \in M(\hat{K})$ such that $\check{a} = \varphi$. The measure a is uniquely determined by the relation $\check{a} = \varphi$.

Proof. We may assume that $\varphi \in P^0(K)$. Let $M_+^0 := \{a \in M(\hat{K}) : \text{a}$ positive, $\|a\| \leq 1\}$. M_+^0 is a compact subset of $M(\hat{K})$ with respect to the weak $*$-topology. Let $(a_i)_{i\in I}$ be a net in M_+^0, which converges in the weak $*$-topology to some $a \in M_+^0$, and $f \in L^1(K)$. Then

$$\int_K f(x)\check{a}_i(x)\,dm(x) = \int_K \int_{\hat{K}} f(x)\alpha(x)\,da_i(\alpha)\,dm(x)$$

$$= \int_{\hat{K}} \int_K f(x)\,\overline{\bar{\alpha}(x)}\,dm(x)\,da_i(\alpha) = \int_{\hat{K}} \hat{f}(\bar{\alpha})\,da_i(\alpha).$$

Hence

$$\lim_i \int_K f(x)\check{a}_i(x)\,dm(x) = \int_{\hat{K}} \hat{f}(\bar{\alpha})\,da(\alpha) = \int_K f(x)\check{a}(x)\,dm(x),$$

and we infer that the mapping $T : M_+^0 \to P^0(K)$, $T(a) = \check{a}$, is continuous, where M_+^0 carries the weak $*$-topology of $M(\hat{K})$ and $P^0(K)$ the weak $*$-topology of $L^\infty(K)$. In particular, the range $T(M_+^0) \subseteq P^0(K)$ is a compact convex subset of $P^0(K)$. For $\alpha \in \hat{K}$ we have $a = \epsilon_\alpha \in M_+^0$ and $T(\epsilon_\alpha) = \check{\epsilon}_\alpha = \alpha$, and the zero measure $a = 0$ is mapped by T to the zero function $0 \in P^0(K)$. By Corollary 2.4.22 it follows that $\mathrm{ex}(P^1(K))$ is exactly $\hat{K}$, and by Proposition 2.4.15 and Theorem 2.4.16 (1) the weak $*$-closure of the convex hull of $\hat{K} \cup \{0\}$ is equal to $P^0(K)$. Therefore $T(M_+^0) = P^0(K)$, and the existence of a positive Borel measure $a \in M(\hat{K})$ such that $\check{a} = \varphi$ is shown. The uniqueness follows by Theorem 3.3.3. ◇

Proposition 3.7.2 *The function $\varphi \in C^b(K) \cap L^1(K)$ is positive definite if, and only if $\hat{\varphi}\,|\mathcal{S}(K) \geq 0$. If $\varphi \in P^b(K) \cap L^1(K)$ then $\hat{\varphi} \in L^1(\hat{K}, \pi)$ and $(\hat{\varphi})^\vee = \varphi$.*

Proof. Let $\varphi \in P^b(K) \cap L^1(K)$. Then $\varphi = \check{a}$, where $a \in M(\hat{K})$, $a \geq 0$. By Corollary 3.3.6 we have $\hat{\varphi} \in L^1(\hat{K}, \pi)$ and $a = \hat{\varphi}\pi$. Theorem 3.3.7 implies $\varphi = (\hat{\varphi})^\vee$, and since $a \geq 0$ we have $\hat{\varphi}|\mathcal{S}(K) \geq 0$. Now assume $\varphi \in C^b(K) \cap L^1(K)$ and $\hat{\varphi}|\mathcal{S}(K) \geq 0$. For $g \in C_c(K)$ we have $\varphi * g \in$

$L^1(K) \cap L^2(K)$. Lemma 3.3.5 (i) implies $(\varphi * g * g^*)^\wedge \in L^1(\hat{K}, \pi)$. Applying Theorem 3.3.7 we obtain

$$0 \leq \int_{\hat{K}} \hat{\varphi}(\alpha)|\hat{g}|^2(\alpha)\, d\pi(\alpha) = \left(\hat{\varphi}|\hat{g}|^2\right)^\vee(e) = ((\varphi * g * g^*)^\wedge)^\vee(e)$$

$$= \varphi * g * g^*(e) = \int_K \overline{\varphi(x)} g * g^*(x)\, dm(x).$$

Now Proposition 2.4.2 implies $\varphi \in P^b(K)$. ◇

Bochner's theorem and Proposition 3.7.2 will be central for the investigation of stochastic processes.

If $\varphi \in P^b(K)$ has compact support it can be written as a convolution of $g \in L^2(K)$ with g^*, where $g^* = \overline{\tilde{g}}$. (Recall a reverse implication in Proposition 2.4.3.)

Proposition 3.7.3 *Let $\varphi \in P^b(K) \cap C_c(K)$. Then $\varphi = g * g^*$ with $g \in L^2(K)$.*

Proof. By Proposition 3.7.2 we have $\hat{\varphi} \in L^1(\hat{K}, \pi) \cap C_0(\hat{K})$ and $\hat{\varphi}|\operatorname{supp} \pi \geq 0$. Put $\psi = \sqrt{\hat{\varphi}}$ on $\mathcal{S}(K) = \operatorname{supp} \pi$ and $\psi = 0$ on $\hat{K} \backslash \mathcal{S}(K)$. Then $\psi \in L^2(\hat{K}, \pi)$, $\psi^2 = \hat{\varphi}$, $\psi = \overline{\psi}$ and for $g = \mathcal{P}^{-1}(\psi) \in L^2(K)$ we have $g = g^*$ and by Proposition 3.3.10 $\varphi = (\hat{\varphi})^\vee = (\psi^2)^\vee = (\mathcal{P}(g)\, \mathcal{P}(g^*))^\vee = g * g^*$. ◇

Now we characterize those $\varphi \in P^b(K)$ with $\varphi = \check{a}$, such that $\operatorname{supp} a \subseteq \mathcal{S}(K)$. The following result is a Bochner-type theorem describing those bounded continuous functions on K which are inverse Fourier-Stieltjes transforms of positive bounded Borel measures $\mu \in M(\mathcal{S}(K))$, $\mathcal{S}(K) = \operatorname{supp} \pi$.

Theorem 3.7.4 *Let $\varphi \in C^b(K)$. Then the following statements are equivalent:*

(i) $\varphi = \check{a}$ *where* $a \in M(\hat{K})$, $a \geq 0$ *and* $\operatorname{supp} a \subseteq \mathcal{S}(K)$.
(ii) $\int_K \overline{\varphi(x)} f(x)\, dm(x) \geq 0$ *for all* $f \in C_c(K)$ *with the property* $\hat{f} \,|\, \mathcal{S}(K) \geq 0$.

Proof. Assume that $\int_K \overline{\varphi(x)} f(x) dm(x) \geq 0$ for all $f \in C_c(K)$ with $\hat{f} \,|\, \mathcal{S}(K) \geq 0$. Define $P_\varphi : C_c(K)^\wedge \,|\, \mathcal{S}(K) \to \mathbb{C}$ by

$$P_\varphi(\hat{f} \,|\, \mathcal{S}(K)) = \int_K \overline{\varphi(x)} f(x)\, dm(x).$$

Note that P_φ is well-defined by Theorem 3.1.8. Obviously, P_φ is a positive functional on $C_c(K)^\wedge|\mathcal{S}(K)$. Choose a family of functions $k_i \in C_c(K)$, $i \in I$ as in Proposition 2.1.6 and define $h_i = k_i * k_i^*$. Then $\lim\limits_i \|h_i * f - f\|_1 = 0$ for each $f \in C_c(K)$ and $\widehat{h_i} \geq 0$, $\|h_i\|_1 \leq 1$. Now let $f \in C_c(K)$ with $\|\hat{f}\|_{\mathcal{S}(K)} \leq 1$. For proving the continuity of P_φ we can assume that $\hat{f}$ is real-valued. Then we have $\widehat{h_i}\left(1 \pm \hat{f}\right) | \mathcal{S}(K) \geq 0$, and hence $P_\varphi\left(\widehat{h_i} \,|\, \mathcal{S}(K)\right) \geq |P_\varphi((h_i * f)^\wedge \,|\, \mathcal{S}(K))|$. Furthermore

$$\lim_i P_\varphi\left(\widehat{h_i} \,|\, \mathcal{S}(K)\right) = \lim_i \int_K \overline{\varphi(x)} h_i(x)\, dm(x) \leq \|\varphi\|_\infty$$

and

$$\lim_i \left|P_\varphi\left((h_i * f)^\wedge \,|\, \mathcal{S}(K)\right)\right|$$

$$= \lim_i \left| \int_K \overline{\varphi(x)} h_i * f(x)\, dm(x) \right| = \left| \int_K \overline{\varphi(x)} f(x)\, dm(x) \right| = \left| P_\varphi\left(\hat{f} \,|\, \mathcal{S}(K)\right) \right|.$$

Thus we have $\left|P_\varphi\left(\hat{f} \,|\, \mathcal{S}(K)\right)\right| \leq \|\varphi\|_\infty$, and so far we have shown that P_φ is a positive, $\|.\|_{\mathcal{S}(K)}$-continuous functional on $C_c(K)^\wedge \,|\, \mathcal{S}(K)$. Therefore there exists a unique positive measure $a \in M(\hat{K})$ with $\operatorname{supp} a \subseteq \mathcal{S}(K)$ and

$$\int_K \overline{\varphi(x)} f(x)\, dm(x) = P_\varphi\left(\hat{f} \,|\, \mathcal{S}(K)\right) = \int_{\hat{K}} \hat{f}(\alpha)\, da(\alpha) = \int_K \overline{\check{a}(x)} f(x)\, dm(x)$$

for every $f \in C_c(K)$. Using the continuity of $\check{a}$ and φ we obtain $\check{a} = \varphi$.

The converse implication follows immediately from

$$\int_K \overline{\varphi(x)} f(x)\, dm(x) = \int_{\hat{K}} \hat{f}(\alpha)\, da(\alpha).$$

◇

Now we can deal in more detail with the problem that $\mathcal{S}(K) = \operatorname{supp} \pi$ is in general not equal to $\hat{K}$, a situation very different to the group case. Examples where $\operatorname{supp} \pi$ is a proper subset of $\hat{K}$ can be found in subsection 4.5.

In subsection 8.2 it is shown that for the Naimark hypergroup K we have $\mathcal{S}(K) \subsetneq \hat{K} \subsetneq \mathcal{X}^b(K)$, see Theorem 8.2.4.

Definition Let $\alpha \in \hat{K}$. We say that the P_2-condition is satisfied in α if for each $\varepsilon > 0$ and every compact subset $C \subseteq K$ there exists some $g \in C_c(K)$ such that $\|g\|_2 = 1$ and

$$\|L_{\tilde{y}} g - \overline{\alpha(y)} g\|_2 < \varepsilon$$

for all $y \in C$.

Now we characterize those $\alpha \in \hat{K}$ that belong to $\mathcal{S}(K) = \operatorname{supp} \pi$.

Theorem 3.7.5 *Let $\alpha \in \hat{K}$. The following conditions are equivalent:*

(i) $\alpha \in \mathcal{S}(K) = \operatorname{supp} \pi$.
(ii) *For all* $f \in C_c(K)$ *with the property* $\hat{f} \mid \mathcal{S}(K) \geq 0$ *one has* $\hat{f}(\alpha) \geq 0$.
(iii) *There exists a net* $(f_i)_{i\in I} \subseteq C_c(K)$ *with norm* $\|f_i\|_2 = 1$ *such that* $f_i * f_i^*$ *converges to* α *uniformly on compact subsets of* K.
(iv) *The* P_2*-condition is satisfied in* α.
(v) *There exists a net* $(g_i)_{i\in I} \subseteq C_c(K)$ *with the property* $\widehat{g_i} \mid \mathcal{S}(K) \geq 0$ *such that* g_i *converges to* α *uniformly on compact subsets of* K.

Proof. Conditions (i) and (ii) are obviously equivalent by Theorem 3.7.4. First of all we show that (i) implies (iii). Choose a compact neighborhood $U \subseteq \hat{K}$ of α such that

$$U \subseteq V\left(\alpha, \frac{\varepsilon}{2}; C\right) = \{\beta \in \hat{K} : |\alpha(x) - \beta(x)| < \varepsilon/2 \quad \text{for all } x \in C\}.$$

Define $h = \chi_U/\pi(U) \in L^1(\hat{K}, \pi)$. Then we have for all $x \in C$

$$|\check{h}(x) - \alpha(x)| = \frac{1}{\pi(U)} \left| \int_U \beta(x)\, d\pi(\beta) - \int_U \alpha(x)\, d\pi(\beta) \right| < \frac{\varepsilon}{2}.$$

For $h^{1/2} = \chi_U/\pi(U)^{1/2}$ exists some $f \in C_c(K)$ such that $\|\hat{f} - h^{1/2}\|_2 < \varepsilon/4$, compare Theorem 3.2.6. Since $\|h^{1/2}\|_2 = 1$ we can assume that $\|\hat{f}\|_2 = \|f\|_2 = 1$. Furthermore we get (compare with Proposition 3.6.6.)

$$\begin{aligned} &\|(f * f^*)^\wedge - h\|_1 \\ &= \||\hat{f}|^2 - h\|_1 \leq \int_{\hat{K}} \left|\hat{f}(\beta) - h^{1/2}(\beta)\right| \cdot \left||\hat{f}|(\beta) + h^{1/2}(\beta)\right| \, d\pi(\beta) \\ &\leq \|\hat{f} - h^{1/2}\|_2 \left(\|\hat{f}\|_2 + \|h^{1/2}\|_2\right) = 2\|\hat{f} - h^{1/2}\|_2 < \varepsilon/2. \end{aligned}$$

Applying Proposition 3.7.2 we obtain for $x \in K$

$$|f * f^*(x) - \check{h}(x)| = \left|((f * f^*)^\wedge)^\vee(x) - \check{h}(x)\right| \leq \|(f * f^*)^\wedge - h\|_1 < \varepsilon/2,$$

and hence $|f * f^*(x) - \alpha(x)| < \varepsilon$ for every $x \in C$.

In order to prove (iii) $\Longrightarrow$ (iv) consider $C \subseteq K$ compact, and let $\varepsilon > 0$. By (iii) exists a function $f \in C_c(K)$ such that $\|f\|_2 = 1$ and

$$|f * f^*(x) - \alpha(x)| < \varepsilon$$

for all $x \in C * C$. We assume that $e \in C$, and $C = \tilde{C}$. Since for all $x, y \in C$

$$|L_y f * f^*(x) - \alpha(y)\alpha(x)| \leq \int_K |f * f^*(z) - \alpha(z)| \, d\omega(y,x)(z) < \varepsilon,$$

and $|f * f^*(x)\alpha(y) - \alpha(x)\alpha(y)| < \varepsilon$, we obtain

$$|L_y f * f^*(x) - f * f^*(x)\alpha(y)| < 2\varepsilon,$$

and hence (using $\|f\|_2 = 1$)

$$\begin{aligned}
&\left| \int_K \overline{L_{\tilde{x}} f(z)}\, [L_y f(z) - \alpha(y) f(z)]\, dm(z) \right| \\
&= \left| \int_K \overline{f(z)}\, [L_x(L_y f)(z) - \alpha(y) L_x f(z)]\, dm(z) \right| \\
&= |L_y f * f^*(x) - \alpha(y) f * f^*(x)| < 2\varepsilon.
\end{aligned}$$

In a similar way we get for $y \in C,\ x \in K$

$$\left| \int_K \overline{\alpha(\tilde{x}) f(z)}\, [L_y f(z) - \alpha(y) f(z)]\, dm(z) \right| = |\alpha(x)||f * f^*(y) - \alpha(y)| < \varepsilon.$$

For $\tilde{x} = y \in C$ we have therefore

$$\|L_y f - \alpha(y) f\|_2^2 = \int_K \overline{[L_y f(z) - \alpha(y) f(z)]}[L_y f(z) - \alpha(y) f(z)]\, dm(z) \le 3\varepsilon.$$

Thus (iii) $\Longrightarrow$ (iv) is shown.

Now assume that the P_2-condition is satisfied in α. We prove that

$$|\hat{f}(\alpha)| \le \sup\{\|f * g\|_2 : g \in L^2(K),\ \|g\|_2 = 1\}$$

for every $f \in C_c(K),\ f \neq 0$. That is $\alpha \in \mathcal{S}(K)$. There exists a function $g \in L^2(K),\ \|g\|_2 = 1$ such that

$$\|L_{\tilde{y}} g - \overline{\alpha(y)} g\|_2 < \varepsilon / \|f\|_1 \qquad \text{for all } y \in \operatorname{supp} f.$$

Since

$$f * g(x) - \hat{f}(\alpha) g(x) = \int_K f(y) \left(L_{\tilde{y}} g(x) - \overline{\alpha(y)} g(x) \right) dm(y),$$

it follows that

$$\|f * g - \hat{f}(\alpha) g\|_2 \le \int_K |f(y)| \cdot \|L_{\tilde{y}} g - \overline{\alpha(y)} g\|_2\, dm(y) < \varepsilon.$$

Thus

$$|\hat{f}(\alpha)| = |\hat{f}(\alpha)| \cdot \|g\|_2 \le \varepsilon + \|f * g\|_2,$$

which implies

$$|\hat{f}(\alpha)| \le \sup\{\|f * g\|_2 : g \in L^2(K),\ \|g\|_2 = 1\}.$$

So far we have shown that the conditions (i), (ii), (iii) and (iv) are equivalent. It is evident that (iii) implies (v). Finally we show (v) $\Longrightarrow$ (ii). Let $f \in C_c(K)$ with $\hat{f} \,|\, \mathcal{S}(K) \geq 0$. There exists a net $(g_i)_{i\in I}$, $g_i \in C_c(K)$ with $\hat{g}_i \,|\, \mathcal{S}(K) \geq 0$, such that g_i converges to α uniformly on $\operatorname{supp} f$. Then $(f * g_i)^\wedge \,|\, \mathcal{S}(K) = \hat{f}\widehat{g_i} \,|\, \mathcal{S}(K) \geq 0$ and

$$\hat{f}(\alpha) = \lim_i \int\limits_K f(x) g_i(\tilde{x})\, dm(x) = \lim_i f{*}g_i(e) = \lim_i \int\limits_{\hat{K}} (f{*}g_i)^\wedge(\beta)\, d\pi(\beta) \geq 0.$$

◇

Remark.

(1) In the case of $\alpha = 1$ the functions f_i of the net $(f_i)_{i\in I}$ with the properties of *(iii)* in Theorem 3.7.5 can be chosen such that $f_i \geq 0$. In fact, given $\epsilon > 0$ let $f_i \in C_c(K)$, $\|f_i\|_2 = 1$ such that $f_i * f_i^*$ converges to 1 uniformly on compact subsets of K. Put $g_i := |f_i| \in C_c(K)$. Then $\|g_i\|_2 = 1$. Furthermore, $0 \leq |f_i * f_i^*(x)| \leq g_i * g_i^*(x)$ and $g_i * g_i^*(x) \leq 1$ for $x \in K$, and hence $0 \leq 1 - g_i{*}g_i^*(x) \leq 1 - |f_i{*}f_i^*(x)| \leq |1 - f_i{*}f_i^*(x)|$. It follows that $g_i * g_i^*$ converges to 1 uniformly on compact subsets of K.

(2) In case of $\alpha = 1$ the functions g in condition P_2 can be chosen to be nonnegative. In fact, the proof of the implementation *(iii)* $\Rightarrow$ *(iv)* shows that functions g in condition P_2 can be selected from the net $(f_i)_{i\in I}$ of *(iii)*. From (1) we know that the f_i can be chosen nonnegative. Therefore we may define that the P_2-condition is satisfied in $\alpha = 1$ if for each $\epsilon > 0$ and every compact subset $C \subseteq K$ there exists some $g \in P^2(K) \cap C_c(K)$, where

$$P^2(K) := \{f \in L^2(K) : f \geq 0,\ \|f\|_2 = 1\},$$

such that $\|L_{\tilde{y}} g - g\|_2 < \epsilon$ for all $y \in C$.

Applying condition (iv) of Theorem 3.7.5 it follows immediately that $\alpha \in \hat{K} \cap L^2(K)$ implies $\alpha \in \mathcal{S}(K) = \operatorname{supp} \pi$. In fact, put $g = \alpha / \|\alpha\|_2$. Then $L_y g = \alpha(y) g$ and hence the P_2-condition is satisfied in α. Independent of Theorem 3.7.5 we can prove the following result.

Proposition 3.7.6 *Let $\alpha \in \hat{K}$. Then $\alpha \in L^2(K)$ if and only if $\pi(\{\alpha\}) > 0$.*

Proof. Let $\alpha \in L^2(K) \cap \hat{K}$. Choose a sequence $(g_n)_{n\in\mathbb{N}}$, $g_n \in C_c(K)$ such that $\lim_{n\to\infty} \|g_n - \alpha\|_2 = 0$. Then $\lim_{n\to\infty} \|L_x g_n - \alpha(x)\alpha\|_2 = 0$ and considering the Plancherel isomorphism we obtain for each $\beta \in \hat{K}$

$$\beta(x)\wp(\alpha)(\beta) = \lim_{n\to\infty} \beta(x) \int_K g_n(y)\overline{\beta(y)}\, dm(y) = \lim_{n\to\infty} \int_K g_n(y)\overline{L_{\tilde{x}}\beta(y)}\, dm(y)$$

$$= \lim_{n\to\infty} \int_K L_x g_n(y)\overline{\beta(y)}\, dm(y) = \alpha(x)\mathcal{P}(\alpha)(\beta).$$

Hence $\mathcal{P}(\alpha)(\beta) = 0$ if $\beta \neq \alpha$. For $\beta = \alpha$ we immediately get $\mathcal{P}(\alpha)(\alpha) = \|\alpha\|_2^2$. The Plancherel-Levitan theorem implies $\pi(\{\alpha\}) = 1/\|\alpha\|_2^2 > 0$. Conversely assume $\pi(\{\alpha\}) > 0$. Consider the delta-function $\epsilon_\alpha(\beta) = \delta_{\alpha,\beta}$. ϵ_α is a non-zero element of $L^2(\hat{K},\pi) \cap L^1(\hat{K},\pi)$. It is easily seen that $(\epsilon_\alpha)^\vee(x) = \pi(\{\alpha\})\alpha(x)$, and Theorem 3.3.9 implies $\alpha = (\epsilon_\alpha)^\vee/\pi(\{\alpha\}) \in L^2(K)$. ◇

8 Positive definiteness on dual spaces

We consider two kinds of positive definiteness on dual spaces of K, a strong one on $\hat{K}$, and a weaker one on $\mathcal{S}(K) \cup \{1\}$, see subsection 3.11.3.

Definition: Let K be a commutative hypergroup. A continuous bounded function $\varphi \in C(\hat{K})$ is called **strongly positive definite** if for each $a \in M(\hat{K})$ with the property $\check{a} \geq 0$, the inequality

$$\int_{\hat{K}} \varphi(\alpha)\, da(\alpha) \geq 0$$

holds. The set of all strongly positive definite functions on $\hat{K}$ is denoted by $SP(\hat{K})$.

Proposition 3.8.1 *Let $\varphi \in SP(\hat{K})$.*

(1) If $a \in M(\hat{K})$ such that $\check{a} \geq 0$, then $(\varphi a)^\vee \geq 0$.
(2) If $a \in M(\hat{K})$ such that $\check{a}$ is real-valued, then $\int_{\hat{K}} \varphi(\alpha)\, da(\alpha)$ is real.
(3) For every $\alpha \in \hat{K}$ we have $\varphi(\bar{\alpha}) = \overline{\varphi(\alpha)}$, and $\bar{\varphi}$ and $Re\,\varphi$ are elements of $SP(\hat{K})$.
(4) If ψ is another element of $SP(\hat{K})$, then $\varphi\psi$ and $c_1\varphi + c_2\psi$ are in $SP(\hat{K})$ for $c_1, c_2 \geq 0$.

Proof.

(1) If $a \in M(\hat{K})$ with $\check{a} \geq 0$, then

$$(\widehat{\epsilon_x} a)^\vee(y) = \int_{\hat{K}} \alpha(y)\, \widehat{\epsilon_x}(\alpha)\, da(\alpha)$$

$$= \int_{\hat{K}} \alpha(y)\, \overline{\alpha(x)}\, da(\alpha) \;=\; \int_{\hat{K}} \int_{K} \alpha(z)\, d\omega(\tilde{x}, y)(z)\, da(\alpha)$$

$$= \int_{\hat{K}} \check{a}(z)\, d\omega(\tilde{x}, y)(z) \;\geq 0 \qquad \text{for every } x \in K.$$

Hence

$$(\varphi a)^{\vee}(\tilde{x}) \;=\; \int_{\hat{K}} \overline{\alpha(x)}\, \varphi(\alpha)\, da(\alpha) \;=\; \int_{\hat{K}} \varphi(\alpha)\, \widehat{\epsilon_{\tilde{x}}}(\alpha)\, da(\alpha) \;\geq 0$$

for all $x \in K$.

(2) Let $\check{a}(x) \in \mathbb{R}$ for any $x \in K$. Then $\|\check{a}\|_\infty \epsilon_1 + a \in M(\hat{K})$ such that $(\|\check{a}\|_\infty \epsilon_1 + a)^{\vee} \geq 0$. Therefore

$$\|\check{a}\|_\infty \varphi(1) \;+\; \int_{\hat{K}} \varphi(\alpha)\, da(\alpha) \;\geq\; 0.$$

Since $\varphi(1) = \int_{\hat{K}} \varphi(\alpha)\, d\epsilon_1(\alpha) \;\geq 0$, it follows that $\int_{\hat{K}} \varphi(\alpha)\, da(\alpha)$ is real-valued.

(3) Since $(\epsilon_\alpha + \epsilon_{\bar{\alpha}})^{\vee} \;= 2\,\mathrm{Re}\ \alpha$, we have $\varphi(\alpha) + \varphi(\bar{\alpha}) \in \mathbb{R}$ by (2). Since $(\epsilon_\alpha - \epsilon_{\bar{\alpha}})^{\vee}/i \;=\; 2\,\mathrm{Im}\ \alpha$, it follows $(\varphi(\alpha) - \varphi(\bar{\alpha}))/i \;\in \mathbb{R}$. Therefore $\varphi(\bar{\alpha}) = \overline{\varphi(\alpha)}$ for each $\alpha \in \hat{K}$. For $a \in M(\hat{K})$ define $\bar{a} \in M(K)$ by

$$\int_{\hat{K}} \sigma(\alpha)\, d\bar{a}(\alpha) \;=\; \int_{\hat{K}} \sigma(\bar{\alpha})\, da(\alpha)$$

for each $\sigma \in C_0(\hat{K})$. Then $(\bar{a})^{\vee}(x) \;=\; a^{\vee}(\tilde{x})$ for all $x \in K$. Hence $(\bar{a})^{\vee} \geq 0$, whenever $\check{a} \geq 0$. It follows

$$\int_{\hat{K}} \overline{\varphi(\alpha)}\, da(\alpha) \;=\; \int_{\hat{K}} \varphi(\bar{\alpha})\, da(\alpha) \;=\; \int_{\hat{K}} \varphi(\alpha)\, d\bar{a}(\alpha) \;\geq\; 0.$$

Thus $\bar{\varphi} \in SP(\hat{K})$ and then $\mathrm{Re}\ \varphi \in SP(\hat{K})$.

(4) The product $\varphi\psi$ is strongly positive definite by (1).

◇

Now we derive a Bochner theorem for strongly positive definite functions on $\hat{K}$.

Theorem 3.8.2 *Let $\varphi \in SP(\hat{K})$. Then there exists a unique positive measure $\nu \in M(K)$ such that $\varphi|S(K) = \hat{\nu}|S(K)$. Conversely, $\hat{\nu} \in SP(\hat{K})$ for every positive measure $\nu \in M(K)$.*

Proof. Let $\nu \in M(K)$ be a positive measure, and let $a \in M(\hat{K})$ such that $\check{a} \geq 0$. Then

$$\int_{\hat{K}} \hat{\nu}(\alpha)\, da(\alpha) = \int_K \check{a}(\tilde{x})\, d\nu(x) \geq 0,$$

i.e. $\hat{\nu} \in SP(\hat{K})$.

Now let $\varphi \in SP(\hat{K})$. If $a \in M(\hat{K})$ such that $\check{a}$ is real-valued, then $\|\check{a}\|_\infty \varphi(1) + \int\limits_{\hat{K}} \varphi(\alpha) da(\alpha) \geq 0$, as we have already noted in the proof of Proposition 3.8.1(2). It follows

$$\left| \int_{\hat{K}} \varphi(\alpha)\, da(\alpha) \right| \leq \varphi(1)\, \|\check{a}\|_\infty.$$

For an arbitrary $a \in M(\hat{K})$ let $a_1 = (a + a^*)/2,\ a_2 = (a - a^*)/(2i)$. By Proposition 3.4.16 we have $\check{a}_1 = \mathrm{Re}\ \check{a}$ and $\check{a}_2 = \mathrm{Im}\ \check{a}$. Hence

$$\left| \int_{\hat{k}} \varphi(\alpha)\, da(\alpha) \right| \leq \varphi(1)\, \|\check{a}_1\|_\infty + \varphi(1)\, \|\check{a}_2\|_\infty \leq 2\varphi(1)\, \|\check{a}\|_\infty.$$

Thus $\phi_0 : M(\hat{K})^\vee \to \mathbb{C},\ \phi_0(\check{a}) := \int\limits_{\hat{K}} \varphi(\alpha) da(\alpha)$ is a sup-norm continuous linear functional. Denote by ϕ the extension to $C_0(K)$ of $\phi_0 | (C_c(\hat{K}))^\vee$. By Riesz's theorem there exists a measure $\nu \in M(K)$ such that

$$\int_{\hat{K}} \varphi(\alpha)\, h(\alpha)\, d\pi(\alpha) = \phi_0(\check{h}) = \int_K \check{h}(x)\, d\tilde{\nu}(x)$$

for each $h \in C_c(\hat{K})$. This measure $\nu \in M(K)$ is positive. In fact, given $f \in C_0(K),\ f \geq 0$ and $\delta > 0$, Proposition 3.3.2(iii) yields a function $h \in C_c(\hat{K})$ such that $\|f - \check{h}\|_\infty < \delta$. We may assume that $\check{h}$ is real-valued. Let $a = \delta\epsilon_1 + h\pi \in M(\hat{K})$. Then $\check{a} \geq 0$, and hence $\phi_0(\check{a}) = \int\limits_{\hat{K}} \varphi(\alpha) da(\alpha) \geq 0$. Obviously,

$$\left| \int_K f(x)\, d\nu(x) - \phi_0(\check{a}) \right| \leq 4\, \varphi(1)\, \delta,$$

and hence $\int\limits_K f(x)\, d\nu(x) \geq 0$. Therefore, ν and also $\tilde{\nu}$ is positive.

For $h \in C_c(\hat{K})$ it follows

$$\int_{\hat{K}} \varphi(\alpha)\, h(\alpha)\, d\pi(\alpha) = \int_K \check{h}(x)\, d\tilde{\nu}(x) = \int_{\hat{K}} \hat{\nu}(\alpha)\, h(\alpha)\, d\pi(\alpha).$$

The continuity of φ and $\hat{\nu}$ implies that $\varphi = \hat{\nu}$ on $\mathrm{supp}\ \pi = \mathcal{S}(K)$. The uniqueness of ν follows by Theorem 3.1.8. ◇

Corollary 3.8.3 *(1) If $\varphi \in SP(\hat{K})$, then $|\varphi(\alpha)| \leq \varphi(1)$ for each $\alpha \in \mathcal{S}(K)$.*

(2) Assume that $\mathcal{S}(K) = \hat{K}$. If $(\varphi_n)_{n\in\mathbb{N}}$ is a sequence of functions in $SP(\hat{K})$ such that φ_n converges uniformly on compact subsets of $\hat{K}$ to a continuous function φ, then φ is strongly positive definite.

Proof.

(1) follows immediately by Theorem 3.8.2.

(2) By (1) and $\varphi_n(1) \to \varphi(1)$ there exists a constant $M \geq 0$ such that $|\varphi_n(\alpha)| \leq M$ for each $n \in \mathbb{N}$ and each $\alpha \in \hat{K}$. Let $a \in M(\hat{K})$ such that $\check{a} \geq 0$, and let $\epsilon > 0$.

Choose a compact subset $C \subseteq \hat{K}$ such that $|a| \Big| \hat{K}\backslash C < \epsilon$. Then

$$\left| \int_{\hat{K}} \varphi(\alpha)\, da(\alpha) - \int_{\hat{K}} \varphi_n(\alpha)\, da(\alpha) \right| \leq 2M\epsilon + \int_C |\varphi(\alpha) - \varphi_n(\alpha)|\, d|a|(\alpha),$$

and $\int\limits_{\hat{K}} \varphi(\alpha) da(\alpha) \geq 0$ follows.

◇

Corollary 3.8.4 *Let $\varphi \in SP(\hat{K}) \cap L^1(\hat{K}, \pi)$. Then $\check{\varphi}$ is nonnegative, $\check{\varphi} \in L^1(K)$ and $(\check{\varphi})^\wedge(\alpha) = \varphi(\alpha)$ for $\alpha \in \mathcal{S}(K)$.*

Proof. By Theorem 3.8.2 we may write $\varphi(\alpha) = \hat{\nu}(\alpha)$ for $\alpha \in \mathcal{S}(K)$, where $\nu \in M^+(K)$. Therefore $\hat{\nu}\pi = \varphi\pi \in M(\hat{K})$, and by Theorem 3.3.6 we obtain that $\check{\varphi} \in L^1(K)$ and $\nu = \check{\varphi}m$. In particular we have $\check{\varphi} \geq 0$. Again by Theorem 3.3.6 it follows that $\varphi\pi = (\check{\varphi}m)^\wedge\pi$. Since $(\check{\varphi})^\wedge$ and φ are continuous functions, $(\check{\varphi})^\wedge(\alpha) = \varphi(\alpha)$ for all $\alpha \in \mathcal{S}(K) = \operatorname{supp} \pi$. ◇

If $\hat{K}$ is a hypergroup under pointwise multiplication, it is interesting to study the relation between strongly positive definite and bounded continuous positive definite functions on $\hat{K}$.

Proposition 3.8.5 *Assume that $\hat{K}$ is a hypergroup under pointwise multiplication. If φ is in $SP(\hat{K})$, then $\widehat{\epsilon_x}\varphi$ is in $P^b(\hat{K})$ for every $x \in K$.*

Proof. Let $c_1, ..., c_n \in \mathbb{C}$ and $\alpha_1, ..., \alpha_n \in \hat{K}$. Denote by $a = \sum\limits_{i=1}^{n} c_i \epsilon_{\alpha_i} \in M(\hat{K})$. Since $(a^*)^\vee(x) = \overline{\check{a}(x)}$, we have

$$(a * a^*)^\vee(x) = \left| \sum_{i=1}^{n} c_i \alpha_i(x) \right|^2 \geq 0.$$

Hence for $x \in K$,

$$\sum_{i=1}^{n}\sum_{j=1}^{n} c_i\bar{c_j}\, \omega(\alpha_i, \bar{\alpha_j})(\widehat{\epsilon_x}\varphi) = \int_{\hat{K}} (\widehat{\epsilon_x}\varphi)(\alpha)\, d(a * a^*)(\alpha)$$

$$= (\varphi(a * a^*))^{\vee}(\tilde{x}) \geq 0$$

by Proposition 3.8.1(1). ◇

If $\hat{K}$ is a hypergroup under pointwise multiplication, we can establish a relationship between the Pontryagin duality and properties of various types of positive definiteness.

Theorem 3.8.6 *Suppose that $\hat{K}$ is a hypergroup under pointwise multiplication. Equivalent are:*

(1) Pontryagin duality holds for K,
(2) $\hat{\hat{K}} \subseteq SP(\hat{K})$,
(3) $P^b(\hat{K}) = SP(\hat{K})$.

Proof. Note that $\mathcal{S}(K) = \hat{K}$, since $\hat{K}$ is a hypergroup. Let $i : K \to \hat{\hat{K}}$ be the embedding map. If $i(K) = \hat{\hat{K}}$, then by Theorem 3.7.1 each $\varphi \in P^b(\hat{K})$ satisfies $\varphi(\alpha) = \check{a}(\alpha)$ for each $\alpha \in \hat{K}$ for some positive measure $a \in M(\hat{\hat{K}})$. Define $\nu \in M(K)$ by $\nu(E) = a(i(E))$ for each Borel set $E \subseteq K$. Then ν is a positive measure on K and $\hat{\nu}(\alpha) = \check{a}(\alpha) = \varphi(\alpha)$ for all $\alpha \in \hat{K}$. Hence $\varphi \in SP(\hat{K})$, i.e. (1) ⇒ (3).

Now suppose that (2) is valid. Let $\sigma \in \hat{\hat{K}}$. Then $\sigma \in SP(\hat{K})$. By Theorem 3.8.2 we know that $\sigma = \hat{\nu}$ for a unique positive measure $\nu \in M(K)$. On the other hand consider the linear functional on $C_c(\hat{K})^{\vee}$,

$$h_\sigma(\check{\varphi}) = \int_{\hat{K}} \sigma(\alpha)\, \varphi(\alpha)\, d\pi(\alpha) = \hat{\varphi}(\bar{\sigma}),$$

where $\varphi \in C_c(\hat{K})$. For $\varphi, \psi \in C_c(\hat{K})$ we have $(\varphi * \psi)^{\wedge}(\bar{\sigma}) = \hat{\varphi}(\bar{\sigma})\hat{\psi}(\bar{\sigma}) = h_\sigma(\check{\varphi})h_\sigma(\check{\psi})$, and $(\varphi * \psi)^{\vee}(x) = \check{\varphi}(x)\check{\psi}(x)$ for all $x \in K$. Hence

$$h_\sigma(\check{\varphi} \cdot \check{\psi}) = h_\sigma((\varphi * \psi)^{\vee}) = (\varphi * \psi)^{\wedge}(\bar{\sigma}) = h_\sigma(\check{\varphi})\, h_\sigma(\check{\psi}),$$

i.e. h_σ is a multiplicative functional on $C_c(\hat{K})^{\vee}$. Since

$$h_\sigma(\check{\varphi}) = \int_{\hat{K}} \hat{\nu}(\alpha)\, \varphi(\alpha)\, d\pi(\alpha) = \int_K \check{\varphi}(\tilde{x})\, d\nu(x),$$

we see that h_σ is continuous on $C_c(\hat{K})^{\vee}$ with respect to the sup-norm. By Proposition 3.3.2(iii) the linear functional h_σ admits a unique continuous extension to $C_0(K)$. This extension is multiplicative, too. Since the structure space $\Delta(C_0(K))$ of the Banach algebra $C_0(K)$ can be identified with K, there exists a point $x \in K$ such that $h_\sigma(\check{\varphi}) = \check{\varphi}(x)$ for every $\varphi \in C_c(\hat{K})$. It follows $\hat{\varphi}(\bar{\sigma}) = \check{\varphi}(x) = \hat{\varphi}(\overline{i(x)})$ for every $\varphi \in C_c(\hat{K})$. Therefore $\sigma = i(x)$ and (2) $\Rightarrow$ (1) is shown. The implication (3) $\Rightarrow$ (2) is obviously true. ◇

Combining Proposition 3.4.13 and Theorem 3.8.6 we get the following.

Corollary 3.8.7 *Let K be discrete and assume that $\hat{K}$ is a hypergroup under pointwise multiplication. Then $SP(\hat{K}) = P^b(\hat{K})$.*

Assuming that K is compact, which is more or less dual to the assumption of Corollary 3.8.7, we derive another relation between $SP(\hat{K})$ and $P^b(\hat{K})$. Whenever K is compact we normalize the Haar measure such that $m(K) = 1$. Denote by $T(K)$ the linear span of $\hat{K}$. The proofs of the following results are similar to that of [305,(30.2)].

Lemma 3.8.8 *Let K be compact and assume that $\hat{K}$ is a hypergroup under pointwise multiplication. Then $T(K)$ is a translation-invariant linear subspace of $C(K)$. Further $T(K)$ is closed under pointwise multiplication and under complex conjugation. $T(K)$ is sup-norm dense in $C(K)$.*

Proof. We show that $T(K)$ is closed under multiplication. Let

$$f = \sum_{i=1}^{n} \lambda_i \alpha_i, \quad g = \sum_{j=1}^{m} \mu_j \beta_j; \qquad \lambda_i, \mu_j \in \mathbb{C}; \;\; \alpha_i, \beta_j \in \hat{K}.$$

Since $\hat{K}$ is a discrete hypergroup, the convolution measures $\omega(\alpha_i, \beta_j) \in M^1(\hat{K})$ are convex combinations of finitely many point measures ϵ_γ, $\gamma \in \hat{K}$, and so $f \cdot g$ is a linear combination of finitely many $\gamma \in \hat{K}$. Since $T(K) = (C_c(\hat{K}))^\vee$, $T(K)$ is sup-norm dense in $C(K)$, see Proposition 3.3.2. The other properties are easily verified. ◇

Lemma 3.8.9 *Let K be compact and assume that $\hat{K}$ is a hypergroup under pointwise multiplication. Let φ be a bounded function on $\hat{K}$, such that $\widehat{\epsilon_x}\varphi \in P^b(\hat{K})$ for each $x \in K$. For $f = \sum_{i=1}^{n} \lambda_i \alpha_i \in T(K)$ define*

$$p_\varphi(f) := \sum_{i=1}^{n} \lambda_i \varphi(\alpha_i).$$

Then $p_\varphi : T(K) \to \mathbb{C}$ is a linear functional, which satisfies

$$p_\varphi(L_x(f\bar{f})) \geq 0$$

for each $f \in T(K)$ and $x \in K$.

Proof. p_φ is a well-defined mapping, since $\hat{K}$ is a linear independent set in $T(K)$. Note that $\hat{K}$ is an orthogonal subset of $L^2(K)$. Further p_φ is obviously linear. Let $f = \sum_{i=1}^{n} \lambda_i \alpha_i \in T(K)$, and write $\alpha_i \bar{\alpha}_j = \sum_{k=1}^{m_{ij}} b_k^{i,j} \gamma_k^{i,j}$, where $b_k^{i,j} \geq 0$, $\gamma_k^{i,j} \in \hat{K}$. Then

$$L_x(f\bar{f}) = \sum_{i,j=1}^{n} \lambda_i \overline{\lambda_j} \sum_{k=1}^{m_{ij}} b_k^{i,j} \gamma_k^{i,j}(x) \gamma_k^{i,j} \in T(K).$$

Hence

$$p_\varphi(L_x(f\bar{f})) = \sum_{i,j=1}^{n} \lambda_i \overline{\lambda_j} \sum_{k=1}^{m_{ij}} b_k^{i,j} \gamma_k^{i,j}(x) \varphi(\gamma_k^{i,j})$$

$$= \sum_{i,j=1}^{m_{ij}} \lambda_i \overline{\lambda_j} \, \omega(\alpha_i, \overline{\alpha_j})(\widehat{\epsilon_{\tilde{x}}} \varphi) \geq 0,$$

since $\widehat{\epsilon_{\tilde{x}}} \varphi$ is positive definite. ◇

Lemma 3.8.10 *Let K be compact and assume that $\hat{K}$ is a hypergroup under pointwise multiplication. Let $\varphi \in l^\infty(\hat{K})$ such that $\widehat{\epsilon_x} \varphi \in P^b(\hat{K})$ for every $x \in K$. Consider some $h \in T(K)$ such that $h \geq 0$, and define for $f \in T(K)$,*

$$p_{\varphi,h}(f) := \int_K p_\varphi(L_{\tilde{x}} f) \, h(x) \, dm(x).$$

Then $p_{\varphi,h}$ is a sup-norm continuous linear functional on $T(K)$, and $p_{\varphi,h}(f) \geq 0$ for each $f \in T(K)$ satisfying $f \geq 0$.

Proof. At first note that $p_{\varphi,h}$ is a linear functional on $T(K)$. Write $\hat{K}$ as a family $(\alpha_i)_{i \in I}$ of orthogonal functions on K. For each $\alpha_i \in \hat{K}$ let $c_i := \int_K |\alpha_i(x)|^2 dm(x)$. Then the function $h \in T(K)$, $h \geq 0$, has the representation

$$h = \sum_{i \in I_h} \lambda_i \alpha_i,$$

where I_h is a finite subset of I, and $\lambda_i \neq 0$. For $f \in T(K)$ we have $f = \sum_{j \in I_f} \mu_j \alpha_j$, $\mu_j \neq 0$, $I_f \subseteq I$ finite. It follows

$$p_{\varphi,h}(f) = \int_K p_\varphi(L_{\tilde{x}} f) \, h(x) \, dm(x) = \sum_{j \in I_f} \mu_j \, \varphi(\alpha_j) \int_K \overline{\alpha_j(x)} \, h(x) \, dm(x)$$

$$= \sum_{j \in I_f} \sum_{i \in I_h} \mu_j \lambda_i \, \varphi(\alpha_j) \int_K \overline{\alpha_j(x)} \, \alpha_i(x) \, dm(x)$$

$$= \sum_{i \in I_f \cap I_h} \mu_i \lambda_i c_i \, \varphi(\alpha_i) \;=\; \sum_{i \in I_f \cap I_h} \mu_i \, p_{\varphi,h}(\alpha_i),$$

and

$$\sum_{i \in I_h} \frac{1}{c_i} \, p_{\varphi,h}(\alpha_i) \int_K f(x) \, \overline{\alpha_i(x)} \, dm(x)$$

$$= \sum_{i \in I_f \cap I_h} \frac{1}{c_i} \, p_{\varphi,h}(\alpha_i) \, \mu_i \int_K |\alpha_i(x)|^2 dm(x)$$

$$= \sum_{i \in I_f \cap I_h} \mu_i p_{\varphi,h}(\alpha_i).$$

Hence we obtain for each $f \in T(K)$,

$$|p_{\varphi,h}(f)| \;=\; \left| \sum_{i \in I_h} \frac{1}{c_i} \, p_{\varphi,h}(\alpha_i) \int_K f(x) \, \overline{\alpha_i(x)} \, dm(x) \right|$$

$$\leq \; M \, \|f\|_\infty$$

where

$$M = \int_K \left| \sum_{i \in I_h} \frac{1}{c_i} \, p_{\varphi,h}(\alpha_i) \, \overline{\alpha_i(x)} \right| \, dm(x).$$

Therefore $p_{\varphi,h}$ is a sup-norm continuous linear functional on $T(K)$. By Lemma 3.8.9 we know that $p_\varphi(L_x(f\bar{f})) \geq 0$ for each $f \in T(K)$ and $x \in K$. It follows that

$$p_{\varphi,h}(f\bar{f}) \geq 0$$

for every $f \in T(K)$. Since $T(K)$ is sup-norm dense in $C(K)$, we infer that $p_{\varphi,h}(f) \geq 0$ for each $f \in T(K)$ such that $f \geq 0$. ◇

Now we consider approximate identities in case that K is compact and commutative. The following proofs proceed as in the group case [305,Theorem 28.53].

Lemma 3.8.11 *Let K be compact and commutative. Let $f, g \in L^2(K)$ and $\varphi = g * f \in C(K) \subseteq L^2(K)$. Then the Plancherel expansion of φ in $L^2(\hat{K}, \pi)$,*

$$\varphi \;=\; \sum_{\alpha \in \hat{K}} \hat{\varphi}(\alpha) \, \pi(\{\alpha\}) \, \alpha$$

converges uniformly on K, and the series converges absolutely.

Proof. The set $\{\alpha \in \hat{K} : \hat{\varphi}(\alpha) \neq 0\}$ is a countable set. Let $\{\alpha_k : k \in \mathbb{N}\} = \{\alpha \in \hat{K} : \hat{\varphi}(\alpha) \neq 0\}$ be arranged in any order. For $n \in \mathbb{N}$ put

$$g_n(x) = \sum_{k=1}^{n} \hat{g}(\alpha_k)\, \alpha_k(x)\, \pi(\{\alpha_k\})$$

and

$$f_n(x) = \sum_{k=1}^{n} \hat{f}(\alpha_k)\, \alpha_k(x)\, \pi(\{\alpha_k\}).$$

Then

$$\|g * f - g_n * f_n\|_\infty \leq \|g - g_n\|_2\, \|f\|_2 + \|g_n\|_2 \|f - f_n\|_2$$

and

$$\begin{aligned} g_n * f_n(x) &= \sum_{j,k=1}^{n} \hat{g}(\alpha_k)\, \hat{f}(\alpha_j)\, \pi(\{\alpha_k\})\, \pi(\{\alpha_j\}) \int_K \alpha_k(y)\, L_{\tilde{y}}\alpha_j(x)\, dm(y) \\ &= \sum_{k=1}^{n} \hat{\varphi}(\alpha_k)\, \alpha_k(x)\, \pi(\{\alpha_k\}) \end{aligned}$$

for all $x \in K$. It follows

$$\varphi(x) = g * f(x) = \sum_{k=1}^{\infty} \hat{\varphi}(\alpha_k)\, \alpha_k(x)\, \pi(\{\alpha_k\})$$

for all $x \in K$, and the series converges uniformly on K. The series converges also absolutely because the series converges to $\varphi(x)$ independently of the enumeration of $\{\alpha \in \hat{K} : \hat{\varphi}(\alpha) \neq 0\}$. ◇

Proposition 3.8.12 *Let K be compact and commutative. Then there exists a net $(h_j)_{j\in J}$ of functions $h_j \in T(K)$, $h_j \geq 0$, $\|h_j\|_1 = 1$ for each $j \in J$, such that*

(1) $\lim_j \|h_j * f - f\|_\infty = 0$ *for all* $f \in C(K)$, *and*

(2) $\lim_j \widehat{h_j}(\alpha) = 1$ *for each* $\alpha \in \hat{K}$.

Proof. Let $(V_i)_{i\in I}$ be a basis of neighbourhoods of $e \in K$. For each $i \in I$ choose $\varphi_i \in P^b(K)$ such that supp $\varphi_i \subseteq V_i$, $\varphi_i \geq 0$, $\|\varphi_i\|_1 = 1$ just as in Proposition 2.3.3, noting that each φ_i has the form $\varphi_i = f_i * \tilde{f}_i$, $f_i \in L^2(K)$, $f_i \geq 0$. By Lemma 3.8.11 we know that

$$\varphi_i(x) = \sum_{k=1}^{\infty} \varphi_i(\alpha_k)\, \alpha_k(x)\, \pi(\{\alpha_k\}),$$

where the series converges uniformly on K and is absolutely convergent. Now, for each $i \in I$ and $n \in \mathbb{N}$ choose $m_n \in \mathbb{N}$ such that

$$\left\| \varphi_i - \sum_{k=1}^{m_n} \widehat{\varphi}_i(\alpha_k)\, \alpha_k\, \pi(\{\alpha_k\}) \right\|_\infty < \frac{1}{2n}$$

and define

$$\psi_{i,n} = \sum_{k=1}^{m_n} \widehat{\varphi}_i(\alpha_k)\, \alpha_k\, \pi(\{\alpha_k\}).$$

We have

$$\overline{\psi_{i,n}(x)} = \sum_{k=1}^{m_n} \widehat{\varphi}_i(\alpha_k)\, \alpha_k(\tilde{x})\, \pi(\{\alpha_k\}) = \psi_{i,n}(\tilde{x}).$$

Put $\kappa_{i,n} = \frac{1}{2}(\psi_{i,n} + \overline{\psi_{i,n}}) + \frac{1}{2n}$. Obviously $\kappa_{i,n} \in T(K)$. Since $\varphi_i \geq 0$ we have $\kappa_{i,n} \geq 0$. Moreover,

$$|1 - \|\kappa_{i,n}\|_1| = |\,\|\varphi_i\|_1 - \|\kappa_{i,n}\|_1| \leq \|\varphi_i - \kappa_{i,n}\|_1 < \frac{1}{n}.$$

Finally, we define

$$h_{(i,n)} := \frac{1}{\|\kappa_{i,n}\|_1}\, \kappa_{i,n}.$$

Consider the net $J := I \times \mathbb{N}$, where $(i_1, n_1) \leq (i_2, n_2)$ if and only if $V_{i_1} \subseteq V_{i_2}$ and $n_1 \leq n_2$. Then $h_j \in T(K)$, $h_j \geq 0$ and $\|h_j\|_1 = 1$, where $j = (i,n) \in J$.

Now let $\varepsilon > 0$ (and $f \in C(K)$, $f \neq 0$). By Proposition 2.3.3 there exists $i_0 \in I$ such that $\|\varphi_i * f - f\|_\infty < \frac{\varepsilon}{2}$ for each $i \geq i_0$. Further there exists $n_0 \in \mathbb{N}$ such that $\|\varphi_i - \kappa_{i,n}\|_\infty < \frac{\varepsilon}{2\|f\|_1}$ for each $i \in I$, $n \geq n_0$. Hence

$$\|\kappa_{i,n} * f - f\|_\infty \leq \|(\kappa_{i,n} - \varphi_i) * f\|_\infty + \|\varphi_i * f - f\|_\infty < \varepsilon$$

for $i \geq i_0$, $n \geq n_0$, and we obtain $\lim_j \|h_j * f - f\|_\infty = 0$, and (1) is proved.

As already shown in Proposition 3.6.7 it follows $\lim_{i \in I} \widehat{\varphi}_i(\alpha) = 1$ for each $\alpha \in \hat{K}$. Hence $\lim_j \widehat{h_j}(\alpha) = 1$ for every $\alpha \in \hat{K}$. ◇

Remark: An elementary modification of the proof of (1) above shows that
$\lim_j \|h_j * f - f\|_p = 0$ for all $f \in L^p(K)$, $1 \leq p < \infty$.

Now we can prove the following Bochner theorem.

Theorem 3.8.13 *Let K be compact, and assume that $\hat{K}$ is a hypergroup under pointwise multiplication. Let $\varphi \in l^\infty(\hat{K})$ such that $\widehat{\epsilon_x}\varphi \in P^b(\hat{K})$ for each $x \in K$. Then there exists a unique positive measure $\nu \in M(K)$ such that $\varphi = \hat{\nu}$. Conversely, $\widehat{\epsilon_x}\hat{\nu} \in P^b(\hat{K})$ for each positive measure $\nu \in M(K)$ and all $x \in K$.*

Proof. Define the linear functional p_φ on $T(K)$ as in Lemma 3.8.9 and define as in Lemma 3.8.10

$$p_{\varphi,h_j}(f) = \int_K p_\varphi(L_{\tilde{x}}f)\, h_j(x)\, dm(x),$$

where $f \in T(K)$ and $h_j \in T(K)$ are the functions of the net $(h_j)_{j\in J}$ introduced in Proposition 3.8.12. If $f \in T(K)$ has the form $f = \sum_{k=1}^{n} \mu_k \alpha_k$, $\mu_k \in \mathbb{C}$, $\alpha_k \in \hat{K}$, then

$$p_{\varphi,h_j}(f) = \sum_{k=1}^{n} \mu_k\, \varphi(\alpha_k) \int_K \overline{\alpha_k(x)}\, h_j(x)\, dm(x) = \sum_{k=1}^{n} \mu_k\, \varphi(\alpha_k) \widehat{h_j}(\alpha_k).$$

By Lemma 3.8.10 we know that $p_{\varphi,h_j}(f) \geq 0$ for each $f \in T(K)$ whenever $f \geq 0$. By Proposition 3.8.12 we get $\lim_j p_{\varphi,h_j}(f) = p_\varphi(f)$ for each $f \in T(K)$. Thus $p_\varphi(f) \geq 0$ for each $f \in T(K)$ whenever $f \geq 0$. Therefore p_φ is a continuous linear functional on $T(K)$ in the sup-norm. In fact, if $f \in T(K)$ is real-valued we have $\|f\|_\infty 1 \pm f \in T(K)$ and $\|f\|_\infty 1 \pm f \geq 0$. Thus $\|f\|_\infty p_\varphi(1) \pm p_\varphi(f) \geq 0$, which implies $p_\varphi(f)$ is real-valued and $|p_\varphi(f)| \leq p_\varphi(1)\, \|f\|_\infty$. For arbitrary $f \in T(K)$ consider Re $f \in T(K)$ and Im $f \in T(K)$, and we obtain $|p_\varphi(f)| \leq 2p_\varphi(1)\, \|f\|_\infty$. $T(K)$ is sup-norm dense in $C(K)$. Hence the sup-norm continuous functional p_φ can be uniquely extended to a sup-norm continuous functional on $C(K)$. Denote this extension still by p_φ. By statement (1) of Proposition 3.8.12 we have also $p_\varphi(f) \geq 0$ for each $f \in C(K)$, $f \geq 0$. By Riesz' theorem there exists a unique positive Borel measure $\nu \in M(K)$ such that

$$p_\varphi(f) = \int_K f(x)\, d\tilde{\nu}(x)$$

for each $f \in C(K)$. In particular

$$\varphi(\alpha) = p_\varphi(\alpha) = \int_K \overline{\alpha(x)}\, d\nu(x) = \hat{\nu}(\alpha)$$

for every $\alpha \in \hat{K}$.

The converse statement holds obviously, since $\widehat{\epsilon_x}\hat{\nu} = (\epsilon_x * \nu)^\wedge$ is positive definite, whenever $\nu \in M(K)$ is positive. ◇

Corollary 3.8.14 *Let K be compact and assume that $\hat{K}$ is a hypergroup under pointwise multiplication. Then $\varphi \in SP(\hat{K})$ if and only if $\widehat{\epsilon_x}\varphi \in P^b(\hat{K})$ for each $x \in K$.*

Corollary 3.8.15 *Let K be compact and assume that $\hat{K}$ is a hypergroup under pointwise multiplication. Then the Pontryagin duality holds for K if and only if for each $\sigma \in \hat{\hat{K}}$ holds $\widehat{\epsilon_x}\sigma \in P^b(\hat{K})$ for every $x \in K$.*

Proof. If $\sigma \in \hat{\hat{K}}$ such that $\widehat{\epsilon_x}\sigma \in P^b(\hat{K})$ for each $x \in K$, then $\sigma \in SP(\hat{K})$ by Corollary 3.8.14. If every $\sigma \in \hat{\hat{K}}$ has this property, Theorem 3.8.6 yields $i(K) = \hat{\hat{K}}$. Conversely, suppose that $i(K) = \hat{\hat{K}}$. If $\sigma \in \hat{\hat{K}}$ and $\sigma = i(y)$ for $y \in K$, then $\widehat{\epsilon_x}\sigma = \omega(x,y)^\wedge$ for each $x \in K$, i.e. $\widehat{\epsilon_x}\sigma \in P^b(\hat{K})$ for every $\sigma \in \hat{\hat{K}}$. ◇

We have studied the strong positive definiteness on $\hat{K}$. It is a useful concept to characterize the Pontryagin duality. For other definitions of positive definiteness on $\hat{K}$ we refer to [96,4.1.35]. A weak positive definiteness condition on $\mathcal{S}(K)$ will be studied in section 3.10.3.

9 Characterization of (inverse) Fourier-Stieltjes transforms

In general it is not easy to decide whether a function $\varphi \in C^b(\hat{K})$ is a Fourier-Stieltjes transform or not. For locally compact abelian groups criteria of Schoenberg or Eberlein characterize Fourier-Stieltjes transforms. These criteria can be extended to commutative hypergroups with some modifications.

Theorem 3.9.1 (Schoenberg) *Let $\varphi \in C^b(\hat{K})$. The following three conditions are equivalent:*

(i) There exists $\mu \in M(K)$ with $\hat{\mu} \,|\, \mathcal{S}(K) = \varphi \,|\, \mathcal{S}(K)$.
(ii) $\varphi\hat{g}$ equals on $\mathcal{S}(K)$ a Fourier transform for each $g \in L^1(K)$.
(iii) For some $M \geq 0$ there is satisfied

$$\left| \int_{\hat{K}} \varphi(\alpha) h(\alpha)\, d\pi(\alpha) \right| \leq M \|\check{h}\|_\infty$$

for every $h \in C_c(\hat{K})$.

Proof. Since $\hat{\mu}\hat{g} = (\mu * g)^\wedge$ and $\mu * g \in L^1(K)$ for every $g \in L^1(K)$, we have immediately that (i) implies (ii). Now assume that (ii) is true. Note that $f \in L^1(K)$ is uniquely determined by the relation

$$\varphi\hat{g} \,|\mathcal{S}(K) = \hat{f} \,|\, \mathcal{S}(K),$$

see Theorem 3.1.8. Thus $\Phi : L^1(K) \to L^1(K)$, $\Phi(g) = f$, where $f \in L^1(K)$ is determined by $\varphi\hat{g} \,|\, \mathcal{S}(K) = \hat{f} \,|\, \mathcal{S}(K)$, is a well-defined linear operator. Moreover this operator is closed, as is easily checked. The closed graph theorem implies some $M \geq 0$ such that $\|\Phi(g)\|_1 \leq M\|g\|_1$ for every $g \in L^1(K)$. Now choose a family $(k_i)_{i\in I}$, $k_i \in C_c(K)$ with $\|k_i\|_1 = 1$, $\operatorname{supp} k_i \to \{e\}$, see Proposition 2.3.3. Obviously $\widehat{k_i} \to 1$ uniformly on compact subsets of $\hat{K}$. Hence we get for each $h \in C_c(\hat{K})$,

$$\int_{\hat{K}} \varphi(\alpha)h(\alpha)\, d\pi(\alpha) = \lim_i \int_{\hat{K}} (\varphi\widehat{k_i})(\alpha)h(\alpha)\, d\pi(\alpha).$$

Writing $f_i = \Phi(k_i)$ we get

$$\int_{\hat{K}} (\varphi\widehat{k_i})(\alpha)h(\alpha)\, d\pi(\alpha) = \int_K f_i(x)\check{h}(\tilde{x})\, dm(x),$$

and then

$$\left|\int_{\hat{K}} (\varphi\widehat{k_i})(\alpha)h(\alpha)\, d\pi(\alpha)\right| \leq \|\check{h}\|_\infty \|f_i\|_1 \leq M\|\check{h}\|_\infty.$$

Therefore $\left|\int_{\hat{K}} \varphi(\alpha)h(\alpha)\, d\pi(\alpha)\right| \leq M\|\check{h}\|_\infty$ is true.

Finally assuming (iii) define a linear functional on $\{\check{h} : h \in C_c(\hat{K})\}$ by

$$\psi(\check{h}) = \int_{\hat{K}} \varphi(\alpha)h(\alpha)\, d\pi(\alpha).$$

By means of assumption (iii) this functional is continuous. It can be uniquely extended to a continuous linear functional on $C_0(K)$, see Proposition 3.3.2(iii). Riesz' representation theorem yields $\mu \in M(K)$ such that for all $h \in C_c(\hat{K})$

$$\int_{\hat{K}} \varphi(\alpha)h(\alpha)\, d\pi(\alpha) = \psi(\check{h}) = \int_K \check{h}(x)\, d\mu(x).$$

Since $\int_K \check{h}(x)\, d\mu(x) = \int_{\hat{K}} \hat{\hat{\mu}}(\alpha)h(\alpha)\, d\pi(\alpha)$, we obtain

$$\varphi \,|\, \operatorname{supp}\pi = \hat{\hat{\mu}} \,|\, \operatorname{supp}\pi.$$

◇

Theorem 3.9.2 *Let $\varphi \in C^b(K)$. The following two conditions are equivalent:*

(i) There is $a \in M(\hat{K})$ with $\check{a} = \varphi$.
(ii) For some $M \geq 0$ there is satisfied

$$\left| \int\limits_K \varphi(x)h(x)\,dm(x) \right| \leq M\|\hat{h}\|_\infty$$

for every $h \in C_c(K)$.

These two conditions hold true, if (iii) is fulfilled, where

(iii) $1 \in \mathcal{S}(K)$ and $\varphi\check{g}$ equals an inverse Fourier transform for each $g \in L^1(\hat{K}, \pi)$.

Proof. Obviously condition (i) implies (ii). Assuming (ii) define a continuous linear functional on $\{\hat{h} : h \in C_c(K)\}$ by

$$\psi(\hat{h}) = \int\limits_K \varphi(x)h(x)\,dm(x).$$

Extending ψ uniquely to $C_0(\hat{K})$ we get $a \in M(\hat{K})$ such that

$$\int\limits_K \varphi(x)h(x)\,dm(x) = \psi(\hat{h}) = \int\limits_{\hat{K}} \hat{h}(\alpha)\,da(\alpha).$$

Putting $\tilde{a}(E) = a(\bar{E})$, where $\bar{E} = \{\bar{\alpha} : \alpha \in E\}$ for $E \subseteq \hat{K}$, we obtain

$$\int\limits_{\hat{K}} \hat{h}(\alpha)\,da(\alpha) = \int\limits_{\hat{K}} \hat{h}(\bar{\alpha})\,d\tilde{a}(\alpha) = \int\limits_K (\tilde{a})^\vee(x)h(x)\,dm(x),$$

i. e. $\varphi = (\tilde{a})^\vee$.

The implication (iii) $\Longrightarrow$ (ii) is shown analoguous as in Theorem 3.9.1. The existence of a family $(k_i)_{i\in I}$, $k_i \in L^1(\hat{K}, \pi)$ with $\|k_i\|_1 = 1$, $\check{k}_i(x) \to 1$ for every $x \in K$ is guaranteed by $1 \in \mathcal{S}(K)$. In fact choose compact neighbourhoods V_i of $1 \in \mathcal{S}(K)$ with $V_i \to \{1\}$, and put $k_i = \chi_{V_i}/\pi(V_i)$. $\diamond$

Remark: If we suppose in Theorem 3.9.2 that $\hat{K}$ is a hypergroup under pointwise multiplication, then condition (iii) is equivalent to (ii) and (i). In fact, we have $\check{a}\check{g} = (a*g)^\vee$ and $a*g \in L^1(\hat{K}, \pi)$ for every $g \in L^1(\hat{K}, \pi)$ and $a \in M(\hat{K})$, i.e. (i) implies (iii).

For locally compact abelian groups Eberlein [202] has given another characterization of Fourier-Stieltjes transforms. A generalization of Eberlein's result to commutative hypergroups is much more involved.

Proposition 3.9.3 *Let $(\varphi_i)_{i\in I}$ be a net of functions $\varphi_i \in C^b(K)$ such that $\varphi_i = \check{a}_i$, where $a_i \in M(\hat{K})$ for every $i \in I$ such that $\sup\limits_i \|a_i\| \le M < \infty$. If $(\varphi_i)_{i\in I}$ converges in the weak $*$-topology to a function $\varphi \in C^b(K)$, then $\varphi = \check{a}$, where $a \in M(\hat{K})$ such that $\|a\| \le M$.*

Proof. Let $h \in L^1(K)$ and define the linear functional σ on $L^1(K)^\wedge \subseteq C_0(\hat{K})$ by

$$\sigma(\hat{h}) := \int_K \varphi(x)\, h(x)\, dm(x) \;=\; \lim_i \int_K \varphi_i(x)\, h(x)\, dm(x)$$

$$= \lim_i \int_{\hat{K}} \hat{h}(\bar{\alpha})\, da_i(\alpha).$$

Hence $|\sigma(\hat{h})| \le M\, \|\hat{h}\|_\infty$. Extending σ to $C_0(\hat{K})$ we get some $a \in M(\hat{K})$ such that $\sigma(\hat{h}) = \int\limits_{\hat{K}} \hat{h}(\alpha) da(\alpha)$, and we obtain

$$\int_{\hat{K}} \hat{h}(\alpha)\, da(\alpha) \;=\; \int_{\hat{K}} \hat{h}(\bar{\alpha})\, d\tilde{a}(\alpha) \;=\; \int_K (\tilde{a})^\vee(x)\, h(x)\, dm(x),$$

hence $\varphi = (\tilde{a})^\vee$. Clearly, $\|a\| \le M$. ◇

Now we construct an approximate identity $(u_j)_{j\in I}$ of $L^1(K)$ such that all $\widehat{u_j}$ have compact support. We have to suppose that $\hat{K}$ is a hypergroup under pointwise multiplication.

Lemma 3.9.4 *Let $\hat{K}$ be a hypergroup under pointwise multiplication. Let $f \in L^1(K)$,*
$\|f\|_1 = 1$, $0 < \varepsilon < 1$. There exists $u \in L^1(K)$ such that $\hat{u} \in C_c(\hat{K})$ and $\|u * f - f\|_1 < \varepsilon$ and $1 - \varepsilon \le \|u\|_1 \le 1 + \varepsilon$.

Proof. By Proposition 2.3.3 there exists $\varphi \in C_c(K)$ such that $\varphi \ge 0$, $\|\varphi\|_1 = 1$ and $\|\varphi * f - f\|_1 < \frac{\varepsilon}{2}$. Consider

$$E := \{h \in L^2(K) :\ \mathcal{P}(h) \in C_c(\hat{K})\}.$$

E is a dense subspace of $L^2(K)$. This follows directly, sincd $\mathcal{P}$ is an isometric isomorphism from $L^2(K)$ onto $L^2(\hat{K},\pi)$, and $C_c(\hat{K})$ is dense in $L^2(\hat{K},\pi)$. Put $\psi \equiv \varphi^{1/2} \in C_c(K)$, and choose $\delta > 0$ such that $\delta(2+\delta) < \frac{\varepsilon}{2}$. Let $h \in E$ such that $\|h - \psi\|_2 < \delta$, and put $u = h^2 \in L^1(K)$. Then $\|\psi\|_2^2 = \|\varphi\|_1 = 1$ and

$$\|\varphi - h^2\|_1 \;=\; \|\psi^2 - h^2\|_1 \;\le\; \|(\psi - h)\psi\|_1 \;+\; \|h(\psi - h)\|_1$$

$$\le\; \|\psi - h\|_2\, \|\psi\|_2 \;+\; \|\psi - h\|_2\, \|h\|_2 \;<\; \delta(2+\delta) \;<\; \frac{\varepsilon}{2}.$$

In particular, $1-\varepsilon \leq \|u\|_1 \leq 1+\varepsilon$. Further,

$$\|u * f - f\|_1 \leq \|\varphi * f - f\|_1 + \|(\varphi - h^2) * f\|_1 < \varepsilon.$$

Finally we prove $\hat{u} \in C_c(\hat{K})$. It is easy to show that $\mathcal{P}(h\bar{\alpha}) = L_\alpha(\mathcal{P}(h))$ for each $\alpha \in \hat{K}$, and $\mathcal{P}(\bar{h})(\beta) = \mathcal{P}(h)(\tilde{\beta})$ for all $\beta \in \hat{K}$. Using Parseval's formula and Proposition 2.2.14 we obtain for $\alpha \in \hat{K}$,

$$\hat{u}(\alpha) = \int_K (h(x)\,\overline{\alpha(x)})\,h(x)\,dm(x) = \int_{\hat{K}} \mathcal{P}(h\bar{\alpha})(\beta)\,\overline{\mathcal{P}(\bar{h})(\beta)}\,d\pi(\beta)$$

$$= \int_{\hat{K}} L_\alpha(\mathcal{P}(h))(\beta)\,\mathcal{P}(h)(\tilde{\beta})\,d\pi(\beta) = \mathcal{P}(h) * \mathcal{P}(h)(\alpha).$$

Since $C_c(\hat{K}) * C_c(\hat{K}) \subseteq C_c(\hat{K})$, it follows $\hat{u} \in C_c(\hat{K})$. ◇

Theorem 3.9.5 *Let $\hat{K}$ be a hypergroup under pointwise multiplication. Then there exists a net $(v_j)_{j\in J}$ of functions $v_j \in L^1(K)$, $\|v_j\|_1 = 1$ and $\widehat{v_j} \in C_c(\hat{K})$ and*

$$\lim_j \|v_j * f - f\|_1 = 0 \qquad \textit{for each } f \in L^1(K).$$

Proof. Let I be the net of Proposition 2.3.3. Direct the net

$$J = \{(i,\varepsilon) : i \in I,\ 0 < \varepsilon < 1\}$$

by $(i_1,\varepsilon_1) \geq (i_2,\varepsilon_2)$ whenever $i_1 \geq i_2$ and $\varepsilon_1 \leq \varepsilon_2$. Consider the functions φ_i of Proposition 2.3.3 and $0 < \varepsilon < 1$. Put $j = (i,\varepsilon) \in J$ and let $f \in L^1(K)$, $f \neq 0$. Use Lemma 3.9.4 to select functions $u_j \in L^1(K)$ such that $\widehat{u_j} \in C_c(\hat{K})$, $1-\varepsilon \leq \|u_j\|_1 \leq 1+\varepsilon$ and $\|u_j * f - f\|_1 < \varepsilon$. Some routine estimates show that the functions $v_j = u_j/\|u_j\|_1$ have the stated properties. ◇

Changing the roles of K and $\hat{K}$ and applying the inverse Fourier transform and the inverse Plancherel transform $\mathcal{P}^{-1}$, the proof of Lemma 3.9.4 shows the following 'dual' result.

Lemma 3.9.6 *Let $\hat{K}$ be a hypergroup under pointwise multiplication. Let $g \in L^1(\hat{K},\pi)$, $\|g\|_1 = 1$, $0 < \varepsilon < 1$. There exists $u \in L^1(\hat{K},\pi)$ such that $\check{u} \in C_c(K)$, $\|u * g - g\|_1 < \varepsilon$ and $1-\varepsilon \leq \|u\|_1 \leq 1+\varepsilon$.*

Consequently we can infer, similar as in Theorem 3.9.5,

Theorem 3.9.7 *Let $\hat{K}$ be a hypergroup under pointwise multiplication. Then there exists a net $(w_i)_{i\in I}$ of functions $w_i \in L^1(\hat{K},\pi)$, $\|w_i\|_1 = 1$ and $\check{w}_i \in C_c(K)$ and $\lim_i \|w_i * g - g\|_1 = 0$ for each $g \in L^1(\hat{K},\pi)$.*

Remark: Note that we do not suppose that the Pontryagin duality holds for K.

Having constructed these approximate identities on $L^1(K)$ and $L^1(\hat{K},\pi)$, respectively, we can generalize Eberlein's theorem to commutative hypergroups K, such that $\hat{K}$ is a hypergroup under pointwise multiplication. Eberlein's condition has to be modified.

Let $\varphi \in C^b(K)$. We say that φ satisfies the **MTE-condition on** K with constant $M \geq 0$, if

$$\left|\sum_{k=1}^{n} \lambda_k L_{\tilde{y}}(\beta\varphi)(x_k)\right| \leq M \sup_{\alpha\in\mathcal{S}(K)} \left|\sum_{k=1}^{n} \lambda_k \alpha(x_k)\right|$$

for each $n \in \mathbb{N}$, $\lambda_1, ..., \lambda_n \in \mathbb{C}$, $x_1, ..., x_n \in K$, $y \in K$, $\beta \in \hat{K}$. (In brief, we write MTE-condition for multiplication-translation-Eberlein condition.)

Proposition 3.9.8 *Let $\varphi \in C^b(K)$. Further let $f \in L^1(K)$ such that $\|f\|_1 \leq 1$, and $g \in L^1(\hat{K},\pi)$ such that $\|g\|_1 \leq 1$. If φ satisfies the MTE-condition on K with constant $M \geq 0$, then $\psi = f * (\check{g}\varphi)$ satisfies*

$$\left|\sum_{k=1}^{n} \lambda_k \psi(x_k)\right| \leq M \sup_{\alpha\in\mathcal{S}(K)} \left|\sum_{k=1}^{n} \lambda_k \alpha(x_k)\right|$$

for each $n \in \mathbb{N}$, $\lambda_1, ..., \lambda_n \in \mathbb{C}$, $x_1, ..., x_n \in K$.

Proof. We obtain

$$\left|\sum_{k=1}^{n} \lambda_k \psi(x_k)\right| = \left|\sum_{k=1}^{n} \lambda_k \int_K f(y)\, L_{\tilde{y}}(\check{g}\varphi)(x_k)\, dm(y)\right|$$

$$= \left|\sum_{k=1}^{n} \lambda_k \int_K f(y) \int_{\hat{K}} g(\beta) \int_K \beta(z)\, \varphi(z)\, d\omega(\tilde{y}, x_k)(z)\, d\pi(\beta)\, dm(y)\right|$$

$$\leq \int_K |f(y)| \int_{\hat{K}} |g(\beta)| \left|\sum_{k=1}^{n} \lambda_k L_{\tilde{y}}(\beta \cdot \varphi)(x_k)\right| d\pi(\beta)\, dm(y)$$

$$\leq M \sup_{\alpha\in\mathcal{S}(K)} \left|\sum_{k=1}^{n} \lambda_k \alpha(x_k)\right|.$$

◇

Proposition 3.9.9 *Let $\hat{K}$ be a hypergroup under pointwise multiplication, and let $\varphi \in C^b(K)$. Let $(f_j)_{j\in J}$ be an approximate identity of $L^1(K)$ such that $\widehat{f_j} \in C_c(\hat{K})$, $\|f_j\|_1 = 1$, and $(g_i)_{i\in I}$ be an approximate identity of $L^1(\hat{K},\pi)$ such that $\check{g}_i \in C_c(K)$, $\|g_i\|_1 = 1$. Then* *(a)* $f_j * (\check{g}_i\varphi) \in L^1(K) \cap C^b(K), \quad (f_j * (\check{g}_i\varphi))^\wedge \in L^1(\hat{K}, \pi) \cap C^b(\hat{K})$ *and* $f_j * (\check{g}_i\varphi) = (f_j * (\check{g}_i\varphi))^{\wedge\vee}$.

(b) $\lim\limits_{i,j} f_j * (\check{g}_i\varphi) = \varphi$ *with respect to the weak* $*$*-topology on* $L^\infty(K)$.

Proof.

(a) Since $\check{g}_i\varphi \in C_c(K)$, we have $f_j * (\check{g}_i\varphi) \in L^1(K) \cap C^b(K)$. Since $\widehat{f_j} \in C_c(\hat{K})$ and $(f_j * (\check{g}_i\varphi))^\wedge = \widehat{f_j}(\check{g}_i\varphi)^\wedge$, it follows $(f_j * (\check{g}_i\varphi))^\wedge \in L^1(\hat{K}, \pi) \cap C^b(\hat{K})$. By the inversion theorem, Theorem 3.3.7, we get $f_j * (\check{g}_i\varphi)(x) = (f_j * (\check{g}_i\varphi))^{\wedge\vee}(x)$ for all $x \in K$.

(b) Let $h \in L^1(K)$. Then

$$\left| \int_K \tilde{h}(x) \, (f_j * (\check{g}_i\varphi)(x) - \varphi(x)) \, dm(x) \right| = |h * (f_j * (\check{g}_i\varphi))(e) - h * \varphi(e)|$$

$$\leq \|h * f_j - h\|_1 \, \|\check{g}_i\varphi\|_\infty + |h * (\check{g}_i\varphi)(e) - h * \varphi(e)| .$$

Since $\|g_i\|_1 = 1$ and $\lim\limits_j \|h * f_j - h\|_1 = 0$, we have to show that

$$\lim_i |h * (\check{g}_i\varphi)(e) - h * \varphi(e)| = 0.$$

Let $\varepsilon > 0$. There exists a compact subset $C \subseteq K$ such that

$$|h * (\check{g}_i\varphi)(e) - h * \varphi(e)| \leq \int_K |\tilde{h}(x)| \, |(\check{g}_i\varphi)(x) - \varphi(x)| \, dm(x)$$

$$\leq \varepsilon + \int_C |\tilde{h}(x)| \, |(\check{g}_i\varphi)(x) - \varphi(x)| \, dm(x).$$

By Proposition 3.3.2(iii) there exists $\psi \in C_c(\hat{K})$ such that $|\varphi(x) - \check{\psi}(x)| < \varepsilon$ for each $x \in C$. It follows

$$|(\check{g}_i\varphi)(x) - \varphi(x)| \leq 2\varepsilon + \left|\check{g}_i(x)\check{\psi}(x) - \check{\psi}(x)\right| \leq 2\varepsilon + \|g_i * \psi - \psi\|_1$$

for all $x \in C$. Since $(g_i)_{i \in I}$ is an approximate identity on $L^1(\hat{K}, \pi)$, we infer that

$$\lim_{i,j} f_j * (\check{g}_i\varphi) = \varphi$$

with respect to the weak $*$-topology.

◇

Proposition 3.9.10 *Let* Φ *be the linear span of* $\{\widehat{\epsilon_x} : x \in K\}$, *and let* Γ *be a compact subset of* $\hat{K}$. *Then* $\Phi|\Gamma$ *is uniformly dense in* $C(\Gamma)$.

Proof. Clearly, the uniform closure $\overline{\Phi|\Gamma}$ is a linear subspace of $C(\Gamma)$. Since $\overline{\widehat{\epsilon_x}(\alpha)} = \widehat{\epsilon_{\tilde{x}}}$, $\overline{\Phi|\Gamma}$ is closed under conjugation. Clearly $\overline{\Phi|\Gamma}$ contains the constant functions and separates points of Γ. $\overline{\Phi|\Gamma}$ is also closed under pointwise products. To prove this it suffices to show that $\widehat{\epsilon_{\tilde{x}}}\widehat{\epsilon_{\tilde{y}}}|\Gamma \in \overline{\Phi|\Gamma}$. We have for $x, y \in K$ and $E = \text{supp}\ \omega(x,y)$,

$$\widehat{\epsilon_{\tilde{x}}}(\alpha)\ \widehat{\epsilon_{\tilde{y}}}(\alpha) \ = \ \alpha(x)\alpha(y) \ = \ \int_E \alpha(z)\ d\omega(x,y)(z)$$

for all $\alpha \in \hat{K}$. Given $\varepsilon > 0$ there is a Borel-partition $E_1, ..., E_m$ of E such that

$$|\alpha(x_1) - \alpha(x_2)| \ < \ \varepsilon \qquad \text{for each } \alpha \in \Gamma,\ x_1, x_2 \in E_j,\ \ j = 1, ..., m.$$

Select $z_j \in E_j$ for each $j = 1, ..., m$. Then for $\alpha \in \Gamma$ we have

$$\left| \alpha(x)\alpha(y) \ - \sum_{j=1}^{m} \alpha(z_j)\ \omega(x,y)(E_j) \right|$$

$$= \left| \sum_{j=1}^{m} \int_{E_j} \alpha(z)\ d\omega(x,y)(z) \ - \ \sum_{j=1}^{m} \int_{E_j} \alpha(z_j)\ d\omega(x,y)(z) \right| \ < \ \varepsilon.$$

Hence we have shown that $\widehat{\epsilon_{\tilde{x}}}\widehat{\epsilon_{\tilde{y}}}|\Gamma$ belongs to $\overline{\Phi|\Gamma}$, i.e. $\overline{\Phi|\Gamma}$ is a Banach algebra. By the Stone-Weierstraß theorem we get $\overline{\Phi|\Gamma} = C(\Gamma)$. ◇

Theorem 3.9.11 *Let $\hat{K}$ be a hypergroup under pointwise multiplication, and let $\varphi \in C^b(K)$. φ satisfies the MTE-condition on K with constant $M \geq 0$ if and only if $\varphi = \check{a}$ for some $a \in M(\hat{K})$ with $\|a\| \leq M$.*

Proof. Assume that $\varphi \in C^b(K)$ satisfies the MTE-condition on K with $M \geq 0$. Denote $\varphi_{i,j} = f_j * (\check{g}_i\varphi)$, where $(f_j)_{j\in J}$ and $(g_i)_{i\in I}$ are approximate identities of $L^1(K)$ and $L^1(\hat{K}, \pi)$, respectively, according to Proposition 3.9.9. By Proposition 3.9.9(a) we know $\varphi_{i,j} = \check{a}_{i,j}$, where $a_{i,j} = \widehat{\varphi_{i,j}} \in L^1(\hat{K}, \pi) \subseteq M(\hat{K})$, and by Proposition 3.9.8 we know that

$$\left| \sum_{k=1}^{n} \lambda_k \varphi_{i,j}(x_k) \right| \ \leq \ M \left\| \sum_{k=1}^{n} \lambda_k \widehat{\epsilon_{\tilde{x}_k}} \right\|_\infty$$

for all $n \in \mathbb{N}$, $\lambda_1, ..., \lambda_n \in \mathbb{C}$, $x_1, ..., x_n \in K$ and each $i \in I$, $j \in J$. Since $a_{i,j} = \widehat{\varphi_{i,j}} = \widehat{f_j}(\check{g}_i\varphi)$, the support of $a_{i,j} \in M(\hat{K})$ is contained in the compact subset $\Gamma_j := \text{supp}\ \widehat{f_j}$ of $\hat{K}$. Hence

$$\|a_{i,j}\| \ = \ \sup\left\{ \left| \int_{\Gamma_j} \psi(\alpha)\ da_{i,j}(\alpha) \right| \ : \ \psi \in C(\Gamma_j),\ \|\psi\|_\infty \leq 1 \right\}.$$

By Proposition 3.9.10 the linear span Φ of $\{\widehat{\epsilon_x} : x \in K\}$ restricted to Γ_j is uniformly dense in $C(\Gamma_j)$. For $\psi \in \Phi|\Gamma_j$, i.e. $\psi = \sum_{k=1}^{n} \lambda_k \widehat{\epsilon_{\widetilde{x_k}}}$, we have

$$\int_K \psi(\alpha)\, da_{i,j}(\alpha) \;=\; \sum_{k=1}^{n} \lambda_k \check{a}_{i,j}(x_k) \;=\; \sum_{k=1}^{n} \lambda_k \varphi_{i,j}(x_k).$$

Therefore, we infer that $\|a_{i,j}\| \leq M$ for each $i \in I$, $j \in J$. Applying Proposition 3.9.9(b) and Proposition 3.9.3 we conclude that there is a Borel measure $a \in M(\hat{K})$ such that $\varphi = \check{a}$ and $\|a\| \leq M$.

Conversely, suppose that $\varphi = \check{a}$, where $a \in M(\hat{K})$ and $\|a\| \leq M$. Clearly, $\beta\check{a} = (\epsilon_\beta * a)^\vee$ for $\beta \in \hat{K}$. Hence

$$\sum_{k=1}^{n} \lambda_k L_{\tilde{y}}(\beta\check{a})(x_k) \;=\; \sum_{k=1}^{n} \lambda_k \int_K (\epsilon_\beta * a)^\vee(z)\, d\omega(\tilde{y}, x_k)(z)$$

$$= \sum_{k=1}^{n} \lambda_k \int_{\hat{K}} \alpha(\tilde{y})\, \alpha(x_k)\, d(\epsilon_\beta * a)(\alpha).$$

Thus,

$$\left|\sum_{k=1}^{n} \lambda_k L_{\tilde{y}}(\beta\check{a})(x_k)\right| \;=\; \left|\int_{\hat{K}} \sum_{k=1}^{n} \lambda_k \alpha(\tilde{y})\, \alpha(x_k)\, d(\epsilon_\beta * a)(\alpha)\right|$$

$$\leq \|\epsilon_\beta * a\| \sup_{\alpha \in \hat{K}} \left|\sum_{k=1}^{n} \lambda_k \alpha(\tilde{y})\, \alpha(x_k)\right| \;\leq\; M \sup_{\alpha \in \hat{K}} \left|\sum_{k=1}^{n} \lambda_k \alpha(x_k)\right|.$$

◊

Of course we have to deal with the impact of the translation by $y \in K$ and the multiplication by $\beta \in \hat{K}$ on the MTE-condition on K. We say that $\varphi \in C^b(K)$ satisfies the **E-condition on** K with constant $M \geq 0$, whenever

$$\left|\sum_{k=1}^{n} \lambda_k \varphi(x_k)\right| \;\leq\; M \sup_{\alpha \in \mathcal{S}(K)} \left|\sum_{k=1}^{n} \lambda_k \alpha(x_k)\right|$$

for $n \in \mathbb{N}$, $\lambda_1, ..., \lambda_n \in \mathbb{C}$, $x_1, ..., x_n \in K$. Clearly, the MTE-condition on K is stronger than the E-condition on K.

Proposition 3.9.12 *Let $\hat{K}$ be a hypergroup under pointwise multiplication. Suppose that $|\text{supp}\, \omega(y,x)| < \infty$ for each $x, y \in K$. If $\varphi \in C^b(K)$ satisfies the E-condition on K with constant $M \geq 0$, then φ satisfies the MTE-condition on K with constant $M \geq 0$.*

Proof. Let $y \in K$, $\beta \in \hat{K}$. Then

$$\omega(\tilde{y}, x_k) = \sum_{j=1}^{m_k} a_{k,j}\, \epsilon_{z_{k,j}},$$

where $z_{k,j} \in K$ and $a_{k,j} \geq 0$, $\sum_{j=1}^{m_k} a_{k,j} = 1$ for $k = 1, ..., n$. Hence

$$\left|\sum_{k=1}^{n} \lambda_k L_{\tilde{y}}(\beta\varphi)(x_k)\right| = \left|\sum_{k=1}^{n} \lambda_k \sum_{j=1}^{m_k} a_{k,j}\beta(z_{k,j})\, \varphi(z_{k,j})\right|$$
$$\leq M \sup_{\alpha\in\hat{K}} \left|\sum_{k=1}^{n} \lambda_k \sum_{j=1}^{m_k} a_{k,j}\beta(z_{k,j})\, \alpha(z_{k,j})\right|.$$

For each $\alpha \in \hat{K}$ we have

$$\left|\sum_{k=1}^{n} \lambda_k \sum_{j=1}^{m_k} a_{k,j}\beta(z_{k,j})\, \alpha(z_{k,j})\right| = \left|\sum_{k=1}^{n} \lambda_k \int_K (\beta\alpha)(z)\, d\omega(\tilde{y}, x_k)(z)\right|$$
$$= \left|\sum_{k=1}^{n} \lambda_k \int_K \int_{\hat{K}} \tau(z)\, d\omega(\beta,\alpha)(\tau)\, d\omega(\tilde{y}, x_k)(z)\right|$$
$$= \left|\int_{\hat{K}} \sum_{k=1}^{n} \lambda_k \tau(\tilde{y})\, \tau(x_k)\, d\omega(\beta,\alpha)(\tau)\right| \leq \sup_{\tau\in\hat{K}} \left|\sum_{k=1}^{n} \lambda_k \tau(\tilde{y})\, \tau(x_k)\right|$$
$$\leq \sup_{\tau\in\hat{K}} \left|\sum_{k=1}^{n} \lambda_k \tau(x_k)\right|.$$

Thus

$$\left|\sum_{k=1}^{n} \lambda_k L_{\tilde{y}}(\beta\varphi)(x_k)\right| \leq M \sup_{\tau\in\hat{K}} \left|\sum_{k=1}^{n} \lambda_k \tau(x_k)\right|.$$

◇

Now we will derive a 'dual' version of Theorem 3.9.11. We apply similar arguments, however we have to check that each step of the proof is valid in the 'dual' version. We recall that we will not assume that the Pontryagin duality holds for K.

Suppose that $\hat{K}$ is a hypergroup under pointwise multiplication. We say that $\varphi \in C^b(K)$ satisfies the **MTE-condition on $\hat{K}$** with constant $M \geq 0$, if

$$\left|\sum_{k=1}^{n} \lambda_k L_{\tilde{\beta}}(\widehat{\epsilon_y}\varphi)(\alpha_k)\right| \leq M \left\|\sum_{k=1}^{n} \lambda_k \alpha_k\right\|_\infty$$

for all $n \in \mathbb{N}$, $\lambda_1, ..., \lambda_n \in \mathbb{C}$, $\alpha_1, ..., \alpha_n \in \hat{K}$, $\beta \in \hat{K}$, $y \in K$.

The dual version of Proposition 3.9.8 is the following result.

Proposition 3.9.13 *Let $\hat{K}$ be a hypergroup under pointwise multiplicaion. Let $\varphi \in C^b(\hat{K})$ and $g \in L^1(\hat{K},\pi)$, $\|g\|_1 \leq 1$ and $f \in L^1(K)$, $\|f\|_1 \leq 1$. If φ satisfies the MTE-condition on $\hat{K}$ with constant $M \geq 0$, then $\psi = g * (\hat{f}\varphi)$ satisfies*

$$\left|\sum_{k=1}^{n} \lambda_k \psi(\alpha_k)\right| \leq M \left\|\sum_{k=1}^{n} \lambda_k\, \alpha_k\right\|_\infty$$

for each $n \in \mathbb{N}$, $\lambda_1, ..., \lambda_n \in \mathbb{C}$, $\alpha_1, ..., \alpha_n \in \hat{K}$.

Proof. A direct calculation shows that

$$\sum_{k=1}^{n} \lambda_k \psi(\alpha_k) = \int_{\hat{K}} g(\beta) \int_K f(y) \left(\sum_{k=1}^{n} \lambda_k L_{\tilde{\beta}}(\widehat{\epsilon_y}\varphi)(\alpha_k)\right) dm(y)\, d\pi(\beta),$$

and the statement follows. ◇

Now we will apply the approximate identities of Theorem 3.9.7 and Theorem 3.9.5.

Proposition 3.9.14 *Let $\hat{K}$ be a hypergroup under pointwise multiplication, and let $\varphi \in C^b(\hat{K})$. Let $(g_i)_{i\in I}$ be an approximate identity of $L^1(\hat{K},\pi)$ such that $\check{g}_i \in C_c(K)$, $\|g_i\|_1 = 1$, and $(f_j)_{j\in J}$ be an approximate identity of $L^1(K)$, such that $\widehat{f_j} \in C_c(\hat{K})$, $\|f_j\|_1 = 1$. Then*

(a) $g_i * (\widehat{f_j}\varphi) \in L^1(\hat{K},\pi) \cap C^b(\hat{K})$, $(g_i * (\widehat{f_j}\varphi))^\vee \in L^1(K) \cap C^b(K)$ and $g_i * (\widehat{f_j}\varphi) = (g_i * (\widehat{f_j}\varphi))^{\vee\wedge}$.

(b) $\lim_{i,j} g_i * (\widehat{f_j}\varphi) = \varphi$ with respect to the weak $*$-topology on $L^\infty(\hat{K},\pi)$.

Proof.

(a) It is obvious how to replace the arguments in the proof of Proposition 3.9.9(a) by their dual statements. The inversion theorem to be applied is Theorem 3.3.8.

(b) Here we also follow the proof of Theorem 3.9.9(b). Proposition 3.3.2(iii) has to be replaced by Theorem 3.2.1.

◇

We have to show two further dual results.

Proposition 3.9.15 *Let $(\psi_i)_{i\in I}$ be a net of functions $\psi_i \in C^b(\hat{K})$ such that $\psi_i|\mathcal{S}(K) = \widehat{\mu_i}|\mathcal{S}(K)$, where $\mu_i \in M(K)$ for every $i \in I$ such that $\sup_i \|\mu_i\| \leq M < \infty$. If $(\psi_i|\mathcal{S}(K))_{i\in I}$ converges in the weak $*$-topology to a function $\psi \in C^b(\mathcal{S}(K))$, then $\psi = \hat{\mu}|\mathcal{S}(K)$, where $\mu \in M(K)$ such that $\|\mu\| \leq M$.*

Proof. We follow the arguments of Proposition 3.9.3. Let $h \in L^1(\hat{K}, \pi)$ and define the linear functional ϑ on $L^1(\mathcal{S}(K), \pi)^\vee$ by

$$\vartheta(\check{h}) := \int_{\mathcal{S}(K)} \psi(\alpha)\, h(\alpha)\, d\pi(\alpha) = \lim_i \int_{\mathcal{S}(K)} \psi_i(\alpha)\, h(\alpha)\, d\pi(\alpha)$$
$$= \lim_i \int_K \check{h}(\tilde{x})\, d\mu_i(x).$$

Thus $|\vartheta(\check{h})| \leq M \, \|\check{h}\|_\infty$. Using Proposition 3.3.2(iii) the extension of ϑ yields a measure $\mu \in M(K)$, such that $\vartheta(\check{h}) = \int\limits_K \check{h}(x) d\mu(x)$ for each $h \in L^1(\hat{K}, \pi)$, and $\|\mu\| \leq M$. Finally

$$\int_K \check{h}(x)\, d\mu(x) = \int_{\hat{K}} \widehat{\tilde{\mu}}(\alpha)\, h(\alpha)\, d\pi(\alpha),$$

i.e. $\widehat{\tilde{\mu}}|\mathcal{S}(K) = \psi|\mathcal{S}(K)$. ◇

Proposition 3.9.16 *Let $\hat{K}$ be a hypergroup under pointwise multiplication, and let F be a compact subset of K. Let $T(K)$ be the linear span of $\hat{K}$. Then $T(K)|F$ is uniformly dense in $C(F)$.*

Proof. The proof is just like that of Proposition 3.9.10. The dual hypergroup structure on $\hat{K}$ is the essential tool. ◇

Having collected the ingredients for the dual version of Theorem 3.9.11 it is routine to prove the following result.

Theorem 3.9.17 *Let $\hat{K}$ be a hypergroup under pointwise multiplication, and let $\varphi \in C^b(\hat{K})$. φ satisfies the MTE-condition on $\hat{K}$ with constant $M \geq 0$ if and only if $\varphi = \hat{\mu}$ for some $\mu \in M(K)$ with $\|\mu\| \leq M$.*

10 Positive characters and a modified convolution

We will show that on every commutative hypergroup K there exists exactly one positive character $\gamma \in \mathcal{S}(K)$. This character replaces the constant character 1 in case 1 is not an element of $\mathcal{S}(K)$.

Let

$$K^* = \{\gamma \in C(K) : \gamma \neq 0,\ \omega(x,y)(\gamma) = \gamma(x)\gamma(y),\ \gamma(\tilde{x}) = \overline{\gamma(x)}$$
$$\text{for all } x, y \in K\}$$

Note that the functions $\gamma \in K^*$ can be unbounded. We have $\hat{K} = K^* \cap C^b(K)$. The functions $\gamma \in K^*$ are called semicharacters. The main results on positive semicharacters are due to Voit, see [691]. The functions $\gamma \in C(K)$, $\gamma \neq 0$ satisfying $\omega(x,y)(\gamma) = \gamma(x)\gamma(y)$ for all $x, y \in K$ are called exponentials, see [647]. Putting $y = e$ in the equation it follows that $\gamma(e) = 1$ for each exponential γ.

Theorem 3.10.1 *Let $\gamma \in K^*$ be a positive semicharacter on K. Then by*

$$\omega_\gamma(x,y) := \frac{1}{\gamma(x)\gamma(y)}\,\gamma\omega(x,y), \qquad x,y \in K,$$

there is defined a convolution on K such that $(K,\omega_\gamma,\tilde{},e)$ is a hypergroup. $(K,\omega_\gamma,\tilde{},e)$ is called modified hypergroup of K.

Proof. Clearly $\omega_\gamma(x,y)$ are probability measures on K, satisfying $\operatorname{supp}\omega_\gamma(x,y) = \operatorname{supp}\omega(x,y)$. To show the weak $*$-continuity of $(x,y) \mapsto \omega_\gamma(x,y)$, $K \times K \to M^1(K)$, we have to check that $(x,y) \mapsto \int_K f(z)\,d\omega_\gamma(x,y)(z)$ is continuous for each $f \in C_c(K)$, since the sup-norm closure of $C_c(K)$ is $C_0(K)$. The mappings $(x,y) \mapsto \frac{1}{\gamma(x)\gamma(y)}$ and $(x,y) \mapsto \int_K f(z)\gamma(z)\,d\omega(x,y)(z)$, where $f \in C_c(K)$, are both continuous. Hence $(x,y) \mapsto \int_K f(z)\,d\omega_\gamma(x,y)(z)$ is continuous. The convolutions of $\mu,\nu \in M(K)$ with respect to ω_γ are denoted by $\mu *_\gamma \nu$. A simple calculation shows the associativity of the convolution ω_γ. Obviously $\omega_\gamma(x,y)^\sim = \omega_\gamma(\tilde{x},\tilde{y})$, and the remaining hypergroup axioms follow from $\operatorname{supp}\omega_\gamma(x,y) = \operatorname{supp}\omega(x,y)$. $\diamond$

Throughout this subsection γ will denote a positive semicharacter on K.

Proposition 3.10.2 *Let $\mu,\nu \in M(K)$ with compact support. Then*

$$\mu *_\gamma \nu \;=\; \gamma\left(\frac{\mu}{\gamma} * \frac{\nu}{\gamma}\right).$$

Proof. At first note that $\frac{\mu}{\gamma}$, $\frac{\nu}{\gamma}$ are elements of $M(K)$. For $f \in C_c(K)$ we have

$$\mu *_\gamma \nu(f) \;=\; \int_K\int_K\int_K f(z)\,d\omega_\gamma(x,y)(z)\,d\mu(x)\,d\nu(y)$$

$$= \int_K\int_K\int_K f(z)\,\gamma(z)\,d\omega(x,y)(z)\,d\frac{\mu}{\gamma}(x)\,d\frac{\nu}{\gamma}(y) \;=\; \frac{\mu}{\gamma} * \frac{\nu}{\gamma}(\gamma f).$$

$\diamond$

Now we consider the relation between dual spaces of K and $K_\gamma := (K,\omega_\gamma,\tilde{},e)$. Denote by $(K_\gamma)^*$ the space of semicharacters on K_γ and by $(K_\gamma)^\wedge$ the space of characters on K_γ. Hence

$$(K_\gamma)^* = \{\alpha \in C(K) : \alpha \neq 0,\ \omega_\gamma(x,y)(\alpha) = \alpha(x)\alpha(y),\ \alpha(\tilde{x}) = \overline{\alpha(x)} \text{ for all } x,y \in K\}$$

and

$$(K_\gamma)^\wedge \;=\; K_\gamma^* \cap C^b(K).$$

Furthermore, let

$$K^*(\gamma) := \left\{ \alpha \in K^* : \sup_{x \in K} \frac{|\alpha(x)|}{\gamma(x)} < \infty \right\}.$$

We equip these spaces with the compact-open topology, see subsection 3.1, as well as K^* and $\hat{K}$.

Proposition 3.10.3 *The mapping $\alpha \mapsto \alpha/\gamma$ is a homeomorphism from K^* onto $(K_\gamma)^*$, and from $K^*(\gamma)$ onto $(K_\gamma)^\wedge$. In particular, $K^*(\gamma) = \{\alpha \in K^* : |\alpha(x)| \le \gamma(x)$ for all $x \in K\}$ and $K^*(\gamma)$ is a locally compact space.*

Proof. Let $\alpha \in K^*$. By the definition of ω_γ it follows directly that $\omega_\gamma(x,y)(\alpha/\gamma) = (\alpha/\gamma)(x)\,(\alpha/\gamma)(y)$. Clearly, $(\alpha/\gamma)(\tilde{x}) = \overline{(\alpha/\gamma)(x)}$. Hence $\alpha/\gamma \in (K_\gamma)^*$. If $\alpha \in K^*(\gamma)$, it follows that $\alpha/\gamma \in (K_\gamma)^\wedge$. In particular $\|\alpha/\gamma\|_\infty = 1$, and hence

$$K^*(\gamma) = \{\alpha \in K^* : |\alpha(x)| \le \gamma(x) \text{ for all } x \in K\}.$$

Since K^*, $K^*(\gamma)$ and $(K_\gamma)^*$, $(K_\gamma)^\wedge$ are equipped with the compact-open topology, both mappings $\alpha \mapsto \alpha/\gamma$ are continuous.

The same arguments show that $\beta \mapsto \beta\gamma$ is a continuous mapping from $(K_\gamma)^*$ to K^* and from $(K_\gamma)^\wedge$ to $K^*(\gamma)$. Therefore $\alpha \mapsto \alpha/\gamma$ is a homeomorphism from K^* onto $(K_\gamma)^*$ and from $K^*(\gamma)$ onto $(K_\gamma)^\wedge$. Since $(K_\gamma)^\wedge$ is a locally compact Hausdorff space, see Theorem 3.1.6, $K^*(\gamma)$ is a locally compact Hausdorff space. $\diamond$

Proposition 3.10.4 *(1) The Haar measure on K_γ is given by $m_\gamma := \gamma^2 m$.*
(2) The mapping $f \mapsto f/\gamma$, $L^2(K) \to L^2(K, m_\gamma)$ is an isometric isomorphism.

Proof.

(1) For $f \in C_c(K)$ we have by Proposition 2.2.13

$$\begin{aligned}
\int_K f(y)\,\gamma^2(y)\,dm(y) &= \int_K (f\cdot\gamma)(y)\,L_{\tilde{y}}\gamma(x)/\gamma(x)\,dm(y) \\
&= \int_K L_x(f\cdot\gamma)(y)\,\gamma(\tilde{y})/\gamma(x)\,dm(y) \\
&= \int_K \int_K \frac{f(z)\gamma(z)}{\gamma(x)\gamma(y)}\,d\omega(x,y)(z)\,\gamma^2(y)\,dm(y) \\
&= \int_K \omega_\gamma(x,y)(f)\,\gamma^2(y)\,dm(y).
\end{aligned}$$

Hence $m_\gamma = \gamma^2 m$ is the Haar measure on K_γ.

(2) follows directly by (1).

◇

At the end of subsection 2.3 the (left-) regular representation $\mu \mapsto L_\mu$ of $M(K)$ on $L^2(K)$ was introduced, $L_\mu(f) = \mu * f$ for $\mu \in M(K)$, $f \in L^2(K)$. The regular representation of $M(K)$ on $L^2(K, m_\gamma)$ is denoted by L_μ^γ, i.e. $L_\mu^\gamma(g) = \mu *_\gamma g$ for $g \in L^2(K, m_\gamma)$. If $\mu \in M(K)$, then μ/γ is in general not a bounded Borel measure on K. Nevertheless the convolution of μ/γ with $f \in L^2(K)$ yields a well-defined element of $L^2(K)$. Moreover, we have the following result.

Proposition 3.10.5 *Let $f \in L^2(K)$, $\mu \in M(K)$. Then for m-almost every $x \in K$ holds*

$$\gamma(x)\, L_\mu^\gamma(f/\gamma)(x) \;=\; \int_K L_{\tilde{y}} f(x)\, d(\mu/\gamma)(y) =: L_{\mu/\gamma}(f)(x).$$

In particular, $L_{\mu/\gamma}$ is a continuous operator on $L^2(K)$ with $\|L_{\mu/\gamma}\| = \|L_\mu^\gamma\|$.

Proof. Let $f \in L^2(K)$. Then $f/\gamma \in L^2(K, m_\gamma)$ and for m-almost all $x \in K$ we have

$$L_\mu^\gamma(f/\gamma)(x) = \mu *_\gamma (f/\gamma)(x) \;=\; \int_K \omega_\gamma(\tilde{y}, x)\, (f/\gamma)\, d\mu(y)$$

$$= \frac{1}{\gamma(x)} \int_K \frac{\omega(\tilde{y}, x)(f)}{\gamma(y)}\, d\mu(y) \;=\; \frac{1}{\gamma(x)} \int_K L_{\tilde{y}} f(x)\, d(\mu/\gamma)(y).$$

Hence

$$\gamma(x) L_\mu^\gamma(f/\gamma)(x) \;=\; \int_K L_{\tilde{y}}\, f(x)\, d\left(\frac{\mu}{\gamma}\right)(y) \quad \text{for } m\text{-almost all } x \in K,$$

and by Proposition 3.10.4 it follows that $L_{\mu/\gamma}(f)$ is an element of $L^2(K)$, and $\|L_{\mu/\gamma}(f)\|_2^2 = \|L_\mu^\gamma(f/\gamma)\|_2^2$. Clearly, $L_{\mu/\gamma}$ is a continuous operator on $L^2(K)$, and $\|L_{\mu/\gamma}\| = \|L_\mu^\gamma\|$. ◇

It is time to give an example of a class of modified hypergroups. Let $K = \mathbb{N}_0$ the polynomial hypergroup generatd by the Chebyshev polynomials of the first kind, $(T_n(x))_{n\in\mathbb{N}_0}$, see subsection 1.2.1. The convolution is given by $\omega(m, n) = \frac{1}{2}\,\epsilon_{|n-m|} + \frac{1}{2}\,\epsilon_{n+m}$. Clearly for each $x \in \mathbb{R}$ the sequence $(\gamma_x(n))_{n\in\mathbb{N}_0}$, where $\gamma_x(n) = T_n(x)$, is a semicharacter of $K = \mathbb{N}_0$. The characters are exactly the sequences γ_x, $x \in [-1, 1]$. (For general results on harmonic analysis of polynomial hypergroups we refer to chapter 4.) The positive semicharacters are the sequences $\gamma_x,\ x > 1$. Fix some $x > 1$ and abbreviate $\gamma_x = \gamma$. Then

$$\omega_\gamma(m,n) \;=\; \frac{\gamma(|n-m|)}{2\gamma(n)\gamma(m)}\,\epsilon_{|n-m|} + \frac{\gamma(n+m)}{2\gamma(n)\,\gamma(m)}\,\epsilon_{n+m}$$

for $n, m \in \mathbb{N}_0$. There exists a unique $a > 0$ such that $x = \cosh(a)$. Then $\gamma(n) = T_n(x) = \cosh(an)$ and we obtain

$$\omega_\gamma(m,n) \;=\; \frac{\cosh(a(n-m))}{2\cosh(an)\cosh(am)}\,\epsilon_{|n-m|} + \frac{\cosh(a(n+m))}{2\cosh(an)\cosh(am)}\,\epsilon_{n+m}.$$

This is the convolution of those polynomial hypergroups generated by cosh-polynomials. They are studied in section 4.5.7.3.

We denote by π_γ the Plancherel measure on $\widehat{K_\gamma}$. Recall that $(K_\gamma)^*$ is the set of all semicharacters on K_γ, and that

$$(K_\gamma)^*\left(\frac{1}{\gamma}\right) \;=\; \left\{\beta \in (K_\gamma)^* : |\beta(x)| \le \frac{1}{\gamma(x)} \text{ for all } x \in K_\gamma,\right\}.$$

Clearly, $\frac{1}{\gamma}$ is a positive semicharacter on K_γ, and $(\omega_\gamma)_{1/\gamma} = \omega$. The Fourier transform on the hypergroup K_γ is denoted by $\hat{\ }^\gamma$. Since for each $\beta \in \widehat{K_\gamma}$ there is a unique $\alpha \in K^*(\gamma)$ such that $\beta = \frac{\alpha}{\gamma}$, we obtain for each $f \in C_c(K_\gamma)$,

$$\hat{f}^\gamma(\beta) \;=\; \int_K \overline{\beta(x)}\, f(x)\, dm_\gamma(x) \;=\; \int_K \overline{\alpha(x)}\, (f\cdot\gamma)(x)\, dm(x) \;=\; (f\cdot\gamma)^\wedge(\alpha).$$

Considering the Fourier-Stieltjes transform on K_γ we have

$$\hat{\mu}^\gamma(\beta) \;=\; \int_K \overline{\alpha(x)}\, d(\mu/\gamma)(x) \;=\; (\mu/\gamma)^\wedge(\alpha)$$

for each $\alpha \in \hat{K},\ \mu \in M_c(K)$. Recall that we write $\mu \in M_c(K)$ if $\mu \in M(K)$ has compact support.

Theorem 3.10.6 *The supports of the Plancherel measures π on $\hat{K}$ and π_γ on $\widehat{K_\gamma}$ satisfy*

$$\mathcal{S}(K) \;=\; supp\ \pi \subseteq \hat{K} \cap K^*(\gamma)$$

and

$$\mathcal{S}(K_\gamma) \;=\; supp\ \pi_\gamma \subseteq \widehat{K_\gamma} \cap (K_\gamma)^*\left(\frac{1}{\gamma}\right).$$

The mapping $\phi(\alpha) := \frac{\alpha}{\gamma}$ is a homeomorphism from $\hat{K} \cap K^*(\gamma)$ onto $\widehat{K_\gamma} \cap (K_\gamma)^*\left(\frac{1}{\gamma}\right)$, and the image measure $\phi(\pi)$ is equal to π_γ.

Proof. We know that $\mathcal{S}(K) = \{\alpha \in \hat{K} : |\hat{\mu}(\alpha)| \leq \|L_\mu\|$ for all $\mu \in M_c(K)\}$, see Proposition 3.1.11. It follows that

$$\mathcal{S}(K) = \{\alpha \in K^* : |\hat{\mu}(\alpha)| \leq \|L_\mu\| \text{ for all } \mu \in M_c(K)\}.$$

(Note that $\|L_x\| \leq 1$.) Now let $\alpha \in \operatorname{supp} \pi \subseteq \hat{K}$. By Proposition 3.10.5 we infer that for each $\mu \in M_c(K)$ holds

$$|(\mu/\gamma)^\wedge(\alpha)| \leq \|L_{\mu/\gamma}\| = \|L^\gamma_\mu\|.$$

Therefore, $\left|\hat{\mu}^\gamma\left(\frac{\alpha}{\gamma}\right)\right| \leq \|L^\gamma_\mu\|$. Using the corresponding characterization of $\mathcal{S}(K_\gamma)$,

$$\mathcal{S}(K_\gamma) = \{\beta \in (K_\gamma)^* : |\hat{\mu}^\gamma(\beta)| \leq \|L^\gamma_\mu\| \text{ for all } \mu \in M_c(K)\},$$

we conclude $\frac{\alpha}{\gamma} \in \mathcal{S}(K_\gamma) \subseteq \widehat{K_\gamma}$, and by Proposition 3.10.3 we have $\alpha \in K^*(\gamma)$. Hence we have shown $\mathcal{S}(K) \subseteq \hat{K} \cap K^*(\gamma)$. Changing the roles of K and K_γ we obtain $\mathcal{S}(K_\gamma) \subseteq \widehat{K_\gamma} \cap (K_\gamma)^*(\frac{1}{\gamma})$. Moreover, by Proposition 3.10.3 we infer that $\phi : \hat{K} \cap K^*(\gamma) \to \widehat{K_\gamma} \cap (K_\gamma)^*(\frac{1}{\gamma})$, $\phi(\alpha) = \frac{\alpha}{\gamma}$, is a homeomorphism. Finally, for $f \in C_c(K)$ we get

$$\int_K |f(x)|^2\, dm_\gamma(x) = \int_K |\gamma(x)\, f(x)|^2\, dm(x) = \int_{\hat{K}} |\widehat{\gamma \cdot f}(\alpha)|^2\, d\pi(\alpha)$$

$$= \int_{\hat{K}} \left|\hat{f}^\gamma\left(\frac{\alpha}{\gamma}\right)\right|^2 d\pi(\alpha) = \int_{\widehat{K_\gamma}} |\hat{f}^\gamma(\beta)|^2\, d\phi(\pi)(\beta).$$

Hence π_γ is the image measure $\phi(\pi)$. ◇

Corollary 3.10.7 *Let $\gamma \in K^*$ be a positive semicharacter on K, and $\alpha \in \mathcal{S}(K)$. Then $|\alpha(x)| \leq \gamma(x)$ for every $x \in K$. In particular, there exists at most one positive character in $\mathcal{S}(K)$.*

Proof. Since $\frac{\alpha}{\gamma} \in \mathcal{S}(K_\gamma) \subseteq \widehat{K_\gamma}$, we have $\frac{|\alpha(x)|}{\gamma(x)} \leq 1$ for all $x \in K$. In particular, if $\beta_1, \beta_2 \in \mathcal{S}(K)$ are positive characters on K, then $\beta_1(x) \leq \beta_2(x) \leq \beta_1(x)$ for all $x \in K$. ◇

Applying Theorem 3.10.6 we get an improvement of Proposition 3.1.2.

Corollary 3.10.8 *Assume $1 \in \mathcal{S}(K)$. Then for each $\alpha \in \hat{K}$, $\alpha \neq 1$, there exists an $x \in K$ such that Re $\alpha(x) < 0$. In particular, if $1 \in \mathcal{S}(K)$, then 1 is the only positive character of $\hat{K}$.*

Proof. First suppose that there is an $x \in K$ such that $\alpha(x) \in \mathbb{C}\backslash\mathbb{R}$. Then there exists some $n \in \mathbb{N}$ such that $\mathrm{Re}\,(\epsilon_x * ... * \epsilon_x)(\alpha) = \mathrm{Re}\,(\alpha(x)^n) < 0$. Since $\epsilon_x * ... * \epsilon_x \in M^1(K)$ there exists $y \in \mathrm{supp}\,(\epsilon_x * ... * \epsilon_x)$ such that $\mathrm{Re}\,\alpha(y) < 0$. Hence it remains to consider the case that $\alpha \in \hat{K}$ is nonnegative. In that case α is a positive character. Indeed, if $\alpha(x) = 0$ we obtain $\omega(x, \tilde{x})(\alpha) = 0$, and from the nonnegativity of α it follows $\alpha(y) = 0$ for each $y \in \mathrm{supp}\,\omega(x, \tilde{x})$. In particular, $\alpha(e) = 0$ contradicting $\alpha \in \hat{K}$. By Corollary 3.10.7 we conclude that $\alpha(x) \geq 1$ for all $x \in K$, i.e. $\alpha = 1$. ◇

Corollary 3.10.9 *If K_γ is the modified hypergroup of K with respect to a positive character $\gamma \in \mathcal{S}(K)$, then $1 \in \mathcal{S}(K_\gamma)$.*

Proof. We only have to recall that π_γ is the image measure $\phi(\pi)$, where ϕ is the homeomorphism from $\mathcal{S}(K)$ onto $\mathcal{S}(K_\gamma)$, $\phi(\alpha) = \frac{\alpha}{\gamma}$. ◇

In Corollary 3.10.7 it is shown that there exists at most one positive character in $\mathcal{S}(K)$. The existence of a positive character is proven in [691,Theorem 2.11].

Theorem 3.10.10 *There exists exactly one positive character α_0 in $\mathcal{S}(K)$, and α_0 is isolated in $\mathcal{S}(K)$ if and only if K is compact.*

For the proof of this we refer to [691, Lemma 2.12 and Theorem 2.11].

11 Duality results based on $\mathcal{S}(K)$

We already know that in general $\mathcal{S}(K)$ or $\hat{K}$ is not equipped with hypergroup structure with respect to pointwise multiplication. Even weaker conditions, such as property (P), see subsection 3.4. may not be satisfied. Nevertheless, without any assumption we are able to construct a translation on $L^2(\mathcal{S}(K), \pi)$. This translation is the basis for further results for general commutative hypergroups.

11.1 *Translation on L^2 and C_0*

We will use the Plancherel isomorphism $\mathcal{P} : L^2(K) \to L^2(\mathcal{S}(K), \pi)$ to define for every $a \in L^\infty(K)$ a convolution operator $M_a \in \mathcal{B}(L^2(\mathcal{S}(K), \pi)$ by

$$M_a(\varphi) := \mathcal{P}(\bar{a}\mathcal{P}^{-1}(\varphi))$$

for every $\varphi \in L^2(\mathcal{S}(K), \pi)$.

M_a is a linear operator and bounded, since $\|M_a(\varphi)\| \leq \|a\|_\infty \|\varphi\|_2$. If $a = 1$ then $M_a = \mathrm{id}$. For proving the following properties of Proposition

3.11.1 below, one can consult the concept of multiplication operators in [185,Example 4.20]. But note that multiplication is performed by elements of $L^\infty(K)$ and not by elements of $L^\infty(\mathcal{S}(K),\pi)$, and note that m and π are not probability measures.

Proposition 3.11.1 *If $a,b \in L^\infty(K)$ then*

$$M_{ab} = M_a \circ M_b \quad \text{and } M_{\bar{a}} = (M_a)^* \quad \text{and } \|M_a\| = \|a\|_\infty.$$

Furthermore $M_a = 0$ if and only if $a = 0$.

Proof. The statements $M_{ab} = M_a \circ M_b$ and $(M_a)^* = M_{\bar{a}}$ are easily shown. Since $\mathcal{P}$ is isometric we get $\|M_a\| = \|a\|_\infty$ from [185,p.88]. Finally, if M_a is the zero-operator, $a\mathcal{P}^{-1}(\varphi)$ is the zero-function in $L^2(K)$ for all $\varphi \in L^2(\mathcal{S}(K),\pi)$, and hence $a = 0$. ◇

Now we restrict the mapping $a \to M_a$, $L^\infty(K) \to \mathcal{B}(L^2(\mathcal{S}(K),\pi))$ to $\mathcal{S}(K) \subseteq L^\infty(K)$, and derive a modul action of $L^1(\mathcal{S}(K),\pi)$ on $L^2(\mathcal{S}(K),\pi)$. A first step is to prove the continuity of $\alpha \mapsto M_\alpha(\varphi)$, $\mathcal{X}^b(K) \to L^2(\mathcal{S}(K),\pi)$ for every $\varphi \in L^2(\mathcal{S}(K),\pi)$.

Lemma 3.11.2 *Let $\varphi \in L^2(\mathcal{S}(K),\pi)$. The mapping $\alpha \mapsto M_\alpha(\varphi)$, $\mathcal{X}^b(K) \to L^2(\mathcal{S}(K),\pi)$ is continuous.*

Proof.

Let $\alpha_0 \in \mathcal{X}^b(K)$, $\varepsilon > 0$. Since $\mathcal{P}^{-1}(\varphi) \in L^2(K)$ there is a compact set $C \subseteq K$ such that

$$\int_{K\setminus C} |\mathcal{P}^{-1}(\varphi)(z)|^2 \, dm(z) < \varepsilon/8.$$

Let $M := \int_C |\mathcal{P}^{-1}(\varphi)(z)|^2 \, dm(z)$ and

$$V(\alpha_0) = \{\alpha \in \mathcal{X}^b(K) : |\alpha(z) - \alpha_0(z)|^2 < \varepsilon/(2M) \text{ for } z \in C\}.$$

Then

$$\begin{aligned}
\|M_\alpha(\varphi) - M_{\alpha_0}(\varphi)\|_2^2 &= \|\bar{\alpha}\mathcal{P}^{-1}(\varphi) - \bar{\alpha}_0\mathcal{P}^{-1}(\varphi)\|_2^2 \\
&= \int_C |\alpha(z) - \alpha_0(z)|^2 \, |\mathcal{P}^{-1}(\varphi)(z)|^2 \, dm(z) \\
&\quad + \int_{K\setminus C} |\alpha(z) - \alpha_0(z)|^2 \, |\mathcal{P}^{-1}(\varphi)(z)|^2 \, dm(z) \\
&< \varepsilon/2 + \varepsilon/2 = \varepsilon,
\end{aligned}$$

if $\alpha \in V(\alpha_0)$. ◇

There are two ways to introduce an action of $L^1(\mathcal{S}(K),\pi)$ on $L^2(\mathcal{S}(K),\pi)$, see (1) and (2) below.

Given $g \in C_c(\mathcal{S}(K))$ and $\varphi \in L^2(\mathcal{S}(K),\pi)$ we use a $L^2(\mathcal{S}(K),\pi)$-valued integral, to define

$$g * \varphi := \int_{\mathcal{S}(K)} g(\alpha) M_{\bar{\alpha}}(\varphi)\, d\pi(\alpha) \ \in L^2(\mathcal{S}(K),\pi).$$

Since $\|M_{\bar{\alpha}}(\varphi)\|_2 \le \|\varphi\|_2$ for each $\alpha \in \mathcal{S}(K)$, we have

$$\|g * \varphi\|_2 \le \int_{\mathcal{S}(K)} |g(\alpha)|\ \|M_{\bar{\alpha}}(\varphi)\|_2\, d\pi(\alpha) \ \le \ \|g\|_1\ \|\varphi\|_2.$$

If $f \in L^1(\mathcal{S}(K),\pi)$ choose a sequence $(g_n)_{n\in\mathbb{N}}$ with $g_n \in C_c(\mathcal{S}(K))$ and $\|f - g_n\|_1 \to 0$ as $n \to \infty$. Now it is easily shown that

$$f * \varphi := \lim_{n\to\infty} g_n * \varphi \ \in L^2(\mathcal{S}(K),\pi) \tag{1}$$

is a well-defined action of $L^1(\mathcal{S}(K),\pi)$ on $L^2(\mathcal{S}(K),\pi)$ with $\|f * \varphi\|_2 \le \|f\|_1\ \|\varphi\|_2$. Another representation of this very weak convolution is: For $f \in L^1(\mathcal{S}(K),\pi)$, $\varphi \in L^2(\mathcal{S}(K),\pi)$ holds

$$f * \varphi \ = \ M_{\overline{\check{f}}}(\varphi) \tag{2}$$

In fact for $g \in C_c(\mathcal{S}(K))$ and $\psi \in L^2(\mathcal{S}(K),\pi)$ we obtain

$$\langle g * \varphi, \psi\rangle \ = \ \int_{\mathcal{S}(K)} g(\alpha)\ \langle M_{\bar{\alpha}}(\varphi), \psi\rangle\ d\pi(\alpha)$$

$$= \int_{\mathcal{S}(K)} g(\alpha) \int_K \alpha(z)\, \mathcal{P}^{-1}(\varphi)(z)\, \overline{\mathcal{P}^{-1}(\psi)(z)}\, dm(z)\, d\pi(\alpha)$$

$$= \int_K \check{g}(z)\, \mathcal{P}^{-1}(\varphi)(z)\, \overline{\mathcal{P}^{-1}(\psi)(z)}\, dm(z) \ = \ \langle M_{\overline{\check{g}}}(\varphi), \psi\rangle.$$

Thus $g * \varphi = M_{\overline{\check{g}}}(\varphi)$ if $g \in C_c(\mathcal{S}(K))$. If $f \in L^1(\mathcal{S}(K),\pi)$ choose again a sequence $(g_n)_{n\in\mathbb{N}}$ with $g_n \in C_c(\mathcal{S}(K))$ and $\|f - g_n\|_1 \to 0$ as $n \to \infty$. Since for $n \to \infty$

$$\|f * \varphi \ - M_{\overline{\check{f}}}(\varphi)\|_2 \ \le \ \|(f - g_n) * \varphi\|_2 \ + \ \|M_{\overline{\check{g}_n} - \overline{\check{f}}}(\varphi)\|_2$$

$$\le \ \Big\|f - g_n\|_1 \ + \ \|\overline{\check{g}_n} - \overline{\check{f}}\|_\infty\Big)\ \|\varphi\|_2 \ \to 0,$$

we get $f * \varphi = M_{\overline{\check{f}}}(\varphi)$, and (2) is shown.

Proposition 3.11.3 *If $\varphi, \psi \in L^1(\mathcal{S}(K), \pi) \cap L^2(\mathcal{S}(K), \pi)$ then $\psi * \varphi = (\check{\psi}\check{\varphi})^\wedge$. In particular $\psi * \varphi \in L^2(\mathcal{S}(K), \pi) \cap C_0(\mathcal{S}(K))$ and*

$$\int_{\mathcal{S}(K)} \psi(\alpha)\, M_{\bar{\alpha}}(\varphi)\, d\pi(\alpha) \;=\; \int_{\mathcal{S}(K)} \varphi(\alpha)\, M_{\bar{\alpha}}(\psi)\, d\pi(\alpha) \qquad (3)$$

Proof. For $\varphi, \psi \in L^1(\mathcal{S}(K), \pi) \cap L^2(\mathcal{S}(K), \pi)$ we have $\mathcal{P}^{-1}(\varphi) = \check{\varphi}$, $\mathcal{P}^{-1}(\psi) = \check{\psi}$ and $\check{\psi}\check{\varphi} \in L^1(K, m)$. Therefore

$$\psi * \varphi \;=\; M_{\overline{\check{\psi}}}(\varphi) \;=\; \mathcal{P}(\check{\psi}\check{\varphi}) \;=\; (\check{\psi}\check{\varphi})^\wedge,$$

especially $\psi * \varphi \in C_0(\mathcal{S}(K))$, and (3) is valid. ◇

Remark: One should note that in Proposition 3.11.3 $\psi * \varphi$ is in general not an element of $L^1(\mathcal{S}(K), \pi)$. If $\hat{K}$ is a hypergroup under pointwise multiplication then $\mathcal{S}(K) = \hat{K}$ and $\psi * \varphi$ is the corresponding convolution in the Banach $*$-algebra $L^1(\mathcal{S}(K), \pi)$.

Although the "convolution" between $L^1(\mathcal{S}(K), \pi)$ and $L^2(\mathcal{S}(K), \pi)$ is a very weak one, it enables us to derive the existence of functions on $\mathcal{S}(K)$, whose inverse transform approximates functions on K in the L^1-norm.

Here we apply M_α to derive the following equivalence result, compare [96,Theorem 2.2.9].

Theorem 3.11.4 *(1) K is discrete if and only if $\mathcal{S}(K)$ is compact.*
(2) K is compact if and only if $\mathcal{S}(K)$ is discrete.

Proof.

(1) We have only to show that the compactness of $\mathcal{S}(K)$ implies that K is discrete. For that it suffices to show that $\{e\}$ is open. At first we prove that $M_\alpha 1 = 1$ π-almost everywhere. In fact for every $f \in C_c(K)$ we see

$$\int_{\mathcal{S}(K)} M_\alpha(1)(\beta)\, \overline{\hat{f}(\beta)}\, d\pi(\beta) \;=\; \int_{\mathcal{S}(K)} 1(\beta)\, \overline{M_{\bar{\alpha}}\hat{f}(\beta)}\, d\pi(\beta)$$

$$= \int_{\mathcal{S}(K)} \overline{\widehat{\alpha f}(\beta)}\, d\pi(\beta) \;=\; \overline{\alpha(e)\, f(e)} \;=\; \overline{f(e)} \;=\; \int_{\mathcal{S}(K)} \overline{\hat{f}(\beta)}\, d\pi(\beta).$$

Since $\{\hat{f} : f \in C_c(K)\}$ is dense in $L^2(\mathcal{S}(K), \pi)$, see Theorem 3.2.6, we have $M_\alpha 1 = 1$ π-almost everywhere. But then $\check{1}(x)\,(M_\alpha 1)^\vee(x) = \alpha(x)\,\check{1}(x)$. Since for $x \neq e$ we can find some $\alpha \in \mathcal{S}(K)$ such that $\alpha(x) \neq 1$, we have $\check{1}(x) = 0$ for $x \neq e$ and $\check{1}(e) = \pi(\mathcal{S}(K))$. The continuity of $\check{1}$ yields that $\{e\}$ is open.

(2) We have only to show that K is compact provided $\mathcal{S}(K)$ is discrete. If $\mathcal{S}(K)$ is discrete, the only positive character α_0 in $\mathcal{S}(K)$ is isolated. By Theorem 3.10.10 it follows that K is compact.

◇

Corollary 3.11.5 *$\mathcal{S}(K)$ is compact if and only if $\hat{K}$ respectively $\mathcal{X}^b(K)$ is compact.*

In subsection 3.4, see Proposition 3.4.10, we have considered the product formula (P) on $\hat{K}$, where all measures $\omega(\alpha,\beta)$ are probability measures, and we found out that the axioms H(2), H(3) and H(6) have to hold in addition to (P) such that $\hat{K}$ is a hypergroup under pointwise multiplication.

Now we introduce a weaker condition than (P) on $\mathcal{S}(K)$ such that a translation can be defined on various function spaces on $\mathcal{S}(K)$. Suppose that there exists $M > 0$ such that for every $\alpha, \beta \in \mathcal{S}(K)$ there exists $\omega(\alpha,\beta) \in M(\mathcal{S}(K))$ such that

$$\alpha(x)\cdot\beta(x) \;=\; \int_{\mathcal{S}(K)} \tau(x)\, d\omega(\alpha,\beta)(\tau) \qquad \text{for all } x \in K \tag{F1}$$

and

$$\|\omega(\alpha,\beta)\| \;\le\; M. \tag{F2}$$

Then we say that $\mathcal{S}(K)$ **satisfies property (F)**. Note that $\omega(\alpha,\beta)(\mathcal{S}(K)) = 1$. Hence the constant M is greater or equal 1.

The proof of the following result is omitted. The proof is nearly identical to that of Proposition 3.4.10.

Proposition 3.11.6 *Suppose that $\mathcal{S}(K)$ satisfies property (F). Then we have*

(1) $(\alpha,\beta) \mapsto \omega(\alpha,\beta)$, $\mathcal{S}(K)\times\mathcal{S}(K) \to M(\mathcal{S}(K))$ is continuous with respect to the weak $*$-topology on $M(\mathcal{S}(K))$.

(2) The canonical extension of ω (see Theorem 1.1.3) satisfies

$$\epsilon_\alpha * \omega(\beta,\gamma) \;=\; \omega(\alpha,\beta) * \epsilon_\gamma$$

for all $\alpha,\beta,\gamma \in \mathcal{S}(K)$.

(3) Defining $\omega(\alpha,\beta)^- \in M(\mathcal{S}(K))$ by

$$\omega(\alpha,\beta)^-(\varphi) \;=\; \omega(\alpha,\beta)(\bar{\varphi})$$

for $\varphi \in C_0(\mathcal{S}(K))$, it holds $\omega(\alpha,\beta)^- = \omega(\bar{\alpha},\bar{\beta})$ for all $\alpha,\beta \in \mathcal{S}(K)$.

Whenever $\mathcal{S}(K)$ satisfies property (F) we can define the translation of any function $\varphi \in C_0(\mathcal{S}(K))$ by

$$L_\alpha\varphi(\beta) = \omega(\alpha,\beta)(\varphi).$$

By Proposition 3.11.6 the function $L_\alpha\varphi$ is continuous and bounded.

Briefly we will see that $L_\alpha\varphi$ is an element of $C_0(\mathcal{S}(K))$.

Theorem 3.11.7 *$\mathcal{S}(K)$ satisfies property (F) if and only if*

$$\|\widehat{\bar{\alpha}f}|\mathcal{S}(K)\|_\infty \le M\,\|\hat{f}|\mathcal{S}(K)\|_\infty$$

for all $f \in L^1(K)$ and $\alpha \in \mathcal{S}(K)$, where M is the constant of (F2).

Proof. Assume that $\mathcal{S}(K)$ satisfies property (F) with constant $M \ge 1$. Clearly $\|L_\alpha\varphi\|_\infty \le M\,\|\varphi\|_\infty$ for all $\varphi \in C_0(\mathcal{S}(K))$, $\alpha \in \mathcal{S}(K)$. Since

$$\begin{aligned} L_\alpha\hat{f}(\beta) &= \int_{\mathcal{S}(K)}\int_K f(x)\,\overline{\tau(x)}\,dm(x)\,d\omega(\alpha,\beta)(\tau) = \int_K f(x)\,\overline{\alpha(x)\beta(x)}\,dm(x) \\ &= \widehat{\bar{\alpha}f}(\beta) \qquad \text{for all } f \in L^1(K),\ \beta \in \mathcal{S}(K), \end{aligned}$$

it follows $\|\widehat{\bar{\alpha}f}|\mathcal{S}(K)\|_\infty \le M\,\|\hat{f}|\mathcal{S}(K)\|_\infty$.

Conversely, suppose this inequality is true for every $f \in L^1(K)$, $\alpha \in \mathcal{S}(K)$. Let $\alpha, \beta \in \mathcal{S}(K)$. The mapping $\hat{f}|\mathcal{S}(K) \mapsto \widehat{\bar{\alpha}f}|\mathcal{S}(K)$ is linear and continuous from the dense subspace $\{\hat{f}|\mathcal{S}(K) : f \in L^1(K)\}$ of $C_0(\mathcal{S}(K))$ into $C_0(\mathcal{S}(K))$. Its unique continuous extension to $C_0(\mathcal{S}(K))$ may be designated by $\varphi \mapsto L_\alpha\varphi$. It satisfies $\|L_\alpha\varphi\|_\infty \le M\,\|\varphi\|_\infty$. Therefore $\varphi \mapsto L_\alpha\varphi(\beta)$ is a continuous linear functional on $C_0(\mathcal{S}(K))$, and Riesz' representation yields $\omega(\alpha,\beta) \in M(\mathcal{S}(K))$ such that $\|\omega(\alpha,\beta)\| \le M$, and $L_\alpha\hat{f}(\beta) = \widehat{\bar{\alpha}f}(\beta)$. Now let $z \in K$. Choose $k_i \in C_c(K)$ such that $k_i \ge 0$, $\|k_i\|_1 = 1$ and $\operatorname{supp} k_i \to \{z\}$. Then we have $\widehat{k_i}(\tau) \to \overline{\tau(z)}$ for every $\tau \in \mathcal{S}(K)$. Since $\omega(\alpha,\beta)$ is a finite regular Borel measure on $\mathcal{S}(K)$ we obtain

$$\lim_i \int_{\mathcal{S}(K)} \widehat{k_i}(\tau)\,d\omega(\alpha,\beta)(\tau) = \int_{\mathcal{S}(K)} \tau(\tilde{z})\,d\omega(\alpha,\beta)(\tau).$$

On the other hand we know that

$$\int_{\mathcal{S}(K)} \widehat{k_i}(\tau)\,d\omega(\alpha,\beta)(\tau) = \widehat{\bar{\alpha}k_i}(\beta) = \int_K k_i(x)\,\overline{\alpha(x)\beta(x)}\,dm(x) \to \overline{\alpha(z)\beta(z)}.$$

Hence we see that

$$\alpha(\tilde{z})\,\beta(\tilde{z}) = \int_{\mathcal{S}(K)} \tau(\tilde{z})\,d\omega(\alpha,\beta)(\tau).$$

◇

Corollary 3.11.8 *Suppose that $\mathcal{S}(K)$ satisfies property (F). Then*

(i) If $\varphi \in C_0(\mathcal{S}(K))$, then $L_\alpha\varphi \in C_0(\mathcal{S}(K))$ for each $\alpha \in \mathcal{S}(K)$.
(ii) If $\varphi \in C_0(\mathcal{S}(K)) \cap L^2(\mathcal{S}(K),\pi)$, then $L_\alpha\varphi \in C_0(\mathcal{S}(K)) \cap L^2(\mathcal{S}(K),\pi)$ and $L_\alpha\varphi = M_\alpha\varphi$ π-almost everywhere.

Proof.

(i) In the proof of Theorem 3.11.7 it is shown that $L_\alpha\hat{f}|\mathcal{S}(K) = \widehat{\bar{\alpha}f}|\mathcal{S}(K) \in C_0(\mathcal{S}(K))$ for each $f \in L^1(K)$, and $\varphi \mapsto L_\alpha\varphi$ is the unique continuous extension to $C_0(\mathcal{S}(K))$ of the mapping $\hat{f}|\mathcal{S}(K) \mapsto \widehat{\bar{\alpha}f}|\mathcal{S}(K)$. Hence $L_\alpha\varphi \in C_0(\mathcal{S}(K))$ for each $\varphi \in C_0(\mathcal{S}(K))$.
(ii) Let $\varphi \in C_0(\mathcal{S}(K))$. By Theorem 3.2.1 there exist $f_n \in L^1(K) \cap L^2(K)$ such that $\|\varphi - \widehat{f_n}|\mathcal{S}(K)\|_\infty \leq \frac{1}{Mn}$. For each $\alpha \in \mathcal{S}(K)$ we get

$$\|L_\alpha\varphi - L_\alpha(\widehat{f_n}|\mathcal{S}(K))\|_\infty \leq \frac{1}{n}.$$

The functions $\widehat{f_n}$ and $\mathcal{P}(f_n)$ coincide as elements of $L^2(\hat{K},\pi)$, see the general remarks below Theorem 3.2.6. Hence

$$M_\alpha(\widehat{f_n}|\mathcal{S}(K)) = \mathcal{P}(\bar{\alpha}f_n)|\mathcal{S}(K) = \widehat{\bar{\alpha}f_n}|\mathcal{S}(K) = L_\alpha(\widehat{f_n}|\mathcal{S}(K)).$$

If $\varphi \in C_0(\mathcal{S}(K)) \cap L^2(\mathcal{S}(K),\pi)$, then it follows that $M_\alpha\varphi = L_\alpha\varphi$ in $L^2(\mathcal{S}(K),\pi)$.

◇

Translation on $C_0(\mathcal{S}(K))$ is a basic tool to define a convolution on $M(\mathcal{S}(K))$. Therefore we consider a property of $\mathcal{S}(K)$ dealing with the existence of bounded translations.

We say that $\mathcal{S}(K)$ satisfies the **bounded dual translation property for** $C_0(\mathcal{S}(K))$, if for each $\alpha \in \mathcal{S}(K)$ there exists a bounded linear operator $L_\alpha \in B(C_0(\mathcal{S}(K)))$ such that

$$L_\alpha(\hat{f}|\mathcal{S}(K)) = \widehat{\bar{\alpha}f}|\mathcal{S}(K) \quad \text{for each } f \in L^1(K) \qquad \text{(duality)}$$

and

$$\sup_{\alpha\in\mathcal{S}(K)} \|L_\alpha\varphi\|_\infty \leq M_\varphi < \infty \quad \text{for all } \varphi \in C_0(\mathcal{S}(K)) \qquad \text{(boundedness).}$$

Corollary 3.11.9 *$\mathcal{S}(K)$ satisfies the bounded dual translation property for $C_0(\mathcal{S}(K))$ if and only if $\mathcal{S}(K)$ satisfies property (F).*

Proof. If (F) holds, then in the proof of Theorem 3.11.7 we have constructed operators $L_\alpha \in \mathcal{B}(C_0(\mathcal{S}(K)))$, such that the duality property and the boundedness property with $M_\varphi = M\,\|\varphi\|$ are satisfied. Conversely, if $\mathcal{S}(K)$ satisfies the bounded dual translation property for $C_0(\mathcal{S}(K))$, then the Theorem of Banach-Steinhaus says that $M_\varphi \leq M\,\|\varphi\|_\infty$, where $M = \sup_{\alpha\in\mathcal{S}(K)} \|L_\alpha\| < \infty$. By Theorem 3.11.7 property (F) follows. ◇

Now it is clear how to define a module-action on $C_0(\mathcal{S}(K))$ and how to define Banach algebras with convolutions. A first step is to prove the continuity of $\alpha \mapsto L_\alpha\varphi,\ \mathcal{S}(K) \to C_0(\mathcal{S}(K))$ for each $\varphi \in C_0(\mathcal{S}(K))$.

Proposition 3.11.10 *Assume that $\mathcal{S}(K)$ satisfies property (F). Let $\varphi \in C_0(\mathcal{S}(K))$. The mapping $\alpha \mapsto L_\alpha\varphi,\ \mathcal{S}(K) \to C_0(\mathcal{S}(K))$ is continuous.*

Proof. Let $\alpha_0 \in \mathcal{S}(K)$, $\varepsilon > 0$. By Theorem 3.2.1 there exists $f \in C_c(K)$ such that $\|\varphi - \hat{f}|\mathcal{S}(K)\|_\infty < \frac{\varepsilon}{3M}$. ($M$ is the constant of property (F).) Let $C = \operatorname{supp} f$ and denote

$$U_{\alpha_0} \;=\; \{\alpha \in \mathcal{S}(K) :\ |\alpha(x) - \alpha(x_0)| < \frac{\varepsilon}{3\|f\|_1} \text{ for all } x \in C\}$$

Then

$$|\widehat{\bar{\alpha}f}(\beta) - \widehat{\bar{\alpha}_0 f}(\beta)| \;\leq\; \int_C |\alpha(x) - \alpha_0(x)|\,|f(x)|\,dm(x) \;<\; \frac{\varepsilon}{3}$$

for every $\beta \in \mathcal{S}(K)$, i.e. $\|L_\alpha(\hat{f}|\mathcal{S}(K)) - L_{\alpha_0}(\hat{f}|\mathcal{S}(K))\|_\infty \;\leq\; \frac{\varepsilon}{3}$. It follows

$$\|L_\alpha\varphi - L_{\alpha_0}\varphi\|_\infty \;\leq\; \|L_\alpha\|\;\|\varphi - \hat{f}|\mathcal{S}(K)\|_\infty + \|L_\alpha(\hat{f}|\mathcal{S}(K)) - L_{\alpha_0}(\hat{f}|\mathcal{S}(K))\|_\infty$$

$$+\; \|L_{\alpha_0}\|\;\|\hat{f}|\mathcal{S}(K) - \varphi\|_\infty \;<\; \varepsilon.$$

◇

We define for $\varphi \in C_0(\mathcal{S}(K))$ and $a \in M(\mathcal{S}(K))$

$$a * \varphi(\alpha) \;:=\; \int_{\mathcal{S}(K)} L_{\tilde{\beta}}\varphi(\alpha)\,da(\beta).$$

Proposition 3.11.11 *Assume that $\mathcal{S}(K)$ satisfies property (F). Let $\varphi \in C_0(\mathcal{S}(K))$, $a \in M(\mathcal{S}(K))$. Then $a * \varphi \in C_0(\mathcal{S}(K))$.*

Proof. By Proposition 3.11.10 we know that for $\alpha_0 \in \mathcal{S}(K)$, $\varepsilon > 0$, there exists a neighbourhood U_{α_0} such that $|L_{\tilde{\beta}}\varphi(\alpha) - L_{\tilde{\beta}}\varphi(\alpha_0)| < \frac{\varepsilon}{\|a\|}$ for all $\alpha \in U_{\alpha_0}$ and all $\beta \in \mathcal{S}(K)$. Then

$$|a * \varphi(\alpha) - a * \varphi(\alpha_0)| \;\leq\; \sup_{\beta\in\mathcal{S}(K)} |L_{\tilde{\beta}}\varphi(\alpha) - L_{\tilde{\beta}}\varphi(\alpha_0)|\;\|a\| \;<\; \varepsilon$$

for $\alpha \in U_{\alpha_0}$, i.e. $a * \varphi \in C^b(\mathcal{S}(K))$. Since $a * \hat{f}(\beta) = \widehat{\check{a}f}(\beta)$ for each $f \in L^1(K)$ and $\beta \in \mathcal{S}(K)$, we see that $a * \hat{f} \in C_0(\mathcal{S}(K))$. Applying Theorem 3.2.1 we get $a * \varphi \in C_0(\mathcal{S}(K))$. ◇

Let $a, b \in M(\mathcal{S}(K))$. Define $a * b \in M(\mathcal{S}(K))$ by

$$a * b(\varphi) := \int_{\mathcal{S}(K)} \int_{\mathcal{S}(K)} L_\alpha \varphi(\beta)\, da(\alpha)\, db(\beta) \;=\; \int_{\mathcal{S}(K)} \tilde{a} * \varphi(\beta)\, db(\beta)$$

for $\varphi \in C_0(\mathcal{S}(K))$. It is easily shown that $\|a * b\| \leq M \|a\| \, \|b\|$. Moreover,

$$(a * b)^\vee(x) \;=\; \int_{\mathcal{S}(K)} \tau(x)\, d(a * b)(\tau) \;=\; \int_{\mathcal{S}(K)} \int_{\mathcal{S}(K)} L_\alpha \widehat{\epsilon_{\tilde{x}}}(\beta)\, da(\alpha)\, db(\beta)$$

$$= \int_{\mathcal{S}(K)} \int_{\mathcal{S}(K)} \alpha(x)\, \beta(x)\, da(\alpha)\, db(\beta) \;=\; \check{a}(x)\, \check{b}(x)$$

for every $x \in K$. Applying Theorem 3.3.3 it follows $(a * b) * c = a * (b * c)$ for $a, b, c \in M(\mathcal{S}(K))$. It follows that $M(\mathcal{S}(K))$ is a commutative algebra with convolution as multiplication. An appropriate involution on $M(\mathcal{S}(K))$ is defined by (see also the end of subsection 3.4)

$$a^*(\varphi) := \int_{\mathcal{S}(K)} \varphi(\alpha)\, da^*(\alpha) \;=\; \overline{\int_{\mathcal{S}(K)} \overline{\varphi(\bar{\alpha})}\, da(\alpha)}$$

for all $\varphi \in C_0(\mathcal{S}(K))$ and $a \in M(\mathcal{S}(K))$. Then

$$\overline{(a^*)^\vee(x)} \;=\; \overline{\int_{\mathcal{S}(K)} \widehat{\epsilon_{\tilde{x}}}(\alpha)\, da^*(\alpha)} \;=\; \int_{\mathcal{S}(K)} \overline{\epsilon_{\tilde{x}}(\bar{\alpha})}\, da(\alpha)$$

$$= \int_{\mathcal{S}(K)} \alpha(x)\, da(\alpha) \;=\; \check{a}(x)$$

for every $x \in K$, i.e. $(a^*)^\vee = \bar{\check{a}}$. It follows

$$((a * b)^*)^\vee \;=\; \overline{(a * b)^\vee} \;=\; \overline{a^\vee}\, \overline{b^\vee} \;=\; (a^*)^\vee\, (b^*)^\vee \;=\; (a^* * b^*)^\vee,$$

and with Theorem 3.3.3 we conclude that $(a * b)^* = a^* * b^*$.

Theorem 3.11.12 *Assume that $\mathcal{S}(K)$ satisfies property (F). Then $M(\mathcal{S}(K))$ is a commutative Banach $*$-algebra.*

Remark:

(1) If the character $\alpha = 1$ is an element of $\mathcal{S}(K)$, then $M(\mathcal{S}(K))$ has a unit ϵ_1. However, in general the constant function 1 is not contained in $\mathcal{S}(K)$.

(2) In general we cannot expect that for $\varphi \in L^1(\mathcal{S}(K),\pi)$ and $\psi \in L^1(\mathcal{S}(K),\pi)$ the measure $\varphi\pi * \psi\pi$ is an element of $L^1(\mathcal{S}(K),\pi)$. Hence $L^1(\mathcal{S}(K),\pi)$ is not a Banach algebra with convolution as multiplication. Whenever $\mathcal{S}(K)$ is compact — this is equivalent to K discrete — $L^1(\mathcal{S}(K),\pi)$ is a Banach $*$-algebra, see subsection 4.13.

Modified hypergroups K_γ of a hypergroup K are investigated in section 3.10, where γ is a positive semicharacter of K. Now we will derive sufficient conditions such that property (F) of $\mathcal{S}(K)$ is transmitted to $\mathcal{S}(K_\gamma)$.

Theorem 3.11.13 *Let γ be a positive semicharacter on K. Assume that $\mathcal{S}(K)$ satisfies property (F). Then $\mathcal{S}(K_\gamma)$ satisfies property (F), provided there exists a constant $M_\gamma > 0$ such that*

$$\|L_{f/\gamma}\| \leq M_\gamma \, \|L_f\| \qquad \textit{for all } f \in C_c(K).$$

Proof. By Theorem 3.10.6 we know that $\phi : \mathcal{S}(K) \to \mathcal{S}(K_\gamma)$, $\phi(\alpha) = \frac{\alpha}{\gamma}$ is a homeomorphism and π_γ is the image measure $\phi(\pi)$. By Theorem 3.11.7 and Proposition 3.1.7 we have to show that there is a constant $M > 0$ such that

$$\|L^\gamma_{\check{\alpha} f/\gamma}\| \leq M \, \|L^\gamma_f\|$$

for all $f \in C_c(K)$ and $\alpha \in \mathcal{S}(K)$. (Note that $C_c(K)$ is dense in $L^1(K_\gamma, m_\gamma)$.)

Property (F) of $\mathcal{S}(K)$ implies in the same manner that there exists a constant $M_1 > 0$ such that $\|L_{\check{\alpha} f}\| \leq M_1 \, \|L_f\|$ for all $f \in C_c(K)$ and $\alpha \in \mathcal{S}(K)$. Proposition 3.10.5 yields

$$\|L^\gamma_{\check{\alpha} f/\gamma}\| = \|L_{\check{\alpha} f/\gamma^2}\| \leq M_1 \, \|L_{f/\gamma^2}\| \leq M_1 \, M_\gamma \, \|L_{f/\gamma}\| = M_1 \, M_\gamma \, \|L^\gamma_f\|$$

for each $f \in C_c(K)$ and $\alpha \in \mathcal{S}(K)$. ◇

Lemma 3.11.14 *Assume that $\mathcal{S}(K)$ satisfies property (F) with constant $M > 0$. Let $a \in M(\mathcal{S}(K))$. Then $\|L_{\check{a} f}\| \leq M \, \|a\| \, \|L_f\|$ for all $f \in L^1(K)$.*

Proof. Applying the bounded translation we obtain for every $f \in L^1(K)$ and $\alpha \in \mathcal{S}(K)$

$$(\check{a} f)^\wedge(\alpha) = \int_{\mathcal{S}(K)} L_{\tilde{\beta}} \hat{f}(\alpha) \, da(\beta) = a * \hat{f}(\alpha).$$

Since $\|L_{\tilde{\beta}} \hat{f}|\mathcal{S}(K)\|_\infty \leq M \, \|\hat{f}|\mathcal{S}(K)\|_\infty$ for each $\beta \in \mathcal{S}(K)$ we get

$$\|a * \hat{f}|\mathcal{S}(K)\|_\infty \leq M \, \|a\| \, \|\hat{f}|\mathcal{S}(K)\|_\infty.$$

It follows

$$\|L_{\check{a}f}\| = \|\widehat{\check{a}f}|\mathcal{S}(K)\|_\infty = \|a * \hat{f}|\mathcal{S}(K)\|_\infty \le M\,\|a\|\,\|L_f\|.$$

◇

Combining Theorem 3.11.13 and Lemma 3.11.14 we have the following result.

Corollary 3.11.15 *Let γ be a positive semicharacter of K such that $\frac{1}{\gamma} = \check{a}$ for some measure $a \in M(\mathcal{S}(K))$. If $\mathcal{S}(K)$ satisfies property (F), then $\mathcal{S}(K_\gamma)$ satisfies property (F), too.*

Consider the polynomial hypergroup $K = \mathbb{N}_0$ generated by the Chebyshev polynomials of the first kind $(T_n(x))_{n\in\mathbb{N}_0}$ and $\gamma(n) = T_n(x)$, where $x > 1$, see the example of subsection 3.10. We know that $\gamma(n) = \cosh(an)$ and K_γ is a polynomial hypergroup generated by cosh-polynomials with parameter $a > 0$. Since

$$\frac{1}{\gamma(n)} = \frac{1}{\cosh(an)} \le \frac{2}{e^{an}},$$

we have $\frac{1}{\gamma} \in l^1(h)$, where $h(0) = 1$, $h(n) = 2$ for $n \in \mathbb{N}$ are the Haar weights. Denote $\varphi(y) = (\frac{1}{\gamma})^\wedge(y)$ for $y \in [-1,1]$. Then $\frac{1}{\gamma} = \check{\varphi}$ by Theorem 3.3.7. It follows that for each polynomial hypergroup K_γ generated by cosh-polynomials the dual space $\mathcal{S}(K_\gamma)$ satisfies property (F).

By Corollary 3.11.15 we know that for each pair $\{\beta_1, \beta_2\} \subseteq \mathcal{S}(K_\gamma)$ exists a Borel measure $\omega_\gamma(\beta_1, \beta_2) \in M(\mathcal{S}(K_\gamma))$ such that

$$\beta_1(x) \cdot \beta_2(x) = \int_{\mathcal{S}(K_\gamma)} \sigma(x)\, d\omega_\gamma(\beta_1, \beta_2)(\sigma)$$

and $\|\omega_\gamma(\beta_1, \beta_2)\| \le M$, provided $\mathcal{S}(K)$ satisfies property (F) and $\frac{1}{\gamma} = \check{a},\ a \in M(\mathcal{S}(K))$.

We have the following relationship between $\omega_\gamma(\beta_1, \beta_2) \in M(\mathcal{S}(K_\gamma))$ and $\omega(\alpha_1, \alpha_2) \in M(\mathcal{S}(K))$ for $\beta_1 = \frac{\alpha_1}{\gamma}$, $\beta_2 = \frac{\alpha_2}{\gamma}$.

Proposition 3.11.16 *Assume that $\mathcal{S}(K)$ satisfies property (F), and that γ is a positive semicharacter on K such that $\frac{1}{\gamma} = \check{a}$ for some $a \in M(\mathcal{S}(K))$. The measures $\omega_\gamma(\beta_1, \beta_2) \in M(\mathcal{S}(K_\gamma))$ are the image measures $\phi(a * \omega(\alpha_1, \alpha_2))$, where $\beta_i = \frac{\alpha_i}{\gamma} = \phi(\alpha_i)$, $\alpha_i \in \mathcal{S}(K)$, $i = 1, 2$.*

Proof. Let $g \in \widehat{\epsilon_{\tilde{x}}}^\gamma|\mathcal{S}(K_\gamma)$, $x \in K$. Since $\widehat{\epsilon_{\tilde{x}}}^\gamma(\frac{\tau}{\gamma}) = \frac{\widehat{\epsilon_{\tilde{x}}}}{\gamma(x)}(\tau)$ for each $\tau \in \mathcal{S}(K)$, we obtain for the image measure

$$\phi(a * \omega(\alpha_1, \alpha_2))(g) = \frac{1}{\gamma(x)} \int_{\mathcal{S}(K)} \widehat{\epsilon_{\tilde{x}}}(\tau)\, d(a * \omega(\alpha_1, \alpha_2))(\tau)$$

$$= \frac{1}{\gamma(x)} \int_{\mathcal{S}(K)} \tau(x)\, d(a * \omega(\alpha_1, \alpha_2))(\tau) = \frac{1}{\gamma(x)} (a * \omega(\alpha_1, \alpha_2))^\vee(x)$$

$$= \frac{1}{\gamma(x)} \check{a}(x)\, (\omega(\alpha_1, \alpha_2))^\vee(x) = \frac{\alpha_1}{\gamma}(x)\, \frac{\alpha_2}{\gamma}(x),$$

i.e.

$$\omega_\gamma\left(\frac{\alpha_1}{\gamma}, \frac{\alpha_2}{\gamma}\right) = \phi(a * \omega(\alpha_1, \alpha_2)).$$

◇

11.2 *Surjectivity of Fourier- and inverse Fourier transforms*

We know already that the range of the Fourier transform on $L^1(K)$ is sup-norm dense in $C_0(\hat{K})$, see Proposition 3.2.1, and that the inverse Fourier transform on $L^1(\mathcal{S}(K), \pi)$ is sup-norm dense in $C_0(K)$, see Proposition 3.3.2(iii). If G is a locally compact abelian group the Fourier transform $^\wedge : L^1(G) \to C_0(\hat{G})$ in onto if and only if G is finite. To extend this result to commutative hypergroups we use the method of C.C.Graham [281]. A central tool is the following result of K.A.Ross.

Theorem 3.11.17 *Let K be a nondiscrete (not necessarily commutative) hypergroup. Then $L^\infty(K) \neq C^b(K)$.*

For a proof we refer to the appendix of [732].

We shall need an analogous statement for the dual space $\mathcal{S}(K)$ with Plancherel measure π.

Proposition 3.11.18 *Let K be a commutative hypergroup such that $\mathcal{S}(K)$ is nondiscrete and metrizable. Then $L^\infty(\mathcal{S}(K), \pi) \neq C^b(\mathcal{S}(K))$.*

Proof. Since $\mathcal{S}(K)$ is nondiscrete there is some $\alpha_0 \in \mathcal{S}(K)$, which is not isolated, and hence there is a sequence $(\alpha_n)_{n\in\mathbb{N}}$ of points $\alpha_n \in \mathcal{S}(K)$ converging to α_0, and $\alpha_n \neq \alpha_m$ for $n \neq m$. Choose open neighbourhoods W_n of α_n such that $\overline{W_n}$ are pointwise disjoint. Put $U = \bigcup_{n=}^{\infty} W_{2n}$ and $V = \bigcup_{n=1}^{\infty} W_{2n-1}$. U and V are open and disjoint and $\alpha_0 \in \bar{U} \cap \bar{V}$. Let f be the characteristic function of U. It is obvious that there is no function $g \in C^b(\mathcal{S}(K))$ such that $g = f$ locally π-almost everywhere. ◇

Theorem 3.11.19 *Let K be a commutative hypergroup such that $\mathcal{S}(K)$ is metrizable. Then the following statements are equivalent.*

(1) The Fourier transform $^\wedge : L^1(K) \to C_0(\hat{K})$ is surjective.
(2) The inverse Fourier transform $^\vee : L^1(\mathcal{S}(K),\pi) \to C_0(K)$ is surjective.
(3) K is finite.

Proof. Obviously K is finite if and only if $\hat{K}$ and $\mathcal{S}(K)$ are finite, and it follows that (3) implies (2) and (1).

In order to show (1) $\Rightarrow$ (3) we use the symbol T instead of $^\wedge$. We know that T is a bounded injective operator from $L^1(K)$ into $C_0(\hat{K})$. The adjoint operator $T^* : M(\hat{K}) \to L^\infty(K)$ is a bounded operator from $M(\hat{K}) \cong C_0(\hat{K})^*$ into $L^\infty(K) \cong L^1(K)^*$, and for $a \in M(\hat{K})$, $f \in L^1(K)$, we have

$$T^*(a)(f) = \int_{\hat{K}} T(f)(\alpha)\, da(\alpha) = \int_{\hat{K}}\int_K f(x)\,\overline{\alpha(x)}\, dm(x)\, da(\alpha)$$

$$= \int_K f(x)\,\check{a}(\tilde{x})\, dm(x) = \int_K f(x)\,\overline{(a^*)^\vee(\tilde{x})}\, dm(x) = \int_K f(x)\,\overline{((a^*)^\vee)^\sim(x)}\, dm(x)$$

that is $T^*(a) = ((a^*)^\vee)^\sim$ in $L^\infty(K)$. Now assume that T is surjective. By the closed graph theorem T is a topological isomorphism from $L^1(K)$ onto $C_0(\hat{K})$. Then $T^* : M(\hat{K}) \to L^\infty(K)$ is also a topological isomorphism. Since $\check{a} \in C^b(K)$ for all $a \in M(\hat{K})$ it follows $L^\infty(K) = C^b(K)$. By Theorem 3.11.17 the hypergroup K has to be discrete and $T^*(a) = ((a^*)^\vee)^\sim$ everywhere on K. Restricting $T^* : M(\hat{K}) \to l^\infty(K)$ to $L^1(\mathcal{S}(K),\pi) = L^1(\hat{K},\pi) \subseteq M(\hat{K})$ we see that $T^*(L^1(\mathcal{S}(K),\pi))$ is a sup-norm closed subspace of $c_0(K)$. Since $L^1(\mathcal{S}(K),\pi)^\vee$ is sup-norm dense in $c_0(K)$, we get $L^1(\mathcal{S}(K),\pi)^\vee = c_0(K)$. Consider now $S : L^1(\mathcal{S}(K),\pi) \to c_0(K)$, $S(\varphi) = \check{\varphi}$. For the adjoint operator $S^* : M(K) \to L^\infty(\mathcal{S}(K),\pi)$ we identify $M(K) \cong c_0(K)^*$ and $L^1(\mathcal{S}(K),\pi)^* \cong L^\infty(\mathcal{S}(K),\pi)$. For $\mu \in M(K)$ and $\varphi \in L^1(\mathcal{S}(K),\pi)$ we have (note that K is discrete),

$$S^*(\mu)(\varphi) = \int_K S(\varphi)(x)\, d\mu(x) = \int_K \int_{\mathcal{S}(K)} \alpha(x)\,\varphi(\alpha)\, d\pi(\alpha)\, d\mu(x)$$

$$= \int_{\mathcal{S}(K)} \varphi(\alpha)\,\hat{\mu}(\tilde{\alpha})\, d\pi(\alpha) \equiv \int_{\mathcal{S}(K)} \varphi(\alpha)\,\overline{(\mu^*)^\wedge(\tilde{\alpha})}\, d\pi(\alpha),$$

that is $S^*(\mu) = ((\mu^*)^\wedge)^\sim$ in $L^\infty(\mathcal{S}(K),\pi)$. Since S is a topological isomorphism, S^* is also a topological isomorphism, and it follows $L^\infty(\mathcal{S}(K),\pi) = C^b(\mathcal{S}(K))$. By Proposition 3.11.18 and by Theorem 3.11.4 this is only possible if K is compact. Therefore K has to be discrete and compact, so K is finite.

In a similar way one can show (2) $\Rightarrow$ (3). Consider $S : L^1(\mathcal{S}(K), \pi) \to C_0(K)$, $S(\varphi) = \check{\varphi}$. Assuming that S is surjective we obtain as before that $S^* : M(K) \to L^\infty(\mathcal{S}(K), \pi)$ is a topological isomorphism, and it follows that K is compact, in particular $\mathcal{S}(K) = \hat{K}$ and $\hat{K}$ is discrete, see Theorem 3.1.9. Moreover, it follows that $S^*(L^1(K))$ is a sup-norm closed subspace of $C_0(\hat{K})$. Since $S^*(\mu) = ((\mu^*)^\wedge)\tilde{}$ for each $\mu \in M(K)$, and $L^1(K)^\wedge$ is sup-norm dense in $C_0(\hat{K})$, it follows that $L^1(K)^\wedge = c_0(\hat{K})$.

Consider $T : L^1(K) \to C_0(\hat{K})$, $T(f) = \hat{f}$. T is a topological isomorphism, $T^* : M(\hat{K}) \to L^\infty(K)$ is a topological isomorphism, too, and $T^*(a) = ((a^*)^\vee)\tilde{}$ for all $a \in M(\hat{K})$. It follows $L^\infty(K) = C^b(K)$, and hence K is discrete. ◇

The proof of Theorem 3.11.19 shows that the following result is also true.

Theorem 3.11.20 *Let K be a commutative hypergroup such that $\hat{K}$ is a hypergroup under pointwise multiplication. Then the conditions (1), (2) and (3) of Theorem 3.11.19 are equivalent.*

11.3 *Bochner theorem on $\mathcal{S}(K)$*

We now introduce positive definite functions on the dual $\mathcal{S}(K)$ of a commutative hypergroup K. Of course, since on the dual no convolution structure is at our disposal, we have to find an appropriate definition of positive definiteness. As dual space we select $\mathcal{S}_1 = \mathcal{S}(K) \cup \{1\}$. We add 1 to $\mathcal{S}(K)$, so that we are able to normalize those functions, which we call positive definite.

Definition. A continuous bounded function $\varphi : \mathcal{S}_1 \to \mathbb{C}$ is called **$T(\mathcal{S}_1)$-positive definite** if for all choices of $n \in \mathbb{N}$; $c_1, ..., c_n \in \mathbb{C}$ and $\alpha_1, ..., \alpha_n \in \mathcal{S}_1$ such that $c_1\alpha_1(x) + ... + c_n\alpha_n(x) \geq 0$ for all $x \in K$, it follows

$$c_1\,\varphi(\alpha_1) + ... + c_n\,\varphi(\alpha_n) \geq 0.$$

Obviously the definition of $T(\mathcal{S}_1)$-positive definiteness can be written in the following way. Denote by

$$T(\mathcal{S}_1) := \text{lin}\,\{\epsilon_\alpha : \alpha \in \mathcal{S}_1\} \subseteq M(\mathcal{S}_1) \subseteq M(\hat{K})$$

the linear span of the point measures of $\alpha \in \mathcal{S}_1$. Then a continuous bounded function $\varphi : \mathcal{S}_1 \to \mathbb{C}$ is $T(\mathcal{S}_1)$-positive definite exactly when

$$\int_{\mathcal{S}_1} \varphi(\alpha)\, da(\alpha) \geq 0 \qquad \text{for all } a \in T(\mathcal{S}_1) \text{ with } \check{a} \geq 0.$$

So we see that our definition is a variant of the various definitions given in [96]. We think our pointwise definition of positive definiteness on $\mathcal{S}_1$ is the appropriate way to characterize positive bounded Borel measures on K.

Proposition 3.11.21 *Let $\varphi : \mathcal{S}_1 \to \mathbb{C}$ be $T(\mathcal{S}_1)$-positive definite. Then the following holds true:*

(i) If $c_1\alpha_1 + ... + c_n\alpha_n \geq 0$ then

$$c_1\,\varphi(\alpha_1)\,\alpha_1 \;+ ... +\; c_n\,\varphi(\alpha_n)\,\alpha_n \;\geq\; 0$$

(ii) If $c_1\alpha_1 + ... + c_n\alpha_n$ is real-valued, then

$$c_1\,\varphi(\alpha_1) \;+ ... +\; c_n\,\varphi_n \;\in\; \mathbb{R}.$$

Proof.

(i) From $c_1\alpha_1(z) + ... + c_n\alpha_n(z) \geq 0$ for all $z \in K$ it follows

$$\begin{aligned} &c_1\,\alpha_1(x)\,\alpha_1(y) \;+ ... +\; c_n\,\alpha_n(x)\,\alpha_n(y) \\ &\quad= \int_K (c_1\alpha_1(z) \;+ ... +\; c_n\alpha_n(z))\; d\omega(x,y)(z) \;\geq 0 \end{aligned}$$

for all $x, y \in K$. Thus $c_1\alpha_1(x)\varphi(\alpha_1) \;+ ... + c_n\alpha_n(x)\varphi(\alpha_n) \geq 0$ for all $x \in K$.

(ii) Since $1(x) = 1 \;\geq 0$ for all $x \in K$ we have $\varphi(1) \geq 0$. For $c_1\alpha_1 + ... + c_n\alpha_n$ real-valued, denote $a := \|c_1\alpha_1 + ... + c_n\alpha_n\|_\infty$. Then

$$a\,1(x) \;+\; c_1\,\alpha_1(x) \;+ ... +\; c_n\,\alpha_n(x) \;\geq 0 \qquad \text{for all } x \in K,$$

which implies $a\varphi(1) + c_1\varphi(\alpha_1) + ... + c_n\varphi(\alpha_n) \;\geq\; 0$. In particular, $c_1\varphi(\alpha_1) + ... + c_n\varphi(\alpha_n) \in \mathbb{R}$.

◇

Proposition 3.11.22 *Let φ and ψ be $T(\mathcal{S}_1)$-positive definite functions on $\mathcal{S}_1$. Then*

(i) $\varphi(1) \geq 0$ and

$$|c_1\,\varphi(\alpha_1) \;+ ... +\; c_n\,\varphi(\alpha_n)| \;\leq\; 2\,\varphi(1)\;\|c_1\alpha_1 \;+ ... +\; c_n\alpha_n\|_\infty$$

for all $n \in \mathbb{N}$, $\;c_1, ..., c_n \in \mathbb{C}$, $\;\alpha_1, ..., \alpha_n \in \mathcal{S}_1$.

(ii) $\lambda_1\varphi + \lambda_2\psi$ is $T(\mathcal{S}_1)$-positive definite for all $\lambda_1, \lambda_2 \geq 0$.

(iii) $\varphi\psi$ is $T(\mathcal{S}_1)$-positive definite.

Proof.

(i) We have already seen that $\varphi(1) \geq 0$. If $c_1\alpha_1 + ... + c_n\alpha_n$ is real-valued, $a := \|c_1\alpha_1 + ... + c_n\alpha_n\|_\infty$, then $a\,1 \pm (c_1\alpha_1 + ... + c_n\alpha_n) \geq 0$ and hence $|c_1\varphi(\alpha_1) + ... + c_n\varphi(\alpha_n)| \leq 2\varphi(1)\,a$.
If $f = c_1\alpha_1 + ... + c_n\alpha_n$ is an arbitrary element of $T(\mathcal{S}_1)$, write $f = f_1 + if_2$, where $f_1 = \frac{1}{2}(f + \overline{f}) \in T(\mathcal{S}_1)$ and $f_2 = \frac{1}{2i}(f - \overline{f}) \in T(\mathcal{S}_1)$. Note that $\overline{\mathcal{S}(K)} = \mathcal{S}(K)$. Then $f_1 = \text{Re}\, f,\ f_2 = \text{Im}\, f$, and we obtain

$$|c_1\varphi(\alpha_1) + ... + c_n\varphi(\alpha_n)| \leq \varphi(1)\,\|f_1\|_\infty + \varphi(1)\,\|f_2\|_\infty \leq 2\,\varphi(1)\,\|f\|_\infty.$$

(ii) follows directly by the definition of $T(\mathcal{S}_1)$-positive definiteness.
(iii) By Proposition 3.11.21 it follows that $c_1\varphi(\alpha_1)\,\alpha_1 + ... + c_n\varphi(\alpha_n)\,\alpha_n \geq 0$, and hence $c_1\varphi(\alpha_1)\,\psi(\alpha_1) + ... + c_n\varphi(\alpha_n)\,\psi(\alpha_n) \geq 0$.

◇

Examples of $T(\mathcal{S}_1)$-positive definite functions are easily found. If $\mu \in M(K)$ is positive, then $\hat{\mu}|\mathcal{S}_1$ is $T(\mathcal{S}_1)$-positive definite. In fact, if $c_1\alpha_1 + ... + c_n\alpha_n \geq 0$, then

$$c_1\,\hat{\mu}(\alpha_1) + ... + c_n\,\hat{\mu}(\alpha_n) = \int_K (c_1\,\alpha_1(\tilde{x}) + ... + c_n\,\alpha(\tilde{x}))\,d\mu(x) \geq 0.$$

The following theorem shows conversely, that every $T(\mathcal{S}_1)$-positive definite function is the Fourier-Stieltjes transform of a positive measure $\mu \in M(K)$.

Theorem 3.11.23 *Let $\varphi : \mathcal{S}_1 \to \mathbb{C}$ be a $T(\mathcal{S}_1)$-positive definite function. Then there exists a unique positive measure $\mu \in M(K)$ such that $\varphi|\mathcal{S}(K) = \hat{\mu}|\mathcal{S}(K)$. Conversely $\hat{\mu}|\mathcal{S}_1$ is a $T(\mathcal{S}_1)$-positive definite function for each positive measure $\mu \in M(K)$.*

Proof. The second statement is just proven above. Now assume φ is $T(\mathcal{S}_1)$-positive definite. Every measure $a \in M(\hat{K})$ compactly supported in $\mathcal{S}(K) \cup \{1\}$ can be approximated in the norm of $M(\hat{K})$ by elements of $T(\mathcal{S}_1)$. In particular, every $h\,d\pi \in M(\hat{K}),\ \ h \in C_c(\hat{K})$, can be approximated by measures from $T(\mathcal{S}_1)$. By Proposition 3.11.22(i) it follows

$$\left|\int_{\hat{K}} \varphi(\alpha)\,h(\alpha)\,d\pi(\alpha)\right| \leq 2\,\varphi(1)\,\|\check{h}\|_\infty$$

for all $h \in C_c(\hat{K})$, and Theorem 3.9.1 implies that there exists $\mu \in M(K)$ with $\hat{\mu}|\mathcal{S}(K) = \varphi|\mathcal{S}(K)$. Note that $\|\mu\| \leq 2\varphi(1)$. From the uniqueness Theorem 3.1.8 follows that $\mu \in M(\hat{K})$ is uniquely determined by $\hat{\mu}|\mathcal{S}(K) = \varphi|\mathcal{S}(K)$. It remains to show that μ is positive. Given $f \in C_0(K),\ \ f \geq 0$

and $\delta > 0$, Proposition 3.3.2(iii) yields a function $h \in C_c(\hat{K})$ such that $\|f - \check{h}\|_\infty < \delta/2$. There exists $b \in T(\mathcal{S}_1)$ with $\|h\,d\pi - b\| < \delta/2$, and so $\|f - \check{b}\|_\infty < \delta$. We may assume that $\check{b}$ is real-valued. Put $a = \delta\epsilon_1 + b$. Since $\check{a} \geq 0$, we get

$$\int_K \check{a}(x)\,d\mu(x) \;=\; \int_{\hat{K}} \varphi(\alpha)\,da(\alpha) \;\geq 0,$$

and by

$$\left|\int_K f(x)\,d\mu(x) \;-\; \int_K \check{a}(x)\,d\mu(x)\right| \;\leq\; 4\,\varphi(1)\,\delta,$$

we see that $\int\limits_K f(x)\,d\mu(x) \geq 0$, i.e. μ is positive. ◇

Corollary 3.11.24 *Let $\varphi : \mathcal{S}_1 \to \mathbb{C}$ be a $T(\mathcal{S}_1)$-positive definite function. Then $\overline{\varphi(\alpha)} = \varphi(\overline{\alpha})$ and $|\varphi(\alpha)| \leq 2\,\varphi(1)$ for all $\alpha \in \mathcal{S}(K)$.*

Proof. Since $\varphi|\mathcal{S}(K) = \hat{\mu}|\mathcal{S}(K)$, $\mu \geq 0$, both statements follow clearly. ◇

12 Invariant means and α-invariant functionals on commutative hypergroups

In this section we will investigate various types of means on commutative hypergroups K, which fulfil certain invariance properties. Since K is commutative, we have not to distinguish between left- and right-variance.

A **mean** on $L^\infty(K)$ is an element $m \in L^\infty(K)^*$ such that $m(1) = 1 = \|m\|$. An **invariant mean** is a mean m which satisfies

$$m(L_x\varphi) \;=\; m(\varphi) \qquad \text{for all } x \in K \quad \text{and } \varphi \in L^\infty(K).$$

A **strongly invariant mean** is a mean m which satisfies

$$m((L_x\varphi)\cdot\psi) \;=\; m(\varphi\cdot(L_{\tilde{x}}\psi)) \qquad \text{for all } x \in K \quad \text{and } \varphi, \psi \in L^\infty(K).$$

The set of all invariant means is denoted by $\mathcal{L}(K)$, and the set of all strongly invariant means is denoted by $\mathcal{SL}(K)$.

Obviously, $\mathcal{SL}(K) \subseteq \mathcal{L}(K)$. $\mathcal{L}(K)$ and $\mathcal{SL}(K)$ are weak $*$-compact convex subsets of $L^\infty(K)^*$. Shortly we will see that the commutativity of K implies that $\mathcal{L}(K)$ is nonempty. $\mathcal{SL}(K)$ can be empty.

Elements $x \in K$ act on $L^\infty(K)^*$ by the map $F \mapsto L_xF$, $L^\infty(K)^* \to L^\infty(K)^*$ where $L_xF(\varphi) = F(L_x\varphi)$ for all $\varphi \in L^\infty(K)$. Note that

$$\|L_xF\| \;=\; \sup\{|F(L_x\varphi)| : \|\varphi\|_\infty \leq 1\} \;\leq\; \|F\|.$$

If $m \in L^\infty(K)^*$ is a mean on $L^\infty(K)$, $L_x m$ is also a mean on $L^\infty(K)$. This follows by a general result for unital C^*-algebras, see e.g. [498,Corollary 3.3.4]. ($L^\infty(K)$ is a unital C^*-algebra.) Hence $\{L_x : x \in K\}$ is a commuting family of continuous and affine mappings (affine denotes the property $L_x(\lambda m_1 + (1-\lambda)m_2) = \lambda L_x(m_1) + (1-\lambda)L_x(m_2)$, $\lambda \in [0,1]$, m_1, m_2 means) from the set of means on $L^\infty(K)$ into itself. The Markov-Kakutani fixed-point theorem yields

Theorem 3.12.1 *Let K be a commutative hypergroup. Then $\mathcal{L}(K) \neq \emptyset$.*

We define

$$J(L^\infty(K)) := \{F \in L^\infty(K)^* : L_x F = F \text{ for all } x \in K\}.$$

Let

$$Bil(L^\infty(K)) = \{B : L^\infty(K) \times L^\infty(K) \to \mathbb{C} : B \text{ bilinear}\}.$$

Each $x \in K$ also determines a map $F \mapsto B^1_{x,F}$, $L^\infty(K)^* \to Bil(L^\infty(K))$ by

$$B^1_{x,F}(\varphi,\psi) = F((L_x\varphi)\cdot\psi) \quad \text{for all } \varphi,\psi \in L^\infty(K),$$

and a second mapping $F \mapsto B^2_{x,F}$, $L^\infty(K)^* \to Bil(L^\infty(K))$ by

$$B^2_{x,F}(\varphi,\psi) = F(\varphi\cdot(L_{\tilde{x}}\psi)) \quad \text{for all } \varphi,\psi \in L^\infty(K).$$

We define

$$\mathcal{S}J(L^\infty(K)) := \{F \in L^\infty(K)^* : B^1_{x,F} = B^2_{x,F} \text{ for all } x \in K\}.$$

Obviously, $J(L^\infty(K))$ and $\mathcal{S}J(L^\infty(K))$ are linear subspaces of $L^\infty(K)^*$ and $0 \in \mathcal{S}J(L^\infty(K)) \subseteq J(L^\infty(K))$.

Proposition 3.12.2 *(i) The linear span*

$$L := span\,\{L_x\varphi - \varphi : \varphi \in L^\infty(K),\ x \in K\}$$

is not dense in $L^\infty(K)$ if and only if $J(L^\infty(K)) \neq \{0\}$.

(ii) The linear span

$$SL := span\,\{L_x\varphi\cdot\psi - \varphi\cdot L_{\tilde{x}}\psi : \varphi,\psi \in L^\infty(K),\ x \in K\}$$

is not dense in $L^\infty(K)$ if and only if $SJ(L^\infty(K)) \neq \{0\}$.

(iii) $SJ(L^\infty(K)) \subsetneq J(L^\infty(K))$ if and only if $\overline{SL} \supsetneq \overline{L}$.

Proof. The proof of (i) follows from the lines of the proof of Proposition 2.1 in [532]. The proofs of (ii) and (iii) also use simple applications of the Hahn-Banach theorem. For the sake of completeness they are given here.

(ii) Assume $\overline{SL} = L^\infty(K)$. If $F \in SJ(L^\infty(K))$, then $B^1_{x,F} = B^2_{x,F}$, i.e. $F(L_x\varphi \cdot \psi - \varphi \cdot L_{\tilde{x}}\psi) = 0$ for all $\varphi, \psi \in L^\infty(K)$ and $x \in K$, and $\overline{SL} = L^\infty(K)$ implies $F = 0$. Conversely, if $\overline{SL} \subsetneq L^\infty(K)$, then there exists some $F \in L^\infty(K)^*$, $F \neq 0$, such that $F|\overline{SL} = 0$. In particular, $F \in SJ(L^\infty(K))$.

(iii) Assume that there is some $F \in J(L^\infty(K))\backslash SJ(L^\infty(K))$. Then $F|\overline{L} = 0$ and there are some $\varphi, \psi \in L^\infty(K)$ such that

$$F(L_x\varphi \cdot \psi - \varphi \cdot L_{\tilde{x}}\psi) \neq 0.$$

Hence $\overline{L} \subsetneq SL$. The inverse implication follows immediately by the Hahn-Banach theorem.

◇

To show the existence or nonexistence of strongly invariant means on K we use another version of the Hahn-Banach theorem, see [546,Lemma 4.27]. Let $SL_{\mathbb{R}}$ be the $\mathbb{R}$-linear span of $\{L_x\varphi\cdot\psi - \varphi\cdot L_{\tilde{x}}\psi : \varphi, \psi \in L^\infty_{\mathbb{R}}(K),\ x \in K\}$, where $L^\infty_{\mathbb{R}}(K)$ is the space of real-valued elements of $L^\infty(K)$. Concerning the relations between means on $L^\infty_{\mathbb{R}}(K)$ and $L^\infty(K)$ we refer to [546, Definition 3.1 and Proposition 3.2].

Proposition 3.12.3 *The following properties are equivalent*

(i) *There exists* $m \in L^\infty_{\mathbb{R}}(K)^*$ *with* $m(1) = 1$, $m \geq 0$ *and* $m|SL_{\mathbb{R}} = 0$.
(ii) $\sup\{\eta(x) : x \in K\} \geq 0$ *for all* $\eta \in SL_{\mathbb{R}}$.

Proof. By the definition of means on $L^\infty_{\mathbb{R}}(K)$ we know that $m(\varphi) \leq \sup\{\varphi(x) : x \in K\}$ for all $\varphi \in L^\infty_{\mathbb{R}}(K)$. Supposing (i) we obtain $0 = m(\eta) \leq \sup\{\eta(x) : x \in K\}$ for all $\eta \in SL_{\mathbb{R}}$. Conversely, assume that (ii) is valid. Let $N(\varphi) = \sup\{\varphi(x) : x \in K\}$ for $\varphi \in L^\infty_{\mathbb{R}}(K)$. Then N is a sublinear functional on $L^\infty_{\mathbb{R}}(K)$ and it dominates the zero functional on $SL_{\mathbb{R}}$. By the Hahn-Banach theorem, see [609], the zero-functional on $SL_{\mathbb{R}}$ can be extended to a linear functional m on $L^\infty_{\mathbb{R}}(K)$ such that $m(\varphi) \leq \sup\{\varphi(x) : x \in K\}$ for all $\varphi \in L^\infty_{\mathbb{R}}(K)$. Then m is a mean on $L^\infty_{\mathbb{R}}(K)$ [546,Proposition 3.2], and $m(\eta) = 0$ for all $\eta \in SL_{\mathbb{R}}$. ◇

Writing $\varphi = L^\infty(K)$ as $\varphi = \varphi_1 + i\varphi_2$, $\varphi_1, \varphi_2 \in L^\infty_{\mathbb{R}}(K)$, and putting $m(\varphi) := m_{\mathbb{R}}(\varphi_1) + i m_{\mathbb{R}}(\varphi_2)$ for a mean $m_{\mathbb{R}}$ on $L^\infty_{\mathbb{R}}(K)$ we obtain a mean on $L^\infty(K)$. If $\varphi, \psi \in L^\infty(K)$ and $\varphi = \varphi_1 + i\varphi_2$, $\psi = \psi_1 + i\psi_2$ with $\varphi_1, \varphi_2, \psi_1, \psi_2 \in L^\infty_{\mathbb{R}}(K)$, then

$$\begin{aligned} L_x\varphi \cdot \psi - \varphi \cdot L_{\tilde{x}}\psi &= (L_x\varphi_1 \cdot \psi_1 - \varphi_1 \cdot L_{\tilde{x}}\psi_1) - (L_x\varphi_2 \cdot \psi_2 - \varphi_2 \cdot L_{\tilde{x}}\psi_2) \\ &\quad + i\,(L_x\varphi_1 \cdot \psi_2 - \varphi_1 \cdot L_{\tilde{x}}\psi_2) + i\,(L_x\varphi_2 \cdot \psi_1 - \varphi_2 \cdot L_{\tilde{x}}\psi_1). \end{aligned}$$

Therefore, if $m_{\mathbb{R}}$ is a mean on $L^\infty_{\mathbb{R}}(K)$ with $m_{\mathbb{R}}|SL_{\mathbb{R}} = 0$, then m is a mean on $L^\infty(K)$ such that $m|SL = 0$. Hence we have the following characterization of the existence of a strongly invariant mean on $L^\infty(K)$.

Theorem 3.12.4 *Let K be a commutative hypergroup. There exists a strongly invariant mean m on $L^\infty(K)$ if and only if* $\sup\{\eta(x) : x \in K\} \geq 0$ *for all $\eta \in SL_{\mathbb{R}}$.*

Strongly invariant means on $L^\infty(K)$ have an interesting separation property.

Proposition 3.12.5 *Suppose that there exists a strongly invariant mean m on $L^\infty(K)$. Let $\alpha, \beta \in \hat{K}$, $\alpha \neq \beta$. Then $m(\alpha \cdot \bar{\beta}) = 0$.*

Proof. Let $x \in K$ such that $\alpha(x) \neq \beta(x)$. Since $L_{\tilde{x}}\bar{\beta} = \beta(x)\bar{\beta}$, we obtain

$$\alpha(x)\, m(\alpha \cdot \bar{\beta}) \;=\; m((L_x\alpha) \cdot \bar{\beta}) \;=\; m(\alpha \cdot (L_{\tilde{x}}\bar{\beta})) \;=\; \beta(x)\, m(\alpha \cdot \bar{\beta}).$$

This is only possible if $m(\alpha \cdot \bar{\beta}) = 0$. ◇

Now fix some $\alpha \in \hat{K}$. We will study linear functionals $m_\alpha \in L^\infty(K)^*$ which have the following properties $m_\alpha(\alpha) = 1$ and

$$m_\alpha(L_x\varphi) \;=\; \alpha(x)\, m_\alpha(\varphi) \qquad \text{for all } x \in K \text{ and } \varphi \in L^\infty(K).$$

We call such functionals $m_\alpha \in L^\infty(K)^*$ an **α-invariant functional** on $L^\infty(K)$. Obviously, an invariant mean m on $L^\infty(K)$ is a 1-invariant functional on $L^\infty(K)$ with the additional property $\|m\| = 1$. Note that we assume throughout that an α-invariant function is a continuous functional.

We will compare the action of K with the action of $L^1(K)$ on $L^\infty(K)$. Consider

$$UC(K) \;=\; \{f \in C^b(K) : x \mapsto L_x f \text{ is continuous from } K \text{ to } L^\infty(K)\},$$

the space of all **bounded uniformly continuous functions** on K. $UC(K)$ is a closed subspace of $C^b(K)$ containing $C_0(K)$.

Remark: There are various ways to define uniform continuity, see [96,Definition 1.2.26]. The choice above is exactly the β-uniform continuity of Bloom-Heyer.

We have already noticed that $f * \varphi \in C^b(K)$, if $f \in L^1(K)$, $\varphi \in L^\infty(K)$, see the remark below Proposition 2.2.13. Applying that $x \mapsto L_x f$, $K \to L^1(K)$, $f \in L^1(K)$, is continuous, one can show the following result.

Proposition 3.12.6 *Let K be a commutative hypergroup. Then*

(1) For $f \in L^1(K)$, $\varphi \in L^\infty(K)$ we have $f * \varphi \in UC(K)$,
(2) $L^1(K) * L^\infty(K) = UC(K)$.

Proof.

(1) We have

$$|f * \varphi(x) - f * \varphi(y)| = \left| \int_K \varphi(z) \left(L_x f(\tilde{z}) - L_y f(\tilde{z})\right) dm(z) \right|$$

$$\leq \|\varphi\|_\infty \|L_x f - L_y f\|_1,$$

and it follows $f * \varphi \in C^b(K)$.
Furthermore we obtain for any $z \in K$ (using the commutativity of K several times),

$$|L_x(f * \varphi)(z) - L_y(f * \varphi)(z)| = |(L_z f) * \varphi(x) - (L_z f) * \varphi(y)|$$

$$\leq \|\varphi\|_\infty \|L_x(L_z f) - L_y(L_z f)\|_1 \leq \|\varphi\|_\infty \|L_z(L_x f) - L_z(L_y f)\|_1$$

$$\leq \|\varphi\|_\infty \|L_x f - L_y f\|_1.$$

Hence

$$\|L_x(f * \varphi) - L_y(f * \varphi)\|_\infty \leq \|\varphi\|_\infty \|L_x f - L_y f\|_1,$$

and (1) is shown.

(2) Note, by (1) the Banach space $UC(K)$ is a $L^1(K)$-module.
Let $\varphi \in UC(K)$ and $\varepsilon > 0$. Choose a neighbourhood V of e with compact closure such that $\|L_x\varphi - \varphi\|_\infty < \varepsilon$ for each $x \in V$, and let $f \in L^1(K)$, $f \geq 0$, supp $f \subseteq V$ and $\|f\|_1 = 1$. Then $\|f * \varphi - \varphi\|_\infty \leq \varepsilon$, and we infer that $L^1(K) * UC(K)$ is norm-dense in $UC(K)$. Since $L^1(K)$ has a bounded approximate unit, by Cohen's factorization theorem [305,(32.50)] we obtain $L^1(K) * UC(K) = UC(K)$. Obviously, it follows $L^1(K) * L^\infty(K) = UC(K)$.

◇

Restricting ourselves to $UC(K) \subseteq L^\infty(K)$ we also consider (strongly) invariant means on $UC(K)$ and α-invariant functionals on $UC(K)$. Clearly a mean m on $UC(K)$ is an element of $UC(K)^*$ such that $m(1) = 1 = \|m\|$.

Proposition 3.12.7 *Let K be a commutative hypergroup.*

(1) If m is a mean on $L^\infty(K)$ such that $m(f * \varphi) = \hat{f}(1)\, m(\varphi)$ for every $\varphi \in L^\infty(K)$, $f \in L^1(K)$, then m is an invariant mean on $L^\infty(K)$.

(2) *If m is an invariant mean on $L^\infty(K)$, then $m(f * \varphi) = \hat{f}(1)\, m(\varphi)$ for every $\varphi \in UC(K)$, $f \in L^1(K)$. In particular, if $g \in L^1(K)$, $g \geq 0$, $\|g\|_1 = 1$, then putting*

$$m^g(\varphi) := m(g * \varphi) \qquad \text{for } \varphi \in L^\infty(K),$$

the functional m^g is also an invariant mean on $L^\infty(K)$.

Proof.

(1) Choose some $g \in L^1(K)$ such that $\hat{g}(1) = 1$. Then

$$m(L_x\varphi) = m(g * L_x\varphi) = m(L_x g * \varphi) = \widehat{L_x g}(1)\, m(\varphi) = m(\varphi)$$

for each $\varphi \in L^\infty(K)$.

(2) Let $\varphi \in UC(K)$ and $f \in C_c(K) \subseteq L^1(K)$. Then $f * \varphi \in UC(K)$ is an $UC(K)$-valued integral,

$$f * \varphi = \int_K L_{\tilde{y}}\varphi\, f(y)\, d\lambda(y).$$

(λ denotes for this part the Haar measure on K.) Hence

$$m(f * \varphi) = \int_K f(y)\, m(L_{\tilde{y}}\varphi)\, d\lambda(y) = \hat{f}(1)\, m(\varphi).$$

Since $C_c(K)$ is dense in $L^1(K)$ we obtain $m(f * \varphi) = \hat{f}(1)\, m(\varphi)$ for every $\varphi \in UC(K)$, $f \in L^1(K)$.

Now let $g \in L^1(K)$, $g \geq 0$, $\|g\|_1 = 1$. We have $m^g(1) = m(g * 1) = \|g\|_1 m(1) = 1$, and for $\varphi \in L^\infty(K)$, $\varphi \geq 0$, it follows $m^g(\varphi) \geq 0$. Therefore m^g is a mean on $L^\infty(K)$. By Proposition 3.12.6 we know that $g * \varphi \in UC(K)$ for each $\varphi \in L^\infty(K)$. it follows

$$m^g(f * \varphi) = m(f * g * \varphi) = \hat{f}(1)\, m(g * \varphi) = \hat{f}(1)\, m^g(\varphi)$$

for every $\varphi \in L^\infty(K)$, $f \in L^1(K)$. By (1) the mean m^g is invariant.

◇

Using similar methods we can show for arbitrary $\alpha \in \hat{K}$.

Proposition 3.12.8 *Let K be a commutative hypergroup, $\alpha \in \hat{K}$.*

(1) *If m_α is a continuous functional on $L^\infty(K)$ such that $m_\alpha(\alpha) = 1$ and $m_\alpha(f * \varphi) = \hat{f}(\alpha)\, m_\alpha(\varphi)$, then m_α is an α-invariant functional on $L^\infty(K)$.*

(2) *If m_α is an α-invariant functional on $L^\infty(K)$, then*

$$m_\alpha(f * \varphi) = \hat{f}(\alpha)\, m_\alpha(\varphi) \qquad \text{for every } \varphi \in UC(K),\ f \in L^1(K).$$

In particular, if $g \in L^1(K)$ with $\hat{g}(\alpha) = 1$, then putting

$$m_\alpha^g(\varphi) := m_\alpha(g * \varphi) \qquad \text{for } \varphi \in L^\infty(K),$$

the functional m_α^g is an α-invariant functional on $L^\infty(K)$.

For strongly invariant means we can derive the following facts.

Proposition 3.12.9 *Let K be a commutative hypergroup. Then*

(1) *If m is a mean on $UC(K)$ such that $m((f * \varphi) \cdot \psi) = m(\varphi \cdot (\tilde{f} * \psi))$ for each $\varphi, \psi \in UC(K)$, $f \in L^1(K)$, then m is a strongly invariant mean on $UC(K)$.*

(2) *Let m be a strongly invariant mean on $UC(K)$. Then*

$$m((f * \varphi) \cdot \psi) = m(\varphi \cdot (\tilde{f} * \psi)) \qquad \text{for all } \varphi, \psi \in UC(K),\ f \in L^1(K).$$

Proof.

(1) Let $(f_i)_{i \in I}$, $f_i \in L^1(K)$, $f_i = \tilde{f}_i$, such that $\lim_i f_i * \varphi = \varphi$ and $\lim_i f_i * \psi = \psi$ in $UC(K)$, see the proof of Proposition 3.12.6(2). For $x \in K$ we have

$$\begin{aligned} m((L_x\varphi) * \psi) &= \lim_i m((f_i * L_x\varphi) \cdot \psi) = \lim_i m((L_x f_i * \varphi) \cdot \psi) \\ &= \lim_i m(\varphi \cdot (\widetilde{L_x f_i} * \psi)) = \lim_i m(\varphi \cdot (L_{\tilde{x}} f_i * \psi)) \\ &= \lim_i m(\varphi \cdot (f_i * L_{\tilde{x}}\psi)) = m(\varphi \cdot (L_{\tilde{x}}\psi)). \end{aligned}$$

(2) Let $f \in C_c(K)$. Then $f * \varphi$ is equal to the $UC(K)$-valued integral $\int_K L_{\tilde{y}}\varphi\, f(y) d\lambda(y)$, λ the Haar measure on K. Obviously $y \mapsto (L_{\tilde{y}}\varphi) \cdot \psi$, $K \to UC(K)$, is a continuous mapping from K into $UC(K)$ and

$$\int_K ((L_{\tilde{y}}\varphi) \cdot \psi)\, f(y)\, d\lambda(y) = \psi \int_K (L_{\tilde{y}}\varphi)\, f(y)\, d\lambda(y) = (f * \varphi) \cdot \psi.$$

Therefore,

$$\begin{aligned} m((f * \varphi) \cdot \psi) &= \int_K m((L_{\tilde{y}}\varphi) \cdot \psi)\, f(y)\, d\lambda(y) \\ &= \int_K m(\varphi \cdot L_{\tilde{y}}\psi)\, \tilde{f}(y)\, d\lambda(y) = m(\varphi \cdot (\tilde{f} * \psi)). \end{aligned}$$

Since $C_c(K)$ is dense in $L^1(K)$, (2) is shown.

◇

For certain $\alpha \in \hat{K}$ we can derive the existence of α-invariant functionals on $L^\infty(K)$ or $UC(K)$ from the existence of strongly invariant means on $L^\infty(K)$ or $UC(K)$.

Theorem 3.12.10 *Let $\alpha \in \hat{K}$. If there exists a strongly invariant mean m on $L^\infty(K)$ or $UC(K)$, such that $m(\alpha\bar{\alpha}) > 0$, then there exists an α-invariant functional m_α on $L^\infty(K)$ or $UC(K)$, respectively.*

Proof. Denote $c_\alpha = m(\bar{\alpha}\alpha)^{-1}$ and define

$$m_\alpha(\varphi) \;=\; c_\alpha\, m(\bar{\alpha}\varphi) \qquad \text{for } \varphi \in L^\infty(K) \quad (\text{or } UC(K)).$$

Then $m_\alpha(\alpha) = 1$ and $m_\alpha(L_x\varphi) = c_\alpha m(\bar{\alpha}L_x\varphi) = c_\alpha m((L_{\tilde{x}}\bar{\alpha})\cdot\varphi) = \alpha(x)c_\alpha m(\bar{\alpha}\varphi) = \alpha(x)m_\alpha(\varphi)$. ◇

Remark:

(1) For polynomial hypergroups we will give a complete characterization of the existence of strongly invariant means. (See subsection 4.8.)
(2) In the next subsection we will investigate relations of invariant means and α-invariant functionals to properties like Reiter's conditions or the existence of approximate units in maximal ideals of $L^1(K)$.

Finally we present a result, which should be compared with Proposition 2.4.3.

Proposition 3.12.11 *Assume that there exists a strongly invariant mean m on $UC(K)$. For every $\psi \in UC(K)$ the function $\varphi(x) = m((L_x\psi)\cdot\bar{\psi})$ is bounded, continuous and positive definite, i.e. $\varphi \in P^b(K)$.*

Proof. We have to prove only that φ is positive definite. Let $x_1, ..., x_n \in K$; $c_1, ..., c_n \in \mathbb{C}$. Applying Proposition 1.1.13 we obtain

$$\sum_{i=1}^{n}\sum_{j=1}^{n} c_i\overline{c_j}\,\omega(x_i,\tilde{x_j})(\varphi) \;=\; \sum_{i=1}^{n}\sum_{j=1}^{n} c_i\overline{c_j}\, m\left(\left(\int_K L_y\psi\, d\omega(x_{i_1},\tilde{x_j})(y)\right)\cdot\bar{\psi}\right)$$

$$= \sum_{i=1}^{n}\sum_{j=1}^{n} c_i\overline{c_j}\, m((L_{\tilde{x_j}}\circ L_{x_i}(\psi))\cdot\bar{\psi})$$

$$= \sum_{i=1}^{n}\sum_{j=1}^{n} c_i\overline{c_j}\, m((L_{x_i}\psi)\cdot(\overline{L_{x_j}\psi})) \;=\; m\left(\left|\sum_{i=1}^{n} c_i L_{x_i}\psi\right|^2\right) \;\geq\; 0.$$

◇

Remark: There exist various amenability properties of the Banach algebras $L^1(K)$ and $M(K)$, for example amenability, weak amenability, φ-amenability and so on. We want to point to the fact that in contrast to the group case, the amenability of K (i.e. the existence of invariant means on $L^\infty(K)$) does not imply the amenability of $L^1(K)$.

13 Reiter's P_1-conditions for commutative hypergroups

Now we discuss Reiter's condition (P_1) and related properties. We will extend the P_1-condition by studying the relation between translation and multiplication with a character $\alpha \in \hat{K}$. Recall Theorem 3.7.5, where it is shown that $\alpha \in \mathcal{S}(K)$ is equivalent to the fact that the P_2-condition is satisfied in α. We will derive several properties of spectral synthesis (e.g. existence of approximate identities in maximal ideals of $L^1(K)$), which are strongly related to P_1-conditions for $\alpha \in \hat{K}$. For commutative groups G we refer to [305], [570] or [608].

The hypergroup K satisfies the condition P_1 if for each $\varepsilon > 0$ and every compact subset $C \subseteq K$ there exists a function $g \in L^1(K)$ with the properties $g \geq 0$, $\|g\|_1 = 1$ and $\|L_y g - g\|_1 < \varepsilon$ for every $y \in C$.

The modified condition P_1 reads as follows.

Definition. We say that the P_1-condition [P_1^*-condition] with bound $M > 0$ is satisfied in $\alpha \in \mathcal{X}^b(K)$ (for the sake of brevity denoted by $P_1(\alpha, M)$ [$P_1^*(\alpha, M)$]) if for each $\varepsilon > 0$ and every compact [finite] subset $C \subseteq K$ there exists some $g \in L^1(K)$ with the following properties

(i) $\mathcal{F}g(\alpha) = 1$
(ii) $\|g\|_1 \leq M$
(iii) $\|L_{\tilde{y}} g - \overline{\alpha(y)}\, g\|_1 < \varepsilon$ for all $y \in C$.

We have the following relation between the conditions P_1 and P_1^*.

Proposition 3.13.1 *The condition $P_1(\alpha, \tilde{M})$ follows from the condition $P_1^*(\alpha, M)$, where the constant $\tilde{M}$ depends only on M and α.*

Proof. Let $s \in L^1(K)$ be a function with $\mathcal{F}s(\alpha) = 1$. Let $C \subseteq K$ be a compact subset and $\delta > 0$. We can choose finitely many open neighbourhoods $V_i = V(y_i)$, $i = 1, ..., m$ in such a way that $C \subseteq \bigcup V_i$ and

$$|\alpha(y) - \alpha(y')| < \frac{\delta}{3M\,\|s\|_1}, \qquad \|L_y s - L_{y'} s\|_1 < \frac{\delta}{3M}$$

for $y, y' \in V_i$. The condition $P_1^*(\alpha, M)$ ensures the existence of a function $g \in L^1(K)$ with $\mathcal{F}g(\alpha) = 1$, $\|g\|_1 \le M$ and $\|L_{\tilde{y}_i} g - \overline{\alpha(y_i)}\, g\|_1 < \frac{\delta}{3\,\|s\|_1}$ for $i = 1, ..., m$. For $y \in C$ there is a set V_j with $y \in V_j$. For $f = g * s$ we have $\mathcal{F}f(\alpha) = 1$, $\|f\|_1 \le M\,\|s\|_1$ and

$$\|L_{\tilde{y}} f - \overline{\alpha(y)}\, f\|_1 \;\le\; \|g\|_1\, \|L_{\tilde{y}} s - L_{\tilde{y}_j} s\|_1 \;+\; \|L_{\tilde{y}_j} g - \overline{\alpha(y_j)}\, g\|_1\, \|s\|_1$$
$$+\; |\overline{\alpha(y_j)} - \overline{\alpha(y)}|\; \|g\|_1\, \|s\|_1 \;<\; \delta.$$

This shows the assertion. ◇

The following proposition will be helpful in proving the main result.

Proposition 3.13.2 *Let $\alpha \in \mathcal{X}^b(K)$ and assume that condition $P_1(\alpha, M)$ holds. Let $\{f_1, ..., f_n\}$ be a finite subset of $I(\alpha) = \{f \in L^1(K) : \mathcal{F}f(\alpha) = 0\}$. Then for every $\varepsilon > 0$ there is a $g \in L^1(K)$ with $\mathcal{F}g(\alpha) = 1$, $\|g\|_1 \le M$ and $\|f_i * g\|_1 < \varepsilon$, $i = 1, ..., n$.*

Proof. Let $\varepsilon > 0$ be given and let $C \subset K$ be a compact set with

$$\int_{K \setminus C} |f_i(y)|\, dm(y) \;<\; \frac{\varepsilon}{4M}, \qquad i = 1, ..., n.$$

Let $\varrho = \max\limits_{i=1,\dots,n} \|f_i\|_1$. Since $P_1(\alpha, M)$ is fulfilled, there is some $g \in L^1(K)$ with $\mathcal{F}g(\alpha) = 1$, $\|g\|_1 \le M$ and

$$\|L_{\tilde{y}} g - \overline{\alpha(y)}\, g\|_1 \;<\; \frac{\varepsilon}{1 + \varrho} \qquad \text{for all } y \in C.$$

So we get

$$\|f_i * g\|_1 \;=\; \int_K \left| \int_K f_i(y)\, L_{\tilde{y}} g(x)\, dm(y) \right| \, dm(x)$$

$$= \int_K \left| \int_K f_i(y)\, L_{\tilde{y}} g(x)\, dm(y) \;-\; \int_K f_i(y)\, \overline{\alpha(y)}\, g(x)\, dm(y) \right| \, dm(x)$$

$$\le \int_K \int_K |f_i(y)| \cdot |L_{\tilde{y}} g(x) - \overline{\alpha(y)}\, g(x)|\, dm(y)\, dm(x)$$

$$= \int_{K \setminus C} |f_i(y)| \cdot \|L_{\tilde{y}} g - \overline{\alpha(y)}\, g\|_1\, dm(y) + \int_C |f_i(y)| \cdot \|L_{\tilde{y}} g - \overline{\alpha(y)}\, g\|_1\, dm(y)$$

$$\le\; 2\, \|g\|_1 \int_{K \setminus C} |f_i(y)|\, dm(y) \;+\; \frac{\varepsilon}{1 + \varrho} \int_C |f_i(y)|\, dm(y)$$

$$\le\; 2M\, \frac{\varepsilon}{4M} \;+\; \frac{\varepsilon}{2(1 + \varrho)}\, \|f_i\| \;<\; \epsilon$$

This inequality is valid for all $i = 1, ..., n$. ◇

By means of Proposition 3.13.2 we are able to give a characterization for the existence of a bounded approximate identity in the maximal ideal $I(\alpha)$, where

$$I(\alpha) := \{f \in L^1(K) : \mathcal{F}f(\alpha) = 0\}.$$

Let A be a commutative Banach algebra. The set of all maximal modular ideals in A is denoted by $\mathrm{Max}(A)$. There is a basic link between $\Delta(A)$ and $\mathrm{Max}(A)$. The mapping $\varphi \mapsto \ker\varphi = \{a \in A : \varphi(a) = 0\}$ is a bijection between $\Delta(A)$ and $\mathrm{Max}(A)$, see for example [356,Theorem 2.1.8].

Thus for $A = L^1(K)$ we have a bijection between $\mathcal{X}^b(K)$ and $\mathrm{Max}(L^1(K))$, $\alpha \mapsto I(\alpha) = \ker h_\alpha = \{f \in L^1(K) : \mathcal{F}f(\alpha) = 0\}$, see section 3.1.

Theorem 3.13.3 *Let $\alpha \in \mathcal{X}^b(K)$. Then $I(\alpha)$ has a bounded approximate identity $(u_\lambda)_{\lambda\in\Lambda}$ with bound M if and only if $P_1(\alpha, M')$ is satisfied, where M' is a constant depending only on M and α.*

Proof. In order to show that there is an approximate identity in the maximal ideal $I(\alpha)$, we have to prove that for every finite set$\{f_1, ..., f_n\} \subset I(\alpha)$ and every $\varepsilon > 0$ there is a $u \in I(\alpha)$ such that $\|f_i - f_i * u\|_1 < \varepsilon$ for $i \in \{1, ..., n\}$. Let $\{f_1, ..., f_n\} \subset I(\alpha)$ be a finite set and $\varepsilon > 0$. By Proposition 3.13.2 there exists a $g \in L^1(K)$ with $\mathcal{F}g(\alpha) = 1, \quad \|g\|_1 \leq M'$ and $\|f_i * g\|_1 \leq \frac{\varepsilon}{2}, \quad i = 1, ..., n$.

Let $(e_\lambda)_{\lambda\in\Lambda}$ be an approximate identity for the Banach algebra $L^1(K)$ with $\|e_\lambda\|_1 = 1$ for all $\lambda \in \Lambda$. There is a $\lambda_0 \in \Lambda$ with

$$\|f_i - f_i * e_\lambda\|_1 < \frac{\varepsilon}{2} \qquad \text{for all } \lambda \geq \lambda_0, \quad i = 1, ..., n.$$

We set $u := e_{\lambda_0} - g * e_{\lambda_0}$. Since

$$\mathcal{F}u(\alpha) = \mathcal{F}e_{\lambda_0}(\alpha) - \mathcal{F}g(\alpha)\,\mathcal{F}e_{\lambda_0}(\alpha) = 0,$$

u is an element of $I(\alpha)$. Furthermore we have for all $i = 1, ..., n$

$$\|f_i - f_i * u\|_1 = \|f_i - f_i * e_{\lambda_0} + f_i * g * e_{\lambda_0}\|_1 \leq \|f_i - f_i * e_{\lambda_0}\|_1 + \|f_i * g\|_1 < \varepsilon$$

and $\|u\|_1 \leq M' + 1$. This shows our assertion.

Conversely, let $(u_\lambda)_{\lambda\in\Lambda}$ be a bounded approximate identity for $I(\alpha)$ with bound $M \geq 0$ and let $s \in L^1(K)$ be a function with $\mathcal{F}s(\alpha) = 1$. We define a net by $g_\lambda := s - s * u_\lambda,\ \ \lambda \in \Lambda$. These functions satisfy $\mathcal{F}g_\lambda(\alpha) = 1$ and $\|g_\lambda\|_1 \leq \|s\|_1(1+M) =: M'$.

Let $C \subset K$ be a given compact subset and $\varepsilon > 0$. By the continuity of the map $K \to L^1(K),\ \ y \mapsto L_y s$ and the continuity of the character $\alpha \in \mathcal{X}^b(K)$ there exist points $y_1, ..., y_n \in C$ and open neighbourhoods $V_i := V(y_i)$ of $y_i,\ \ i = 1, ..., n$ with $C \subseteq \bigcup_{i=1}^n V_i$ such that

$$\|L_{\tilde{y}} s - L_{\tilde{y}_k} s\|_1 < \frac{\varepsilon}{3(1+M)}, \quad |\alpha(y) - \alpha(y_k)| < \frac{\varepsilon}{3M'}$$
$$\text{for all } y \in V_k,\ \ 1 \leq k \leq n.$$

For any $y \in K$ we set $f_y := L_{\tilde{y}} s - \overline{\alpha(y)}\, s$. Since $\mathcal{F} f_y(\alpha) = 0$ this function is an element of $I(\alpha)$. Moreover, we have

$$L_{\tilde{y}} g_\lambda \ - \ \overline{\alpha(y)}\, g_\lambda \ = \ f_y \ - \ f_y * u_\lambda.$$

Since $(u_\lambda)_{\lambda \in \Lambda}$ is an approximate identity for $I(\alpha)$ there is a $\lambda_0 \in \Lambda$ with

$$\|f_{y_k} \ - \ f_{y_k} * u_{\lambda_0}\|_1 \ < \ \frac{\varepsilon}{3}$$

for all $k = 1, ..., n$. This gives us

$$\|L_{\tilde{y}_k} g_{\lambda_0} \ - \ \overline{\alpha(y_k)}\, g_{\lambda_0}\|_1 \ < \ \frac{\varepsilon}{3}$$

for all $k = 1, ..., n$. Applying the triangle inequality we end up with

$$\|L_{\tilde{y}} g_{\lambda_0} \ - \ \overline{\alpha(y)}\, g_{\lambda_0}\|_1 \ < \ \epsilon \qquad \text{for all } y \in C.$$

This completes the proof. ◇

Now we will give the relation between the modified Reiter condition $P_1(\alpha, M)$ and the existence of α-invariant functionals m_α on $L^\infty(K)$.

Theorem 3.13.4 *Let $\alpha \in \hat{K}$. Then the condition $P_1(\alpha, M)$ is satisfied if and only if there exists $m_\alpha \in (L^\infty(K))^*$ with*

(i) $m_\alpha(\alpha) = 1$,
(ii) $\|m_\alpha\| \ \leq \ M$,
(iii) $m_\alpha(L_y f) \ = \ \alpha(y)\, m_\alpha(f)$ *for all* $f \in L^\infty(K),\ \ y \in K$.

Proof. Assume the condition $P_1(\alpha, M)$ is satisfied. For $\varepsilon > 0$ and a compact set $C \subset K$ let $g \in L^1(K)$ according to $P_1(\alpha, M)$. We define the functional $m_{\varepsilon,C}$ on $L^\infty(K)$ by the rule

$$m_{\varepsilon,C}(f) \ = \ \int_K f(\tilde{x})\, g(x)\, dm(x).$$

We have $m_{\varepsilon,C}(\alpha) = 1$ and

$$\|m_{\varepsilon,C}\| \ \leq \ \|g\|_1 \ \leq \ M.$$

Hence the functionals $m_{\varepsilon,C}$ are uniformly bounded. Moreover for $y \in C$ we have

$$m_{\varepsilon,C}(L_y f) = \int_K L_y f(\tilde{x})\, g(x)\, dm(x) = \int_K f(\tilde{x})\, L_y g(x)\, dm(x).$$

Thus

$$|m_{\varepsilon,C}(L_y f) - \alpha(y)\, m_{\varepsilon,C}(f)| \leq \int_K |f(x)| \cdot |L_y g(x) - \alpha(y) g(x)|\, m(x)$$

$$\leq \|f\|_\infty \, \|L_y g - \alpha(y) g(x)\|_1 \leq \varepsilon \, \|f\|_\infty.$$

The family of functionals $m_{\varepsilon,C}$ form a net, where the indices (ε, C) are partially ordered by

$$(\varepsilon, C) \prec (\varepsilon', C') \quad \text{if} \quad \varepsilon' \leq \varepsilon, \;\; C \subset C'.$$

Let m_α be an accumulation point of this net. Then $\|m_\alpha\| \leq M$ and $m_\alpha(\alpha) = 1$. Moreover

$$m_\alpha(L_y f) = \alpha(y)\, m_\alpha(f).$$

Conversely assume that an α-invariant functional m_α exists. Since m_α belongs to the second dual of $L^1(K)$ by the Goldstine theorem [189,p.424] there is a net $(f_i)_{i\in I}$ $*$-weakly convergent to m_α such that $\|f_i\|_1 \leq M$. In particular we have $\hat{f}_i(\alpha) \to m_\alpha(\alpha)$. Since $m_\alpha(\alpha) = 1$ we can assume $\hat{f}_i(\alpha) = 1$. For any $y \in K$ and $f \in L^\infty(K)$ we have

$$\int_K L_{\tilde{y}} f_i(x)\, \overline{f(x)}\, dm(x) = \int_K f_i(x)\, \overline{L_y f(x)}\, dm(x) \to m_\alpha(L_y f) = \alpha(y)\, m_\alpha(f).$$

Therefore

$$\int_K \left(L_{\tilde{y}} f_i(x) - \overline{\alpha(y)}\, f_i(x) \right) \overline{f(x)}\, dm(x) \to 0.$$

Fix $y_1, y_2, ..., y_m \in K$. Let $F_{k,i} = L_{\tilde{y}_k} f_i - \overline{\alpha(y_k)}\, f_i$. The m-tuple

$$\mathbf{F}_i = (F_{i,1}, F_{i,2}, .., F_{m,i})$$

forms a net weakly convergent to 0 in the product space $L^1(K) \times \cdots \times L^1(K)$. By [189, Corollary 14, p.422] there is a sequence of convex combinations of $\mathbf{F}_i$ convergent to 0 in norm. Hence for every $\varepsilon > 0$ there is a function $g \in L^1(K)$, a convex linear combination of f_i's, such that $\hat{g}(\alpha) = 1$, $\|g\|_1 \leq M$ and

$$\|L_{\tilde{y}_i} g - \overline{\alpha(y_i)}\, g\|_1 < \varepsilon \qquad \text{for } i = 1, ..., m.$$

Now our assertion follows from Proposition 3.13.1. ◇

Till now we have shown that for $\alpha \in \hat{K}$ the following conditions are equivalent:

(i) There exists an α-invariant functional $m_\alpha \in L^\infty(K)^*$ with $\|m_\alpha\| \le M$.
(ii) The Reiter condition $P_1(\alpha, M)$ is satisfied with bound $M \ge 1$.
(iii) $I(\alpha)$ has a bounded approximate identity $(u_\lambda)_{\lambda\in\Lambda}$.

We will add a fourth and fifth property to the three equivalence conditions (i), (ii) and (iii).

Denote by

$$J(\alpha, M) := \{f \in L^1(K) : \hat{f}(\alpha) = 1, \ \|f\|_1 \le M\}.$$

Obviously Reiter's $P_1(\alpha, M)$-condition is equivalent to the existence of a net $(g_i)_{i\in I}$, $g_i \in J(\alpha, M)$ such that $\|L_x g_i - \alpha(x) g_i\|_1 \to 0$ uniformly on compacta.

Proposition 3.13.5 *Let $\alpha \in \hat{K}$ and let $(g_i)_{i\in I}$ be a net in $J(\alpha, M)$ such that*
$\|L_x g_i - \alpha(x) g_i\|_1 \to 0$ uniformly on compacta. Then

$$\|\mu * g_i - \hat{\mu}(\alpha) g_i\|_1 \to 0 \qquad \text{for all } \mu \in M(K).$$

Proof. Since $\|g_i\|_1 \le M$ and $\|\mu * g\|_1 \le \|\mu\| \, \|g\|_1$ for $g \in L^1(K)$, we can suppose that $\mu \in M(K)$ has compact support C. Using Fubini's theorem we have for $g \in L^1(K)$

$$\|\mu * g - \hat{\mu}(\alpha) g\|_1 \le \int_K \int_K |L_{\tilde{x}} g(y) - \alpha(\tilde{x}) g(y)| \, d\mu(x) \, dm(y)$$

$$= \int_C \|L_{\tilde{x}} g - \alpha(\tilde{x}) g\|_1 \, d\mu(x),$$

and since $\sup\{ \|L_{\tilde{x}} g_i - \alpha(\tilde{x}) g\|_1 : x \in C\} \to 0$ the required result follows. ◇

For $\alpha \in \hat{K}$ and $M > 0$ denote by

$$T(\alpha, M) := \left\{ \sum_{i=0}^{n} \beta_i L_{\tilde{y}_i} : n \in \mathbb{N}_0;\ \beta_0, ..., \beta_n \in \mathbb{C},\ \sum_{i=0}^{n} |\beta_i| \le M : \right.$$

$$\left. y_0, ..., y_n \in K,\ \sum_{i=0}^{n} \beta_i \, \overline{\alpha(y_i)} = 1 \right\} \subseteq \mathcal{B}(L^1(K)).$$

Proposition 3.13.6 *Let $\alpha \in \hat{K}$ and $g \in L^1(K)$ such that $\hat{g}(\alpha) = 1$. Then $g * f \in \overline{T(\alpha, M)(f)} \subseteq L^1(K)$ for every $f \in L^1(K)$, where $M = \|g\|_1 + 1$.*

Proof. Let $\varepsilon > 0$ and put $\delta = \min\{\frac{1}{2}, \frac{\varepsilon}{2\|f\|_1+\|g\|_1}\}$. There exists a compact subset $C \subseteq K$ such that $\int_{K\setminus C} |g(y)|\, dm(y) \le \delta$. Since $x \mapsto L_x f,\ K \to L^1(K)$ is continuous, there exist neighbourhoods $V(y_i)$ of $y_i \in C,\ i = 1, ..., n$, such that $C \subseteq \bigcup_{i=1}^{n} V(y_i)$ and

$$\|L_{\tilde{y}} f - L_{\tilde{y_i}} f\|_1 < \delta \quad \text{and } |\alpha(\tilde{y}) - \alpha(\tilde{y_i})| < \delta/\|g\|_1 \quad \text{for } y \in V(y_i).$$

We put $y_0 = e,\ A_0 = K\setminus C,\ A_1 = V(y_1) \cap C$ and inductively $A_i = V(y_i) \cap \left(C\setminus \bigcup_{j=1}^{i-1} A_j\right)$ for $i = 2, ..., n$. For $i \in \{1, ..., n\}$ put

$$\beta_i := \int_{A_i} g(y)\, dm(y) \quad \text{and } \beta_0 := 1 - \sum_{i=1}^{n} \beta_i \alpha(\tilde{y_i}).$$

Then $\sum_{i=0}^{n} \beta_i\, \overline{\alpha(y_i)} = 1$, and

$$|\beta_0| = \left| \hat{g}(\alpha) - \sum_{i=1}^{n} \beta_i \cdot \overline{\alpha(y_i)} \right|$$

$$= \left| \int_{A_0} g(y)\, \overline{\alpha(y)}\, dm(y) + \sum_{i=1}^{n} \int_{A_i} g(y) \left(\overline{\alpha(y)} - \overline{\alpha(y_i)}\right) dm(y) \right|$$

$$\le \delta + \sum_{i=1}^{n} \int_{A_i} |g(y)|\, |\alpha(y) - \alpha(y_i)|\, dm(y) \le 2\delta \le 1.$$

Hence

$$\sum_{i=0}^{n} |\beta_i| \le 1 + \sum_{i=1}^{n} \int_{A_i} |g(y)|\, dm(y) \le 1 + \|g\|_1 =: M.$$

For m-almost every $x \in K$ we obtain

$$\left| g * f(x) - \sum_{i=1}^{n} \beta_i\, L_{\tilde{y_i}} f(x) \right| \le \int_{A_0} |g(y)\, |L_{\tilde{y}} f(x) - \beta_0 f(x)|\, dm(y)$$

$$+ \sum_{i=1}^{n} \int_{A_i} |g(y)|\, |L_{\tilde{y}} f(x) - L_{\tilde{y_i}} f(x)\,|\, dm(y).$$

Using Fubini's theorem we get

$$\left\| g * f - \sum_{i=0}^{n} \beta_i L_{\tilde{y_i}} f \right\|_1 \le \|f\|_1(\delta + |\beta_0|\, \delta) + \delta\, \|g\|_1 \le \delta(2\|f\|_1 + \|g\|_1) \le \varepsilon.$$

◇

For $f \in L^1(K)$ we define

$$d_{\alpha,M}(f) := d(0, T(\alpha, M)(f)) \;=\; \inf\{\, \|h\|_1 : h \in T(\alpha, M)(f)\}.$$

If

$$A = \sum_{i=0}^{n} \beta_i \, L_{\tilde{y}_i} \;\in T(\alpha, M),$$

then

$$\widehat{A(f)}(\alpha) \;=\; \sum_{i=0}^{n} \beta_i \, \overline{\alpha(y_i)} \, \hat{f}(\alpha) \;=\; \hat{f}(\alpha),$$

hence $|\hat{f}(\alpha)| \;=\; |\widehat{A(f)}(\alpha)| \;\leq\; \|A(f)\|_1$. Therefore for each $\alpha \in \hat{K}$, $M > 0$ and $f \in L^1(K)$ we have

$$|\hat{f}(\alpha)| \;\leq\; d_{\alpha,M}(f). \qquad (1)$$

If $P_1(\alpha, M)$ is satisfied, we will obtain that $d_{\alpha,M+1}(f) \leq M\,|\hat{f}(\alpha)|$.

Proposition 3.13.7 *Suppose that the Reiter condition $P_1(\alpha, M)$ is satisfied. Then $d_{\alpha,M+1}(f) \leq M\,|\hat{f}(\alpha)|$ for all $f \in L^1(K)$.*

Proof. By Proposition 3.13.5 condition $P_1(\alpha, M)$ implies that there exists a net $(g_i)_{i\in I}$ in $J(\alpha, M)$ such that $\|f * g_i - \hat{f}(\alpha) g_i\|_1 \to 0$. In particular we have $\|f * g_i\|_1 \to |\hat{f}(\alpha)|\;\|g_i\|_1$. By Proposition 3.13.6 we know that $f * g_i \in \overline{T(\alpha, M+1)(f)}$ for each $i \in I$. Therefore we obtain $d_{\alpha,M+1}(f) \leq M\,|\hat{f}(\alpha)|$. ◇

Combining (1) with this result we have for each $f \in L^1(K)$

$$|\hat{f}(\alpha)| \;\leq\; d_{\alpha,M+1}(f) \;\leq\; M\,|\hat{f}(\alpha)| \;\leq\; (M+1)\,|\hat{f}(\alpha)|$$

whenever $P_1(\alpha, M)$ is satisfied.

Definition. We say that the **Glicksberg-Reiter property** $G(\alpha, M)$ with bound $M \geq 1$ is satisfied in $\alpha \in \hat{K}$, if

$$|\hat{f}(\alpha)| \;\leq\; d_{\alpha,M}(f) \;\leq\; M\,\hat{f}(\alpha) \qquad (2)$$

for every $f \in L^1(K)$.

Proposition 3.13.7 says that $P_1(\alpha, M)$ implies $G(\alpha, M+1)$. Conversely we show the following result.

Proposition 3.13.8 *Suppose that the Glicksberg-Reiter property $G(\alpha, M)$ is satisfied. Then $I(\alpha)$ has a bounded approximative identity $(u_i)_{i\in I}$ with $\|u_i\|_1 \leq M+1$ for all $i \in I$.*

Proof. Let $\varepsilon > 0$ and $f \in I(\alpha)$. Since $G(\alpha, M)$ is satisfied we have $d_{\alpha,M}(f) = 0$, and hence there exists $A = \sum_{i=0}^{n} \beta_i L_{\tilde{y}_i} \in T(\alpha, M)$ such that $\|A(f)\|_1 < \varepsilon$. Let $\mu = \sum_{i=0}^{n} \beta_i \varepsilon_{y_i}$. Then $\hat{\mu}(\alpha) = 1$, $\|\mu\| \leq M$ and $\|\mu * f\|_1 < \varepsilon$. Let $(e_i)_{i \in I}$ be an approximate identity in $L^1(K)$ with $\|e_i\|_1 = 1$, and put $u_i := e_i - e_i * \mu \in L^1(K)$. Then $\widehat{u_i}(\alpha) = 0$, $\|u_i\|_1 \leq M + 1$ and

$$\|u_i * f - f\|_1 \leq \|e_i * f - f\|_1 + \|e_i * \mu * f\| \leq \|e_i * f - f\| + \varepsilon.$$

Hence $(u_i)_{i \in I}$ is an approximate identity in $I(\alpha)$ with bound $M + 1$. ◇

There is a second equivalent formulation of the Glicksberg-Reiter property $G(\alpha, M)$.

Definition. We say that the **Glicksberg-Reiter property** $\tilde{G}(\alpha, M)$ with bound $M \geq 1$ is satisfied in $\alpha \in \hat{K}$, if $d_{\alpha,M}(f) = 0$ for all $f \in I(\alpha)$.

Obviously $G(\alpha, M)$ implies $\tilde{G}(\alpha, M)$.

Proposition 3.13.9 *Suppose that $\tilde{G}(\alpha, M)$ is satisfied. Then $G(\alpha, M')$ is satisfied with a bound $M' > M$.*

Proof. Let $\varepsilon > 0$, $f \in L^1(K)$ such that $\hat{f}(\alpha) \neq 0$. Choose $g \in L^1(K)$ such that $\hat{g}(\alpha) = 1$. Then $f - \hat{f}(\alpha)g \in I(\alpha)$, and condition $\tilde{G}(\alpha, M)$ yields $A \in T(\alpha, M)$ such that $\|A(f - \hat{f}(\alpha)g)\|_1 < \varepsilon$. Hence

$$\begin{aligned} \varepsilon > \|A(f) - \hat{f}(\alpha)\, A(g)\|_1 &\geq \|A(f)\|_1 - |\hat{f}(\alpha)|\, \|A(g)\|_1 \\ &\geq \|A(f)\|_1 - |\hat{f}(\alpha)|\, \|g\|_1\, M, \end{aligned}$$

i.e. $\|A(f)\|_1 < M' \, |\hat{f}(\alpha)| + \varepsilon$, where $M' = M\, \|g\|_1$. Since $\|g\|_1 \geq |\hat{g}(\alpha)| = 1$, we have $M' \geq M$. As $\varepsilon > 0$ can be chosen arbitrarily small, we have $\|A(f)\|_1 \leq M' \, |\hat{f}(\alpha)|$ and then

$$|\hat{f}(\alpha)| \leq d_{\alpha,M'}(f) \leq M' \, |\hat{f}(\alpha)|.$$

◇

In [357] for arbitrary Banach algebras A the property of φ-amenability for a homomorphism φ from A onto $\mathbb{C}$ is introduced. For $A = L^1(K)$ this property is strongly related to the existence of α-invariant functionals on $L^\infty(K)$.

We collect the above results.

Corollary 3.13.10 *Let $\alpha \in \hat{K}$. Equivalent are*

(1) The Reiter condition $P_1(\alpha, M)$ is satisfied with bound $M \geq 1$.

(2) $I(\alpha)$ *has a bounded approximate identity* $(u_i)_{i\in I}$ *with bound* $\tilde{M} \geq 1$.
(3) *There exists an* α*-invariant functional* $m_\alpha \in L^\infty(K)^*$ *with* $\|m_\alpha\| \leq M$.
(4) *The Glicksberg-Reiter property* $G(\alpha, M')$ *is satisfied with bound* $M' \geq 1$.
(5) *The Glicksberg-Reiter property* $\tilde{G}(\alpha, M'')$ *is satisfied with bound* $M'' \geq M' \geq 1$.

Remark:

(i) A short glimpse on the proofs show that there is a simple, direct relation between the different bounds $M, M', M'', \tilde{M}$ and $\|m_\alpha\|$ depending on the choice of $f \in L^1(K)$ with $\hat{f}(\alpha) = 1$.
(ii) The statements (1) and (2) are equivalent even for $\alpha \in \mathcal{X}^b(K)$.
(iii) It is routine to derive that for $g \in L^1(K)$ the following equivalence is true:

$$\hat{g}(1) = 1 = \|g\|_1 \iff g \geq 0 \quad \text{and} \quad \|g\|_1 = 1.$$

Hence, for the special case $\alpha = 1$ and $M = 1$ the Reiter condition $P(1,1)$ is the following property:
For every $\varepsilon > 0$ and every compact subset $C \subseteq K$ there exists some $g \in P^1(K) := \{f \in L^1(K) : f \geq 0, \|f\|_1 = 1\}$ such that $\|L_{\tilde{y}}g - g\|_1 < \varepsilon$ for all $y \in C$. This property is usually called Reiter's condition P_1.

We now investigate a modification of the Reiter condition (P_1), which is related to the existence of strongly invariant means, see subsection 3.12.

Definition. We say that K satisfies the **strong Reiter condition** (SP_1^*) if for every $\varepsilon > 0$ and every finite subset $F \subseteq K$ and finite subset $\Phi \subseteq L^\infty(K)$ there exists some $f \in P^1(K) = \{f \in L^1(K) : f \geq 0, \|f\|_1 = 1\}$ such that

$$\|L_x(\varphi \cdot f) - (L_x\varphi) \cdot f\|_1 < \varepsilon$$

for all $x \in F$ and $\varphi \in \Phi$.

Proposition 3.13.11 *If there exists a strongly invariant mean on* $L^\infty(K)$*, then* K *satisfies* (SP_1^*).

Proof. Using the Goldstine theorem [189,p.424] we know that the embedding of the unit ball $B \subseteq L^1(K)$ is dense in the unit ball $C \subseteq L^\infty(K)^*$ with respect to the weak $*$-topology. Hence, the existence of a strongly invariant mean $m \in L^\infty(K)^*$ yields a net $(f_j)_{j\in I}$, $f_j \in B$ such that

$$\int_K f_j(x)\, \psi(x)\, dm(x) \longrightarrow m(\psi) \quad \text{for all } \psi \in L^\infty(K).$$

Since $m(1) = 1$ and m is positive we can choose the f_j from $P^1(K)$. For any $y \in K$ and $\varphi, \psi \in L^\infty(K)$ we have

$$\int_K L_y(\varphi \cdot f_j)(x)\, \psi(x)\, dm(x) = \int_K f_j(x)\, \varphi(x)\, L_{\tilde{y}}\psi(x)\, dm(x) \;\to\; m(\varphi \cdot L_{\tilde{y}}\psi)$$

and

$$\int_K f_j L_y\varphi(x)\, \psi(x)\, dm(x) \;\to\; m(L_y\varphi \cdot \psi).$$

Hence,

$$\int_K (L_y(\varphi \cdot f_j)(x) - (f_j \cdot L_y\varphi)(x))\, \psi(x)\, dm(x) \;\to\; 0$$

for all $\psi \in L^\infty(K)$. For $F = \{y_1, ..., y_m\}$ and $\Phi = \{\varphi_1, ..., \varphi_n\}$ define

$$g_{k,l,j} \;=\; L_{y_k}(\varphi_l f_j) - f_j(L_{y_k}\varphi_l).$$

Then the net $g_j = (g_{1,1,j}, ..., g_{m,n,j}) \in L^1(K)^{mn} = L^1(K) \times \cdots \times L^1(K)$ converges to zero in the product space $L^1(K)^{mn}$ with respect to the weak topology. Since for convex subsets of $L^1(K)^{mn}$ the norm-closure coincides with the weak closure, a convex combination of the g_j converges to zero in the norm of $L^1(K)^{mn}$, see [189,p.422]. Hence, there exists some $f \in P^1(K)$, a convex combination of the functions f_j, such that

$$\|L_x(\varphi \cdot f) \;-\; (L_x\varphi) \cdot f\|_1 \;<\; \varepsilon$$

for all $x \in F$, $\varphi \in \Phi$. ◇

Proposition 3.13.12 *If K satisfies (SP_1^*), then there exists a strongly invariant mean m on $L^\infty(K)$.*

Proof. Let $\varepsilon > 0$, $F \subseteq K$ finite and $\Phi \subseteq L^\infty(K)$ finite. Then there exists $f \in P^1(K)$ such that $\|L_x(\varphi f) - (L_x\varphi) \cdot f\|_1 < \varepsilon$ for all $x \in F$ and $\varphi \in \Phi$. The function f determines a linear functional $m_{\varepsilon,F,\Phi} \in L^\infty(K)^*$ by

$$m_{\varepsilon,F,\Phi}(\psi) \;=\; \int_K \psi(y)\, f(y)\, dm(y) \qquad \text{for } \psi \in L^\infty(K).$$

This functional is positive and satisfies $\|m_{\varepsilon,F,\Phi}\| = 1$.

In particular all $m_{\varepsilon,F,\Phi}$ are elements of a weak $*$-compact subset of $L^\infty(K)^*$. For every $\varphi, \psi \in L^\infty(K)$ and $x \in K$ we have

$$m_{\varepsilon,F,\Phi}(\varphi \cdot L_{\tilde{x}}\psi) = \int_K \varphi(y)\, L_{\tilde{x}}\psi(y)\, f(y)\, dm(y) = \int_K L_x(\varphi f)(y)\, \psi(y)\, dm(y),$$

which follows by Proposition 2.2.5. Therefore,

$$|m_{\varepsilon,F,\Phi}(\varphi \cdot L_{\tilde{x}}\psi) \;-\; m_{\varepsilon,F,\Phi}(\psi \cdot L_x\varphi)|$$

$$= \left| \int_K (L_x(\varphi f)(y) - f(y)\, L_x\varphi(y))\, \psi(y)\, dm(y) \right|$$

$$\leq \|\psi\|_\infty \, \|L_x(\varphi f) - (L_x\varphi) \cdot f)\|_1 < \varepsilon \quad \text{for all } \varphi \in \Phi \text{ and } x \in F,$$

and all $\psi \in L^\infty(K)$.

Defining $(\varepsilon_1, F_1, \Phi_1) < (\varepsilon_2, F_2, \Phi_2)$ whenever $\varepsilon_2 < \varepsilon_1$, $F_1 \subseteq F_2$, $\Phi_1 \subseteq \Phi_2$, we get a partial order. With respect to this partial order, the functionals $m_{\varepsilon,F,\Phi}$ form a net. This net has an accumulation point $m \in L^\infty(K)^*$ satisfying $\|m\| = m(1) = 1$ and $m(\varphi \cdot (L_{\tilde{x}}\psi)) = m((L_x\varphi) \cdot \psi)$ for all $\varphi, \psi \in L^\infty(K)$ and $x \in K$. ◇

Corollary 3.13.13 *Let K be a commutative hypergroup. K satisfies (SP_1^*) if and only if there exists a strongly invariant mean on $L^\infty(K)$.*

The function $f \in P^1(K)$ in the (SP_1^*)-condition can even be chosen to be continuous with compact support.

Lemma 3.13.14 *If K satisfies (SP_1^*), then the function f in the definition of (SP_1^*) can be chosen from $f \in P^1(K) \cap C_c(K)$.*

Proof. Let $\varepsilon > 0$, $F \subseteq K$ finite, $\Phi \subseteq L^\infty(K)$ finite, and denote $M = \max\{ \|\varphi\|_\infty : \varphi \in \Phi\}$. Then there exists $g \in P^1(K)$ such that $\|L_x(\varphi \cdot g) - (L_x\varphi) \cdot g\|_1 < \varepsilon/3$ for all $x \in F$, $\varphi \in \Phi$. Therefore, there is some $f \in P^1(K) \cap C_c(K)$ such that $\|g - f\|_1 < \varepsilon/(3M)$. It follows that

$$\|L_x(\varphi \cdot f) - (L_x\varphi) \cdot f\|_1 \leq \|L_x(\varphi \cdot f) - L_x(\varphi \cdot g)\|_1$$
$$+ \|L_x(\varphi \cdot g) - (L_x\varphi) \cdot g\|_1 + \|(L_x\varphi) \cdot g - (L_x\varphi) \cdot f\|_1 < \varepsilon.$$

◇

Restricting from $L^\infty(K)$ to $UC(K)$ we may replace the finiteness of $F \subseteq K$ in (SP_1^*) by compactness. In fact we can show

Theorem 3.13.15 *Assume that (SP_1^*) is satisfied. For $\varepsilon > 0$, $C \subseteq K$ compact and $\Phi \subseteq UC(K)$ finite there exists some $f \in P^1(K) \cap C_c(K)$ such that*

$$\|L_x(\varphi \cdot f) - (L_x\varphi) \cdot f\|_1 < \varepsilon \quad \textit{for all } x \in C,\ \varphi \in \Phi.$$

Proof. Let $M = \max\{ \|\varphi\|_\infty : \varphi \in \Phi\}$. Choose $g \in P^1(K) \cap C_c(K)$. There exist $x_1, ..., x_n \in K$ and open neighbourhoods U_{x_i} of x_i, $i = 1, ..., n$, such that

$$C \subseteq \bigcup_{i=1}^n U_{x_i} \text{ and } \|L_x g - L_y g\|_1 < \frac{\varepsilon}{3M} \text{ and } \|L_x\varphi - L_y\varphi\|_\infty < \frac{\varepsilon}{3}$$

for all $x, y \in U_{x_i}$, and every $\varphi \in \Phi$. Condition (SP_1^*) yields $f \in P^1(K) \cap C_c(K)$ such that

$$\|L_{x_i}(\varphi \cdot f) - (L_{x_i}\varphi) \cdot f\|_1 < \frac{\varepsilon}{3} \quad \text{for all } \varphi \in \Phi,\ i = 1, ..., n.$$

Now let $x \in C$. Then $x \in U_{x_i}$ for some $i = 1, ..., n$, and we obtain

$$\|L_x(\varphi \cdot f) * g - L_{x_i}(\varphi \cdot f) * g\|_1 = \|(\varphi \cdot f) * L_x g - (\varphi \cdot f) * L_{x_i} g\|_1$$

$$\leq \|\varphi \cdot f\|_1 \, \|L_x g - L_{x_i} g\|_1 < \frac{\varepsilon}{3}.$$

Hence,

$$\|L_x(\varphi \cdot f) * g - ((L_x\varphi) \cdot f) * g\|_1 \leq \|L_x(\varphi \cdot f) * g - L_{x_i}(\varphi \cdot f) * g\|_1$$

$$+ \|L_{x_i}(\varphi \cdot f) * g - ((L_{x_i}\varphi) \cdot f) * g\|_1 + \|((L_{x_i}\varphi) \cdot f) * g - ((L_x\varphi) \cdot f) * g\|_1 < \varepsilon.$$

Choosing $g \in P^1(K) \cap C_c(K)$ from an approximate identity in $L^1(K)$ it follows that $\|L_x(\varphi \cdot f) - (L_x\varphi) \cdot f\|_1 < \varepsilon$ for all $x \in C$ and $\varphi \in \Phi$. ◇

(Strongly) invariant means, α-invariant functionals and Reiter's condition for polynomial hypergroups are studied in subsections 4.8 and 4.9, including many examples and counter-examples.

Chapter 4

Fourier analysis on polynomial hypergroups

In subsection 1.2.1 we have seen that each orthogonal polynomial sequence $(R_n(x))_{n\in\mathbb{N}_0}$ with degree $R_n = n$, $R_n(1) = 1$ and nonnegative coefficients $g(m, n; k)$ generates a commutative hypergroup on $\mathbb{N}_0$. We briefly studied some examples, the Chebyshev hypergroup, the cosh-hypergroup, and the whole class of hypergroups generated by the ultraspherical polynomials. The stochastic processes studied in Ch.6 are indexed by $\mathbb{N}_0$ and enjoy a weak stationarity condition related to a polynomial hypergroup on $\mathbb{N}_0$. So in this section we deal with problems of harmonic analysis on polynomial hypergroups and we extend the list of examples considerably.

1 Dual spaces

Suppose that $(R_n(x))_{n\in\mathbb{N}_0}$ generates a polynomial hypergroup on $\mathbb{N}_0$, see Theorem 1.2.2. The Haar measure – up to a multiplicative constant — is given by the weights

$$h(n) := m(\{n\}) = g(n, n; 0)^{-1} = \left(\int_{\mathbb{R}} R_n^2(x)\, d\pi(x)\right)^{-1} = \mu_n^{-1},$$

compare Theorem 2.1.1.

The translation of a sequence $a = (a(n))_{n\in\mathbb{N}_0}$ writes as

$$L_m\, a(n) = \sum_{k=|m-n|}^{m+n} g(m, n; k)\, a(k)$$

and for $a, b \in l^1(h) := l^1(\mathbb{N}_0, h)$ we have

$$\sum_{n=0}^{\infty} L_m\, a(n)\, b(n)\, h(n) = \sum_{n=0}^{\infty} a(n)\, L_m\, b(n)\, h(n),$$

see Theorem 2.2.10.

Concerning the Banach algebras $l^1(h)$ and $M(\mathbb{N}_0)$ one has to notify that $M(\mathbb{N}_0)$ is exactly

$$l^1 = \{a = (a(n))_{n\in\mathbb{N}_0} : \sum_{n=0}^{\infty} |a(n)| < \infty\},$$

and hence the map $a \mapsto ah$, $l^1(h) \to l^1 = M(\mathbb{N}_0)$ is an isometric isomorphism from the Banach algebra $l^1(h)$ onto l^1. The involution in $l^1(h)$ is given by $a^*(n) = \overline{a(n)}$. The modul action of l^1 on $l^p(h) = l^p(\mathbb{N}_0, h)$, $1 \le p < \infty$, is given by

$$a * b\,(m) = \sum_{n=0}^{\infty} L_n\, b(m)\, a(n),$$

where $a \in l^1$, $b \in l^p(h)$.

The convolution of $a \in l^p(h)$ and $b \in l^q(h)$, $\frac{1}{p} + \frac{1}{q} = 1$, is given by

$$a * b\,(m) = \sum_{n=0}^{\infty} L_n\, b(m)\, a(n)\, h(n),$$

and $(a * b(m))_{m\in\mathbb{N}_0}$ is a sequence converging to zero.

The dual spaces of the polynomial hypergroups can easily be characterized by means of the product formula (L) of Lemma 1.2.1. In fact, define for $z \in \mathbb{C}$

$$\alpha_z : \mathbb{N}_0 \longrightarrow \mathbb{C}, \qquad \alpha_z(n) = R_n(z)$$

and let

$$D := \{z \in \mathbb{C} : (R_n(z))_{n\in\mathbb{N}_0} \text{ is bounded}\} \qquad \text{and} \qquad D_s := D \cap \mathbb{R}.$$

Theorem 4.1.1 *Suppose that $(R_n(x))_{n\in\mathbb{N}_0}$ generates a polynomial hypergroup on $\mathbb{N}_0$. Then*

(i) $\sup\{|R_n(z)| : n \in \mathbb{N}_0\} = \|\alpha_z\|_\infty = 1$ *for every $z \in D$.*
In particular, $D = \{z \in \mathbb{C} : |R_n(z)| \le 1$ for all $n \in \mathbb{N}_0\}$,

(ii) $\mathcal{X}^b(\mathbb{N}_0) = \{\alpha_z : z \in D\}$ *and* $\hat{\mathbb{N}}_0 = \{\alpha_x : x \in D_s\}$,

(iii) the mapping $D \to \mathcal{X}^b(\mathbb{N}_0)$, $z \mapsto \alpha_z$ and $D_s \to \hat{\mathbb{N}}_0$, $x \mapsto \alpha_x$ are homeomorphisms. The inverse mapping is given by $\alpha \mapsto a_0\alpha(1) + b_0$. (Recall that $R_1(x) = \frac{1}{a_0}(x - b_0)$.)
In particular, D and D_s are compact subsets of $\mathbb{C}$ and $\mathbb{R}$, respectively. Furthermore $D_s \subseteq [1 - 2a_0, 1]$.

Proof.

(i) The characters $\alpha \in \mathcal{X}^b(\mathbb{N}_0)$ are exactly the bounded sequences $\alpha \in l^\infty$ different from zero and enjoying

$$\alpha(n)\,\alpha(m) \;=\; \omega(n,m)(\alpha) \;=\; \sum_{k=|n-m|}^{n+m} g(n,m;k)\,\alpha(k).$$

Hence for every $z \in D$ by formula (L) we have $\alpha_z \in \mathcal{X}^b(\mathbb{N}_0)$, and for $x \in D_s$ we get $\alpha_x \in \hat{\mathbb{N}}_0$. Since each character $\alpha \in \mathcal{X}^b(\mathbb{N}_0)$ fulfils $\|\alpha\|_\infty = 1 = \alpha(0)$ the assertion (i) follows.

(ii) We have already shown that $\alpha_z \in \mathcal{X}^b(\mathbb{N}_0)$ for all $z \in D$. Conversely let $\alpha \in \mathcal{X}^b(\mathbb{N}_0)$. Put $z_0 = a_0\alpha(1) + b_0$. Then $\alpha(1) = R_1(z_0)$ and by induction we get $\alpha(n) = R_n(z_0) = \alpha_{z_0}(n)$ for every $n \in \mathbb{N}_0$. Of course $z_0 \in D$, and assertion (ii) is shown.

(iii) The mapping $z \mapsto \alpha_z$ is bijective and the inverse $\alpha_z \mapsto z = a_0\alpha_z(1)+b_0$ is continuous. Since $\mathcal{X}^b(\mathbb{N}_0)$ is compact, see Theorem 3.1.9, $z \mapsto \alpha_z$ is a homeomorphism. Considering the polynomial $R_1(x)$ the inclusion $D_s \subseteq [1-2a_0, 1]$ follows by means of (i).

◇

In the following we shall identify $\mathcal{X}^b(\mathbb{N}_0)$ and $\hat{\mathbb{N}}_0$ with the corresponding compact subsets D and D_s of the complex plane and the real line, respectively. Hence the Fourier transform of $d = (d(n))_{n\in\mathbb{N}_0} \in l^1(h)$ is an orthogonal expansion

$$\hat{d}(x) \;=\; \sum_{k=0}^{\infty} d(k)\,R_k(x)\,h(k), \qquad x \in D_s,$$

or $\mathcal{F}d(z) = \sum_{k=0}^{\infty} d(k)\,R_k(z)h(k), \;\; z \in D.$

To determine $\mathcal{S}(\mathbb{N}_0) \subseteq \hat{\mathbb{N}}_0$ we consider the translation operator $L_1 \in B(l^2(h))$. Since $L_{n+1} = \frac{1}{a_n}(L_1 \circ L_n - b_nL_n - c_nL_{n-1})$ where a_n, b_n, c_n are the recurrence coefficients, we obtain that the commutative C^*-algebra $C^*_\lambda(\mathbb{N}_0) = \{L_a : a \in l^1(h)\}^-$ of subsection 3.1 is generated by L_1. Now it is well-known, see [609,Theorem 11.19], that $\mathcal{S}(\mathbb{N}_0) \cong \Delta(C^*_\lambda(\mathbb{N}_0))$ is homeomorphic to the spectrum $\sigma(L_1)$ of the Hilbert space operator $L_1 \in \mathcal{B}(l^2(h))$ via the mapping $\alpha \mapsto k_\alpha(L_1)$, $\mathcal{S}(\mathbb{N}_0) \to \mathbb{C}$, see Theorem 3.1.10.

Now $k_\alpha(L_1) = \frac{1}{h(1)}\,\hat{\epsilon}_1(\alpha) = R_1(x)$, where $\alpha = \alpha_x$ or $x = a_0\alpha(1) + b_0$ according to the proof of Theorem 4.1.1(ii) and $\epsilon_1(k) = \delta_{1,k}$. The action of the translation operator L_1 on $l^2(h)$ can be represented by an infinite matrix with respect to the canonical basis $(\epsilon_k)_{k\in\mathbb{N}_0}$, where $\epsilon_k(n) = \delta_{k,n}$.

$$\begin{pmatrix} 0 & 1 & 0 & 0 & & \\ c_1 & b_1 & a_1 & 0 & & 0 \\ 0 & c_2 & b_2 & a_2 & & \\ & & \ddots & \ddots & \ddots & \\ & 0 & & & & \end{pmatrix} \cong L_1.$$

We have the following result:

Theorem 4.1.2 *Suppose that $(R_n(x))_{n\in\mathbb{N}_0}$ generates a polynomial hypergroup on $\mathbb{N}_0$. Then*

(i) $a_0\sigma(L_1) + b_0 \subseteq D_s$ *and* $\mathcal{S}(\mathbb{N}_0) = \{\alpha_x : x \in \sigma(L_1)\}$. *The mapping* $x \mapsto \alpha_x,\ \sigma(L_1) \to \mathcal{S}(\mathbb{N}_0)$ *is a homeomorphism.*

(ii) $\max\{|y| : y \in \sigma(L_1)\} \leq \|L_1\| \leq 1$.

(iii) $y \in \sigma(L_1)$ *if and only if* $y = R_1(x),\ x \in D_s$ *is an approximate eigenvalue of* L_1, *i.e. for every* $\varepsilon > 0$ *there exists some* $g \in l^2(h),\ \|g\|_2 = 1$ *such that* $\|L_1 g - R_1(x)g\|_2 < \varepsilon$.

Proof. (i) is already shown. Statements (ii) and (iii) are known results of the spectrum of a bounded operator on a Hilbert space. $\diamond$

The theorem of Plancherel-Levitan (Theorem 3.2.6) yields the orthogonalization measure π of $(R_n(x))_{n\in\mathbb{N}_0}$. Indeed we have

Theorem 4.1.3 *Suppose that $(R_n(x))_{n\in\mathbb{N}_0}$ generates a polynomial hypergroup on $\mathbb{N}_0$. Then there exists a unique probability measure π on D_s such that*

$$\sum_{n=0}^{\infty} |d(n)|^2\, h(n) = \int_{D_s} |\hat{d}(x)|^2\, d\pi(x)$$

for every $d \in l^1(h)$. The measure π is the orthogonalization measure of $(R_n(x))_{n\in\mathbb{N}_0}$. For the support of π the following identities are valid:

$$\text{supp}\ \pi = a_0\sigma(L_1) + b_0 = \{x \in D_s : |\hat{d}(x)| \leq \|L_d\| \text{ for all } d \in l^1(h)\}$$

Proof. We apply the mapping $\alpha \mapsto a_0\alpha(1) + b_0,\ \widehat{\mathbb{N}_0} \to D_s$ and take the image measure of the Plancherel measure. Using the polarization identity it is easy to see that this image measure is the orthogonalization measure of the polynomial sequence $(R_n(x))_{n\in\mathbb{N}_0}$. Note that π is unique, since supp π is compact. $\diamond$

We will write $S = \text{supp}\ \pi = a_0\sigma(L_1) + b_0$. In the sequel we use $D \cong \mathcal{X}^b(\mathbb{N}_0),\ D_s \cong \widehat{\mathbb{N}_0}$ and $S \cong \mathcal{S}(\mathbb{N}_0)$ as dual spaces, when $K = \mathbb{N}_0$

is a polynomial hypergroup. The polynomial hypergroup $\mathbb{N}_0$ is completely determined by the recurrence coefficients a_n, b_n, c_n, $n \in \mathbb{N}$, and thus $\sigma(L_1)$, $\widehat{\mathbb{N}_0}$ and $\mathcal{X}^b(\mathbb{N}_0)$ are uniquely determined. The sets $S = \text{supp}\ \pi$, D_s and D depend on the choice of $a_0 > 0, b_0 \in \mathbb{R}$ for $R_1(x) = \frac{1}{a_0}(x - b_0)$. In case of $b_n = 0$, $n \in \mathbb{N}$ (the symmetric case) we always choose $a_0 = 1, b_0 = 0$.

For $K = \mathbb{N}_0$ discrete, $\hat{K} = D_s$ compact, the inversion theorems 3.3.7 and 3.3.8 are not very surprising. Concerning results on approximation by polynomial expansion with $(R_n(x))_{n\in\mathbb{N}_0}$ we refer to subsection 4.14. For $a \in M(D_s) = C(D_s)^*$ we define the **inverse Fourier-Stieltjes transform** by

$$\check{a}(n) = \int_{D_s} R_n(x)\, da(x), \qquad n \in \mathbb{N}_0.$$

Obviously $L^1(D_s, \pi)$ can be viewed as closed linear subspace of $M(D_s)$ via the identification $\varphi \mapsto \varphi\pi$, $L^1(D_s, \pi) \to M(D_s)$. So the **inverse Fourier transform** of $\varphi \in L^1(D_s, \pi)$ is given by

$$\check{\varphi}(n) = \int_{D_s} R_n(x)\, \varphi(x)\, d\pi(x) \qquad \text{for } n \in \mathbb{N}_0.$$

Theorem 4.1.4 *Suppose that $(R_n(x))_{n\in\mathbb{N}_0}$ generates a polynomial hypergroup on $\mathbb{N}_0$. Then the following inversion formulas hold:*

(i) Let $d \in l^1(h)$. Then for every $n \in \mathbb{N}_0$ is true

$$d(n) = \int_{D_s} \hat{d}(x)\, R_n(x)\, d\pi(x)$$

(ii) Let $\varphi \in L^1(D_s, \pi)$ such that $\check{\varphi} \in l^1(h)$. Then for π-almost every $x \in D_s$ is true

$$\varphi(x) = \sum_{n=0}^{\infty} \check{\varphi}(n)\, R_n(x)\, h(n)$$

If φ is continuous, then the equality holds for all $x \in$ supp $\pi = S$.

An important result heavily relating on the Banach algebraic structure of $l^1(h)$ and its image under the Fourier transformation in $C(D_s)$ is Wiener's theorem. If we restrict to $S \subseteq D_s$ we get

$$A(S) := \{\hat{d}\,|\,S : \ d \in l^1(h)\}$$

a self-adjoint subalgebra of $C(S)$. Recall that $(d * c)^\wedge = \hat{d}\hat{c}$ and $(d^*)^\wedge = \overline{\hat{d}}$. Applying the Stone-Weierstraß theorem we see that $A(S)$ is sup-norm-dense in $C(S)$. We can impose another norm on $A(S)$,

$$\|\hat{d}\,|S\|_A = \sum_{n=0}^{\infty} |d(n)|\, h(n)$$

the norm of $l^1(h)$. By the uniqueness theorem $\| \, \|_A$ is a norm on $A(S)$. Applying the convolution on $L^2(S,\pi)$ defined in subsection 3.11.1 we get the following result, which generalizes a theorem of M.Riesz, compare [415,Theorem 15.1] .

Theorem 4.1.5 *Let $\varphi \in C(S)$. Then $\varphi \in A(S)$ if and only if there exists $\psi, \varrho \in L^2(S,\pi)$ such that $\varphi = \psi * \varrho$. Moreover*

$$\|\varphi\|_A \le \|\psi\|_2 \, \|\varrho\|_2.$$

Proof. If $\varphi = \psi * \varrho$, $\psi, \varrho \in L^2(S,\pi)$, then $\check{\psi}, \check{\varrho} \in l^2(h)$ and $\psi * \varrho = (\check{\psi}\, \check{\varrho})^\wedge$, see Proposition 3.11.3, that means $\varphi \in A(S)$.
Conversely if $\varphi \in A(S)$, set $a(n) = |\check{\varphi}(n)|^{1/2}$ and $b(n) = a_n e^{it_n}$, where $t_n \in [0, 2\pi[$ is chosen such that $\check{\varphi}(n) = |\check{\varphi}(n)|\, e^{it_n}$. Evidently $a = (a(n))_{n\in\mathbb{N}_0}$, $b = (b(n))_{n\in\mathbb{N}_0} \in l^2(h)$. Let $\psi = \mathcal{P}(a)$, $\varrho = \mathcal{P}(b) \in L^2(S,\pi)$. Then $\check{\psi} = a$, $\check{\varrho} = b$ and $\check{\varphi} = ab \in l^1(h)$. By Proposition 3.11.3, we have $\varphi = (\check{\varphi})^\wedge = \psi * \varrho$. The norm-inequality follows by

$$\|\varphi\|_A = \sum_{n=0}^{\infty} |\check{\psi}(n)\, \check{\varrho}(n)|\, h(n) \le \|\check{\psi}\|_2 \, \|\check{\varrho}\|_2 = \|\psi\|_2 \, \|\varrho\|_2.$$

$\diamond$

Remark: If $\varphi : S \to \mathbb{C}$ has the form $\varphi = \psi * \overline{\psi}$, $\psi \in L^2(S,\pi)$, then $\check{\varphi} = \check{\psi}\check{\overline{\psi}} \in l^1(h)$. Extending φ to $S_1 = S \cup \{1\}$ by defining $\varphi : S_1 \to \mathbb{C}$ by $\varphi = (\check{\varphi})^\wedge | S_1$, Theorem 3.11.23 yields that $\varphi : S_1 \to \mathbb{C}$ is $T(S_1)$-positive definite.

For Wiener's theorem we have to take into account the dual space $D \subseteq \mathbb{C}$, since D is homeomorphic to $\Delta(l^1(h))$. Hence we consider $A(D) := \{\mathcal{F}d : d \in l^1(h)\}$, where $\mathcal{F}$ is the Fourier transform of d on $D \cong \mathcal{X}^b(\mathbb{N}_0) \cong \Delta(l^1(h))$.

Theorem 4.1.6 (Wiener)
If $\varphi \in A(D)$ satisfies $\varphi(z) \neq 0$ for every $z \in D$, then $1/\varphi \in A(D)$.

Proof. At first we show a result true for any commutative Banach algebra $\mathcal{B}$ with unit $e \in \mathcal{B}$: An element $b \in \mathcal{B}$ is invertible if and only if $h(b) \neq 0$ for all $h \in \Delta(\mathcal{B})$. In fact, if $b \in \mathcal{B}$ is invertible, then $1 = h(e) = h(bb^{-1}) = h(b)\, h(b^{-1})$. Therefore $h(b) \neq 0$ for every $h \in \Delta(\mathcal{B})$. For the converse implication we apply the bijection between the set of all maximal ideals of $\mathcal{B}$ and the structure space $\Delta(\mathcal{B})$, given by $h \to \ker h$. Now assume that $h(b) \neq 0$ for every $h \in \Delta(\mathcal{B})$, but b not invertible in $\mathcal{B}$. Then $\mathcal{B}b$ is an ideal in $\mathcal{B}$, and hence there exists some $h \in \Delta(\mathcal{B})$ such that $\mathcal{B}b \subseteq \ker h$.

But then $0 = h(eb) = h(b)$, a contradiction.
Now consider $\mathcal{B} \cong l^1(h)$, which is isometric isomorphic to $(A(D), \| \ \|_A)$. By the foregoing result the function $\varphi \in A(D)$ is an invertible element of the Banach algebra $A(D)$. Hence there exists $\psi \in A(D)$ such that $\varphi\psi = 1$. ◇

2 Comparison of dual spaces

The basic dual objects $\mathcal{S}(\mathbb{N}_0)$, $\hat{\mathbb{N}}_0$ and $\mathcal{X}^b(\mathbb{N}_0)$ are homeomorphic to (and are identified with) the compact subsets $S = \text{supp}\ \pi$, $D_s = \{x \in \mathbb{R} : |R_n(x)| \leq 1 \text{ for all } n \in \mathbb{N}_0\}$ and $D = \{z \in \mathbb{C} : |R_n(z)| \leq 1 \text{ for all } n \in \mathbb{N}_0\}$, respectively, see Theorem 4.1.1 and Theorem 4.1.3. The homeomorphisms are given by $\alpha \mapsto a_0\alpha(1) + b_0$, $\alpha \in \mathcal{S}$, $\alpha \in \hat{\mathbb{N}}_0$ and $\alpha \in \mathcal{X}^b(\mathbb{N}_0)$, respectively.

Consider the Banach $*$-algebra $l^1(h)$ and define for $d \in l^1(h)$

$$r(d) := \lim_{n\to\infty} \|d^n\|_1^{1/n},$$

where $d^n = d^{n-1} * d$, $d^0 = \epsilon_0$. Further set

$$\sigma(d) := \{\lambda \in \mathbb{C} : d - \lambda\epsilon_0 \text{ is not invertible in } l^1(h)\}$$

and set for $d \in l^1(h)$

$$s(d) := \sup\{|\hat{d}(x)| : x \in D_s\}.$$

Then $\|L_d\|$, $s(d)$, $r(d)$ and $\|d\|_1$ are submultiplicative norms on $l^1(h)$ and

$$\|L_d\| \leq s(d) \leq r(d) \leq \|d\|_1.$$

We begin by recalling some results from section 3.4 for the special case of polynomial hypergroups $K = \mathbb{N}_0$.

Proposition 4.2.1 *Suppose $(R_n(x))_{n\in\mathbb{N}_0}$ generates a polynomial hypergroup. Then*

(1) We have $S = \text{supp}\ \pi = D_s$ if and only if $\|L_d\| = s(d)$ for every $d \in l^1(h)$.
(2) We have $D_s = D$ if and only if $s(d) = r(d)$ for all $d \in l^1(h)$.

Corollary 4.2.2 *We have $S = D_s = D$ if and only if $\|L_d\| = r(d)$ for every $d \in l^1(h)$.*

Remark: The equality $D_s = D$ is equivalent to $l^1(h)$ being a symmetric Banach $*$-algebra.

Growth properties give sufficient conditions for $S = D_s = D$. For polynomial hypergroups the growth conditions are given with respect to the Haar weights $h(n)$.

Proposition 4.2.3 *The polynomial hypergroup $\mathbb{N}_0$ is of*

(1) polynomial growth if and only if there is a $\beta \geq 0$ such that $h(n) = O(n^\beta)$ as $n \to \infty$,
(2) subexponential growth if and only if for every $\varepsilon > 0$ there exists a constant $K = K(\varepsilon) > 0$ such that $h(n) \leq K(1+\varepsilon)^n$ for all $n \in \mathbb{N}_0$.

Proof.

(1) Assume that $h(n) = O(n^\beta)$, $n \to \infty$. Let $C \subseteq \mathbb{N}_0$ be finite, and $n_0 = \max C$. Since $C^n \subseteq \{0, 1, .., nn_0\}$, it follows

$$h(C^n) = O\left((nn_0 + 1)(nn_0)^\beta\right) = O(n^{\beta+1}) \qquad \text{as } n \to \infty.$$

Conversely, consider $C = \{1\}$. Since $n \in \{1\}^n$, the polynomial growth of $\mathbb{N}_0$ implies $h(n) \leq h(\{1\}^n) = O(n^\alpha)$, $n \to \infty$.

(2) Assume that for $\varepsilon > 0$ there is $K > 0$ such that $h(n) \leq K(1+\varepsilon)^n$. Let $C \subseteq \mathbb{N}_0$ be finite, $\delta > 0$ and $n_0 = \max C$. There exists some $\varepsilon > 0$ such that $\sum_{k=0}^{n_0} (1+\varepsilon)^k \leq 1 + \delta$. Since $C^n \subseteq \{0, ..., nn_0\}$ we get by the assumption

$$h(C^n) \leq \sum_{k=0}^{nn_0} h(k) \leq K \sum_{k=0}^{nn_0} (1+\varepsilon)^k$$

$$= K \left(1 + \sum_{i=0}^{n-1} (1+\varepsilon)^{n_0 i} \sum_{j=1}^{n_0} (1+\varepsilon)^j\right) \leq K \left(1 + (1+\delta) \sum_{i=0}^{n-1} ((1+\varepsilon)^{n_0})^i\right)$$

$$\leq K \left(1 + \sum_{i=1}^{n} (1+\delta)^i\right) \leq M(C, \delta)\, (1+\delta)^n,$$

wherein we may put $M(C, \delta) = K \, \frac{1+\delta}{\delta}$.

The converse implication follows immediately by considering $C = \{1\}$.

◇

We state Theorem 3.4.6 explicitly for polynomial hypergroups.

Theorem 4.2.4 *If the polynomial hypergroup* $\mathbb{N}_0$ *is of subexponential growth, then* $S = D_s = D$.

Our next topic is to derive an equivalent condition to $1 \in \operatorname{supp} \pi = S$ where π is the orthogonalization measure of $(R_n(x))_{n\in\mathbb{N}_0}$. Since $a_0 + b_0 = 1$, we have $1 \in S = a_0\sigma(L_1) + b_0$ exactly when $1 \in \sigma(L_1)$. Recall that $\sigma(L_1)$ is the spectrum of the translation operator $L_1 \in \mathcal{B}(l^2(h))$.

Theorem 4.2.5 *Suppose* $(R_n(x))_{n\in\mathbb{N}_0}$ *generates a polynomial hypergroup such that* $1 \notin S$, *then* $\left\{\frac{H(n)}{a_n h(n)} : n \in \mathbb{N}_0\right\}$ *is bounded, where* $H(n) = \sum_{k=0}^{n} h(k)$.

Proof. Given $n \in \mathbb{N}_0$ denote by χ_n the sequence with $\chi_n(k) = 1$ for $k = 0, ..., n$ and $\chi_n(k) = 0$ for $k = n+1, n+2, ...$ An easy computation shows that

$$(\mathrm{id} - L_1)(\chi_n)(k) = 0 \quad \text{for all } k \in \mathbb{N}_0 \backslash \{n, n+1\}$$

and

$$(\mathrm{id} - L_1)(\chi_n)(n) = a_n, \qquad (\mathrm{id} - L_1)(\chi_n)(n+1) = -c_{n+1}.$$

Hence

$$\|(\mathrm{id} - L_1)(\chi_n)\|_2^2 = a_n^2\, h(n) + c_{n+1}^2 h(n+1) = a_n\,(a_n + c_{n+1})\, h(n).$$

Since $1 \notin \sigma(L_1)$, there exists $A = (\mathrm{id} - L_1)^{-1} \in \mathcal{B}(l^2(h))$. It follows

$$\|A \circ (\mathrm{id} - L_1)(\chi_n)\|_2^2 = \|\chi_n\|_2^2 = \sum_{k=0}^{n} h(k) = H(n),$$

and

$$\|A \circ (\mathrm{id} - L_1)(\chi_n)\|_2^2 \le \|A\|^2\, \|(\mathrm{id} - L_1)(\chi_n)\|_2^2$$

$$= \|A\|^2\, a_n h(n)\,(a_n + c_{n+1}) \le 2\,\|A\|^2\, a_n h(n).$$

Hence $H(n) \le 2\,\|A\|^2\, a_n h(n)$ for all $n \in \mathbb{N}_0$, so that $\left\{\frac{H(n)}{a_n h(n)} : n \in \mathbb{N}_0\right\}$ is bounded. ◇

In order to prove the converse implication we begin with finding the solution $\alpha = (\alpha(n))_{n\in\mathbb{N}_0}$ of $(\mathrm{id} - L_1)(\alpha) = \epsilon_0$.

Lemma 4.2.6 *A sequence* $\alpha = (\alpha(n))_{n\in\mathbb{N}_0}$ *satisfies* $(id - L_1)(\alpha) = \epsilon_0$ *exactly when* $\alpha(n+1) = \alpha(0) - \sum_{k=0}^{n} \frac{1}{a_k h(k)}$ *for all* $n \in \mathbb{N}_0$.

Proof. We have $(\mathrm{id}-L_1)(\alpha)(0) = 1$ if and only if $\alpha(1) - \alpha(0) = -1$. For $n \geq 1$ we see that

$$(\mathrm{id} - L_1)(\alpha)(n) = \alpha(n) - (a_n\alpha(n+1) + b_n\alpha(n) + c_n\alpha(n-1)) = 0$$

is equivalent to

$$a_n\,(\alpha(n+1) - \alpha(n)) = c_n\,(\alpha(n) - \alpha(n-1)).$$

By iteration,

$$\alpha(n+1)-\alpha(n) = \frac{c_n}{a_n}\,(\alpha(n)-\alpha(n-1)) = \frac{c_n c_{n-1}\cdots c_1}{a_n a_{n-1}\cdots a_1}\cdot\frac{-1}{a_0} = \frac{-1}{a_n h(n)},$$

and the statement follows. ◇

Next we have to study whether the sequence $\alpha = (\alpha(n))_{n\in\mathbb{N}_0}$ of Lemma 4.2.6 is a member of $l^2(h)$. We will apply some facts about the weighted Cesaro operator $C \in \mathcal{B}(l^2(h))$. For any sequence $\gamma = (\gamma(n))_{n\in\mathbb{N}_0}$ define $C(\gamma) = (C(\gamma)(n))_{n\in\mathbb{N}_0}$ by

$$C(\gamma)(n) = \frac{1}{H(n)}\sum_{k=0}^{n}\gamma(k)\,h(k).$$

A result due to Hardy says that C is a bounded linear operator on $l^2(h)$, and $\|C\| \leq 2$, see [290,9.10]. We shall call $C \in \mathcal{B}(l^2(h))$ the weighted Cesaro operator. The adjoint operator $C^* \in \mathcal{B}(l^2(h))$ satisfies

$$C^*\gamma(n) = \sum_{k=n}^{\infty}\gamma(k)\,\frac{h(k)}{H(k)}.$$

Definition. *We say that* $(R_n(x))_{n\in\mathbb{N}_0}$ *satisfies* **condition (B)** *if* $\{\frac{H(n)}{a_n h(n)} : n \in \mathbb{N}_0\}$ *is bounded.*

Lemma 4.2.7 *Assume that* $(R_n(x))_{n\in\mathbb{N}_0}$ *satisfies condition (B). Then*

$$\sum_{k=0}^{\infty}\frac{1}{a_k\,h(k)} < \infty.$$

Proof. By property (B) we know that there is some $K > 0$ such that

$$\sum_{k=0}^{n}\left(\frac{1}{a_k h(k)}\right)^2 h(k) \leq K\sum_{k=0}^{n}\left(\frac{1}{H(k)}\right)^2 h(k)$$

for each $n \in \mathbb{N}_0$. Since $C(\epsilon_0) = \left(\frac{1}{H(n)}\right)_{n\in\mathbb{N}_0}$, Hardy's result on weighted Cesaro operators implies $\left(\frac{1}{H(n)}\right)_{n\in\mathbb{N}_0} \in l^2(h)$. Hence $\left(\frac{1}{a_n h(n)}\right)_{n\in\mathbb{N}_0} \in l^2(h)$, which is $\sum_{k=0}^{\infty}\frac{1}{a_k^2 h(k)} < \infty$. Finally $a_k^2 \leq a_k$ yields $\sum_{k=0}^{\infty}\frac{1}{a_k h(k)} < \infty$. ◇

Following Lemma 4.2.6 the sequence $\alpha = (\alpha(n))_{n\in\mathbb{N}_0}$ is further on defined by

$$\alpha(n) = \sum_{k=n}^{\infty} \frac{1}{a_k h(k)} \qquad \text{for } n \in \mathbb{N}_0, \qquad (i)$$

(provided the series converges).

In order to show that $\alpha \in l^2(h)$ whenever (B) holds, we apply the adjoint weighted Cesaro operator $C^* \in \mathcal{B}(l^2(h))$. Define a sequence $\beta = (\beta(n))_{n\in\mathbb{N}_0}$ by

$$\beta(n) = \frac{H(n)}{a_n (h(n))^2} \qquad \text{for } n \in \mathbb{N}_0. \qquad (ii)$$

Note that $C^*\beta(n) = \alpha(n)$.

Lemma 4.2.8 *Assume that $(R_n(x))_{n\in\mathbb{N}_0}$ satisfies condition (B). Then $(\beta(n))_{n\in\mathbb{N}_0}$ is an element of $l^2(h)$.*

Proof. We have $\beta(n) \leq K\frac{1}{h(n)}$ for all $n \in \mathbb{N}_0$. Hence we have to show that $\left(\frac{1}{h(n)}\right)_{n\in\mathbb{N}_0} \in l^2(h)$. Applying Lemma 4.2.7 it follows

$$\sum_{k=0}^{\infty} \frac{1}{(h(k))^2}\, h(k) \leq \sum_{k=0}^{\infty} \frac{1}{a_k h(k)} < \infty.$$

◇

Since $C^*\beta = \alpha$, Lemma 4.2.8 yields

Proposition 4.2.9 *Assume that $(R_n(x))_{n\in\mathbb{N}_0}$ satisfies condition (B). Then the sequence $(\alpha(n))_{n\in\mathbb{N}_0}$ given by (i) is the unique solution in $l^2(h)$ of the equation $(id{-}L_1)(\alpha) = \epsilon_0$.*

Now it is easy to determine $\alpha^m \in l^2(h)$ such that

$$(\mathrm{id} - L_1)(\alpha^m) = \frac{\epsilon_m}{h(m)}$$

for $m \in \mathbb{N}$. By the following Proposition 4.2.10(3) we get $\alpha^m = L_m\alpha/h(m)$.

Proposition 4.2.10 *Assume that $(R_n(x))_{n\in\mathbb{N}_0}$ satisfies condition (B). The operators $L_m \in \mathcal{B}(l^2(h))$ act as follows for each $m \in \mathbb{N}$:*

(1) $L_m\epsilon_0 = \epsilon_m$,

(2) $L_m\alpha(k) = \alpha(m)$ *for* $k = 0, 1, ..., m$ *and*
$L_m\alpha(k) = \alpha(k)$ *for* $k = m+1, m+2, \dots$,

(3) $(id - L_1)(L_m\alpha) = L_m \circ (id{-}L_1)(\alpha) = L_m\epsilon_0 = \epsilon_m$.

Proof. The recurrence relation (1) of subsection 1.2.1 implies that the operators $L_n \in \mathcal{B}(l^2(h))$ satisfy

$$L_1 \circ L_n = a_n L_{n+1} + b_n L_n + c_n L_{n-1}, \qquad n \in \mathbb{N}, \qquad (*)$$

(with $L_0 = \mathrm{id}$).

(1) Obviously $L_1 \epsilon_0 = \epsilon_1$. Assume that $L_m \epsilon_0 = \epsilon_m$ and $L_{m-1}\epsilon_0 = \epsilon_{m-1}$ is already shown. Then by $(*)$

$$a_m L_{m+1}\epsilon_0 = L_1 \epsilon_m - b_m \epsilon_m - c_m \epsilon_{m-1} = a_m \epsilon_{m+1}.$$

(2) For $m = 1$ we have

$$L_1 \alpha = \alpha - \epsilon_0 = (\alpha(1), \alpha(1), \alpha(2), \alpha(3), ...).$$

We use again induction and assume that the statement holds for m and $m-1$, $m \geq 1$. Then for $k = 0, ..., m-1$ we have

$$L_{m+1}\alpha(k) = \frac{1}{a_m} (\alpha(m) - b_m \alpha(m) - c_m \alpha(m-1))$$

$$= \alpha(m) + \frac{c_m}{a_m} (\alpha(m) - \alpha(m-1)) = \alpha(m) - \frac{c_m}{a_m} \frac{1}{a_{m-1} h(m-1)}$$

$$= \alpha(m) - \frac{1}{a_m h(m)} = \alpha(m+1).$$

For $k = m$ it follows

$$L_{m+1}\alpha(m) = \frac{1}{a_m} (a_m \alpha(m+1) + b_m \alpha(m) + c_m \alpha(m))$$

$$- \frac{b_m}{a_m} \alpha(m) - \frac{c_m}{a_m} \alpha(m) = \alpha(m+1).$$

For $k = m+1, m+2, ...$ we have

$$L_{m+1}\alpha(k) = \frac{1}{a_m} (a_k \alpha(k+1) + b_k \alpha(k) + c_k \alpha(k-1))$$

$$- \frac{b_m}{a_m} \alpha(k) - \frac{c_m}{a_m} \alpha(k)$$

$$= \frac{1}{a_m} \left[a_k \left(\alpha(k) - \frac{1}{a_k h(k)} \right) + b_k \alpha(k) + c_k \left(\alpha(k) + \frac{1}{a_{k-1} h(k-1)} \right) \right]$$

$$- \frac{b_m}{a_m} \alpha(k) - \frac{c_m}{a_m} \alpha(k)$$

$$= \frac{1}{a_m} \left[\alpha(k) - \frac{1}{h(k)} + \frac{1}{h(k)} \right] - \frac{b_m}{a_m} \alpha(k) - \frac{c_m}{a_m} \alpha(k) = \alpha(k).$$

(3) Obviously L_m commutes with $\mathrm{id} - L_1$. Hence

$$(\mathrm{id} - L_1)(L_m\alpha) = L_m(\mathrm{id} - L_1)(\alpha) = L_m\epsilon_0 = \epsilon_m.$$

$\diamond$

For fixed $m \in \mathbb{N}$ define the sequence β^m by

$$\beta^m(k) = 0 \qquad \text{for } k = 0, ..., m-1 \quad \text{and}$$

$$\beta^m(k) = \frac{H(k)}{a_k(h(k))^2} \qquad \text{for } k = m, m+1, \ldots .$$

By the proof of Lemma 4.2.8 we know $\beta^m \in l^2(h)$ whenever (B) is satisfied. Moreover,

$$C^*(\beta^m)(n) = \sum_{k=m}^{\infty} \frac{\beta^m(k)h(k)}{H(k)} = \sum_{k=m}^{\infty} \frac{1}{a_k h(k)} = \alpha(m) \qquad \text{for } n \le m,$$

$$C^*(\beta^m)(n) = \sum_{k=n}^{\infty} \frac{1}{a_k h(k)} = \alpha(n) \qquad \text{for } n > m.$$

Proposition 4.2.10(2) says that $C^*(\beta^m) = L_m\alpha$. We put $\beta^0 = \beta$, where β is defined in (ii).

Now we can combine the results above to determine $(\mathrm{id} - L_1)^{-1}$, provided (B) is satisfied. We define $\varphi = (\varphi(k))_{k\in\mathbb{N}_0}$ by

$$\varphi(k) = \frac{H(k)^2}{a_k(h(k))^2} \qquad \text{for } k \in \mathbb{N}_0. \qquad (iii)$$

If $\frac{H(k)}{a_k h(k)} \le K$ for all $k \in \mathbb{N}_0$, then $\varphi(k) \le K^2$ for all $k \in \mathbb{N}_0$.

Hence, if (B) holds, the multiplication with $\varphi \in l^\infty$ defines a bounded operator M_φ on $l^2(h)$, where $M_\varphi(\gamma)(n) = \varphi(n)\gamma(n),\ \gamma \in l^2(h)$.

Theorem 4.2.11 *Assume that $(R_n(x))_{n\in\mathbb{N}_0}$ satisfies condition (B). Then $C^* \circ M_\varphi \circ C \in \mathcal{B}(l^2(h))$ is the inverse of id$-L_1$, where φ is the sequence of (iii).*

Proof. We know that $C(\epsilon_m)(k) = 0$ for $k = 0, ..., m-1$ and $C(\epsilon_m)(k) = \frac{1}{H(k)}$ for $k = m, m+1, ...$, and so $M_\varphi \circ C(\epsilon_m) = \beta^m$. Hence $C^* \circ M_\varphi \circ C(\epsilon_m) = L_m(\alpha)$. In particular

$$(\mathrm{id} - L_1) \circ (C^* \circ M_\varphi \circ C)(\epsilon_m) = \epsilon_m.$$

Furthermore,

$$C^* \circ M_\varphi \circ C \circ (\mathrm{id} - L_1)(\epsilon_0) = C^* \circ M_\varphi \circ C(\epsilon_0 - \epsilon_1)$$

$$= \alpha - L_1(\alpha) = (\mathrm{id} - L_1)(\alpha) = \epsilon_0$$

and for $m \geq 1$

$$C^* \circ M_\varphi \circ C \circ (\mathrm{id} - L_1)(\epsilon_m) = (C^* \circ M \circ C)\,(\epsilon_m - (a_m \epsilon_{m+1} + b_m \epsilon_m + c_m \epsilon_{m-1}))$$

$$= L_m(\alpha) - (a_m L_{m+1}(\alpha) + b_m L_m(\alpha) + c_m L_{m-1}(\alpha))$$

$$= L_m(\alpha) - L_1 \circ L_m(\alpha) = (\mathrm{id} - L_1)\,(L_m(\alpha)) = \epsilon_m.$$

Since $C^* \circ M_\varphi \circ C$ is a bounded linear operator on $l^2(h)$ we obtain

$$C^* \circ M_\varphi \circ C \circ (\mathrm{id} - L_1) = \mathrm{id} = (\mathrm{id} - L_1) \circ C^* \circ M_\varphi \circ C,$$

i.e. $(\mathrm{id} - L_1)^{-1} = C^* \circ M_\varphi \circ C.$ ◇

Recalling that $S = \sigma(L_1)$ we have proven:

Corollary 4.2.12 *Suppose $(R_n(x))_{n\in\mathbb{N}_0}$ generates a polynomial hypergroup. Then $1 \notin S$ if and only if $(R_n(x))_{n\in\mathbb{N}_0}$ satisfies condition (B).*

3 Nevai class and dual spaces

Now we investigate polynomial hypergroups generated by orthogonal polynomials $R_n(x)$, which belong to the Nevai class. We assume throughout that the orthogonalization measure π is a probability measure, and we suppose that $(R_n(x))_{n\in\mathbb{N}_0}$ generates a polynomial hypergroup on $\mathbb{N}_0$. We recall the recurrence relation of the $R_n(x)$, compare (1) and (2) of subsection 1.2.1:

$$R_1(x)\,R_n(x) = a_n R_{n+1}(x) + b_n R_n(x) + c_n R_{n-1}(x), \qquad n \in \mathbb{N}, \tag{1}$$

$$R_0(x) = 1, \qquad R_1(x) = \frac{1}{a_0}\,(x - b_0), \tag{2}$$

where $a_0 > 0,\ b_0 \in \mathbb{R},\ a_0 + b_0 = 1$ and $a_n > 0,\ b_n \geq 0,\ c_n > 0,\ a_n + b_n + c_n = 1$ for each $n \in \mathbb{N}$.

Consider the orthonormal version $(p_n(x))_{n\in\mathbb{N}_0}$ of $(R_n(x))_{n\in\mathbb{N}_0}$. It satisfies the recurrence relation

$$x p_n(x) = \alpha_n p_{n+1}(x) + \beta_n p_n(x) + \gamma_n p_{n-1}(x), \qquad n \in \mathbb{N}, \tag{3}$$

$$p_0(x) = 1, \qquad p_1(x) = \frac{1}{\alpha_0}\,(x - \beta_0), \tag{4}$$

where $\alpha_0, \beta_0 \in \mathbb{R}$ and $\alpha_n, \beta_n, \gamma_n \in \mathbb{R}$ for each $n \in \mathbb{N}$.

The recurrence coefficients of $p_n(x)$ are

$$\alpha_0 = \sqrt{c_1}\, a_0, \qquad \beta_0 = b_0 \tag{5}$$

and

$$\alpha_n = a_n \frac{p_n(1)}{p_{n+1}(1)}\, a_0 = \sqrt{c_{n+1}a_n}\, a_0, \qquad \beta_n = b_n a_0 + b_0,$$

$$\gamma_n = c_n \frac{p_n(1)}{p_{n-1}(1)}\, a_0 = \sqrt{c_n a_{n-1}}\, a_0 \qquad \text{for } n \in \mathbb{N}. \tag{6}$$

We suppose now that the recurrence coefficients a_n, b_n, c_n are convergent. Write $a = \lim_{n\to\infty} a_n$, $b = \lim_{n\to\infty} b_n$, $c = \lim_{n\to\infty} c_n$. Then $a, b, c \in [0,1]$ and $a+b+c = 1$. It follows that $\lim_{n\to\infty} \beta_n = ba_0 + b_0$, $\lim_{n\to\infty} \alpha_n = \lim_{n\to\infty} \sqrt{ca}\, a_0 = \lim_{n\to\infty} \gamma_n$. Using the terminology of Nevai [509, §3.1, Definition 6] we have for the orthogonalization measure π

$$\pi \in M(ba_0 + b_0, 2\sqrt{ca}\, a_0).$$

Applying a theorem of Blumenthal, see [509, §3.3, Theorem 7], we have the following result.

Theorem 4.3.1 *Suppose that $(R_n(x))_{n\in\mathbb{N}_0}$ generates a polynomial hypergroup on $\mathbb{N}_0$, and assume that the recurrence coefficients a_n, b_n, c_n of (1) and (2) are convergent. Then*

$$S = supp\, \pi = A \cup B, \qquad A \cap B = \emptyset,$$

where

$$A = [ba_0 + b_0 - 2\sqrt{ca}\, a_0,\, ba_0 + b_0 + 2\sqrt{ca}\, a_0],$$

$a = \lim_{n\to\infty} a_n$, $b = \lim_{n\to\infty} b_n$, $c = \lim_{n\to\infty} c_n$, *and B is at most denumerable, isolated, and the only two possible limit points of B are the two endpoints of A.*

It is an easy calculation (using $a_0 + b_0 = 1$ and $a + b + c = 1$) to derive for the two endpoints of A:

$$ba_0 + b_0 + 2\sqrt{ca}\, a_0 = 1 - (\sqrt{a} - \sqrt{c})^2\, a_0$$

and

$$ba_0 + b_0 - 2\sqrt{ca}\, a_0 = 1 - (\sqrt{a} + \sqrt{c})^2\, a_0.$$

Definition. We say that the polynomial hypergroup on $\mathbb{N}_0$ generated by $(R_n(x))_{n\in\mathbb{N}_0}$ belongs to the **Nevai class** with R_n-parameters a, c, if the recurrence coefficients a_n, c_n are convergent with $a = \lim_{n\to\infty} a_n$, $c = \lim_{n\to\infty} c_n$. We call a_0 the **starting parameter** for $(R_n(x))_{n\in\mathbb{N}_0}$.

Throughout we will use the decomposition $S = A \cup B$ of Theorem 4.3.1. By Theorems 4.1.1 and 4.1.3 we know that changing the starting parameter a_0 leads only to an affine changing of S, D_s and D.

Proposition 4.3.2 *Suppose $\mathbb{N}_0$ is a polynomial hypergroup belonging to the Nevai class with R_n-parameters a, c. Then the following is true:*

(i) $0 \leq c \leq \frac{1}{2}$,
(ii) $1 \in S =$ *supp* π *if and only if* $a = c$,
(iii) $[-1, 1] = A$ *if and only if* $a = c$ *and* $a_0 = \frac{1}{2a}$.

Proof.

(i) We show that $c > \frac{1}{2}$ is not possible. We know that $g(n, n; 0) < 1$ for each $n \in \mathbb{N}$, and hence $h(n) = g(n, n; 0)^{-1} > 1$ for each $n \in \mathbb{N}$.
If $c > \frac{1}{2}$ choose c_0 such that $\frac{1}{2} < c_0 < c$. Then there exists $n_0 \in \mathbb{N}$ such that $c_0 \leq c_n$ for $n \geq n_0$. Then $a_n \leq 1 - c_0 < \frac{1}{2}$ for $n \geq n_0$. Then $h(n) = \frac{a_1 \cdots a_{n-1}}{c_1 \cdots c_n}$ tends to zero, which contradicts $h(n) > 1$.

(ii) If $a = c$ we have $1 \in A \subseteq S$. Conversely, suppose $1 \in S \backslash A$. Then 1 is an isolated point of S. Then there exists $f \in C(S)$ such that $f(1) = 1$ and $f|S \backslash \{1\} = 0$. Then

$$\check{f}(n) = \int_S f(x)\, R_n(x)\, d\pi(x) = \pi(\{1\}) > 0$$

for each $n \in \mathbb{N}_0$. But by Proposition 3.3.2(ii) $\check{f}(n)$ has to converge to zero. Therefore $1 \in S$ yields $1 \in A$, which means $a = c$.

(iii) Obviously, $a = c$ if and only if 1 is the right endpoint of A. In case of $a = c$ we have $a_0 = \frac{1}{2a}$ exactly when -1 is the left endpoint of A.

◇

Now we will deal with polynomial hypergroups generated by $(R_n(x))_{n\in\mathbb{N}_0}$ with $b_n = 0$ for each $n \in \mathbb{N}_0$. In that case π is a probability measure with support in $[-1, 1]$. Moreover π is symmetric, i.e. $\pi_- = \pi$, where $\pi_-(A) = \pi(-A)$ for each Borel set A in $[-1, 1]$. In fact, since $R_n(x) = (-1)^n R_n(-x)$, it follows immediately that π_- is also an orthogonalization measure of $(R_n(x))_{n\in\mathbb{N}_0}$. The orthogonalization measure of $(R_n(x))_{n\in\mathbb{N}_0}$ has compact support, and thus is uniquely determined, see [138], i.e. $\pi_- = \pi$.

Theorem 4.3.3 *Let $c \in [0,1]$. Suppose $\mathbb{N}_0$ is a polynomial hypergroup belonging to the Nevai class with R_n-parameters $a = 1-c$ and c. As starting parameter for $(R_n(x))_{n\in\mathbb{N}_0}$ we take $a_0 = 1$.*

(i) The only possible values of c are $0 < c \le \frac{1}{2}$.
(ii) For $c \in]0, \frac{1}{2}]$ we have

$$S = supp\ \pi = [-2\sqrt{c(1-c)},\, 2\sqrt{c(1-c)}\,],$$

$$D_s = [-1,1], \qquad \textit{and}$$

$$D = \{z \in \mathbb{C} : |z - 2\sqrt{c(1-c)}| + |z + 2\sqrt{c(1-c)}| \le 2\}.$$

Proof.

(i) Note that $c > \frac{1}{2}$ is not possible, which is already shown in Proposition 4.3.2(i).

(ii) Consider $0 < c \le \frac{1}{2}$ and let $d = 2\sqrt{c(1-c)}$. By Theorem 4.3.1 we have $S = A \cup B,\ A \cap B = \emptyset,\ A = [-d,d] \cup B,\ B$ a countable subset of $[-1,1]$ with possible limit points $-d$ and d. We show that $B \subseteq [-d,d]$. If B is not a subset of $[-d,d],\ \alpha_0 = \max S > 0$ is an isolated point of S.
Note that π is symmetric. By the separating property of the zeros of $R_n(x)$ (see [138]), we infer that $R_n(\alpha_0) > 0$ for all $n \in \mathbb{N}_0$. Now define $Q_n(x) := R_n(\alpha_0 x)/R_n(\alpha_0)$. It is easily shown that $(Q_n(x))_{n\in\mathbb{N}_0}$ generates a polynomial hypergroup on $\mathbb{N}_0$. The support of the corresponding orthogonalization measure contains 1 as an isolated point. As already shown in the proof of Proposition 4.3.2(ii), this is impossible, and $B \subseteq [-d,d]$ is shown.

(iii) Now we prove $D = \{z \in \mathbb{C} : |z-d| + |z+d| \le 2\}$, which implies also $D_s = [-1,1]$. Bear in mind that $E := \{z \in \mathbb{C} : |z-d| + |z+d| \le 2\}$ is an ellipse with the focus points $-d$ and d, and with the boundary $\{(\cos t,\ (1-2c)\sin t) : t \in [0, 2\pi[\,\}$. Consider the Joukovsky function

$$\phi : \mathbb{C}\backslash\{0\} \to C, \qquad \phi(u) = \frac{1}{u} + c(1-c)\,u,$$

and set

$$M_1 = \{u \in \mathbb{C} : \frac{1}{1-c} \le |u| \le \frac{2}{d}\}, \qquad M_2 = \{u \in \mathbb{C} : 0 < |u| < \frac{1}{1-c}\}.$$

We have $\phi(M_1) = E$ and $\phi(M_2) = \mathbb{C}\backslash E$. Furthermore, ϕ maps the circle $\{u \in \mathbb{C} : |u| = \frac{2}{d}\}$ onto $[-d,d]$, and the circle $\{u \in \mathbb{C} : |u| =$

$\frac{1}{1-c}\}$ onto the boundary of E. Further, by Poincare's theorem, see [509, §4.1, Theorem 13], we have for $z \in \mathbb{C}\backslash[-d,d]$

$$\frac{z}{d} + \sqrt{\left(\frac{z}{d}\right)^2 - 1} = \lim_{n\to\infty} \frac{p_n(z)}{p_{n-1}(z)} = \lim_{n\to\infty} \frac{\sqrt{h(n)}\,R_n(z)}{\sqrt{h(n-1)}\,R_{n-1}(z)}.$$

Hence

$$\lim_{n\to\infty} \frac{R_n(z)}{R_{n-1}(z)} = \frac{1}{2(1-c)}\,(z + \sqrt{z^2 - d^2}).$$

(We take the branch of the square root above such that $|z + \sqrt{z^2-d^2}| > 1$, whenever $z \in \mathbb{C}\backslash[-d,d]$.) For any $z \in \mathbb{C}\backslash[-d,d]$ there is a $u \in \mathbb{C}$ with $|u| < \frac{2}{d}$ such that $\phi(u) = z$. An easy calculation yields

$$\frac{1}{2(1-c)}\,(\phi(u) + \sqrt{\phi(u)^2 - d^2}\,) = \frac{1}{u(1-c)}.$$

For $z \in \mathbb{C}\backslash E$ we have $\phi(u) = z$, where $0 < |u| < \frac{1}{1-c}$. Therefore,

$$\lim_{n\to\infty} \frac{|R_n(z)|}{|R_{n-1}(z)|} = \frac{1}{|u|\,(1-c)} > 1,$$

and z cannot be an element of D, whenever $z \in \mathbb{C}\backslash E$. Note that statement (ii) is proved completely for the case $c = \frac{1}{2}$. in fact, for $c = \frac{1}{2}$ the results above yield

$$[-1,1] = S \subseteq D_s \subseteq [-1,1],$$

and $D = D_s$, since $D \subseteq \mathbb{R}$. Continuing our proof for $0 < c < \frac{1}{2}$, let $z \in \overset{\circ}{E}\backslash[-d,d]$. There exists $u \in \mathbb{C}$ such that $\frac{1}{1-c} < |u| < \frac{2}{d}$ and $\phi(u) = z$, ($\overset{\circ}{E}$ is the ellipse without its boudary). Hence

$$\lim_{n\to\infty} \frac{|R_n(z)|}{|R_{n-1}(z)|} = \frac{1}{|u|\,(1-c)} < 1,$$

and z is an element of D, whenever $z \in \overset{\circ}{E}\backslash[-d,d]$. As D is closed we obtain $D = E$.

It remains to prove that $c = 0$ is not possible. The arguments used in (ii) yield supp $\pi = S = \{0\}$, whenever $c = 0$. This contradicts $|\text{supp}\ \pi| = \infty$.

◇

Many examples of polynomial hypergroups belonging to the Nevai class we will meet in section 4.5.

4 Nonnegativity of linearization coefficients

In section 1.2.1 we have already met the Chebyshev hypergroup, the cosh-hypergroup and the hypergroup generated by ultraspherical polynomials. Having investigated the dual spaces of polynomial hypergroups it is time to present a list of polynomial hypergroups and derive their properties concerning the dual spaces S, D_s and D. In this paragraph we prove general results concerning the nonnegativity of the linearization coefficients $g(m,n;k)$. Recall that the orthogonal polynomials $R_n(x)$, that possibly generate a polynomial hypergroup on $\mathbb{N}_0$, are defined by the recurrence relations (1) and (2) of subsection 1.2.1, where $a_n > 0$, $c_n > 0$, $b_n \in \mathbb{R}$ for $n \in \mathbb{N}$ and $a_0 > 0$, $b_0 \in \mathbb{R}$. To get a hypergroup structure on $\mathbb{N}_0$ we have to suppose, in addition, that $R_n(1) = 1$ (which is equivalent to $a_n + b_n + c_n = 1$, $n \in \mathbb{N}$ and $a_0 + b_0 = 1$) and – the most important property – that all linearization coefficients $g(m,n;k)$ are nonnegative, see Theorem 1.2.2.

The proof of Theorem 4.4.1 below follows from the lines of the proof of [18,Theorem 5.2], which uses the monic version of the orthogonal polynomial sequence. We notify that the condition $R_n(1) = 1$ is not used.

Theorem 4.4.1 *Let $(R_n(x))_{n\in\mathbb{N}_0}$ be an orthogonal polynomial sequence with recurrence coefficients $a_n > 0$, $c_n > 0$, $b_n \in \mathbb{R}$; $n \in \mathbb{N}$. If*

$$b_{n+1} \ge b_n, \quad n \in \mathbb{N}, \quad \text{and} \quad c_2 a_1 \ge c_1, \quad c_{n+1}a_n \ge c_n a_{n-1}, \quad n = 2,3,..., \tag{1}$$

then all the linearization coefficients $g(m,n;k)$ are nonnegative.

Proof. We use induction on m. By symmetry we may assume $1 \le m \le n$. Then

$$\begin{aligned}
a_m\, R_{m+1}(x)\, R_n(x) &= (R_1(x)R_m(x) - b_m R_m(x) - c_m R_{m-1}(x))\, R_n(x)\\
&= R_m(x)\,(a_n R_{n+1}(x) + b_n R_n(x) + c_n R_{n-1}(x)) - b_m R_m(x)R_n(x)\\
&\quad - c_m R_{m-1}(x)R_n(x)\\
&= a_n R_m(x)R_{n+1}(x) + (b_n - b_m)\, R_m(x)R_n(x) + c_n R_m(x)R_{n-1}(x)\\
&\quad - c_m R_{m-1}(x)R_n(x)\\
&= a_n R_m(x)R_{n+1}(x) + (b_n - b_m)\, R_m(x)R_n(x)\\
&\quad + (c_n a_{n-1} - c_m a_{m-1})\, R_{m-1}(x)R_n(x)\, \frac{1}{a_{m-1}}\\
&\qquad + \frac{c_n}{a_{m-1}}\,(a_{m-1}R_m(x)R_{n-1}(x) - a_{n-1}R_{m-1}(x)R_n(x)),
\end{aligned} \tag{2}$$

where we have to set $a_0 = 1$ (only for the proof). The first three terms on the right-hand side of (2) have nonnegative linearization coefficients.

So we have to show that $a_{m-1}R_m(x)R_{n-1}(x) - a_{n-1}R_{m-1}(x)R_n(x)$ has nonnegative linearization coefficients. To prove this, write

$$\Delta_{m,n}(x) = a_m R_{m+1}(x)R_n(x) - a_n R_m(x)R_{n+1}(x),$$

and apply a second induction. In fact, by (2) $\Delta_{m,n}(x)$ has nonnegative linearization coefficients if $\Delta_{m-1,n-1}(x)$ has this property. Since

$$\begin{aligned}\Delta_{0,n-m}(x) &= R_1(x)R_{n-m}(x) - a_{n-m}R_{n-m+1}(x) \\ &= b_{n-m}R_{n-m}(x) + c_{n-m}R_{n-m-1}(x)\end{aligned}$$

in case of $n > m$, and $\Delta_{0,0}(x) = 0$, we obtain that $R_{m+1}(x)R_n(x)$ has nonnegative linearization coefficients. ◇

Remark: The conditions of Theorem 4.4.1 cover many of the cases. However, for ultraspherical polynomials $R_n^{(\alpha)}(x)$ the sequence $(c_{n+1}a_n)_{n\in\mathbb{N}}$ is nondecreasing if and only if $\alpha \geq \frac{1}{2}$. So the case $-\frac{1}{2} \leq \alpha < \frac{1}{2}$ is not covered by Theorem 4.4.1. The drawback of Theorem 4.4.1 is the fact that, independent of the normalization of the $R_n(x)$, one ends up with a monotonicity criterion which involves actually the recurrence coefficients of the monic version. R.Szwarc ([648] and [649]) has proven a general result where directly the monotonicity of the recurrence coefficients is applied.

To derive Szwarc's result we begin with some observations concerning renormalizations of orthogonal polynomials. Let $(P_n(x))_{n\in\mathbb{N}_0}$ be an orthogonal polynomial sequence with a known recurrence relation

$$x\,P_n(x) = \alpha_n P_{n+1}(x) + \beta_n P_n(x) + \gamma_n P_{n-1}(x), \qquad n \in \mathbb{N}, \tag{3}$$

and

$$P_0(x) = 1, \qquad P_1(x) = \frac{x}{\alpha_0} - \frac{\beta_0}{\alpha_0}, \tag{4}$$

where $\alpha_n > 0$, $\gamma_n > 0$, $\beta_n \in \mathbb{R}$ for $n \in \mathbb{N}$, and $\alpha_0 > 0$ and $\beta_0 < 1$. A simple modification of (3) is

$$P_1(x)\,P_n(x) = \frac{\alpha_n}{\alpha_0}\,P_{n+1}(x) + \frac{\beta_n - \beta_0}{\alpha_0}\,P_n(x) + \frac{\gamma_n}{\alpha_0}\,P_{n-1}(x), \qquad n \in \mathbb{N}.$$

A renormalization of $(P_n(x))_{n\in\mathbb{N}_0}$ is given by $Q_n(x) = \kappa_n P_n(x)$, where

$$\kappa_0 = 1, \qquad \kappa_n = \frac{\alpha_0\cdots\alpha_{n-1}}{\gamma_1\cdots\gamma_n}, \qquad n \in \mathbb{N}. \tag{5}$$

The orthogonal polynomials $Q_n(x)$ satisfy

$$x\,Q_n(x) = \gamma_{n+1}Q_{n+1}(x) + \beta_n Q_n(x) + \alpha_{n-1}Q_{n-1}(x), \qquad n \in \mathbb{N}, \tag{6}$$

and

$$Q_0(x) = 1, \qquad Q_1(x) = \frac{x}{\gamma_1} - \frac{\beta_0}{\gamma_1}. \tag{7}$$

The normalization of $(P_n(x))_{n\in\mathbb{N}}$ that can lead to hypergroups is $R_n(x) = \frac{1}{P_n(1)}\, P_n(x)$. For that we assume that $P_n(1) > 0$ for all $n \in \mathbb{N}_0$. Then $R_n(x)$ satisfy (1) and (2) of subsection 1.2.1 with

$$a_0 \;=\; 1 - \beta_0, \qquad b_0 \;=\; \beta_0, \tag{8}$$

and for $n \in \mathbb{N}$

$$a_n \;=\; \frac{P_{n+1}(1)}{P_n(1)(1-\beta_0)}\,\alpha_n, \quad b_n \;=\; \frac{\beta_n - \beta_0}{1-\beta_0}, \quad c_n \;=\; \frac{P_{n-1}(1)}{P_n(1)(1-\beta_0)}\,\gamma_n. \tag{9}$$

Then $a_n > 0$, $c_n > 0$ and $a_n + b_n + c_n = 1$ if $n \in \mathbb{N}$. To guarantee that $b_n \geq 0$ we have to assume in (3) that $\beta_n \geq \beta_0$.

If we can prove that if $2 \leq m \leq n$ and

$$P_m(x)\, P_n(x) \;=\; \sum_{k=n-m}^{n+m} c(m,n;k)\, P_k(x)$$

with

$c(m,n;k) \geq 0$ (or equivalently, $Q_m(x)Q_n(x) \;=\; \sum_{k=n-m}^{n+m} d(m,n;k)\, Q_k(x)$ with $d(m,n;k) \geq 0$) then

$$g(m,n;k) \;=\; c(m,n;k)\,\frac{P_k(1)}{P_m(1)P_n(1)} \;=\; d(m,n;k)\,\frac{\kappa_m\kappa_n P_k(1)}{\kappa_k P_m(1)P_n(1)} \;\geq\; 0.$$

Therefore $(R_n(x))_{n\in\mathbb{N}_0}$ will generate a polynomial hypergroup on $\mathbb{N}_0$, whenever $R_n(x) = P_n(x)/P_n(1)$, where $P_n(x)$ satisfies (3), (4) with $P_n(1) > 0$ with $\alpha_n > 0$, $\gamma_n > 0$, $\beta_n \geq \beta_0$ for $n \in \mathbb{N}$, $\alpha_0 > 0$, $\beta_0 < 1$ and $c(m,n;k) \geq 0$ (or equivalently, $d(m,n;k) \geq 0$).

For concrete examples we will use the polynomials $P_n(x)$ defined by (3), (4). In the proof of the Theorem below we use $Q_n(x)$ of (6), (7).

Let $u = (u(m,n))_{m,n\in\mathbb{N}_0}$ be a matrix indexed by $\mathbb{N}_0 \times \mathbb{N}_0$. Define operators T_1 and T_2 acting on the matrices by

$$\begin{aligned} T_1\, u(m,n) \;&=\; \gamma_{m+1}u(m+1,n) + \beta_m u(m,n) + \alpha_{m-1}u(m-1,n), \\ &\qquad m \in \mathbb{N},\ \ n \in \mathbb{N}_0, \\ T_1\, u(0,n) \;&=\; \gamma_1 u(1,n) + \beta_0 u(0,n), \qquad n \in \mathbb{N}_0, \end{aligned}$$

and

$$\begin{aligned} T_2\, u(m,n) \;&=\; \gamma_{n+1}u(m,n+1) + \beta_n u(m,n) + \alpha_{n-1}u(m,n-1), \\ &\qquad m \in \mathbb{N}_0,\ \ n \in \mathbb{N} \\ T_2\, u(m,0) \;&=\; \gamma_1 u(m,1) + \beta_0 u(m,0), \qquad m \in \mathbb{N}_0, \end{aligned}$$

and finally set

$$H\, u \;=\; T_1 u \;-\; T_2 u.$$

If $u(m,n) = Q_m(x)Q_n(x)$ for some $x \in \mathbb{R}$, then $T_1 u(m,n) = x\,u(m,n)$ and $T_2 u(m,n) = x\,u(m,n)$. Hence $Hu = 0$.

For $k \in \mathbb{N}_0$ consider

$$u_k(m,n) := d(m,n;k) = \int_{\mathbb{R}} Q_m(x)Q_n(x)Q_k(x)\,d\pi(x) \Big/ \int_{\mathbb{R}} Q_k^2(x)\,d\pi(x), \qquad m,n \in \mathbb{N}_0$$

It follows that $Hu_k = 0$ for every $k \in \mathbb{N}_0$. Moreover

$$u_k(m,0) = d(m,0;k) = \begin{cases} 1 & \text{if } k = m \\ 0 & \text{if } k \neq m \end{cases}.$$

Hence the u_k's are solutions of the following discrete boundary value problem.

Proposition 4.4.2 *The orthogonal polynomials $Q_n(x)$ possess nonnegative linearization coefficients $d(m,n;k)$ if and only if all solutions $u = (u(m,n))_{m,n\in\mathbb{N}_0}$ of the discrete boundary value problem*

$$H\,u = 0, \qquad u(m,0) \geq 0, \ \ m \in \mathbb{N}_0 \tag{10}$$

satisfy $u(m,n) \geq 0$ for all $m \geq n \geq 0$.

Proof. We have already seen that $d(.,.;k)$ are solutions of (10). Since $d(m,n;k) = d(n,m;k)$ the nonnegativity of all solutions implies the nonnegativity of the linearization coefficients $d(m,n;k)$.

Conversely suppose that all coefficients $d(m,n;k)$ are nonnegative. Let $u(m,n)$ be a solution of (10) and put $v(m,n) = \sum_{k=0}^{\infty} u(k,0)d(m,n;k)$. Then $Hv = 0$, $v(m,0) = u(m,0) \geq 0$. By the uniqueness of the solution of (10) it follows $v(m,n) = u(m,n) \geq 0$. ◇

An inner product on the space of the matrices is given by

$$\langle u, v\rangle = \sum_{m,n=0}^{\infty} u(m,n)\,\overline{v(m,n)}\,.$$

For simplicity put $v(m,-1) = 0 = v(-1,n)$ and $\alpha_{-1} = 0 = \gamma_{-1}$. The adjoint operator H^* of H acts as

$$\begin{aligned} H^* v(m,n) &= \alpha_m v(m+1,n) + \beta_m v(m,n) + \gamma_m v(m-1,n) \\ &\quad -\alpha_n v(m,n+1) - \beta_n v(m,n) - \gamma_n v(m,n-1) \end{aligned} \tag{11}$$

if $m,n \in \mathbb{N}_0$. (H^* corresponds to the "translation" induced by $(P_n(x))_{n\in\mathbb{N}_0}$, whereas H corresponds to that of $(Q_n(x))_{n\in\mathbb{N}_0}$.)

For $m \geq n \geq 1$ let $\Delta_{m,n}$ denote the "triangle" in $\mathbb{N}_0 \times \mathbb{N}_0$ defined by

$$\Delta_{m,n} := \{(i,j) \in \mathbb{N}_0 \times \mathbb{N}_0 : 0 \leq j \leq i,\ |m-i| < n-j\}.$$

The three edges of the triangle $\Delta_{m,n}$ are $(0, m-n+1)$, $(0, m+n-1)$ and $(m, n-1)$.

Proposition 4.4.3 (Szwarc)
The discrete boundary value problem (10) admits nonnegative solutions if for all $m \geq n \geq 1$ there exists a matrix $v_{m,n}(i,j)$ such that

(1) supp $v_{m,n} \subseteq \Delta_{m,n}$
(2) $H^* v_{m,n}(m,n) < 0$
(3) $H^* v_{m,n}(i,j) \geq 0$ for $(i,j) \neq (m,n)$.

Proof. Let $u(m,n)$ be a solution of (10). We use induction on n to show $u(m,n) \geq 0$. The boundary value condition says $u(m,0) \geq 0$ for all $m \in \mathbb{N}_0$. Suppose $u(m,j) \geq 0$ for all $0 \leq j \leq n, \quad m \in \mathbb{N}_0$. At first we observe that the support of $H^* v_{m,n}$ is contained in the increased triangle with edges $(0, m-n)$, $(0, m+n)$, (m,n),

$$\operatorname{supp}(H^* v_{m,n}) \subseteq \{(i,j) \in \mathbb{N}_0 \times \mathbb{N}_0 : 0 \leq j \leq i, \ |m-i| \leq n-j\}.$$

Then for any $m \in \mathbb{N}_0$

$$0 = \langle Hu, v_{m,n+1} \rangle = \langle u, H^* v_{m,n+1} \rangle$$

$$= u(m,n+1)\, H^* v_{m,n+1}(m,n+1) + \sum_{j \leq n} u(i,j)\, H^* v_{m,n+1}(i,j).$$

Hence

$$-H^* v_{m,n+1}(m,n+1)\, u(m,n+1) = \sum_{j \leq n} u(i,j)\, H^* v_{m,n+1}(i,j),$$

and by (2), (3) and the induction assumption we conclude $u(m,n+1) \geq 0$. ◇

Now we have to come up with a suitable choice of $v_{m,n}$, where we must take into account the special form of action of H^*, see (11). The appropriate choice is

$$v_{m,n}(i,j) = \begin{cases} 1 & \text{for } (i,j) \in \Delta_{m,n}, \ (m+n)-(i+j) \text{ odd} \\ 0 & \text{otherwise} \end{cases}.$$

The direct calculation shows

$$H^* v_{m,n}(i,j) = \begin{cases} -\gamma_n & \text{if } (i,j) = (m,n) \\ \beta_i - \beta_j & \text{if } (i,j) \in \Delta_{m,n}, \ (m+n)-(i+j) \text{ odd} \\ \gamma_i + \alpha_i - \gamma_j - \alpha_j & \text{if } (i,j) \in \Delta_{m,n}, \ (m+n)-(i+j) \text{ even} \\ \gamma_i - \gamma_j & \text{if } (i,j) = (m+n-k,k), \ k = 0,\dots,n-1 \\ \alpha_i - \gamma_j & \text{if } (i,j) = (m-n+k,k), \ k = 0,\dots,n-1 \end{cases}.$$

In order to fulfil the assumptions of Proposition 4.4.3 it suffices that the recurrence coefficients satisfy corresponding monotonicity criteria. More precisely we have shown:

Theorem 4.4.4 (Szwarc)
The orthogonal polynomials $P_n(x)$ have nonnegative linearization coefficients $c(m,n;k)$ whenever the recurrence coefficients $\alpha_n > 0,\ \gamma_n > 0,\ \beta_n \in \mathbb{R},\ n \in \mathbb{N}_0\ (\gamma_0 = 0)$ of (3), (4) satisfy:

$$(\gamma_n)_{n\in\mathbb{N}_0},\ (\beta_n)_{n\in\mathbb{N}_0},\ (\gamma_n+\alpha_n)_{n\in\mathbb{N}_0} \ \text{are nondecreasing,}$$

$$\text{and } \gamma_n \le \alpha_n \text{ for all } n \in \mathbb{N}. \tag{12}$$

Proof. The conditions of (12) guarantee that the assumptions of Proposition 4.4.3 are fulfilled. ◇

In the case of even orthogonalization measures π, R.Szwarc has derived another sufficient criterion for nonnegativity of the linearization coefficients by splitting assumptions according to the parity of n. The proof of this result consists in considering an adapted boundary value problem, and then to show that the matrices $v_{m,n}$ (from above) guarantee that the solutions are nonnegative. More precisely we state without proof, see [649].

Theorem 4.4.5 (Szwarc)
Suppose the orthogonal polynomials $P_n(x)$ are symmetric (i.e. $\beta_n = 0$ for all $n \in \mathbb{N}_0$). The linearization coefficients $c(m,n;k)$ are nonnegative whenever the recurrence coefficients $\alpha_n > 0,\ \gamma_n > 0,\ n \in \mathbb{N}_0\ (\gamma_0 = 0)$ of (3), (4) satisfy

$$(\gamma_{2n})_{n\in\mathbb{N}_0},\ (\gamma_{2n+1})_{n\in\mathbb{N}_0},\ (\gamma_{2n}+\alpha_{2n})_{n\in\mathbb{N}_0},\ (\gamma_{2n+1}+\alpha_{2n+1})_{n\in\mathbb{N}_0}$$

$$\text{are nondecreasing, and } \gamma_n \le \alpha_n \text{ for all } n \in \mathbb{N}. \tag{13}$$

It remains to show under which conditions $(P_n(x))_{n\in\mathbb{N}}$ can be normalized to $R_n(x)$ such that $R_n(1) = 1$ and $a_n > 0,\ c_n > 0,\ b_n \in \mathbb{R}$ for $n \in \mathbb{N}_0\ (c_0 = 0)$, see (8), (9). As already mentioned this is possible if $\alpha_n > 0,\ \gamma_n > 0,\ \beta_n \ge \beta_0,\ n \in \mathbb{N},\ \ \alpha_0 > 0,\ \beta_0 < 1$ and $P_n(1) > 0$ for $n \in \mathbb{N}_0$. Thus it is important to check $P_n(1) > 0$ and better to determine $P_n(1)$ explicitly.

For that reason we restrict ourselves to the case of even orthogonalization measures π, i.e. $\beta_n = 0$ for all $n \in \mathbb{N}_0$.

Proposition 4.4.6 *Let $(P_n(x))_{n\in\mathbb{N}_0}$ be defined by (3), (4), and suppose that $\alpha_n + \gamma_n = 1,\ \ \gamma_n \ne 1$ for every $n \in \mathbb{N}$, and set $\gamma_0 = 1 - \alpha_0$. Then*

$$P_n(1) = 1 + \sum_{k=1}^{n} \frac{\gamma_0 \cdots \gamma_{k-1}}{(1-\gamma_0)\cdots(1-\gamma_{k-1})} \qquad \text{for} \quad n \in \mathbb{N}. \tag{14}$$

Proof. We apply induction. If $n = 1$, then $1 + \frac{\gamma_0}{1-\gamma_0} = \frac{1}{\alpha_0} = P_1(1)$. Suppose that (14) is true for n. Then

$$\begin{aligned}
P_{n+1}(1) &= \frac{1}{1-\gamma_n}\left(P_n(1) - d_n P_{n-1}(1)\right) \\
&= \frac{1}{1-\gamma_n}\left(1 + \sum_{k=1}^{n} \frac{\gamma_0 \cdots \gamma_{k-1}}{(1-\gamma_0)\cdots(1-\gamma_{k-1})}\right) \\
&\quad - \frac{\gamma_n}{1-\gamma_n}\left(1 + \sum_{k=1}^{n-1} \frac{\gamma_0 \cdots \gamma_{k-1}}{(1-\gamma_0)\cdots(1-\gamma_{k-1})}\right) \\
= \left(1 + \sum_{k=1}^{n-1} \frac{\gamma_0 \cdots \gamma_{k-1}}{(1-\gamma_0)\cdots(1-\gamma_{k-1})}\right) &+ \frac{1}{(1-\gamma_n)} \cdot \frac{\gamma_0 \cdots \gamma_{n-1}}{(1-\gamma_0)\cdots(1-\gamma_{n-1})} \\
&= \left(1 + \sum_{k=1}^{n-1} \frac{\gamma_0 \cdots \gamma_{k-1}}{(1-\gamma_0)\cdots(1-\gamma_{k-1})}\right) \\
&\quad + \left(\frac{\gamma_0 \cdots \gamma_{n-1}}{(1-\gamma_0)\cdots(1-\gamma_{n-1})} + \frac{\gamma_0 \cdots \gamma_n}{(1-\gamma_0)\cdots(1-\gamma_n)}\right) \\
&= 1 + \sum_{k=1}^{n+1} \frac{\gamma_0 \cdots \gamma_{k-1}}{(1-\gamma_0)\cdots(1-\gamma_{k-1})}.
\end{aligned}$$

◇

Remark: The assumption $\alpha_n + \gamma_n = 1$ for $n \in \mathbb{N}$ nearly meets (besides $\gamma_0 = 0$) the condition of $P_n(1) = 1$ for all $n \in \mathbb{N}_0$. Formula (14) reveals the effect of $\gamma_0 \neq 0$.

Often the recurrence coefficients in (3), (4) can be written as

$$\gamma_n = \frac{\lambda_n}{\lambda_n + \mu_{n+1}} \quad \text{and} \quad \alpha_n = \frac{\mu_{n+1}}{\lambda_n + \mu_{n+1}}, \quad n \in \mathbb{N}_0, \tag{15}$$

where $(\lambda_n)_{n\in\mathbb{N}_0}$ and $(\mu_n)_{n\in\mathbb{N}_0}$ are two sequences of positive numbers up to λ_0, for which we only assume that $\lambda_0 \neq -\mu_1$. Then formula (14) writes as

$$P_n(1) = 1 + \sum_{k=1}^{n} \frac{\lambda_0 \cdots \lambda_{k-1}}{\mu_1 \cdots \mu_k}, \qquad n \in \mathbb{N}. \tag{16}$$

For general $\alpha_n > 0$, $\beta_n = 0$, $n \in \mathbb{N}_0$, $\gamma_n > 0$, $n \in \mathbb{N}$ we can use the concept of chain sequences to check whether $P_n(x)$ can be normalized such that $R_n(x)$ has the desired recurrence coefficients of (1), (2) in subsection 1.2.1, i.e. $a_n > 0$, $c_n > 0$, $a_n + c_n = 1$ for $n \in \mathbb{N}$, and $a_0 = 1$, $b_0 = 0$.

Definition: A sequence $(\sigma_n)_{n\in\mathbb{N}}$ is a positive **chain sequence**, if there is a sequence $(\gamma_n)_{n\in\mathbb{N}_0}$ such that $0 \leq \gamma_0 < 1$, $0 < \gamma_n < 1$ and $\sigma_n = \gamma_n(1 - \gamma_{n-1})$ for $n \in \mathbb{N}$. The sequence $(\gamma_n)_{n\in\mathbb{N}_0}$ is called **parameter sequence** for $(\sigma_n)_{n\in\mathbb{N}}$. A parameter sequence $(c_n)_{n\in\mathbb{N}_0}$ is called minimal if $c_0 = 0$.

A systematic development of the theory of chain sequences can be found in [742] or [138].

Proposition 4.4.7 *The orthogonal polynomial sequence $(P_n(x))_{n\in\mathbb{N}_0}$ with recurrence relation (3), (4) (with $\beta_n = 0$) can be normalized to $R_n(x) = k_n P_n(x)$ such that (1), (2) in subsection 1.2.1 hold with $a_0 = 1$, $b_n = 0$ for $n \in \mathbb{N}_0$ and $a_n = 1 - c_n$, $0 < c_n < 1$ for $n \in \mathbb{N}$ if and only if $(\sigma_n)_{n\in\mathbb{N}}$ is a positive chain sequence, where $\sigma_n = \gamma_n \alpha_{n-1}$ for $n \in \mathbb{N}$.*

Proof. The recurrence coefficients satisfy

$$a_n = \frac{k_n}{k_{n+1}}\,\alpha_n, \quad n \in \mathbb{N}_0 \qquad \text{and} \quad c_n = \frac{k_n}{k_{n-1}}\,\gamma_n, \quad n \in \mathbb{N}, \; c_0 = 0.$$

Hence $\sigma_n = \gamma_n \alpha_{n-1} = c_n a_{n-1} = c_n(1 - c_{n-1})$, and the equivalence statement follows. ◇

The sequence $(c_n)_{n\in\mathbb{N}_0}$ (with $c_0 = 0$) is the minimal parameter sequence for $(\gamma_n \alpha_{n-1})_{n\in\mathbb{N}}$. The normalization coefficients $k_n = \frac{1}{P_n(1)}$, $n \in \mathbb{N}$, are given by

$$k_n = \prod_{k=0}^{n-1} \frac{\alpha_k}{1 - c_k}. \tag{17}$$

If the recurrence coefficients α_n, γ_n have the form of (15), there is a sufficient criterion for $\sigma_n = \gamma_n \alpha_{n-1}$ being a chain sequence.

Proposition 4.4.8 *Suppose σ_n is given as*

$$\sigma_n = \frac{\lambda_n \mu_n}{(\lambda_{n-1} + \mu_n)(\lambda_n + \mu_{n+1})}, \qquad n \in \mathbb{N},$$

where $\lambda_n, \mu_n > 0$ if $n \in \mathbb{N}$, and $\lambda_0 < 0$, $\lambda_0 \neq -\mu_1$. If

$$0 < \lambda_n - \lambda_{n-1} \le \mu_n - \mu_{n-1}, \qquad n \in \mathbb{N},$$

(with $\mu_0 = 0$), then $(\sigma_n)_{n\in\mathbb{N}}$ is a chain sequence.

Proof. We construct the minimal parameter sequence $(c_n)_{n\in\mathbb{N}_0}$ recursively by showing

$$0 < c_n \le \frac{\mu_n}{\lambda_n + \mu_{n+1}} < 1, \qquad n \in \mathbb{N}, \tag{18}$$

where c_{n+1} is defined by the recurrence $c_{n+1} = \frac{\sigma_{n+1}}{1-c_n}$. With $c_0 = 0$ and $c_1 = \sigma_1$ we see that

$$0 < c_1 = \frac{\lambda_1 \mu_1}{(\lambda_0 + \mu_1)(\lambda_1 + \mu_2)} \le \frac{\mu_1}{\lambda_1 + \mu_2} < 1.$$

Suppose $c_1, ..., c_n$ are constructed and fulfil (18). Then

$$0 < \frac{\lambda_n + \mu_{n+1} - \mu_n}{\lambda_n + \mu_{n+1}} \le 1 - c_n < 1,$$

and hence

$$c_{n+1} = \frac{\sigma_{n+1}}{1-c_n} = \frac{\lambda_{n+1}\mu_{n+1}}{(\lambda_n+\mu_{n+1})(\lambda_{n+1}+\mu_{n+2})(1-c_n)}$$
$$\leq \frac{\lambda_{n+1}\mu_{n+1}}{(\lambda_{n+1}+\mu_{n+2})(\lambda_n+\mu_{n+1}-\mu_n)} \leq \frac{\mu_{n+1}}{\lambda_{n+1}+\mu_{n+2}} < 1.$$

◇

There is another useful result on positive chain sequences.

Theorem 4.4.9 *Let $(\sigma_n)_{n\in\mathbb{N}}$ be a positive chain sequence and $\sigma_n = c_n(1-c_{n-1})$, where $(c_n)_{n\in\mathbb{N}}$ is the minimal parameter sequence. If $\lim_{n\to\infty} \sigma_n = \sigma$, then $0 \leq \sigma \leq \frac{1}{4}$ and*

$$\lim_{n\to\infty} c_n = \frac{1}{2}\left(1 - \sqrt{1-4\sigma}\right).$$

In particular, if $c_n(1-c_{n-1}) \to \frac{1}{4}$, then $c_n \to \frac{1}{2}$.

For the proof of Theorem 4.4.9 we refer to [138, Ch.III, Theorem 6.4].

5 Examples of polynomial hypergroups

We begin by the ultraspherical polynomials to see how the results of section 4.4 fit into this class. Recall that we know the linearization coefficients in that case explicitly, see subsection 1.2.1.

5.1 *Ultraspherical polynomials*

The usual recurrence formula of ultraspherical polynomials is

$$(n+1)\,P_{n+1}(x) = 2x(n+\alpha+\frac{1}{2})\,P_n(x) - (2\alpha+n)\,P_{n-1}(x), \qquad n\in\mathbb{N}$$

and

$$P_0(x) = 1, \qquad P_1(x) = (2\alpha+1)x, \qquad (\alpha > -1)$$

see [641,(4.7.17)] (we changed the parameters to $\lambda = \alpha + \frac{1}{2}$). Hence the coefficients in (3), (4) in subsection 4.4 are

$$\alpha_n = \frac{n+1}{2n+2\alpha+1}, \quad \beta_n = 0, \quad \gamma_n = \frac{n+2\alpha}{2n+2\alpha+1},$$
$$n\in\mathbb{N}, \text{ and } \alpha_0 = \frac{1}{2\alpha+1}.$$

Since $P_n(1) = \frac{(2\alpha+1)_n}{n!}$, see [641,(4.7.3)], we have the recurrence formula for the polynomials $R_n^{(\alpha,\alpha)}(x)$

$$x\cdot R_n^{(\alpha,\alpha)}(x) = \frac{n+2\alpha+1}{2n+2\alpha+1}\,R_{n+1}^{(\alpha,\alpha)}(x) + \frac{n}{2n+2\alpha+1}\,R_{n-1}^{(\alpha,\alpha)}(x),$$
$$n\in\mathbb{N}, \qquad (i)$$

and

$$R_0^{(\alpha,\alpha)}(x) \;=\; 1, \qquad R_1^{(\alpha,\alpha)}(x) \;=\; x. \qquad\qquad (ii)$$

It is easy to check that $c_n = \frac{n}{2n+2\alpha+1}$ is nondecreasing if and only if $\alpha \geq -\frac{1}{2}$. By Theorem 4.4.4 it follows that $(R_n^{(\alpha,\alpha)})_{n\in\mathbb{N}_0}$ generates a polynomial hypergroup on $\mathbb{N}_0$ whenever $\alpha \geq -\frac{1}{2}$, which belongs to the Nevai class with R_n-parameters $a = \frac{1}{2}$, $c = \frac{1}{2}$ and starting parameter $a_0 = 1$. The explicit form of the linearization coefficients $g(m, n; k)$ we have already written down. The Haar weights are

$$h(n) \;=\; \frac{(2n+2\alpha+1)(2\alpha+1)_n}{(2\alpha+1)n!}, \qquad n \in \mathbb{N}_0, \qquad\qquad (iii)$$

the leading coefficient of $R_n^{(\alpha,\alpha)}(x)$, i.e. $R_n^{(\alpha,\alpha)}(x) = \varrho_n x^n + ...$, is easily calculated as

$$\varrho_n \;=\; \frac{(2\alpha+n+1)_n}{2^n(\alpha+1)_n}, \qquad n \in \mathbb{N}_0.$$

As is well-known the orthogonalization measure, i.e. the Plancherel measure, is

$$d\pi(x) \;=\; c_\alpha \chi_{[-1,1]}(1-x^2)^\alpha\, dx, \quad \text{where} \quad c_\alpha \;=\; \frac{\Gamma(2\alpha+2)}{2^{2\alpha+1}(\Gamma(\alpha+1))^2}. \qquad (iv)$$

Note that $c_{-1/2} = \frac{1}{\pi}$, $c_0 = \frac{1}{2}$ and $c_{1/2} = \frac{2}{\pi}$.

Since the polynomial hypergroup is of polynomial growth, $h(n) = O(n^{2\alpha+1})$ as $n \to \infty$, it follows by Theorem 4.2.4 that $[-1, 1] = \operatorname{supp} \pi = S = D_s = D$.

Within the ultraspherical polynomials there are well-known examples.

1.a. **Chebyshev polynomials of the first kind**

For $\alpha = -\frac{1}{2}$ we get the Chebyshev polynomials of the first kind,

$$T_n(x) \;=\; R_n^{(-\frac{1}{2},-\frac{1}{2})}(x) \;=\; \cos(nt), \qquad x \;=\; \cos t.$$

Special features of the corresponding hypergroup on $\mathbb{N}_0$ are:

$$g(m, n, n-m) \;=\; \frac{1}{2} \;=\; g(m, n, n+m), \qquad 1 \leq m \leq n$$

and

$$g(m, n, k) \;=\; 0 \quad \text{for } 1 \leq m \leq n, \;\; k = n-m+1, .., n+m-1.$$

The Haar weights are $h(0) = 1$, $h(n) = 2$ for $n \in \mathbb{N}$.

1.b. **Chebyshev polynomials of the second kind**

For $\alpha = \frac{1}{2}$ we get the Chebyshev polynomials of the second kind

$$U_n(x) \;=\; R_n^{(\frac{1}{2},\frac{1}{2})}(x) \;=\; \frac{\sin((n+1)t)}{(n+1)\sin t}, \qquad x \;=\; \cos t$$

The linearization coefficients are for $1 \le m \le n$

$$g(m,n;k) \;=\; \frac{k+1}{(n+1)(m+1)}$$

$$\text{if } k = n-m, n-m+2, ..., n+m-2, n+m,$$

and

$$g(m,n;k) \;=\; 0 \quad \text{if } k = n-m+1, n-m+3, ..., n+m-1.$$

The Haar weights are $h(0) = 1,\;\; h(n) = (n+1)^2$.

The ultraspherical polynomials belong to the larger class of Jacobi polynomials. The class of symmetric Jacobi polynomial sequences are exactly the ultraspherical polynomial sequences.

5.2 *Jacobi polynomials*

Let $\alpha, \beta \in \mathbb{R}$ such that $\alpha \ge \beta > -1,\;\; \alpha + \beta + 1 \ge 0$. Put for $n \in \mathbb{N}$

$$a_n \;=\; \frac{2(n+\alpha+\beta+1)(n+\alpha+1)(\alpha+\beta+2)}{(2n+\alpha+\beta+2)(2n+\alpha+\beta+1)\,2(\alpha+1)}$$

$$b_n \;=\; \frac{\alpha-\beta}{2(\alpha+1)}\left[1 - \frac{(\alpha+\beta+2)(\alpha+\beta)}{(2n+\alpha+\beta+2)(2n+\alpha+\beta)}\right]$$

$$c_n \;=\; \frac{2n(n+\beta)(\alpha+\beta+2)}{(2n+\alpha+\beta+1)(2n+\alpha+\beta)\,2(\alpha+1)}. \qquad (i)$$

Further set

$$a_0 \;=\; \frac{2(\alpha+1)}{\alpha+\beta+2}, \qquad b_0 \;=\; \frac{\beta-\alpha}{\alpha+\beta+2}. \qquad (ii)$$

One can check that $a_n > 0,\; c_n > 0,\; b_n \ge 0$ and $a_n + b_n + c_n = 1$ if $n \in \mathbb{N}$. Obviously $a_0 > 0$ and $a_0 + b_0 = 1$. The orthogonal polynomials $R_n^{(\alpha,\beta)}(x)$ defined by the coefficients of (i), (ii) are Jacobi polynomials, see [641]. The whole work to check that the linearization coefficients $g(m,n;k)$ are nonnegative (if $\alpha \ge \beta > -1,\; \alpha+\beta+1 \ge 0$) is done in [252,Theorem]. The region for the parameters α, β such that $g(m,n;k) \ge 0$ can be extended slightly, see [253]. In fact, the linearization coefficients are nonnegative if and only if $(\alpha, \beta) \in V$, where $V = \{(\alpha,\beta) \in \mathbb{R}^2 : \alpha \ge \beta > -1$ and

$a(a+5)(a+3)^2 \geq (a^2-7a-24)b^2$ with $a = \alpha+\beta+1,\ b = (\alpha-\beta)\}$. In particular if $\alpha \geq \beta > -1,\ \alpha+\beta+1 \geq 0$ then $(\alpha,\beta) \in V$.

So $(R_n^{(\alpha,\beta)}(x))_{n\in\mathbb{N}_0}$ generates a polynomial hypergroup on $\mathbb{N}_0$ whenever $(\alpha,\beta) \in V$. This polynomial hypergroup belongs to the Nevai class with R_n-parameter $a = c = \frac{1}{2}\,\frac{\alpha+\beta+2}{2(\alpha+1)}$ and starting parameter $a_0 = \frac{1}{2a}$. (We refer to Proposition 4.3.2(iii).)

For $\alpha = \beta$ we get the ultraspherical polynomials.

The Haar weights are

$$h(n) \;=\; \frac{(2n+\alpha+\beta+1)(\alpha+\beta+1)_n(\alpha+1)_n}{(\alpha+\beta+1)n!(\beta+1)_n}, \qquad n \in \mathbb{N}_0. \qquad (iii)$$

The leading coefficients of $R_n^{(\alpha,\beta)}(x) = \varrho_n x^n + \ldots$ are given by

$$\varrho_n \;=\; \frac{(\alpha+\beta+n+1)_n}{2^n(\alpha+1)_n}, \qquad n \in \mathbb{N}_0.$$

The orthogonalization measure is

$$d\pi(x) \;=\; c_{\alpha,\beta}\chi_{[-1,1]}(1-x)^\alpha(1+x)^\beta\, dx, \qquad (iv)$$

where

$$c_{\alpha,\beta} \;=\; \frac{\Gamma(\alpha+\beta+2)}{2^{\alpha+\beta+1}\Gamma(\alpha+1)\Gamma(\beta+1)}.$$

The generated polynomial hypergroup is of polynomial growth. In fact, $h(n) = O(n^{2\alpha+1})$ as $n \to \infty$, independently of β. Hence Theorem 4.2.4 implies $[-1,1] = \operatorname{supp}\,\pi = S = D_s = D$.

5.3 *Generalized Chebyshev polynomials*

Let $\alpha > -1,\ \beta > -1$. The coefficients of the recurrence relation are

$$a_n \;=\; \begin{cases} \dfrac{k+\alpha+\beta+1}{2k+\alpha+\beta+1} & \text{if } n = 2k,\ \ k \in \mathbb{N} \\[2ex] \dfrac{k+\alpha+1}{2k+\alpha+\beta+2} & \text{if } n = 2k+1,\ \ k \in \mathbb{N}_0 \end{cases}$$

$$c_n \;=\; \begin{cases} \dfrac{k}{2k+\alpha+\beta+1} & \text{if } n = 2k,\ \ k \in \mathbb{N} \\[2ex] \dfrac{k+\beta+1}{2k+\alpha+\beta+2} & \text{if } n = 2k+1,\ \ k \in \mathbb{N}_0, \end{cases} \qquad (i)$$

$b_n = 0$ if $n \in \mathbb{N}_0$ and $a_0 = 1$. The corresponding orthogonal polynomials are the generalized Chebyshev polynomials $T_n^{(\alpha,\beta)}(x)$. They are related to the Jacobi polynomials $R_k^{(\alpha,\beta)}(x)$ by

$$T_n^{(\alpha,\beta)}(x) = \begin{cases} R_k^{(\alpha,\beta)}(2x^2-1) & \text{if } n = 2k, \ k \in \mathbb{N} \\ x\, R_k^{(\alpha,\beta+1)}(2x^2-1) & \text{if } n = 2k+1, \ k \in \mathbb{N}_0, \end{cases} \qquad (ii)$$

see [138,Ch.V,(2.40)]. It is straightforward to check that the assumptions of Theorem 4.4.5 are fulfilled if and only if $\alpha \geq \beta > -1, \ \ \alpha+\beta+1 \geq 0$. Hence $(T_n^{(\alpha,\beta)}(x))_{n\in\mathbb{N}_0}$ generates a polynomial hypergroup on $\mathbb{N}_0$ whenever $\alpha \geq \beta > -1, \ \ \alpha+\beta+1 \geq 0$. These hypergroups belong to the Nevai class with R_n-parameters $a = c = \frac{1}{2}$ and starting parameter $a_0 = 1$. In particular, by Theorem 4.3.3 we infer that $[-1,1] = \text{supp}\ \pi = S = D_s = D$.

The Haar weights are $h(0) = 1$ and

$$h(n) = \begin{cases} \dfrac{(2k+\alpha+\beta+1)(\alpha+\beta+1)_k(\alpha+1)_k}{(\alpha+\beta+1)k!(\beta+1)_k} & \text{if } n = 2k, \ k \in \mathbb{N} \\[2ex] \dfrac{(2k+\alpha+\beta+2)(\alpha+\beta+2)_k(\alpha+1)_k}{k!(\beta+1)_{k+1}} & \text{if } n = 2k+1, \ k \in \mathbb{N}_0. \end{cases} \qquad (iii)$$

The orthogonalization measure is

$$d\pi(x) = d_{\alpha,\beta}\chi_{[-1,1]}(1-x^2)^\alpha\, |x|^{2\beta+1}\, dx, \qquad (iv)$$

where $d_{\alpha,\beta} = \frac{\Gamma(\alpha+\beta+2)}{\Gamma(\alpha+1)\Gamma(\beta+1)}$.

5.4 *Associated ultraspherical polynomials*

Fix $\alpha \geq -\frac{1}{2}, \ \ \nu > 0$. The associated ultraspherical polynomials $P_n(x)$ satisfy the recurrence relation

$$(n+\nu+1)\, P_{n+1}(x) = 2x\, (n+\nu+\alpha+\frac{1}{2})\, P_n(x) - (2\alpha+n+\nu)\, P_{n-1}(x), \quad n \in \mathbb{N}$$

and

$$P_0(x) = 1, \qquad P_1(x) = \frac{2\alpha+2\nu+1}{\nu+1}\, x,$$

see for example [124]. They are obtained by replacing in the recurrence coefficients of ultraspherical polynomials the "n" by "$n+\nu$". The coefficients corresponding to $\alpha_n, \beta_n, \gamma_n$ in (3), (4) in section 4.4 are

$$\alpha_n = \frac{n+\nu+1}{2n+2\nu+2\alpha+1}, \quad \beta_n = 0, \quad \gamma_n = \frac{n+\nu+2\alpha}{2n+2\nu+2\alpha+1}, \quad n \in \mathbb{N},$$

and

$$\alpha_0 = \frac{\nu+1}{2\nu+2\alpha+1}, \qquad \beta_0 = 0.$$

We shall use (16) to determine $P_n(1)$. To do this put

$$\lambda_n = n+\nu+2\alpha \text{ for } n\in\mathbb{N}_0 \quad \text{and } \mu_n = n+\nu \text{ for } n\in\mathbb{N}.$$

Then

$$\gamma_n = \frac{\lambda_n}{\lambda_n+\mu_{n+1}} \quad \text{and } \alpha_n = \frac{\mu_{n+1}}{\lambda_n+\mu_{n+1}} \quad \text{for } n\in\mathbb{N}_0.$$

Note that $\lambda_0 \neq -\mu_1$ since $\nu > 0$. By formula (16) we have

$$P_n(1) = 1+\sum_{k=1}^{n}\frac{(2\alpha+\nu)_k}{(1+\nu)_k} = \frac{(2\alpha+\nu)_{n+1}-(\nu)_{n+1}}{2\alpha\,(\nu+1)_n}, \qquad n\in\mathbb{N},$$

where the second equality is formula (7.1.1) of [289,p.151]. Denote by

$$\sigma_n = \frac{\lambda_n\mu_n}{(\lambda_{n-1}+\mu_n)(\lambda_n+\mu_{n+1})} = \frac{(n+\nu+2\alpha)(n+\nu)}{(2n+2\nu+2\alpha+1)(2n+2\nu+2\alpha-1)}, \quad n\in\mathbb{N}.$$

Obviously $(\sigma_n)_{n\in\mathbb{N}}$ is a positive chain sequence, and $(\gamma_n)_{n\in\mathbb{N}_0}$ is a parameter sequence for $(\sigma_n)_{n\in\mathbb{N}}$, provided $\gamma_0 \geq 0$, i.e. $2\alpha+\nu \geq 0$. Proposition 4.4.8 yields that $(\sigma_n)_{n\in\mathbb{N}}$ is also a chain sequence when $2\alpha+\nu<0$. According to Proposition 4.4.7 we have the recurrence formula for the polynomials $R_n^{(\alpha)}(x;\nu) = P_n(x)/P_n(1)$,

$$R_1^{(\alpha)}(x;\nu)\,R_n^{(\alpha)}(x;\nu) = (1-c_n)\,R_{n+1}^{(\alpha)}(x;\nu) + c_n\,R_{n-1}^{(\alpha)}(x;\nu), \quad n\in\mathbb{N}$$

and $R_0^{(\alpha)}(x;\nu)=1$, $R_1^{(\alpha)}(x;\nu)=x$, where

$$\begin{aligned} c_n &= \frac{P_{n-1}(1)(n+\nu+2\alpha)}{P_n(1)(2n+2\nu+2\alpha+1)} \\ &= \frac{(\nu+n)(2\alpha+\nu)_{n+1}-(n+\nu+2\alpha)(\nu)_{n+1}}{(2n+2\nu+2\alpha+1)\left[(2\alpha+\nu)_{n+1}-(\nu)_{n+1}\right]} \qquad (i) \end{aligned}$$

for $n\in\mathbb{N}$, satisfy $0<c_n<1$.

For the positivity of the linearization coefficients we observe that $\sigma_n \leq \sigma_{n+1}$ for $n\in\mathbb{N}$ if and only if $\alpha\geq\frac{1}{2}$. Hence, Theorem 4.4.1 yields a polynomial hypergroup structure on $\mathbb{N}_0$ if $\alpha\geq\frac{1}{2}$. For the case $-\frac{1}{2}\leq\alpha<\frac{1}{2}$ we suppose $2\alpha+\nu\geq 0$. Then $\lambda_0\geq 0$ and $\lambda_n\leq\mu_{n+1}$ for $n\in\mathbb{N}$ and $(\lambda_n/\mu_{n+1})_{n\in\mathbb{N}}$ is nondecreasing. This implies that $\gamma_n\leq\alpha_n$ for $n\in\mathbb{N}$ and $(\gamma_n)_{n\in\mathbb{N}}$ is nondecreasing. Thus Theorem 4.4.4 implies that $(R_n^{(\alpha)}(x;\nu))_{n\in\mathbb{N}_0}$ generates a polynomial hypergroup. For the case $2\alpha+\nu<0$ we use Theorem 4.4.4 again, now with the recurrence coefficients c_n and $a_n = 1-c_n$. A tedious calculation shows that $c_n\leq c_{n+1}$. Furthermore, we have $\sigma_n\to\frac{1}{4}$, and hence the minimal parameter sequence $(c_n)_{n\in\mathbb{N}_0}$ converges towards $\frac{1}{2}$. Thus $c_n\leq\frac{1}{2}\leq 1-c_n = a_n$, and hence the assumptions of Theorem 4.4.4 are satisfied.

Summarizing, $(R_n^{(\alpha)}(x,\nu))_{n\in\mathbb{N}_0}$ (with c_n defined by (i)) generates a polynomial hypergroup, whenever $\alpha\geq-\frac{1}{2}$, $\nu>0$, and the hypergroups belong to the Nevai class with R_n-parameters $a = c = \frac{1}{2}$ and starting parameter $a_0 = 1$. By Theorem 4.3.3 we have $[-1,1] = \text{supp}\,\pi = S = D_s = D$.

We can calculate the Haar weights $h(n)$ from the coefficients α_n, γ_n and also from λ_n, μ_n. In fact, we have for $n \in \mathbb{N}$

$$h(n) = \frac{\alpha_0 \cdots \alpha_{n-1}}{\gamma_1 \cdots \gamma_n} (P_n(1))^2$$

and

$$h(n) = \frac{\mu_1 \cdots \mu_n (\lambda_n + \mu_{n+1})}{\lambda_1 \cdots \lambda_n (\lambda_0 + \mu_1)} (P_n(1))^2.$$

Therefore we have for the associated ultraspherical polynomials

$$h(n) = \frac{(2n + 2\alpha + 2\nu + 1)}{4\alpha^2(2\alpha + 2\nu + 1)(\nu + 1)_n(2\alpha + \nu + 1)_n} ((2\alpha + \nu)_{n+1} - (\nu)_{n+1})^2. \quad (ii)$$

The orthogonalization measure is determined in [548]. See also [124]. We have $d\pi(x) = f(x)\,\chi_{[-1,1]}dx$ with

$$f(\cos t) = \frac{(\sin t)^{2\alpha}}{|{}_2F_1(\frac{1}{2} - \alpha, \nu; \nu + \alpha + \frac{1}{2}; e^{2it})|^2}, \qquad 0 \le t \le \pi \qquad (iii)$$

up to a multiplicative constant.

5.5 *Pollaczek polynomials*

Fix $\alpha > -\frac{1}{2}$, $\mu \ge 0$. Define the orthogonal polynomials $P_n(x)$ by

$$(n+1)\, P_{n+1}(x) = 2x\,(n+\alpha+\mu+\frac{1}{2})\, P_n(x) - (n+2\alpha)\, P_{n-1}(x), \qquad n \in \mathbb{N}$$

and

$$P_0(x) = 1, \qquad P_1(x) = (2\alpha + 2\mu + 1)\,x.$$

These orthogonal polynomials were first studied by Pollaczek [548]. See also [138] or [349]. We call them Pollaczek polynomials. The recursion coefficients $\alpha_n, \beta_n, \gamma_n$ in (3), (4) in section 4.4 are

$$\alpha_n = \frac{n+1}{2n + 2\alpha + 2\mu + 1}, \quad \beta_n = 0, \quad \gamma_n = \frac{n + 2\alpha}{2n + 2\alpha + 2\mu + 1}, \quad n \in \mathbb{N},$$

and

$$\alpha_0 = \frac{1}{2\alpha + 2\mu + 1}, \qquad \beta_0 = 0.$$

For $\mu = 0$ we obtain the ultraspherical polynomials, see subsection 4.5.1.

Using the recurrence relation of the Laguerre polynomials $L_n^{(\alpha)}(x)$, see [641,(5.1.10)], we obtain

$$P_n(1) = L_n^{(2\alpha)}(-2\mu) = (2\alpha+1)_n \sum_{k=0}^{n} \frac{(2\mu)^k}{(2\alpha+1)_k(n-k)!k!},$$

where the second equality is (5.1.6) of [641].

The recurrence formula for the polynomials $R_n(x;\alpha,\mu) = P_n(x)/P_n(1)$ is

$$R_1(x;\alpha,\mu)\, R_n(x;\alpha,\mu) = (1-c_n)\, R_{n+1}(x;\alpha,\mu) + c_n\, R_{n-1}(x;\alpha,\mu), \qquad n \in \mathbb{N}$$

and

$$R_0(x;\alpha,\mu) = 1, \qquad R_1(x;\alpha,\mu) = x,$$

where

$$c_n = \frac{P_{n-1}(1)}{P_n(1)}\, \gamma_n$$

$$= \frac{n}{2n+2\alpha+2\mu+1} \left(\sum_{k=0}^{n-1} \binom{n-1}{k} \frac{(2\mu)^k}{(2\alpha+1)_k} \right) \left(\sum_{k=0}^{n} \binom{n}{k} \frac{(2\mu)^k}{(2\alpha+1)_k} \right)^{-1} \qquad (i)$$

for $n \in \mathbb{N}$.

By the recursion formula for the $P_n(x)$ we obtain $1 - c_n = \frac{P_{n+1}(1)}{P_n(1)}\, \alpha_n$. Hence, $0 < c_n < 1$ holds for $n \in \mathbb{N}$ and $(1-c_{n-1})c_n = \alpha_{n-1}\gamma_n$. Therefore $(\sigma_n)_{n\in\mathbb{N}}$, where $\sigma_n = \alpha_{n-1}\gamma_n$, is a positive chain sequence.

If $\alpha \geq \frac{1}{2}$, we obtain $\sigma_{n+1} \geq \sigma_n$ for $n \in \mathbb{N}$. Theorem 4.4.1 yields that $(R_n(x;\alpha,\mu))_{n\in\mathbb{N}_0}$ generates a polynomial hypergroup on $\mathbb{N}_0$, whenever $\alpha \geq \frac{1}{2}$, $\mu \geq 0$. If $-\frac{1}{2} < \alpha \leq \frac{1}{2}$, we have $\gamma_n \leq \alpha_n$, $\gamma_n \leq \gamma_{n+1}$ and $\alpha_n + \gamma_n \leq \alpha_{n+1} + \gamma_n$ for $n \in \mathbb{N}$. However, $\alpha_0 \leq \alpha_1 + \gamma_1$ is not satisfied for some $\mu \geq 0$ if $-\frac{1}{2} < \alpha < 0$. A transformation of random walk polynomials (see subsection 4.5.9) enables us to derive that a polynomial hypergroup is determined, if $-\frac{1}{2} < \alpha < 0$ and $0 \leq \mu < \alpha + \frac{1}{2}$. Summarizing, $(R_n(x;\alpha,\mu))_{n\in\mathbb{N}_0}$ generates a polynomial hypergroup on $\mathbb{N}_0$, provided $\alpha \geq 0$, $\mu \geq 0$ or $-\frac{1}{2} < \alpha < 0$, $0 \leq \mu < \alpha + \frac{1}{2}$. Applying asymptotic properties of Laguerre polynomials, see Theorem 8.22.3 of [641], we obtain

$$\frac{P_{n-1}(1)}{P_n(1)} = \frac{L_{n-1}^{(2\alpha)}(-2\mu)}{L_n^{(2\alpha)}(-2\mu)} \to 1 \qquad \text{as } n \to \infty.$$

Hence $\lim_{n\to\infty} c_n = \frac{1}{2}$ and the generated polynomial hypergroup belongs to the Nevai class with $a = c = \frac{1}{2}$ and $a_0 = 1$. In particular, we have $[-1,1] = \operatorname{supp} \pi = S = D_s = D$.

The Haar weights are given by

$$h(n) = \frac{(2n+2\alpha+2\mu+1)(2\alpha+1)_n}{(2\alpha+2\mu+1)\, n!} \left(\sum_{k=0}^{n} \binom{n}{k} \frac{(2\mu)^k}{(2\alpha+1)_k} \right)^2 . \qquad (ii)$$

The orthogonalization measure is (up to a multiplicative constant)

$$d\pi(x) = f(x)\, \chi_{[-1,1]}\, dx$$

$$f(\cos t) = (\sin t)^{2\alpha} \left| \Gamma(\alpha + \frac{1}{2} + i\mu \cot t) \right|^2 \exp((2t-\pi)\,\mu \cot t), \quad 0 \le t \le \pi, \qquad (iii)$$

see Pollaczek [548].

5.6 *Associated Pollaczek polynomials*

Fix $\alpha > -\frac{1}{2}$, $\mu \ge 0$, $\nu \ge 0$. Define the associated Pollaczek polynomials $P_n(x) = P_n^{(\nu)}(x;\alpha,\mu)$ by

$$(n+\nu+1)\, P_{n+1}(x) = 2x\,(n+\nu+\alpha+\mu+\frac{1}{2})\, P_n(x) - (n+\nu+2\alpha)\, P_{n-1}(x), \quad n \in \mathbb{N}$$

and

$$P_0(x) = 1, \qquad P_1(x) = \frac{2\alpha+2\nu+2\mu+1}{\nu+1}\, x.$$

These polynomials are the associated ones of those considered in subsection 4.5.5. They were first studied by Pollaczek [548], see also [138,p.185], and we call them associated Pollaczek polynomials. The recursion coefficients $\alpha_n, \beta_n, \gamma_n$ in (3) and (4) in section 4.4 are

$$\alpha_n = \frac{n+\nu+1}{2n+2\nu+2\alpha+2\mu+1}, \quad \beta_n = 0,$$

$$\gamma_n = \frac{n+\nu+2\alpha}{2n+2\nu+2\alpha+2\mu+1}, \quad n \in \mathbb{N}$$

and

$$\alpha_0 = \frac{\nu+1}{2\alpha+2\nu+2\mu+1}, \qquad \beta_0 = 0.$$

The value of $P_n(1)$ is given by $L_n^{(2\alpha)}(-2\mu;\nu)$, where $L_n^{(\alpha)}(x;\nu)$ denotes the associated Laguerre polynomials investigated in [38] and [349]. The associated Laguerre polynomials $L_n^{(\alpha)}(x;\nu)$ are generated by

$$(n+\nu+1)\, L_{n+1}^{(\alpha)}(x;\nu) = (-x+2n+2\nu+\alpha+1)\, L_n^{(\alpha)}(x;\nu) - (n+\nu+\alpha)\, L_{n-1}^{(\alpha)}(x;\nu),$$

$$L_0^{(\alpha)}(x;\nu) = 1, \qquad L_1^{(\alpha)}(x;\nu) = (-x+2\nu+\alpha+1)/(\nu+1).$$

Hence we see that

$$P_n(1) = L_n^{(2\alpha)}(-2\mu;\nu).$$

Furthermore, the leading coefficient of $L_n^{(\alpha)}(x;\nu)$ is $(-1)^n/(\nu+1)_n$ as is easily derived from their recurrence relation.

Since the zeros of the associated Laguerre polynomials lie in $]0,\infty[$ positivity of $L_n^{(2\alpha)}(-2\mu;\nu)$ follows. Now set

$$c_n = \frac{L_{n-1}^{(2\alpha)}(-2\mu;\nu)}{L_n^{(2\alpha)}(-2\mu;\nu)}\,\gamma_n, \qquad n\in\mathbb{N}. \qquad (i)$$

Applying the recursion formula of $L_n^{(2\alpha)}(x;\nu)$ we obtain

$$1-c_n = \frac{L_{n+1}^{(2\alpha)}(-2\mu;\nu)}{L_n^{(2\alpha)}(-2\mu;\nu)}\,\alpha_n, \qquad n\in\mathbb{N}. \qquad (ii)$$

Hence we see that $0 < c_n < 1$ and $(1-c_{n-1})c_n = \alpha_{n-1}\gamma_n$. Therefore, setting $\sigma_n = \alpha_{n-1}\gamma_n$, we know $(\sigma_n)_{n\in\mathbb{N}}$ is a positive chain sequence and $(\gamma_n)_{n\in\mathbb{N}_0}$ is a parameter sequence for $(\sigma_n)_{n\in\mathbb{N}}$. If $\alpha \geq \frac{1}{2}$ then $\sigma_n = \alpha_{n-1}\gamma_n$ are nondecreasing and Theorem 4.4.1 yields a polynomial hypergroup. For $-\frac{1}{2} < \alpha \leq \frac{1}{2}$ we have $\gamma_n \leq \alpha_n$, $\gamma_n \leq \gamma_{n+1}$ and $\alpha_n+\gamma_n \leq \alpha_{n+1}+\gamma_{n+1}$ for $n\in\mathbb{N}$. However, $\alpha_0 \leq \alpha_1+\gamma_1$ only if $0\leq\alpha\leq\frac{1}{2}$. In this case Theorem 4.4.4 yields a polynomial hypergroup.

Summarizing we have the following result: The orthogonal polynomials $R_n(x) = R_n^{(\nu)}(x;\alpha,\mu) = P_n^{(\nu)}(x;\alpha,\mu)/P_n^{(\nu)}(1;\alpha,\mu)$ generate a polynomial hypergroup on $\mathbb{N}_0$, provided $\alpha\geq 0,\ \mu\geq 0,\ \nu\geq 0$.

The Haar weights are given by

$$h(n) = \frac{(2n+2\nu+2\alpha+2\mu+1)(\nu+1)_n}{(2\alpha+2\nu+2\mu+1)(\nu+2\alpha+1)_n}\left(L_n^{(2\alpha)}(-2\mu;\nu)\right)^2. \qquad (iii)$$

The orthogonalization measure is determined by Pollaczek in [548]. Up to a multiplicative constant it is

$$f(\cos t) = (\sin t)^{2\alpha}\left|\Gamma(\alpha+\frac{1}{2}+\nu+i\,\mu\,\cot t)\right|^2 \exp((2t-\pi)\,\mu\,\cot t)$$

$$\times\left|{}_2F_1(\frac{1}{2}-\alpha+i\,\cot t,\,\nu;\,\alpha+\frac{1}{2}+\nu+i\,\mu\,\cot t;\,e^{2it})\right|^{-2}, \qquad 0\leq t\leq\pi. \qquad (iv)$$

It follows that $[-1,1] = \mathrm{supp}\,\pi = S = D_s$.

5.7 *Orthogonal polynomials with constant monic recursion formula*

Fix $\gamma>0,\ \alpha>0$ Consider monic orthogonal polynomials determined by

$$x\,\phi_n(x;\gamma,\alpha) = \phi_{n+1}(x;\gamma,\alpha) + \mu_n\,\phi_{n-1}(x;\gamma,\alpha), \qquad n\in\mathbb{N}$$

and

$$\phi_0(x;\gamma,\alpha) = 1, \qquad \phi_1(x;\gamma,\alpha) = x,$$

where $\mu_n = \gamma$ for $n = 2, 3, \ldots$ and $\mu_1 = \alpha\gamma$.

Note that $\phi_n(x; \frac{1}{4}, 2)$ are the monic Chebyshev polynomials of the first kind, $\frac{1}{2^{n-1}} T_n(x)$, and $\phi_n(x; \frac{1}{4}, 1)$ are the monic Chebyshev polynomials of the second kind, i.e. $\phi_n(x; \frac{1}{4}, 1) = \frac{n+1}{2^n} R_n^{(\frac{1}{2},\frac{1}{2})}(x)$. Notice also that the recursion formula for the $\phi_n(x;\gamma,\alpha)$ determines an orthogonal polynomial sequence as long as $\gamma > 0$ and $\alpha > 0$, see [138, Theorem 4.4, Ch.1].

Linearizing the products, write

$$\phi_m(x;\gamma,\alpha)\, \phi_n(x;\gamma,\alpha) = \sum_{k=n-m}^{n+m} d(m,n;k)\, \phi_k(x;\gamma,\alpha).$$

Applying the recursion formulas (vii) and (ix) of Section 1.2.1 we obtain by induction on m, where $1 \leq m \leq n$,

$$d(m,n;n-m+k) = \begin{cases} 1 & k = 2m \\ \alpha\gamma^m & k = 0,\ m = n \\ \gamma^m & k = 0,\ m < n \\ \gamma^{m-1-\frac{k}{2}}(2\gamma - \alpha\gamma) & k = 2, 4, \ldots, 2m-2 \\ 0 & k \text{ odd} \end{cases}.$$

We use now a slight renormalization of the $\phi_n(x;\gamma,\alpha)$. Moreover, we restrict $0 < \alpha \leq 2$ to have all $d(m,n;k) \geq 0$. Combined with $0 < \gamma \leq \frac{1}{4}$ we know that the support of the orthogonalization measure is contained in $[-1,1]$, see below.

Let $0 < \gamma < \frac{1}{4}$ and write $\omega = \sqrt{1-4\gamma}$. Set

$$\gamma_n = \frac{1}{2}(1-\omega) \quad \text{and} \quad \alpha_n = 1 - \gamma_n = \frac{1}{2}(1+\omega), \qquad n \in \mathbb{N}$$

and

$$\alpha_0 = \frac{\alpha}{2}(1+\omega) = \alpha_1\alpha$$

and define the orthogonal polynomials $P_n(x) = P_n(x;\gamma,\alpha)$ by

$$P_n(x) = \frac{1}{\alpha_1^n\alpha}\, \phi_n(x;\gamma,\alpha) \qquad \text{for } n \in \mathbb{N}, \quad \text{and } P_0(x) = 1.$$

Then we have

$$x\, P_n(x) = \alpha_n P_{n+1}(x) + \gamma_n P_{n-1}(x), \qquad n \in \mathbb{N}_0$$

and

$$P_0(x) = 1, \qquad P_1(x) = \frac{x}{\alpha_0}.$$

For that we have to have in mind that $\gamma_n \alpha_n = \gamma_1 \alpha_1 = \gamma$.

Now we can proceed as in the examples before. In order to obtain $P_n(1)$ we use Proposition 4.4.6 and obtain

$$P_n(1) = 1 + \frac{1-\alpha_0}{\alpha_0} \sum_{k=0}^{n-1} \left(\frac{\gamma_1}{\alpha_1}\right)^k = 1 + \frac{1-\alpha_0}{\alpha_0} \frac{((1+\omega)^n - (1-\omega)^n)}{2\omega\,(1+\omega)^{n-1}}$$

$$= \frac{(2-2\alpha)\,[(1+\omega)^n - (1-\omega)^n] + \alpha\,\left[(1+\omega)^{n+1} - (1-\omega)^{n+1}\right]}{2\alpha\omega\,(1+\omega)^n}, \quad n \in \mathbb{N}.$$

The latter equality follows by direct calculation. By Wall's comparison test, see [138, Ch.3, Theorem 5.7], the sequence $(\mu_n)_{n\in\mathbb{N}}$ is a chain sequence. The minimal parameter sequence for $(\mu_n)_{n\in\mathbb{N}}$ is determined by the values of $\phi_n(1;\gamma,\alpha) = \frac{\alpha}{2^n}\,(1+\omega)^n P_n(1)$ in the following way: $c_1 = \mu_1 = \alpha\gamma$, and for $n = 2, 3, \ldots$

$$c_n = \frac{\phi_{n-1}(1)}{\phi_n(1)}\,\mu_n = \frac{\phi_{n-1}(1)}{\phi_n(1)}\,\gamma$$

$$= 2\gamma\,\frac{(2-2\alpha)\,\left[(1+\omega)^{n-1} - (1-\omega)^{n-1}\right] + \alpha\,[(1+\omega)^n - (1-\omega)^n]}{(2-2\alpha)\,[(1+\omega)^n - (1-\omega)^n] + \alpha\,[(1+\omega)^{n+1} - (1-\omega)^{n+1}]}.$$

(Use the recursion formula of the $\phi_n(x;\gamma,\alpha)$ to show that $c_n(1-c_{n-1}) = \mu_n$ for $n \in \mathbb{N}$, where $c_0 = 0$.)

The polynomials

$$R_n(x) = R_n(x;\gamma,\alpha) = \frac{P_n(x;\gamma,\alpha)}{P_n(1;\gamma,\alpha)} = \frac{\phi_n(x;\gamma,\alpha)}{\phi_n(1;\gamma,\alpha)}$$

generate a polynomial hypergroup on $\mathbb{N}_0$, provided $0 < \gamma \leq \frac{1}{4}$ and $0 < \alpha \leq 2$. (The case $\gamma = \frac{1}{4}$ will be considered in 7.1 below.)

The three-term recursion coefficients are the c_n from above and $a_n = 1 - c_n$, $b_n = 0$. The linearization coefficients $g(m,n;k)$ are

$$g(m,n;k) = d(m,n,k)\,\frac{\phi_k(1;\gamma,\alpha)}{\phi_m(1;\gamma,\alpha)\,\phi_n(1;\gamma,\alpha)}.$$

Before we continue to determine the orthogonalization measure and the Haar weights of these polynomial hypergroups we consider some subclasses.

5.7.1 4.5.7.1 Geronimus polynomials

We begin with the boundary case $\gamma = \frac{1}{4}$, which was excluded above. In that case we proceed as above. Let $\phi_n(x) = \phi_n(x; \frac{1}{4}, \alpha)$ and put $P_n(x) = P_n(x; \frac{1}{4}, \alpha) = \frac{2^n}{\alpha}\, \phi_n(x; \frac{1}{4}, \alpha)$.

We have $P_n(1) = 1 + \frac{2-\alpha}{\alpha}\, n,\;\; \phi_n(1) = \frac{1}{2^n}\, (\alpha + (2-\alpha)n)$ and

$$c_1 \;=\; \frac{\alpha}{4}, \qquad c_n \;=\; \frac{\alpha + (2-\alpha)(n-1)}{2\,(\alpha + (2-\alpha)n)}, \qquad n = 2, 3, \ldots .$$

Using Theorem 4.4.4 it follows that the polynomials $R_n(x) = R_n(x; \frac{1}{4}, \alpha) = \frac{1}{P_n(1)}\, P_n(x)$ generate a polynomial hypergroup on $\mathbb{N}_0$ for $0 < \alpha \le 2$. These polynomials are called Geronimus polynomials, see [270]. Their hypergroup structure is studied in [405]. (In the notation of [405] we have $\alpha = \frac{4}{a}$.) The Geronimus polynomials have an interesting property. They can be represented as linear combinations of Chebyshev polynomials of first and second kind, i.e.

$$R_n(x) \;=\; \frac{2(\alpha-1)}{\alpha + (2-\alpha)n}\, T_n(x) \;+\; \frac{(2-\alpha)(n+1)}{\alpha + (2-\alpha)n}\, R_n^{(\frac{1}{2},\frac{1}{2})}(x).$$

This follows directly from $\phi_n(x; \frac{1}{4}, \alpha) \;=\; (\alpha - 1)\phi_n(x; \frac{1}{4}, 2) + (2 - \alpha)\phi_n(x; \frac{1}{4}, 1)$. In fact, the monic polynomials $(\alpha - 1)\phi_n(x; \frac{1}{4}, 2) + (2 - \alpha)\phi_n(x; \frac{1}{4}, 1)$ fulfil exactly the recursion formula of $\phi_n(x; \frac{1}{4}, \alpha)$. Conversely, if $Q_n(x) = \tau\, \phi_n(x; \frac{1}{4}, 2) + \sigma\, \phi_n(x; \frac{1}{4}, 1)$ with $\tau + \sigma = 1$ and $\tau > -1$ then $Q_n(x) = \phi_n(x; \frac{1}{4}, \alpha)$ with $\alpha = \tau + 1$.

5.7.2 4.5.7.2 Polynomials connected with homogeneous trees

Another interesting subclass is among these polynomials.
Let $q \ge 1$ and put $\gamma = \frac{q}{(q+1)^2}$ and $\alpha = \frac{q+1}{q}$. Then $0 < \gamma \le \frac{1}{4}$ and $0 < \alpha \le 2$. Hence the polynomials $R_n(x; \gamma, \alpha) = R_n(x; q)$ with $\gamma = \frac{q}{(q+1)^2}$ and $\alpha = \frac{q+1}{q}$ generate for each $q \ge 1$ a polynomial hypergroup on $\mathbb{N}_0$. Furthermore we have in that case

$$\omega \;=\; \sqrt{1 - 4\gamma} \;=\; \frac{q-1}{q+1},$$

and for the coefficients of the polynomials $P_n(x; \gamma, \alpha)$ we obtain $\alpha_0 = 1,\;\; \gamma_n = \frac{1}{q+1},\;\; \alpha_n = \frac{q}{q+1}$. Therefore, $P_n(1; \gamma, \alpha) = 1$, which implies $R_n(x; q) = P_n(x; \gamma, \alpha)$. The polynomials $R_n(x; q)$ satisfy

$$x \cdot R_n(x; q) \;=\; \frac{q}{q+1}\, R_{n+1}(x; q) \;+\; \frac{1}{q+1}\, R_{n-1}(x; q)$$

and $R_0(x; q) = 1,\; R_1(x; q) = x$.

The polynomials $R_n(x;q)$ are connected with homogeneous trees of degree q, $q \in \mathbb{N}$. They are also called Cartier-Dunau polynomials, see e.g. [127],[11],[12],[13],[413,(6)] and [447]. The explicit form of the linearization coefficients $g(m,n;k)$ is rather simple:

$$g(m,n;n-m) = \frac{1}{(q+1)\,q^{m-1}}, \qquad g(m,n;n+m) = \frac{q}{q+1}$$

and

$$g(m,n;n+m-2k) = \frac{q-1}{(q+1)\,q^k} \qquad \text{for } k = 1,2,...,m-1,$$

$$g(m,n;n+m-k) = 0 \qquad \text{for } k = 1,3,...,2m-1,$$

see [405,3(d)].

5.7.3 *4.5.7.3 Cosh-polynomials*

For $\alpha = 2$ and $0 < \gamma \le \frac{1}{4}$ we get an orthogonal polynomial sequence which has rather simple linearization coefficients. In fact, from above we get for the monic versions

$$d(m,n;n-m+k) = \begin{cases} 1 & k = 2m \\ 2\gamma^m & k = 0,\ m = n \\ \gamma^m & k = 0,\ m < n \\ 0 & \text{else,} \end{cases}$$

where $1 \le m \le n$.

For the polynomials $\phi_n(x;\gamma,2)$ and $P_n(x;\gamma,2)$ we have $\alpha_0 = (1+\omega)$, and

$$P_n(1;\gamma,2) = \frac{(1+\omega)^n + (1-\omega)^n}{2\,(1+\omega)^n},$$

$$\phi_n(1;\gamma,2) = \frac{1}{2^{n-1}}\,(1+\omega)^n\,P_n(1;\gamma,2) = \frac{1}{2^n}\,((1+\omega)^n + (1-\omega)^n).$$

A parameter transformation yields a very nice representation of the recurrence coefficients a_n and c_n. Choose the unique parameter $a \ge 0$ such that

$$\cosh(a) = \frac{1}{2\sqrt{\gamma}} = \frac{1}{\sqrt{1-\omega^2}}.$$

A rather tedious calculation shows that

$$c_n = \frac{\phi_{n-1}(1;\gamma,2)}{\phi_n(1;\gamma,2)}\,\mu_n = \frac{\cosh(a(n-1))}{2\cosh(an)\,\cosh(a)}. \qquad (*)$$

This can be also shown in a rather simple way. In fact, define $R_n(x)$ by

$$x\,R_n(x) \;=\; a_n\,R_{n+1}(x) + c_n\,R_{n-1}(x), \qquad n\in\mathbb{N},$$

$R_0(x)=1,\; R_1(x)=x$, and

$$c_n \;=\; \frac{\cosh(a(n-1))}{2\cosh(an)\,\cosh(a)}, \qquad a_n \;=\; 1-c_n \;=\; \frac{\cosh(a(n+1))}{2\cosh(an)\,\cosh(a)}.$$

The monic version $\phi_n(x)$ of $R_n(x)$ is $\phi_n(x) = \left(\prod_{k=1}^{n-1} a_k\right) R_n(x)$. It is easily shown that $\phi_n(x)$ satisfy

$$x\,\phi_n(x) \;=\; \phi_{n+1}(x) + \mu_n\,\phi_{n-1}(x), \qquad \phi_0(x) \;=\; 1, \quad \phi_1(x) \;=\; x$$

with $\mu_n = \frac{1}{4(\cosh(a))^2}$ for $n=2,3,...$ and $\mu_1 = \frac{1}{2(\cosh(a))^2}$, and so the equality $(*)$ is shown.

From the linearization coefficients $d(m,n;k)$ we can also derive that for $1\le m\le n$

$$g(m,n;n+m) \;=\; \frac{\cosh(a(n+m))}{2\cosh(an)\,\cosh(am)},$$

$$g(m,n;n-m) \;=\; \frac{\cosh(a(n-m))}{2\cosh(an)\,\cosh(am)}$$

and $g(m,n;k)=0$ for $k=n-m+1,...,n+m-1$.

We return now to the general case of orthogonal polynomials with constant monic recursion formula, and we determine the orthogonalization measure and the Haar weights. A simple change of variables yields the orthogonalization measure. In fact, set

$$Q_n(x;\gamma,\alpha) \;=\; \frac{1}{(2\sqrt{\gamma})^n}\,\phi_n(2\gamma x;\gamma,\alpha).$$

Then the $Q_n(x;\gamma,\alpha)$ satisfy the recurrence relation of the $\phi_n(x;\frac{1}{4},\alpha)$. Hence, $Q_n(x;\gamma,\alpha) = \phi_n(x;\frac{1}{4},\alpha)$. The orthogonalization measure of the Geronimus polynomials $\phi_n(x;\frac{1}{4},\alpha)$ is known, [138,p.204] or [270]. Hence the support of the orthogonalization measure is $[-2\sqrt{\gamma},2\sqrt{\gamma}]$ and the measure is given by $d\pi(x)=f(x)\,dx$, where

$$f(x) \;=\; \frac{\sqrt{4\gamma-x^2}}{\alpha^2\gamma-(\alpha-1)x^2} \qquad \text{for } x\in[-2\sqrt{\gamma},2\sqrt{\gamma}],$$

up to a multiplicative constant. In case of $\gamma=\frac{1}{4}$ the Haar weights are $h(0)=1,\; h(1)=\frac{4}{\alpha}$ and $h(n)=\alpha\,(1+\frac{2-\alpha}{\alpha}\,n)^2$ for $n=2,3,\ldots$.

If $0 < \gamma < \frac{1}{4}$ we have $h(0) = 1$, $h(1) = \frac{1}{\alpha\gamma}$ and

$$h(n) = \frac{\left[(2-2\alpha)\left((1+\omega)^n - (1-\omega)^n\right) + \alpha\left((1+\omega)^{n+1} - (1-\omega)^{n+1}\right)\right]^2}{\alpha\,\gamma^n\,4^{n+1}\,(1-4\gamma)}$$

for $n = 2, 3, \ldots$.

Using the results of section 4.3 we get the following results: If $\gamma = \frac{1}{4}$, then $[-1,1] = \operatorname{supp}\pi = S = D_s = D$. In case of $0 < \gamma < \frac{1}{4}$, $[-2\sqrt{\gamma}, 2\sqrt{\gamma}] = \operatorname{supp}\pi = S$, $D_s = [-1,1]$ and $D = \{z \in \mathbb{C} : |z - 2\sqrt{\gamma}| + |z + 2\sqrt{\gamma}| \leq 2\}$.

5.8 *Karlin-McGregor polynomials*

This class of orthogonal polynomials generates polynomial hypergroups, for which the dual spaces fulfil interesting properties. Karlin-McGregor polynomials were first considered in [363]. They are defined by the recurrence relation (1), (2) of subsection 1.2.1, where the coefficients are

$$a_n = \begin{cases} \dfrac{\alpha-1}{\alpha} & \text{if } n = 2k+1, \quad k \in \mathbb{N}_0 \\[2ex] \dfrac{\beta-1}{\beta} & \text{if } n = 2k, \quad k \in \mathbb{N}, \end{cases}$$

$b_n = 0$, $c_n = 1 - a_n$ for $n \in \mathbb{N}$, and $a_0 = 1$, $b_0 = 0$. For $\alpha = \beta = 2$ the Karlin-McGregor polynomials are the Chebyshev polynomials of the first kind. The Karlin-McGregor polynomials $R_n(x) = R_n(x; \alpha, \beta)$ generate a polynomial hypergroup on $\mathbb{N}_0$, whenever $\alpha, \beta \geq 2$. In fact, Theorem 4.4.5 implies immediately that the linearization coefficients $g(m,n;k)$ are non-negative if $\alpha, \beta \geq 2$. (So we assume that $\alpha, \beta \geq 2$.) Applying the recursion formulas (vii) and (ix) of subsection 1.2.1 we can determine the coefficients $g(m,n;k)$ explicitly. By induction on m we obtain

$$g(m,n;m+n) = \begin{cases} \dfrac{\beta-1}{\beta} & \text{if } n \text{ and } m \text{ even} \\[2ex] \dfrac{\alpha-1}{\alpha} & \text{if } n \text{ or } m \text{ odd} \end{cases},$$

$$g(m,n;m+n-2k)$$
$$= \begin{cases} \left(\frac{1}{(\alpha-1)(\beta-1)}\right)^{k/2} \frac{\beta-2}{\beta} & k \text{ even; } n \text{ or } m \text{ even} \\ \left(\frac{1}{(\alpha-1)(\beta-1)}\right)^{k/2} \frac{\alpha-2}{\alpha} & k \text{ even; } n \text{ and } m \text{ odd} \\ \left(\frac{\alpha}{\beta(\alpha-1)}\right)^{(k+1)/2} \left(\frac{\beta}{\alpha(\beta-1)}\right)^{(k-1)/2} \frac{\alpha-2}{\alpha} & k \text{ odd; } n \text{ or } m \text{ even} \\ \left(\frac{\alpha}{\beta(\alpha-1)}\right)^{(k-1)/2} \left(\frac{\beta}{\alpha(\beta-1)}\right)^{(k+1)/2} \frac{\beta-2}{\beta} & k, n \text{ and } m \text{ odd} \end{cases}$$

for $1 \leq k \leq m-1,\ m \geq 2$,

$$g(m,n;n-m) = \begin{cases} \left(\frac{1}{(\alpha-1)(\beta-1)}\right)^{m/2} \frac{\beta-1}{\beta} & m \text{ even} \\ \left(\frac{1}{(\alpha-1)(\beta-1)}\right)^{(m-1)/2} \frac{1}{\beta} & m \text{ odd; } n \text{ even} \\ \left(\frac{1}{(\alpha-1)(\beta-1)}\right)^{(m-1)/2} \frac{1}{\alpha} & m \text{ and } n \text{ odd} \end{cases}$$

for $m \geq 2$, and $g(m,n;m+n-k) = 0$ for k odd.

The Haar weights are $h(0) = 1$ and

$$h(n) = \begin{cases} ((\alpha-1)(\beta-1))^{n/2} \frac{\beta}{\beta-1} & n \text{ even} \\ ((\alpha-1)(\beta-1))^{(n-1)/2}\ \alpha & n \text{ odd.} \end{cases}$$

In case of $\alpha = \beta \geq 2$ the Karlin-McGregor polynomials are the polynomials connected with homogeneous trees studied in 4.5.7.2. The parameter $\alpha = \beta \geq 2$ of the Karlin-McGregor polynomials and the parameter $q \geq 1$ of the polynomials connected with homogeneous trees are related by $\alpha = q+1$.

In order to determine $R_n(z)$ for $z \in \mathbb{C}$ we use an approach which fits very well for this class of polynomials. In fact we observe that

$$z^2\, R_{2n}(z) = r\, R_{2n+2}(z) + s\, R_{2n}(z) + t\, R_{2n-2}(z)$$

beginning with $R_0(z) = 1,\ R_2(z) = \frac{\alpha}{\alpha-1} z^2 - \frac{1}{\alpha-1}$, where

$$r = \frac{(\alpha-1)(\beta-1)}{\alpha\,\beta}, \quad s = \frac{(\alpha-1)+(\beta-1)}{\alpha\,\beta}, \quad t = \frac{1}{\alpha\,\beta}.$$

Now we can apply the method of difference equations with constant coefficients to calculate $R_{2n}(z)$ and then $R_{2n+1}(z)$. It is well known that

$$R_{2n}(z) = c\lambda_1^n + d\lambda_2^n,$$

where

$$\lambda_{1,2} = \frac{(z^2-s) \pm \sqrt{(z^2-s)^2 - 4rt}}{2r}, \quad \text{if } (z^2-s)^2 \neq 4rt$$

and

$$R_{2n}(z) = \lambda^n (1+nd), \quad \text{whenever } (z^2-s)^2 = 4rt.$$

Assuming $\alpha \geq 2$, $\beta \geq 2$ we see that $(R_n(x))_{n\in\mathbb{N}_0}$ is bounded for each $x \in [-1,1]$. Hence

$$D_s = [-1,1].$$

The orthogonalization measure π is determined in example (iii), section 2 of [363]. We begin by considering $x = 0$. We have $R_n(0) = 0$ for n odd, and

$$R_n(0) = \left(\frac{-1}{\alpha-1}\right)^{n/2}$$

for n even. Hence we obtain

$$\sum_{n=0}^{\infty} R_n^2(0)\, h(n) = 1 + \frac{\beta}{\beta-1}\sum_{k=1}^{\infty}\left(\frac{\beta-1}{\alpha-1}\right)^k.$$

If $\alpha > \beta \geq 2$ it follows that $\sum_{n=0}^{\infty} R_n^2(0)\, h(n) = \frac{\alpha}{\alpha-\beta}$. Applying Theorem 4.1.3 we conclude that $\pi(\{0\}) > 0$. This means that π has a discrete part in $x = 0$ whenever $\alpha > \beta \geq 2$. The measure π is supported by S,

$$S = \operatorname{supp} \pi = \begin{cases} [-v,-w] \cup [w,v] & \text{for } \alpha \leq \beta, \\ \{0\} \cup [-v,-w] \cup [w,v] & \text{for } \alpha > \beta, \end{cases}$$

where

$$v = \frac{\sqrt{\alpha-1}+\sqrt{\beta-1}}{\sqrt{\alpha\beta}}, \qquad w = \frac{|\sqrt{\alpha-1}-\sqrt{\beta-1}|}{\sqrt{\alpha\beta}},$$

see [363].

Concerning D we refer to [544,Theorem 4.1]. Using the results of [544,Proposition 3.2] it follows that

$$D = \begin{cases} \{\lambda \in \mathbb{C} : \lambda^2 \in E_1\} & \text{for } \beta \geq \alpha \\ \{0\} \cup \{\lambda \in \mathbb{C} : \lambda^2 \in E_1\} & \text{for } \beta < \alpha, \end{cases}$$

where E_1 is the (maybe degenerate) ellipse in $\mathbb{C}$,

$$E_1 = \{x+iy \in \mathbb{C} : \frac{(x-r)^2}{(1-r)^2} + \frac{y^2}{(1-s)^2} \leq 1\}$$

with $r = \frac{\alpha+\beta-2}{\alpha\beta}$, $s = \frac{\alpha+\beta}{\alpha\beta}$.

5.9 *Random walk polynomials*

We will continue to random walk polynomials which describe the states of a linear-growth birth and death process. Such random walk polynomials are studied in [30,§6].

Fix $a \geq 1,\ b \geq 0$. The random walk polynomials $R_n(x;a,b)$ are determined by the recurrence formula,

$$x\,R_n(x;a,b) \;=\; a_n R_{n+1}(x;a,b) \;+\; c_n R_{n-1}(x;a,b), \qquad n \in \mathbb{N},$$

$$R_0(x;a,b) \;=\; 1, \qquad R_1(x;a,b) \;=\; x, \qquad\qquad (i)$$

where

$$a_n \;=\; \frac{an+b}{(a+1)n+b}, \qquad c_n \;=\; \frac{n}{(a+1)n+b}.$$

For $a = 1$ the polynomials $R_n(x;1,b)$ are the ultraspherical polynomials $R_n^{(\alpha,\alpha)}(x)$, where $\alpha = \frac{1}{2}(b-1)$. For $b = 0$ the polynomials coincide with the polynomials $R_n(x;q)$ connected with homogeneous trees, where $q = a$. Obviously, the recurrence coefficients a_n, c_n satisfy the assumptions of Theorem 4.4.4, if $a \geq 1,\ b \geq 0$. Hence $(R_n(x;a,b))_{n\in\mathbb{N}_0}$ generates a polynomial hypergroup on $\mathbb{N}_0$.

In order to find the dual spaces and the orthogonalization measure we carry out the following transformation, see [30].

For $a > 1,\ b > 0$ set

$$Q_n(x;a,b) \;=\; \frac{(\sqrt{a})^n\,(b/a)_n}{n!}\;R_n\left(\frac{2\sqrt{a}}{a+1}\,x;a,b\right).$$

Then formulae (i) become

$$(n+1)\,Q_{n+1}(x;a,b) \;=\; 2x\left(n+\frac{b}{a+1}\right)\,Q_n(x;a,b)$$

$$-\,(n+\frac{b}{a}-1)\,Q_{n-1}(x;a,b), \qquad\qquad (ii)$$

$$Q_0(x;a,b) \;=\; 1, \qquad Q_1(x;a,b) \;=\; \frac{2b}{a+1}\,x.$$

This recurrence relation is related to that of Pollaczek polynomials. In fact, with

$$\alpha \;=\; \frac{1}{2}\left(\frac{b}{a}-1\right) \quad \text{and } \mu \;=\; b\left(\frac{1}{a+1}\;-\;\frac{1}{2a}\right)$$

we obtain $Q_n(x;a,b) = P_n(x;\alpha,\mu)$ for $n \in \mathbb{N}_0$, where $P_n(x;\alpha,\mu)$ are the Pollaczek polynomials, see subsection 4.5.5.

Applying the results of Theorem 4.3.3 it follows for the hypergroup generated by the random walk polynomials $R_n(x;a,b)$, $a > 1$, $b > 0$ that

$$S = \text{supp}\,\pi = \left[-2\,\frac{\sqrt{a}}{a+1}, 2\,\frac{\sqrt{a}}{a+1}\right], \qquad D_s = [-1,1]$$

and

$$D = \{z \in \mathbb{C} : |z - \frac{\sqrt{a}}{a+1}| + |z + 2\frac{\sqrt{a}}{a+1}| \leq 2\}.$$

The orthogonalization measure π can be established from the corresponding one for Pollaczek polynomials: $d\pi(x) = f(x)\chi_{[-1,1]}dx$, where (up to a multiplicative constant)

$$f(x) = s(x)^{(\frac{b}{a}-1)} \left|\Gamma\left(\frac{1}{2}\frac{b}{a} + i\,\frac{b(a-1)}{2a}\,\frac{x}{s(x)}\right)\right|^2$$

$$\times \exp\left((2\arccos x - \pi)\,\frac{b(a-1)}{2a}\,\frac{x}{s(x)}\right),$$

with

$$s(x) = \sqrt{4a - x^2(a+1)^2} \quad \text{for } -2\,\frac{\sqrt{a}}{a+1} \leq x \leq 2\,\frac{\sqrt{a}}{a+1}.$$

We refer also to [30,(6.59),(6.60)]. The Haar weights are

$$h(n) = \frac{a^n(\frac{b}{a})_n\,((a+1)n+b)}{b\,n!}, \qquad n \in \mathbb{N}_0.$$

5.10 *Grinspun polynomials and some generalizations*

(a) Altering only the first recurrence coefficients c_1 and a_1 of the Chebyshev polynomials $T_n(x)$ of the first kind leads us to the Grinspun polynomials.

Fix $a \in \mathbb{R}$, $a > 1$ and define

$$c_1 = \frac{1}{a}, \quad a_1 = \frac{a-1}{a} \quad \text{and } c_n = a_n = \frac{1}{2} \quad \text{for } n = 2,3,\ldots.$$

The symmetric orthogonal polynomials $R_n(x;a)$ defined by these recurrence coefficients are studied by Grinspun, see [138, Ch.VI,(13.10)]. They are orthogonal on $[-1,1]$ with respect to $d\pi(x) = 1/[(1+\nu x^2)\sqrt{1-x^2}]\,dx$, $\nu = a^2 - 2a$, and belong to the larger class of Bernstein-Szegö polynomials. In fact, under the notation of [641,Ch.2.6] we have

$$\varrho(x) = 1 + \nu x^2 = |h(e^{i\theta})|^2,$$

where

$$h(e^{i\theta}) = \frac{a}{2} + \frac{a-2}{2}\, e^{2i\theta}, \qquad x = \cos\theta.$$

Applying the formulas (vii), (viii) and (ix) of subsection 1.2.1, one can show inductively that

$$g(2,n;n-2) = \begin{cases} \frac{1}{2(a-1)} & \text{if } n=2 \\ \frac{a}{4(a-1)} & \text{if } n \geq 3\,, \end{cases} \qquad g(2,n;n) = \begin{cases} \frac{3(a-2)}{4(a-1)} & \text{if } n=2 \\ \frac{a-2}{2(a-1)} & \text{if } n \geq 3, \end{cases}$$

and $g(2,n;n+2) = \frac{a}{4(a-1)}$. For $3 \leq m \leq n$ it follows

$$g(m,n;n-m) = \begin{cases} \frac{1}{2(a-1)} & \text{if } n=m \\ \frac{a}{4(a-1)} & \text{if } n > m, \end{cases}$$

$$g(m,n;n-m+2) = \begin{cases} \frac{a-2}{2(a-1)} & \text{if } n=m \\ \frac{a-2}{4(a-1)} & \text{if } n > m \end{cases}$$

and $g(m,n;n+m-2k) = 0$ if $k = 2,3,...,m-2$, and

$$g(m,n;n+m-2) = \frac{a-2}{4(a-1)}, \qquad g(m,n;n+m) = \frac{a}{4(a-1)},$$

and obviously $g(m,n;n+m-k) = 0$ for $m \geq 2$ and $k = 1,3,...,2m-1$. The coefficients $g(m,n;k)$ are all nonnegative if and only if $a \geq 2$. In particular, $(R_n(x;a))_{n\in\mathbb{N}_0}$ generates a polynomial hypergroup when $a \geq 2$. This follows also by Theorem 4.4.4.

The Haar weights are

$$h(0) = 1, \quad h(1) = a \quad \text{and } h(n) = 2(a-1) \quad \text{for } n \geq 2.$$

Applying induction it follows directly that $R_0(x;a) = 1$, $R_1(x;a) = x$ and

$$R_n(x;a) = \frac{a}{2(a-1)}\, T_n(x) + \frac{a-2}{2(a-1)}\, T_{n-2}(x) \quad \text{for } n = 2,3,\ldots.$$

Using the notation of section 4.6 in advance we have $T_n \overset{d}{\geq} R_n(\cdot;a)$.

(b) Let $a \geq 2$ and $b \geq 2$. Altering the first and second recurrence coefficients c_1, a_1 and c_2, a_2 we get a generalization of the Grinspun polynomials by setting

$$c_1 = \frac{1}{a}, \quad a_1 = \frac{a-1}{a}, \quad c_2 = \frac{1}{b}, \quad a_2 = \frac{b-1}{b}$$

and

$$c_n = a_n = \frac{1}{2} \qquad \text{for } n = 3, 4, \ldots.$$

Let $R_n(x; a, b)$ be the symmetric orthogonal polynomials defined by these coefficients. Theorem 4.4.5 implies that $(R_n(x))_{n\in\mathbb{N}_0}$ generates a polynomial hypergroup on $\mathbb{N}_0$. By Theorem 4.3.3 it follows $[-1, 1] = S = D_s = D$. The Haar weights are $h(0) = 1$, $h(1) = a$, $h(2) = b(a-1)$, $h(n) = 2(b-1)(a-1)$ for $n \geq 3$.
Define connection coefficients

$$d_{2,2} = \frac{a}{2(a-1)}, \quad d_{2,0} = \frac{a-2}{2(a-1)},$$

$$d_{3,3} = \frac{ab}{4(a-1)(b-1)}, \quad d_{3,1} = \frac{3ab - 4(a+b-1)}{4(a-1)(b-1)} = 1 - d_{3,3}$$

and

$$d_{4,4} = d_{3,3}, \quad d_{4,2} = d_{2,0}, \quad d_{4,0} = \frac{a(b-2)}{4(a-1)(b-1)} = d_{2,2} - d_{3,3}.$$

It follows

$$R_0(x; a, b) = T_0(x) = 1, \quad R_1(x; a, b) = T_1(x) = x,$$

$$R_2(x; a, b) = d_{2,2}T_2(x) + d_{2,0}T_0(x), \quad R_3(x; a, b) = d_{3,3}T_3(x) + d_{3,1}T_1(x),$$

and applying induction we obtain

$$R_n(x; a, b) = d_{4,4}T_n(x) + d_{4,2}T_{n-2}(x) + d_{4,0}T_{n-4}(x) \qquad \text{for } n = 4, 5, \ldots.$$

In particular, $T_n \overset{d}{\geq} R_n(\cdot; a, b)$.

(c) Let $a \geq 2$, $b \geq 2$ and $c \geq 2$, and put

$$c_1 = \frac{1}{a}, \quad a_1 = \frac{a-1}{a}, \quad c_2 = \frac{1}{b}, \quad a_2 = \frac{b-1}{b}, \quad c_3 = \frac{1}{c}, \quad a_3 = \frac{c-1}{c}$$

and $c_n = a_n = \frac{1}{2}$ for $n = 4, 5, \ldots$, and let $(R_n(x; a, b, c))_{n\in\mathbb{N}_0}$ be the corresponding symmetric orthogonal polynomials. Applying Theorem 4.4.5 we know that $(R_n(x; a, b, c))_{n\in\mathbb{N}_0}$ generates a polynomial hypergroup on $\mathbb{N}_0$ provided $a \geq c$. Moreover, by Theorem 4.3.3 it follows $[-1, 1] = S = D_s = D$. The Haar weights are $h(0) = 1$, $h(1) = a$, $h(2) = b(a-1)$, $h(3) = c(b-1)(a-1)$, $h(n) = 2(c-1)(b-1)(a-1)$.
We will show now that $T_n \overset{d}{\geq} R_n(\cdot; a, b, c)$ is valid for $a, b, c \geq 2$. Proving this a lot of (elementary) calculations is involved.

Define connection coefficients

$$f_{2,2} = \frac{a}{2(a-1)}, \quad f_{2,0} = \frac{a-2}{2(a-1)}; \quad f_{3,3} = f_{2,2}\,\frac{b}{2(b-1)}, \quad f_{3,1} = 1 - f_{3,3}.$$

By (b) we obtain

$$R_0(x;a,b,c) \;=\; T_0(x) \;=\; 1, \quad R_1(x;a,b,c) \;=\; T_1(x) \;=\; x,$$

$$R_2(x;a,b,c) = f_{2,2}T_2(x) + f_{2,0}T_0(x) \text{ and } R_3(x;a,b,c) = f_{3,3}T_3(x) + f_{3,1}T_1(x).$$

The recurrence formula of $xR_3(x;a,b,c)$ yields

$$R_4(x;a,b,c) \;=\; f_{4,4}T_4(x) + f_{4,2}T_2(x) + f_{4,0}T_0(x)$$

with

$$f_{4,4} \;=\; f_{3,3}\,\frac{c}{2(c-1)}, \quad f_{4,2} \;=\; \frac{c}{2(c-1)} - \frac{1}{c-1}\,f_{2,2} \;=\; \frac{ac-(a+c)}{2(a-1)(c-1)}$$

and

$$f_{4,0} \;=\; \frac{c}{2(c-1)}\,f_{3,1} \;-\; \frac{1}{c-1}\,f_{2,0}.$$

Using that

$$\frac{c}{2}\,(1-f_{3,3}) - f_{2,0} \;\geq\; \left(\frac{c}{2}-1\right) \;+\; \left(f_{3,3} - \frac{c}{2}\,f_{3,3}\right) \;=\; \frac{c-2}{2}\,f_{3,1}$$

we get $f_{4,0} \geq \frac{1}{c-1}\frac{c-2}{2} f_{3,1} \geq 0$. Therefore we have $f_{4,4} \geq 0,\ f_{4,2} \geq 0,\ f_{4,0} \geq 0$. Continuing this way by considering $xR_4(x;a,b,c)$ we obtain

$$R_5(x;a,b,c) \;=\; f_{5,5}T_5(x) + f_{5,3}T_3(x) + f_{5,1}T_1(x)$$

with $f_{5,5} = f_{4,4},\ f_{5,3} = f_{4,4} + f_{4,2} - f_{3,3},\ f_{5,1} = f_{4,2} + 2f_{4,0} - f_{3,1}$. Using $f_{2,2} \leq 1$ we get $f_{5,3} \geq \frac{c-2}{2(c-1)} f_{3,1}$. Applying that

$$f_{4,2} \;=\; f_{4,4} + \frac{c}{2(c-1)}\,f_{3,1} \;-\; \frac{1}{c-1}\,f_{2,2}$$

we finally obtain

$$f_{5,1} \;=\; \frac{1}{c-1}\left(2f_{3,1} + \frac{a-2}{2(a-1)} + \frac{c-2}{2}\right) \;\geq\; 0.$$

Thus we have $f_{5,5} \geq 0,\ f_{5,3} \geq 0,\ f_{5,1} \geq 0$.
For $R_6(x;a,b,c)$ we get the representation

$$R_6(x;a,b,c) \;=\; f_{6,6}T_6(x) + f_{6,4}T_4(x) + f_{6,2}T_2(x) + f_{6,0}T_0(x)$$

with $f_{6,6} = f_{5,5},\ f_{6,4} = f_{5,3},\ f_{6,2} = f_{4,0},\ f_{6,0} = f_{3,3} - f_{4,4}$. Applying induction we obtain for $n = 6, 7, \ldots$

$$R_n(x;a,b,c) \;=\; f_{6,6}T_n(x) + f_{6,4}T_{n-2}(x) + f_{6,2}T_{n-4}(x) + f_{6,0}T_{n-6}(x)$$

and $T_n \overset{d}{\geq} R_n(\cdot;a,b,c)$ is shown.

5.11 *Continuous q-ultraspherical polynomials*

We will show that continuous q-ultraspherical polynomials with parameters $-1 < q < 1$, $-1 < \beta < 1$ generate polynomial hypergroups. We divide this class into three subclasses with the goal to derive many of their properties with elementary calculations.

The continuous q-ultraspherical polynomials $C_n(x;\beta|q)$ are given by

$$C_0(x;\beta|q) = 1, \quad C_1(x;\beta|q) = 2x(1-\beta)/(1-q)$$

and

$$x\,C_n(x;\beta|q) = \alpha_n C_{n+1}(x;\beta|q) + \gamma_n C_{n-1}(x;\beta|q) \qquad \text{for } n \in \mathbb{N} \qquad (1)$$

with

$$\alpha_n = \frac{1-q^{n+1}}{2(1-\beta q^n)}, \quad n \in \mathbb{N}_0, \quad \text{and } \gamma_n = \frac{1-\beta^2 q^{n-1}}{2(1-\beta q^n)}, \quad n \in \mathbb{N},$$

see [349,(13.2.12)] or [120,(1.1)].

Define the renormed version $R_n(x;\beta|q)$ of $C_n(x;\beta|q)$ by the recurrence relation (1) and (2) of subsection 1.2.1 with

$$c_1 = \gamma_1\alpha_0, \quad a_1 = 1 - c_1, \quad b_1 = 0,$$

$$c_n = \gamma_n\alpha_{n-1}/a_{n-1}, \quad a_n = 1 - c_n, \quad b_n = 0 \quad \text{for } n = 2,3,... \qquad (2)$$

and

$$a_0 = 1, \quad b_0 = c_0 = 0.$$

We have to check that $c_n > 0$, $a_n > 0$.

(a) We consider $0 < q < 1$, $-1 < \beta < 1$.
We prove by induction on $n \in \mathbb{N}$

$$0 < c_n < \frac{1-q^n}{2(1-\beta q^n)} \quad \text{and} \quad \frac{1-\beta^2 q^n}{2(1-\beta q^n)} < a_n < 1. \qquad (3)$$

In fact, for $n = 1$ we have $c_1 = \frac{(1-q)(1+\beta)}{4(1-\beta q)}$. Hence $0 < c_1 < \frac{1-q}{2(1-\beta q)}$ and thus

$$1 > a_1 > 1 - \frac{1-q}{2(1-\beta q)} = \frac{1+q(1-2\beta)}{2(1-\beta q)} > \frac{1-\beta^2 q}{2(1-\beta q)},$$

since $q > 0$. By the induction assumption we have $\frac{1}{a_n} < \frac{2(1-\beta q^n)}{1-\beta^2 q^n}$. Hence

$$0 < c_{n+1} = \gamma_{n+1}\alpha_n/a_n < \frac{1-q^{n+1}}{2(1-\beta q^{n+1})}$$

and therefore

$$1 > a_{n+1} > 1 - \frac{1-q^{n+1}}{2(1-\beta q^{n+1})} = \frac{1-q^{n+1}(2\beta-1)}{2(1-\beta q^{n+1})} > \frac{1-\beta^2 q^{n+1}}{2(1-\beta q^{n+1})}.$$

Obviously, by (3) it follows $c_n > 0,\ a_n > 0$ for $n \in \mathbb{N}$.
Denote

$$\kappa_n := \prod_{k=0}^{n-1} a_k / \prod_{k=0}^{n-1} \alpha_k \qquad \text{for } n \in \mathbb{N}$$

and write (for the moment)

$$P_0(x;\beta|q) = 1, \quad P_1(x;\beta|q) = \kappa_1 R_1(x;\beta|q)$$

and $P_n(x;\beta|q) = \kappa_n R_n(x;\beta|q)$. Since $a_{n-1}c_n = \alpha_{n-1}\gamma_n$ and $\frac{\kappa_n}{\kappa_{n+1}} = \frac{\alpha_n}{a_n}$, the comparison of the recurrence relations shows that $P_n(x;\beta|q) = C_n(x;\beta|q)$, the continuous q-ultraspherical polynomials defined by (1). Note that it follows immediately that $C_n(1;\beta|q) = 1/\kappa_n > 0$, and

$$a_n = \alpha_n \frac{C_{n+1}(1;\beta|q)}{C_n(1;\beta|q)}, \qquad c_n = \gamma_n \frac{C_{n-1}(1;\beta|q)}{C_n(1;\beta|q)}. \tag{4}$$

An explicit representation of the continuous q-ultraspherical polynomials is

$$C_n(x;\beta|q) = \sum_{k=0}^{n} \frac{(\beta;q)_k(\beta;q)_{n-k}}{(q;q)_k(q;q)_{n-k}} e^{i(n-2k)\theta}, \quad x = \cos\theta, \tag{5}$$

where $(a;q)_0 = 1$, $(a;q)_n = (1-a)(1-aq)\cdots(1-aq^{n-1})$, $n \in \mathbb{N}\cup\{\infty\}$, are the q-shifted factorials, see [349,(13.2.1)]. In particular we have

$$C_n(1;\beta|q) = \sum_{k=0}^{n} \frac{(\beta;q)_k(\beta;q)_{n-k}}{(q;q)_k(q;q)_{n-k}}.$$

The linearization formula of the products $C_m(x;\beta|q)\cdot C_n(x;\beta|q)$ was already calculated in 1894 by Rogers [580,p.29]. We also refer to [120,Theorem 1]. An explicit formula (with a sketch of the proof) can be found in [349,(13.3.10)]. Another proof (based on basic hypergeometric series) is contained in [261,(8.5)]. The coefficients are products of factors which are positive when $-1 < \beta < 1$ and $-1 < q < 1$. Note that the explicit determination of the linearization coefficients of $C_m(x;\beta|q)\cdot C_n(x;\beta|q)$ is rather involved. Restricting the parameters to $0 < q < 1$, $-1 < \beta < q$, a direct calculation shows that $c_2a_1 \geq a_1$, $c_{n+1}a_n \geq c_n a_{n-1}$ for $n = 2,3,\ldots$. Then already Theorem 4.4.1 implies that the linearization coefficients are positive. Using

$C_n(1;\beta|q) = 1/\kappa_n > 0$ we can conclude till now, that $(R_n(x;\beta|q))_{n\in\mathbb{N}_0}$ generates a polynomial hypergroup on $\mathbb{N}_0$ when $0 < q < 1$, $-1 < \beta < 1$. We collect some properties of these polynomial hypergroups. The sequence $(\sigma_n)_{n\in\mathbb{N}}$, $\sigma_n := c_n a_{n-1} = \gamma_n \alpha_{n-1}$ is a chain sequence, and $(c_n)_{n\in\mathbb{N}_0}$ is its minimal parameter sequence. Since $\lim_{n\to\infty} \sigma_n = \frac{1}{4}$, it follows by Theorem 4.4.9 that

$$\lim_{n\to\infty} c_n = \frac{1}{2} = \lim_{n\to\infty} a_n.$$

By Theorem 4.3.3 we obtain that $[-1,1] = S = D_s = D$.
Applying (4) successively one gets for the Haar weights

$$h(n) = h(n-1)\,\frac{a_{n-1}}{c_n} = \frac{\alpha_0\cdot\alpha_1\cdots\alpha_{n-1}}{\gamma_1\cdots\gamma_n}\, C_n^2(1;\beta|q).$$

Concerning the orthogonalization measure π we refer to [349,Theorem 13.2.1]. There it is shown that supp $\pi = [-1,1]$ and $d\pi(x) = \frac{\omega(x)}{\sqrt{1-x^2}}\,dx$ with

$$\omega(\cos\theta) = \frac{(e^{2i\theta};q)_\infty(e^{-2i\theta};q)_\infty}{(\beta e^{2i\theta};q)_\infty(\beta e^{-2i\theta};q)_\infty}. \tag{6}$$

For the special case $\beta = q^{\alpha+\frac{1}{2}}$, $\alpha > -\frac{1}{2}$, denote the recurrence coefficients of $C_n(x;\beta|q)$ by $\alpha_n(q)$ and $\gamma_n(q)$. An elementary calculation shows that

$$\lim_{q\to 1} \alpha_n(q) = \frac{n+1}{2n+2\alpha+1} \quad\text{and}\quad \lim_{q\to 1}\gamma_n(q) = \frac{n+2\alpha}{2n+2\alpha+1}.$$

These limits are exactly the recurrence coefficients of the ultraspherical polynomials $P_n(x)$ of subsection 4.5.1. It follows that

$$\lim_{q\to 1} R_n(x;q^{\alpha+\frac{1}{2}}|q) = R_n^{(\alpha,\alpha)}(x),$$

the normalized ultraspherical polynomials.
An interesting subclass of the continuous q-ultraspherical polynomials is obtained if β is zero. The continuous q-Hermite polynomials $H_n(x|q)$ are defined by

$$H_0(x|q) = 1, \quad H_1(x|q) = 2x$$

and

$$2x\,H_n(x|q) = H_{n+1}(x|q) + (1-q^n)\,H_{n-1}(x|q), \qquad n\in\mathbb{N},$$

see [349, (13.1.1) and (13.1.2)]. It is clear that

$$H_n(x|q) = (q;q)_n\, C_n(x;0|q).$$

(b) Now we consider $q = 0$, $-1 < \beta < 1$. The monic version $\phi_n(x;\beta)$ of $C_n(x;\beta|0)$ ($C_n(x;\beta|0)$ defined by (1) above) satisfies the recursion formula

$$x\,\phi_n(x;\beta) = \phi_{n+1}(x;\beta) + \mu_n\,\phi_{n-1}(x;\beta), \tag{7}$$
$$\phi_0(x;\beta) = 1, \quad \phi_1(x;\beta) = x,$$

where $\mu_1 = \frac{1}{4}(1+\beta)$ and $\mu_n = \frac{1}{4}$ for $n = 2, 3, \ldots$. In fact the monic version is given by

$$\phi_n(x;\beta) = \frac{1}{2^n(1-\beta)}\,C_n(x;\beta|0) \qquad \text{for } n \in \mathbb{N},$$

and

$$\phi_0(x;\beta) = 1 = C_0(x;\beta|0).$$

It is easily shown that $\phi_n(x;\beta)$ satisfy (7). In particular, $\phi_n(x;\beta)$ are the monic Geronimus polynomials studied in subsection 4.5.7.1 (with $\gamma = \frac{1}{4}$ and $\alpha = 1+\beta$). From section 4.5.7 we know that $(R_n(x;\beta|0))_{n\in\mathbb{N}_0}$ generates a polynomial hypergroup and $[-1,1] = S = D_s = D$, $d\pi(x) = f(x)dx$ with

$$f(x) = \frac{\sqrt{1-x^2}}{\frac{1}{4}(1+\beta^2) - \beta x^2}$$

(up to a multiplicative constant), and

$$h(0) = 1, \quad h(1) = \frac{4}{1+\beta}, \quad h(n) = (1+\beta)\left(1 + \frac{1-\beta}{1+\beta}\,n\right)^2$$
$$\text{for } n = 2, 3, \ldots.$$

(c) Let $-1 < q < 0$, $-1 < \beta < 1$. Now we have to apply results based on basic hypergeometric series. Since $C_n(1;\beta|q) = \sum_{k=0}^{n} \frac{(\beta;q)_k(\beta;q)_{n-k}}{(q;q)_k(q;q)_{n-k}}$ also is valid for $-1 < q < 0$ (see [261,(7.4.2)]), $C_n(1;\beta|q)$ is positive. Hence by (4) the recurrence coefficients c_n and a_n are positive. As already noted in part (a), Roger's linearization formula (see [261,(8.5.1)]) yields that $(R_n(x;\beta|q))_{n\in\mathbb{N}_0}$ generates a polynomial hypergroup. We refer also to Theorem 2(i) of [659], where the positivity of the linearization coefficients is shown. Other properties of the polynomial hypergroup, where $-1 < q < 0$, are derived exactly as in (a). In particular, $[-1,1] = S = D_s = D$ is true and π is given by (6). A remarkable property of the chain sequence $(\sigma_n)_{n\in\mathbb{N}}$,

$$\sigma_n = a_{n-1}c_n = \alpha_{n-1}\gamma_n = \frac{1}{4} + \frac{1}{4}\,\frac{q^{n-1}(\beta-q)(1-\beta)}{(1-\beta q^{n-1})(1-\beta q^n)},$$

is, that the coefficients are oscillating around $\frac{1}{4}$, whenever $-1 < q < 0$.

5.12 *Little q-Legendre polynomials*

Fix a parameter $0 < q < 1$. The little q-Legendre polynomials $R_n(x;q)$ are defined by $a_0 = \frac{1}{q+1}$, $b_0 = \frac{q}{q+1}$ and for $n \in \mathbb{N}$

$$a_n = q^n \, \frac{(1+q)(1-q^{n+1})}{(1-q^{2n+1})(1+q^{n+1})},$$
$$b_n = \frac{(1-q^n)(1-q^{n+1})}{(1+q^n)(1+q^{n+1})},$$
$$c_n = q^n \, \frac{(1+q)(1-q^n)}{(1-q^{2n+1})(1+q^n)},$$

and by the recurrence relations (1) and (2) of subsection 1.2.1. Note that $a_n + b_n + c_n = 1$ for each $n \in \mathbb{N}$.

$(R_n(x;q))_{n\in\mathbb{N}_0}$ generates a polynomial hypergroup on $\mathbb{N}_0$, see [385]. The Haar weights are given by $h(n) = \frac{1-q^{2n+1}}{(1-q)q^n}$, $n \in \mathbb{N}_0$, and satisfy

$$\lim_{n\to\infty} \frac{h(n)}{h(n+1)} = \frac{1}{q} > 1.$$

That means $h(n)$ is of exponential growth. Nevertheless,

$$\{1\} \cup \{1-q^k : k \in \mathbb{N}_0\} = S = D_s = D.$$

Furthermore $\pi(\{1-q^k\}) = q^k(1-q)$, $k \in \mathbb{N}_0$ and $\pi(\{1\}) = 0$, see [515]. Note that the polynomials $R_n(x;q)$ are the polynomials $R_n(1-x)$ studied in section 3 of [515].

We refer to Theorem 4.2.4, and note that subexponential growth of $h(n)$ is not a necessary condition for $S = D_s = D$. Furthermore we point to example (2) below Theorem 3.5.3. There it is shown that the polynomial hypergroup generated by the little q-Legendre polynomials has a D-regular algebra $l^1(h)$. This is a remarkable fact in the context of the statement of Theorem 3.5.13.

Final Remark: Of course there exist further classes of orthogonal polynomial sequences containing $(R_n(x))_{n\in\mathbb{N}_0}$ which generate polynomial hypergroups on $\mathbb{N}_0$. For example certain associated continuous q-ultraspherical polynomials (see [614] or Askey-Wilson polynomials (see [615]). Moreover, we will meet further examples in the following sections.

6 Connection coefficients and property (T)

Let $(R_n(x))_{n\in\mathbb{N}_0}$ be an orthogonal polynomial sequence generating a polynomial hypergroup on $\mathbb{N}_0$. Furthermore let $(P_n(x))_{n\in\mathbb{N}_0}$ be an arbitrary sequence of (not necessarily orthogonal) polynomials $P_n(x)$ with degree $P_n = n$ and $P_n(1) = 1$. Then one can write

$$P_n(x) \;=\; \sum_{k=0}^{n} c(n,k)\, R_k(x).$$

The unique coefficients $c(n,k)$ are called **connection coefficients**, compare [15]. If all connection coefficients $c(n,k)$ are nonnegative, we shall briefly write

$$R_n \overset{d}{\geq} P_n.$$

Setting $x = 1$, we see that $1 = \sum_{k=0}^{n} c(n,k)$, and hence if $z \in D$, i.e. $|R_n(z)| \leq 1$ for all $n \in \mathbb{N}_0$, then $|P_n(z)| \leq 1$ for all $n \in \mathbb{N}_0$.

Theorem 4.6.1 *Assume that the orthogonal polynomial sequence $(R_n(x))_{n\in\mathbb{N}_0}$ generates a polynomial hypergroup on $\mathbb{N}_0$, and let $(P_n(x))_{n\in\mathbb{N}_0}$ be a sequence of polynomials with degree $P_n = n$ and $P_n(1) = 1$. Suppose that $R_n \overset{d}{\geq} P_n$. If $\varphi \in C(D_s)$ such that $\varphi(x) = \sum_{k=0}^{\infty} d(k)\, P_k(x)$ for all $x \in D_s$, where $\sum_{k=0}^{\infty} |d(k)| < \infty$, then $\varphi \in A(S)$, that is*

$$\varphi(x) \;=\; \hat{e}(x) \;=\; \sum_{k=0}^{\infty} e(k)\, R_k(x)\, h(k)$$

for all $x \in S = supp\, \pi$, where $e = (e(k))_{k\in\mathbb{N}_0} \in l^1(h)$.

If in addition $d(k) \geq 0$ for all $k \in \mathbb{N}_0$, then $\varphi|S_1$ is $T(S_1)$-positive definite, where $S_1 = S \cup \{1\}$.

Proof. For every $\mu \in M(D_s)$ we have

$$\left|\int_{D_s} \varphi(x)\, d\mu(x)\right| \;\leq\; M \sup_{n\in\mathbb{N}_0} \left|\int_{D_s} P_n(x)\, d\mu(x)\right|,$$

where $M := \sum_{k=0}^{\infty} |d(k)| < \infty$. Moreover,

$$\left|\int_{D_s} P_n(x)\, d\mu(x)\right| \;\leq\; \sum_{k=0}^{n} c(n,k) \left|\int_{D_s} R_k(x)\, d\mu(x)\right|$$

$$\leq\; \sup_{n\in\mathbb{N}_0} \left|\int_{D_s} R_n(x)\, d\mu(x)\right| \;=\; \|\check{\mu}\|_\infty.$$

Now consider $\mu \in M(D_s)$ of the form $d\mu = f\,d\pi,\ f \in C(D_s)$. We have

$$\left| \int_{D_s} \varphi(x)\, f(x)\, d\pi(x) \right| \le M\, \|\check{f}\|_\infty,$$

and Theorem 3.9.1 implies $\varphi(x) = \hat{e}(x)$ for all $x \in S$, where $e = (e(n))_{n\in\mathbb{N}_0} \in l^1(h)$.

If in addition $d(k) \ge 0$ for all $k \in \mathbb{N}_0$, take $x_1,...,x_n \in S_1 = S \cup \{1\}$; $c_1,...,c_n \in \mathbb{C}$ such that $c_1R_k(x_1)+...+c_nR_k(x_n) \ge 0$ for all $k \in \mathbb{N}_0$. Then for any $k \in \mathbb{N}_0$

$$c_1P_k(x_1) + ... + c_nP_k(x_n) = \sum_{l=0}^{k} c(k,l)\,(c_1R_l(x_1) + ... + c_nR_l(x_n)) \ge 0$$

Hence

$$c_1\varphi(x_1) + ... + c_n\varphi(x_n) = \sum_{k=0}^{\infty} d(k)\,(c_1P_k(x_1) + ... + c_nP_k(x_n)) \ge 0,$$

and the second statement is shown, too. ◇

Examples:

(1) Consider the ultraspherical polynomials $R_n^{(\alpha,\alpha)}$, $\alpha \ge -1/2$, see subsection 1.2.1. In that case we have $\operatorname{supp}\pi = D_s = D = [-1,1]$. In [18,(7.34)] it is shown that $R_n^{(\alpha,\alpha)} \overset{d}{\ge} R_n^{(\gamma,\gamma)}$ if $\gamma \ge \alpha \ge -1/2$. Therefore we have $A(D_\gamma) \subseteq A(D_\alpha)$, where

$$A(D_\alpha) := \{\varphi \in C([-1,1]) : \varphi(x) = \sum_{k=0}^{\infty} d(k)\, R_k^{(\alpha,\alpha)}(x)\, h_\alpha(k),$$

$$\sum_{k=0}^{\infty} |d(k)|\, h_\alpha(k) < \infty\}.$$

($h_\alpha(n)$ are the corresponding Haar weights, $h_\alpha(n) = \dfrac{(2n+2\alpha+1)\,(2\alpha+1)_n}{(2\alpha+1)\,n!}$.)

Moreover, the polynomial hypergroup $\mathbb{N}_0$ generated by $R_n^{(\alpha,\alpha)}(x)$ has the property that $[-1,1] \cong \hat{\mathbb{N}}_0$ is a hypergroup under pointwise multiplication, see subsection 1.2.2, which fulfils the Pontryagin duality, see Proposition 3.4.13. Applying Theorem 3.11.23 the notions of $T(S_1)$-positive definiteness and the positive definiteness on the dual hypergroup $D_s = S_1 = [-1,1]$ coincide.
Denote by $P(D_\alpha)$ the set of all continuous functions on $[-1,1]$, which are positive definite with respect to the dual hypergroup on $[-1,1]$

generated by $R_n^{(\alpha,\alpha)}(x)$. The second statement of Theorem 4.1.4 yields $P(D_\gamma) \subseteq P(D_\alpha)$, if $\gamma \geq \alpha \geq -1/2$.
In case of $R_n(x) = T_n(x)$, the Chebyshev polynomials of the first kind, the relation $T_n \overset{d}{\geq} P_n$ corresponds to the **condition (T)** which will be studied below.

(2) An interesting case occurs for $P_n(x) = (R_1(x))^n$. (This means $P_n(x) = x^n$ whenever the $R_n(x)$ are symmetric.) If we already know the representation $(R_1(x))^n = \sum_{k=0}^{n} c(n,k)\, R_k(x)$ we get easily

$$(R_1(x))^{n+1} = \sum_{k=0}^{n+1} c(n+1,k)\, R_k(x),$$

where

$$\begin{aligned} c(n+1,0) &= c(n,1)\, c_1 \\ c(n+1,1) &= c(n,2)\, c_2 \;+\; c(n,1)\, b_1 \;+\; c(n,0) \\ c(n+1,k) &= c(n,k+1)\, c_{k+1} \;+\; c(n,k)\, b_k \;+\; c(n,k-1)\, a_{k-1} \end{aligned}$$

for $k = 2, ..., n-1$

$$\begin{aligned} c(n+1,n) &= c(n,n)\, b_n \;+\; c(n,n-1)\, a_{n-1} \\ c(n+1,n+1) &= c(n,n)\, a_n. \end{aligned}$$

By induction we obtain immediately that $R_n \overset{d}{\geq} P_n$, where $P_n(x) = (R_1(x))^n$. Notify that $(R_n(x))_{n\in\mathbb{N}_0}$ is any orthogonal polynomial sequence that generates a polynomial hypergroup on $\mathbb{N}_0$. By Theorem 4.6.1 we know that each $\varphi \in C(D_s)$, $\varphi(x) = \sum_{k=0}^{\infty} d(k)\,(R_1(x))^k$, $\sum_{k=0}^{\infty} |d(k)| < \infty$, is an element of $A(S)$. $\varphi|S_1$ is $T(S_1)$-positive definite, if in addition $d(k) \geq 0$ for all $k \in \mathbb{N}_0$.

(3) Consider the Grinspun polynomials and their generalizations of subsection 4.5.10 (a), (b) and (c). We know already that they satisfy property (T).

Using Theorem 3.9.2 one can easily establish a dual version of Theorem 4.6.1 characterizing those sequences $\varphi = (\varphi(n))_{n\in\mathbb{N}_0}$ that are inverse Fourier-Stieltjes transforms. Again, let $(R_n(x))_{n\in\mathbb{N}_0}$ be an orthogonal polynomial sequence generating a polynomial hypergroup on $\mathbb{N}_0$. Let $(P_n(x))_{n\in\mathbb{N}_0}$ be an arbitrary sequence of polynomials $P_n(x)$ with degree

$P_n = n$ and $P_0(x) = 1$. We write $R_n \overset{c}{\geq} P_n$ if for every $x \in D_s$ there exists $\mu_x \in M^1(D_s)$ such that

$$P_n(x) = \int_{D_s} R_n(y)\, d\mu_x(y)$$

for all $n \in \mathbb{N}_0$. Note that μ_x is uniquely determined by the above equation. Obviously $|P_n(x)| \leq 1$ for all $x \in D_s$ and $n \in \mathbb{N}_0$.

Theorem 4.6.2 *Assume that $(R_n(x))_{n\in\mathbb{N}_0}$ generates a polynomial hypergroup, and let $(P_n(x))_{n\in\mathbb{N}_0}$ be a sequence of polynomials with degree $P_n = n$ and $P_0(x) = 1$. Suppose $R_n \overset{c}{\geq} P_n$. Let*

$$\varphi(n) = \int_{D_s} P_n(x)\, db(x), \qquad n \in \mathbb{N}_0,$$

where $b \in M(D_s)$. Then there exists some $a \in M(D_s)$ such that

$$\varphi(n) = \int_{D_s} R_n(x)\, da(x) = \check{a}(n).$$

If in addition b is a regular positive Borel measure, then φ is a positive definite sequence.

Proof. For $g \in C_c(\mathbb{N}_0)$ we have

$$\left| \sum_{n\in\mathbb{N}_0} \varphi(n)\, g(n)\, h(n) \right| \leq \|b\| \sup_{x\in D_s} \left| \sum_{n\in\mathbb{N}_0} P_n(x)\, g(n)\, h(n) \right|.$$

Moreover,

$$\left| \sum_{n\in\mathbb{N}_0} P_n(x)\, g(n)\, h(n) \right| = \int_{D_s} \left| \sum_{n\in\mathbb{N}_0} R_n(y)\, g(n)\, h(n) \right| d\mu_x(y)$$

$$\leq \sup_{y\in D_s} \left| \sum_{n\in\mathbb{N}_0} R_n(y)\, g(n)\, h(n) \right| = \|\hat{g}\|_\infty,$$

and hence

$$\left| \sum_{n\in\mathbb{N}_0} \varphi(n)\, g(n)\, h(n) \right| \leq \|b\|\, \|\hat{g}\|_\infty.$$

Theorem 3.9.2 implies $\varphi = \check{a}$ for some $a \in M(D_s)$. To show the second statement, note that

$$\omega(n,m)(\varphi) = \int_{D_s} \sum_{k=|n-m|}^{n+m} g(n,m;k)\, P_k(x)\, db(x)$$

$$= \int_{D_s} \int_{D_s} \sum_{k=|n-m|}^{n+m} g(n,m;k)\, R_k(y)\, d\mu_x(y)\, db(x)$$

$$= \int_{D_s} \int_{D_s} R_n(y)\, R_m(y)\, d\mu_x(y)\, db(x)$$

If b is a regular positive Borel measure, let $\lambda_1, ..., \lambda_m \in \mathbb{C}$ and $n_1, ..., n_m \in \mathbb{N}_0$. Then

$$\sum_{i,j=1}^{m} \lambda_i\, \overline{\lambda}_j\, \omega(n_i, n_j)(\varphi) = \int_{D_s} \int_{D_s} \sum_{i,j=1}^{m} \lambda_i\, \overline{\lambda}_j\, R_{n_i}(y)\, R_{n_j}(y)\, d\mu_x(y)\, db(x)$$

$$= \int_{D_s} \int_{D_s} \left| \sum_{i=1}^{m} \lambda_i\, R_{n_i}(y) \right|^2 d\mu_x(y)\, db(x) \geq 0.$$

◇

Examples:

(1) Consider the ultraspherical polynomials $R_n^{(\alpha,\alpha)}$, $\alpha \geq -1/2$, see subsection 1.2.2. It is shown in [22] that

$$R_n^{(\alpha,\alpha)} \overset{c}{\geq} R_n^{(\gamma,\gamma)} \qquad \text{if } \gamma \geq \alpha \geq -\frac{1}{2}.$$

Therefore, if $\varphi(n) = \int_{-1}^{1} R_n^{(\gamma,\gamma)}(x)\, db(x)$, $b \in M([-1,1])$, then

$$\varphi(n) = \int_{-1}^{1} R_n^{(\alpha,\alpha)}(x)\, da(x) \qquad \text{for some } a \in M([-1,1]).$$

Moreover, it follows by Theorem 4.6.2 that $(\varphi(n))_{n\in\mathbb{N}_0}$ is positive definite with respect to $R_n^{(\alpha,\alpha)}(x)$, whenever $(\varphi(n))_{n\in\mathbb{N}_0}$ is positive definite with respect to $R_n^{(\gamma,\gamma)}(x)$.

(2) To apply Theorem 4.6.2, let $R_n(x) = R_n^{(\alpha,\alpha)}(x)$, $\alpha > -\frac{1}{2}$, the ultraspherical polynomials, once again, and let $P_n(x) = \dfrac{2\alpha+1}{2n+2\alpha+1}\, x^n$. Now we derive that

$$P_n(x) = \int_{D_s} R_n^{(\alpha,\alpha)}(y)\, d\mu_x(y) \qquad \text{for some } \mu_x \in M^+([-1,1]).$$

Applying the generating function of $C_n^{\lambda}(x) = R_n^{(\alpha,\alpha)}(x)\, \dfrac{(2\alpha+1)_n}{n!}$, $\lambda = \alpha + \dfrac{1}{2}$, $\alpha > -\dfrac{1}{2}$

$$\sum_{n=0}^{\infty} C_n^{\lambda}(x)\, t^n = (1 - 2xt + t^2)^{-\lambda},$$

see [18,(3.16)], we obtain

$$\sum_{n=0}^{\infty} \left(\frac{2\alpha+1}{2n+2\alpha+1}\, t^n \right) R_n^{(\alpha,\alpha)}(x)\, h(n) \;=\; (1-2xt+t^2)^{-\alpha-\frac{1}{2}} \qquad (\dot{I})$$

for all $x \in [-1,1],\ 0 \le t < 1$. Setting for $0 \le t < 1,\ x \in [-1,1]$

$$f_t(x) \;=\; (1-2xt+t^2)^{-\alpha-\frac{1}{2}},$$

we see $f_t \in C([-1,1]),\ f_t \ge 0$ and

$$\frac{2\alpha+1}{2n+2\alpha+1}\, t^n \;=\; \check{f}_t(n) \qquad (\dot{I}\dot{I})$$

for every $n \in \mathbb{N}_0$. Moreover,

$$\|f_t\|_1 \;=\; \int_{-1}^{1} f_t(x)\, d\pi(x) \;=\; \check{f}_t(0) \;=\; 1$$

Considering the limit case $t \to 1-$, we set $f_1(x) \;=\; (2-2x)^{-\alpha-\frac{1}{2}}$ for $x \in [-1,1[$. Using

$$c_{\alpha,\beta} := \int_{-1}^{1} (1-x)^{\alpha}\, (1+x)^{\beta}\, dx \;=\; 2^{(\alpha+\beta+1)}\, \frac{\Gamma(\alpha+1)\,\Gamma(\beta+1)}{\Gamma(\alpha+\beta+2)}$$

we get

$$\|f_1\|_1 = \int_{-1}^{1} (2-2x)^{-\alpha-\frac{1}{2}}\, d\pi(x) = c_{\alpha,\alpha}^{-1}\, 2^{-\alpha-\frac{1}{2}} \int_{-1}^{1} (1-x)^{-\frac{1}{2}}\, (1+x)^{\alpha}\, dx$$

$$= 2^{-\alpha-\frac{1}{2}}\, \frac{c_{-\frac{1}{2},\alpha}}{c_{\alpha,\alpha}} \;=\; \frac{\Gamma(\frac{1}{2})\,\Gamma(2\alpha+2)}{\Gamma(\alpha+\frac{3}{2})\,\Gamma(\alpha+1)\, 2^{2\alpha+1}}.$$

Legendre's duplication formula for the gamma function gives

$$\|f_1\|_1 \;=\; \frac{\Gamma(\frac{1}{2})\, 2^{2\alpha+\frac{3}{2}}}{\sqrt{2\pi}\, 2^{2\alpha+1}} \;=\; 1.$$

So we see that $\lim_{t\to 1-} f_t(x) \;=\; f_1(x)$ for π-almost every $x \in [-1,1]$ and $\|f_t\|_1 \;=\; \|f_1\|_1$ for all $0 \le t < 1$. By a convergence theorem of Riesz it follows $\lim_{t\to 1-} \|f_t - f_1\|_1 \;=0$. Hence

$$\check{f}_1(n) \;=\; \lim_{t\to 1-} \check{f}_t(n) \;=\; \frac{2\alpha+1}{2n+2\alpha+1}\,.$$

It is worthwile to formulate this result explicitly.

Proposition 4.6.3 *Let $\alpha > -\frac{1}{2}$. The inverse Fourier transform of $f_1(x) = (2-2x)^{-\alpha-\frac{1}{2}}$, $x \in [-1,1[$, with respect to $R_n^{(\alpha,\alpha)}(x)$ is given by*

$$\check{f}_1(n) = \frac{2\alpha+1}{2n+2\alpha+1}. \qquad (\dot{I}\dot{I}\dot{I})$$

In particular, the sequence $(\frac{2\alpha+1}{2n+2\alpha+1})_{n\in\mathbb{N}_0}$ is positive definite with respect to the polynomial hypergroup generated by $(R_n^{(\alpha,\alpha)}(x))_{n\in\mathbb{N}_0}$.

Since $R_n^{(\alpha,\alpha)}(-x) = (-1)^n R_n^{(\alpha,\alpha)}(x)$, it follows from (i)

$$\frac{2\alpha+1}{2n+2\alpha+1}\,(-t)^n = \check{f}_{-t}(n) \qquad (\dot{I}V)$$

for all $0 \le t < 1$, $n \in \mathbb{N}_0$. Clearly, the parameter space for t is now extended to $t \in]-1,1[$, and $f_t(x) = (1-2xt+t^2)^{-\alpha-\frac{1}{2}}$. In the same way we get from (ii):

$$\check{f}_{-1}(n) = \frac{2\alpha+1}{2n+2\alpha+1}\,(-1)^n,$$

where $f_{-1}(x) = (2+2x)^{-\alpha-\frac{1}{2}}$, $x \in]-1,1]$.

Combining (i), (ii), (iii), and (iv), we see that

$$P_n(t) = \int_{-1}^{1} R_n^{(\alpha,\alpha)}(x)\, f_t(x)\, d\pi(x),$$

for each $t \in [-1,1]$, $n \in \mathbb{N}_0$. Moreover, $f_t\pi \in M^1([-1,1])$, and so $R_n^{(\alpha,\alpha)} \overset{c}{\ge} P_n$. Thus we can apply Theorem 4.6.2 and get the following result on moment sequences:

Proposition 4.6.4 *Let $\alpha > -\frac{1}{2}$. If $\psi(n) = \int\limits_{-1}^{1} x^n db(x)$, $b \in M([-1,1])$, then*

$$\frac{2\alpha+1}{2n+2\alpha+1}\,\psi(n) = \int_{-1}^{1} R_n^{(\alpha,\alpha)}(x)\, da(x)$$

for some $a \in M([-1,1])$. Moreover, if $(\psi(n))_{n\in\mathbb{N}_0}$ is a moment sequence defined by a positive regular bounded Borel measure on $[-1,1]$, then $(\frac{2\alpha+1}{2n+2\alpha+1}\,\psi(n))_{n\in\mathbb{N}_0}$ is positive definite with respect to $(R_n^{(\alpha,\alpha)}(x))_{n\in\mathbb{N}_0}$.

Of course we have to derive a sufficient criterion for $R_n \overset{d}{\ge} P_n$, whenever $(R_n(x))_{n\in\mathbb{N}_0}$ generates a polynomial hypergroup on $\mathbb{N}_0$, and $(P_n(x))_{n\in\mathbb{N}_0}$ is an arbitrary sequence of polynomials with degree $P_n = n$ and $P_n(1) = 1$. We suppose $P_n(1) = 1$, so that Theorem 4.6.1 can be applied. Essential for the following result is a theorem of Askey, see [17,Theorem 1]. Assume $R_n(x)$ are defined by the recurrence relations (1) and (2) of subsection 1.2.1, where $a_0 > 0$, $b_0 \in \mathbb{R}$, $a_0+b_0 = 1$ and $a_n > 0$, $b_n \geq 0$, $c_n > 0$, $a_n+b_n+c_n = 1$ for each $n \in \mathbb{N}_0$. Assume that $P_n(x)$ are also defined by a recurrence relation of the form

$$P_1(x)\, P_n(x) = \alpha_n P_{n+1}(x) + \beta_n P_n(x) + \gamma_n P_{n-1}(x), \quad n \in \mathbb{N}, \qquad (i)$$

$$P_0(x) = 1, \qquad P_1(x) = \frac{1}{\alpha_0}\,(x - \beta_0) \qquad (ii)$$

where $\alpha_n > 0$, $\beta_n \in \mathbb{R}$ for $n \in \mathbb{N}_0$, $\gamma_n \geq 0$ and $\alpha_n + \beta_n + \gamma_n = 1$ for $n \in \mathbb{N}$. $\alpha_n + \beta_n + \gamma_n = 1$ yields $P_n(1) = 1$. Note that $\gamma_n = 0$ is allowed.

We refer to Example (2) below Theorem 4.6.1, where $\alpha_0 = a_0$ and $\beta_0 = b_0$, $\alpha_n = 1$, $\beta_n = 0$, $\gamma_n = 0$ for $n \in \mathbb{N}$. For the proof we have to consider monic versions $\tilde{R}_n(x)$ and $\tilde{P}_n(x)$ of $(R_n(x))_{n\in\mathbb{N}_0}$ and $(P_n(x))_{n\in\mathbb{N}_0}$, i.e. $\tilde{R}_n(x) = x^n + ...$ and $\tilde{P}_n(x) = x^n + ...$ Hence put

$$\tilde{R}_0(x) = \tilde{P}_0(x) = 1, \quad \tilde{R}_1(x) = a_0 R_1(x), \quad \tilde{P}_1(x) = \alpha_0 P_1(x)$$

and for $n \geq 2$

$$\tilde{R}_n(x) = a_0^n \prod_{k=1}^{n-1} a_k R_n(x), \quad \tilde{P}_n(x) = \alpha_0^n \prod_{k=1}^{n-1} \alpha_k P_n(x).$$

Then $(\tilde{R}_n(x))_{n\in\mathbb{N}_0}$ and $(\tilde{P}_n(x))_{n\in\mathbb{N}_0}$ satisfy the following recursion formulas

$$x\, \tilde{R}_n(x) = \tilde{R}_{n+1}(x) + \tilde{b}_n \tilde{R}_n(x) + \tilde{c}_n \tilde{R}_{n-1}(x) \qquad (iii)$$

and

$$x\, \tilde{P}_n(x) = \tilde{P}_{n+1}(x) + \tilde{\beta}_n \tilde{P}_n(x) + \tilde{\gamma}_n \tilde{P}_{n-1}(x) \qquad (iv)$$

for $n \in \mathbb{N}_0$ with $\tilde{R}_{-1}(x) = 0 = \tilde{P}_{-1}(x)$, $\tilde{b}_0 = b_0$, $\tilde{\beta}_0 = \beta_0$, $\tilde{b}_n = a_0 b_n + b_0$, $\tilde{\beta}_n = \alpha_0\beta_n + \beta_0$ for $n \in \mathbb{N}$, and $\tilde{c}_1 = a_0^2 c_1$, $\tilde{\gamma}_1 = \alpha_0^2\gamma_1$, $\tilde{c}_n = a_0^2 a_{n-1} c_n$, $\tilde{\gamma}_n = \alpha_0^2\alpha_{n-1}\gamma_n$ for $n \geq 2$.

Now we can apply Theorem 1 of [17] and obtain:

If $\tilde{\beta}_n \leq \tilde{b}_k$ for $n \in \mathbb{N}_0$ and $k = 0, 1, ..., n$, $\tilde{\gamma}_1 \leq \tilde{c}_1$ and $\tilde{\gamma}_n \leq \tilde{c}_k$ for $n \geq 2$ and $k = 1, ..., n$,

then

$$\tilde{P}_n(x) = \sum_{k=0}^{n} d(n,k)\, \tilde{R}_k(x) \quad \text{with } d(n,k) \geq 0.$$

(Note that the proof of Askey is also correct when we extend the assumption $\gamma_n > 0$ to $\gamma_n \geq 0$.) Since the polynomials $R_n(x)$ and $P_n(x)$ are scaled by positive constants we have the following result.

Theorem 4.6.5 *Let $(R_n(x))_{n\in\mathbb{N}_0}$ generate a polynomial hypergroup and let $(P_n(x))_{n\in\mathbb{N}_0}$ be a polynomial sequence defined by (i) and (ii) above. If the defining coefficients a_n, b_n, c_n of $(R_n(x))_{n\in\mathbb{N}_0}$ and $\alpha_n, \beta_n, \gamma_n$ of $(P_n(x))_{n\in\mathbb{N}_0}$ fulfil the inequalities*

$$\beta_0 \le b_0, \quad \alpha_0\beta_n + \beta_0 \le b_0 \quad \textit{and} \quad \alpha_0\beta_n + \beta_0 \le a_0 b_k + b_0 \qquad (v)$$

for $n \in \mathbb{N}$ and $k = 1, ..., n$; and

$$\alpha_0^2\gamma_1 \le a_0^2 c_1, \qquad \alpha_0^2\alpha_{n-1}\gamma_n \le a_0^2 c_1 \qquad (vi,a)$$

and

$$\alpha_0^2\alpha_{n-1}\gamma_n \le a_0^2 a_{k-1} c_k \qquad (vi,b)$$

for $n \ge 2$ and $k = 2, ..., n$, then

$$P_n(x) = \sum_{k=0}^{n} c(n,k)\, R_k(x) \quad \textit{with} \quad c(n,k) \ge 0, \quad \textit{i.e.}\ R_n \overset{d}{\ge} P_n.$$

Theorem 4.6.5 has many applications. We will examine some examples.

(1) Consider first the **ultraspherical polynomials**. Let $-\frac{1}{2} \le \alpha \le \beta$, and $R_n^{(\alpha,\alpha)}(x)$, $R_n^{(\beta,\beta)}(x)$ the corresponding ultraspherical polynomials. We write $\tilde{c}_n(\alpha)$ and $\tilde{c}_n(\beta)$ for the recurrence coefficients of the monic versions of $R_n^{(\alpha,\alpha)}(x)$ and $R_n^{(\beta,\beta)}(x)$, respectively. We have

$$\tilde{c}_1(\alpha) = \frac{1}{2\alpha+3} \quad \text{and} \quad \tilde{c}_n(\alpha) = \frac{(n+2\alpha)\,n}{(2n+2\alpha-1)(2n+2\alpha+1)} \quad \text{for } n \ge 2.$$

It is easy to check that $\tilde{c}_n(\alpha) \ge \tilde{c}_n(\beta)$ and $\tilde{c}_1(\alpha) \ge \cdots \ge \tilde{c}_n(\alpha)$ for each $n \in \mathbb{N}$, provided that $-\frac{1}{2} \le \alpha \le \beta \le \frac{1}{2}$. Thus Theorem 4.6.5 tells us $R_n^{(\alpha,\alpha)} \overset{d}{\ge} R_n^{(\beta,\beta)}$ for $-\frac{1}{2} \le \alpha \le \beta \le \frac{1}{2}$. (From [18,(7.34)] we know that $R_n^{(\alpha,\alpha)} \overset{d}{\ge} R_n^{(\beta,\beta)}$ for $-\frac{1}{2} \le \alpha \le \beta$.)

(2) Now consider the **Pollaczek polynomials** $R_n(x;\beta,\mu)$, see subsection 4.5.5, and their relation to the ultraspherical polynomials $R_n^{(\alpha,\alpha)}(x)$. We write $\tilde{c}_n(\alpha)$ for the recurrence coefficients of the monic version of $R_n^{(\alpha,\alpha)}(x)$, and $\tilde{c}_n(\beta,\mu)$ for that of the monic version of $R_n(x;\beta,\mu)$, $\beta > -\frac{1}{2}$, $\mu \ge 0$. we have

$$\tilde{c}_1(\beta,\mu) = \frac{1+2\beta}{(2\beta+2\mu+1)(2\beta+2\mu+3)}$$

and

$$\tilde{c}_n(\beta,\mu) = \frac{(n+2\beta)\,n}{(2n+2\beta+2\mu-1)(2n+2\beta+2\mu+1)} \quad \text{for } n \ge 2.$$

Obviously $\tilde{c}_n(\alpha) \ge \tilde{c}_n(\beta) \ge \tilde{c}_n(\beta,\mu)$ and $\tilde{c}_1(\alpha) \ge \cdots \ge \tilde{c}_n(\alpha)$ for $n \in \mathbb{N}$, whenever $-\frac{1}{2} \le \alpha \le \beta \le \frac{1}{2}$. From Theorem 4.6.5 we have $R_n^{(\alpha,\alpha)} \overset{d}{\ge} R_n(\cdot;\beta,\mu)$ for $-\frac{1}{2} \le \alpha \le \beta \le \frac{1}{2}$, and $\mu \ge 0$.

(3) Consider the **associated ultraspherical polynomials** $R_n^{(\alpha)}(x;\nu)$, see subsection 4.5.4. We begin with analyzing the relation between $R_n^{(\alpha)}(x;\nu)$ and $R_n^{(1/2)}(x;\nu)$. Write $\tilde{c}_n(\alpha,\nu)$ for the recurrence coefficients of the monic version of $R_n^{(\alpha)}(x;\nu)$, $\alpha \geq -\frac{1}{2}$, $\nu > 0$. We have

$$\tilde{c}_1(\alpha,\nu) = \frac{(1+\nu)(1+\nu+2\alpha)}{(2\alpha+2\nu+1)(2\alpha+2\nu+3)}$$

and

$$\tilde{c}_n(\alpha,\nu) = \frac{(n+\nu)(n+\nu+2\alpha)}{(2n+2\alpha+2\nu-1)(2n+2\alpha+2\nu+1)} \qquad \text{for } n \geq 2.$$

We point to two special cases: $\alpha = -\frac{1}{2}$ and $\alpha = \frac{1}{2}$. It is easy to check that $R_n^{(-1/2)}(x;\nu) = T_n(x)$ and $R_n^{(1/2)}(x;\nu) = U_n(x)$ for each $\nu > 0$. For $\alpha \geq \frac{1}{2}$, $\nu > 0$ it follows

$$\tilde{c}_n(\alpha,\nu) \leq \frac{1}{4} = \tilde{c}_n\left(\frac{1}{2},\nu\right) = \ldots = \tilde{c}_1\left(\frac{1}{2},\nu\right).$$

From Theorem 4.6.5 we obtain $U_n \overset{d}{\geq} R_n^{(\alpha)}(\cdot,\nu)$ for $\alpha \geq \frac{1}{2}$, $\nu > 0$.
Now we analyze the relation between $R_n^{(\alpha)}(x;\nu)$ and $R_n^{(\alpha)}(x;\mu)$ for $-\frac{1}{2} < \alpha < \frac{1}{2}$ and $0 < \nu \leq \mu$. It is straightforward to prove that $\tilde{c}_n(\alpha,\nu) \geq \tilde{c}_n(\alpha,\mu)$ for each $n \in \mathbb{N}$. Since $\tilde{c}_{n+1}(\alpha,\nu) = \tilde{c}_n(\alpha,\nu+1) \leq \tilde{c}_n(\alpha,\nu)$, Theorem 4.6.5 implies $R_n^{(\alpha)}(\ ;\nu) \overset{d}{\geq} R_n^{(\alpha)}(\ ;\mu)$ for $-\frac{1}{2} < \alpha < \frac{1}{2}$, $0 < \nu \leq \mu$.

(4) Consider the **Geronimus polynomials** $R_n(x;a)$, where $R_n(x;a)$ are the polynomials $R_n(x;\frac{1}{4},\alpha)$ with $a = \frac{4}{\alpha}$ of subsection 4.5.7.1. They generate a polynomial hypergroup on $\mathbb{N}_0$ for every $a \geq 2$. The recurrence coefficients $\tilde{c}_n(a)$ of the monic version of $R_n(x;a)$ are $\tilde{c}_1(a) = \frac{1}{a}$ and $\tilde{c}_n(a) = \frac{1}{4}$ for $n \geq 2$. Therefore by Theorem 4.6.5 we obtain $R_n(\cdot;a) \overset{d}{\geq} R_n(\cdot,b)$ for $2 \leq a \leq 4$ and $a \leq b$.

In addition to Theorem 4.6.5 there are further results on connection coefficients, see for example R.Szwarc [650]. We also refer to [18,Lecture 7] and [349,Ch.9.1]. For further applications we refer to a result for Jacobi polynomials, $R_n^{(\alpha,\beta)} \overset{d}{\geq} R_n^{(\gamma,\beta)}$ if $\gamma \geq \alpha > -1$, $\beta > -1$, see [18,(733)].

An important property of orthogonal polynomials $P_n(x)$ is **condition (T)**, that is $T_n \overset{d}{\geq} P_n$, where $T_n(x) = R_n^{(-1/2,-1/2)}(x)$ are the Chebyshev polynomials of the first kind. By Theorem 4.6.5 we have immediately a sufficient criterion for $(P_n(x))_{n\in\mathbb{N}_0}$ satisfying condition (T).

Corollary 4.6.6 *Let $(P_n(x))_{n\in\mathbb{N}_0}$ be an orthogonal polynomial sequence defined by (i) and (ii) with coefficients $\alpha_n, \beta_n, \gamma_n$. If $\beta_0 \leq 0$, $\alpha_0\beta_n + \beta_0 \leq 0$ for $n \in \mathbb{N}$, and $\alpha_0^2\gamma_1 \leq \frac{1}{2}$, $\alpha_0^2\alpha_{n-1}\gamma_n \leq \frac{1}{4}$ for $n \geq 2$, then $(P_n(x))_{n\in\mathbb{N}_0}$ satisfies condition (T).*

In case that $(R_n(x))_{n\in\mathbb{N}_0}$ generates a polynomial hypergroup on $\mathbb{N}_0$, which belongs to the Nevai class, a result by Nevai can be used to derive condition (T). We will use the notation of section 4.3.

Theorem 4.6.7 *Let $(R_n(x))_{n\in\mathbb{N}_0}$ be an orthogonal polynomial sequence defined as in section 4.3 by (1) and (2) with coefficients a_n, b_n, c_n, which are convergent. Let $a = \lim_{n\to\infty} a_n$, $b = \lim_{n\to\infty} b_n$, $c = \lim_{n\to\infty} c_n$ and write $\alpha = ba_0 + b_0$ and $\beta = 2\sqrt{caa_0}$. (That means the orthogonalization measure π of $(R_n(x))_{n\in\mathbb{N}_0}$ belongs to $M(\alpha,\beta)$.) Then for each $n \in \mathbb{N}$, $k \in \mathbb{N}_0$ we have*

$$\lim_{m\to\infty} g(m+n,n;k)\,\frac{(h(m)\,h(m+n))^{\frac{1}{2}}}{h(k)}$$
$$= \frac{1}{\pi}\int_{\alpha-\beta}^{\alpha+\beta} R_k(x)\,T_n\left(\frac{x-\alpha}{\beta}\right)(\beta^2-(x-\alpha)^2)^{-\frac{1}{2}}\,dx$$

Proof. Setting $f(x) = R_k(x)$ the assertion follows by [509, §4.2, Theorem 13]. ◇

For $\alpha \in \mathbb{R}$, $\beta > 0$, define $Q_n(x) = T_n\left(\frac{x-\alpha}{\beta}\right)$. Obviously, the $Q_n(x)$ are polynomials orthogonal with respect to $d\mu(x) = (\beta^2-(x-\alpha)^2)^{-\frac{1}{2}}dx$ on $[\alpha-\beta,\alpha+\beta]$. The recurrence relations of the corresponding monic orthogonal polynomials $\tilde{Q}_n(x)$ are

$$\tilde{Q}_0(x) = 1, \qquad \tilde{Q}_1(x) = x-\alpha$$

$$x\,\tilde{Q}_n(x) = \tilde{Q}_{n+1}(x) + \alpha\tilde{Q}_n(x) + \gamma_n\tilde{Q}_{n-1}(x) \qquad \text{for } n \in \mathbb{N},$$

where $\gamma_1 = \frac{\beta^2}{2}$, $\gamma_n = \frac{\beta^2}{4}$ for $n \geq 2$.

By Corollary 4.6.6 we infer that $Q_n(x) = \sum_{k=0}^{n} d(n,k)\,T_k(x)$ with $d(n,k) \geq 0$, whenever $\alpha \leq 0$, $0 < \beta \leq 1$.

Theorem 4.6.8 *Let $(R_n(x))_{n\in\mathbb{N}_0}$ be an orthogonal polynomial sequence as in Theorem 4.6.7, which generates a polynomial hypergroup on $\mathbb{N}_0$ (viz. the generated polynomial hypergroup belongs to the Nevai class with R_n-parameters a, c and starting parameter a_0). If $\alpha \leq 0$ and $0 < \beta \leq 1$, then $(R_n(x))_{n\in\mathbb{N}_0}$ satisfies condition (T).*

Proof. The linearization coefficients $g(m,n;k)$ of $(R_n(x))_{n\in\mathbb{N}_0}$ are nonnegative. Applying Theorem 4.6.7 and $T_n \overset{d}{\geq} Q_n$, the connection coefficients $c(n,k)$ of $R_n(x) = \sum_{k=0}^{n} c(n,k)\,T_k(x)$ are nonnegative. ◇

7 Growth condition (H) and Følner condition

We have already investigated means and strongly invariant means on commutative hypergroups in section 3.12. For polynomial hypergroups the growth of the Haar weights $h(n)$ is essential for the representation of translation-invariant means. We obtain such means as cluster points of averages on certain sequences which we call summing sequences. We refer to the construction of Følner sequences in the group case.

Definition 1: A sequence $(A_n)_{n\in\mathbb{N}_0}$, where $A_n \subseteq \mathbb{N}_0$ for all $n \in \mathbb{N}_0$ is called **summing sequence** on the polynomial hypergroup $K = \mathbb{N}_0$, if it satisfies

(i) A_n are finite subsets of $\mathbb{N}_0$,
(ii) $\bigcup_{n\in\mathbb{N}_0} A_n = \mathbb{N}_0$,
(iii) $A_n \subseteq A_{n+1}$ for every $n \in \mathbb{N}_0$,
(iv) $\lim_{n\to\infty} \frac{h((L_k A_n)\Delta A_n)}{h(A_n)} = 0$ for every $k \in \mathbb{N}$. (Δ denotes the symmetric difference of sets and $L_k A_n = \{k\} * A_n$.)

The growth condition for $h(n)$ which we have in mind is **property (H)**, that is

$$\lim_{n\to\infty} \frac{h(n)}{H(n)} = 0,$$

where $H(n) := \sum_{k=0}^{n} h(k)$.

It is straightforward to show that property (H) is satisfied whenever $h(n)$ is of polynomial growth. We derive now a sufficient criterion for (H), which can be easily checked.

Theorem 4.7.1 *Suppose $(R_n(x))_{n\in\mathbb{N}_0}$ generates a polynomial hypergroup. Put $\sigma_0 = 0$, $\sigma_n = h(n)/H(n-1)$ for $n \in \mathbb{N}$. Then*

(1) $\sigma_n \to \varrho$, $\varrho > 0$, *if and only if* $\frac{a_{n-1}}{c_n} \to 1 + \varrho$ *as* $n \to \infty$,

(2) $\sigma_n \to 0$ *and* $\frac{\sigma_n}{\sigma_{n-1}} \to 1$ *if and only if* $\frac{a_{n-1}}{c_n} \to 1$ *as* $n \to \infty$.

Proof. By induction, we obtain

$$H(n) = \sum_{k=0}^{n} h(k) = \prod_{k=0}^{n} (1+\sigma_k).$$

In particular, it follows $h(n) = \left(\prod_{k=0}^{n-1}(1+\sigma_k)\right)\sigma_n$, and hence

$$\frac{a_{n-1}}{c_n} = \frac{h(n)}{h(n-1)} = \frac{\sigma_n}{\sigma_{n-1}}\,(1+\sigma_{n-1}).$$

Now $\sigma_n \to \varrho$, $\varrho > 0$ implies $\frac{a_{n-1}}{c_n} \to 1+\varrho$. If $\sigma_n \to 0$ and $\frac{\sigma_n}{\sigma_{n-1}} \to 1$ we get then $\frac{a_{n-1}}{c_n} \to 1$.

To show the converse implications in (1) and (2), denote by $q_n = \frac{c_n}{a_{n-1}}$, and let $\varepsilon > 0$. Assuming $\frac{a_{n-1}}{c_n} \to 1+\varrho$ with $\varrho \geq 0$, there exists $m \in \mathbb{N}$ such that $\frac{1}{1+\varrho} - \varepsilon \leq q_{m+n} \leq \frac{1}{1+\varrho} + \varepsilon$ for all $n \in \mathbb{N}_0$.

We consider at first the case $\varrho = 0$ and suppose $0 < \varepsilon < 1$. Since

$$0 < q_{m+n}\sigma_{m+n} = \frac{q_{m+n}h(m+n)}{H(m+n-1)} = \frac{h(m+n-1)}{H(m+n-1)}$$

$$\leq \frac{h(m+n-1)}{\sum\limits_{k=m}^{m+n-1} h(k)}$$

$$= \frac{h(m+n-1)}{h(m+n-1)(1+q_{m+n-1}+q_{m+n-1}q_{m+n-2}+\ldots+q_{m+n-1}\cdots q_{m+1})}$$

$$= \frac{1}{1+(1-\varepsilon)+\ldots+(1-\varepsilon)^{n-1}} = \frac{\varepsilon}{1-(1-\varepsilon)^n},$$

and $0 < \varepsilon < 1$ was arbitrary, it follows $q_n\sigma_n \to 0$, and so $\sigma_n \to 0$.

Now suppose $\varrho > 0$. Since $\frac{a_{n-1}}{c_n} \to 1+\varrho > 1$ and $h(n) = h(n-1)\,\frac{a_{n-1}}{c_n}$, it follows $h(n) \to \infty$ as $n \to \infty$. Moreover,

$$\frac{1}{q_{m+n}\sigma_{m+n}} = \frac{H(m-1)}{h(m+n-1)} + \frac{\sum\limits_{k=m}^{m+n-1} h(k)}{h(m+n-1)}$$

$$\geq \frac{H(m-1)}{h(m+n-1)} + 1 + \left(\frac{1}{1+\varrho} - \varepsilon\right) + \ldots + \left(\frac{1}{1+\varrho} - \varepsilon\right)^{n-1}.$$

Hence we get

$$\liminf_{n\to\infty} \frac{1}{q_n\sigma_n} \geq \frac{1}{1-\frac{1}{1+\varrho}} = \frac{1+\varrho}{\varrho},$$

and so

$$\liminf_{n\to\infty} \frac{1}{\sigma_n} = \liminf_{n\to\infty} \frac{1}{\sigma_n q_n} \lim_{n\to\infty} q_n \geq \frac{1}{\varrho},$$

which implies $\limsup_{n\to\infty} \sigma_n \le \varrho$. Similarly we obtain

$$\frac{1}{q_{m+n}\sigma_{m+n}} \le \frac{H(m-1)}{h(m+n-1)} + 1 + \left(\frac{1}{1+\varrho} + \varepsilon\right) + \ldots + \left(\frac{1}{1+\varrho} + \varepsilon\right)^{n-1},$$

which implies

$$\limsup_{n\to\infty} \frac{1}{q_n\sigma_n} \le \frac{1}{1-\frac{1}{1+\varrho}} = \frac{1+\varrho}{\varrho},$$

and then $\liminf_{n\to\infty} \sigma_n \ge \varrho$. Finally we have shown $\lim_{n\to\infty} \sigma_n = \varrho$. ◇

Corollary 4.7.2 *Suppose $(R_n(x))_{n\in\mathbb{N}_0}$ generates a polynomial hypergroup. If $\frac{a_{n-1}}{c_n} \to 1$ as $n \to \infty$, then property (H) is satisfied.*

For $m \in \mathbb{N}_0$ write $Q_m(n) := \frac{1}{h(n+m)} H(n)$. The following identities are obvious

$$Q_m(n) = Q_{m-1}(n+1) - \frac{h(n+1)}{h(n+m)} \qquad (1)$$

$$Q_m(n) = Q_{m-1}(n) \frac{h(n+m-1)}{h(n+m)} = \cdots = Q_1(n) \frac{h(n+1)}{h(n+m)}$$

$$= Q_0(n) \frac{h(n)}{h(n+m)}. \qquad (2)$$

Lemma 4.7.3 *The numbers $Q_m(n)$ satisfy the following recurrence formula,*

$$Q_m(n)\left(1 + \frac{1}{Q_1(n)}\right) = Q_{m-1}(n+1). \qquad (3)$$

Proof. By (1) and (2) it follows

$$Q_m(n) = Q_{m-1}(n+1) - \frac{h(n+1)}{h(n+m)} = Q_{m-1}(n+1) - \frac{Q_m(n)}{Q_1(n)}$$

and hence $Q_m(n)\,(1 + \frac{1}{Q_1(n)}) = Q_{m-1}(n+1)$. ◇

The important asymptotic property (H) will be used subsequently several times.

Proposition 4.7.4 *Suppose that property (H) is satisfied. Then*

$$\lim_{n\to\infty} \frac{h(n+m)}{H(n)} = \lim_{n\to\infty} \frac{1}{Q_m(n)} = 0 \qquad \textit{for all } m \in \mathbb{N}_0.$$

Proof. We apply induction on m. For $m = 0$ the statement is just property (H). Suppose the statement is shown for $k = 0, \ldots, m-1$. Then by formula (3) it follows immediately that $Q_m(n) \to \infty$ as $n \to \infty$. ◇

A natural candidate for a summing sequence is $(S_n)_{n\in\mathbb{N}_0}$, where $S_n = \{0, 1, \ldots, n\}$. Obviously $(S_n)_{n\in\mathbb{N}_0}$ satisfies (i) – (iii) of the Definition 1 of summing sequences. We will call $(S_n)_{n\in\mathbb{N}_0}$ the **canonical sequence**.

Theorem 4.7.5 *The canonical sequence $(S_n)_{n\in\mathbb{N}_0}$ is a summing sequence if and only if property (H) is satisfied.*

Proof. For $k \le n$ we have

$$\{n+k\} \subseteq (L_k S_n)\,\Delta S_n \subseteq \{n+1, n+2, ..., n+k\}.$$

Hence

$$\frac{h(n+k)}{h(S_n)} \le \frac{h((L_k S_n)\,\Delta S_n)}{h(S_n)} \le \frac{h(n+1)+\cdots+h(n+k)}{h(S_n)}$$

$$= \frac{1}{Q_1(n)} + \cdots + \frac{1}{Q_k(n)}.$$

If property (H) is satisfied, Proposition 4.7.4 implies that $(S_n)_{n\in\mathbb{N}_0}$ is a summing sequence. Conversely, if $(S_n)_{n\in\mathbb{N}_0}$ is a summing sequence, the first inequality with $k = 1$ yields $0 \le \frac{h(n+1)}{H(n+1)} \le \frac{h(n+1)}{h(S_n)} \to 0$. Hence property (H) is satisfied. $\diamond$

There are other sequences of sets than the canonical sequence $(S_n)_{n\in\mathbb{N}_0}$, which are summing sequences. An example is constructed as follows.

Proposition 4.7.6 *Let $(A_n)_{n\in\mathbb{N}_0}$ be a sequence of sets $A_n \subseteq \mathbb{N}_0$ with*

(i) $A_n \subseteq A_{n+1}$

(ii) $\bigcup_{n\in\mathbb{N}_0} A_n = \mathbb{N}_0$

(iii) $A_n = S_n \backslash I_n$, *where $(S_n)_{n\in\mathbb{N}_0}$ is the canonical sequence and*

$$|I_n| \le C \qquad \textit{for all } n \in \mathbb{N}_0.$$

If property (H) is satisfied and if the Haar weights are nondecreasing, then $(A_n)_{n\in\mathbb{N}_0}$ is a summing sequence.

Proof. We have

$$h((L_k A_n)\,\Delta A_n) \le h((L_k S_n)\,\Delta S_n) + h(L_k I_n) + h(I_n)$$

$$\le h((L_k S_n)\,\Delta S_n) + C \cdot (2k+2) \cdot h(n+k).$$

By Proposition 4.7.4 and Theorem 4.7.5 and $\lim_{n\to\infty} \frac{h(S_n)}{h(A_n)} = 1$ it follows that $(A_n)_{n\in\mathbb{N}_0}$ is a summing sequence. $\diamond$

Remarks:

(1) Summing sequences play an important role in forming averages and deriving ergodic theorems for actions of locally compact groups on measure spaces, see e.g. [285].

(2) An example of a sequence $(A_n)_{n\in\mathbb{N}_0}$ satisfying (i), (ii), (iii), but not condition (iv), where the polynomial hypergroup is the Chebyshev hypergroup on $\mathbb{N}_0$, is the following:

$$A_n := \{0,1,...,2n\} \cup \{2n+1, 2n+3, ..., 4n-1\}.$$

Then $L_1 A_n = \{0,1,...,2n,2n+1\} \cup \{2n+2, 2n+4, ..., 4n\}$, and

$$(L_1 A_n)\,\Delta A_n = \{2n+2, 2n+3, ..., 4n\}.$$

Since $h(0) = 1$, $h(n) = 2$ for $n \in \mathbb{N}$, we get $h(A_n) = 1 + 6n$ and $h((L_1 A_n)\,\Delta A_n) = 4n - 2$. Hence

$$\lim_{n\to\infty} \frac{h((L_1 A_n)\,\Delta A_n)}{h(A_n)} = \frac{2}{3}.$$

Note that the Chebyshev hypergroup satisfies property (H).

Proposition 4.7.7 *If $\frac{h(n)}{H(n)} \geq C > 0$ for all $n \in \mathbb{N}_0$, then there does not exist any summing sequence. In particular, there cannot exist any summing sequence whenever $h(n)$ grow exponentially.*

Proof. Let $(A_n)_{n\in\mathbb{N}_0}$ be a sequence of sets satisfying (i), (ii) and (iii) of Definition 1. Put $m_n := \max A_n$. Then $m_n \leq m_{n+1}$, $\lim_{n\to\infty} m_n = \infty$ and $m_n + 1 \in (L_1 A_n)\,\Delta A_n$. Therefore,

$$\frac{h((L_1 A_n)\,\Delta A_n)}{h(A_n)} \geq \frac{h(m_n+1)}{h(A_n)} \geq \frac{h(m_n+1)}{H(m_n+1)} \geq C > 0,$$

and (iv) cannot hold. ◇

Our general goal is to construct translation invariant means on $l^p(h)$ as limits of elements in the dual space $(l^p(h))^*$. We will use the characteristic functions χ_{A_n} of $A_n \subseteq \mathbb{N}_0$. In contrast to the group case we have to keep in mind that $L_k\chi_{A_n}$ is no more a characteristic function. We only know that $\operatorname{supp} L_k\chi_{A_n} = \{k\} * A_n$.

Definition 2: A sequence $(A_n)_{n\in\mathbb{N}_0}$, where $A_n \subseteq \mathbb{N}_0$ for all $n \in \mathbb{N}_0$ is said to satisfy the **Følner condition** (F_p), $1 \leq p < \infty$, if

(i) A_n are finite subsets of $\mathbb{N}_0$,

(ii) $\bigcup_{n\in\mathbb{N}_0} A_n = \mathbb{N}_0$,

(iii) $A_n \subseteq A_{n+1}$ for every $n \in \mathbb{N}_0$,

(iv) $\displaystyle\lim_{n\to\infty} \frac{\|L_k\chi_{A_n} - \chi_{A_n}\|_p^p}{h(A_n)} = 0 \qquad$ for all $k \in \mathbb{N}$.

Remark: If $K = G$ is a group, condition (iv) of Definition 1 coincides with condition (iv) of Definition 2 in case $p = 1$.

Lemma 4.7.8 *If* $\lim_{n\to\infty} \frac{\|L_1\chi_{A_n}-\chi_{A_n}\|_p^p}{h(A_n)} = 0$, *then*

$$\lim_{n\to\infty} \frac{\|L_k\chi_{A_n} - \chi_{A_n}\|_p^p}{h(A_n)} = 0 \quad \textit{for every } k \in \mathbb{N}.$$

Proof. The translation operators $L_k \in B(l^p(h))$ satisfy the three-term recurrence relation

$$L_k L_1 = L_1 L_k = a_k L_{k+1} + b_k L_k + c_k L_{k-1}$$

for $k \in \mathbb{N}$. Moreover, the operator norms fulfil $\|L_k\| \leq 1$ for each $k \in \mathbb{N}$. We apply induction with respect to k. Assume that property (iv) of the Følner condition (F_p) is fulfilled for k and $k-1$. Then

$$\begin{aligned}\|L_k L_1\chi_{A_n} - \chi_{A_n}\|_p &\leq \|L_1 L_k\chi_{A_n} - L_1\chi_{A_n}\|_p + \|L_1\chi_{A_n} - \chi_{A_n}\|_p \\ &\leq \|L_k\chi_{A_n} - \chi_{A_n}\|_p + \|L_1\chi_{A_n} - \chi_{A_n}\|_p,\end{aligned}$$

and hence (using $a_k + b_k + c_k = 1$),

$$\begin{aligned}\|L_{k+1}\chi_{A_n} - \chi_{A_n}\|_p &= \|\frac{1}{a_k}(L_k L_1\chi_{A_n} - b_k L_k\chi_{A_n} - c_k L_{k-1}\chi_{A_n}) - \chi_{A_n}\|_p \\ &\leq \|\frac{1}{a_k}(L_k L_1\chi_{A_n} - \chi_{A_n})\|_p + \|\frac{b_k}{a_k}(L_k\chi_{A_n} - \chi_{A_n})\|_p \\ &\quad + \|\frac{c_k}{a_k}(L_{k-1}\chi_{A_n} - \chi_{A_n})\|_p \\ &\leq \frac{1}{a_k}\left(\|L_k\chi_{A_n} - \chi_{A_n}\|_p + \|L_1\chi_{A_n} - \chi_{A_n}\|_p\right) \\ &\quad + \frac{b_k}{a_k}\|L_k\chi_{A_n} - \chi_{A_n}\|_p + \frac{c_k}{a_k}\|L_{k-1}\chi_{A_n} - \chi_{A_n}\|_p.\end{aligned}$$

By the assumptions it follows $\lim_{n\to\infty} \frac{\|L_{k+1}\chi_{A_n}-\chi_{A_n}\|_p^p}{h(A_n)} = 0.$ ◇

Theorem 4.7.9 *The canonical sequence* $(S_n)_{n\in\mathbb{N}_0}$ *satisfies the Følner condition* (F_1) *if and only if* $\frac{a_n h(n)}{H(n)} \to 0$.

Proof. By Lemma 4.7.8 it is sufficient to consider (iv) of Definition 2 only for $k = 1$. Using the recurrence relation we have

$$L_1\chi_{S_n}(k) - \chi_{S_n}(k) = a_n + b_n + c_n - 1 = 0$$

$$\text{for } k = 0, 1, ..., n-1 \text{ and } k = n+2, n+3, ...,$$

and

$$L_1\chi_{S_n}(n) - \chi_{S_n}(n) = b_n + c_n - 1 = -a_n, \quad L_1\chi_{S_n}(n+1) - \chi_{S_n}(n+1) = c_{n+1}.$$

Hence

$$\frac{\|L_1\chi_{S_n} - \chi_{S_n}\|_1}{h(S_n)} = \frac{1}{H(n)}(a_n h(n) + c_{n+1}h(n+1)) = \frac{2a_n h(n)}{H(n)}.$$

◇

Corollary 4.7.10 *(1) If property (H) holds or if $a_n \to 0$, then $(S_n)_{n\in\mathbb{N}_0}$ satisfies (F_1).*
(2) If there exist $c > 0$, $n_0 \in \mathbb{N}_0$, such that $a_n \geq c$ or $c_n \geq c$ for all $n \geq n_0$, and if the canonical sequence $(S_n)_{n\in\mathbb{N}_0}$ satisfies (F_1) then the hypergroup fulfils property (H).

Proof.

(1) is obviously true by Theorem 4.7.9.
(2) If $(S_n)_{n\in\mathbb{N}_0}$ satisfies (F_1), then

$$0 = \lim_{n\to\infty} \frac{\|L_1\chi_{S_n} - \chi_{S_n}\|_1}{H(n)} = \lim_{n\to\infty} \frac{2a_n h(n)}{H(n)} = \lim_{n\to\infty} \frac{2c_{n+1}h(n+1)}{H(n)} \geq \lim_{n\to\infty} \frac{2c_{n+1}h(n+1)}{H(n+1)}.$$

Since at least one of the sequences $(a_n)_{n\in\mathbb{N}_0}$, $(c_n)_{n\in\mathbb{N}_0}$ is bounded away from zero, property (H) follows.

◇

We know that $1 \in S$ if and only if $\{\frac{H(n)}{a_n h(n)} : n \in \mathbb{N}_0\}$ is unbounded, see Corollary 4.2.12. Therefore the following conclusion is true.

Corollary 4.7.11 *The canonical sequence $(S_n)_{n\in\mathbb{N}_0}$ has a subsequence $(S_{n_k})_{k\in\mathbb{N}_0}$ which fulfils the Følner condition (F_1) if and only if $1 \in S$.*

Now we investigate how the Følner conditions (F_p) depend on p.

Proposition 4.7.12 *Suppose that the canonical sequence $(S_n)_{n\in\mathbb{N}_0}$ satisfies (F_p) for some $p \geq 1$. Then it satisfies (F_q) for all $q \geq p$.*

Proof. We have

$$\|L_1\chi_{S_n} - \chi_{S_N}\|_q^q = a_n^q\, h(n) + c_{n+1}^q\, h(n+1) = a_n^q\, h(n) + a_n c_{n+1}^{q-1}\, h(n) = h(n)\,(a_n^q + a_n c_{n+1}^{q-1}).$$

Because of $0 < a_n c_n < 1$ for $n \in \mathbb{N}$ we obtain

$$\|L_1\chi_{S_n} - \chi_{S_n}\|_q^q \leq \|L_1\chi_{S_n} - \chi_{S_n}\|_p^p \qquad \text{for } 1 \leq p \leq q.$$

◇

We will write for $1 < p < \infty$ $f_n = \frac{\chi_{S_n}}{H(n)^{1/p}}$ and $g_n = \frac{\chi_{S_n}}{H(n)} = f_n^p$. Obviously $\|f_n\|_p = 1$, $\|g_n\|_1 = 1$ and $\|L_1 g_n\|_1 = 1$.

Lemma 4.7.13 *Suppose that the canonical sequence $(S_n)_{n\in\mathbb{N}_0}$ satisfies (F_p), $p > 1$. Then*

$$\lim_{n\to\infty} \|(L_1 f_n)^p - g_n\|_1 = 0.$$

Proof. We will use the inequality $|y^p - x^p| \leq p\,|y-x|(x+y)^{p-1}$ for all $x, y \geq 0$, which follows by the elementary mean-value theorem. We have

$$\|(L_1 f_n)^p - f_n^p\|_1 = \sum_{k=0}^{\infty} |(L_1 f_n)^p(k) - f_n^p(k)|\; h(k)$$

$$\leq p \sum_{k=0}^{\infty} |L_1 f_n(k) - f_n(k)|\; [L_1 f_n(k) + f_n(k)]^{p-1}\, h(k).$$

Since $(L_1 f_n + f_n)^{p-1} \in l^{\frac{p}{p-1}}(h)$, Hölder's inequality implies

$$\|(L_1 f_n)^p - f_n^p\|_1 \leq p\,\|L_1 f_n - f_n\|_p\, \|L_1 f_n + f_n\|_p^{p-1}.$$

By $\|L_1 f_n\|_p \leq \|f_n\|_p = 1$ we obtain

$$\|(L_1 f_n)^p - g_n\|_1 \leq p\, 2^{p-1}\, \|L_1 f_n - f_n\|_p.$$

◇

Proposition 4.7.14 *Suppose that the canonical sequence $(S_n)_{n\in\mathbb{N}_0}$ satisfies (F_p), $p > 1$. Then $(S_n)_{n\in\mathbb{N}_0}$ satisfies (F_1).*

Proof. Let $\varepsilon > 0$. By Lemma 4.7.13 there exists $n_0 \in \mathbb{N}$ such that $\|(L_1 f_n)^p - g_n\|_1 < \frac{\varepsilon}{2}$ for all $n \geq n_0$. We know $(L_1 f)^p(k) \leq L_1(\,|f|^p)(k)$ for every $k \in \mathbb{N}_0$, see Proposition 2.2.3. In particular, $L_1 g_n(k) \geq (L_1 f_n)^p(k)$. We obtain

$$\|(L_1 f_n)^p - L_1 g_n\|_1 = \sum_{k=0}^{\infty} |L_1 g_n(k) - (L_1 f_n)^p(k)|\; h(k)$$

$$= \sum_{k=0}^{\infty} (L_1 g_n(k) - (L_1 f_n)^p(k))\; h(k) = \|L_1 g_n\|_1 - \|(L_1 f_n)^p\|_1$$

$$= \|g_n\|_1 - \|(L_1 f_n)^p\|_1 \leq \|(L_1 f_n)^p - g_n\|_1 < \frac{\varepsilon}{2} \quad \text{for all } n \geq n_0.$$

Finally we obtain

$$\|L_1 g_n - g_n\|_1 \leq \|L_1 g_n - (L_1 f_n)^p\|_1 + \|(L_1 f_n)^p - g_n\|_1 < \frac{\varepsilon}{2} + \frac{\varepsilon}{2} = \varepsilon$$

for all $n \geq n_0$. ◇

Combining the results above we have a dichotomy-result.

Theorem 4.7.15 *The canonical sequence satisfies (F_p) for all $1 \leq p < \infty$ or for none.*

8 Strongly invariant means on polynomial hypergroups

We continue the investigations of subsection 3.12 for the special case of polynomial hypergroups $K = \mathbb{N}_0$, starting by deriving sufficient conditions for the existence of strongly invariant means from growth properties. Recall that a mean m on l^∞ is an element $m \in (l^\infty)^*$ such that $m(1) = 1 = \|m\|$. For the proof of the following two results we use summing sequences $(A_n)_{n\in\mathbb{N}_0}$. Based on the pairing

$$\langle \varphi, f\rangle = \sum_{n=0}^{\infty} \varphi(n)\, f(n)\, h(n) \qquad \text{for } \varphi \in l^\infty,\ f \in l^1(h),$$

one has $l^1(h)^* \cong l^\infty$. Because of that we can isometrically embed $l^1(h)$ into $(l^\infty)^*$ via the mapping $f \mapsto j(f)$, where $j(f)(\varphi) = \langle \varphi, f\rangle$ for $f \in l^1(h)$, $\varphi \in l^\infty$.

We write $P^1(h) := \{f \in l^1(h) : f \geq 0,\ \|f\|_1 = 1\}$. Obviously,

$$P^1(h) = \overline{\mathrm{co}\,\{\epsilon_n : \mathbb{N}_0\}}$$

(norm closure of the convex hull of the point measures ϵ_n). Moreover, $j(P^1(h))$ is weak $*$-dense in the set of all means on l^∞, see [532,Proposition 0.1]. From this fact it follows immediately that $|m(\varphi)| \leq m(|\varphi|)$ for all $\varphi \in l^\infty$. Moreover, Proposition 0.1 of [532] says that $m \in (l^\infty)^*$ is a mean if and only if $m(1) = 1$ and $m(\varphi) \geq 0$ whenever $\varphi \geq 0$ in l^∞. We recall that there exist invariant means on l^∞, see Theorem 3.12.1.

Proposition 4.8.1 *A mean m on l^∞ is strongly invariant if*

$$m((L_1\varphi)\psi) = m(\varphi(L_1\psi)) \qquad \textit{for all } \varphi, \psi \in l^\infty.$$

Proof. We use induction on n. Assume that $m((L_k\varphi)\psi) = m(\varphi(L_k\psi))$ for $k = n-1$ and $k = n$. Then

$$m((L_{n+1}\varphi)\psi) = \frac{1}{a_n}\, m((L_1 \circ L_n\varphi)\psi) - \frac{b_n}{a_n}\, m((L_n\varphi)\psi) - \frac{c_n}{a_n}\, m((L_{n-1}\varphi)\psi)$$

$$= \frac{1}{a_n}\, m(\varphi(L_n \circ L_1\psi)) - \frac{b_n}{a_n}\, m(\varphi(L_n\psi)) - \frac{c_n}{a_n}\, m(\varphi(L_{n-1}\psi))$$

$$= m(\varphi(L_{n+1}\psi)).$$

◇

We consider now the canonical sequence $(S_n)_{n\in\mathbb{N}_0}$, where $S_n = \{0, 1, ..., n\}$ and denote by

$$f_n := \frac{1}{h(S_n)}\, \chi_{S_n} = \frac{1}{H(n)}\, \chi_{S_n} \in P^1(h).$$

Proposition 4.8.2 *For each $\varphi, \psi \in l^\infty$ and $n \in \mathbb{N}_0$ we have*

$$|j(f_n)((L_1\varphi)\psi) - j(f_n)(\varphi(L_1\psi))|$$

$$= \frac{1}{H(n)} |(\psi(n)\,\varphi(n+1) - \psi(n+1)\varphi(n))\, a_n h(n)|.$$

Proof. We have

$$j(f_n)((L_1\varphi)\psi) - j(f_n)(\varphi(L_1\psi))$$

$$= \frac{1}{H(n)} \sum_{k=0}^{\infty} \chi_{S_n}(k)\,((L_1\varphi)(k)\,\psi(k) - \varphi(k)\,(L_1\psi)(k))\,h(k)$$

$$= \frac{1}{H(n)} \sum_{k=0}^{\infty} (L_1(\chi_{S_n} \cdot \psi)(k) - \chi_{S_n}(k)\,(L_1\psi)(k))\,\varphi(k)\,h(k),$$

where we have used that

$$\sum_{k=0}^{\infty} L_1\varphi(k)\,(\chi_{S_n} \cdot \psi)(k)\,h(k) = \sum_{k=0}^{\infty} \varphi(k)\,L_1(\chi_{S_n} \cdot \psi)(k)\,h(k),$$

compare Theorem 2.2.6. Since

$$L_1(\chi_{S_n} \cdot \psi)(k) - \chi_{S_n}(k)\,(L_1\psi)(k) = \begin{cases} 0 & k = 0, \ldots, n-1, \\ & \quad n+2, n+3, \ldots \\ -a_n\psi(n+1) & k = n \\ c_{n+1}\psi(n) & k = n+1 \end{cases},$$

we get

$$|j(f_n)\,((L_1\varphi)\psi) - j(f_n)\,(\varphi(L_1\psi))|$$

$$= \frac{1}{H(n)} |-a_n\psi(n+1)\,\varphi(n)\,h(n) + c_{n+1}\psi(n)\,\varphi(n+1)\,h(n+1)|$$

$$= \frac{1}{H(n)} |(\psi(n)\,\varphi(n+1) - \psi(n+1)\,\varphi(n))\,a_n h(n)|.$$

◇

Combining Proposition 4.8.1 and Proposition 4.8.2 we obtain

Theorem 4.8.3 *If $\frac{a_n h(n)}{H(n)} \to 0$, then the weak $*$-cluster points of $(j(f_n))_{n\in\mathbb{N}_0}$ are strongly invariant means on l^∞. In particular, there exist strongly invariant means on l^∞, whenever $\frac{a_n h(n)}{H(n)} \to 0$.*

For the last statement in Theorem 4.8.3 we have used that each net in a compact space has a convergent subnet.

Applying Theorem 4.7.9 we have the following result.

Corollary 4.8.4 *If $(S_n)_{n\in\mathbb{N}_0}$ satisfies the Følner condition F_1, then the weak $*$-cluster points of $(j(f_n))_{n\in\mathbb{N}_0}$ are strongly invariant means on l^∞.*

Now we will derive an interesting characterization of the existence of strongly invariant means on l^∞. We shall use the result of Theorem 3.12.4. We begin by deriving a more accessible description of the linear space

$$SL := \text{ span } \{L_n\varphi\cdot\psi - \varphi\cdot L_n\psi : n\in\mathbb{N}_0,\ \varphi,\psi\in l^\infty\}.$$

Lemma 4.8.5

$$SL \;=\; \textit{span } \{L_1\varphi\cdot\psi - \varphi\cdot L_1\psi : \varphi,\psi\in l^\infty\}.$$

Proof. We will use induction on n and again the identity

$$L_{n+1} \;=\; \frac{1}{a_n}\, L_1\circ L_n \;-\; \frac{b_n}{a_n}\, L_n \;-\; \frac{c_n}{a_n}\, L_{n-1}. \qquad (*)$$

Suppose that $L_k\varphi_1\cdot\psi_1 - \varphi_1\cdot L_k\psi_1$ are elements of

$$SL_1 := \text{ span } \{L_1\varphi_2\cdot\psi_2 - \varphi_2\cdot L_1\psi_2 : \varphi_2,\psi_2\in l^\infty\}$$

for $k = n-1, n$ and for all $\varphi_1,\psi_1\in l^\infty$. From $(*)$ and $L_n\circ L_1 = L_1\circ L_n$ we obtain for all $\varphi_1,\psi_1\in l^\infty$,

$$L_{n+1}\varphi_1\cdot\psi_1 - \varphi_1\cdot L_{n+1}\psi_1 \;=\; \left[L_1\left(\frac{L_n\varphi_1}{a_n}\right) - b_n\,\frac{L_n\varphi_1}{a_n} - c_n\,\frac{L_{n-1}\varphi_1}{a_n}\right]\psi_1$$

$$-\,\varphi_1\left[L_n\left(\frac{L_1\psi_1}{a_n}\right) - b_n\,\frac{L_n\psi_1}{a_n} - c_n\,\frac{L_{n-1}\psi_1}{a_n}\right].$$

By the assumption we have

$$L_{n-1}\varphi_1\cdot\psi_1 - \varphi_1\cdot L_{n-1}\psi_1 \in SL_1, \quad L_n\varphi_1\cdot\psi_1 - \varphi_1\cdot L_n\psi_1 \in SL_1.$$

Moreover,

$$L_1\left(\frac{L_n\varphi_1}{a_n}\right)\cdot\psi_1 - \frac{L_n\varphi_1}{a_n}\cdot L_1\psi_1 \in SL_1$$

and by the assumption

$$\varphi_1\cdot L_n\left(\frac{L_1\psi_1}{a_n}\right) - L_n\varphi_1\cdot\frac{L_1\psi_1}{a_n} \in SL_1.$$

Hence we have

$$L_1\left(\frac{L_n\varphi_1}{a_n}\right)\cdot\psi_1 - \varphi_1\cdot L_n\left(\frac{L_1\psi_1}{a_n}\right) \;=\; \left(L_1\left(\frac{L_n\varphi_1}{a_n}\right)\cdot\psi_1 - \frac{L_n\varphi_1}{a_n}\cdot L_1\psi_1\right)$$

$$-\left(\varphi_1\cdot L_n\left(\frac{L_1\psi_1}{a_n}\right) - L_n\varphi_1\cdot\frac{L_1\psi_1}{a_n}\right) \in SL_1.$$

Therefore Lemma 4.8.5 is shown. ◇

Let $\varphi, \psi \in l^\infty$. Then

$$\begin{aligned} L_1\varphi(n)\,\psi(n) - \varphi(n)\,L_1\psi(n) &= a_n(\varphi(n+1)\,\psi(n) - \psi(n+1)\varphi(n)) \\ &\quad - c_n(\varphi(n)\,\psi(n-1) - \psi(n)\,\varphi(n-1)) \\ &= a_n\omega(n) - c_n\omega(n-1), \end{aligned} \tag{1}$$

where $\omega(n) = \varphi(n+1)\psi(n) - \psi(n+1)\varphi(n)$ for all $n \in \mathbb{N}_0$. Note that we set $a_0 = 1$ and $c_0 = 0$.

It is important to point out that for any sequence $\omega \in l^\infty$ there are $\varphi, \psi \in l^\infty$ such that $\omega(n) = \varphi(n+1)\psi(n) - \psi(n+1)\varphi(n)$. This can be easily shown by setting

$$\varphi(2k) = \begin{cases} 0 & \text{for } k \text{ even} \\ 1 & \text{for } k \text{ odd} \end{cases}, \qquad \varphi(2k+1) = \begin{cases} \omega(2k) & \text{for } k \text{ even} \\ -\omega(2k+1) & \text{for } k \text{ odd} \end{cases} \tag{2}$$

and

$$\psi(2k) = \begin{cases} 1 & \text{for } k \text{ even} \\ 0 & \text{for } k \text{ odd} \end{cases}, \qquad \psi(2k+1) = \begin{cases} \omega(2k+1) & \text{for } k \text{ even} \\ -\omega(2k) & \text{for } k \text{ odd} \end{cases}. \tag{3}$$

Therefore, there is again a simplification.

Lemma 4.8.6 *We have*

$$\textit{span}\,\{L_1\varphi \cdot \psi - \varphi \cdot L_1\psi : \varphi, \psi \in l^\infty\} = \{L_1\varphi \cdot \psi - \varphi \cdot L_1\psi : \varphi, \psi \in l^\infty\}.$$

Proof. Let $\varphi_1, \psi_1, \varphi_2, \psi_2 \in l^\infty$ and $n \in \mathbb{N}_0$. By (1) we have

$$(L_1\varphi_1 \cdot \psi_1 - \varphi_1 \cdot L_1\psi_1 + L_1\varphi_2 \cdot \psi_2 - \varphi_2 \cdot L_1\psi_2)(n)$$

$$= a_n\omega_1(n) - c_n\omega_1(n-1) + a_n\omega_2(n) - c_n\omega_2(n-1) = a_n\omega(n) - c_n\omega(n-1)$$

with $\omega_1, \omega_2 \in l^\infty$ and $\omega = \omega_1 + \omega_2$. Finally, in view of (1), (2) and (3) there exist $\varphi, \psi \in l^\infty$ such that

$$L_1\varphi_1 \cdot \psi_1 - \varphi_1 \cdot L_1\psi_1 + L_1\varphi_2 \cdot \psi_2 - \varphi_2 \cdot L_1\psi_2 = L_1\varphi \cdot \psi - \varphi \cdot L_1\psi.$$

◇

Now, by Theorem 3.12.4, we have the following result for polynomial hypergroups.

Theorem 4.8.7 *There exists a strongly invariant mean m on l^∞ exactly when*

$$\sup_{n\in\mathbb{N}_0} (L_1\varphi(n)\,\psi(n) - \varphi(n)\,L_1\psi(n)) \geq 0 \qquad \textit{for all } \varphi, \psi \in l^\infty_{\mathbb{R}}.$$

Proof. By Lemmas 4.8.5 and 4.8.6 it follows that

$$\sup_{n\in\mathbb{N}_0} \left(L_1\varphi(n)\,\psi(n) - \varphi(n)\,L_1\psi(n)\right) \geq 0 \quad \text{for all } \varphi,\psi \in l^\infty_\mathbb{R}$$

is equivalent to $\sup_{n\in\mathbb{N}_0} \eta(n) \geq 0$ for all $\eta \in SL_\mathbb{R}$, and Theorem 3.12.4 yields the statement. ◇

Consequently, the nonexistence of a strongly invariant mean on l^∞ is characterized as follows.

Proposition 4.8.8 *There does not exist a strongly invariant mean on l^∞ if and only if there exist $\delta > 0$ and $\omega \in l^\infty_\mathbb{R}$ such that*

$$a_n\omega(n) - c_n\omega(n-1) \leq -\delta \qquad \textit{for all } n \in \mathbb{N}_0.$$

By Theorem 4.8.3 we know already that there exists a strongly invariant mean on l^∞ whenever $\frac{a_n h(n)}{H(n)} \to 0$. We point out that this mean m is a weak $*$-cluster point of $(j(f_n))_{n\in\mathbb{N}_0}$.

A small improvement is obtained by an application of Proposition 4.8.8.

Proposition 4.8.9 *If the set $\{\frac{H(n)}{a_n h(n)} : n \in \mathbb{N}_0\}$ is unbounded, then there exists a strongly invariant mean on l^∞.*

Proof. Suppose that there does not exist a strongly invariant mean on l^∞. By Proposition 4.8.8 there are $\omega \in l^\infty_\mathbb{R}$ and $\delta > 0$ such that

$$a_n\omega(n) + \delta \leq c_n\omega(n-1) \qquad \text{for all } n \in \mathbb{N}.$$

Since $c_n = a_{n-1}h(n-1)/h(n)$ it follows that

$$h(n)\,a_n\omega(n) + h(n)\,\delta \leq a_{n-1}h(n-1)\,\omega(n-1) \quad \text{for all } n \in \mathbb{N}.$$

Adding $h(n-1)\delta$ to both sides of the inequality, we obtain for all $n \geq 2$,

$$\begin{aligned} h(n)\,a_n\omega(n) + h(n)\,\delta + h(n-1)\,\delta &\leq a_{n-1}h(n-1)\,\omega(n-1) + h(n-1)\,\delta \\ &\leq a_{n-2}h(n-2)\,\omega(n-2). \end{aligned}$$

Iterating this step it follows that

$$h(n)\,a_n\omega(n) + \delta\sum_{k=1}^{n} h(k) \leq a_1h(1)\,\omega(1) + \delta\,h(1) \leq \omega(0) \leq -\delta$$

for all $n \in \mathbb{N}$. Hence we obtain for all $n \in \mathbb{N}_0$

$$\frac{h(n)a_n}{H(n)}\,\omega(n) + \delta \leq 0.$$

However, there exists a subsequence such that $h(n_k)a_{n_k}/H(n_k) \to 0$ as $k \to \infty$, contradicting $\omega \in l^\infty_\mathbb{R}$ and $\delta > 0$. Hence there exists a strongly invariant mean on l^∞. ◇

Proposition 4.8.10 *Assume that there exists a strongly invariant mean m on l^∞. Then $\{\frac{H(n)}{a_n h(n)} : n \in \mathbb{N}_0\}$ is unbounded.*

Proof. Suppose that $\{\frac{H(n)}{a_n h(n)} : n \in \mathbb{N}_0\}$ is bounded. Putting $\omega(n) = \frac{H(n)}{a_n h(n)}$ for all $n \in \mathbb{N}_0$ we get a sequence $\omega \in l^\infty_{\mathbb{R}}$ with the property

$$a_n\omega(n) - c_n\omega(n-1) = \frac{H(n)}{h(n)} - \frac{H(n-1)c_n}{h(n-1)a_{n-1}} = \frac{1}{h(n)}\,(H(n) - H(n-1)) = 1.$$

For this sequence ω choose $\varphi, \psi \in l^\infty_{\mathbb{R}}$ according to (2) and (3). Then we obtain $1 = m(1) = m(L_1\varphi \cdot \psi - \varphi \cdot L_1\psi) = 0$, a contradiction. ◇

Corollary 4.8.11 *For polynomial hypergroups $K = \mathbb{N}_0$ are equivalent*

(1) There exists a strongly invariant mean on l^∞.
(2) $\{\frac{H(n)}{a_n h(n)} : n \in \mathbb{N}_0\}$ is an unbounded set.
(3) $1 \in S$.
(4) The P_2-condition is satisfied in 1.

Proof. We have only to combine Proposition 4.8.9, Proposition 4.8.10 with Theorem 3.7.5 and Corollary 4.2.12. ◇

Examples

(i) On orthogonal polynomials defined by homogeneous trees or on cosh-polynomials there does not exist a strongly invariant mean. (Excluding $\gamma = \frac{1}{4}$.)
(ii) On little q-Legendre polynomials there exist strongly invariant means. Note that a_n tends to zero.
(iii) On Jacobi polynomials, generalized Chebyshev polynomials, Pollaczek polynomials and others exist strongly invariant means.

9 Reiter's P_1-conditions for polynomial hypergroups

As a matter of course we have to deal with Reiter's P_1-condition for polynomial hypergroups. We refer to subsection 3.13 for all the results shown for general commutative hypergroups. We begin with a detailed investigation for which $z \in D$ the condition $P_1(z, M)$ holds. We already know that for $z = 1$ condition $P_1(1, 1)$ is valid. For polynomial hypergroups condition $P_1(z, M)$ is the following property.

Proposition 4.9.1 *Assume that $(R_n(x))_{n\in\mathbb{N}_0}$ generates a polynomial hypergroup on $\mathbb{N}_0$, and let $z \in D$. Then the condition $P_1(z, M)$ is satisfied if and only if for every $\epsilon > 0$ there exists $g \in l^1(h)$ such that*

(i) $\mathcal{F}g(z) = 1$,
(ii) $\|g\|_1 \leq M$,
(iii) $\|L_1 g - R_1(z)g\|_1 < \epsilon$.

Proof. The crucial point is that we can restrict to the translation L_1, which follows immediately from

$$L_{n+1} = \frac{1}{a_n} L_1 \circ L_n - \frac{b_n}{a_n} L_n - \frac{c_n}{a_n} L_{n-1}.$$

◇

Given $z \in D$ define $g_n \in l^1(h)$ by

$$g_n(k) = \frac{1}{H(n,z)} R_k(z) \quad \text{for } 0 \leq k \leq n \quad \text{and } g_n(k) = 0 \quad \text{for } k \geq n+1,$$

where $H(n,z) := \sum_{k=0}^{n} |R_k(z)|^2 \, h(k)$. We have for each $n \in \mathbb{N}_0$

$$\mathcal{F}g_n(z) = \frac{1}{H(n,z)} \sum_{k=0}^{n} R_k(z)\, \overline{R_k(z)}\, h(k) = 1,$$

and

$$\|g_n\|_1 = \frac{1}{H(n,z)} \sum_{k=0}^{n} |R_k(z)|\, h(k).$$

Furthermore,

$$L_1 g_n(k) - R_1(z) g_n(k) = \begin{cases} 0 & k = 0, ..., n-1, n+2, n+3, ... \\ -a_n R_{n+1}(z) & k = n \\ c_{n+1} R_n(z) & k = n+1. \end{cases}$$

It follows

$$\|L_1 g_n - R_1(z) g_n\|_1 = \frac{a_n h(n)}{H(n,z)} \left(|R_n(z)| + |R_{n+1}(z)| \right).$$

Theorem 4.9.2 *Assume that $(R_n(x))_{n\in\mathbb{N}_0}$ generates a polynomial hypergroup on $\mathbb{N}_0$, and let $z \in D$. Then the condition $P_1(z,M)$ is satisfied if there exist a subsequence $n_1 < n_2 < \cdots < n_m < \cdots$ such that*

(i) there is a constant $M_z \geq 1$ satisfying

$$\frac{\sum_{k=0}^{n_m} |R_k(z)|\, h(k)}{\sum_{k=0}^{n_m} |R_k(z)|^2\, h(k)} \leq M_z \qquad \textit{for all } m \in \mathbb{N}, \qquad \textit{and}$$

(ii)

$$\lim_{m\to\infty} \frac{a_{n_m} h(n_m)}{H(n_m, z)} \left(|R_{n_m}(z)| + |R_{n_m+1}(z)| \right) = 0.$$

Now we will establish a negative result for $P_1(x,M)$, showing that for many $x \in D_s$ condition $P_1(x,M)$ does not hold. We apply the equivalence of $P_1(x,M)$ and the existence of α_x-invariant functionals on l^∞, see Theorem 3.13.4. Recall that $\alpha_z(n) = R_n(z)$.

Theorem 4.9.3 *Assume that $(R_n(x))_{n\in\mathbb{N}_0}$ generates a polynomial hypergroup on $\mathbb{N}_0$, and let $x \in D_s$. If $\pi(\{x\}) = 0$ and $R_n(x) \to 0$, then α_x-invariant functionals on l^∞ do not exist. In particular, $P_1(x,M)$ is not satisfied, whenever $\pi(\{x\}) = 0$ and $R_n(x) \to 0$.*

Proof. Suppose that an α_x-invariant functional m_x on l^∞ exists. Since $L_k(\epsilon_0) = \epsilon_k/h(k)$ we have

$$m_x(\epsilon_k) \;=\; h(k)\, R_k(x)\, m_x(\epsilon_0). \qquad (*)$$

Let $\sigma_k :=$ sign $(R_k(x))$. Then we obtain for each $n \in \mathbb{N}$

$$M \;\geq\; \left| m_x\left(\sum_{k=0}^{n} \sigma_k \epsilon_k\right)\right| \;=\; |m_x(\epsilon_0)| \sum_{k=0}^{n} |R_k(x)|\, h(k)$$

$$\geq\; |m_x(\epsilon_0)| \sum_{k=0}^{n} |R_k(x)|^2\, h(k).$$

If $m_x(\epsilon_0) \neq 0$ we have the estimate

$$\sum_{k=0}^{\infty} |R_k(x)|^2\, h(k) \;\leq\; \frac{M}{|m_x(\epsilon_0)|}.$$

In particular $\alpha_x \in l^2(h)$. By Proposition 3.7.6 this is only possible if $\pi(\{x\}) > 0$. Therefore $m_x(\epsilon_0) = 0$, and by $(*)$ we get $m_x(\epsilon_k) = 0$ for each $k \in \mathbb{N}_0$.

Let α_x^n denote the truncated sequence defined by

$$\alpha_x^n(k) \;=\; \begin{cases} 0 & \text{for } k \leq n \\ R_k(x) & \text{for } k > n. \end{cases}$$

Then

$$m_x(\alpha_x) \;=\; \sum_{k=0}^{n} R_k(x)\, m_x(\epsilon_k) + m_x(\alpha_x^n) \;=\; m_x(\alpha_x^n)$$

for each $n \in \mathbb{N}_0$. It follows

$$|m_x(\alpha_x)| \;=\; |m_x(\alpha_x^n)| \;\leq M \sup_{k>n} |R_k(x)|.$$

Since $R_n(x) \to 0$ we infer that $m_x(\alpha_x) = 0$ in contradiction to $m_x(\alpha_x) = 1$. ◇

In the proof of the above theorem we show that $\pi(\{x\}) = 0$ implies $m_x(\epsilon_k) = 0$ for each $k \in \mathbb{N}_0$. This fact implies the following result.

Proposition 4.9.4 *Let $x \in D_s$ such that $\pi(\{x\}) = 0$. If there exists an α_x-invariant functional m_x on l^∞, then $m_x|c_0 = 0$ where c_0 is the subspace of l^∞ consisting of sequences tending to zero.*

By Proposition 4.9.4 we obtain another result about the nonexistence of α_x-invariant functionals on l^∞.

Corollary 4.9.5 *Assume that π is a continuous measure, i.e. $\pi(\{x\}) = 0$ for every $x \in D_s$. If $\sum_{n=0}^{\infty} \frac{1}{h(n)}$ is convergent, then an α_x-invariant functional on l^∞ does not exist for π-almost every $x \in D_s$. In particular, $P_1(x, M)$ is not satisfied for π-almost every $x \in D_s$.*

Proof. We have

$$\int_{D_s} \sum_{n=0}^{\infty} R_n^2(x)\, d\pi(x) = \sum_{n=0}^{\infty} \frac{1}{h(n)} < \infty.$$

Hence $\sum_{n=0}^{\infty} R_n^2(x)$ is convergent π-almost everywhere. Thus $R_n(x) \to 0$ π-almost everywhere, and by Theorem 4.9.3 the statement follows. ◇

A useful tool to study the asymptotic behaviour of certain summation processes is the function

$$\mu(y) := \limsup_{n\to\infty} \frac{H(n,y)}{H(n)} \equiv \lim_{m\to\infty} \frac{1}{H(n_m)} \sum_{k=0}^{n_m} R_k^2(y)\, h(k) \qquad \text{for } y \in D_s$$

and the set

$$\mathcal{T} := \{y \in D_s : \mu(y) > 0\}.$$

($(n_m)_{m\in\mathbb{N}}$ is a subsequence in $\mathbb{N}_0$ depending on y.) We call μ the **structural function** of D_s. We will use μ also in subsection 4.10.

Proposition 4.9.6 *Suppose that $x \in \mathcal{T}$ with*

$$\lim_{m\to\infty} \frac{1}{H(n_m)} \sum_{k=0}^{n_m} R_k^2(x)\, h(k) = \mu(x) > 0,$$

and assume that $\{\frac{H(n_m)}{a_{n_m} h(n_m)} : m \in \mathbb{N}\}$ is unbounded. Then $P_1(x, M)$ is satisfied.

Proof. Since $x \in \mathcal{T}$ we have

$$\frac{1}{H(n_m,x)} \sum_{k=0}^{n_m} |R_k(x)|\, h(k) \leq \frac{H(n_m)}{H(n_m,x)} \to \frac{1}{\mu(x)} \qquad \text{as } m \to \infty.$$

Hence there exists $M \geq 1$ such that $\frac{H(n_m)}{H(n_m,x)} \leq M$ for all $m \in \mathbb{N}$. In particular,

$$\frac{1}{H(n_m,x)} \sum_{k=0}^{n_m} |R_k(x)|\, h(k) \;\leq\; M \qquad \text{for each } m \in \mathbb{N}_0.$$

Since $\frac{a_{n_m} h(n_m)}{H(n_m,x)} \leq M \frac{a_{n_m} h(n_m)}{H(n_m)}$ and $\{\frac{H(n_m)}{a_{n_m} h(n_m)} : m \in \mathbb{N}\}$ is unbounded, it follows

$$\lim_{m\to\infty} \frac{a_{n_m} h(n_m)}{H(n_m,x)} \left(|R_{n_m}(x)| + |R_{n_m+1}(x)| \right) \;=\; 0.$$

By Theorem 4.9.2 condition $P_1(x,M)$ is satisfied. ◇

Obviously, $1 \in \mathcal{T}$ is valid. Moreover, if π is symmetric (i.e. the recurrence coefficients b_n are zero), we know that $R_n(-1) = (-1)^n$, and so $-1 \in \mathcal{T}$.

The next theorem, based strongly on a result of Nevai, allows to characterize the subset $\mathcal{T}$, provided π satisfies certain conditions.

Theorem 4.9.7 *Suppose that $(R_n(x))_{n\in\mathbb{N}_0}$ generates a polynomial hypergroup on $\mathbb{N}_0$ and that $D_s = [-1,1]$. Further assume that the Radon-Nikodym derivative π' of π satisfies Szegö's condition*

$$\int_{-1}^{1} \frac{\ln(\pi'(x))}{\sqrt{1-x^2}}\, dx \;>\; -\infty.$$

Let $x \in]-1,1[$. If the measure π is absolutely continuous in a neighbourhood of x, π' is continuous in x and $\pi'(x) > 0$, then

(i) $x \in \mathcal{T}$, provided $\limsup\limits_{n\to\infty} \frac{n+1}{H(n)} > 0$,

(ii) $x \notin \mathcal{T}$, provided $\lim\limits_{n\to\infty} \frac{n+1}{H(n)} = 0$.

Proof. Theorem 4.5.2 of [510] (see also [509, 6.2., Theorem 35]) yields for those $x \in]-1,1[$, fulfilling the assumptions above,

$$\lim_{n\to\infty} \frac{n+1}{H(n,x)} \;=\; C\,\pi'(x)\,\sqrt{1-x^2} \;>\; 0.$$

Hence, if $\lim\limits_{m\to\infty} \frac{n_m+1}{H(n_m)} > 0$, we get $\lim\limits_{m\to\infty} \frac{H(n_m,x)}{H(n_m)} > 0$, i.e. $x \in \mathcal{T}$. Moreover, if $\lim\limits_{n\to\infty} \frac{n+1}{H(n)} = 0$, we have $\lim\limits_{m\to\infty} \frac{H(n_m,x)}{H(n_m)} = 0$ for each subsequence $(n_m)_{m\in\mathbb{N}}$, i.e. $x \notin \mathcal{T}$. ◇

Corollary 4.9.8 *Suppose that $(R_n(x))_{n\in\mathbb{N}_0}$ generates a polynomial hypergroup on $\mathbb{N}_0$ so that $D_s = [-1,1]$ and $d\pi(x) = w(x)dx$, $w(x)$ a positive continuous function on $]-1,1[$. If $h(n) \leq C$ for all $n \in \mathbb{N}_0$, then $]-1,1[\subseteq \mathcal{T}$. In particular, $P_1(x,M)$ is satisfied for each $x \in]-1,1]$.*

Proof. Using the boundedness of $h(n)$, Theorem 2(6) of [472] yields that $(w(x)\sqrt{1-x^2})^{-1}$ is essentially bounded. That means, π' satisfies Szegö's condition. Since $\frac{n+1}{H(n)} \geq \frac{1}{C}$ for each $n \in \mathbb{N}_0$, Theorem 4.9.7 yields that $]-1,1[\subseteq \mathcal{T}$, and hence $]-1,1] \subseteq \mathcal{T}$. Moreover, $\frac{H(n)}{a_n h(n)} \geq \frac{n+1}{C}$. By Proposition 4.9.6 it follows that $P_1(x,M)$ is satisfied for each $x \in]-1,1]$. $\diamond$

By Corollary 4.9.8 we obtain that for the Chebyshev polynomials of the first kind $(T_n(x))_{n\in\mathbb{N}_0}$ Reiter's condition $P_1(x,M)$ is satisfied for each $x \in [-1,1] = D_s$.

Whereas $P_1(x,M)$ is satisfied for all $x \in D_s = [-1,1]$ if $R_n^{(-\frac{1}{2},-\frac{1}{2})}(x) = T_n(x)$, $P_1(x,M)$ is not satisfied for all $x \in]-1,1[$, when the polynomial hypergroup is generated by the Jacobi polynomials $R_n^{(\alpha,\beta)}(x)$, $\alpha \geq \beta > -1$ and $\alpha+\beta+1 \geq 0$ (but not $\alpha = -\frac{1}{2} = \beta$). In fact by [641, (4.1.1) and (8.21.18)] we know that $|R_n^{(\alpha,\beta)}(x)| = O(n^{-\alpha-\frac{1}{2}})$ for each $x \in]-1,1[$. Hence $R_n^{(\alpha,\beta)}(x) \to 0$ for each $x \in]-1,1[$ provided $\alpha > -\frac{1}{2}$. By Theorem 4.9.3 condition $P_1(x,M)$ is not satisfied for each $x \in]-1,1[$. In case of $\alpha = \beta > -\frac{1}{2}$, condition $P_1(-1,M)$ is satisfied, since (i) and (ii) of Theorem 4.9.2 hold for $z=-1$. Finally, if $\alpha > \beta$ we have $R_n(-1) \to 0$ as $n \to \infty$, see [641, (4.11) and (4.14)]. Hence $P(-1,M)$ is not fulfilled for $x=-1$, whenever $\alpha > \beta$. For $x=1$ Reiter's condition $P_1(1,M)$ is always satisfied. So we have a complete description for those $x \in [-1,1]$ enjoying or not enjoying $P_1(x,M)$ for Jacobi polynomials $R_n^{(\alpha,\beta)}(x)$, $\alpha \geq \beta > 1$ and $\alpha+\beta+1 \geq 0$.

To have an example for which $h(n) \to \infty$ and $P_1(x,M)$ is valid for an interior point $x \in D_s$, we consider the generalized Chebyshev polynomials $T_n^{(\alpha,\beta)}(x)$, $\alpha \geq \beta > -1$, $\alpha+\beta+1 \geq 0$, see subsection 4.5.3, and select $x = 0 \in D_s$. For symmetric polynomials $R_n(x)$ the recurrence relations yield $R_{2k+1}(0) = 0$ and

$$R_{2k}(0) = (-1)^k \frac{c_1}{a_1}\frac{c_3}{a_3}\cdots\frac{c_{2k-1}}{a_{2k-1}} \qquad \text{for } k = 0,1,\ldots.$$

For the generalized Chebyshev polynomials $T_n^{(\alpha,\beta)}(x)$ it follows $T_{2k}^{(\alpha,\beta)}(0) = \frac{(\beta+1)_k}{(\alpha+1)_k}$. To keep the calculation simple we restrict to the case $\alpha=\beta=0$, in particular $T_{2k}^{(0,0)}(0) = 1$. By the identities (iii) of subsection 4.5.3 we obtain for the corresponding Haar weights $h(n) = n+1$ and $H(n) = \frac{(n+1)(n+2)}{2}$. It follows

$$\sum_{j=0}^{2k} (T_j^{(0,0)}(0))^2\, h(j) = \sum_{l=0}^{k} h(2l) = (k+1)^2$$

and

$$\frac{1}{H(2k)} \sum_{j=0}^{2k} (T_j^{(0,0)}(0))^2 \, h(j) \;=\; \frac{k+1}{2k+1}.$$

Moreover,

$$\frac{1}{H(2k+1)} \sum_{j=0}^{2k+1} (T_j^{0,0)}(0))^2 \, h(j) \;=\; \frac{k+1}{2k+3}.$$

Hence for the structural function μ we have

$$\mu(0) \;=\; \lim_{n\to\infty} \frac{H(n,0)}{H(n)} \;=\; \frac{1}{2}.$$

In addition $\frac{H(n)}{a_n h(n)} = n+1$ for n even and $\frac{H(n)}{a_n h(n)} = n+2$ for n odd. By Proposition 4.9.6 the Reiter condition $P_1(0, M)$ is satisfied.

10 Positive definite sequences and mean ergodic theorems

In subsection 4.1 we have already met positive definite sequences. By Bochner's theorem a bounded sequence $(\varphi(n))_{n\in\mathbb{N}_0}$ is positive definite (with respect to $(R_n(x))_{n\in\mathbb{N}_0}$) if and only if there is a unique bounded positive Borel measure $a \in M(D_s)$ such that $\check{a} = \varphi$. By Proposition 3.7.2 an $l^1(h)$-convergent sequence $(\varphi(n))_{n\in\mathbb{N}_0} \in l^1(h)$ is positive definite if and only if $\hat{\varphi}(x) = \sum_{k=0}^{\infty} \varphi(k)\, R_k(x)\, h(k) \geq 0$ for all $x \in S$.

We add some further results to these characterizations of positive definite sequences.

Proposition 4.10.1 *Suppose that $(R_n(x))_{n\in\mathbb{N}_0}$ generates a polynomial hypergroup on $\mathbb{N}_0$. Let $(\varphi(n))_{n\in\mathbb{N}_0}$ be a bounded positive definite sequence, and assume that $\sum_{k=0}^{p} c_k R_k(x)$, $x \in D_s$, is a non-negative polynomial of arbitrary degree p. Then the*

$$\psi(n) \;=\; \sum_{k=0}^{p} L_n\, \varphi(k)\, c_k$$

form a bounded positive definite sequence $(\psi(n))_{n\in\mathbb{N}_0}$.

Proof. Let $\varphi = \check{a}$, $a \in M(D_s)$ positive. Then

$$L_n\, \varphi(k) \;=\; \int_{D_s} R_n(x)\, R_k(x)\, da(x).$$

Setting $f(x) = \sum_{k=0}^{p} c_k R_k(x)$,

$$(fa)^\vee(n) = \int_{D_s} R_n(x)\, f(x)\, da(x) = \sum_{k=0}^{p} c_k \int_{D_s} R_k(x)\, R_n(x)\, da(x)$$

$$= \sum_{k=0}^{p} L_n\, \varphi(k)\, c_k = \psi(n).$$

Since fa is a positive measure, $(\psi(n))_{n\in\mathbb{N}_0}$ is positive definite. ◇

Theorem 4.10.2 *Suppose that $(R_n(x))_{n\in\mathbb{N}_0}$ generates a polynomial hypergroup on $\mathbb{N}_0$, and let $(\varphi(n))_{n\in\mathbb{N}_0}$ be a bounded sequence. Then $(\varphi(n))_{n\in\mathbb{N}_0}$ is positive definite if and only if $\sum_{k=0}^{p} c_k\varphi(k) \geq 0$ for all non-negative polynomials $\sum_{k=0}^{p} c_k R_k(x) \geq 0,\ x \in D_s$ of arbitrary degree p.*

Proof. Assume that $(\varphi(n))_{n\in\mathbb{N}_0}$ is positive definite. By Proposition 4.10.1 $(\psi(n))_{n\in\mathbb{N}_0}$ is positive definite, where $\psi(n) = \sum_{k=0}^{p} L_n\varphi(k)\, c_k$. In particular, $0 \leq \psi(0) = \sum_{k=0}^{p} c_k\varphi(k)$.

Conversely, let $\lambda_0, ..., \lambda_n \in \mathbb{C}$. Consider the non-negative polynomial

$$\left|\sum_{k=0}^{n} \lambda_k\, R_k(x)\right|^2 = \sum_{i,j=0}^{n} \lambda_i\, \overline{\lambda}_j\, R_i(x)\, R_j(x)$$

$$= \sum_{i,j=0}^{n} \lambda_i\, \overline{\lambda}_j \sum_{l=|i-j|}^{i+j} g(i,j;l)\, R_l(x).$$

By the assumption follows

$$0 \leq \sum_{i,j=0}^{n} \lambda_i\, \overline{\lambda}_j \sum_{l=|i-j|}^{i+j} g(i,j;l)\, \varphi(l) = \sum_{i,j=0}^{n} \lambda_i\, \overline{\lambda}_j\, L_i\, \varphi(j),$$

and so $(\varphi(n))_{n\in\mathbb{N}_0}$ is positive definite. ◇

Remark: This characterization of positive definiteness on $K = \mathbb{N}_0$ is a "predual" version of the positive definiteness given in section 3.11.3.

Let $(\varphi(n))_{n\in\mathbb{N}_0}$ be a bounded positive definite sequence with representing bounded positive Borel measure $a \in M(D_s)$ such that $\varphi = \check{a}$. Consider the Hilbert space $L^2(D_s, a)$, and denote the scalar product in

$L^2(D_s, a)$ by $\langle \,.\,, \,.\, \rangle_a$ and the norm by $\| \,.\, \|_{2,a}$. Obviously, $R_n \in L^2(D_s, a)$ and $\langle R_n, R_k \rangle_a = L_n\varphi(k)$. We come now to a mean ergodic theorem for positive definite sequences. Prior to deriving these results we formulate the so-called Christoffel-Darboux formula, see [641].

Proposition 4.10.3 *Let $(R_n(x))_{n\in\mathbb{N}_0}$ be an orthogonal polynomial sequence generating a polynomial hypergroup on $\mathbb{N}_0$. The following identities hold for $n \in \mathbb{N}$:*

(1)

$$\sum_{k=0}^{n} R_k(x)\, R_k(y)\, h(k) \;=\; a_0\, a_n\, h(n)\, \frac{R_{n+1}(x)R_n(y) - R_n(x)R_{n+1}(y)}{x-y}$$

for $x, y \in \mathbb{R}$, $x \neq y$.

(2)

$$\sum_{k=0}^{n} R_k^2(x)\, h(k) \;=\; a_0\, a_n\, h(n)\, \left[R'_{n+1}(x)\, R_n(x) \;-\; R_{n+1}(x)\, R'_n(x)\right]$$

for all $x \in \mathbb{R}$.

(3)

$$\left(\sum_{k=0}^{n} h(k)\right) / h(n) \;=\; a_0\, a_n\, (R'_{n+1}(1) \;-\; R'_n(1)).$$

Proof. (3) follows from (2), and (2) follows from (1). But (1) is the Christoffel-Darboux formula for that normalization we use, i.e. $R_n(1) = 1$. $\diamond$

As in the previous section we shortly write $H(n) = \sum_{k=0}^{n} h(k)$.

Consider the mean functions defined by

$$D_n(x) := \frac{1}{H(n)} \sum_{k=0}^{n} R_k(x)\, h(k), \qquad x \in D_s.$$

$(D_n(x))_{n\in\mathbb{N}_0}$ is the normalized Dirichlet kernel with respect to $(R_n(x))_{n\in\mathbb{N}_0}$.

By Proposition 4.10.3(1) we have for $x \in D_s$, $x \neq 1$, $n \in \mathbb{N}_0$,

$$D_n(x) \;=\; \frac{1}{H(n)} \sum_{k=0}^{n} R_k(x)\, h(k) \;=\; \frac{h(n)}{H(n)}\, a_0\, a_n\, \frac{R_{n+1}(x) - R_n(x)}{x-1}.$$

Now we can prove a mean ergodic theorem.

Theorem 4.10.4 *Let $(R_n(x))_{n\in\mathbb{N}_0}$ generate a polynomial hypergroup on $\mathbb{N}_0$, and let $\varphi = (\varphi(n))_{n\in\mathbb{N}_0}$ be a bounded positive definite sequence, and $\varphi = \check{a},\ \ a \in M(D_s)$ positive.*

If $\frac{a_n h(n)}{H(n)}$ converges to zero, or if $R_n(x) \to 0$ for all $x \in D_s\backslash\{1\}$, then D_n converges towards $\chi_{\{1\}}$ in $L^2(D_s, a)$.

Proof. For $x \in D_s\backslash\{1\}$ we have

$$|D_n(x)| \ \le\ \frac{h(n)a_n a_0\left(|R_{n+1}(x)| + |R_n(x)|\right)}{H(n)\ |x-1|}.$$

Hence, by the assumptions the mean functions $D_n(x)$ converge pointwise to $\chi_{\{1\}}(x)$ for all $x \in D_s$. By Lebesgue's theorem of dominated convergence it follows

$$\lim_{n\to\infty} \|D_n - \chi_{\{1\}}\|_{2,a} \ =\ 0.$$

◇

Remark:

(1) Assuming in Theorem 4.10.4 that a subsequence $(\frac{a_{n_k} h(n_k)}{H(n_k)})_{k\in\mathbb{N}}$ converges to zero, we obtain that

$$\lim_{k\to\infty} \|D_{n_k} - \chi_{\{1\}}\|_{2,a} \ =\ 0.$$

(2) There are some examples where

$$\frac{h(n)}{H(n)} \longrightarrow \kappa > 0, \qquad\quad a_n \to \alpha > 0$$

and $b_n = 0$ for all $n \in \mathbb{N}_0$. Then $R_n(-1) = (-1)^n$ and $D_{2n}(-1) \to \kappa a_0 \alpha$ and $D_{2n-1}(-1) \to -\kappa a_0 \alpha$. Hence $D_n(x)$ does not converge in $x = -1$. But if we suppose that $R_n(x) \to 0$ for all $x \in D_s\backslash\{-1, 1\} =]-1, 1[$, and assume that the positive measure $a \in M(D_s)$ has no point mass in $x = -1$, i.e. $a(\{-1\}) = 0$, then the proof of Theorem 4.10.4 reveals that $\lim\limits_{n\to\infty} \|D_n - \chi_{\{1\}}\|_{2,a} = 0$ holds true.

From the mean ergodic theorem we obtain an interesting result on bounded positive definite sequences.

Corollary 4.10.5 *Suppose $(R_n(x))_{n\in\mathbb{N}_0}$ generates a polynomial hypergroup on $\mathbb{N}_0$, and assume that it satisfies one of the conditions of Theorem 4.10.4. For any bounded positive definite sequence $(\varphi(n))_{n\in\mathbb{N}_0}$, the limits*

(1)

$$\lim_{n\to\infty} \frac{1}{(H(n))^2} \sum_{i,j=0}^{n} L_i\varphi(j)\, h(i)\, h(j)$$

(2)

$$\lim_{n\to\infty} \frac{1}{H(n)} \sum_{k=0}^{n} L_m\varphi(k)\, h(k), \qquad m \in \mathbb{N}_0,$$

exist and are equal to $a(\{1\})$*, where* a *is the bounded positive Borel measure* $a \in M(D_s)$ *with* $\check{a} = \varphi$.

Proof.

(1) Obviously,

$$\| \frac{1}{H(n)} \sum_{k=0}^{n} R_k\, h(k) \|_{2,a}^2 = \frac{1}{(H(n))^2} \sum_{i,j=0}^{n} L_i\varphi(j)\, h(i)\, h(j).$$

Then the existence of the limit in (1) follows directly from Theorem 4.10.4. Moreover, we see that the limit is equal to $\|\chi_{\{1\}}\|_{2,a}^2 = a(\{1\})$.

(2) Since

$$\langle \frac{1}{H(n)} \sum_{k=0}^{n} R_k\, h(k), R_m\rangle_a = \frac{1}{H(n)} \sum_{k=0}^{n} L_m\, \varphi(k)\, h(k),$$

the limit of (2) exists for any $m \in \mathbb{N}_0$, and is equal to $\langle \chi_{\{1\}}, R_m\rangle_a = a(\{1\})$.

$\diamond$

We mention an additional fact concerning the above limits.

Corollary 4.10.6 *Let* $(R_n(x))_{n\in\mathbb{N}_0}$ *generate a polynomial hypergroup on* $\mathbb{N}_0$*, and assume it satisfies one of the two conditions of Theorem 4.10.4. For any bounded positive definite sequence* $(\varphi(n))_{n\in\mathbb{N}_0}$*, the limits of (1) and (2) in Corollary 4.10.5 are equal to*

$$\inf \{ \sum_{i,j=1}^{n} c_i\, \bar{c}_j\, L_{m_i}\varphi(m_j) : n \in \mathbb{N};\ c_1, ..., c_n \in \mathbb{C},\ c_1 + ... + c_n = 1; \\ m_1, ..., m_n \in \mathbb{N}_0\}$$

Proof. Let $c_1 + ... + c_n = 1$. Then for any $x \in D_s$

$$\chi_{\{1\}}(x) \le |c_1\, R_{m_1}(x) + ... + c_n\, R_{m_n}(x)|^2,$$

and so

$$a(\{1\}) = \|\chi_1\|_{2,a}^2 \le \|c_1\, R_{m_1} + ... + c_n\, R_{m_n}\|_{2,a}^2 \\ = \sum_{i,j=1}^{n} c_i\, \bar{c}_j\, L_{m_i}\varphi(m_j).$$

By Corollary 4.10.5 it follows that $a(\{1\})$ is equal to the greatest lower bound of $\sum_{i,j=1}^{n} c_1 \bar{c}_j L_{m_i}\varphi(m_j)$, wherein $c_1 + ... + c_n = 1$. ◇

We have already mentioned in section 2.4 that the product of two positive definite functions is in general not positive definite. Nevertheless, we can get further information on a bounded positive definite sequence $(\varphi(n))_{n\in\mathbb{N}_0}$ by studying $(R_n(y)\varphi_n)_{n\in\mathbb{N}_0}$, $y \in D_s$, which is no longer positive definite, in general. For a fixed $y \in D_s$, $y \neq 1$, consider

$$D_{n,y}(x) = \frac{1}{H(n)} \sum_{k=0}^{n} R_k(y)\, R_k(x)\, h(k), \qquad x \in D_s.$$

If for example $(R_n(x))_{n\in\mathbb{N}_0}$ satisfies condition (H), then $D_{n,y}(x) \to 0$ as $n \to \infty$, for all $x \in D_s$, $x \neq y$; compare Proposition 4.10.3(1).

The case $x = y$ is more delicate. Of course, since $(D_{n,y}(y))_{n\in\mathbb{N}_0}$ is a bounded sequence, there exists a convergent subsequence. However, in order to apply the mean ergodic theorems below we have to study the convergence of $(D_{n,y}(y))$ and, if possible, determine the value of the limit. We begin with an easy example.

Example. Let $R_n(x) = T_n(x)$ be the Chebyshev polynomials of the first kind, see subsection 1.2.1. The Haar weights are $h(0) = 1$ and $h(n) = 2$ for $n \in \mathbb{N}$, and hence $H(n) = 2n+1$ for $n \in \mathbb{N}$. Since $T_n(x) = \cos(nt)$ where $x = \cos t$, we have $T_n'(x) = n\,\frac{\sin(nt)}{\sin t}$.

Let $x \in]-1,1[$. Applying formula (2) of Proposition 4.10.3 we get

$$D_{n,x}(x) = \frac{1}{2n+1}\left[\frac{(n+1)\sin((n+1)t)}{\sin t}\cos(nt) - \frac{n\sin(nt)}{\sin t}\cos((n+1)t)\right]$$

$$= \frac{1}{2n+1}\left[n + \frac{\sin((n+1)t)}{\sin t}\cos(nt)\right].$$

since $t \in]0,\pi[$ we know that $\left(\frac{\sin((n+1)t)}{(2n+1)\sin t}\right)_{n\in\mathbb{N}_0}$ converges towards zero, and we get $\lim_{n\to\infty} D_{n,x}(x) = \frac{1}{2}$. Obviously, if $x = 1$ or $x = -1$ then $\lim_{n\to\infty} D_{n,x}(x) = 1$.

The structural function

$$\mu(y) = \limsup_{n\to\infty} D_{n,y}(y) = \limsup_{n\to\infty} \frac{H(n,y)}{H(n)}$$

and the set $\mathcal{T} = \{y \in D_s : \mu(y) > 0\}$ are already introduced in subsection 4.9.

Since $0 \leq D_{n,y}(y) \leq 1$ we have two cases:

(1) If $y \notin \mathcal{T}$, then $\lim\limits_{n\to\infty} D_{n,y}(y) \;=\; \mu(y) = 0$.
(2) If $y \in \mathcal{T}$, then $\limsup\limits_{n\to\infty} D_{n,y}(y) \;=\; \lim\limits_{m\to\infty} D_{n_m,y}(y) \;=\; \mu(y) > 0$.

Theorem 4.10.7 *Assume that $(R_n(x))_{n\in\mathbb{N}_0}$ generates a polynomial hypergroup on $\mathbb{N}_0$ and $\frac{a_n h(n)}{H(n)}$ converges to zero. Let $y \in D_s\backslash\{1\}$, and let $\varphi = (\varphi(n))_{n\in\mathbb{N}_0}$ be a bounded positive definite sequence and $\varphi = \check{a}$, $a \in M(D_s)$ positive. if*

(1) $\mu(y) = 0$, *then* $\lim\limits_{n\to\infty} \|D_{n,y}\|_{2,a} \;=\; 0$,
(2) $\mu(y) > 0$, *then* $\limsup\limits_{n\to\infty} \|D_{n,y} - \mu(y)\,\chi_{\{y\}}\|_{2,a} \;=\; 0$.

Proof. We know by the assumption and (1) of Proposition 4.10.3, that $D_{n,y}(x)$ converges pointwise to zero for all $x \in D_s$, if $y \notin \mathcal{T}$, and that $D_{n_m,y}(x)$ converges pointwise to $\mu(y)\,\chi_{\{y\}}(x)$ for all $x \in D_s$, if $y \in \mathcal{T}$, where $(n_m)_{m\in\mathbb{N}}$ is a subsequence in $\mathbb{N}_0$ depending on y. By the dominated convergence theorem the statements (1) and (2) follow. ◇

Corollary 4.10.8 *Assume that $(R_n(x))_{n\in\mathbb{N}_0}$ generates a polynomial hypergroup on $\mathbb{N}_0$ and $\frac{a_n h(n)}{H(n)}$ converges to zero. Let $y \in D_s\backslash\{1\}$, and let $\varphi = (\varphi(n))_{n\in\mathbb{N}_0}$ be a bounded positive definite sequence and $\varphi = \check{a}$, $a \in M(D_s)$ positive. If*

(1) $\mu(y) = 0$, *then*

$$\lim_{n\to\infty} \frac{1}{H(n)^2} \sum_{i,j=0}^{n} R_i(y)\,R_j(y)\;L_i\varphi(j)\,h(i)\,h(j) \;=\; 0$$

and

$$\lim_{n\to\infty} \frac{1}{H(n)} \sum_{k=0}^{n} R_k(y)\;L_l\varphi(k)\;h(k) \;=\; 0 \qquad \textit{for each } l \in \mathbb{N}_0,$$

(2) $\mu(y) > 0$, *then*

$$\lim_{m\to\infty} \frac{1}{H(n_m)^2} \sum_{i,j=0}^{n_m} R_i(y)\,R_j(y)\;L_i\varphi(j)\,h(j) \;=\; \mu(y)^2\;a(\{y\})$$

and

$$\lim_{m\to\infty} \frac{1}{H(n_m)} \sum_{k=0}^{n_m} R_k(y)\;L_l\varphi(k)\,h(k) \;=\; \mu(y)\,R_l(y)\;a(\{y\}), \qquad l \in \mathbb{N}_0,$$

where

$$\mu(y) \;=\; \lim_{m\to\infty} D_{n_m,y}(y) \;=\; \limsup_{n\to\infty} D_{n,y}(y).$$

Proof. Obviously,

$$\|D_{n,y}\|_{2,a}^2 = \frac{1}{H(n)^2} \sum_{i,j=0}^{n} R_i(y)\, R_j(y)\, L_i\varphi(j)\, h(i)\, h(j),$$

and

$$\langle D_{n,y}, R_l \rangle_a = \frac{1}{H(n)} \sum_{k=0}^{n} R_k(y)\, L_l\varphi(k)\, h(k).$$

By Theorem 4.10.7 the statements follow. ◇

The results above suggest that we can find the discrete part of the positive measure $a \in M(D_s)$, where $\check{a} = \varphi$, φ a bounded positive definite sequence. However, we have to assume that the sequences $(D_{n,y}(y))_{n\in\mathbb{N}_0}$ converge for each $y \in D_s$. That means $\limsup_{n\to\infty} D_{n,y}(y) = \lim_{n\to\infty} D_{n,y}(y)$ for every $y \in D_s$.

Corollary 4.10.9 *Suppose that $(R_n(x))_{n\in\mathbb{N}_0}$ generates a polynomial hypergroup on $\mathbb{N}_0$ and $\frac{a_n h(n)}{H(n)}$ converges to zero. Assume that $(D_{n,y}(y))_{n\in\mathbb{N}_0}$ converges for every $y \in D_s$. Let $(\varphi(n))_{n\in\mathbb{N}_0}$ be a bounded positive definite sequence.*

(1) If $y_1, y_2 \in D_s$, $y_1 \neq y_2$, then

$$\lim_{n\to\infty} \frac{1}{H(n)^2} \sum_{i,j=0}^{n} R_i(y_1)\, R_j(y_2)\, L_i\varphi(j)\, h(i)\, h(j) = 0.$$

(2) There exists at most a countable infinite set of values of $y \in D_s$ for which

$$\Phi(y) := \lim_{n\to\infty} \frac{1}{H(n)} \sum_{k=0}^{n} R_k(y)\, \varphi(k)\, h(k)$$

differs from zero.

Proof.

(1) By Theorem 4.10.7 we know that D_{n,y_1} and D_{n,y_2} converge in $L^2(D_s, a)$ to $\mu(y_1)\chi_{\{y_1\}}$ and $\mu(y_2)\chi_{\{y_2\}}$, respectively. But $\chi_{\{y_1\}}$ and $\chi_{\{y_2\}}$ are orthogonal in $L^2(D_s, a)$, and hence statement (1) is shown.
(2) Since for any $a \in M(D_s)$ there exist at most countable infinite $y \in D_s$ with $a(\{y\}) \neq 0$.

◇

We call

$$\Phi(y) = \lim_{n\to\infty} \frac{1}{H(n)} \sum_{k=0}^{n} R_k(y)\,\varphi(k)\,h(k)$$

the **mean value function** on D_s of $\varphi = (\varphi(n))_{n\in\mathbb{N}_0}$, provided the limits exist.

Theorem 4.10.10 *Assume $(R_n(x))_{n\in\mathbb{N}_0}$ generates a polynomial hypergroup on $\mathbb{N}_0$ and $\frac{a_n h(n)}{H(n)}$ converges to zero and for $y \in D_s\backslash\{1\}$ assume that $(D_{n,y}(y))_{n\in\mathbb{N}_0}$ converges. Let $(\varphi(n))_{n\in\mathbb{N}_0}$ be a bounded positive definite sequence and let Φ be its mean value function. Suppose that $\{y \in D_s : \Phi(y) \neq 0\} \subseteq \mathcal{T}$ and denote by $y_1, y_2, \ldots$ these (at most countable) values with $\Phi(y) \neq 0$. Put*

$$\psi(n) := \sum_{i=1}^{\infty} R_n(y_i)\,\Phi(y_i)\,\frac{1}{\mu(y_i)}.$$

Then

(1) the series $\sum_{i=1}^{\infty} \Phi(y_i)\,\frac{1}{\mu(y_i)}$ is convergent and $\sum_{i=1}^{\infty} \Phi(y_i)\,\frac{1}{\mu(y_i)} \leq \varphi(0)$.
The series $\sum_{i=1}^{\infty} R_n(y_i)\,\Phi(y_i)\,\frac{1}{\mu(y_i)}$ converges uniformly in $n \in \mathbb{N}_0$,

(2) the sequence $(\psi(n))_{n\in\mathbb{N}_0}$ is bounded positive definite.

Proof. Let $a \in M(D_s)$ such that $\check{a} = \varphi$, and write $a = a_c + a_d$, where a_d is the discrete part of a. Then

$$\sum_{i=1}^{\infty} \Phi(y_i)\,\frac{1}{\mu(y_i)} = \sum_{i=1}^{\infty} a(\{y_i\}) = \|a_d\| \leq \|a\| = \varphi(0),$$

and the statements of (1) follow easily. Also assertion (2) is obvious, since $\psi(n) = \check{a}_d(n)$, and a_d is a positive Borel measure on D_s. ◇

For a detailed analysis of the discrete part of any measure $a \in M(D_s)$ we refer to subsection 4.12.

Let us continue with a brief list of positive definite sequences.

(1) Choose some $n \in \mathbb{N}_0$ and consider the positive measure $a \in M(D_s)$ given by $da(x) = (R_n(x))^2 d\pi(x)$. Since

$$(R_n(x))^2 = \sum_{k=0}^{2n} g(n,n;k)\,R_k(x),$$

we get

$$\int_{D_s} R_m(x)\, da(x) = \sum_{k=0}^{2n} g(n,n;k) \int_{D_s} R_k(x)\, R_m(x)\, d\pi(x)$$
$$= g(n,n;k)\, h(k)^{-1}$$

for $m = 0, ..., 2n$, and

$$\int_{D_s} R_m(x)\, da(x) \;=\; 0 \qquad \text{for } m = 2n+1, ...$$

Putting $\varphi(k) = g(n,n;k)h(k)^{-1}$ for $k = 0, ..., 2n$ and $\varphi(k) = 0$ for $k = 2n+1, ...$, the sequence $(\varphi(k))_{k\in\mathbb{N}_0}$ is positive definite.

(2) Fix again $n \in \mathbb{N}_0$, and consider the positive measure $a \in M(D_s)$ given by $da(x) = (\sum_{k=0}^{n} R_k(x)h(k))^2 d\pi(x)$. Denote by χ_n the sequence defined by $\chi_n(k) = 1$ for $k = 0, ..., n$ and $\chi_n(k) = 0$ for $k = n+1, ...$. Then

$$\left(\sum_{k=0}^{n} R_k(x)\, h(k)\right)^2 \;=\; (\hat{\chi}_n(x))^2 \;=\; (\chi_n * \chi_n)^\wedge(x),$$

and hence

$$\int_{D_s} R_m(x)\, da(x) \;=\; (\chi_n * \chi_n)^{\wedge\vee}(m) \;=\; \chi_n * \chi_n(m).$$

Putting $\varphi(k) = \chi_n * \chi_n(k)$ for $k = 0, ..., 2n$ and $\varphi(k) = 0$ for $k = 2n+1, ...$, the sequence $(\varphi(k))_{k\in\mathbb{N}_0}$ is positive definite. This sequence is used to construct the Fejér approximation process, see subsection 4.14. It is a particular case of positive definite sequences of the form $a * a^*,\ a \in l^2(h)$.

(3) Let $\alpha > -\frac{1}{2}$, and consider the polynomial hypergroup generated by the ultraspherical polynomials $R_n^{(\alpha,\alpha)}(x)$. In Proposition 4.6.3 we have shown that $(\varphi(k))_{k\in\mathbb{N}_0}$ is positive definite, where $\varphi(k) = \frac{2\alpha+1}{2k+2\alpha+1}$. Appealing to Proposition 4.6.4 we obtain that $(\varphi(k))_{k\in\mathbb{N}_0}$ is positive definite, wherein $\varphi(k) = \frac{2\alpha+1}{2k+2\alpha+1}\, x^k,\ \ x \in [-1,1]$. Here we choose in Proposition 4.6.4 for $b \in M([-1,1])$ the point measure in $x \in [-1,1]$. Selecting for b the Lebesgue measure restricted to $[0,1]$, we see that $\left(\frac{2\alpha+1}{(2k+2\alpha+1)(k+1)}\right)_{k\in\mathbb{N}_0}$ is positive definite.

(4) Let $\alpha \geq \beta > -1,\ \ \alpha+\beta+1 \geq 0$, and consider the polynomial hypergroup generated by the Jacobi polynomials $R_n^{(\alpha,\beta)}(x)$, compare subsection 4.5.2 below. Fix some $n \in \mathbb{N}_0$ and define by

$$a_{n,k} \;=\; \frac{n!\,\Gamma(n+\alpha+\beta+2)}{(n-k)!\,\Gamma(n+k+\alpha+\beta+2)} \qquad \text{for } k = 0, ..., n.$$

From equality (2.30) in [18] and (4.3.4) in [641] we can derive

$$A_n(x) := \sum_{k=0}^{n} a_{n,k}\, R_k(x)\, h(k)$$
$$= \frac{\Gamma(n+\alpha+\beta+2)}{\Gamma(n+\beta+2)\,\Gamma(\alpha+1)\,2^{\alpha+\beta+1}} \left(\frac{1+x}{2}\right)^n \geq 0$$

for all $x \in D_s = [-1,1]$. We have

$$\int_{-1}^{1} R_m(x)\, A_n(x)\, d\pi(x) \;=\; a_{n,m} \qquad \text{for } m = 0, ..., n.$$

Putting $\varphi(k) = a_{n,k}$ for $k = 0, ..., n$, and $\varphi(k) = 0$ for $k = n+1, ...$, we see that $(\varphi(k))_{k\in\mathbb{N}_0}$ is positive definite.

The next examples are very important. They are the exponentials of quadratic forms on the polynomial hypergroup $\mathbb{N}_0$. We call a complex-valued sequence $(q(n))_{n\in\mathbb{N}_0}$ an **additive sequence** with respect to $(R_n(x))_{n\in\mathbb{N}_0}$, if

$$L_m q(n) \;=\; q(n) + q(m)$$

for all $n, m \in \mathbb{N}_0$. (We refer to subsection 4.13.3, where quadratic forms on commutative hypergroups are studied. Also see [647,2.2].)

Theorem 4.10.11 *Let $(R_n(x))_{n\in\mathbb{N}_0}$ generate a polynomial hypergroup on $\mathbb{N}_0$. A sequence $(q(n))_{n\in\mathbb{N}_0}$ is an additive sequence with respect to $(R_n(x))_{n\in\mathbb{N}_0}$ if and only if there exists a complex number c such that*

$$q(n) \;=\; c\, R_n'(1)$$

for all $n \in \mathbb{N}_0$.

Proof. Differentiating the recurrence formulas (1) and (2) of subsection 1.2.1 and setting $x = 1$, we see that $s(n) := a_0 R_n'(1)$ satisfies

$$s(0) \;=\; 0, \qquad s(1) \;=\; 1 \qquad \text{and}$$
$$1 + s(n) \;=\; a_n s(n+1) + b_n s(n) + c_n s(n-1) \qquad \text{for } n \in \mathbb{N}.$$

Conversely, if $(q(n))_{n\in\mathbb{N}_0}$ is additive, it follows

$$q(0) \;=\; 0, \qquad c := q(1) \qquad \text{and}$$
$$c + q(n) \;=\; a_n q(n+1) + b_n q(n) + c_n q(n-1) \qquad \text{for } n \in \mathbb{N}.$$

Comparing both recursion formulas it is clear that $(q(n))_{n\in\mathbb{N}_0}$ is additive if and only if $q(n) = cR_n'(1)$. ◇

For the Jacobi polynomials $R_n^{(\alpha,\beta)}(x)$ the additive sequences are

$$q(n) \;=\; c\, \frac{n(n+1+\alpha+\beta)}{\alpha+\beta+2}, \qquad c \in \mathbb{C}.$$

This follows immediately from the recurrence relation of the $R_n^{(\alpha,\beta)}(x)$. Additive sequences with respect to other orthogonal polynomial systems can be found in subsection 4.13.3.

Theorem 4.10.12 *Let $(R_n(x))_{n\in\mathbb{N}_0}$ generate a polynomial hypergroup on $\mathbb{N}_0$. Assume that $1 \in D_s$ is not isolated in D_s, and that for $x, y \in D_s$ the product sequence $(R_n(x)R_n(y))_{n\in\mathbb{N}_0}$ is positive definite. Given a nonnegative additive sequence $(q(n))_{n\in\mathbb{N}_0}$ on $\mathbb{N}_0$ with respect to $(R_n(x))_{n\in\mathbb{N}_0}$, then the sequence $(\exp(-tq(n)))_{n\in\mathbb{N}_0}$, $t > 0$, is positive definite.*

Proof. By the assumption the sequences $((R_n(x))^m)_{n\in\mathbb{N}_0}$ are positive definite for each $x \in D_s$ and $m \in \mathbb{N}$. Using the exponential power series it follows that $(\exp(tR_n(x)))_{n\in\mathbb{N}_0}$ is positive definite for each $x \in D_s$ and $t > 0$. Hence $(\exp(-t(1 - R_n(x)))_{n\in\mathbb{N}_0}$ is positive definite, too. If $x \neq 1$ replace t by $t(1-x)^{-1}$ obtaining that $(\exp(-t(1-R_n(x))(1-x)^{-1}))_{n\in\mathbb{N}_0}$ is positive definite for each $x \in D_s$. Since 1 is not isolated in D_s, $R_n'(1)$ is the limit of $(1 - R_n(x))(1-x)^{-1}$, where x tends in D_s towards 1. By Theorem 2.4.9 the (pointwise) limit of positive definite sequences is positive definite. Theorem 4.10.11 yields that $(\exp(-tq(n)))_{n\in\mathbb{N}_0}$ is positive definite for any $t > 0$. ◇

Remark:

The assumption that the product sequences $(R_n(x)R_n(y))_{n\in\mathbb{N}_0}$ are positive definite is satisfied for Jacobi polynomials or generalized Chebyshev polynomials, see subsection 4.13.1. The assumption that 1 is not isolated in D_s holds true if supp $\pi = D_s$, as the following short argument shows: Assume that $1 \in D_s =$ supp π is an isolated point of D_s. Then $\pi(\{1\}) > 0$, and there is $f \in C(D_s)$ such that $f(1) = 1$ and $f|D_s\backslash\{1\} = 0$. This implies $\hat{f}(n) = \int_{D_s} f(x)R_n(x)\,d\pi(x) = \pi(\{1\})$ in contradiction to $\hat{f}(n) \to 0$ as $n \to \infty$.

11 Boundedness of Haar weights and the Turan determinants

We have seen that the boundedness of $\{h(n) : n \in \mathbb{N}_0\}$ allows to derive many properties of the corresponding polynomial hypergroup, see for example Theorem 4.2.4 or Corollary 4.9.8. We add some further results induced by the boundedness of the Haar weights. Furthermore we introduce the Turan determinants and conjugate orthogonal polynomials and analyze their properties.

We begin by deriving examples of polynomial hypergroups on $\mathbb{N}_0$ with bounded Haar weights $h(n)$.

Proposition 4.11.1 *Let $(R_n(x))_{n\in\mathbb{N}_0}$ be a symmetric orthogonal polynomial sequence with recurrence coefficients c_k, a_k, $k \in \mathbb{N}$, such that $0 < c_k \leq$*

$c_{k+1} \leq \frac{1}{2}$, $a_k + c_k = 1$ *and* $\lim\limits_{k\to\infty} c_k = \frac{1}{2}$. *If* $\sum\limits_{k=1}^{\infty} \frac{a_k - c_k}{c_k}$ *is a convergent series, then the polynomial hypergroup on* $\mathbb{N}_0$ *generated by* $(R_n(x))_{n\in\mathbb{N}_0}$ *has Haar weights* $h(n)$, $n \in \mathbb{N}_0$, *nondecreasing and* $\lim\limits_{n\to\infty} h(n) = M < \infty$.

Proof. $(R_n(x))_{n\in\mathbb{N}_0}$ generates a polynomial hypergroup on $\mathbb{N}_0$ by Theorem 4.4.4, and $(h(n))_{n\in\mathbb{N}_0}$ is nondecreasing, since $\frac{a_n}{c_{n+1}} \geq 1$. Since $c_n h(n) = \frac{a_1 a_2 \cdots a_{n-1}}{c_1 c_2 \cdots c_{n-1}}$ it suffices to show that

$$\ln\left(\frac{a_1 a_2 \cdots a_n}{c_1 c_2 \cdots c_n}\right) = \sum_{k=1}^{n} \ln\left(\frac{a_k}{c_k}\right)$$

is convergent as $n \to \infty$. We have $\ln(\frac{a_k}{c_k}) \leq \frac{a_k}{c_k} - 1 = \frac{a_k - c_k}{c_k}$, and $\lim\limits_{n\to\infty} h(n) < \infty$ follows whenever $\sum\limits_{k=1}^{\infty} \frac{a_k - c_k}{c_k}$ converges. ◇

Examples of polynomial hypergroups with bounded Haar weights.

Let $\beta > 1$ and $\alpha \geq 0$. Put

$$c_k = \frac{k^\beta}{2k^\beta + \alpha}, \quad a_k = 1 - c_k.$$

Then $0 < c_k \leq c_{k+1} \leq \frac{1}{2}$ and $\lim\limits_{k\to\infty} c_k = \frac{1}{2}$. Furthermore, $(a_k - c_k)/c_k = \frac{\alpha}{k^\beta}$ and $\sum\limits_{k=1}^{\infty} \frac{a_k - c_k}{c_k}$ is convergent. By Proposition 4.11.1 the generated polynomial hypergroup on $\mathbb{N}_0$ has bounded Haar weights. Note that the case $\beta = 1$ leads to ultraspherical polynomials, which have unbounded Haar weights whenever $\alpha > 0$, see subsection 4.5.1. Applying Theorem 4.3.3 we know that $S = D_s = D = [-1, 1]$.

Remark: We refer to a result due to Korous, see [641, Theorem 7.1.3], which might be used to derive further examples of orthogonal polynomial sequences with bounded Haar weights.

Theorem 4.11.2 *Let* $(R_n(x))_{n\in\mathbb{N}_0}$ *be a symmetric orthogonal polynomial sequence generating a polynomial hypergroup. Suppose that the Haar weights* $h(n)$ *are bounded by* $M \geq 1$.

(a) If the recurrence coefficients a_n *satisfy* $\frac{1}{2} \leq a_n < 1$ *for* $n \geq n_0$, $n_0 \in \mathbb{N}$, *then* $\sum\limits_{n=n_0}^{\infty} (1 - \frac{c_n}{a_n}) < \infty$, *in particular* $a_n \to \frac{1}{2}$ *and* $c_n \to \frac{1}{2}$. *Moreover,* $[-1, 1] = S = D_s = D$. *If, in addition, the recurrence coefficients satisfy* $\frac{1}{2} \leq a_{n+1} \leq a_n < 1$ *for* $n \geq n_0$, *then* $n(a_n - \frac{1}{2}) \to 0$ *and* $n(\frac{1}{2} - c_n) \to 0$.

(b) *If* $\frac{1}{2} \leq a_{n+1} \leq a_n < 1$ *for* $n \geq n_0$, *then* $d\pi(x) = g(x)\, d\lambda(x)$, *where* λ *is the Lebesgue measure on* $[-1,1]$, *and* g *is a continuous function with* $g(x) > 0$ *for all* $x \in]-1,1[$.

Proof.

(a) Below, let $n \geq n_0$. Put $v_n := 1 - \frac{c_n}{a_n}$. By $a_n + c_n = 1$ it follows $0 \leq v_n < 1$. The products

$$p_n := \prod_{k=n_0}^{n} (1 - v_k)$$

satisfy $p_{n_0} \geq p_{n_0+1} \geq \cdots \geq p_n > 0$ and

$$p_n = \frac{c_{n_0}}{a_{n_0}} \cdots \frac{c_n}{a_n} = \frac{h(n_0-1)\, a_{n_0-1}}{h(n)\, a_n}.$$

Hence $(p_n)_{n \geq n_0}$ is a convergent sequence bounded below by $\frac{2h(n_0-1)a_{n_0-1}}{M} > 0$. Let $p = \lim_{n\to\infty} p_n$. It follows

$$0 < p \leq p_n = \prod_{k=n_0}^{n} (1 - v_k) \leq \exp\left(-\sum_{k=n_0}^{n} v_k\right) \qquad \text{for all } n \in \mathbb{N}.$$

If $\sum\limits_{k=n_0}^{\infty} v_k = \infty$, then $\exp\left(-\sum\limits_{k=n_0}^{n} v_k\right)$ tends to zero as $n \to \infty$, a contradiction. Hence $\sum\limits_{k=n_0}^{\infty} v_k < \infty$ and so $v_n \to 0$, i.e. $a_n \to \frac{1}{2}$ and $c_n \to \frac{1}{2}$. By Theorem 4.3.3 follows $S = D_s = D = [-1,1]$. If $\frac{1}{2} \leq a_{n+1} \leq a_n < 1$ then the sequence $(v_n)_{n \geq n_0}$ is not increasing, tending to zero. By [374, 80, Satz] we obtain $nv_n \to 0$, i.e. $n(a_n - \frac{1}{2}) \to 0$ and $n(c_n + \frac{1}{2}) \to 0$.

(b) By the monotonicity of the a_n we have $1 \leq c_n + a_{n-1}$. Hence $\frac{1}{4} - a_{n-1}c_n \leq (a_{n-1} - \frac{1}{2})(\frac{1}{2} - c_n)$. By (a) it follows $n^2(1 - 4a_{n-1}c_n) \to 0$ as $n \to \infty$. In particular $\sum\limits_{n=1}^{\infty} |1 - 4a_{n-1}c_n| < \infty$. By [688,Theorem 6] we obtain that π is absolutely continuous on $]-1,1[$ and $g(x) = \pi'(x)$ is continuous and positive on $]-1,1[$. We have to check that π contains no point measures in $x = \pm 1$. This is not possible, since for each $f \in C(S)$ the sequence $(\check{f}(n))_{n \in \mathbb{N}_0}$ has to be an element of $c_0(\mathbb{N}_0)$, see Proposition 3.3.2, and $R_n(1) = 1$, $R_n(-1) = (-1)^n$.

◇

We want to state part of Theorem 4.11.2 explicitly.

Corollary 4.11.3 *If $(R_n(x))_{n\in\mathbb{N}_0}$ generates a polynomial hypergroup with bounded Haar weights, then $S = D_s = D$. If in addition $(R_n(x))_{n\in\mathbb{N}_0}$ is symmetric and $\frac{1}{2} \le a_n < 1$ for $n \ge n_0$, then $[-1,1] = S = D_s = D$.*

Proof. The first statement follows by Theorem 4.2.4, the second statement is contained in Theorem 4.11.2(a). ◇

Remark: For symmetric polynomial hypergroups it is easy to derive that $R_n(0) = 0$ for $n = 2k+1$, $k \in \mathbb{N}_0$, and

$$R_{2k}(0) = (-1)^k \frac{c_1}{a_1}\frac{c_3}{a_3}\cdots\frac{c_{2k-1}}{a_{2k-1}}, \qquad k \in \mathbb{N}. \tag{1}$$

Hence the condition $0 \in D_s$ is equivalent to

$$c_1 c_3 \cdots c_{2k-1} \le a_1 a_3 \cdots a_{2k-1} \qquad \text{for all } k \in \mathbb{N}. \tag{2}$$

Regarding only $k = 1$ and $k = 2$, we have the following result, which is interesting in connection with Corollary 4.11.3.

Proposition 4.11.4 *Let $(R_n)_{n\in\mathbb{N}_0}$ be a symmetric orthogonal polynomial sequence generating a polynomial hypergroup. If $0 \in D_s$, then necessarily $a_1 \ge \frac{1}{2}$ and $a_1 + a_3 \ge 1$.*

Proof. We have only to note that $c_1 \le a_1$ is equivalent to $a_1 \ge \frac{1}{2}$, and $c_1c_3 \le a_1a_3$ is equivalent to $a_1 + a_3 \ge 1$. ◇

Proposition 4.11.5 *Let $(R_n(x))_{n\in\mathbb{N}_0}$ be a symmetric orthogonal polynomial sequence generating a polynomial hypergroup. Suppose that the Haar weights $h(n)$ are bounded. If the recurrence coefficients a_n satisfy $\frac{1}{2} \le a_n < 1$ for $n \ge n_0$, $n_0 \in \mathbb{N}$, then there exists $c > 0$ such that $|R_{2k}(0)| \ge c$ for each $k \in \mathbb{N}$.*

Proof. Denote by

$$p_1(k) := \frac{c_1c_3\cdots c_{2k-1}}{a_1a_3\cdots a_{2k-1}} \quad \text{and } p_2(k) := \frac{c_2c_4\cdots c_{2k}}{a_2a_4\cdots a_{2k}}$$

for $k \in \mathbb{N}$. We may assume $n_0 = 2k_0$. Let $h(n) \le M$ and $M_0 = p_2(k_0)$. By the assumptions we obtain $p_2(k) \le M_0$ for $k \ge k_0$, and hence

$$|R_{2k}(0)| = p_1(k) = \frac{1}{h(2k)a_{2k}p_2(k)} \ge \frac{1}{MM_0}$$

for all $k \ge k_0$. It follows that there is some $c > 0$ such that $|R_{2k}(0)| \ge c$ for all $k \in \mathbb{N}$. ◇

Corollary 4.11.6 *Let $(R_n(x))_{n\in\mathbb{N}_0}$ be a symmetric orthogonal polynomial sequence generating a polynomial hypergroup. Suppose that the Haar weights $h(n)$ are bounded, and the recurrence coefficients a_n satisfy $\frac{1}{2} \le a_n < 1$ for $n \ge n_0$, $n_0 \in \mathbb{N}$. Then $0 \in \mathcal{T}$. In particular, the condition $P_1(0, M)$ is satisfied.*

Proof. We can assume that $n_0 = 2k_0 - 1$, $k_0 \in \mathbb{N}$. For $k \ge k_0$ we have $a_{2k-1} \ge \frac{1}{2}$ and $c_{2k} \le \frac{1}{2}$. Hence $h(2k) = h(2k-1)\,\frac{a_{2k-1}}{c_{2k}} \ge h(2k-1)$, i.e. $h(2k-1) + h(2k) \le 2h(2k)$ for $k \ge k_0$. We have to show that

$$\lim_{n\to\infty} \frac{1}{H(n)} \sum_{k=0}^{n} R_k^2(0)\, h(k) \;>\; 0.$$

Having in mind that $h(n) \ge 1$ it is enough to show

$$\lim_{k\to\infty} \left(\sum_{l=k_0}^{k} R_{2l}^2(0)\, h(2l) \right) \Big/ \sum_{l=k_0}^{k} \left(h(2l) + h(2l-1)\right) \;>\; 0.$$

We have

$$\left(\sum_{l=k_0}^{k} R_{2l}^2(0)\, h(2l) \right) \Big/ \sum_{l=k_0}^{k} \left(h(2l) + h(2l-1)\right)$$

$$\ge \frac{1}{2} \left(\sum_{l=k_0}^{k} R_{2l}^2(0)\, h(2l) \right) \Big/ \sum_{l=k_0}^{k} h(2l) \;\ge\; \frac{c^2}{2} \;>\; 0,$$

where c is the positive constant of Proposition 4.10.5. It follows

$$\lim_{n\to\infty} \frac{1}{H(n)} \sum_{k=0}^{n} R_k^2(0)\, h(k) \;>\; 0,$$

and $0 \in \mathcal{T}$ is shown. The additional statement follows by Proposition 4.9.6. ◇

An interesting tool for investigating the behaviour of the sequences $(R_n(x))_{n\in\mathbb{N}_0}$, $x \in D_s$, are the Turan determinants of $(R_n(x))_{n\in\mathbb{N}_0}$. The Turan determinants are defined by

$$\Delta_n(x) \;=\; R_n^2(x) - R_{n-1}(x) R_{n+1}(x) \tag{3}$$

for $n \in \mathbb{N}$, $x \in D_s$. Obviously $\Delta_n(1) = 0$, and whenever $(R_n(x))_{n\in\mathbb{N}_0}$ is symmetric, then $\Delta_n(-1) = 0$. If $R_n(x) = T_n(x)$ are the Chebyshev polynomials of the first kind, the corresponding Turan determinants are

$$\Delta_n(x) \;=\; \frac{1}{2}\,(1 + T_{2n}(x)) \;-\; \frac{1}{2}\,(T_2(x) + T_{2n}(x)) \;=\; \frac{1}{2}\,(1 - T_2(x))$$

for $x \in [-1,1]$ and $n \in \mathbb{N}$, i.e. $\Delta_n(x) = \sin^2 t$ for $x = \cos t$ and each $n \in \mathbb{N}$. If $\Delta_n(x) \geq 0$ for every $n \in \mathbb{N}$, $x \in D_s$, we can derive various properties of the polynomial hypergroup generated by $(R_n(x))_{n\in\mathbb{N}_0}$. Note that the inequality of the Turan determinants depends on the normalization of the orthogonal polynomials. Here we will use the normalization $R_n(1) = 1$.

The following results valid for symmetric polynomial hypergroups are based on general results due to [663] and [73].

Proposition 4.11.7 *Let $(R_n(x))_{n\in\mathbb{N}_0}$ be a symmetric orthogonal polynomial sequence, and assume $0 < a_n, c_n < 1$, $a_n + c_n = 1$. Then*

$$\text{(i)}\ \Delta_n(x) = R_n^2(x) + \frac{c_n}{a_n} R_{n-1}^2(x) - \frac{x}{a_n} R_{n-1}(x)\,R_n(x), \qquad n \in \mathbb{N},$$

$$\text{(ii)}\ \Delta_{n-1}(x) = R_{n-1}^2(x) + \frac{a_{n-1}}{c_{n-1}} R_n^2(x) - \frac{x}{c_{n-1}} R_{n-1}(x)\,R_n(x) \text{ and}$$

$$\text{(iii)}\ (a_{n-1} - c_{n-1})\,a_n\Delta_n(x) - (a_n - c_n)\,c_{n-1}\Delta_{n-1}(x)$$
$$= (c_n - c_{n-1})\,(R_{n-1}^2(x) + R_n^2(x) - 2x\,R_{n-1}(x)\,R_n(x)) \quad \text{for } n \geq 2.$$

Proof. (i) and (ii) : By the recurrence relations we obtain

$$\Delta_n(x) = R_n^2(x) - R_{n-1}(x)\left(\frac{1}{a_n}\,x\,R_n(x) - \frac{c_n}{a_n} R_{n-1}(x)\right)$$
$$= R_n^2(x) + \frac{c_n}{a_n} R_{n-1}^2(x) - \frac{x}{a_n} R_{n-1}(x)\,R_n(x)$$

and

$$\Delta_{n-1}(x) = R_{n-1}^2(x) - R_n(x)\left(\frac{1}{c_{n-1}}\,x\,R_{n-1}(x) - \frac{a_{n-1}}{c_{n-1}} R_n(x)\right)$$
$$= R_{n-1}^2(x) + \frac{a_{n-1}}{c_{n-1}} R_n^2(x) - \frac{x}{c_{n-1}}\,x\,R_{n-1}(x)\,R_n(x).$$

(iii) Multiply the first equation by $a_n(a_{n-1} - c_{n-1})$ and the second equation by $-c_{n-1}(a_n - c_n)$, and add both resulting equations. We obtain

$$(a_{n-1} - c_{n-1})\,a_n\Delta_n(x) - (a_n - c_n)\,c_{n-1}\Delta_{n-1}(x)$$
$$= (c_n a_{n-1} - c_{n-1}a_n)\,(R_n^2(x) + R_{n-1}^2(x))$$
$$-(a_{n-1} - a_n - c_{n-1} + c_n)\,x\,R_{n-1}(x)\,R_n(x).$$

Finally $a_n + c_n = 1$ for $n \in \mathbb{N}$ gives equation (iii). ◇

Theorem 4.11.8 *Let $(R_n(x))_{n\in\mathbb{N}_0}$ be a symmetric orthogonal polynomial sequence generating a polynomial hypergroup on $\mathbb{N}_0$ such that $\frac{1}{2} \leq a_{n+1} \leq a_n < 1$ for each $n \in \mathbb{N}$. Then $\Delta_n(x) > 0$ for $-1 < x < 1$ and each $n \in \mathbb{N}$.*

Proof. At first note that by the assumption $(c_n)_{n\in\mathbb{N}}$ converges to some c, $0 < c \le \frac{1}{2}$ and $S = \operatorname{supp}\pi = [-2\sqrt{c(1-c)}, 2\sqrt{c(1-c)}]$, $D_s = [-1,1]$, see Theorem 4.3.3. The orthogonal polynomials $R_{n-1}(x)$ and $R_n(x)$ have no common zeros, see [138]. Hence by an elementary consideration it follows that $R_{n-1}^2(x) + R_n^2(x) - 2xR_{n-1}(x)R_n(x) > 0$ for $-1 < x < 1$, $n \in \mathbb{N}$. Assume that $c_n < a_n$ for each $n \in \mathbb{N}$. By Proposition 4.11.7(iii) we obtain that $\Delta_n(x) > 0$ for $-1 < x < 1$ provided that $\Delta_1(x) > 0$ for $-1 < x < 1$. Since $\Delta_1(x) = R_1^2(x) - R_2(x) = \frac{c_1}{a_1}(1-x^2)$, we have $\Delta_1(x) > 0$ for $-1 < x < 1$. Finally suppose that $a_n = c_n = \frac{1}{2}$ for some $n \in \mathbb{N}$. Let $n_0 \in \mathbb{N}$ be the smallest n with this property. Then $a_n = c_n = \frac{1}{2}$ for all $n \ge n_0$, and we obtain by Proposition 4.11.7(i) that $\Delta_n(x) = R_n^2(x) + R_{n-1}^2(x) - 2xR_{n-1}(x)R_n(x) > 0$ for $n \ge n_0$, $-1 < x < 1$. For the n smaller than n_0 the consideration of the first case yields $\Delta_n(x) > 0$ for $-1 < x < 1$. ◇

There is a particular relation between $\Delta_n(0)$ and $h(n)$. Since $R_{2k+1}(0) = 0$, we have

$$\Delta_{2n}(0) = R_{2n}^2(0) = \left(\frac{c_1 c_3 \cdots c_{2n-1}}{a_1 a_3 \cdots a_{2n-1}}\right)^2.$$

It follows

$$\Delta_{2n}(0)\, h(2n)\, a_{2n} = \left(\frac{c_1 c_3 c_5 \cdots c_{2n-1}}{c_2 c_4 c_6 \cdots c_{2n}}\right)\left(\frac{a_2 a_4 a_6 \cdots a_{2n}}{a_1 a_3 a_5 \cdots a_{2n-1}}\right).$$

Furthermore,

$$\Delta_{2n+1}(0) = -R_{2n}(0)\, R_{2n+2}(0) = \left(\frac{c_1 c_3 \cdots c_{2n-1}}{a_1 a_3 \cdots a_{2n-1}}\right)^2 \frac{c_{2n+1}}{a_{2n+1}},$$

and hence $\Delta_{2n+1}(0)h(2n+1)a_{2n+1} = \Delta_{2n}(0)h(2n)a_{2n}$.

Proposition 4.11.9 *Let $(R_n(x))_{n\in\mathbb{N}_0}$ be a symmetric orthogonal polynomial sequence generating a polynomial hypergroup on $\mathbb{N}_0$. Then*

$$\Delta_{2n+1}(0)\, h(2n+1)\, a_{2n+1} = \Delta_{2n}(0)\, h(2n)\, a_{2n}.$$

If $\frac{1}{2} \le a_{k+1} \le a_k < 1$ for each $k \in \mathbb{N}$, then

$$\frac{c_1}{a_1} \le \Delta_{2n}(0)\, h(2n)\, a_{2n} \le 1 \qquad \textit{for all } n \in \mathbb{N}.$$

Proof. Since $c_{2k-1} \le c_{2k}$ and $a_{2k} \le a_{2k-1}$ it follows $\Delta_{2n}(0)h(2n)a_{2n} \le 1$. Since $c_{2k+1} \ge c_{2k}$ and $a_{2k+1} \le a_{2k}$ and $a_{2k} \ge \frac{1}{2}$, we obtain

$$\Delta_{2n}(0)\, h(2n)\, a_{2n} = \frac{c_1 a_2 c_3 a_4 \cdots c_{2n-1} a_{2n}}{c_2 a_1 c_4 a_3 \cdots c_{2n} a_{2n-1}}$$

$$\geq \frac{c_1 a_2 c_2 a_3 c_3 a_4 \cdots c_{2n-1} a_{2n}}{c_2 a_1 c_3 a_2 c_4 a_3 \cdots c_{2n} a_{2n-1}} = \frac{c_1 a_{2n}}{a_1 c_{2n}} \geq \frac{c_1}{a_1}.$$

◇

If there exists a lower bound $c_x > 0$ for the Turan determinants $\Delta_n(x)$, $x \in]-1,1[$ fixed, then $x \in \mathcal{T}$. In fact, we can show the following results.

Theorem 4.11.10 *Let $(R_n(x))_{n\in\mathbb{N}_0}$ be a symmetric orthogonal polynomial sequence generating a polynomial hypergroup on $\mathbb{N}_0$ such that $\frac{1}{2} \leq a_n < 1$ for $n \in \mathbb{N}$. Let $x \in]-1,1[$ and suppose that there exists some $c_x > 0$ such that $\Delta_n(x) \geq c_x$ for all $n \in \mathbb{N}$. Then $x \in \mathcal{T}$. If in addition property (H) holds, then condition $P_1(x, M)$ is satisfied.*

Proof. At first we note that for each $n \in \mathbb{N}$

$$\max\{|R_{n-1}(x)|,\ |R_n(x)|,\ |R_{n+1}(x)|\} \geq \sqrt{c_x/2}.$$

In fact, if $\max\{|R_{n-1}(x)|, |R_n(x)|, |R_{n+1}(x)|\} < \sqrt{c_x/2}$ for some $n \in \mathbb{N}$, then

$$R_n^2(x) + |R_{n-1}(x)|\,|R_{n+1}(x)| < c_x.$$

Since $\Delta_n(x) \leq R_n^2(x) + |R_{n-1}(x)|\,|R_{n+1}(x)|$, we have a contradiction to $\Delta_n(x) \geq c_x$ for each $n \in \mathbb{N}$.

Consider the triples $\{3k, 3k+1, 3k+2\}$, $k \in \mathbb{N}_0$. Each triple determined by $k \in \mathbb{N}_0$ contains at least one $n_k \in \{3k, 3k+1, 3k+2\}$ such that $R_{n_k}^2(x) \geq c_x/2$. Select for each $k \in \mathbb{N}_0$ such an n_k and note that $n_{k+1} - n_k \leq 5$. Since $h(n) \leq h(n+1)$, we have $h(n_k + 1) + ... + h(n_{k+1}) \leq 5h(n_{k+1})$. It follows

$$\sum_{j=0}^{n_k} R_j^2(x)\, h(j) / \sum_{j=0}^{n_k} h(j) \geq \sum_{l=0}^{k} R_{n_l}^2(x)\, h(n_l) / \sum_{j=0}^{n_k} h(j)$$

$$\geq \frac{1}{5} \sum_{l=0}^{k} R_{n_l}^2(x)\, h(n_l) / \sum_{l=0}^{k} h(n_l) \geq \frac{c_x}{10} > 0,$$

and $x \in \mathcal{T}$ is proven. If $\frac{h(n)}{H(n)} \to 0$, then $P_1(x, M)$ follows by Proposition 4.9.6. ◇

Since $R_n(1) = 1$ and $R_n(-1) = (-1)^n$ we infer that $R_{n+2}(x) - R_n(x)$ is a polynomial with zeros $x = 1$ and $x = -1$. Hence $R_{n+2}(x) - R_n(x)$ is divisible by $x^2 - 1$ and

$$Q_n(x) = \frac{R_{n+2}(x) - R_n(x)}{x^2 - 1} = \frac{R_{n+2}(x) - R_n(x)}{a_1(R_2(x) - R_0(x))} \tag{4}$$

is a polynomial of degree n. If $P(x)$ is a polynomial of degree smaller than n, we have

$$\int_S Q_n(x)\, P(x)\, (1-x^2)\, d\pi(x) \;=\; \int_S (R_n(x) - R_{n+2}(x))\, P(x)\, d\pi(x) \;=\; 0.$$

Hence the polynomials $Q_n(x)$ are orthogonal with respect to $d\pi^*(x) = \frac{1}{a_1}(1-x^2)d\pi(x)$. Note that π^* is a probability measure, however $Q_0(x) = \frac{1}{a_1}$. In particular, we have to normalize the $Q_n(x)$, in case we are interested whether these polynomials generate a polynomial hypergroup. Since $\operatorname{supp}\pi^* \subseteq [-1,1]$ we have $Q_n(1) \neq 0$ and we can put $S_n(x) := \frac{Q_n(x)}{Q_n(1)}$.

An extension of the Christoffel–Darboux formula, see Proposition 4.10.3, is a useful tool to get more information about $(S_n(x))_{n\in\mathbb{N}_0}$. For example the connection coefficients between $R_n(x)$ and $S_n(x)$ can be explicitly determined.

Proposition 4.11.11 *Let $(R_n(x))_{n\in\mathbb{N}_0}$ be a symmetric orthogonal polynomial sequence generating a polynomial hypergroup on $\mathbb{N}_0$. The following identities hold:*

(1) $$\sum_{k=0}^{[n/2]} R_{n-2k}(x)\, R_{n-2k}(y)\, h(n-2k) \;=\; a_n a_{n+1} h(n)$$
$$\frac{R_{n+2}(x)R_n(y) - R_n(x)R_{n+2}(y)}{x^2-y^2}$$
for $x, y \in \mathbb{R}$, $x \neq y$ and $x \neq -y$. In particular,

$$\sum_{k=0}^{[n/2]} R_{n-2k}(x)\, h(n-2k) \;=\; a_n a_{n+1} h(n)\, \frac{R_{n+2}(x) - R_n(x)}{x^2-1}$$

for $x \neq 1$.

(2) $$\sum_{k=0}^{[n/2]} R_{n-2k}^2(x) h(n-2k) = a_n a_{n+1} h(n) \frac{R'_{n+2}(x)R_n(x) - R'_n(x)R_{n+2}(x)}{2x}$$
for $x \neq 0$. In particular,

$$\sum_{k=0}^{[n/2]} h(n-2k) \;=\; a_n a_{n+1} h(n)\, \frac{1}{2}\, (R'_{n+2}(1) - R'_n(1)),$$

and

$$\sum_{k=0}^{[n/2]} R_{n-2k}^2(0)\, h(n-2k) \;=\; a_n a_{n+1} h(n)$$

$$\frac{1}{2}(R''_{n+2}(0)\, R_n(0) \;-\; R''_n(0)\, R_{n+2}(0)).$$

Proof.

(1) By the recurrence relations of $(R_n(x))_{n\in\mathbb{N}_0}$ we obtain for each $m \in \mathbb{N}_0$

$$x^2 R_m(x) = a_m a_{m+1} R_{m+2}(x) + a_m c_{m+1} R_m(x) + c_m a_{m-1} R_m(x) + c_m c_{m-1} R_{m-2}(x), \qquad (*)$$

(with $c_0 = 0,\ a_{-1} = 0,\ a_0 = 1$).

For fixed $n \in \mathbb{N}_0$ it follows for $x \neq y,\ x \neq -y$

$$R_{n+2}(x)\, R_n(y) \;-\; R_n(x)\, R_{n+2}(y) \;=\; \frac{R_n(x) R_n(y)}{a_n a_{n+1}}\,(x^2 - y^2) + \frac{c_{n-1}c_n}{a_n a_{n+1}}\,[R_n(x)\, R_{n-2}(y) \;-\; R_{n-2}(x) R_n(y)],$$

i.e.

$$a_n a_{n+1} h(n)\, \frac{R_{n+2}(x)R_n(y) - R_n(x)R_{n+2}(y)}{x^2 - y^2} \;=\; h(n)\, R_n(x)\, R_n(y) + c_{n-1}c_n h(n)\, \frac{R_n(x)R_{n-2}(y) - R_{n-2}(x)R_n(y)}{x^2 - y^2}.$$

Applying $c_{n-1}c_n h(n) = a_{n-2}a_{n-1}h(n-2)$ and $(*)$ for $m = n-2$ it follows

$$a_n a_{n+1} h(n)\, \frac{R_{n+2}(x)R_n(y) - R_n(x)R_{n+2}(y)}{x^2 - y^2}$$
$$= h(n)\, R_n(x)\, R_n(y) \;+\; h(n-2)\, R_{n-2}(x)R_{n-2}(y) + a_{n-4}a_{n-3}h(n-4)\, \frac{R_{n-2}(x)R_{n-4}(y) - R_{n-4}(x)R_{n-2}(y)}{x^2 - y^2}.$$

By iteration, the first equation of (1) follows.

(2) The first equation of (2) is obtained by $y \to x$.

◇

Corollary 4.11.12 *Let $(R_n(x))_{n\in\mathbb{N}_0}$ be a symmetric orthogonal polynomial sequence generating a polynomial hypergroup on $\mathbb{N}_0$. Then*

$$Q_n(1) \;=\; \frac{1}{a_n a_{n+1} h(n)} \sum_{k=0}^{[n/2]} h(n-2k) \;=\; \frac{1}{2}(R'_{n+2}(1) - R'_n(1)) \;>\; 0.$$

Example. Consider $R_n(x) = T_n(x) = R^{(-\frac{1}{2},-\frac{1}{2})}(x)$, the Chebyshev polynomials of the first kind. The corresponding polynomials $Q_n(x)$ are the Chebyshev polynomials of the second kind. Hence $S_n(x) = Q_n(x)/Q_n(1) = R_n^{(+\frac{1}{2},+\frac{1}{2})}(x)$, see subsection 4.5.1. Since $T'_n(1) = n^2$, Corollary 4.11.12 gives $Q_n(1) = 2(n+1)$ and so

$$Q_n(x) \;=\; 2\, \frac{\sin((n+1)t)}{\sin t}, \qquad x = \cos t.$$

This example suggests itself to call $(S_n(x))_{n\in\mathbb{N}_0}$ the **conjugate** orthogonal polynomial sequence of $(R_n(x))_{n\in\mathbb{N}_0}$.

We begin by noting that $S_n(x) = \sum\limits_{k=0}^{n} c(n,k)R_k(x)$ is such that the connection coefficients $c(n,k)$ are nonnegative, i.e. $R_n \overset{d}{\geq} S_n$, see subsection 4.6. This follows easily from Proposition 4.11.11 and Corollary 4.11.12.

Furthermore one can add further equations to those of Corollary 4.10.9.

Proposition 4.11.13 *Let $(R_n(x))_{n\in\mathbb{N}_0}$ be a symmetric orthogonal polynomial sequence generating a polynomial hypergroup on $\mathbb{N}_0$. Then*

$$Q_n(1) = \frac{a_n h(n) + H(n)}{2a_n a_{n+1} h(n)} = \frac{1}{2a_{n+1}}\left(1 + \frac{H(n)}{a_n h(n)}\right).$$

Proof. Since $c_n h(n) = a_{n-1}h(n-1)$ we have

$$a_n h(n) = h(n) - c_n h(n) = h(n) - a_{n-1}h(n-1).$$

Continuing the calculation we get

$$a_n h(n) = h(n) - h(n-1) + \cdots \pm h(0).$$

Hence

$$a_n h(n) + H(n) = 2\sum_{k=0}^{[n/2]} h(n-2k)$$

and

$$Q_n(1) = \frac{a_n h(n) + H(n)}{2a_n a_{n+1} h(n)}.$$

$\diamond$

The recurrence relations of $(R_n(x))_{n\in\mathbb{N}_0}$ yield that

$$xQ_n(x) = a_{n+2}Q_{n+1}(x) + c_n Q_{n-1}(x), \qquad n \in \mathbb{N} \tag{5}$$

with $Q_0(x) = \frac{1}{a_1}$ and $Q_1(x) = \frac{1}{a_1 a_2}\, x$.

Hence the recurrence relations of $(S_n(x))_{n\in\mathbb{N}_0}$ are

$$xS_n(x) = \alpha_n S_{n+1}(x) + \gamma_n S_{n-1}(x) \tag{6}$$

with

$$\alpha_n = a_{n+2}\frac{Q_{n+1}(1)}{Q_n(1)}, \qquad \gamma_n = c_n\frac{Q_{n-1}(1)}{Q_n(1)}.$$

Obviously, $\alpha_0 = 1$, $\gamma_1 = c_1 a_2$ and $\alpha_n + \gamma_n = 1$.

Since $\frac{\alpha_{n+1}}{\alpha_{n-1}} = \frac{Q_{n-1}(1)}{Q_n(1)}$, we obtain

$$1 = \alpha_n + \gamma_n = \alpha_n + c_n \frac{Q_{n-1}(1)}{Q_n(1)} = \alpha_n + \frac{c_n a_{n+1}}{\alpha_{n-1}}.$$

In particular, we have the following relation between the recurrence coefficients c_n, a_n of $R_n(x)$ and γ_n, α_n of $S_n(x)$:

$$\gamma_n \alpha_{n-1} = c_n a_{n+1}, \qquad n \in \mathbb{N}. \tag{7}$$

Applying Theorem 4.4.1 we can derive a sufficient condition such that the conjugate orthogonal polynomial sequence $(S_n(x))_{n\in\mathbb{N}_0}$ generates a polynomial hypergroup on $\mathbb{N}_0$.

Theorem 4.11.14 *Let $(R_n(x))_{n\in\mathbb{N}_0}$ be a symmetric orthogonal polynomial sequence (with recurrence coefficients c_n, a_n) generating a polynomial hypergroup on $\mathbb{N}_0$. If the sequence $(c_n a_{n+1})_{n\in\mathbb{N}}$ is increasing, then $(S_n(x))_{n\in\mathbb{N}_0}$ generates a polynomial hypergroup on $\mathbb{N}_0$. Moreover, the sequence $(\gamma_n)_{n\in\mathbb{N}_0}$ is strictly increasing.*

Proof. By (7) the sequence $(\gamma_n \alpha_{n-1})_{n\in\mathbb{N}}$ is increasing. By Theorem 4.4.1 $(S_n(x))_{n\in\mathbb{N}_0}$ generates a polynomial hypergroup on $\mathbb{N}_0$. Moreover $(c_n a_{n+1})_{n\in\mathbb{N}}$ is a chain sequence. According to Theorem 5.6 of [138,Ch.III] the minimal parameter sequence $(\gamma_n)_{n\in\mathbb{N}_0}$ is strictly increasing. ◇

Remark: The property that $(c_n a_{n+1})_{n\in\mathbb{N}}$ is increasing is equivalent to the property that for $n \in \mathbb{N}$

$$c_{n+1} - c_n \geq \frac{c_{n+1}}{a_{n+1}} (c_{n+2} - c_{n+1}) \tag{8}$$

is valid.

In fact, using $a_{n+2} = c_{n+1} + a_{n+1} - c_{n+2}$, it follows immediately that $c_n a_{n+1} \leq c_{n+1} a_{n+2}$ holds exactly when $c_{n+1} - c_n \geq \frac{c_{n+1}}{a_{n+1}} (c_{n+2} - c_{n+1})$.

Examples.

(1) **Ultraspherical polynomials** $R_n^{(\alpha,\alpha)}(x)$, $\alpha \geq -\frac{1}{2}$.
From subsection 4.5.1 we know already that for $R_n(x) = R^{(\alpha,\alpha)}(x)$ the conjugate polynomials are exactly $S_n(x) = R_n^{(\alpha+1,\alpha+1)}(x)$. It is worthwhile to mention that $(c_n a_{n+1})_{n\in\mathbb{N}}$ is an increasing sequence for each $\alpha \geq -\frac{1}{2}$, which can be easily checked using (8), and that $(c_n a_{n-1})_{n\in\mathbb{N}}$ is not increasing for $-\frac{1}{2} \leq \alpha < \frac{1}{2}$, see the Remark below Theorem 4.4.1.

(2) **Geronimus polynomials** $R_n(x; \frac{1}{4}, \alpha)$, $0 < \alpha \leq 2$, see subsection 4.5.7.1.

The recurrence coefficients c_n, a_n of $R_n(x; \frac{1}{4}, \alpha)$ are for $n \in \mathbb{N}$

$$c_n = \frac{\alpha + (2-\alpha)(n-1)}{2(\alpha + (2-\alpha)n)} \quad \text{and} \quad a_n = \frac{\alpha + (2-\alpha)(n+1)}{2(\alpha + (2-\alpha)n)}.$$

An elementary calculation shows that for each $n \in \mathbb{N}$

$$\frac{c_{n+2} - c_{n+1}}{c_{n+1} - c_n} = \frac{\alpha + (2-\alpha)n}{\alpha + (2-\alpha)(n+2)} = \frac{c_{n+1}}{a_{n+1}}$$

is valid. Hence inequality (8) holds and Theorem 4.11.14 implies that the conjugate Geronimus polynomials $S_n(x; \frac{1}{4}, \alpha)$, $0 < \alpha \leq 2$, generate a polynomial hypergroup on $\mathbb{N}_0$.

(3) **Polynomials connected with homogeneous trees** $R_n(x; q)$, $q \geq 1$, see subsection 4.5.7.2.
The recurrence coefficients c_n, a_n, of $R_n(x; q)$ are $c_n = \frac{1}{q+1}$, $a_n = \frac{q}{q+1}$ for $n \in \mathbb{N}$. We have $c_n a_{n+1} = \frac{q}{(q+1)^2}$, and we obtain from Theorem 4.11.14 that the conjugate polynomials $S_n(x; q)$ generate a polynomial hypergroup on $\mathbb{N}_0$.

Finally we consider the family of orthogonal polynomials $(R_n(x))_{n \in \mathbb{N}_0}$ introduced below Proposition 4.11.1.

Lemma 4.11.15 *Put* $f(x) = \frac{x^\beta}{2x^\beta + \alpha}$ *and* $g(x) = \frac{(x+1)^\beta + \alpha}{2(x+1)^\beta + \alpha}$ *and define* $h(x) = f(x)g(x)$ *for* $x \geq 1$. *Then* h *is a monotone increasing function on* $[1, \infty[$.

Proof. Since $f'(x) = \frac{\alpha\beta x^{\beta-1}}{(2x^\beta+\alpha)^2}$ and $g'(x) = \frac{-\alpha\beta(x+1)^{\beta-1}}{(2(x+1)^\beta+\alpha)^2}$, we have to show that

$$h'(x) = \frac{\alpha\beta x^{\beta-1}[(x+1)^\beta + \alpha]}{(2x^\beta + \alpha)^2[2(x+1)^\beta + \alpha]} - \frac{\alpha\beta(x+1)^{\beta-1}x^\beta}{[2(x+1)^\beta + \alpha]^2(2x^\beta + \alpha)}$$

is nonnegative for each $x \geq 1$. Hence we have to show that

$$\frac{x^{\beta-1}[(x+1)^\beta + \alpha]}{2x^\beta + \alpha} \geq \frac{(x+1)^{\beta-1}x^\beta}{2(x+1)^\beta + \alpha} \qquad (*)$$

holds true for $x \geq 1$. Inequality $(*)$ is equivalent to

$$\left[(x+1)^\beta + \alpha\right]\left[2(x+1)^\beta + \alpha\right] \geq x(x+1)^{\beta-1}\left[2x^\beta + \alpha\right]. \qquad (**)$$

Since $(x+1)^{2\beta} \geq x^{\beta+1}(x+1)^{\beta-1}$ and $(x+1)^\beta \geq x(x+1)^{\beta-1}$ for $x \geq 1$, inequality $(**)$ is obviously true. ◇

(4) The class of polynomial hypergroups with **bounded Haar weights** introduced below Proposition 4.11.1.
Let $\beta > 1$ and $\alpha \geq 0$. Define $R_n(x)$ by the recurrence coefficients $c_n = \frac{n^\beta}{2n^\beta+\alpha}$, $a_n = \frac{n^\beta+\alpha}{2n^\beta+\alpha}$ for $n \in \mathbb{N}$. We know already that $(R_n(x))_{n\in\mathbb{N}_0}$ generates a polynomial hypergroup on $\mathbb{N}_0$ with bounded Haar weights. We have $c_n = f(n)$ and $a_{n+1} = g(n)$ with the functions f and g of Lemma 4.11.15. Hence $(c_n a_{n+1})_{n\in\mathbb{N}}$ is an increasing sequence. Applying Theorem 4.11.14 the conjugate polynomials $S_n(x)$ generate a polynomial hypergroup on $\mathbb{N}_0$.

Now we go back to the Turan determinants $\Delta_n(x)$ of a symmetric polynomial sequence $(R_n(x))_{n\in\mathbb{N}_0}$. Using the conjugate polynomials $Q_n(x)$ we get more information about $\Delta_n(x)$.

Proposition 4.11.16 *Let $(R_n(x))_{n\in\mathbb{N}_0}$ be a symmetric orthogonal polynomial sequence. For $n \in \mathbb{N}$ and $-1 < x < 1$ holds*

$$\frac{\Delta_n(x)}{1-x^2} = c_n a_n Q_{n-1}^2(x) - c_{n-1}a_{n+1}Q_{n-2}(x)Q_n(x). \tag{9}$$

Proof. We have

$$R_{n+1}(x) - xR_n(x) = c_n(R_{n+1}(x) - R_{n-1}(x)) = c_n(x^2-1)\, Q_{n-1}(x)$$

and

$$xR_n(x) - R_{n-1}(x) = a_n(R_{n+1}(x) - R_{n-1}(x)) = a_n(x^2-1)\, Q_{n-1}(x).$$

It follows

$$\begin{aligned}
&(x^2-1)^2\left[c_n a_n Q_{n-1}^2(x) - c_{n-1}a_{n+1}Q_{n-2}(x)\, Q_n(x)\right] \\
&= (R_{n+1}(x) - xR_n(x))(xR_n(x) - R_{n-1}(x)) \\
&\quad -(R_n(x) - xR_{n-1}(x))(xR_{n+1}(x) - R_n(x)) \\
&= (1-x^2)(R_n^2(x) - R_{n-1}(x)R_{n+1}(x)) = (1-x^2)\Delta_n(x).
\end{aligned}$$

◇

Theorem 4.11.17 (Berg, Szwarc)
Let $(R_n(x))_{n\in\mathbb{N}_0}$ be a symmetric orthogonal polynomial sequence with $\frac{1}{2} \leq a_{n+1} \leq a_n < 1$ for all $n \in \mathbb{N}$ and $(c_n a_{n+1})_{n\in\mathbb{N}}$ increasing. Then

$$\Delta_n(x) \geq \frac{2}{h(n)}\frac{c_1 a_2}{a_1}(1-x^2) \quad \textit{for } -1 < x < 1,\ n \in \mathbb{N}.$$

Proof. Put $D_n(x) = a_n Q_{n-1}^2(x) - a_{n+1} Q_{n-2}(x) Q_n(x)$. By Proposition 4.11.16 it follows

$$\frac{\Delta_n(x)}{1-x^2} \geq c_{n-1}\, D_n(x) \qquad \text{for } -1 < x < 1.$$

Applying the recurrence relation (5) for $Q_n(x)$ we obtain

$$\begin{aligned} D_n(x) &= a_n Q_{n-1}^2(x) - Q_{n-2}(x)\,(xQ_n(x) - c_{n-1}Q_{n-2}(x)) \\ &= a_n Q_{n-1}^2(x) + c_{n-1}Q_{n-2}^2(x) - xQ_{n-2}(x)\,Q_{n-1}(x) \end{aligned}$$

and

$$\begin{aligned} D_{n-1}(x) &= a_{n-1}Q_{n-2}^2(x) - a_n Q_{n-1}(x)\left(\frac{1}{c_{n-2}}\, xQ_{n-2}(x) - \frac{a_n}{c_{n-2}}\, Q_{n-1}(x)\right) \\ &= a_{n-1}Q_{n-2}^2(x) + \frac{a_n^2}{c_{n-2}}\, Q_{n-1}^2(x) - \frac{a_n}{c_{n-2}}\, xQ_{n-2}(x)\, Q_{n-1}(x). \end{aligned}$$

Subtracting the second equation from the first, we obtain

$$D_n(x) - \frac{c_{n-2}}{a_n}\, D_{n-1}(x) = \frac{c_{n-1}a_n - c_{n-2}a_{n-1}}{a_n}\, Q_{n-2}^2(x) \geq 0$$

for $n \geq 3$, since $(c_n a_{n+1})_{n\in\mathbb{N}}$ is increasing. Iterating

$$D_n(x) \geq \frac{c_{n-2}}{a_n}\, D_{n-1}(x),$$

it follows

$$D_n(x) \geq \frac{c_1 \cdot c_2 \cdots c_{n-2}}{a_3 \cdot a_4 \cdots a_n}\, D_2(x).$$

Finally,

$$D_2(x) = a_2 Q_1^2(x) + \frac{c_1}{a_1^2} - \frac{xQ_1(x)}{a_1} = \frac{c_1}{a_1^2}.$$

Therefore,

$$\begin{aligned} \Delta_n(x) &\geq c_{n-1}\, \frac{c_1 \cdots c_{n-2}}{a_3 \cdots a_n}\, \frac{c_1}{a_1^2}\,(1-x^2) = \frac{1}{h(n)}\, \frac{c_1 a_2}{a_1 c_n a_n}\,(1-x^2) \\ &\geq \frac{2}{h(n)}\, \frac{c_1 a_2}{a_1}\,(1-x^2) \end{aligned}$$

for each $x \in]-1,1[$, $n \in \mathbb{N}$. ◇

Remark. By Proposition 4.11.9 we know that

$$\Delta_{2n}(0) \leq \frac{1}{h(2n)a_{2n}} \leq \frac{2}{h(2n)}$$

and

$$\Delta_{2n+1}(0) \leq \frac{1}{h(2n+1)a_{2n+1}} \leq \frac{2}{h(2n+1)}.$$

Hence Theorem 4.11.17 implies $\Delta_n(x) \geq \Delta_n(0)\, \frac{c_1 a_2}{a_1}\,(1-x^2)$ for each $x \in]-1,1[$, $n \in \mathbb{N}$.

Corollary 4.11.18 *Let $(R_n(x))_{n\in\mathbb{N}_0}$ be symmetric and generate a polynomial hypergroup with bounded Haar weights. If $\frac{1}{2} \le a_{n+1} \le a_n < 1$ for all $n \in \mathbb{N}$ and $(c_n a_{n+1})_{n\in\mathbb{N}}$ is increasing, then $[-1,1] = \mathcal{T}$ and Reiter's condition $P_1(x,M)$ is satisfied for each $x \in [-1,1]$.*

Proof. By Proposition 4.11.2 and Corollary 4.11.3 we have $c_n \to \frac{1}{2}$, $a_n \to \frac{1}{2}$ and $[-1,1] = S = D_s = D$. Theorem 4.11.17 and Theorem 4.11.10 yield $x \in \mathcal{T}$ for every $x \in [-1,1]$. Finally Reiter's condition $P_1(x,M)$ is satisfied by Theorem 4.9.6 for each $x \in [-1,1]$. ◇

Remark. We have already shown that for the class of polynomial hypergroups on $\mathbb{N}_0$ with recurrence coefficients

$$c_n = \frac{n^\beta}{2n^\beta + \alpha}, \qquad a_n = \frac{n^\beta + \alpha}{2n^\beta + \alpha},$$

$\beta > 1$, $\alpha \ge 0$, the assumptions of Corollary 4.11.18 are fulfilled.

Altering only the first recurrence coefficients c_1 and a_1 of the Chebyshev polynomials $T_n(x)$ of the first kind leads us to another class of polynomial hypergroups. It is interesting to study the effect of changing c_1 and a_1 concerning the Turan determinants or the conjugate polynomials.

(5) **Grinspun polynomials**

Fix $a \in \mathbb{R}$, $a \ge 2$ and define

$$c_1 = \frac{1}{a}, \quad a_1 = \frac{a-1}{a} \quad \text{and } c_n = a_n = \frac{1}{2} \quad \text{for } n = 2,3,\ldots.$$

The symmetric orthogonal polynomials $R_n(x)$ defined by these recurrence coefficients are studied in 4.5.10 (a).

We can easily determine the Turan determinants of $(R_n(x))_{n\in\mathbb{N}_0}$,

$$\Delta_1(x) = \frac{1}{a}\,(1 - R_2(x))$$

and

$$\Delta_n(x) = \frac{1}{2(a-1)}\left(1 + \frac{a-4}{2}\,R_2(x) - \frac{a-2}{2}\,R_4(x)\right) \quad \text{for each } n \ge 2.$$

Finally we use formula (7) to derive the recurrence coefficients γ_n, α_n of the conjugate polynomial sequence $(S_n(x))_{n\in\mathbb{N}_0}$ of $(R_n(x))_{n\in\mathbb{N}_0}$, see (6). By induction it follows immediately

$$\gamma_n = \frac{(a-1)(n-1)+1}{2(a-1)n+2}, \qquad \alpha_n = \frac{(a-1)(n+1)+1}{2(a-1)n+2} \quad \text{for } n \in \mathbb{N}.$$

The Grinspun polynomials $(R_n(x))_{n\in\mathbb{N}_0}$ generate a polynomial hypergroup on $\mathbb{N}_0$, and the assumptions of Corollary 4.11.18 are satisfied.

12 A Wiener Theorem

A theorem of Norbert Wiener characterizes the discrete part a_d of a complex Borel measure $a \in M(\mathbb{T})$ on the torus group $\mathbb{T}$,

$$\sum_{z\in\mathbb{T}} |\, a(\{z\})\,|^2 \;=\; \lim_{n\to\infty} \frac{1}{2n+1} \sum_{k=-n}^{n} |\hat{a}(k)|^2,$$

where $\hat{a}$ is the Fourier-Stieltjes transform of a, see [282,p.415]. In particular, $a_d = 0$ if and only if $\lim_{n\to\infty} \frac{1}{2n+1} \sum_{k=-n}^{n} |\hat{a}(k)|^2 = 0$. We will present analogous results for polynomial hypergroups.

Let $(R_n(x))_{n\in\mathbb{N}_0}$ generate a polynomial hypergroup on $\mathbb{N}_0$. Consider

$$D_{n,y}(x) \;=\; \frac{1}{H(n)} \sum_{k=0}^{n} R_k(y)\; R_k(x)\; h(k) \qquad \text{for } x, y \in D_s,$$

compare subsection 4.10, the structural function $\mu(y) = \limsup_{n\to\infty} D_{n,y}(y)$ and $\mathcal{T} = \{y \in D_s : \mu(y) > 0\}$. The following result extends Theorem 4.10.10.

Theorem 4.12.1 *Suppose that $(R_n(x))_{n\in\mathbb{N}_0}$ generates a polynomial hypergroup on $\mathbb{N}_0$ and $\frac{a_n h(n)}{H(n)}$ converges to zero. Let $a \in M(D_s)$ with discrete part a_d. The following conditions are equivalent:*

(i) $\mathcal{T} \cap \mathit{supp}\; a_d \;=\; \emptyset.$

(ii) $\lim_{n\to\infty} \frac{1}{H(n)} \sum_{k=0}^{n} |\check{a}(k)|^2\; h(k) \;=\; 0$

Proof. Denote by $\Delta = \{(x,x) : x \in D_s\}$ the diagonal of $D_s \times D_s$. We have

$$\frac{1}{H(n)} \sum_{k=0}^{n} |\check{a}(k)|^2\; h(k) \;=\; \int_{D_s} \int_{D_s} D_{n,x}(y)\; da(x)\; d\bar{a}(y)$$

$$= \int_{\Delta} D_{n,x}(y)\; da \times \bar{a}(x,y) + \int_{(D_s\times D_s)\setminus\Delta} D_{n,x}(y)\; da \times \bar{a}(x,y). \qquad (*).$$

Since $\frac{a_n h(n)}{H(n)} \to 0$, we obtain by Proposition 4.10.3(1) that $\lim_{n\to\infty} |D_{n,x}(y)| = 0$ for all $(x,y) \in (D_s \times D_s)\setminus\Delta$. Obviously $|D_{n,x}(y)| \le 1$ for all $(x,y) \in D_s \times D_s$. Therefore, applying Lebesgue's theorem of dominated convergence, it follows

$$\lim_{n\to\infty} \int_{(D_s\times D_s)\setminus\Delta} D_{n,x}(y)\; da \times \bar{a}(x,y) \;=\; 0.$$

For the first integral in (*) we obtain by means of Fubini's theorem [304,Theorem 14.25],

$$\int_{\Delta} D_{n,x}(y)\, da \times \bar{a}(x,y) \;=\; \int_{\Delta} D_{n,x}(x)\, \bar{a}(\{x\})\, da(x)$$

$$=\; \sum_{x \in \operatorname{supp} a_d} D_{n,x}(x)\; |a(\{x\})|^2 .$$

Hence

$$\frac{1}{H(n)} \sum_{k=0}^{n} |\breve{a}(k)|^2\, h(k) \;=\; \sum_{x \in \operatorname{supp} a_d} D_{n,x}\; |a(\{x\})|^2 \;+\; \varrho_n,$$

where $\varrho_n \to 0$ as $n \to \infty$. Since $D_{n,x}(x) \geq 0$, condition (ii) implies (i). Conversely, Lebesgue's theorem of dominated convergence applied to the discrete measure

$$\nu \;=\; \sum_{x \in \operatorname{supp} a_d} |a(\{x\})|^2\, \epsilon_x$$

yields that (i) implies (ii). Recall that $D_s \backslash \mathcal{T} = \{x \in D_s : \mu(x) = \lim\limits_{n\to\infty} D_{n,x}(x) = 0\}$. ◇

A drawback of Theorem 4.12.1 is the fact that $\mathcal{T}$ might be a small subset of D_s. Indeed, for the ultraspherical polynomials $R_n^{(\alpha,\alpha)}(x)$, $\alpha \geq -\frac{1}{2}$, see subsection 4.5.1, we know that $h(n) = O(n^{2\alpha+1})$. Hence $\frac{a_n h(n)}{H(n)} \to 0$ as $n \to \infty$. However, Theorem 4.9.7 yields that $]-1,1[\, \cap\, \mathcal{T} = \emptyset$ whenever $\alpha > -\frac{1}{2}$, whereas $]-1,1[\, \subseteq \mathcal{T}$ for $\alpha = -\frac{1}{2}$. It follows that $\mathcal{T} = \{-1,1\}$ in case of $\alpha > -\frac{1}{2}$ and $\mathcal{T} = [-1,1]$ for $\alpha = -\frac{1}{2}$. Some further examples are

1. Jacobi polynomials $R_n^{(\alpha,\beta)}(x)$, $\alpha \geq \beta > -1$, $\alpha + \beta + 1 \geq 0$.
   For $\alpha = \beta$ we get the ultraspherical polynomials, and as just noted: $\mathcal{T} = [-1,1]$ for $\alpha = \beta = -\frac{1}{2}$ and $\mathcal{T} = \{-1,1\}$ for $\alpha = \beta > -\frac{1}{2}$. The asymptotic properties of $R_n^{(\alpha,\beta)}(x)$, $x \in D_s = [-1,1]$ yield that $\mathcal{T} = \{1\}$ if $\alpha > \beta$, $\alpha > -\frac{1}{2}$.
2. Symmetric orthogonal polynomials with bounded Haar weights.
   Applying Corollary 4.11.18 we know that for the polynomial hypergroups with recurrence coefficients

$$c_n \;=\; \frac{n^\beta}{2n^\beta + \alpha}, \qquad a_n \;=\; \frac{n^\beta + \alpha}{2n^\beta + \alpha}, \qquad \beta > 1,\ \alpha \geq 0,$$

   $\mathcal{T} = [-1,1]$ is true. The same is true for the Grinspun polynomials, see subsection 4.5.10 (a).

A direct extension of Theorem 4.12.1 allows to replace $\mathcal{T}$ by a bigger set. But the measures $a \in M(D_s)$ have to fulfil a certain property.

Fix some $\kappa > 0$ and define

$$\mathcal{T}_\kappa := \{x \in D_s : (n^\kappa D_{n,x}(x))_{n\in\mathbb{N}_0} \text{ is unbounded or } \limsup_{n\to\infty} n^\kappa D_{n,x}(x) > 0\}.$$

Theorem 4.12.2 *Suppose that $(R_n(x))_{n\in\mathbb{N}_0}$ generates a polynomial hypergroup on $\mathbb{N}_0$ and $\frac{n^\kappa a_n h(n)}{H(n)}$ converges to zero, where $\kappa > 0$. Let $a \in M(D_s)$ such that there exists some $a \times \bar{a}$- integrable majorant of all the functions $n^\kappa D_{n,y}(x)$ on $D_s \times D_s$. The following conditions are equivalent.*

(i) $\mathcal{T}_\kappa \cap \mathit{supp}\, a_d = \emptyset$.

(ii) $\lim\limits_{n\to\infty} \dfrac{n^\kappa}{H(n)} \sum\limits_{k=0}^{n} |\check{a}(k)|^2\, h(k) = 0.$

Proof. We have (see the proof of Theorem 4.12.1)

$$\frac{n^\kappa}{H(n)} \sum_{k=0}^{n} |\check{a}(k)|^2\, h(k)$$

$$= \int_\Delta n^\kappa\, D_{n,x}(y)\, da \times \bar{a}(x,y) + \int_{(D_s\times D_s)\setminus\Delta} n^\kappa\, D_{n,x}(y)\, da \times \bar{a}(x,y).$$

Since $\frac{n^\kappa a_n h(n)}{H(n)} \to 0$ we get $\lim\limits_{n\to\infty} n^\kappa\, |D_{n,x}(y)| = 0$ for all $(x,y) \in (D_s \times D_s)\setminus\Delta$. By the assumption on $a \in M(D_s)$ there exists some $a\times\bar{a}$-integrable majorant of $n^\kappa |D_{n,x}(y)|$ and it follows

$$\lim_{n\to\infty} \int_{(D_s\times D_s)\setminus\Delta} n^\kappa\, D_{n,x}(y)\, da \times \bar{a}(x,y) = 0$$

and then

$$\frac{n^\kappa}{H(n)} \sum_{k=0}^{n} |\check{a}(k)|^2\, h(k) = \sum_{\kappa\in \mathrm{supp}\, a_d} n^\kappa\, D_{n,x}(x)\, |a(\{x\})|^2 + \varrho_n,$$

where $\varrho_n \to 0$. Now, clearly (ii) implies (i). Conversely, if (i) holds, we know that $\lim\limits_{n\to\infty} n^\kappa D_{n,x}(x) = 0$ for each $x \in \mathrm{supp}\, a_d$, and (ii) is valid. ◇

We consider the ultraspherical polynomials $R_n^{(\alpha,\alpha)}(x)$, $\alpha \geq -\frac{1}{2}$. Since $|R_n^{(\alpha,\alpha)}(x)| = O(n^{-\alpha-\frac{1}{2}})$ for each $x \in\,]-1,1[$ and $h(n) = O(n^{2\alpha+1})$, we obtain that $D_{n,x}(x) = O(n^{-2\alpha-1})$ for each $x \in\,]-1,1[$. It follows that $\mathcal{T}_\kappa = [-1,1]$ for $\kappa \geq 2\alpha+1$. Furthermore we have $\lim\limits_{n\to\infty} \frac{n^\kappa a_n h(n)}{H(n)} = 0$ if $0 \leq \kappa < 1$. Hence Theorem 4.12.2 can be applied as follows: If $-\frac{1}{2} \leq \alpha < 0$ then supp $a_d = \emptyset$ if and only if

$$\lim_{n\to\infty} \frac{n^\kappa}{H(n)} \sum_{k=0}^{n} |\check{a}(k)|^2\, h(k) = 0 \qquad \text{for } \kappa = 2\alpha+1.$$

13 Homogeneous Banach spaces on S

13.1 *Dual structure on S*

Before studying the approximation of functions on $S = \operatorname{supp} \pi \subseteq D_s$ by orthogonal expansions in section 4.14 we introduce so-called homogeneous Banach spaces on S. For this class of Banach spaces several approximation procedures can be applied, see section 4.14.

In subsection 3.11.1 we have already investigated the property (F) of $\mathcal{S}(K)$ for general commutative hypergroups K. We recall the main results for the special case of polynomial hypergroups $K = \mathbb{N}_0$.

Property (F) of $S = \operatorname{supp} \pi$ holds true, if for every $x, y \in S$ there exists a regular real-valued Borel measure $\mu_{x,y} \in M(S)$ such that

$$R_n(x)\, R_n(y) \;=\; \int_S R_n(z)\, d\mu_{x,y}(z) \quad \text{and} \quad \|\mu_{x,y}\| \;\leq\; M$$

for some $M \geq 1$, M independent of $x, y \in S$.

Theorem 4.13.1 *S satisfies property (F) if and only if*

$$\|\widehat{\alpha_x f}|S\,\|_\infty \;\leq\; M\; \|\hat{f}|S\,\|_\infty$$

for all $f \in l^1(h)$, $x \in S$, where $\alpha_x(n) = R_x(n)$ and M is the constant of property (F).

Whenever S satisfies property (F) the translation of $\varphi \in C(S)$ by some $x \in S$ is defined by

$$L_x\varphi(y) \;=\; \mu_{x,y}(\varphi) \;=\; \int_S \varphi(z)\, d\mu_{x,y}(z) \quad \text{for } y \in S. \tag{1}$$

$L_x\varphi$ is an element of $C(S)$. Moreover, the mapping $x \mapsto L_x\varphi$, $S \to C(S)$, is continuous. We define for $\mu \in M(S)$ and $\varphi \in C(S)$ a module action (i.e. a convolution operator)

$$\mu * \varphi(y) \;=\; \int_S L_x\varphi(y)\, d\mu(x). \tag{2}$$

By Proposition 3.11.11 we know already that $\mu * \varphi \in C(S)$.

Note that we have just used that S is compact. In particular, $C(S)$ is a subset of $L^1(S, \pi)$, and so we can consider the inverse Fourier transform of $\varphi \in C(S)$. The following result can be shown directly and shows that the inverse Fourier transform transfers the module action to a multiplication operator.

Proposition 4.13.2 *Assume that S satisfies property (F). Let $\varphi \in C(S)$, $y \in S$ and $\mu \in M(S)$. Then*

(i) $(L_y\varphi)^\vee(n) = R_n(y)\, \check{\varphi}(n),$

(ii) $(\mu * \varphi)^\vee(n) = \check{\mu}(n)\, \check{\varphi}(n)$

for all $n \in \mathbb{N}_0$.

Proof.

(i) Let $f \in l^1(h)$. Then $L_y\hat{f}(x) = \sum_{k=0}^{\infty} f(k)\, R_k(x) R_k(y)\, h(k)$ and therefore,

$$(L_y\hat{f})^\vee(n) = \int_S L_y\hat{f}(x)\, R_n(x)\, d\pi(x)$$

$$= \sum_{k=0}^{\infty} f(k)\, R_k(y)\, h(k) \int_S R_k(x)\, R_n(x)\, d\pi(x) = f(n)\, R_n(y).$$

Given $\varepsilon > 0$ there exists $f \in l^1(h)$ such that $\|\varphi - \hat{f}\|_\infty < \frac{\varepsilon}{M}$. Since $|L_y\varphi(x) - L_y\hat{f}(x)| \leq \|\varphi - \hat{f}\|_\infty\, \|\mu_{x,y}\|$, we obtain

$$\|L_y\varphi - L_y\hat{f}\|_\infty \leq M\, \|\varphi - \hat{f}\|_\infty < \varepsilon.$$

Finally,

$$|(L_y\varphi)^\vee(n) - R_n(y)\check{\varphi}(n)| \leq \|L_y\varphi - L_y\hat{f}\|_\infty + |R_n(y)|\, \|\hat{f} - \varphi\|_\infty,$$

and (i) follows.

(ii)

$$(\mu * \varphi)^\vee(n) = \int_S \mu * \varphi(x)\, R_n(x)\, d\pi(x)$$

$$= \int_S \int_S L_y\varphi(x)\, d\mu(y)\, R_n(x)\, d\pi(x)$$

$$= \int_S R_n(y)\, \check{\varphi}(n)\, d\mu(y) = \check{\mu}(n)\, \check{\varphi}(n),$$

where we have applied Lebesgue's theorem of dominated convergence and (i).

◇

Since S is compact each $\varphi \in C(S)$ is an element of $L^1(S, \pi)$ and $\varphi(x)\, d\pi(x) \in M(S)$. Thus we introduce a dual convolution on $C(S)$ by setting

$$\varphi * \psi(y) := \int_S L_y\psi(x)\, \varphi(x)\, d\pi(x).$$

Theorem 4.13.3 *Assume that S satisfies property (F). Given $\varphi, \psi \in C(S)$ we have*

$$\int_S L_y\psi(x)\, \varphi(x)\, d\pi(x) \;=\; \varphi * \psi(y) \;=\; \psi * \varphi(y) \;=\; \int_S L_y\varphi(x)\, \psi(x)\, d\pi(x).$$

Furthermore, $\varphi * \psi \in C(S)$.

Proof. We know already that the functions $\varphi * \psi$ and $\psi * \varphi$ are continuous. Since $(\mu_{x,y})^\vee(n) = R_n(x)R_n(y) = (\mu_{y,x})^\vee(n)$ for each $n \in \mathbb{N}$, it follows $\mu_{x,y} = \mu_{y,x}$. Calculating the inverse Fourier transform, Proposition 4.13.2 yields

$$\int_S \varphi * \psi(y)\, R_n(y)\, d\pi(y) \;=\; \int_S \int_S L_y\psi(x)\, \varphi(x)\, d\pi(x)\, R_n(y)\, d\pi(y)$$

$$= \int_S \int_S L_x\psi(y)\, R_n(y)\, d\pi(y)\, \varphi(x)\, d\pi(x) \;=\; \int_S \check{\psi}(n)\, R_n(x)\, \varphi(x)\, d\pi(x)$$

$$= \check{\psi}(n)\, \check{\varphi}(n).$$

In the same way we get

$$\int_S \psi * \varphi(y)\, R_n(y)\, d\pi(y) \;=\; \check{\psi}(n)\, \check{\varphi}(n).$$

Thus $(\varphi * \psi)\pi = (\psi * \varphi)\pi$, and then $\varphi * \psi = \psi * \varphi$ by the continuity of both functions. ◇

Corollary 4.13.4 *Assume that S satisfies property (F). For $\varphi \in C(S)$, $\mu \in M(S)$ holds true*

$$\|\mu * \varphi\|_\infty \;\leq\; M\, \|\mu\|\, \|\varphi\|_\infty,$$

where M is the constant (independent of φ, μ) of condition (F).

In particular for $\varphi, \psi \in C(S)$ is valid

$$\|\varphi * \psi\|_\infty \;\leq\; M\, \|\varphi\|_1\, \|\psi\|_\infty \qquad \text{and} \qquad \|\varphi * \psi\|_\infty \;\leq\; M\, \|\psi\|_1\, \|\varphi\|_\infty.$$

In order to introduce translations and convolutions on $L^p(S,\pi)$ one could follow the lines of section 2.2. We will use throughout the denseness of $C(S)$ in $L^p(S,\pi)$. Obviously, the convolution of $\mu, \nu \in M(S)$ is given by

$$\mu * \nu(\varphi) \;=\; \int_S \int_S L_y\varphi(x)\, d\mu(y)\, d\nu(x) \;=\; \int_S \mu * \varphi(x)\, d\nu(x) \qquad (3)$$

for each $\varphi \in C(S)$. Since

$$|\mu * \nu(\varphi)| \;\leq\; \int_S |\mu * \varphi(x)|\, d|\nu|(x) \;\leq\; M\, \|\mu\|\, \|\nu\|\, \|\varphi\|_\infty,$$

$\mu * \nu \in M(S)$ and $\|\mu * \nu\| \;\leq\; M\, \|\mu\|\, \|\nu\|$. Note that $\mu * \nu = \nu * \mu$, since $L_y\varphi(x) = L_x\varphi(y)$.

Proposition 4.13.5 *Assume that S satisfies property (F). Let $\varphi \in C(S)$ and $\mu \in M(S)$. Then $(\mu * \varphi)\,\pi = \mu * \varphi\pi$.*

Proof. For $\psi \in C(S)$ it follows by Lebesgue's theorem of dominated convergence and Theorem 4.13.3,

$$(\mu*\varphi)\,\pi(\psi) = \int_S (\mu*\varphi)(x)\,\psi(x)\,d\pi(x) = \int_S \int_S L_y\varphi(x)\,d\mu(y)\,\psi(x)\,d\pi(x)$$

$$= \int_S \int_S L_y\varphi(x)\,\psi(x)\,d\pi(x)\,d\mu(y) = \int_S \int_S L_y\psi(x)\,\varphi(x)\,d\pi(x)\,d\mu(y)$$

$$= \int_S \int_S L_y\psi(x)\,d\mu(y)\,\varphi(x)\,d\pi(x) = \mu * \varphi\pi(\psi).$$

◇

Corollary 4.13.6 *Assume that S satisfies property (F). If $y \in S$, $\mu \in M(S)$ and $\varphi \in C(S)$, then $L_y\varphi \in C(S)$ and $\mu * \varphi \in C(S)$. Furthermore,*

$$\|L_y\varphi\|_1 \le M\,\|\varphi\|_1 \quad \text{and} \quad \|\mu * \varphi\|_1 \le M\,\|\mu\|\,\|\varphi\|_1,$$

where M is the constant of condition (F).

Proof. We know already that $L_y\varphi$ and $\mu * \varphi$ are continuous. By Proposition 4.13.5 we get

$$\|\mu * \varphi\|_1 = \|(\mu * \varphi)\pi\| = \|\mu * \varphi\pi\| \le M\,\|\mu\|\,\|\varphi\pi\|$$

$$= M\,\|\mu\|\,\|\varphi\|_1,$$

and since $\epsilon_y * \varphi = L_y\varphi$, we get also $\|L_y\varphi\|_1 \le M\,\|\varphi\|_1$. ◇

Now we can extend the translations and convolutions from $C(S)$ to $L^p(S,\pi)$, $1 \le p \le \infty$.

Proposition 4.13.7 *Assume that S satisfies property (F). Let $\varphi \in L^1(S,\pi)$, $y \in S$ and $\mu \in M(S)$.*

(1) There is a unique $L_y\varphi \in L^1(S,\pi)$ which agrees with definition (1) when $\varphi \in C(S)$. Moreover, $(L_y\varphi)\pi = \epsilon_y * \varphi\pi$ and $\|L_y\varphi\|_1 \le M\,\|\varphi\|_1$.

(2) There is a unique $\mu*\varphi \in L^1(S,\pi)$ which agrees with definition (2) when $\varphi \in C(S)$. Moreover, $(\mu * \varphi)\pi = \mu * \varphi\pi$ and $\|\mu * \varphi\|_1 \le M\,\|\mu\|\,\|\varphi\|_1$.

(3) $y \mapsto L_y\varphi$, $S \to L^1(S,\pi)$ is continuous, and

$$\mu * \varphi = \int_S L_y\varphi\,d\mu(y),$$

the $L^1(S,\pi)$-integral of $y \mapsto L_y\varphi$.

Proof.

(1) Choose a sequence $(\varphi_n)_{n\in\mathbb{N}}$ in $C(S)$ with $\|\varphi_n - \varphi\|_1 \to 0$. By Corollary 4.13.6 we have that $(L_y\varphi_n)_{n\in\mathbb{N}}$ is a Cauchy sequence in $L^1(S,\pi)$. Let the limit of $(L_y\varphi_n)_{n\in\mathbb{N}}$ be denoted by $L_y\varphi$. It is easily shown that $L_y\varphi \in L^1(S,\pi)$ is well-defined, and $L_y\varphi$ agrees with the element defined in (1) when φ is continuous. By Proposition 4.13.5 we have $(L_y\varphi_n)\pi = \epsilon_y * \varphi_n\pi$ for every $n \in \mathbb{N}$. Since $\|\epsilon_y*\varphi_n\pi - \epsilon_y*\varphi\pi\| \le M\ \|\varphi_n - \varphi\|_1 \to 0$ as $n \to \infty$, it follows

$$(L_y\varphi)\pi \;=\; \lim_{n\to\infty} (L_y\varphi_n)\pi \;=\; \lim_{n\to\infty} \epsilon_y * \varphi_n\pi \;=\; \epsilon_y * \varphi\pi.$$

$\|L_y\varphi\|_1 \le M\ \|\varphi\|_1$ is now obvious.

(2) is shown in the same way as part (1) by replacing ϵ_y by $\mu \in M(S)$.

(3) At first consider $\psi \in C(S)$, $x_0 \in S$, $\varepsilon > 0$. By the compactness of S we get a neighbourhood U of x_0 such that

$$|L_x\psi(y) - L_{x_0}\psi(y)| \;<\; \frac{\varepsilon}{3}$$

for all $y \in S$ and $x \in U$. Hence we have $\|L_x\psi - L_{x_0}\psi\|_1 < \frac{\varepsilon}{3}$.
For $\varphi \in L^1(S,\pi)$ choose $\psi \in C(S)$ such that $\|\varphi - \psi\|_1 < \frac{\varepsilon}{3M}$, and choose for that ψ a neighbourhood U of x_0 as above. For $x \in U$ holds

$$\|L_x\varphi - L_{x_0}\varphi\|_1 \;\le\; \|L_x\varphi - L_x\psi\|_1 + \|L_x\psi - L_{x_0}\psi\|_1 + \|L_{x_0}\psi - L_{x_0}\varphi\|_1 \;<\varepsilon.$$

To show $\mu * \varphi = \int\limits_S L_y\varphi\, d\mu(y)$ we note that this equality is pointwise valid whenever $\varphi \in C(S)$. If $\varphi \in L^1(S,\pi)$ choose $\varphi_n \in C(S)$ such that $\|\varphi - \varphi_n\|_1 \to 0$. Then

$$\|\int_S (L_y\varphi - L_y\varphi_n)\, d\mu(y)\|_1 \;\le\; \int_S \|L_y\varphi - L_y\varphi_n\|_1\, d|\mu|(y) \;\longrightarrow\; 0$$

and we get the equality for $\varphi \in L^1(S,\pi)$, too.

◇

Now we can extend the translation by $y \in S$ and the convolution by $\mu \in M(S)$ to actions on $L^p(S,\pi)$ for $1 \le p \le \infty$.

Corollary 4.13.8 *Assume that S satisfies property (F). Let $1 \le p \le \infty$ and $\varphi \in L^p(S,\pi)$, $y \in S$ and $\mu \in M(S)$. There exist unique $L_y\varphi \in L^p(S,\pi)$ and $\mu * \varphi \in L^p(S,\pi)$ which agree with definition (1) and (2), respectively. Moreover, $\|L_y\varphi\|_p \le M\ \|\varphi\|_p$, $\|\mu * \varphi\|_p \le M\ \|\mu\|\ \|\varphi\|_p$.*

Proof. Let $p = \infty$ and $\varphi \in L^\infty(S,\pi)$. Define $L_y\varphi$ as the unique element of $L^\infty(S,\pi)$ satisfying

$$\int_S L_y\varphi(x)\,\psi(x)\,d\pi(x) = \int_S \varphi(x)\,L_y\psi(x)\,d\pi(x)$$

for all $\psi \in L^1(S,\pi)$, where $L_y\psi$ is defined by Proposition 4.13.7(1). Obviously,

$$\|L_y\varphi\|_\infty = \sup_{\|\psi\|_1\le 1}\left|\int_S \varphi(x)\,L_y\psi(x)\,d\pi(x)\right| \le \sup_{\|\psi\|_1\le 1}\|L_y\psi\|_1\,\|\varphi\|_\infty \le M\,\|\varphi\|_\infty.$$

Similarly we get $\mu * \varphi \in L^\infty(S,\pi)$ and $\|\mu * \varphi\|_\infty \le M\,\|\mu\|\,\|\varphi\|_\infty$. For $1 < p < \infty$ we can apply the Riesz-Thorin interpolation theorem. Given $L_y \in B(L^1(S,\pi))$ and $L_y \in B(L^\infty(S,\pi))$ then $L_y \in B(L^p(S,\pi))$ and $\|L_y\varphi\|_p \le M\,\|\varphi\|_p$. In the same way we get an operator $\varphi \mapsto \mu * \varphi,\ L^p(S,\pi) \to L^p(S,\pi)$ with $\|\mu * \varphi\|_p \le M\,\|\mu\|\,\|\varphi\|_p$. ◇

Corollary 4.13.9 *Assume that S satisfies property (F). Let $1 \le p < \infty,\ \varphi \in L^p(S,\pi)$. Then $y \mapsto L_y\varphi,\ S \to L^p(S,\pi)$ is continuous and*

$$\mu * \varphi = \int_S L_y\varphi\,d\mu(y),$$

the $L^p(S,\pi)$-integral of $y \mapsto L_y\varphi$.

Proof. Let $\psi \in C(S),\ x_0 \in S,\ \varepsilon > 0$. There exists a neighbourhood of U of x_0 such that

$$|L_x\psi(y) - L_{x_0}\psi(y)|^p < \varepsilon^p$$

for all $y \in S$ and $x \in U$. It follows that $\|L_x\psi - L_{x_0}\psi\|_p < \varepsilon$ for all $x \in U$. Given $\varphi \in L^p(S,\pi)$ choose $\psi \in C(S)$ such that $\|\varphi - \psi\|_p < \frac{\varepsilon}{M}$. By Corollary 4.13.8 we know that for any $x \in S$ it holds $\|L_x(\varphi - \psi)\|_p < \varepsilon$ and therefore $\|L_x\varphi - L_{x_0}\varphi\|_p < 3\varepsilon$ for all $x \in U$, where U is a neighbourhood of x_0 chosen as above for the function ψ.

Finally, by Proposition 4.13.7(3) it follows that $\mu * \varphi$ is equal to $\int_S L_y\varphi\,d\mu(y)$ π-almost everywhere on S. Since $L^p(S,\pi) \subseteq L^1(S,\pi)$ for $1 < p \le \infty$ this implies $\mu * \varphi = \int_S L_y\varphi\,d\mu(y)$. ◇

In (3) the convolution $\mu * \nu$ of $\mu,\nu \in M(S)$ is defined. This convolution obeys the associativity law.

Lemma 4.13.10 *Assume that S satisfies property (F). For $x, y, z \in S$ we have*

$$\mu_{x,y} * \epsilon_z = \epsilon_x * \mu_{y,z}.$$

Proof. For each $n \in \mathbb{N}_0$ we have

$$\mu_{x,y} * \epsilon_z(R_n) = \int_S \int_S L_u(R_n)(v)\, d\mu_{x,y}(u)\, d\epsilon_z(v)$$

$$= \int_S \int_S R_n(u)\, R_n(v)\, d\mu_{x,y}(u)\, d\epsilon_z(v) = (R_n(x)R_n(y))R_n(z)$$

$$= R_n(x)(R_n(y)R_n(z)) = \epsilon_x * \mu_{y,z}(R_n).$$

Since the closure of the linear span of the polynomials $R_n(x)$, $n \in \mathbb{N}_0$, is dense in $C(S)$, it follows $\mu_{x,y} * \epsilon_z = \epsilon_x * \mu_{y,z}$. ◇

Proposition 4.13.11 *Assume that S satisfies property (F).*

(1) For $x, y, z \in S$, $\varphi \in C(S)$ holds $\mu_{x,y} * \varphi(z) = \mu_{y,z} * \varphi(x)$.
(2) For $\mu, \nu, \omega \in M(S)$ holds $(\mu * \nu) * \omega = \mu * (\nu * \omega)$.

Proof.

(1) By definition we have $\mu_{x,y} * \varphi(z) = \mu_{x,y} * \epsilon_z(\varphi) = \mu_{x,y}(L_z\varphi)$, and Lemma 4.13.10 implies

$$\mu_{x,y} * \varphi(z) = \mu_{x,y} * \epsilon_z(\varphi) = \epsilon_x * \mu_{y,z}(\varphi) = \mu_{y,z} * \varphi(x).$$

(2) $$(\mu * \nu) * \omega(\varphi) = \int_S \int_S L_z\varphi(u)\, d(\mu * \nu)(u)\, d\omega(z)$$

$$= \int_S \int_S \int_S \mu_{x,y}(L_z\varphi)\, d\mu(x)\, d\nu(y)\, d\omega(z)$$

$$= \int_S \int_S \int_S \mu_{y,z}(L_x\varphi)\, d\mu(x) d\nu(y)\, d\omega(z)$$

$$= \int_S \int_S L_x\varphi(v)\, d\mu(x)\, d(\nu * \omega)(v) = \mu * (\nu * \omega)(\varphi),$$

holds for each $\varphi \in C(S)$, where we used the statement of (1).

◇

For each $\mu \in M(S)$ we define $\|\mu\|_0 = M\,\|\mu\|$. Then $\|\cdot\|_0$ is a norm on $M(S)$ which is equivalent to the usual norm $\|\cdot\|$ on $M(S)$. For $f \in L^1(S,\pi)$ we define $\|f\|_0 = M\,\|f\|_1 = M\,\|f\pi\|$.

Applying Proposition 4.13.7 and Proposition 4.13.11 it follows

Theorem 4.13.12 *Assume that S satisfies property (F). With the convolution $*$ and the norm $\|\cdot\|_0$ the Banach space $M(S)$ is a Banach algebra. $L^1(S,\pi)$ is a closed ideal in $M(S)$. In particular, $L^1(S,\pi)$ is a Banach algebra with norm $\|\cdot\|_0$.*

Remark. Setting

$$\mu^*(\varphi) = \overline{\int_S \overline{\varphi(x)}\, d\mu(x)} \qquad \text{for } \varphi \in C(S),$$

an appropriate involution is defined on $M(S)$ such that $M(S)$ and $L^1(S,\pi)$ are Banach $*$-algebras.

We present now examples of $(R_n(x))_{n\in\mathbb{N}_0}$ generating a polynomial hypergroup on $\mathbb{N}_0$ such that S satisfies property (F). Using positive semicharacters γ of a polynomial hypergroup $\mathbb{N}_0$ such that S satisfies property (F) and $\frac{1}{\gamma} = \check{a}$ for some measure $a \in M(S)$, we can deduce many further examples of polynomial hypergroups $\mathbb{N}_0 = K_\gamma$ such that the dual space $S(K_\gamma)$ satisfies property (F). We refer to Corollary 3.11.15 and Proposition 3.11.16.

Remark. If $(R_n(x))_{n\in\mathbb{N}_0}$ generates a polynomial hypergroup with property (P), see Proposition 3.4.10, then S satisfies property (F). In that case the measures $\mu_{x,y}$ are probability measures and exist, moreover, for all $x, y \in D_s$.

Examples.

(1) **Jacobi polynomials** $(R_n^{(\alpha,\beta)}(x))_{n\in\mathbb{N}_0}$, see subsection 4.5.2. If the pair of parameters (α,β) is a member of J_s, where

$$J_s = \{(\alpha,\beta) : \alpha \geq \beta \geq -1 \text{ and } (\beta \geq -\frac{1}{2} \text{ or } \alpha+\beta \geq 0)\}$$

we know that $(R_n^{(\alpha,\beta)}(x))_{n\in\mathbb{N}_0}$ generates a polynomial hypergroup on $\mathbb{N}_0$ such that the Pontryagin duality is valid, see subsection 4.15. Hence, $[-1,1] = S = D_s = D$ and the measure $\mu_{x,y}$ generate a hypergroup on S. (Note that J_s is a proper subset of V, where V is defined in subsection 4.5.2.) The special case of $\alpha = \beta$, $\alpha \geq -\frac{1}{2}$ is already mentioned in subsection 1.2.2.

(2) **Generalized Chebyshev polynomials** $(T_n^{(\alpha,\beta)}(x))_{n\in\mathbb{N}_0}$, see subsection 4.5.3. We know that $(T_n^{(\alpha,\beta)}(x))_{n\in\mathbb{N}_0}$ generates a polynomial hypergroup on $\mathbb{N}_0$ whenever $\alpha \geq \beta > -1$, $\alpha+\beta+1 \geq 0$. Applying Theorem 1(iii) of [398] we obtain that S satisfies property (F), if $\alpha \geq \beta > -1$, $\alpha+\beta+1 > 0$ or $\alpha = \beta = -\frac{1}{2}$. Note even that property (P) is valid whenever $\alpha > \beta \geq -\frac{1}{2}$ or $\alpha = \beta = -\frac{1}{2}$. See also section 4.15 on dual convolution. Moreover, $[-1,1] = S = D_s = D$.

(3) **Cosh-polynomials** $(R_n(x))_{n\in\mathbb{N}_0}$
Fix $a \geq 0$ and define $(R_n(x))_{n\in\mathbb{N}_0}$, as in subsection 4.5.7.3. The polynomials $R_n(x)$ generate a polynomial hypergroup. Determine $0 < \gamma \leq \frac{1}{4}$

by $\cosh(a) = \frac{1}{2\sqrt{\gamma}}$. Then $S = [-2\sqrt{\gamma}, 2\sqrt{\gamma}]$ and Corollary 3.11.15 implies that S satisfies property (F) for each $a \geq 0$. See also section 4.15.

13.2 *Homogeneous Banach spaces on S*

Throughout this subsection we assume that $(R_n(x))_{n\in\mathbb{N}_0}$ generates a polynomial hypergroup such that S satisfies **property (F)**. In particular the translation $L_x \in B(L^1(S,\pi))$ for $x \in S$ is a well-defined operator. M will always be the smallest constant with $\|\mu_{x,y}\| \leq M$ for all $x, y \in S$, see subsection 4.13.1. Note that $M \geq 1$. Following the lines of [364] we introduce homogeneous Banach spaces on S.

Definition *We call a linear subspace B of $L^1(S,\pi)$ a homogeneous Banach space on S with respect to $(R_n(x))_{n\in\mathbb{N}_0}$, if it is endowed with a norm $\|\cdot\|_B$ such that*

(B1) $R_n|S \in B$ *for all* $n \in \mathbb{N}_0$.
(B2) B *is complete with respect to* $\|\cdot\|_B$ *and* $\|\cdot\|_1 \leq \|\cdot\|_B$.
(B3) For every $\varphi \in B$, $x \in S$ *we have* $L_x\varphi \in B$ *and* $\|L_x\varphi\|_B \leq M\,\|\varphi\|_B$.
(B4) For every $\varphi \in B$ *the map* $x \mapsto L_x\varphi$, $S \to B$ *is continuous.*

We know already that $B = L^p(S,\pi)$, $1 \leq p < \infty$ and $B = C(S)$ are homogeneous Banach spaces on S with respect to $(R_n(x))_{n\in\mathbb{N}_0}$.

Proposition 4.13.13 *Let B be a homogeneous Banach space on S with respect to $(R_n(x))_{n\in\mathbb{N}_0}$. Then for each $\psi \in B$ and $\varphi \in L^1(S,\pi)$, $\varphi * \psi$ is an element of B and $\|\varphi * \psi\|_B \leq M\,\|\varphi\|_1\,\|\psi\|_B$.*

Proof. Let $\varphi \in C(S)$. Considering the B-valued integral we have for $\psi \in B$,

$$\varphi * \psi = \int_S \varphi(x)\, L_x\psi\, d\pi(x) \in B$$

and $\|\varphi * \psi\|_B \leq M\,\|\varphi\|_1\,\|\psi\|_B$. If $\varphi \in L^1(S,\pi)$ choose a sequence $(\varphi_n)_{n\in\mathbb{N}}$, $\varphi_n \in C(S)$, such that $\|\varphi - \varphi_n\|_1 \to 0$. ◇

Corollary 4.13.14 *Let B be a homogeneous Banach space on S with respect to $(R_n(x))_{n\in\mathbb{N}_0}$. B is a Banach algebra with convolution as multiplication and norm $\|\cdot\|_0 = M\,\|\cdot\|_B$.*

Proof. We have only to note that $B \subseteq L^1(S,\pi)$ and $\|\cdot\|_1 \leq \|\cdot\|_B$. ◇

Definition *We call a homogeneous Banach space on S with respect to $(R(x))_{n\in\mathbb{N}_0}$* **character-invariant***, if for every $\varphi \in B$, $n \in \mathbb{N}_0$ we have $R_n|S \cdot \varphi \in B$ and*
$\|R_n|S \cdot \varphi\|_B \leq \|\varphi\|_B$.

Proposition 4.13.15 *Let B be a character-invariant homogeneous Banach space on S with respect to $(R_n(x))_{n\in\mathbb{N}_0}$. Then for each $\psi \in B$ and $d \in l^1(h)$, $\hat{d}|S \cdot \psi$ is an element of B and $\|\hat{d}|S \cdot \psi\|_B \leq \|d\|_1 \, \|\psi\|_B$.*

Proof. For every $n \in \mathbb{N}_0$ the partial sum $\varphi_n = \left(\sum\limits_{k=0}^{n} d(k)R_k h(k)\right)|S \cdot \psi$ is an element of B, and for $n > m$ we have

$$\|\varphi_n - \varphi_m\|_B \leq \left(\sum_{k=m+1}^{n} |d(k)| \, h(k)\right) \|\psi\|_B.$$

Now it is clear that $(\varphi_n)_{n\in\mathbb{N}_0}$ converges in B to $\hat{d}|S \cdot \psi$ and $\|\hat{d}|S \cdot \psi\|_B \leq \|d\|_1 \, \|\psi\|_B$. ◇

13.2.1 *4.13.2.1 Wiener algebra $A(S)$*

Denote by $A(S) = \{\varphi \in C(S) : \check{\varphi} \in l^1(h)\}$ and set $\|\varphi\|_A = \|\check{\varphi}\|_1$. We have already studied $A(S)$ in section 4.1, see Theorem 4.1.5. Note that by Theorem 3.3.8,

$$\{\varphi \in C(S) : \check{\varphi} \in l^1(h)\} = \{\hat{d}|S : d \in l^1(h)\}.$$

Theorem 4.13.16 *$A(S)$ with norm $\|\cdot\|_A$ is a homogeneous Banach space on S with respect to $(R_n(x))_{n\in\mathbb{N}_0}$. $A(S)$ is also character-invariant.*

Proof. Obviously $A(S)$ is a linear space, and $\|\cdot\|_A$ is a norm by the uniqueness theorem, Theorem 3.1.8. Since $\check{R}_m(n) = h(n)^{-1}\delta_{m,n}$, condition (B1) is fulfilled. Since $\varphi \mapsto \check{\varphi}$, $A(S) \to l^1(h)$, is an isometric isomorphism from $A(S)$ onto $l^1(h)$, we get that $(A(S), \|\cdot\|_A)$ is complete. Since $\|\varphi\|_A = \|\check{\varphi}\|_1 \geq \|\check{\varphi}\|_2 = \|\varphi\|_2 \geq \|\varphi\|_1$, we see that (B2) is valid. For $\varphi \in A(S)$, $x \in S$, we know $L_x\varphi \in C(S)$, $(L_x\varphi)^\vee(n) = R_n(x)\check{\varphi}(n)$, and hence $L_x\varphi \in A(S)$, and $\|L_x\varphi\|_A \leq \|\varphi\|_A \leq M\,\|\varphi\|_A$.
To show (B4) let $x_0 \in S$, $\varepsilon > 0$. There exists $N \in \mathbb{N}$ and $\psi \in A(S)$ such that $\|\varphi - \psi\|_A < \frac{\varepsilon}{4M}$ and $\check{\psi}(n) = \check{\varphi}(n)$ for $n = 0, 1, ..., N$ and $\check{\psi}(n) = 0$ for $n \geq N+1$. Furthermore there is some $\delta > 0$ with $\|L_x\psi - L_{x_0}\psi\|_A < \frac{\varepsilon}{2}$ for all $x \in S$ such that $|x - x_0| < \delta$. It follows $\|L_x\varphi - L_{x_0}\varphi\|_A < \varepsilon$ for all $x \in S$ with $|x - x_0| < \delta$. Finally, $(R_n|S \cdot \varphi)^\vee = L_n\check{\varphi}$, where L_n is the

translation operator on $l^1(h)$, and we conclude that $R_n|S \cdot \varphi \in A(S)$ and $\|R_n|S \cdot \varphi\|_A \leq \|\varphi\|_A$. ◇

Proposition 4.13.17 *$A(S)$ is the smallest character-invariant homogeneous Banach space on S with respect to $(R_n(x))_{n\in\mathbb{N}_0}$. That is: If B is a character-invariant homogeneous Banach space on S with respect to $(R_n(x))_{n\in\mathbb{N}_0}$, then $A(S) \subseteq B$ and $\|\cdot\|_B \leq c\,\|\cdot\|_A$, $c > 0$.*

Proof. For $\varphi \in A(S)$ and any $n \in \mathbb{N}$ put $\varphi_n = \sum_{k=0}^{n} \check{\varphi}(k) R_k h(k)$. Then $\varphi_n \in B$, and for $n > m$,

$$\|\varphi_n - \varphi_m\|_B \leq \sum_{k=m+1}^{n} |\check{\varphi}(k)|\ \|R_k\|_B\, h(k).$$

Since B is character-invariant, we have $\|R_k\|_B = \|R_k R_0\|_B \leq \|R_0\|_B$. Putting $c = \|R_0\|_B$ we obtain that $(\varphi_n)_{n\in\mathbb{N}_0}$ converges in B to φ, and $\|\varphi\|_B \leq c\,\|\check{\varphi}\|_1 = c\,\|\varphi\|_A$. ◇

13.2.2 *4.13.2.2 The $A^p(S)$-algebras*

Let $1 \leq p < \infty$ and denote by

$$A^p(S) = \{\varphi \in L_1(S,\pi) : \check{\varphi} \in l^p(h)\},$$

and set $\|\varphi\|^p := \|\varphi\|_1 + \|\check{\varphi}\|_p$. We follow the investigations of [401] for locally compact abelian groups, see also [400,Ch.6]. However, we have to have in mind that S bears only a weak hypergroup structure. Note that $A^1(S)$ is the Wiener algebra $A(S)$ with a different but equivalent norm. In fact, $\|\check{\varphi}\|_1 \geq \|\check{\varphi}\|_2 = \|\varphi\|_2 \geq \|\varphi\|_1$, and hence

$$\|\varphi\|_A = \|\check{\varphi}\|_1 \leq \|\varphi\|^1 = \|\check{\varphi}\|_1 + \|\varphi\|_1 \leq 2\,\|\varphi\|_A.$$

Proposition 4.13.18 *Let $1 \leq p < \infty$. $A^p(S)$, equipped with $\|\cdot\|^p$, is a Banach space.*

Proof. Obviously, $\|\cdot\|^p$ is a norm. $A^p(S)$ is also complete. Let $(\varphi_n)_{n\in\mathbb{N}}$ be a Cauchy sequence. Then $(\varphi_n)_{n\in\mathbb{N}}$ and $(\check{\varphi}_n)_{n\in\mathbb{N}}$ are Cauchy sequences in $L^1(S,\pi)$ and $l^p(h)$, respectively. Hence there exist $\varphi \in L^1(S,\pi)$ and $\psi \in l^p(h)$ such that $\|\varphi_n - \varphi\|_1 \to 0$ and $\|\check{\varphi}_n - \psi\|_p \to 0$. Then $\|\check{\varphi}_n - \check{\varphi}\|_\infty \to 0$, and $\|\check{\varphi} - \psi\|_\infty \leq \|\check{\varphi} - \check{\varphi}_n\|_\infty + \|\check{\varphi}_n - \psi\|_p \to 0$. Hence $\check{\varphi} = \psi \in l^p(h)$, and $\|\varphi_n - \varphi\|^p \to 0$. ◇

Theorem 4.13.19 *Let $1 \leq p < \infty$. $A^p(S)$ with norm $\|\cdot\|^p$ is a homogeneous Banach space on S with respect to $(R_n(x))_{n\in\mathbb{N}_0}$. $A^p(S)$ is also character-invariant.*

Proof. Obviously, $A^p(S)$ satisfies properties (B1) and (B2). Since $(L_x\varphi)^\vee(n) = R_n(x)\check{\varphi}(n)$, we have $L_x\varphi \in A^p(S)$ for each $x \in S$, $\varphi \in A^p(S)$. Furthermore,

$$\|L_x\varphi\|^p = \|L_x\varphi\|_1 + \left(\sum_{k=0}^{\infty} |R_k(x)\,\check{\varphi}(k)|^p\, h(k)\right)^{1/p}$$

$$\leq M\,\|\varphi\|_1 + \|\check{\varphi}\|_p \leq M\,\|\varphi\|^p,$$

and (B3) is shown. Let $\varphi \in A^p(S)$, $x_0 \in S$ and $\varepsilon > 0$. Choose $N \in \mathbb{N}$ such that $\sum\limits_{k=N+1}^{\infty} |\check{\varphi}(k)|^p\, h(k) < (\frac{\varepsilon}{4})^p$, and put $\psi = \sum\limits_{k=0}^{N} \check{\varphi}(k) R_k h(k)$. Then $\|\check{\varphi} - \check{\psi}\|_p < \frac{\varepsilon}{4}$. Since $(L_y\psi)^\vee(n) = R_n(y)\check{\psi}(n)$ for any $y \in S$, $n \in \mathbb{N}_0$, there exists some $\delta > 0$ such that $\|(L_x\psi)^\vee - (L_{x_0}\psi)^\vee\|_p < \frac{\varepsilon}{2}$ for all $x \in S$ with $|x - x_0| < \delta$. It follows $\|(L_x\varphi)^\vee - (L_{x_0}\varphi)^\vee\|_p < \varepsilon$ for all $x \in S$ with $|x - x_0| < \delta$. Since $L^1(S,\pi)$ is a homogeneous Banach space there is some $\eta > 0$ such that $\|L_x\varphi - L_{x_0}\varphi\|^p < 2\varepsilon$ for all $x \in S$ with $|x - x_0| < \eta$, and (B4) is shown. $A^p(S)$ is character-invariant, too. In fact, $(R_n|S\cdot\varphi)^\vee = L_n\check{\varphi}$, L_n the translation operator on $l^p(h)$. Hence $R_n|S\cdot\varphi \in A^p(S)$ and

$$\|R_n|S\cdot\varphi\|^p \leq \|\varphi\|_1 + \|L_n\check{\varphi}\|_p \leq \|\varphi\|^p.$$

◇

We have the following relations between the A^p-algebras, which can be easily checked. $A^1(S) = A(S)$, $A^2(S) = L^2(S,\pi)$ and for $1 \leq p \leq q < \infty$ holds $A^p(S) \subseteq A^q(S)$.

13.2.3 *4.13.2.3 Beurling algebra $A_*(S)$*

Denote by

$$A_*(S) = \{\varphi \in A(S) : \sum_{k=0}^{\infty} \left(\sup_{l\geq k} |\check{\varphi}(l)|\right) h(k) < \infty\}.$$

$A_*(S)$ is a linear space. Setting $\tilde{\varphi}(k) := \sup\limits_{l\geq k} |\check{\varphi}(l)|$ we define

$$\|\varphi\|_{A_*} = \sum_{k=0}^{\infty} \tilde{\varphi}(k)\, h(k).$$

$\|\cdot\|_{A_*}$ is a norm on $A_*(S)$ and clearly

$$\|\varphi\|_{A_*} \geq \|\varphi\|_A \geq \|\varphi\|_1 \qquad \text{for all } \varphi \in A_*(S).$$

Proposition 4.13.20 *Assume that the Haar weights $h(n)$ are increasing. Then $A_*(S)$ with norm $\|\cdot\|_{A_*}$ is a Banach space.*

Proof. We have to show that $A_*(S)$ is complete. Let $(\varphi_n)_{n\in\mathbb{N}}$ be a Cauchy sequence in $A_*(S)$. Then $(\varphi_n)_{n\in\mathbb{N}}$ is a Cauchy sequence in $A(S)$. Put $\varphi = \lim\limits_{n\to\infty} \varphi_n \in A(S)$. Given $\varepsilon > 0$ and $M \in \mathbb{N}$ there exists $N_0 \in \mathbb{N}$ such that

$$\|\varphi_n - \varphi_m\|_{A_*} < \frac{\varepsilon}{2} \qquad \text{for } n, m \geq N_0,$$

and for $\delta := \frac{\varepsilon}{2(M+1)} > 0$ there exists $N_M \geq N_0$ such that

$$\|\check{\varphi}h - \check{\varphi}_n h\|_\infty \leq \|\varphi - \varphi_n\|_A < \delta \qquad \text{for } n \geq N_M.$$

We have

$$\sum_{k=0}^{M} \left(\sup_{l\geq k} |\check{\varphi}(l) - \check{\varphi}_n(l)| \right) h(k)$$

$$\leq \sum_{k=0}^{M} \left(\sup_{l\geq k} |\check{\varphi}(l) - \check{\varphi}_{N_M}(l)| \right) h(k) + \sum_{k=0}^{\infty} \left(\sup_{l\geq k} |\check{\varphi}_{N_M}(l) - \check{\varphi}_n(l)| \right) h(k).$$

Applying the monotonicity of $(h(n))_{n\in\mathbb{N}_0}$ we get

$$\sum_{k=0}^{M} \left(\sup_{l\geq k} |\check{\varphi}(l) - \check{\varphi}_{N_M}(l)| \right) h(k) \leq \sum_{k=0}^{M} \sup_{l\geq k} |\check{\varphi}(l) - \check{\varphi}_{N_M}(l)|\, h(l)$$

$$\leq \sum_{k=0}^{M} \|\check{\varphi}h - \check{\varphi}_{N_M} h\|_\infty \leq (M+1)\,\delta = \frac{\varepsilon}{2}.$$

Furthermore we have for any $n \geq N_0$

$$\sum_{k=0}^{\infty} \left(\sup_{l\geq k} |\check{\varphi}_{N_M}(l) - \check{\varphi}_n(l)| \right) h(k) < \frac{\varepsilon}{2}.$$

So we have shown that for each $M \in \mathbb{N}$ and $n \geq N_0$

$$\sum_{k=0}^{M} \left(\sup_{l\geq k} |\check{\varphi}(l) - \check{\varphi}_n(l)| \right) h(k) < \varepsilon.$$

This yields $\varphi - \varphi_n \in A_*(S)$ and $\|\varphi - \varphi_n\|_{A_*} \leq \varepsilon$ for all $n \geq N_0$, which implies $\varphi \in A_*(S)$ and $\|\varphi - \varphi_n\|_{A_*} \to 0$ as $n \to \infty$. ◇

Proposition 4.13.21 *Assume that the Haar weights $h(n)$ are increasing. Then the linear subspace $A_{*,c}(S) = \{\varphi \in A_*(S) : \check{\varphi}(k) \neq 0$ at most for finitely many $k \in \mathbb{N}_0\}$ is dense in $A_*(S)$.*

Proof. Let $\varphi \in A_*(S)$, $\varepsilon > 0$. There is an $N \in \mathbb{N}_0$ such that $\sum_{k=N}^{\infty} \tilde{\varphi}(k)\, h(k) < \frac{\varepsilon}{2}$. Define $\psi \in A_{*,c}(S)$ by setting $\check{\psi}(k) = \check{\varphi}(k)$ for $k = 0, ..., 2N-1$ and $\check{\psi}(k) = 0$ for $k = 2N, 2N+1, \ldots$. Then $\|\varphi - \psi\|_{A_*} = S_1 + S_2$, where

$$S_1 = \sum_{k=0}^{N} \left(\sup_{l \geq k} |\check{\varphi}(l) - \check{\psi}(l)| \right) h(k)$$

and

$$S_2 = \sum_{k=N+1}^{\infty} \left(\sup_{l \geq k} |\check{\varphi}(l) - \check{\psi}(l)| \right) h(k).$$

Since $|\check{\varphi}(l) - \check{\psi}(l)| = 0$ for $l = 0, ..., 2N-1$ and $|\check{\varphi}(l) - \check{\psi}(l)| = |\check{\varphi}(l)|$ for $l \geq 2N$, we obtain for every $k = 0, ..., 2N-1$

$$\sup_{l \geq k} |\check{\varphi}(l) - \check{\psi}(l)| = \sup_{l \geq 2N} |\check{\varphi}(l)| = \tilde{\varphi}(2N),$$

in particular $S_1 = \sum_{k=0}^{N} \tilde{\varphi}(2N)\, h(k)$. Since $\tilde{\varphi}(2N) \leq ... \leq \tilde{\varphi}(N+1) \leq \tilde{\varphi}(N)$ and since the weights $h(n)$ are increasing, we get

$$S_1 \leq \sum_{k=0}^{N} \tilde{\varphi}(k+N)\, h(k+N) = \sum_{k=N}^{2N} \tilde{\varphi}(k)\, h(k) < \frac{\varepsilon}{2}.$$

Considering S_2 we have

$$S_2 = \sum_{k=N+1}^{2N-1} \tilde{\varphi}(2N)\, h(k) + \sum_{k=2N}^{\infty} \tilde{\varphi}(k)\, h(k)$$

$$\leq \sum_{k=N+1}^{2N-1} \tilde{\varphi}(k)\, h(k) + \sum_{k=2N}^{\infty} \tilde{\varphi}(k)\, h(k) < \frac{\varepsilon}{2}.$$

So $A_{*,c}(S)$ is dense in $A_*(S)$. ◇

Theorem 4.13.22 *Assume that the Haar weights $h(n)$ are increasing. Then $A_*(S)$ with norm $\|\cdot\|_{A_*}$ is a homogeneous Banach space on S with respect to $(R_n(x))_{n \in \mathbb{N}_0}$.*

Proof. Obviously, $R_n|S \in A_*(S)$ and $\|\varphi\|_{A_*} \geq \|\varphi\|_1$ for $\varphi \in A_*(S)$. Furthermore $(L_x\varphi)\tilde{}(k) = \sup_{l \geq k} |R_l(x)\check{\varphi}(l)| \leq \tilde{\varphi}(k)$ for $x \in S$, and so (B3) holds true. It remains to show (B4). Given $\varphi \in A_*(S)$, $x_0 \in S$ and $\varepsilon > 0$, choose $\psi \in A_{*,c}(S)$ such that $\|\varphi - \psi\|_{A_*} < \frac{\varepsilon}{4}$. Let N be the smallest integer such that $\check{\psi}(n) = 0$ for all $n \geq N+1$. Since

$$(L_x\psi - L_{x_0}\psi)\tilde{}(k) = \sup_{l \geq k} |R_l(x) - R_l(x_0)|\, |\check{\psi}(l)|$$

$$\leq \sup_{l \leq N} |R_l(x) - R_l(x_0)|\, \tilde{\psi}(0),$$

there exists $\delta > 0$ such that

$$(L_x\psi - L_{x_0}\psi)\tilde{}(k)\, h(k) < \frac{\varepsilon}{2(N+1)}$$

for $k = 0, .., N$ and $x \in S$ with $|x - x_0| < \delta$. Hence $\|L_x\psi - L_{x_0}\psi\|_{A_*} < \frac{\varepsilon}{2}$ for all $x \in S$, $|x - x_0| < \delta$. It follows

$$\|L_x\varphi - L_{x_0}\varphi\|_{A_*} < \frac{\varepsilon}{2} + 2\,\|\varphi - \psi\|_{A_*} < \varepsilon \qquad \text{for } x \in S,\ |x - x_0| < \delta.$$

◇

13.2.4 *4.13.2.4 The algebra $U_B(S)$ of uniform convergent Fourier series*

For $\varphi \in L^1(S, \pi)$ denote by

$$\mathcal{D}_n\varphi(x) = \sum_{k=0}^{n} \check{\varphi}(k)\, R_k(x)\, h(k)$$

for $x \in S$, $n \in \mathbb{N}_0$, the n-th Fourier partial sum for $R_k(x)$.

Let B be a homogeneous Banach space on S with respect to $(R_n(x))_{n\in\mathbb{N}_0}$. Set

$$U_B(S) = \{\varphi \in B : \lim_{n\to\infty} \|\mathcal{D}_n\varphi - \varphi\|_B = 0\},$$

$$\|\varphi\|_U := \sup_{n\in\mathbb{N}_0} \|\mathcal{D}_n\varphi\|_B.$$

Note that $R_k \in B$ implies $\mathcal{D}_n\varphi \in B$ for every $\varphi \in L^1(S,\pi)$. If $\varphi \in U_B(S)$ then $(\mathcal{D}_n\varphi)_{n\in\mathbb{N}_0}$ converges in B, and so $\sup_{n\in\mathbb{N}_0} \|\mathcal{D}_n\varphi\|_B < \infty$. If $\|\varphi\|_U = 0$ then $\mathcal{D}_n\varphi = 0$ for all $n \in \mathbb{N}_0$, and hence $\check{\varphi}(k) = 0$ for all $k \in \mathbb{N}_0$. The uniqueness theorem yields $\varphi = 0$ in $L^1(S,\pi)$. The other properties of a norm are easily shown.

Theorem 4.13.23 *Let B be a homogeneous Banach space on S with respect to $(R_n(x))_{n\in\mathbb{N}_0}$ satisfying $\sup_{n\in\mathbb{N}_0} \|R_n\|_B < \infty$. Then $U_B(S)$ with norm $\|\cdot\|_U$ is a homogeneous Banach space on S with respect to $(R_n(x))_{n\in\mathbb{N}_0}$.*

Proof. Since $\mathcal{D}_n R_k = R_k$ for $n \geq k$ we have $R_k \in U_B(S)$ for each $k \in \mathbb{N}_0$. By $\|\varphi\|_1 \leq \|\varphi\|_B \leq \|\varphi - \mathcal{D}_n\varphi\|_B + \|\mathcal{D}_n\varphi\|_B \leq \|\varphi - \mathcal{D}_n\varphi\|_B + \|\varphi\|_U$ it follows $\|\varphi\|_1 \leq \|\varphi\|_U$ for each $\varphi \in U_B(S)$. $U_B(S)$ is also complete. In fact, if $(\varphi_n)_{n\in\mathbb{N}}$ is a Cauchy sequence in $U_B(S)$, $\varepsilon > 0$, then there is $N_1 \in \mathbb{N}$ such that $\sup\limits_{n\in\mathbb{N}_0} \|\mathcal{D}_n\varphi_k - \mathcal{D}_n\varphi_l\|_B < \varepsilon$ for all $k, l \geq N_1$. Since $\|\varphi\|_B \leq \|\varphi\|_U$ the sequence $(\varphi_n)_{n\in\mathbb{N}}$ converges in B. Let $\varphi = \lim\limits_{n\to\infty} \varphi_n \in B$, and $N_2 \in \mathbb{N}$ such that $\|\varphi - \varphi_n\|_B < \varepsilon$ for every $n \geq N_2$. Then

$$|\check{\varphi}_k(m) - \check{\varphi}(m)| \leq \|\varphi_k - \varphi\|_1 \leq \|\varphi_k - \varphi\|_B < \varepsilon$$

for all $k \geq N_2$ and $m \in \mathbb{N}_0$. Let $C = \sup\limits_{n\in\mathbb{N}_0} \|R_n\|_B < \infty$. For each $n \in \mathbb{N}$ there exists $K_n \in \mathbb{N}$ such that

$$\|\mathcal{D}_n\varphi_k - \mathcal{D}_n\varphi\|_B \leq C \left(\sum_{m=0}^{n} |\check{\varphi}_k(m) - \check{\varphi}(m)|\, h(m) \right) < \varepsilon$$

for all $k \geq K_n$. We can assume $K_n \geq N := \max\{N_1, N_2\}$, and obtain for $k \geq N$ and each $n \in \mathbb{N}_0$

$$\begin{aligned} \|\mathcal{D}_n(\varphi - \varphi_k) - (\varphi - \varphi_k)\|_B &\leq \|\mathcal{D}_n(\varphi - \varphi_{K_n})\|_B + \|\mathcal{D}_n\varphi_{K_n} - \mathcal{D}_n\varphi_k\|_B \\ &\quad + \|\varphi - \varphi_k\|_B < 3\varepsilon. \end{aligned}$$

Finally $\varphi_N \in U_B(S)$ yields some $n_0 \in \mathbb{N}$ such that $\|\mathcal{D}_n\varphi_N - \varphi_N\|_B < \varepsilon$ for all $n \geq n_0$, and we get finally

$$\|\mathcal{D}_n\varphi - \varphi\|_B \leq \|\mathcal{D}_n(\varphi - \varphi_N) - (\varphi - \varphi_N)\|_B + \|\mathcal{D}_n\varphi_N - \varphi_N\|_B < 4\varepsilon$$

for $n \geq n_0$. Hence $\varphi \in U_B(S)$ and

$$\|\mathcal{D}_n\varphi - \mathcal{D}_n\varphi_k\|_B \leq \|\mathcal{D}_n(\varphi - \varphi_k) - (\varphi - \varphi_k)\|_B + \|\varphi - \varphi_k\|_B < 4\varepsilon$$

for $k \geq N$ implies $\varphi = \lim\limits_{k\to\infty} \varphi_k$ in $U_B(S)$. So far (B1) and (B2) are proven.

Since for $x, y \in S$

$$\mathcal{D}_n L_x\varphi(y) = \sum_{k=0}^{n} \check{\varphi}(k)\, R_k(x)\, R_k(y)\, h(k) = L_x\mathcal{D}_n\varphi(y),$$

and hence

$$\|\mathcal{D}_n L_x\varphi - L_x\varphi\|_B = \|L_x(\mathcal{D}_n\varphi - \varphi)\|_B \leq M\, \|\mathcal{D}_n\varphi - \varphi\|_B$$

for $\varphi \in U_B(S)$, it follows $L_x\varphi \in U_B(S)$. We have also

$$\|L_x\varphi\|_U = \sup_{n\in\mathbb{N}_0} \|\mathcal{D}_n L_x\varphi\|_B \leq M \sup_{n\in\mathbb{N}_0} \|\mathcal{D}_n\varphi\|_B = M\, \|\varphi\|_U,$$

and (B3) is shown. (Note that the constant M of condition (B3) for the homogeneous Banach space B, is also the constant for $U_B(S)$.)

It remains to prove that $x \mapsto L_x\varphi,\ S \to U_B(S)$ is continuous. Let $\varphi \in U_B(S)$, $x_0 \in S$ and $\varepsilon > 0$. There exists $N \in \mathbb{N}_0$ with $\|\mathcal{D}_n\varphi - \varphi\|_B < \frac{\varepsilon}{3M}$ for all $n \geq N$, and there exists $\delta_1 > 0$ such that

$$\|L_x\varphi - L_{x_0}\varphi\|_B < \frac{\varepsilon}{3} \qquad \text{for every } x \in S \text{ with } |x - x_0| < \delta_1.$$

Hence

$$\|\mathcal{D}_n L_x\varphi - \mathcal{D}_n L_{x_0}\varphi\|_B = \|L_x\mathcal{D}_n\varphi - L_{x_0}\mathcal{D}_n\varphi\|_B$$

$$\leq \|L_x(\mathcal{D}_n\varphi) - L_x\varphi\|_B + \|L_x\varphi - L_{x_0}\varphi\|_B + \|L_{x_0}\varphi - L_{x_0}\mathcal{D}_n\varphi\|_B < \varepsilon$$

for all $n \geq N$ and $x \in S,\ |x - x_0| < \delta_1$. There exists also $\delta_2 > 0$ such that

$$\|\mathcal{D}_n L_x\varphi - \mathcal{D}_n L_{x_0}\varphi\|_B \leq C\left(\sum_{k=0}^{n} |(L_x\varphi)\check{}(k) - (L_{x_0}\varphi)\check{}(k)|\, h(k)\right) < \varepsilon$$

for all $n = 0, ..., N$ and $x \in S$ with $|x - x_0| < \delta_2$. Hence we have shown $\|L_x\varphi - L_{x_0}\varphi\|_U \leq \varepsilon$ for all $x \in S$ with $|x - x_0| < \min\{\delta_1, \delta_2\}$. $\diamond$

Remark: If $B = A_*(S)$ the assumption $\sup_{n\in\mathbb{N}_0} \|R_n\|_B < \infty$ in Theorem 4.13.23 is, in general, not satisfied. In fact, using the notation of subsection 4.13.2.3 we have $\tilde{R_n}(k) = \frac{1}{h(n)}$ for $k = 0, ..., n$, and $\tilde{R_n}(k) = 0$ for $k = n + 1, n + 2, \ldots$. Hence $\|R_n\|_{A_*} = \frac{1}{h(n)} \sum_{k=0}^{n} h(k) = \frac{H(n)}{h(n)}$, and so $\sup_{n\in\mathbb{N}_0} \|R_n\|_B = \infty$, whenever $(R_n(x))_{n\in\mathbb{N}_0}$ fulfils property (H), see section 4.7.

13.2.5 *4.13.2.5 Lipschitz algebra $\mathrm{lip}_B(\lambda)$*

Let B be a homogeneous Banach space on S with respect to $(R_n(x))_{n\in\mathbb{N}_0}$. We assume in this subsection that $1 \in S$. This implies that 1 is not an isolated point of S, see the proof of the Remark at the end of section 4.10. Choose $\lambda \in]0, 1[$ and denote by

$$\mathrm{lip}_B(\lambda) = \{\varphi \in B : \lim_{x\to 1-} \frac{\|\varphi - L_x\varphi\|_B}{(1-x)^\lambda} = 0\},$$

$$\mathrm{Lip}_B(\lambda) = \{\varphi \in B : \sup_{x\in S\setminus\{1\}} \frac{\|\varphi - L_x\varphi\|_B}{(1-x)^\lambda} < \infty\}$$

and

$$\|\varphi\|_L = \|\varphi\|_B + \sup_{x\in S\setminus\{1\}} \frac{\|\varphi - L_x\varphi\|_B}{(1-x)^\lambda} \qquad \text{for } \varphi \in \mathrm{Lip}_B(\lambda).$$

Proposition 4.13.24 *Let B be a homogeneous Banach space on S with respect to $(R_n(x))_{n\in\mathbb{N}_0}$ and let $0<\lambda<1$. Then $Lip_B(\lambda)$ with norm $\|\cdot\|_L$ is a Banach space and $lip_B(\lambda)$ is a closed subspace.*

Proof. Obviously $\|\cdot\|_L$ is a norm on $\mathrm{Lip}_B(\lambda)$. We have to show the completeness of $\mathrm{Lip}_B(\lambda)$ and $\mathrm{lip}_B(\lambda)$ with respect to $\|\cdot\|_L$. If $(\varphi_n)_{n\in\mathbb{N}}$ is a Cauchy sequence in $\mathrm{Lip}_B(\lambda)$ and $\varepsilon>0$, then there exists $N\in\mathbb{N}$ such that

$$\sup_{x\in S\backslash\{1\}} \frac{\|(\varphi_n-\varphi_m)-L_x(\varphi_n-\varphi_m)\|_B}{(1-x)^\lambda} < \frac{\varepsilon}{4}$$

for $m,n\geq N$. Since $(\varphi_n)_{n\in\mathbb{N}}$ is a Cauchy sequence in B, there exists $\varphi\in B$ such that $\lim_{n\to\infty}\|\varphi-\varphi_n\|_B=0$. In particular, given $x\in S\backslash\{1\}$ there is $N(x)\in\mathbb{N}$, $N(x)\geq N$ such that

$$\|\varphi-\varphi_n\|_B \leq \frac{\varepsilon}{8M}\,(1-x)^\lambda$$

for all $n\geq N(x)$. Considering an $x\in S\backslash\{1\}$, we get

$$\frac{\|(\varphi-\varphi_n)-L_x(\varphi-\varphi_n)\|_B}{(1-x)^\lambda}$$

$$\leq \frac{\|(\varphi-\varphi_{N(x)})-L_x(\varphi-\varphi_{N(x)})\|_B}{(1-x)^\lambda} + \frac{\|(\varphi_{N(x)}-\varphi_n)-L_x(\varphi_{N(x)}-\varphi_n)\|_B}{(1-x)^\lambda}$$

$$\leq \frac{2M\,\|\varphi_{N(x)}-\varphi\|_B}{(1-x)^\lambda} + \sup_{y\in S\backslash\{1\}} \frac{\|(\varphi_{N(y)}-\varphi_n)-L_y(\varphi_{N(y)}-\varphi_n)\|_B}{(1-x)^\lambda}$$

$$< \frac{\varepsilon}{4}+\frac{\varepsilon}{4}=\frac{\varepsilon}{2} \qquad \text{for all } n\geq N.$$

This is valid for every $x\in S\backslash\{1\}$. If we can show $\varphi\in\mathrm{Lip}_B(\lambda)$, it follows $\lim_{n\to\infty}\|\varphi-\varphi_n\|_L=0$. Since $\varphi_N\in\mathrm{Lip}_B(\lambda)$ we get

$$\frac{\|\varphi-L_x\varphi\|_B}{(1-x)^\lambda} \leq \frac{\|(\varphi-\varphi_N)-L_x(\varphi-\varphi_N)\|_B}{(1-x)^\lambda} + \frac{\|\varphi_N-L_x\varphi_N\|_B}{(1-x)^\lambda},$$

and $\varphi\in\mathrm{Lip}_B(\lambda)$ follows. Moreover, if all $\varphi_n\in\mathrm{lip}_B(\lambda)$, then $\varphi\in\mathrm{lip}_B(\lambda)$. In fact, then there exists $\delta>0$ such that

$$\frac{\|\varphi_N-L_x\varphi_N\|_B}{(1-x)^\lambda} < \frac{\varepsilon}{2}$$

for all $x\in S\backslash\{1\}$ with $|x-1|<\delta$. Then the inequality above yields

$$\frac{\|\varphi-L_x\varphi\|_B}{(1-x)^\lambda} < \varepsilon$$

for $x\in S\backslash\{1\}$ with $|x-1|<\delta$, i.e. $\varphi\in\mathrm{lip}_B(\lambda)$. ◇

Theorem 4.13.25 *Let B be a homogeneous Banach space on S with respect to $(R_n(x))_{n\in\mathbb{N}_0}$ and let $0<\lambda<1$. Then $lip_B(\lambda)$ with norm $\|\cdot\|_L$ is a homogeneous Banach space on S with respect to $(R_n(x))_{n\in\mathbb{N}_0}$.*

Proof. We begin by proving $R_n \in \mathrm{lip}_B(\lambda)$ for $n\in\mathbb{N}$. We have

$$\frac{\|R_n - L_x R_n\|_B}{(1-x)^\lambda} = \frac{|1-R_n(x)|}{(1-x)^\lambda}\,\|R_n\|_B = |(1-x)^{1-\lambda}P(x)|\;\|R_n\|_B$$

$$\leq C\,(1-x)^\lambda, \qquad \text{where } P(x) = \frac{1-R_n(x)}{1-x}, \quad x\neq 1.$$

Since $0<\lambda<1$ we have $R_n \in \mathrm{lip}_B(\lambda)$. (B2) holds obviously. For $\varphi \in \mathrm{lip}_B(\lambda)$ and $x,y\in S$ it follows

$$\|L_y\varphi - L_x(L_y\varphi)\|_B \leq M\,\|\varphi - L_x\varphi\|_B,$$

and hence $L_y\varphi \in \mathrm{lip}_B(\lambda)$ and $\|L_y\varphi\|_L \leq M\,\|\varphi\|_L$, that is (B3).

It remains to derive (B4). If $\varphi\in\mathrm{lip}_B(\lambda)$, $y_0\in S$ and $\varepsilon>0$, then there is a $\delta>0$ such that

$$\frac{\|\varphi - L_x\varphi\|_B}{(1-x)^\lambda} \leq \frac{\varepsilon}{2M} \qquad \text{for all } x\in S\backslash\{1\} \quad \text{with } |x-1|<\delta,$$

and an $\eta>0$ such that

$$\|L_y\varphi - L_{y_0}\varphi\|_B < \frac{\varepsilon\delta^\lambda}{2(1+M)} \qquad \text{for all } y\in S \quad \text{with } |y-y_0|<\eta.$$

We obtain for any $y\in S$

$$\frac{\|(L_y\varphi - L_{y_0}\varphi) - L_x(L_y\varphi - L_{y_0}\varphi)\|_B}{(1-x)^\lambda}$$

$$\leq \frac{\|L_y(\varphi - L_x\varphi)\|_B + \|L_{y_0}(\varphi - L_x\varphi)\|_B}{(1-x)^\lambda}$$

$$\leq \frac{2M\,\|\varphi - L_x\varphi\|_B}{(1-x)^\lambda} < \varepsilon \qquad (i)$$

for all $x\in S\backslash\{1\}$ with $|x-1|<\delta$.

If $x\in S$, $|x-1|\geq\delta$ then $\frac{1}{(1-x)^\lambda} \leq \frac{1}{\delta^\lambda}$, and hence

$$\frac{\|(L_y\varphi - L_{y_0}\varphi) - L_x(L_y\varphi - L_{y_0}\varphi)\|_B}{(1-x)^\lambda}$$

$$\leq \frac{\|L_y\varphi - L_{y_0}\varphi\|_B + \|L_x(L_y\varphi - L_{y_0}\varphi)\|_B}{(1-x)^\lambda}$$

$$\leq \frac{(1+M)\,\|L_y\varphi - L_{y_0}\varphi\|_B}{(1-x)^\lambda} < \varepsilon. \qquad (ii)$$

Combining inequalities (i) and (ii) we have

$$\sup_{x\in S\setminus\{1\}} \frac{\|(L_y\varphi - L_{y_0}\varphi) - L_x(L_y\varphi - L_{y_0}\varphi)\|_B}{(1-x)^\lambda} \leq \varepsilon$$

for all $y \in S$, $|y - y_0| < \eta$. Hence the continuity of $y \mapsto L_y\varphi$, $S \to \mathrm{lip}_B(\lambda)$ follows. ◇

At the end of subsection 4.13.3 (Theorem 4.13.37), we give further examples of homogeneous Banach spaces on S. They are based on a certain differential operator, which will be investigated in subsection 4.13.3.

13.3 *Quadratic forms and a differential operator*

Before we investigate another homogeneous Banach space we introduce an unbounded operator in $L^2(S,\pi)$. This operator can be defined for every polynomial hypergroup without any assumption on a dual structure on S. If a bounded product formula (F) holds for $R_n(x)$, the operator is a differential operator, and in case of Jacobi polynomials this operator is the Jacobi differential operator, see for example [744].

Let K be a commutative hypergroup. A continuous function $q : K \to \mathbb{C}$ is called a **quadratic form**, if

$$\omega(x,y)(q) \;+\; \omega(x,\tilde{y})(q) \;=\; 2\,(q(x) + q(y)) \qquad (1)$$

for all $x, y \in K$.

Obviously a quadratic form q satisfies

$$q(e) \;=\; 0, \qquad q(\tilde{x}) \;=\; q(x) \qquad \text{and}$$

$$\mu * \nu(q) \;+\; \mu * \tilde{\nu}(q) \;=\; 2\,(\nu(q)\mu(K) + \mu(q)\nu(K)) \qquad (2)$$

for all $\mu, \nu \in M(K)$ with compact support.

Proposition 4.13.26 *Let q be a quadratic form, and let $\mu \in M(K)$ have compact support. Then*

$$\mu^n(q) \;=\; n^2\mu(K)^{n-1}\mu(q) \;-\; (n(n-1)/2)\,\mu(K)^{n-2}\mu * \tilde{\mu}(q)$$

for each $n \in \mathbb{N}, n \geq 2$, where $\mu^n = \mu * \cdots * \mu$, the n-fold convolution of μ.

Proof. We prove the statement by induction on n. By (2) we obtain

$$\mu^{n+1}(q) \;=\; 2\left(\mu^n(q)\mu(K) + \mu(K)^n\mu(q)\right) \;-\; \mu^n * \tilde{\mu}(q)$$

and

$$\mu^n * \tilde{\mu}(q) \;=\; \frac{1}{2}\left(\mu^{n-1} * (\mu * \tilde{\mu})(q) \;+\; \mu^{n-1} * (\mu * \tilde{\mu})(q)\right)$$

$$=\; \mu^{n-1}(q)\mu(K)^2 \;+\; \mu(K)^{n-1}\mu * \tilde{\mu}(q).$$

Hence by the induction assumption

$$\mu^{n+1}(q) \;=\; 2\mu(K)^n\mu(q)$$

$$+\, 2\mu(K)\,\left(n^2\mu(K)^{n-1}\mu(q) \;-\; (n(n-1)/2)\,\mu(K)^{n-2}\mu * \tilde{\mu}(q)\right)$$

$$-\,\mu(K)^2\,\left((n-1)^2\mu(K)^{n-2}\mu(q) \;-\; ((n-1)(n-2)/2)\,\mu(K)^{n-3}\mu * \tilde{\mu}(q)\right)$$

$$-\,\mu(K)^{n-1}\mu * \tilde{\mu}(q)$$

$$=\; \mu(K)^n(n+1)^2\mu(q) \;-\; \mu(K)^{n-1}\,((n+1)n/2)\,\mu * \tilde{\mu}(q).$$

◇

Corollary 4.13.27 *Let q be a quadratic form. Then for each $x \in K$,*

$$\lim_{n\to\infty} \frac{\epsilon_x^n(q)}{n^2} \;=\; q(x) - \frac{1}{2}\,\omega(x,\tilde{x})(q).$$

In particular, if q is nonnegative (i.e. $q(x) \geq 0$ for all $x \in K$) then

$$\omega(x,\tilde{x})(q) \;\leq\; 2q(x) \;\leq\; \omega(x,x)(q).$$

Proof. Consider $\mu = \epsilon_x$ and apply Proposition 4.13.26. ◇

Corollary 4.13.28 *The only bounded quadratic form q on K is $q = 0$.*

Proof. Let $x \in K$. Considering $\mu = \epsilon_x^n * \epsilon_{\tilde{x}}^n$ and $\nu = \epsilon_x * \epsilon_{\tilde{x}} = \omega(x,\tilde{x})$, equation (2) yields

$$\epsilon_x^{n+1} * \epsilon_{\tilde{x}}^{n+1}(q) \;=\; \epsilon_x^n * \epsilon_{\tilde{x}}^n(q) \;+\; \epsilon_x * \epsilon_{\tilde{x}}(q).$$

Therefore $\epsilon_x^n * \epsilon_{\tilde{x}}^n(q) \;=\; n \cdot \epsilon_x * \epsilon_{\tilde{x}}(q)$. The boundedness of q implies that $\omega(x,\tilde{x})(q) = \epsilon_x * \epsilon_{\tilde{x}}(q) = 0$. Hence $\lim_{n\to\infty} \frac{\epsilon_x^n(q)}{n^2} = q(x)$. Again by the boundedness of q we get $q(x) = 0$. ◇

Now we characterize the quadratic forms defined on polynomial hypergroups $K = \mathbb{N}_0$. For every hypergroup K with the property $\tilde{x} = x$ for each $x \in K$, equation (1) reduces to

$$\omega(x,y)(q) \;=\; q(x) + q(y). \tag{3}$$

Hence a quadratic form on the polynomial hypergroup $\mathbb{N}_0$ is exactly an additive sequence with respect to $(R_n(x))_{n\in\mathbb{N}_0}$, see section 4.10.

Theorem 4.10.11 tells us that the additive sequences $(q(n))_{n\in\mathbb{N}_0}$ with respect to $(R_n(x))_{n\in\mathbb{N}_0}$ are exactly of the form

$$q(n) \;=\; c\, R_n'(1), \tag{4}$$

$c \in \mathbb{C}$.

Using formula (3) of Proposition 4.10.3 we have

$$a_0\,(R_{n+1}'(1) \;-\; R_n'(1)) \;=\; \frac{H(n)}{h(n)\,a_n},$$

and it follows

Proposition 4.13.29 *Let $\mathbb{K} = \mathbb{N}_0$ be a polynomial hypergroup. Then the additive sequences on $K = \mathbb{N}_0$ are given by $q(0) = 0$ and*

$$q(n) \;=\; c\cdot\sum_{k=0}^{n-1}\frac{H(k)}{h(k)\,a_k}, \qquad c\in\mathbb{C}, \quad \textit{for } n\in\mathbb{N}.$$

Examples:

(1) Consider the polynomial hypergroup generated by Jacobi polynomials $R_n^{(\alpha,\beta)}(x)$. We know already (see section 4.10):

$$q(n) \;=\; c\,\frac{n(n+\alpha+\beta+1)}{\alpha+\beta+2}, \qquad c\in\mathbb{C}. \tag{5}$$

(2) For the generalized Chebyshev polynomials $T_n^{(\alpha,\beta)}(x)$, see 4.5.3, we obtain:

If $n = 2k$, then

$$(T_n^{(\alpha,\beta)})'(x) \;=\; 4x\cdot(R_k^{(\alpha,\beta)})'(2x^2-1)$$

$$=\; \frac{2k(k+\alpha+\beta+1)}{\alpha+1}\; x\, R_{k-1}^{(\alpha+1,\beta+1)}(2x^2-1)$$

$$=\; \frac{n(n+2\alpha+2\beta+2)}{2(\alpha+1)}\; T_{n-1}^{(\alpha+1,\beta)}(x),$$

and if $n = 2k+1$, then

$$(T_n^{(\alpha,\beta)})'(x) = R_k^{(\alpha,\beta+1)}(2x^2-1) + 4x^2 \cdot (R_k^{(\alpha,\beta+1)})'(2x^2-1)$$

$$= T_{n-1}^{(\alpha,\beta+1)}(x) + 4x^2 \frac{k(k+\alpha+\beta+2)}{2(\alpha+1)} R_{k-1}^{(\alpha+1,\beta+2)}(2x^2-1)$$

$$= T_{n-1}^{(\alpha,\beta+1)}(x) + 4x \frac{k(k+\alpha+\beta+2)}{2(\alpha+1)} T_{n-2}^{(\alpha+1,\beta+1)}(x)$$

$$= T_{n-1}^{(\alpha,\beta+1)}(x) + \frac{(n-1)(n+2\alpha+2\beta+3)}{2(\alpha+1)} x\, T_{n-2}^{(\alpha+1,\beta+1)}(x).$$

Therefore,

$$q(n) = (T_n^{(\alpha,\beta)})'(1) = c\,\frac{n(n+2\alpha+2\beta+2)}{2(\alpha+1)} \qquad \text{if } n = 2k,\ k \in \mathbb{N}, \tag{6}$$

and

$$q(n) = (T_n^{(\alpha,\beta)})'(1) = c\left(1 + \frac{(n-1)(n+2\alpha+2\beta+3)}{2(\alpha+1)}\right) \qquad \text{if } n = 2k+1,\ k \in \mathbb{N}_0.$$

(3) The Geronimus polynomials $R_n(x) = R_n(x; \frac{1}{4}, \alpha)$, see 4.5.7.1 generate a polynomial hypergroup on $\mathbb{N}_0$ for $0 < \alpha \le 2$. Since they can be represented by

$$R_n(x) = \frac{2(\alpha-1)}{\alpha+(2-\alpha)n} T_n(x) + \frac{(2-\alpha)(n+1)}{\alpha+(2-\alpha)n} R_n^{(\frac{1}{2},\frac{1}{2})}(x),$$

we have

$$R_n'(x) = \frac{2(\alpha-1)n^2}{\alpha+(2-\alpha)n} R_n^{(\frac{1}{2},\frac{1}{2})}(x) + \frac{(2-\alpha)n(n+1)(n+2)}{3(\alpha+(2-\alpha)n)} R_n^{(\frac{3}{2},\frac{3}{2})}(x),$$

and hence

$$q(n) = c\,R_n'(1) = c\,\frac{n[6(\alpha-1)+(2-\alpha)(n+1)(n+2)]}{3(\alpha+(2-\alpha)n)}. \tag{7}$$

(4) Finally we consider the orthogonal polynomials $R_n(x;q)$ connected with homogeneous trees, see 4.5.7.2. Consider $q > 1$. Then $a_n = \frac{q}{q+1}$, $h(0) = 1$ and $h(n) = (q+1)q^{n-1}$ for $n \in \mathbb{N}$. Furthermore,

$$H(n) = \frac{(q+1)q^n - 2}{q-1},$$

and

$$\frac{H(n)}{h(n)\,a_n} = \frac{q+1}{q-1}\left(1 - \frac{2}{(q+1)q^n}\right) \qquad \text{for } n \in \mathbb{N}.$$

By Proposition 4.13.29 we obtain

$$q(n) = c\,R_n'(1;q) = \frac{c}{q-1}\left[n(q+1) - 2\,\frac{q-q^{1-n}}{q-1}\right]. \tag{8}$$

Now we introduce an unbounded operator in $L^2(\mathcal{S},\pi)$. Write $s(n) = a_0 R_n'(1)$. Denote by $D(A) = \mathrm{lin}\{R_n : n \in \mathbb{N}_0\}$ the linear span of the R_n in $L^2(S,\pi)$, and define $A : D(A) \to L^2(S,\pi)$ by

$$A(R_n) \;=\; (1+s(n))\, R_n \tag{9}$$

for $n \in \mathbb{N}_0$. Therefore, if $f \;=\; \sum\limits_{k=0}^{n} \alpha_k R_k \in D(A)$, then

$$A(f) \;=\; f \;+\; \sum_{k=0}^{n} \alpha_k \; s(k)\; R_k \;=\; f \;+\; a_0 \sum_{k=0}^{n} \alpha_k \; R_k'(1)\; R_k,$$

Proposition 4.13.30 *Let $K = \mathbb{N}_0$ be a polynomial hypergroup. Then $A : D(A) \to L^2(S,\pi)$ is a symmetric positive definite operator with point spectrum $\sigma_p(A) = \{1+s(n) : n \in \mathbb{N}_0\}$ and corresponding eigenfunctions $R_n, n \in \mathbb{N}_0$.*

Proof. If $f \;=\; \sum\limits_{k=0}^{n} \alpha_k R_k,\; g \;=\; \sum\limits_{l=0}^{n} \beta_l R_l \;\in D(A)$, then

$$\langle Af, g\rangle \;=\; \langle f, g\rangle \;+\; \sum_{k=0}^{n} s(k)\; \alpha_k \overline{\beta}_k\, \frac{1}{h(k)} \;=\; \langle f, Ag\rangle,$$

where we used that $s(k)$ is real-valued. The $s(k)$ are strictly increasing. Hence

$$\langle Af, f\rangle \;=\; \sum_{k=0}^{n} |\alpha_k|^2 \,(1+s(k))\, \frac{1}{h(k)} \;\geq\; \sum_{k=0}^{n} |\alpha_k|^2 \, \frac{1}{h(k)} \;=\; \|f\|_2^2.$$

Obviously $1+s(n)$ are eigenvalues of A with eigenfunctions R_n, respectively. If $\lambda \in \mathbb{C}$ is an eigenvalue of A with eigenfunction $f \;=\; \sum\limits_{k=0}^{n} \alpha_k R_k,\; f \neq 0$, then

$$0 \;=\; (A - \lambda\,\mathrm{id})\, f \;=\; \sum_{k=0}^{n} (1+s(k)-\lambda)\; \alpha_k R_k.$$

The R_k are linearly independent. Hence $(1+s(k)-\lambda)\alpha_k = 0$ for $k = 0, ..., n$. Since at least one α_k is different from zero, we have then $\lambda = 1+s(k)$. ⋄

Proposition 4.13.31 *The operator $A : D(A) \to L^2(S,\pi)$ is essentially self-adjoint. Its closure $\overline{A} = A^{**} : D(\overline{A}) \to L^2(S,\pi)$ fulfils*

(i) $D(\overline{A}) \;=\; \{f \in L^2(S,\pi) : \; \sum\limits_{k=0}^{\infty} (1+s(k))^2 \; |\check{f}(k)|^2 \; h(k) \; < \infty\}$

(ii) $\sigma(\overline{A}) = \sigma_p(\overline{A}) = \sigma_p(A) = \{1 + s(k) : k \in \mathbb{N}_0\}$ *with corresponding eigenfunctions* R_k.

Proof. By [673,Satz 21.4] the operator A is essentially self-adjoint. The statements (i) and (ii) are part of [673,Satz 21.1]. ◇

Now we introduce the operator we are interested in. (The operator A was introduced, since A has positive eigenvalues.) Denote by

$$D(B) := \{f \in L^2(S,\pi) : \sum_{k=0}^{\infty} s(k)^2 \, |\check{f}(k)|^2 \, h(k) < \infty\}$$

and

$$B : D(B) \to L^2(S,\pi), \qquad Bf := \overline{A}f - f, \quad f \in D(B).$$

Obviously $D(B) = D(\overline{A})$. Moreover, we have:

Proposition 4.13.32 *The operator* $B : D(B) \to L^2(S,\pi)$ *is self-adjoint and explicitly given by*

$$Bf = \sum_{k=0}^{\infty} s(k) \, \check{f}(k) \, R_k \, h(k), \qquad f \in D(B) = D(\overline{A}). \tag{10}$$

Furthermore, $B = \overline{A - id}$ *holds true.*

Proof. By [673,Lemma 17.2] it follows

$$B^* = (\overline{A} - \mathrm{id})^* = \overline{A}^* - \mathrm{id} = \overline{A} - \mathrm{id} = B,$$

i.e. B is self-adjoint. If $f \in D(B)$, $k \in \mathbb{N}_0$, we get

$$(Bf)^{\vee}(k) = \langle (\overline{A} - \mathrm{id})f, R_k \rangle = \langle f, \overline{A}R_k \rangle - \langle f, R_k \rangle$$

$$= (1 + s(k)) \, \check{f}(k) - \check{f}(k) = s(k) \, \check{f}(k),$$

which means that $Bf = \sum_{k=0}^{\infty} s(k) \, \check{f}(k) \, R_k \, h(k)$ in $L^2(S,\pi)$.

Finally note that $A-\mathrm{id}$ is closable, since $A-\mathrm{id} \subseteq \overline{A}-\mathrm{id}$. Moreover, $B^* = (\overline{A}-\mathrm{id})^* = \overline{A}^* - \mathrm{id} = A^* - \mathrm{id} = (A-\mathrm{id})^*$. Therefore,

$$B = B^{**} = (A - \mathrm{id})^{**} = \overline{(A - \mathrm{id})}.$$

◇

Now we will show that for many polynomial hypergroups $K = \mathbb{N}_0$ the operator B is a differential operator. We have to recall the translation operator $M_x \in \mathcal{B}(L^2(S,\pi))$ for each $x \in S$, see subsection 3.11.1. For $x \in D$ and $f \in L^2(S,\pi)$ define

$$M_x f = \mathcal{P}(\alpha_x \cdot \check{f}).$$

We know already that $x \mapsto M_x f$, $D \to L^2(S,\pi)$ is a continuous mapping. Furthermore $(M_x f)^\vee = \alpha_x \cdot \check{f}$ for each $x \in D, f \in L^2(S,\pi)$. In case that S satisfies property (F) and $S = D$, see examples (1) and (2) at the end of subsection 4.13.1, then M_x is exactly the operator L_x, compare Corollary 4.13.8. For the polynomials connected with homogeneous trees, see example (3), M_x is equal to L_x on S, which is a proper subset of D. The same is true for the example we can derive from Corollary 3.11.15 and Proposition 3.11.16.

We have to restrict ourselves to polynomial hypergroups $K = \mathbb{N}_0$ such that two conditions are satisfied:

(1)

$$\text{There exists } x_0 < 1 \text{ such that } [x_0, 1] \subseteq D_s. \qquad (\alpha)$$

(2) For each $n \in \mathbb{N}_0$ holds

$$\sup_{x\in[x_0,1]} |R_n'(x)| \le R_n'(1). \qquad (\beta)$$

We emphasize that for the following construction of the differential operator ∂_1 we do not need property (F) of S.

Proposition 4.13.33 *Assume that $(R_n(x))_{n\in\mathbb{N}_0}$ fulfils conditions (α) and (β). If $f \in D(B)$, then*

$$\lim_{x\to 1-} \left\| \frac{f - M_x f}{1 - R_1(x)} - Bf \right\|_2 = 0.$$

Proof. Let $x_0 \le x < 1$. It follows by Plancherel's theorem

$$\left\| \frac{f - M_x f}{1 - R_1(x)} - Bf \right\|_2^2 = \left\| \sum_{k=0}^{\infty} \left(\frac{1 - R_k(x)}{1 - R_1(x)} - a_0 R_k'(1) \right) \check{f}(k)\, R_k\, h(k) \right\|_2^2$$

$$= \sum_{k=0}^{\infty} a_0^2 \left| \frac{1 - R_k(x)}{1 - x} - R_k'(1) \right|^2 |\check{f}(k)|^2\, h(k)$$

$$= a_0^2 \sum_{k=0}^{\infty} |R_k'(\varrho_k(x)) - R_k'(1)|^2\, |\check{f}(k)|^2\, h(k),$$

where $x < \varrho_k(x) < 1$. By property (β) we have

$$|R_k'(\varrho_k(x)) - R_k'(1)| \leq 2R_k'(1).$$

Now a routine application of Lebesgue's theorem of dominated convergence yields the statement. ◇

Condition (α) is satisfied for many examples, in particular if $D_s = [-1,1]$, see section 4.5.

Condition (β) holds whenever property (T) holds, see section 4.6. The polynomials $R_n(x)$ fulfil property (T) if $R_n(x) = \sum_{k=0}^{n} c(n,k)T_k(x)$ with $c(n,k) \geq 0$ for all $n \in \mathbb{N}_0$.

Proposition 4.13.34 *If $(R_n(x))_{n\in\mathbb{N}_0}$ fulfils property (T), then condition (β) is satisfied.*

Proof. For the Chebyshev polynomials of the first kind $T_n(x) = R_n^{(-\frac{1}{2},-\frac{1}{2})}(x)$ we have $T_n'(x) = n^2 U_{n-1}(x)$, where $U_n(x) = R_n^{(\frac{1}{2},\frac{1}{2})}(x)$ are the Chebyshev polynomials of the second kind. Hence

$$R_n'(x) = \sum_{k=1}^{n} c(n,k)\, k^2\, U_{k-1}(x),$$

and so

$$\sup_{x\in[-1,1]} |R_n'(x)| = \sum_{k=0}^{n} c(n,k)\, k^2 = \sum_{k=0}^{n} c(n,k)\, T_k'(1) = R_n'(1).$$

◇

Definition. *Assume that $(R_n(x))_{n\in\mathbb{N}_0}$ fulfils conditions (α) and (β). Denote by*

$$H_2^{(1)}(S) = \{f \in L^2(S,\pi) : \lim_{x\to 1-} \frac{f - M_x f}{1 - R_1(x)} \text{ exists in } L^2(S,\pi)\}$$

and

$$\partial_1 : H_2^{(1)}(S) \to L^2(S,\pi), \qquad \partial_1 f = \lim_{x\to 1-} \frac{f - M_x f}{1 - R_1(x)}. \tag{11}$$

We call ∂_1 the differential operator (of first order) with respect to $(R_n(x))_{n\in\mathbb{N}_0}$.

Theorem 4.13.35 *Assume that $(R_n(x))_{n\in\mathbb{N}_0}$ fulfils conditions (α) and (β). Then for the differential operator $\partial_1 : H_2^{(1)}(S) \to L^2(S,\pi)$ we have*

(i) $H_2^{(1)}(S) = D(B) = D(\overline{A})$.

(ii)

$$\partial_1 f = \lim_{x\to 1-} \frac{f - M_x f}{1 - R_1(x)} = \sum_{k=0}^{\infty} s(k)\, \check{f}(k)\, R_k\, h(k) = B(f) = \overline{A} f - f$$

for every $f \in H_2^{(1)}(S)$.

(iii) ∂_1 *is self-adjoint.*

(iv) $\sigma(\partial_1) = \sigma_p(\partial_1) = \{s(k) : k \in \mathbb{N}_0\}$ *with corresponding eigenfunctions* R_k.

Proof. We have only to show that $H_2^{(1)}(S) = D(B)$ and ∂_1 coincides with B given by (10). The other statements are already shown above.

By Proposition 4.13.33 we have $D(B) \subseteq H_2^{(1)}(S)$. Since $(A-\mathrm{id})(R_n) = \partial_1 R_n$, it follows $A-\mathrm{id} \subseteq \partial_1$, and therefore $\partial_1^* \subseteq (A-\mathrm{id})^* = B^*$, as we have already noted in the proof of Proposition 4.13.32. Since ∂_1 is a symmetric operator, we obtain $\partial_1 \subseteq \partial_1^* \subseteq B^* = B$ by Proposition 4.13.32. So we have shown $H_2^{(1)} = D(B)$ and $\partial_1 = B$. ◇

Remark: Jacobi polynomials fulfil a differential equation

$$\frac{d}{dx}\left(w^{(\alpha,\beta)}(x)(1-x^2)\,\frac{d}{dx}\,R_n^{(\alpha,\beta)}(x)\right)$$
$$= -n(n+\alpha+\beta+1)\, w^{(\alpha,\beta)}(x)\, R_n^{(\alpha,\beta)}(x),$$

see [641,Theorem 4.2.1], where $d\pi^{(\alpha,\beta)}(x) = w^{(\alpha,\beta)}(x)\,dx$. This means that in the Jacobi case ∂_1 coincides with the classical Jacobi differential operator. We point out that the operator ∂_1 defined by the translation operator M_x is an appropriate generalization of the Jacobi differential operator for many orthogonal polynomial sequences $(R_n(x))_{n\in\mathbb{N}_0}$.

Theorem 4.13.36 *Assume that* S *satisfies property (F) and* $1 \in S$. *Furthermore assume that* (β) *is valid. Put*

$$\|\varphi\|_{2,1} := \|\varphi\|_2 + \|\partial_1\varphi\|_2 \qquad \text{for } \varphi \in H_2^{(1)}(S).$$

With this norm $H_2^{(1)}(S)$ *is a homogeneous Banach space on* S.

Proof. Since we assume $1 \in S \subseteq D_s$, condition (α) is satisfied. $H_2^{(1)}(S)$ is complete. In fact, if $(\varphi_n)_{n\in\mathbb{N}}$ is a Cauchy sequence in $H_2^{(1)}(S)$, then φ_n and $\partial_1\varphi_n$ converge in $L^2(S,\pi)$ towards $\varphi \in L^2(S,\pi)$ and $\psi \in L^2(S,\pi)$, respectively. Since ∂_1 is a closed operator, it follows that $\partial_1\varphi = \psi$. It is routine to show that $(\varphi_n)_{n\in\mathbb{N}}$ converges in $H_2^{(1)}(S)$ towards $\varphi \in H_2^{(1)}(S)$. Obviously $\|\sigma\|_1 \le \|\sigma\|_2 \le \|\sigma\|_{2,1}$ for all $\sigma \in H_2^{(1)}(S)$, and (B2) is shown. Condition (B1) is satisfied since $\partial_1 R_n = s(n)R_n$. For $\varphi \in H_2^{(1)}(S), x \in S$ we have

$$(L_x\partial_1\varphi)^\vee(n) = R_n(x)\,(\partial_1\varphi)^\vee(n) = R_n(x)\,s(n)\,\check{\varphi}(n) = (\partial_1 L_x\varphi)^\vee(n)$$

for all $n \in \mathbb{N}_0$. Hence $L_x \partial_1 \varphi = \partial_1 L_x \varphi$, and thus $L_x \varphi \in H_2^{(1)}(S)$ and $\|L_x \varphi\|_{2,1} \leq M \|\varphi\|_{2,1}$. It remains to show (B4). If $\varphi \in H_2^{(1)}(S)$, $x_0 \in S$, $\varepsilon > 0$, there exists $\delta > 0$ such that $\|L_x \varphi - L_{x_0} \varphi\|_2 < \varepsilon/2$ and $\|\partial_1 L_x \varphi - \partial_1 L_{x_0} \varphi\|_2 < \varepsilon/2$ for all $x \in S$ with $|x - x_0| < \delta$. Hence $\|L_x \varphi - L_{x_0} \varphi\|_{2,1} < \varepsilon$ for all $x \in S$, $|x - x_0| < \delta$. ◇

14 Orthogonal expansions

14.1 *Triangular schemes and approximate identities*

Approximation methods of periodic functions by Fourier series have a long history and there is a lot of fundamental work on this topic. In this subsection we will concentrate on orthogonal Fourier expansions on $S = \operatorname{supp} \pi$, where $(R_n(x))_{n \in \mathbb{N}_0}$ constitute the basic functions. We assume that the $R_n(x)$ generate a polynomial hypergroup on $\mathbb{N}_0$. (We would like to point out that several of the following results concerning orthogonal expansions are valid also for some orthogonal polynomial sequences which do not generate polynomial hypergroups on $\mathbb{N}_0$.)

We present methods of approximation in Banach spaces of functions on S with norm $\|\cdot\|_B$ such that $R_n|S \in B$ for all $n \in \mathbb{N}_0$ and such that the linear span of $\{R_n|S : n \in \mathbb{N}_0\}$ is dense in B. Moreover we assume throughout that $\|\cdot\|_1 \leq \|\cdot\|_B$. In particular, $C(S)$ and $L^p(S, \pi)$, $1 \leq p < \infty$, belong to this general class of Banach spaces. The first results are valid for these Banach spaces without any assumption on S. For the following main results (e.g. Theorem 4.14.10) we have to suppose that S satisfies property (F).

For $n \in \mathbb{N}_0$ let $a_{n,k}$, $k = 0, ..., n$ be a triangular matrix of complex numbers. Sometimes we consider only a subsequence n_m of $\mathbb{N}_0$, for example, $n = 2m$ for the Fejér approximation. Another example of this kind are the de la Vallée-Poussin kernels. Obviously the following results are valid for these kernels, too.

We define the polynomials

$$A_n(x) = \sum_{k=0}^{n} a_{n,k} \, R_k(x) \, h(k), \qquad x \in S. \tag{1}$$

We call $(A_n(x))_{n \in \mathbb{N}_0}$ a kernel. In the sequel $(A_n(x))_{n \in \mathbb{N}_0}$ will be the kernel determined by the triangular scheme $a_{n,k}$. We also may identify A_n with a continuous linear operator from B into B by

$$A_n \varphi(x) = \sum_{k=0}^{n} a_{n,k} \, \check{\varphi}(k) \, R_k(x) \, h(k) \tag{2}$$

for $x \in S$ and $\varphi \in B$, where $(B, \|\cdot\|_B)$ is an arbitrary Banach space of functions contained in $L^1(S, \pi)$.

The weight coefficients $a_{n,k}$ have to be chosen appropriately to guarantee concrete features of the approximation process. The most important feature is described by the following definition.

Definition: We say the kernels $(A_n)_{n \in \mathbb{N}_0}$ are an **approximate identity** with respect to B, if

$$\lim_{n \to \infty} \|A_n \varphi - \varphi\|_B = 0$$

for all $\varphi \in B$.

Testing this property for $\varphi = R_m \in B$ we see that $A_n R_m = a_{n,m}$ if $m \leq n$ and $A_n R_m = 0$ if $m > n$. Hence

$$\lim_{n \to \infty} a_{n,m} = 1 \qquad \text{for each } m \in \mathbb{N}_0 \tag{3}$$

are necessary conditions, so that $(A_n)_{n \in \mathbb{N}_0}$ is an approximate identity.

In order to use below the well-known Banach-Steinhaus theorem we have to show that the linear span of $\{R_n | S : n \in \mathbb{N}_0\}$ is dense in B. If $B = C(S)$, this follows immediately by the theorem of Stone-Weierstraß, see Theorem 3.2.1. If $B = L^p(S, \pi)$, $1 \leq p < \infty$, the linear span of $\{R_n | S : n \in \mathbb{N}_0\}$ is dense in B, since $C(S)$ is dense in $L^p(S, \pi)$ and $\|\cdot\|_1 \leq \|\cdot\|_p \leq \|\cdot\|_\infty$ for $1 \leq p < \infty$.

A direct consequence of the Banach-Steinhaus theorem is the following result.

Theorem 4.14.1 *Let $B = C(S)$ or $B = L^p(S, \pi)$, $1 \leq p < \infty$. We have $\lim_{n \to \infty} \|A_n \varphi - \varphi\|_B = 0$ for all $\varphi \in B$ if and only if (3) is true, and*

$$\|A_n \varphi\|_B \leq C \, \|\varphi\|_B \qquad \text{for all } \varphi \in B, \ n \in \mathbb{N}_0 \tag{4}$$

are satisfied.

Denote by $\|A_n\|^B$ the operator norm of $A_n : B \to B$, $A_n(\varphi) = A_n \varphi$. For $x, y \in S$ we introduce

$$A_n(x, y) = \sum_{k=0}^{n} a_{n,k} \, R_k(x) \, R_k(y) \, h(k). \tag{5}$$

Proposition 4.14.2 *Let $B = C(S)$ or $B = L^p(S, \pi)$, $1 \leq p < \infty$. Denote by $\|A_n(x, \cdot)\|_1 = \int_S |A_n(x, y)| \, d\pi(y)$. Then*

(1) $\|A_n\|^B \leq \sup_{x \in S} \|A_n(x, \cdot)\|_1$ *for any B.*

(2) $\|A_n\|^{L^1(S,\pi)} = \|A_n\|^{C(S)} \geq \sup_{x\in S} \|A_n(x,\cdot)\|_1$.

Proof. Evidently, $A_n\varphi(x) = \int_S A_n(x,y)\,\varphi(y)\,d\pi(y)$.

(1) If $B = C(S)$ then $\|A_n\varphi\|_\infty \leq \sup_{x\in S} \|A_n(x,\cdot)\|_1\,\|\varphi\|_\infty$ and statement (1) follows for $B = C(S)$.
Now let $B = L^p(S,\pi)$, $1 \leq p < \infty$. For $\frac{1}{p} + \frac{1}{q} = 1$ Hölder's inequality implies

$$\left(\int_S |\varphi(y)|\,|A_n(x,y)|\,d\pi(y)\right)^p$$
$$\leq \|A_n(x,\cdot)\|_1^{p/q} \int_S |\varphi(y)|^p\,|A_n(x,y)|\,d\pi(y).$$

Using $A_n(x,y) = A_n(y,x)$ and Fubini's theorem we obtain

$$\|A_n\varphi\|_p^p \leq \left(\sup_{x\in S} \|A_n(x,\cdot)\|_1\right)^{p/q} \int_S |\varphi(y)|^p \int_S |A_n(x,y)|\,d\pi(x)\,d\pi(y)$$
$$\leq \left(\sup_{x\in S} \|A_n(x,\cdot)\|_1\right)^p \|\varphi\|_p^p.$$

(2) Recalling a well-known equality for the L^1-norm, we obtain

$$\|A_n(x,\cdot)\|_1 = \sup_{\|\varphi\|_\infty=1} \left|\int_S \varphi(y)\,A_n(x,y)\,d\pi(y)\right| = \sup_{\|\varphi\|_\infty=1} |A_n\varphi(x)|$$
$$\leq \sup_{\|\varphi\|_\infty=1} \|A_n\varphi\|_\infty = \|A_n\|^{C(S)}.$$

Furthermore,

$$\|A_n\|^{L^1(S,\pi)} = \sup_{\|\psi\|_1=1} \|A_n(\psi)\|_1 = \sup_{\|\psi\|_1=1}\ \sup_{\|\varphi\|_\infty=1} \left|\int_S \varphi(y)\,A_n\psi(y)\,d\pi(y)\right|$$
$$= \sup_{\|\varphi\|_\infty=1}\ \sup_{\|\psi\|_1=1} \left|\int_S A_n\,\varphi(y)\,\psi(y)\,d\pi(y)\right| = \sup_{\|\varphi\|_\infty=1} \|A_n\varphi\|_\infty = \|A_n\|^{C(S)}.$$

Note that $\int_S A_n\varphi(y)\,\psi(y)\,d\pi(y) = \int_S \varphi(y)\,A_n\psi(y)\,d\pi(y)$ was applied, an identity easily to be checked.

◇

Corollary 4.14.3 *Let* $B = C(S)$ *or* $B = L^p(S,\pi)$, $1 \leq p < \infty$. *The following holds true for* A_n*:*

(1) If (3) holds and if

$$\sup_{n\in\mathbb{N}_0}\sup_{x\in S}\|A_n(x,\cdot)\|_1 \;<\; \infty, \tag{6}$$

then $\lim_{n\to\infty}\|A_n\varphi-\varphi\|_B = 0$ *for all* $\varphi\in B$.

(2) For $B = L^1(S,\pi)$ *and* $B = C(S)$ *we have* $\lim_{n\to\infty}\|A_n\varphi-\varphi\|_B = 0$ *for all* $\varphi\in B$ *if and only if (3) and (6) hold.*

Proof. Combine the statements of Theorem 4.14.1 and Proposition 4.14.2. ◇

Corollary 4.14.3 shows that $(A_n)_{n\in\mathbb{N}_0}$ is an approximate identity with respect to $L^1(S,\pi)$ if and only if it is an approximate identity with respect to $C(S)$. Moreover, if $(A_n)_{n\in\mathbb{N}_0}$ is an approximate identity with respect to $C(S)$, then it is also with respect to $L^p(S,\pi)$, $1<p<\infty$.

Condition (3) becomes more handsome if the operator A_n is positive. An operator A from $C(S)$ into $C(S)$ is called positive, if $\varphi\geq 0$ implies $A\varphi\geq 0$.

Proposition 4.14.4 *Assume that the operators* $A_n : C(S)\to C(S)$ *are positive, and* $1\in S$. *If* $\lim_{n\to\infty} a_{n,0} = \lim_{n\to\infty} a_{n,1} = 1$, *then*

$$\lim_{n\to\infty} a_{n,k} \;=\; 1$$

for all $k\in\mathbb{N}_0$.

Proof. Since the A_n are positive operators and $A_nR_k = a_{n,k}R_k$, the coefficients $a_{n,k}$ are real-valued. For the Dirichlet kernel $\mathcal{D}_k(x) = \sum\limits_{i=0}^{k} R_i(x)h(i)$, $x\in D_s$, we have $|\mathcal{D}_k(x)|\leq\sum\limits_{i=0}^{k} h(i)$ and

$$(1-R_1(x))\,\mathcal{D}_k(x) \;=\; a_k\,h(k)\,(R_k(x)-R_{k+1}(x)),$$

see Proposition 4.10.3(1). Hence for all $x\in D_s$,

$$-(1-R_1(x))\,\frac{\sum_{i=0}^{k}h(i)}{a_k\,h(k)} \;\leq\; R_k(x)-R_{k+1}(x) \;\leq\; (1-R_1(x))\,\frac{\sum_{i=0}^{k}h(i)}{a_k\,h(k)}$$

If $n>k$ it follows by the positivity of the operator A_n

$$-(a_{n,0}-a_{n,1}R_1(x))\,\frac{\sum_{i=0}^{k}h(i)}{a_k\,h(k)} \;\leq\; a_{n,k}R_k(x)-a_{n,k+1}R_{k+1}(x)$$

$$\leq\; (a_{n,0}-a_{n,1}R_1(x))\,\frac{\sum_{i=0}^{k}h(i)}{a_k\,h(k)},$$

for all $x \in S$. Putting $x = 1$ we obtain

$$|a_{n,k} - a_{n,k+1}| \le \frac{\sum_{i=0}^{k} h(i)}{a_k\, h(k)}\, |a_{n,0} - a_{n,1}|,$$

and it follows $\lim_{n\to\infty} a_{n,k} = 1$ for all $k \in \mathbb{N}_0$. ◇

One should note that in general $A_n(x) \ge 0$ for all $x \in S$ does not imply that $A_n : C(S) \to C(S)$ is positive. However, the following result is true.

Proposition 4.14.5 *Suppose that the coefficients $a_{n,k}$ are real-valued. Then the operator $A_n : C(S) \to C(S)$ is positive if and only if*

$$A_n(x,y) \;=\; \sum_{k=0}^{n} a_{n,k}\, R_k(x)\, R_k(y)\, h(k) \;\ge\; 0 \qquad \text{for all } x,y \in S. \qquad (*)$$

In particular, if S satisfies property (F) with positive measures $\mu_{x,y} \in M^+(S)$, then $A_n : C(S) \to C(S)$ is positive if and only if

$$A_n(z) \;=\; \sum_{k=0}^{n} a_{n,k}\, R_k(z)\, h(k) \;\ge\; 0 \qquad \text{for all } z \in S.$$

Proof. Assume that $(*)$ is valid, then

$$A_n\varphi(x) = \sum_{k=0}^{n} a_{n,k}\, \check{\varphi}(k)\, R_k(x)\, h(k)$$

$$= \int_S \sum_{k=0}^{n} a_{n,k}\, R_k(x)\, R_k(y)\, h(k)\, \varphi(y)\, d\pi(y) \;\ge\; 0$$

if $\varphi \ge 0$. Conversely, suppose that $(*)$ does not hold. Then there exist $x_0, y_0 \in S$, $\delta < 0$ and an open set U with $y \in U$ such that $\sum_{k=0}^{n} a_{n,k} R_k(x_0) R_k(y) h(k) < \delta$ for each $y \in U$. Choose $V \subseteq U$ compact with $\pi(V) > 0$. By Urysohn's lemma there is a continuous function $g : \mathbb{R} \to [0,1]$ with $g(y) = 1$ for all $y \in V$ and $g(y) = 0$ for all $y \in \mathbb{R}\backslash U$, For $\varphi = g|S \in C(S)$ it follows

$$A_n\varphi(x_0) \;=\; \int_S \sum_{k=0}^{n} a_{n,k}\, R_k(x_0)\, R_k(y)\, h(k)\, \varphi(y)\, d\pi(y) \;<\; \pi(V)\,\delta \;<\; 0.$$

This contradicts the positivity of the operator A_n. ◇

Because of the results of Corollary 4.14.3 our goal now is to derive conditions such that $\|A_n(x,\cdot)\|_1$ is bounded uniformly in $x \in S$. Mainly we will use that S satisfies property (F).

Proposition 4.14.6 *Suppose that S satisfies property (F). If the kernels $(A_n(x))_{n\in\mathbb{N}_0}$ satisfy $\sup_{n\in\mathbb{N}_0} \|A_n\|_1 < \infty$, then $\sup_{n\in\mathbb{N}_0} \sup_{x\in S} \|A_n(x,\cdot)\|_1 < \infty$. Moreover, $A_n\varphi = A_n * \varphi$ for all $\varphi \in C(S)$, where the convolution $*$ is the one defined in subsection 4.13.1.*

Proof. Let $C := \sup_{n\in\mathbb{N}_0} \|A_n\|_1$. For any $\varphi \in C(S)$ we obtain

$$\begin{aligned} A_n\varphi(x) &= \sum_{k=0}^{n} a_{n,k}\, \check{\varphi}(k)\, R_k(x)\, h(k) \\ &= \int_S \sum_{k=0}^{n} a_{n,k}\, \varphi(y)\, R_k(y)\, R_k(x)\, h(k)\, d\pi(y) \\ &= \int_S \int_S A_n(z)\, \varphi(y)\, d\mu_{x,y}(z)\, d\pi(y) \\ &= \int_S \varphi(y)\, L_y A_n(x)\, d\pi(y) = A_n * \varphi(x) \end{aligned}$$

for all $x \in S$. Hence $\|A_n\varphi\|_\infty \leq M\,\|A_n\|_1\,\|\varphi\|_\infty \leq MC\,\|\varphi\|_\infty$ by Corollary 4.13.4. By Proposition 4.14.2 follows $\sup_{x\in S} \|A_n(x,\cdot)\|_1 \leq MC$ for all $n \in \mathbb{N}_0$. ◇

In the proof above we have used property (F) of S to obtain a boundedness condition for the double sums $A_n(x,y)$. It is a remarkable fact that the validity of property (F) of S is equivalent to a boundedness condition of **triple sums.**

For a triangular matrix $a_{n,k}$, $k = 0, ..., n$, define

$$A_n(x,y,z) = \sum_{k=0}^{n} a_{n,k}\, R_k(x)\, R_k(y)\, R_k(z)\, h(k) \tag{7}$$

for $x, y, z \in S$ and denote

$$\|A_n(x,y,\cdot)\|_1 = \int_S |A_n(x,y,z)|\, d\pi(y).$$

Theorem 4.14.7 *If* $\lim_{n\to\infty} a_{n,k} = 1$ *for all* $k \in \mathbb{N}_0$ *and*

$$\sup_{n\in\mathbb{N}_0} \sup_{x,y\in S} \|A_n(x,y,\cdot)\|_1 = M < \infty,$$

then S *satisfies property (F).*

Proof. Let $P(S)$ be the linear space of polynomials on S. Put $u_k = R_k(x)R_k(y)$ for fixed $x, y \in S$ and define a linear functional $L : P(S) \to \mathbb{C}$ by

$$L(Q) = \sum_{k=0}^{m} v_k\, u_k\, h(k), \qquad \text{when } Q(z) = \sum_{k=0}^{m} v_k\, R_k(z)\, h(k).$$

Since $\check{Q}(k) = v_k$ for $k = 0, 1, ..., m$ and $\check{Q}(k) = 0$ for $k \geq m+1$, we obtain

$$|L(Q)| = \left|\sum_{k=0}^{\infty} \check{Q}(k)\, u_k\, h(k)\right| = \lim_{n\to\infty} \left|\sum_{k=0}^{n} \check{Q}(k)\, u_k\, h(k)\right|$$

$$= \lim_{n\to\infty} \left| \int_S Q(z) \sum_{k=0}^{n} a_{n,k}\, R_k(z)\, R_k(x)\, R_k(y)\, h(k)\, d\pi(z) \right| \le M\, \|Q\|_\infty.$$

Thus L is continuous with respect to $\|\cdot\|_\infty$. The space $P(S)$ is dense in $C(S)$. Hence by $L(\varphi) = \lim_{m\to\infty} L(Q_m)$ for $\lim_{m\to\infty} Q_m = \varphi$, a continuous linear functional is defined on $C(S)$, which fulfills $\|L\| \le M$ and $L(R_k) = u_k = R_k(x)R_k(y)$.

Riesz's representation theorem yields a regular real-valued Borel measure $\mu_{x,y}$ on S such that

$$L(\varphi) = \int_S \varphi(z)\, d\mu_{x,y}(z) \qquad \text{for all } \varphi \in C(S) \quad \text{and } \|\mu_{x,y}\| \le M.$$

◇

A converse result is valid, too.

Proposition 4.14.8 *Assume that* $\sup_{n\in\mathbb{N}_0} \sup_{u\in S} \|A_n(u,\cdot)\|_1 = M_1 < \infty$. *If* S *satisfies property (F), then*

$$\sup_{n\in\mathbb{N}_0} \sup_{x,y\in S} \|A_n(x,y,\cdot)\|_1 < \infty.$$

Proof. For $x, y \in S$ and $n \in \mathbb{N}_0$ we obtain by Fubini's theorem

$$\|A_n(x,y,\cdot)\|_1 = \int_S \left| \sum_{k=0}^{n} a_{n,k}\, R_k(z) \int_S R_k(u)\, d\mu_{x,y}(u)\, h(k) \right| d\pi(z)$$

$$\le \int_S \int_S \left| \sum_{k=0}^{n} a_{n,k}\, R_k(u)\, R_k(z)\, h(k) \right| d\pi(z)\, d\mu_{x,y}(u)$$

$$= \int_S \|A_n(u,\cdot)\|_1\, d\mu_{x,y}(u) \le M_1 \cdot M,$$

where M is the constant determined by property (F). ◇

From now on we will assume that S satisfies property (F), and so we can derive results which are valid for homogeneous Banach spaces B on S. Note that $C(S)$ and $L^p(S,\pi)$, $1 \le p < \infty$ are homogeneous Banach spaces on S.

Lemma 4.14.9 *Let* S *satisfy property (F) and let* B *be a homogeneous Banach space on* S. *Let* $\varphi \in B$. *The* B*-valued integral*

$$\int_S L_x\varphi\, R_k(x)\, d\pi(x)$$

is equal to $\check{\varphi}(k) R_k$ *in* B.

Proof. Since $B \subseteq L^1(S,\pi)$ and $\|\cdot\|_1 \le \|\cdot\|_B$ the B-valued integral is the same as the $L^1(S,\pi)$-integral. By Proposition 4.13.7(3) we know that

$$\int_S L_x\varphi \, R_k(x)\, d\pi(x) \;=\; (R_k\pi) * \varphi.$$

Choose a sequence $(\varphi_n)_{n\in\mathbb{N}}$, $\varphi_n \in C(S)$, such that $\|\varphi_n - \varphi\|_1 \to 0$. Applying Theorem 4.13.3 we infer that

$$\int_S L_x\varphi_n(y)\, R_k(x)\, d\pi(x) \;=\; \int_S \varphi_n(y)\, L_y R_k(x)\, d\pi(x) \;=\; R_k(y)\, \check{\varphi}_n(k)$$

for each $y \in S$. Obviously $R_k(y)\check{\varphi}_n(k)$ converges towards $R_k(y)\check{\varphi}(k)$ and

$$\left\| \int_S (L_x\varphi - L_x\varphi_n)\, R_k(x)\, d\pi(x) \right\|_1 \;\le\; \int_S \|L_x\varphi - L_x\varphi_n\|_1\, d\pi(x) \;\longrightarrow 0 \qquad \text{as } n \to \infty.$$

Hence $\int\limits_S L_x\varphi\, R_k(x)\, d\pi(x) = \check{\varphi}(k) R_k$. ◇

Theorem 4.14.10 *Suppose that S satisfies property (F). Let B be a homogeneous Banach space on S such that the linear span of $\{R_n|S : n \in \mathbb{N}_0\}$ is dense in B. Then*

$$\lim_{n\to\infty} \|A_n\varphi - \varphi\|_B \;=\; 0 \qquad \textit{for all } \varphi \in B$$

if and only if (3) is true and

$$\|A_n * \varphi\|_B \;\le\; C\, \|\varphi\|_B \qquad \textit{for all } \varphi \in B,\; n \in \mathbb{N}_0.$$

Proof. Using B-valued integration, Lemma 4.14.9 yields

$$A_n\varphi \;=\; \sum_{k=0}^{n} a_{n,k}\, \check{\varphi}(k)\, R_k\, h(k) \;=\; \sum_{k=0}^{n} a_{n,k} \left(\int_S L_x\varphi\, R_k(x)\, d\pi(x) \right) h(k)$$

$$=\; \int_S L_x\varphi\, A_n(x)\, d\pi(x) \;=\; A_n * \varphi.$$

Now the stated equivalence is again exactly the Banach-Steinhaus theorem. ◇

Corollary 4.14.11 *Suppose that S satisfies property (F). Let B be a homogeneous Banach space on S such that the linear span of $R_n|S$ is dense in B. If $\lim\limits_{n\to\infty} a_{n,k} = 1$ and $\sup\limits_{n\to\infty} \|A_n\|_1 < \infty$, then*

$$\lim_{n\to\infty} \|A_n\varphi - \varphi\|_B \;=\; 0 \qquad \textit{for each } \varphi \in B.$$

Proof. We have only to apply Proposition 4.13.13 and Theorem 4.14.10. ◇

A strong property for kernels $(A_n(x))_{n\in\mathbb{N}_0}$ is their summability.

Definition: Let $(R_n(x))_{n\in\mathbb{N}_0}$ generate a polynomial hypergroup on $\mathbb{N}_0$ such that $1 \in S$. A kernel $(A_n(x))_{n\in\mathbb{N}_0}$ is called a **summability kernel** if

(1) $\int\limits_S A_n(x)\, d\pi(x) = (A_n)^\vee(0) = a_{n,0} = 1$ for each $n \in \mathbb{N}_0$,

(2) $\int\limits_S |A_n(x)|\, d\pi(x) = \|A_n\|_1 \leq M$ for each $n \in \mathbb{N}_0$,

(3) For each $0 < \delta < 1$,

$$\lim_{n\to\infty} \int_{S\setminus[\delta,1]} |A_n(x)|\, d\pi(x) = 0.$$

Following the lines of the proof of Lemma 2.2 and Theorem 2.11 of Katznelson [364] one obtains:

Theorem 4.14.12 *Let $(R_n(x))_{n\in\mathbb{N}_0}$ generate a polynomial hypergroup on $\mathbb{N}_0$ such that $1 \in S$ and property (F) is satisfied. Let B be a homogeneous Banach space on S. If $(A_n(x))_{n\in\mathbb{N}_0}$ is a summability kernel, then the B-valued integrals $\int\limits_S A_n(x)\, L_x\varphi\, d\pi(x)$ satisfy*

$$\left\| \varphi - \int_S A_n(x)\, L_x\varphi\, d\pi(x) \right\|_B \longrightarrow 0 \qquad \textit{as } n \to \infty$$

for every $\varphi \in B$.

We already noted in the proof of Theorem 4.14.10 that the B-valued integral $\int\limits_S A_n(x) L_x\varphi\, d\pi(x)$ is equal to $A_n * \varphi$. Hence Theorem 4.14.12 says that $(A_n(x))_{n\in\mathbb{N}_0}$ is an approximate identity with respect to B, whenever $(A_n(x))_{n\in\mathbb{N}_0}$ is a summability kernel.

Corollary 4.14.13 *Let $(R_n(x))_{n\in\mathbb{N}_0}$ generate a polynomial hypergroup on $\mathbb{N}_0$ such that $1 \in S$ and property (F) is satisfied. Assume there exists a summability kernel $(A_n(x))_{n\in\mathbb{N}_0}$. Then the linear span of $\{R_n|S : n \in \mathbb{N}_0\}$ is dense in every homogeneous Banach space on S.*

Remark: We refer to Proposition 2.3.3 and Proposition 3.6.7, where we have constructed approximate identities $(\varphi_i)_{i\in I}$, $\varphi_i \in C_c(K)$, where K is a hypergroup. The construction of approximate identities $(A_n(x))_{n\in\mathbb{N}_0}$, $A_n \in C(S)$, is much more involved, since on the one hand S is in general not a hypergroup, and on the other hand $A_n(x) =$

$\sum_{k=0}^{n} a_{n,k} R_k(x) h(k)$ are completely determined by the triangular scheme $a_{n,k} = (A_n)^\vee(k)$.

14.2 *Construction of approximate identities*

The simplest way to define approximate identities would be to choose $a_{n,k} = 1$ for $k = 0, ..., n$. This leads to the so-called Dirichlet kernel $\mathcal{D}_n(x)$,

$$\mathcal{D}_n(x) = \sum_{k=0}^{n} R_k(x)\, h(k).$$

Only in few cases (depending on the polynomial hypergroup or on the Banach space B), the Dirichlet kernel is an approximate identity. We postpone the discussion of the Dirichlet kernel to the end of this subsection.

14.2.1 *4.14.2.1 Fejér-type kernels*

There are more or less two types of kernels, which are related to the trigonometric Fejér kernel of Fourier series on the torus $\mathbb{T}$. The trigonometric Fejér kernel is given by

$$F_n(z) = \sum_{k=-n}^{n} \left(1 - \frac{|k|}{n+1}\right) z^k = \frac{1}{n+1} \left(\frac{\sin\left((n+1)\frac{t}{2}\right)}{\sin \frac{t}{2}}\right)^2, \quad z = e^{it} \in \mathbb{T}$$

and the trigonometric Dirichlet kernel is given by

$$\mathcal{D}_n(z) = \sum_{k=-n}^{n} z^k = \frac{\sin\left((2n+1)\frac{t}{2}\right)}{\sin \frac{t}{2}}, \qquad z = e^{it} \in \mathbb{T}.$$

Hence

$$F_{2n}(z) = \frac{1}{2n+1}\, \mathcal{D}_n^2(z) = \frac{1}{\int_0^{2\pi} \mathcal{D}_n^2(e^{it})\, dt}\, \mathcal{D}_n^2(z).$$

This leads us to define a Fejér-type kernel $(F_{2n}(x))_{n\in\mathbb{N}_0}$ by setting

$$F_{2n}(x) := \frac{1}{\int_S \mathcal{D}_n^2(y)\, d\pi(y)}\, \mathcal{D}_n^2(x), \qquad x \in S. \qquad (F1)$$

For $n \in \mathbb{N}_0$ consider the sequence χ_n defined by

$$\chi_n(k) = \begin{cases} 1 & \text{for } k = 0, ..., n \\ 0 & \text{for } k = n = 1, ... \end{cases}$$

In order to investigate the Fejér (or lateron the Vallée Poussin) kernels we calculate the convolution products $\chi_n * \chi_m$.

Lemma 4.14.14 *Let $n, m \in \mathbb{N}_0$ with $n \geq m$. For the sequence $\chi_n * \chi_m$ we have*

$$\begin{aligned}
&(i)\ \chi_n * \chi_m(k) = \sum_{j=0}^{m} h(j) \quad \text{for } k = 0, 1, ..., n-m, \\
&(ii)\ \chi_n * \chi_m(k) = \sum_{j=0}^{m} \sum_{l=|k-j|}^{\min\{n,k+j\}} g(j,k;l)\, h(j) \\
&\quad \text{for } k = n-m+1, ..., n+m-1 \\
&(iii)\ \chi_n * \chi_m(n+m) = g(n+m, m; n)\, h(m) \\
&(iv)\ \chi_n * \chi_m(k) = 0 \quad \text{for } k = n+m+1, ...
\end{aligned}$$

Proof. By the definition of the convolution we obtain

$$\chi_n * \chi_m(k) = \sum_{j=0}^{m} \sum_{l=|k-j|}^{k+j} g(j,k;l)\, \chi_n(l)\, h(j)$$

If $0 \leq k \leq n-m$ and $0 \leq j \leq m$, we have $0 \leq k+j \leq n$ and hence

$$\sum_{l=|k-j|}^{k+j} g(j,k;l)\, \chi_n(l) = \sum_{l=|k-j|}^{k+j} g(j,k;l) = 1.$$

This shows (i). Also (ii), (iii) and (iv) follow immediately. See also subsection 1.2.1. ◇

The case $n = m$ in Lemma 4.14.14 is summarized in the following statement.

Lemma 4.14.15 *Let $n \in \mathbb{N}_0$. For the sequence $\chi_n * \chi_n$ we have*

$$\begin{aligned}
&(i)\ \chi_n * \chi_n(0) = \sum_{j=0}^{n} h(j) = H(n) \\
&(ii)\ \chi_n * \chi_n(k) = \sum_{j=0}^{n} \sum_{l=|k-j|}^{\min\{n,k+j\}} g(j,k;l)\, h(j) \\
&\quad \text{for } k = 1, ..., 2n-1. \\
&(iii)\ \chi_n * \chi_n(2n) = g(n, 2n; n)\, h(n) \\
&(iv)\ \chi_n * \chi_n(k) = 0 \quad \text{for } k = 2n+1, ...
\end{aligned}$$

Finally formula (ii) of Lemma 4.14.15 yields

Proposition 4.14.16 *Let $n \in \mathbb{N}_0$ and $k \leq n$. Then*

$$\frac{\chi_n * \chi_n(k)}{H(n)} = 1 - \frac{1}{H(n)} \left(\sum_{j=n+1-k}^{n} \sum_{l=n+1}^{k+j} g(j,k;l)\, h(j) \right).$$

Proof. By $\sum_{l=|k-j|}^{k+j} g(j,k;l) = 1$ we obtain for $k \leq n$ from Lemma 4.14.15 (ii)

$$\frac{\chi_n * \chi_n(k)}{H(n)} = \frac{1}{H(n)} \left(\sum_{j=0}^{n} \sum_{l=|k-j|}^{k+j} g(j,k;l)\, h(j) - \sum_{j=n+1-k}^{n} \sum_{l=n+1}^{k+j} g(j,k;l)\, h(j) \right)$$

$$= 1 - \frac{1}{H(n)} \left(\sum_{j=n+1-k}^{n} \sum_{l=n+1}^{k+j} g(j,k;l)\, h(j) \right).$$

◇

Since $\mathcal{D}_n^2(x) = (\chi_n * \chi_n)^\wedge(x)$ and $\int_S \mathcal{D}_n^2(y)\, d\pi(y) = \chi_n * \chi_n(0) = H(n)$, it follows that

$$F_{2n}(x) = \sum_{k=0}^{2n} a_{2n,k}\, R_k(x)\, h(k),$$

where

$$a_{2n,k} = \frac{1}{H(n)}\, \chi_n * \chi_n(k).$$

Proposition 4.14.17 *Let $(R_n(x))_{n\in\mathbb{N}_0}$ generate a polynomial hypergroup and suppose that property (H) is satisfied. Then the coefficient $a_{2n,k}$ of the Fejér-type kernel $(F_{2n}(x))_{n\in\mathbb{N}_0}$ satisfy $\lim_{n\to\infty} a_{2n,k} = 1$ for each $k \in \mathbb{N}_0$.*

Proof. Note that $a_{2n,0} = 1$. Using that $g(j,k;l)h(j) = g(l,k;j)h(l)$, see (vi) of subsection 1.2.1, and applying Proposition 4.7.4, we obtain directly by Proposition 4.14.16 that $\lim_{n\to\infty} a_{2n,k} = 1$ for each $k \in \mathbb{N}_0$. ◇

Remarks:

(1) We recall that property (H) implies $1 \in S$, see Corollary 4.2.12. Furthermore, we recall that 1 cannot be an isolated point of S. Otherwise there would exist $\varphi \in C(S)$ such that $\varphi(1) = 1$ and $\varphi|S\backslash\{1\} = 0$. Then $\check{\varphi}(n) = \pi(\{1\}) > 0$ for all $n \in \mathbb{N}_0$. But $\check{\varphi}(n) \to 0$ as $n \to \infty$ by Proposition 3.3.2(ii).

(2) Since $F_{2n}(x) \geq 0$ it follows $\|F_{2n}\|_1 = a_{2n,0} = 1$. For Chebyshev polynomials of the first kind we obtain $a_{2n,k} = 1 - \frac{k}{2n+1}$.

(3) In order to define the Fejér-type kernels also for odd indices, one can put

$$a_{2n+1,k} = \frac{L_1(\chi_n * \chi_n)(k)}{H(n)}.$$

Since

$$L_1(\chi_n{*}\chi_n)(k) \;=\; a_k\,(\chi_n{*}\chi_n)(k{+}1) + b_k\,(\chi_n{*}\chi_n)(k) + c_k\,(\chi_n{*}\chi_n)(k{-}1)$$

it follows $\lim\limits_{n\to\infty} a_{2n+1,k} = 1$.

(4) For the special case of the polynomial hypergroup generated by the Chebyshev polynomials of the first kind, we can use the trigonometric approximation of functions on $\mathbb{T} = \{e^{it} : t \in [0, 2\pi[\ \}$. For that case there are many proposals for triangular schemes $\alpha_{n,k}$, $k = -n, ..., n$, see [125] or [415]. Any triangular scheme $\alpha_{n,k}$ symmetric in k is a candidate for an approximation based on the polynomials $T_n(x)$. The essential criteria for the trigonometric approximation are $\lim\limits_{n\to\infty} \alpha_{n,k} = 1$ and $\sup\limits_{n\in\mathbb{N}_0} \|\mathfrak{A}_n\|_1 < \infty$, where

$$\mathfrak{A}_n(e^{it}) \;=\; \sum_{k=-n}^{n} \alpha_{n,k}\, e^{ikt}.$$

We define $a_{n,k} = \alpha_{n,k}$ for $k = 0, ..., n$. Applying Proposition 4.14.6 we have to check that $\sup\limits_{n\in\mathbb{N}_0} \|A_n\|_1 < \infty$, where

$$A_n(y) \;=\; \sum_{k=0}^{n} a_{n,k}\, T_k(y)\, h(k).$$

The normalized orthogonalization measure of $(T_n(x))_{n\in\mathbb{N}_0}$ is $d\pi(x) = \frac{1}{\pi}\frac{dx}{\sqrt{1-x^2}}$. It follows

$$\|A_n\|_1 = \int_{-1}^{1} \left|\sum_{k=0}^{n} a_{n,k}\, T_k(x)\, h(k)\right| d\pi(x) = \frac{1}{\pi}\int_0^{\pi} \left|\sum_{k=-n}^{n} \alpha_{n,k}\, \cos kt\right| dt$$

$$= \frac{1}{2\pi}\int_0^{2\pi} \left|\sum_{k=-n}^{n} \alpha_{n,k}\, e^{ikt}\right| dt \;=\; \|\mathfrak{A}_n\|_1.$$

Examples:

(i) **Fejér kernel**
Let $a_{n,k} = 1 - \frac{k}{n+1}$, $k = 0, ..., n$. For even $n \in \mathbb{N}$ the triangular scheme coincides with that constructed above, see Remark (2).

(ii) **Bochner-Riesz kernel**
Let $a_{n,k} = 1 - (\frac{k}{n+1})^2$. The trigonometric kernel $\mathfrak{A}_n(e^{it})$ can be represented by Jacobi polynomials $R_n^{(\alpha,\beta)}(\cos t)$, $\alpha = \frac{3}{2}$, $\beta = -\frac{1}{2}$ (up to multiplicative constants), see [415,Ch.9].

There are many further examples, e.g. Rogosinski kernel, Fejér-Korovkin kernel or generalized Dirichlet kernels, see [336,section 4]. It should be mentioned that the Bochner-Riesz kernel belongs to the class of generalized Dirichlet kernels.

Theorem 4.14.18 *Let $(R_n(x))_{n\in\mathbb{N}_0}$ generate a polynomial hypergroup satisfying property (H). Assume that S satisfies property (F). Then $(F_{2n}(x))_{n\in\mathbb{N}_0}$ is an approximate identity with respect to $L^1(S,\pi)$.*

Proof. Since $\|F_{2n}\|_1 = 1$ for each $n \in \mathbb{N}_0$, Proposition 4.14.6 yields

$$\sup_{n\in\mathbb{N}_0}\sup_{x\in S} \|F_{2n}(x,\cdot)\|_1 < \infty.$$

Now combine Proposition 4.14.17 and Corollary 4.14.3. ◇

Applying Theorem 4.14.10 it follows

Corollary 4.14.19 *Let B be a homogeneous Banach space on S, such that the linear span of $\{R_n|S : n \in \mathbb{N}_0\}$ is dense in B. If the assumptions of Theorem 4.14.18 are valid, then $(F_{2n}(x))_{n\in\mathbb{N}_0}$ is an approximate identity with respect to B.*

We derive now another Fejér-type kernel from the following formula valid for the trigonometric Fejér kernel: We refer to the paper [516] of Obermaier.

$$F_n(z) = \frac{1}{n+1}\sum_{k=0}^{n} \mathcal{D}_k(z) = 1 + \sum_{k=1}^{n}\left(1 - \frac{k}{n+1}\right) 2\cos(nt)$$

$$= \frac{1}{n+1}\,\frac{1-\cos((n+1)t)}{1-\cos t} \qquad \text{for } z = e^{it},\ n \in \mathbb{N}_0.$$

Transferring this formula to orthogonal polynomial sequences $(R_n(x))_{n\in\mathbb{N}_0}$ the Christoffel-Darboux formula is very useful.

Proposition 4.14.20 *Let $(R_n(x))_{n\in\mathbb{N}_0}$ be an orthogonal polynomial sequence generating a polynomial hypergroup on $\mathbb{N}_0$. Then*

$$\frac{1-R_{n+1}(x)}{1-R_1(x)} = \sum_{k=0}^{n} b_{n,k}\, R_k(x)\, h(k) \qquad \textit{for all } n \in \mathbb{N}_0,$$

where $b_{0,0} = 1$ and

$$b_{n,k} := \frac{1}{h(k)a_k} + \frac{1}{h(k+1)a_{k+1}} + \frac{1}{h(n)a_n}$$

for $n \geq 1$ and $k = 1,...,n$, and $b_{n,0} = 1 + b_{n,1}$.

In particular, $b_{n,k} = b_{n,0} - b_{k-1,0}$ for $k = 0,...,n$, where $b_{-1,0} = 0$.

Proof. Write

$$Q_n(x) = \frac{1 - R_{n+1}(x)}{1 - R_1(x)}.$$

$Q_n(x)$ is a polynomial of degree n. By Proposition 4.10.3(1) we obtain

$$\mathcal{D}_n(x) = \sum_{k=0}^{n} R_k(x)\, h(k) = a_0 a_n h(n)\, \frac{R_{n+1}(x) - R_n(x)}{x - 1}$$

$$= a_n h(n)\, \frac{R_n(x) - R_{n+1}(x)}{1 - R_1(x)} \qquad \text{for } n \geq 1.$$

Therefore,

$$Q_n(x) = Q_{n-1}(x) + \frac{1}{h(n)a_n}\, \mathcal{D}_n(x) \qquad \text{for } n \geq 1,$$

which implies

$$Q_n(x) = \mathcal{D}_0(x) + \frac{1}{h(1)a_1}\, \mathcal{D}_1(x) + \cdots + \frac{1}{h(n)a_n}\, \mathcal{D}_n(x)$$

for each $n \in \mathbb{N}_0$. It follows that for $n \geq 1$,

$$b_{n,k} = \sum_{l=k}^{n} \frac{1}{h(l)a_l} \qquad \text{for } k = 1, ..., n, \quad \text{and } b_{n,0} = 1 + b_{n,1}.$$

In particular,

$$b_{n,k} = b_{n,0} - b_{k-1,0}.$$

◇

According to Proposition 4.14.20 we define the modified Fejér kernel $(\mathcal{F}_n(x))_{n\in\mathbb{N}_0}$ by $\mathcal{F}_0(x) = 1$ and for $n \geq 1$

$$\mathcal{F}_n(x) = \frac{1}{b_{n,0}} \sum_{k=0}^{n} b_{n,k}\, R_k(x)\, h(k) = \sum_{k=0}^{n} \left(1 - \frac{b_{k-1,0}}{b_{n,0}}\right) R_k(x)\, h(k)$$

$$= \sum_{k=0}^{n} a_{n,k}\, R_k(x)\, h(k), \qquad (F2)$$

where

$$a_{n,k} := 1 - \frac{b_{k-1,0}}{b_{n,0}} \qquad \text{for } k = 0, ..., n$$

and

$$b_{n,k} := \sum_{l=k}^{n} \frac{1}{h(l)a_l} \qquad \text{for } k = 1, ..., n.$$

Note that $a_{n,0} = 1$ and $a_{n,1} = 1 - \frac{1}{b_{n,0}}$.

Hence $\lim\limits_{n\to\infty} a_{n,1} = 1$ is equivalent to $\lim\limits_{n\to\infty} b_{n,0} = \sum\limits_{l=1}^{\infty} \frac{1}{h(l)a_l} = \infty$.

Using Corollary 4.2.12 and Lemma 4.2.7 we infer that a necessary condition for $\lim\limits_{n\to\infty} a_{n,1} = 1$ is that $1 \in S$. Obviously $\lim\limits_{n\to\infty} a_{n,k} = 1$ for each $k \geq 2$ exactly if $\lim\limits_{n\to\infty} a_{n,1} = 1$. Since $b_{n,0} > 0$ and $\frac{1-R_{n+1}(x)}{1-R_1(x)} \geq 0$, applying Proposition 4.14.20 we obtain $\mathcal{F}_n(x) \geq 0$ for each $x \in S$. Therefore $\|\mathcal{F}_n\|_1 = 1$ for each $n \in \mathbb{N}_0$.

Now we can conclude (in a similar way as in Theorem 4.14.18):

Theorem 4.14.21 *Let $(R_n(x))_{n\in\mathbb{N}_0}$ generate a polynomial hypergroup satisfying*
$\sum\limits_{k=1}^{\infty} \frac{1}{h(k)a_k} = \infty$. *Assume that S satisfies property (F). Then $(\mathcal{F}_n(x))_{n\in\mathbb{N}_0}$ is an approximate identity with respect to $L^1(S,\pi)$. Moreover, $(\mathcal{F}_n(x))_{n\in\mathbb{N}_0}$ is a summability kernel.*

Proof. It remains to prove condition (3) of summability kernels. By Proposition 4.14.20 we have

$$\mathcal{F}_n(x) = \frac{1}{b_{n,0}} \frac{1 - R_{n+1}(x)}{1 - R_1(x)}.$$

Given $0 < \delta < 1$ we obtain

$$\int_{S\setminus[\delta,1]} |\mathcal{F}_n(x)|\, d\pi(x) \leq \frac{2}{b_{n,0}} \int_{S\setminus[\delta,1]} \frac{1}{1 - R_1(x)}\, d\pi(x)$$

$$\leq \frac{2}{b_{n,0}} \frac{1}{1 - R_1(\delta)} \longrightarrow 0 \quad \text{as } n \to \infty.$$

◇

Corollary 4.14.22 *Let B be a homogeneous Banach space on S. If the assumptions of Theorem 4.14.21 are valid, then span$\{R_n|S : n \in \mathbb{N}_0\}$ is dense in B and $(\mathcal{F}_n(x))_{n\in\mathbb{N}_0}$ is an approximate identity with respect to B.*

Examples:

14.2.2 *(1) Jacobi polynomials $R_n^{(\alpha,\beta)}(x)$*

If $(\alpha,\beta) \in J$, where $J = \{(\alpha,\beta) : \alpha \geq \beta > -1 \text{ and } \alpha + \beta \geq -1\}$, then $(R_n^{(\alpha,\beta)}(x))_{n\in\mathbb{N}_0}$ generates a polynomial hypergroup on $\mathbb{N}_0$ and $S = D_s = D = [-1,1]$ satisfies property (F). Note that if $(\alpha,\beta) \in J_s$, where

$J_s = \{(\alpha, \beta) : \alpha \geq \beta > -1 \text{ and } (\beta \geq -\frac{1}{2} \text{ or } \alpha + \beta \geq 0)\} \subseteq J$, then $(R_n^{(\alpha,\beta)}(x))_{n\in\mathbb{N}_0}$ generates a polynomial hypergroup satisfying the Pontryagin duality. (See subsection 4.15.)

Hence Theorem 4.14.18 or Corollary 4.14.19 yield that $(F_{2n}(x))_{n\in\mathbb{N}_0}$ is an approximate identity with respect to B for Jacobi polynomials, provided $(\alpha, \beta) \in J$. Note that property (H) is satisfied for these polynomial hypergroups. (Use Corollary 4.7.2.) Furthermore we would like to point out that Proposition 4.14.5 and Proposition 4.14.4 imply that for $(\alpha, \beta) \in J_s$ we have to show only $a_{2n,1} \to 1$ as n tends to infinity.

Concerning the modified Fejér kernel $(\mathcal{F}_n(x))_{n\in\mathbb{N}_0}$ for $R_n^{(\alpha,\beta)}(x)$, $(\alpha, \beta) \in J$, the condition $\sum_{k=1}^{\infty} \frac{1}{h(k)\, a_k} = \infty$ is satisfied exactly when $-\frac{1}{2} \leq \alpha \leq 0$. Note that $h(n) = O(n^{2\alpha+1})$.

14.2.3 *(2) Generalized Chebyshev polynomials* $T_n^{(\alpha,\beta)}(x)$

If $(\alpha \geq \beta > -1,\ \alpha + \beta + 1 > 0)$ or $\alpha = \beta = -\frac{1}{2}$ we know that $S = D_s = D = [-1, 1]$ satisfies property (F). Property (H) is fulfilled. (Use Corollary 4.7.2.) Hence $(F_{2n}(x))_{n\in\mathbb{N}_0}$ is then an approximate identity with respect to B. Notify that property (P) is valid if $\alpha > \beta \geq \frac{1}{2}$ or $\alpha = \beta = -\frac{1}{2}$. (See also subsection 4.15.)

For the modified Fejér kernel $(\mathcal{F}_n(x))_{n\in\mathbb{N}_0}$ for $T_n^{(\alpha,\beta)}(x)$ with property (F), the condition $\sum_{k=1}^{\infty} \frac{1}{h(k)\, a_k} = \infty$ is satisfied exactly when $-\frac{1}{2} \leq \alpha \leq 0$, since $h(n) = O(n^{2\alpha+1})$.

14.2.4 *4.14.2.2 De la Vallée-Poussin kernels*

A special property of the following de la Vallée-Poussin kernel is to reproduce polynomials up to a certain degree. The Fejér-type kernel $F_{2n}(x))_{n\in\mathbb{N}_0}$ reproduces only the constant functions. In order to define $F_{2n}(x)$ we used $\mathcal{D}_n^2(x)$, see (F1), and investigated the sequences $\chi_n * \chi_n$. Now we use the products $\mathcal{D}_n(x)\mathcal{D}_m(x)$ and investigate $\chi_n * \chi_m$ for $n \geq m$. We will focus on the case $n = 2m$, compare [364, Ch.I, 2.13].

We define

$$V_m^n(x) := \frac{1}{\int_S \mathcal{D}_n(y)\, \mathcal{D}_m(y)\, d\pi(y)}\, \mathcal{D}_n(x)\, \mathcal{D}_m(x) \qquad (V1)$$

for $x \in S$ and $n, m \in \mathbb{N}_0,\ n \geq m$.

Since $\mathcal{D}_n(x)\mathcal{D}_m(x) = (\chi_n * \chi_m)^\wedge(x)$ and

$$\int_S \mathcal{D}_n(y)\, \mathcal{D}_m(y)\, d\pi(y) \;=\; \chi_n * \chi_m(0),$$

it follows

$$V_m^n(x) \;=\; \sum_{k=0}^{n+m} v_{m,k}^n\, R_k(x)\, h(k)$$

where

$$v_{m,k}^n \;=\; \frac{\chi_n * \chi_m(k)}{\chi_n * \chi_m(0)} \;=\; \frac{\chi_n * \chi_m(k)}{H(m)},$$

see Lemma 4.14.14.

Proposition 4.14.23 *For the de la Vallée-Poussin kernel $(V_m^n(x))_{n,m\in\mathbb{N}_0}$, $n \geq m$, we have $V_m^n P = P$ for each polynomial P with $\deg P \leq n-m$.*

Proof. Let $P(x) = \sum_{j=0}^{n-m} d_j R_j(x)$. Then $\check{P}(k) = \frac{d_k}{h(k)}$ for $k = 0, 1, \ldots,$ $n-m$ and $\check{P}(k) = 0$ for $k = n-m+1, \ldots$. By Lemma 4.14.14(i) we obtain $v_{m,k}^n = 1$ for $k = 0, 1, \ldots, n-m$. Hence

$$V_m^n\, P(x) \;=\; \sum_{k=0}^{n+m} v_{m,k}^n\, \check{P}(k)\, R_k(x)\, h(k) \;=\; \sum_{k=0}^{n-m} d_k\, R_k(x) \;=\; P(x).$$

◇

Proposition 4.14.24 *For the de la Vallée-Poussin kernel $(V_m^n(x))_{n,m\in\mathbb{N}_0}$, $n \geq m$, we have*

$$\|V_m^n\|_1 \;\leq\; \frac{H(n)}{H(m)} \qquad \textit{for all } n, m \in \mathbb{N}_0,\ n \geq m.$$

Proof. Using $2\mathcal{D}_n(x)\mathcal{D}_m(x) = \mathcal{D}_n^2(x) + \mathcal{D}_m^2(x) - (\mathcal{D}_n(x) - \mathcal{D}_m(x))^2$ we obtain

$$2\int_S |\mathcal{D}_n(x)\mathcal{D}_m(x)|\, d\pi(x)$$

$$\leq\; \int_S \mathcal{D}_n^2(x)\, d\pi(x) \;+\; \int_S \mathcal{D}_m^2(x)\, d\pi(x) \;+\; \int_S (\mathcal{D}_n(x) - \mathcal{D}_m(x))^2\, d\pi(x)$$

$$=\; H(n) \;+\; H(m) \;+\; \sum_{j=m+1}^{n} h(j) \;=\; 2\, H(n).$$

Hence

$$\|V_m^n\|_1 = \frac{1}{H(m)} \int_S |\mathcal{D}_n(x)\mathcal{D}_m(x)|\, d\pi(x) \le \frac{H(n)}{H(m)}.$$

$\diamond$

The coefficients $v_{m,k}^n$ are zero for $k \ge m+n+1$. If $n = m$ we get the Fejér-type kernels $F_{2m}(x)$ i.e. $F_{2m}(x) = V_m^m(x)$. An interesting case is $n = 2m$. By Proposition 4.14.23 we have $v_{m,k}^{2m} = 1$ for $k = 0, 1, ..., m$, i.e. $V_m^{2m} R_k = R_k$ for $k = 0, 1, ..., m$. Moreover we have the following result.

Theorem 4.14.25 *Let $(R_n(x))_{n\in\mathbb{N}_0}$ generate a polynomial hypergroup on $\mathbb{N}_0$. Assume that S satisfies property (F) and assume that $h(n) \sim n^\gamma$ for some $\gamma > 0$. Then $(V_m^{2m}(x))_{m\in\mathbb{N}_0}$ is an approximate identity with respect to $L^1(S,\pi)$.*

Proof. We know $v_{m,k}^{2m} = 1$ for $k = 0, 1, ..., m$. In particular $\lim_{m\to\infty} v_{m,k}^{2m} = 1$ for each $k \in \mathbb{N}_0$. By Proposition 4.14.6 it remains to show that $\|V_m^{2m}\|_1 \le M$ for all $m \in \mathbb{N}_0$. Since

$$\frac{H(2m)}{H(m)} \sim \frac{(2m)^{\gamma+1}}{m^{\gamma+1}},$$

Proposition 4.14.24 implies the boundedness of $\|V_m^{2m}\|_1$. $\diamond$

Corollary 4.14.26 *Let B be a homogeneous Banach space on S such that the linear span of $\{R_n|S : n \in \mathbb{N}_0\}$ is dense in B. If the assumptions of Theorem 4.14.25 are valid, then $(V_m^{2m}(x))_{m\in\mathbb{N}_0}$ is an approximate identity with respect to B.*

As examples for which the assumptions of Theorem 4.14.25 are satisfied we refer to the subsection 4.14.2.1 of Fejér-type kernels:

(1) Jacobi polynomials $R_n^{(\alpha,\beta)}(x)$ with $(\alpha,\beta) \in J$.
(2) Generalized Chebyshev polynomials $T_n^{(\alpha,\beta)}(x)$ with ($\alpha \ge \beta > -1$, $\alpha + \beta + 1 > 0$) or $\alpha = \beta = -\frac{1}{2}$.

Finally we shortly describe another class of de la Vallée-Poussin kernels. For trigonometric polynomials a de la Vallée-Poussin kernel is defined by

$$V_n(t) = \frac{(n!)^2}{(2n)!}(\cos t + 1)^n = 1 + \sum_{k=1}^{n} v_{n,k} \cos(kt)$$

with

$$v_{n,k} = \frac{(n!)^2}{(n-k)!(n+k)!} \qquad \text{for } t \in [-\pi,\pi],$$

see [125,Section2.5.2].

An explicit extension of this formula for Jacobi polynomials $R_n^{(\alpha,\beta)}(x)$ (and the history of this sum) is due to Askey [18,(2.30)].

Applying formula (4.3.4) of Szegö's book [641] we get

$$\frac{\Gamma(n+\alpha+\beta+2)}{\Gamma(n+\beta+2)\,\Gamma(\alpha+1)\,2^{\alpha+\beta+1}}\left(\frac{1+x}{2}\right)^n = \sum_{k=0}^{n} v_{n,k}\, R_k^{(\alpha,\beta)}(x)\, h(k)$$

for all $x \in [-1,1]$, $n \in \mathbb{N}_0$, with the coefficients

$$v_{n,k} = \frac{n!\,\Gamma(n+\alpha+\beta+2)}{(n-k)!\,\Gamma(n+k+\alpha+\beta+2)}. \qquad (*)$$

(See also example (4) of subsection 4.10.) Hence we define for the polynomial hypergroup generated by Jacobi polynomials

$$\mathcal{V}_n(x) = \sum_{k=0}^{n} v_{n,k}\, R_n^{(\alpha,\beta)}(x)\, h(k) \qquad (V2)$$

with the coefficients $v_{n,k}$ of $(*)$.

The asymptotic properties of the Gamma function imply that $\lim_{n\to\infty} v_{n,k} = 1$ for each $k \in \mathbb{N}_0$. Since $\mathcal{V}_n(x) \geq 0$ for $x \in [-1,1]$, Theorem 4.14.18 or Corollary 4.14.19 yield that $(\mathcal{V}_n(x))_{n\in\mathbb{N}_0}$ is an approximate identity with respect to B if $(\alpha,\beta) \in J$.

In [514] Obermaier generalized this access to construct approximate identities. In particular, for kernel functions

$$\mathcal{V}_n(x) := \left(\frac{1+x}{2}\right)^n = \sum_{k=0}^{n} v_{n,k}\, R_k(x)\, h(k)$$

sufficient conditions are derived such that $\lim_{n\to\infty} v_{n,k} = 1$ for $k \in \mathbb{N}_0$, see [514,Theorem 3.1]. It is shown that for generalized Chebyshev polynomials $T_n^{(\alpha,\beta)}(x)$ the kernels $\mathcal{V}_n(x))_{n\in\mathbb{N}_0}$ are approximate identities with respect to B if $(\alpha \geq \beta > -1,\ \alpha+\beta+1 > 0)$ or $\alpha = \beta = -\frac{1}{2}$. see [514,Example 4.2].

Remark: Further examples of approximate identities based on triangular schemes $a_{n,k}$ are obtained in [336]. There the orthogonal polynomials $R_n(x)$, $n \in \mathbb{N}_0$, belong to the class of Bernstein-Szegö polynomials. Property (F) is not used. The construction of the approximate identities is derived from known results for Chebyshev polynomials of the first kind.

14.2.5 *4.14.2.3 Dirichlet kernel*

The Dirichlet kernel is defined by $\mathcal{D}_n(x) = \sum_{k=0}^{n} R_k(x)h(k)$, i.e. $a_{n,k} = 1$ for $k = 0, 1, ..., n$, and

$$\mathcal{D}_n(x, y) = \sum_{k=0}^{n} R_k(x)\, R_k(y)\, h(k) \qquad \text{for } x, y \in S.$$

Convergence with respect to $L^1(S, \pi)$ or $C(S)$ is determined by

$$\sup_{n\in\mathbb{N}_0} \sup_{x\in S} \|\mathcal{D}_n(x, \cdot)\|_1 < \infty,$$

see Corollary 4.14.3.

We derive a necessary property of $(R_n(x))_{n\in\mathbb{N}_0}$ such that $\sup_{n\in\mathbb{N}_0} \sup_{x\in S} \|\mathcal{D}_n(x, \cdot)\|_1 < \infty$. A result of Maté, Névai and Totik [472] will be an essential tool.

The orthonormal version of $R_n(x)$ is denoted by $p_n(x)$, i.e. $p_n(x) = \sqrt{h(n)}\, R_n(x)$.

Theorem 4.14.27 *Assume that $(R_n(x))_{n\in\mathbb{N}_0}$ generates a polynomial hypergroup, and suppose $S = [-1, 1]$ and $\pi' > 0$ almost everywhere. (π' is the absolute continuous part of π.) If $\sup_{n\in\mathbb{N}_0} \sup_{x\in S} \|\mathcal{D}_n(x, \cdot)\|_1 < \infty$, then the Haar weights $h(n)$ are bounded.*

Proof. Let $\sup_{x\in S} \|\mathcal{D}_n(x, \cdot)\|_1 \leq M$ for every $n \in \mathbb{N}_0$. Since

$$R_n(x)\, R_n(y)\, h(n) = \mathcal{D}_n(x, y) - \mathcal{D}_{n-1}(x, y) \qquad \text{for } n \geq 1,$$

we get

$$\begin{aligned} \|p_n\|_\infty\, \|p_n\|_1 &= \sup_{x\in S} \sqrt{h(n)}\, |R_n(x)| \int_S |R_n(y)|\, \sqrt{h(n)}\, d\pi(y) \\ &\leq \|\mathcal{D}_n(x, \cdot)\|_1 + \|\mathcal{D}_{n-1}(x, \cdot)\| \leq 2M. \end{aligned}$$

Now we apply [472,Theorem 2(7)] with $p = 1$ and the function $g(x) = 1$ for $x \in [-1, 1]$. it follows $\liminf_{n\to\infty} \|p_n\|_1 > 0$. Since $\|p_n\|_\infty\, \|p_n\|_1 \leq 2M$ and $\|R_n\|_\infty = 1$, it follows that $h(n)$ is bounded. ◇

Remark: Theorem 4.14.27 applied to Jacobi polynomials $R_n^{(\alpha,\beta)}(x)$, $(\alpha, \beta) \in J\backslash\{(-\frac{1}{2}, -\frac{1}{2})\}$, shows that $(\mathcal{D}_n(x))_{n\in\mathbb{N}_0}$ is not an approximate identity with respect to $L^1(S, \pi)$ or $C(S)$. For the case $\alpha = \beta = -\frac{1}{2}$ (the Chebyshev polynomials of the first kind) we refer to [568] and [456], where it is shown that $\|\mathcal{D}_n(1,\cdot)\|_1 \sim \ln(n)$. Hence $(\mathcal{D}_n(x))_{n\in\mathbb{N}_0}$ is not an approximate identity with respect to $L^1(S,\pi)$ or $C(S)$ for each $(\alpha,\beta)\in J$.

Given $1 < p < \infty$ the following result for Jacobi polynomials $R_n^{(\alpha,\beta)}(x)$, $(\alpha,\beta)\in J$, is true:

If $p \in]\frac{4(\alpha+1)}{2\alpha+3}, \frac{4(\alpha+1)}{2\alpha+1}[$, then $(\mathcal{D}_n(x))_{n\in\mathbb{N}_0}$ is an approximate identity with respect to $L^p(S,\pi)$ and if $p \notin]\frac{4(\alpha+1)}{2\alpha+3}, \frac{4(\alpha+1)}{2\alpha+1}[$, then there exists a function $\varphi\in L^p(S,\pi)$ such that $\mathcal{D}_n\varphi$ does not converge. For these convergence and nonconvergence results we refer to [549], [494] and [188]. The operator norm of $\mathcal{D}_n$ we will meet again in section 4.16.

14.3 *Selective approximate identities*

We have seen that the approximation procedures in norm depend on the condition that S satisfies property (F). Now we will focus on the convergence at points $y\in S$ for all functions $\varphi\in C(S)$. We will achieve positive results without requiring property (F). We assume throughout that $(R_n(x))_{n\in\mathbb{N}_0}$ generates a polynomial hypergroup on $\mathbb{N}_0$.

Let $y\in S$. For $n\in\mathbb{N}_0$ let $b_{n,k}^y$, $k=0,...,n$, be a triangular matrix of complex numbers and define

$$B_n^y(x) = \sum_{k=0}^n b_{n,k}^y \, R_k(x)\, h(k) \qquad \text{for } x\in S.$$

Definition 1. We say that the sequence $(B_n^y)_{n\in\mathbb{N}_0}$ is a **selective approximate identity** with respect to $y\in S$, if

$$\lim_{n\to\infty}\int_S B_n^y(x)\,\varphi(x)\,d\pi(x) = \varphi(y) \qquad \text{for all } \varphi\in C(S).$$

If $(A_n)_{n\in\mathbb{N}_0}$ is an approximate identity with respect to the Banach space $C(S)$ according to subsection 4.14.1, then $b_{n,k}^y = a_{n,k}R_k(y)$ generates a selective identity with respect to y. In fact, we have then

$$B_n^y(x) = \sum_{k=0}^n a_{n,k}\, R_k(y)\, R_k(x)\, h(k) = A_n(x,y)$$

and we get

$$A_n\varphi(y) = \sum_{k=0}^n b_{n,k}^y\, \check{\varphi}(k)\, h(k) = \int_S B_n^y(x)\,\varphi(x)\,d\pi(x),$$

and so $\int\limits_S B_n^y(x)\varphi(x)d\pi(x)$ converges to $\varphi(y)$, since $(A_n)_{n\in\mathbb{N}_0}$ is an approximate identity with respect to $C(S)$.

This suggests to define for $\varphi \in C(S)$,

$$B_n^y \varphi := \int_S B_n^y(x)\, \varphi(x)\, d\pi(x) \;=\; \sum_{k=0}^{n} b_{n,k}^y\, \check{\varphi}(k)\, h(k).$$

B_n^y is a continuous linear functional on $C(S)$ and for the norm of this functional we have

$$\|B_n^y\| \;=\; \int_S |B_n^y(x)|\, d\pi(x) \;=\; \|B_n^y\|_1$$

by Riesz representation theorem.

The following characterization of selective approximate identities is not surprising, compare Theorem 4.14.1.

Theorem 4.14.28 *Let $y \in S$. Then $(B_n^y)_{n\in\mathbb{N}_0}$ is a selective approximate identity with respect to y if and only if*

(1) $\lim_{n\to\infty} b_{n,k}^y \;=\; R_k(y)$ *for all $k \in \mathbb{N}_0$, and*
(2) $\|B_n^y\|_1 \le C$ *for all $n \in \mathbb{N}_0$.*

Proof. Suppose that (1) and (2) hold. Let

$$Q(x) \;=\; \sum_{k=0}^{m} v_k\, R_k(x)\, h(k)$$

be an arbitrary polynomial of degree m. For $n \ge m$ we have $B_n^y Q = \sum_{k=0}^{m} b_{n,k}^y\, v_k\, h(k)$. By (1) it follows that $\lim_{n\to\infty} B_n^y Q = Q(y)$. Now let $\varphi \in C(S)$ and $\varepsilon > 0$ be arbitrary. Choose a polynomial Q such that $\|Q - \varphi\|_\infty < \varepsilon$. By (2) it follows

$$\begin{aligned} |B_n^y\varphi - \varphi(y)| \;&\le\; |B_n^y\varphi - B_n^y Q| \;+\; |B_n^y Q - Q(y)| \;+\; |Q(y) - \varphi(y)| \\ &<\; C\varepsilon \;+\; |B_n^y Q - Q(y)| \;+\; \varepsilon. \end{aligned}$$

Hence, for sufficiently large n we get $|B_n^y\varphi - \varphi(y)| \;<\; (C+2)\,\varepsilon$, and it is shown that $(B_n^y)_{n\in\mathbb{N}_0}$ is a selective approximate identity with respect to y.

Conversely assume that $(B_n^y)_{n\in\mathbb{N}_0}$ is a selective approximate identity with respect to y. It follows that $\sup\{\, |B_n^y\varphi| \,:\, n \in \mathbb{N}_0\} < \infty$ for every $\varphi \in C(S)$. By the Banach-Steinhaus theorem we get (2), and by $B_n^y R_k = b_{n,k}^y$ for $n \ge k$ condition (1) follows. ◇

Corollary 4.14.29 *Let $y \in S$ and assume that $B_n^y(x) \ge 0$ for all $x \in S$, $n \in \mathbb{N}_0$. Then $(B_n^y)_{n\in\mathbb{N}_0}$ is a selective approximate identity with respect to y if and only if*

$$\lim_{n\to\infty} b_{n,k}^y \;=\; R_k(y) \qquad \textit{for all } k \in \mathbb{N}_0.$$

Proof. It is sufficient to prove that in case of $B_n^y(x) \geq 0$ for all $n \in \mathbb{N}_0$, $x \in S$ the convergence condition $\lim_{n\to\infty} b_{n,k}^y = R_k(y)$ for all $k \in \mathbb{N}_0$ implies the boundedness condition (2) of Theorem 4.14.28. We have for $n \in \mathbb{N}$

$$\|B_n^y\|_1 = \int_S \sum_{k=0}^n b_{n,k}^y \, R_k(x) \, d\pi(x) = b_{n,0}^y,$$

and

$$\lim_{n\to\infty} b_{n,0}^y = R_0(y) = 1.$$

Hence $(\|B_n^y\|_1)_{n\in\mathbb{N}_0}$ is bounded. ◇

We start with an example with convergence at the point $y = 1$.

Consider the modified Fejér kernels $(\mathcal{F}_n(x))_{n\in\mathbb{N}_0}$ defined by (F2) in subsection 4.14.2. Using the facts collected before Theorem 4.14.20 we obtain by Corollary 4.14.29

Proposition 4.14.30 *Assume that* $\sum_{k=1}^{\infty} \frac{1}{h(k)a_k} = \infty$. *Then* $1 \in S$ *and* $(\mathcal{F}_n(x))_{n\in\mathbb{N}_0}$ *is a selective approximate identity with respect to* $y = 1$.

If $b_{n,k}^y$ is continuous in y, we can add to the pointwise approximation of Theorem 4.14.28 a global approximation procedure on S.

Theorem 4.14.31 *Assume that* $y \mapsto b_{n,k}^y$ *is a continuous mapping from* S *to* $\mathbb{C}$ *for all* $n, k \in \mathbb{N}_0$. *Denote by* B_n *the linear operator from* $C(S)$ *into* $C(S)$, *which is defined by*

$$B_n\varphi(y) := \int_S B_n^y(x) \, \varphi(x) \, d\pi(x) = B_n^y \varphi.$$

Then $\lim_{n\to\infty} \|B_n\varphi - \varphi\|_\infty = 0$ *for all* $\varphi \in C(S)$ *if and only if*

(1) $\lim_{n\to\infty} b_{n,k}^y = R_k(y)$ **uniformly** *on* S *for all* $k \in \mathbb{N}_0$, *and*
(2) $\|B_n\varphi\|_\infty \leq C \, \|\varphi\|_\infty$ *for all* $\varphi \in C(S)$, $n \in \mathbb{N}_0$.

Proof. Since

$$B_n\varphi(y) = B_n^y \varphi = \sum_{k=0}^n b_{n,k}^y \, \check{\varphi}(k) \, h(k),$$

we see that $B_n\varphi \in C(S)$ for each $\varphi \in C(S)$. The proof of this equivalence statement is quite similar to the proof of Theorem 4.14.28. We only have to incorporate that the convergence of $b_{n,k}^y$ to $R_k(y)$ is uniform with respect to $y \in S$ as $n \to \infty$. ◇

The following construction of selective approximate identities is strongly related to that of Fejér-type kernels, see subsection 4.14.2. Consider $b_{n,k}^y = R_k(y)$ and define

$$\mathcal{D}_n^y(x) = \sum_{k=0}^{n} R_k(y)\, R_k(x)\, h(k), \tag{1}$$

the Dirichlet kernel with respect to $y \in S$. In order to construct the Fejér-type kernel with respect to $y \in S$, we have to consider $(\mathcal{D}_n^y(x))^2$. Obviously

$$(\mathcal{D}_n^y(x))^2 = \sum_{k=0}^{2n} \mu_{2n,k}^y\, R_k(x)\, h(k), \tag{2}$$

where the coefficients $\mu_{2n,k}^y$ are uniquely determined. For $n \in \mathbb{N}_0$ denote

$$\chi_n^y(k) = \begin{cases} R_k(y) & \text{for } k = 0, ..., n \\ 0 & \text{for } k = n+1, ... \end{cases}$$

We have $\mathcal{D}_n^y(x) = (\chi_n^y)^\wedge(x)$ and therefore $(\mathcal{D}_n^y(x))^2 = (\chi_n^y * \chi_n^y)^\wedge(x)$ for all $x \in S$. Define

$$F_{2n}^y(x) = (\mathcal{D}_n^y(x))^2 / \int_S (\mathcal{D}_n^y(x))^2 d\pi(x). \tag{3}$$

Note that

$$\int_S (\mathcal{D}_n^y(x))^2\, d\pi(x) = \chi_n^y * \chi_n^y(0) = \sum_{j=0}^{n} R_j^2(y)\, h(j),$$

and

$$\chi_n^y * \chi_n^y(k) = \sum_{j=0}^{n} L_k\, \chi_n^y(j)\, R_j(y)\, h(j)$$

$$= \sum_{j=0}^{n} \sum_{l=|k-j|}^{\min\{n,k+j\}} g(k,j;l)\, R_l(y)\, R_j(y)\, h(j)$$

for $k = 1, ..., 2n-1$,

$$\chi_n^y * \chi_n^y(2n) = g(n,2n;n)\, R_n(y)\, R_n(y)\, h(n),$$

and

$$\chi_n^y * \chi_n^y(k) = 0 \quad \text{for } k = 2n+1, ...$$

Compare with the formulas of Lemma 4.14.15.

Since $g(k,j;l) = 0$ for $l > k+j$, we have

$$\mu^y_{2n,k} = \sum_{j=0}^{n} \sum_{l=|k-j|}^{n} g(k,j;l)\, R_l(y)\, R_j(y)\, h(j). \tag{4}$$

In particular,

$$\mu^y_{2n,0} = \chi^y_n * \chi^y_n(0) = \sum_{j=0}^{n} R_j^2(y)\, h(j),$$

and we obtain for the Fejér-type kernel $F^y_{2n}(x)$ of (3)

$$F^y_{2n}(x) = \sum_{k=0}^{2n} a^y_{2n,k}\, R_k(x)\, h(k) \tag{5}$$

with coefficients

$$a^y_{2n,k} = \frac{\mu^y_{2n,k}}{\mu^y_{2n,0}}.$$

Definition 2. Let $y \in S$. We say that property (H_y) holds, if

$$\lim_{n\to\infty} \frac{R_n^2(y)\, h(n)}{\sum_{k=0}^{n} R_k^2(y)\, h(k)} = 0.$$

If $y = 1 \in S$, property (H_y) is nothing else but property (H), which is studied in subsection 4.7. Studying consequences of (H_y), we have to consider that $R_k(y)$ can be zero. It is worthwhile to mention that the case $R_k(y) = 0 = R_{k+1}(y)$ is not possible. This follows immediately from

$$R'_{k+1}(y)\, R_k(y) - R'_k(y)\, R_{k+1}(y) > 0,$$

see the Christoffel-Darboux formula in Proposition 4.10.3(2).

Subsequently we will use the notation $H(n,y) := \sum\limits_{k=0}^{n} R_k^2(y)\, h(k)$.

Proposition 4.14.32 *Let $y \in S$ and suppose that property (H_y) holds. Then*

$$\lim_{n\to\infty} \frac{R^2_{n+m}(y)\, h(n+m)}{H(n,y)} = 0 \qquad \textit{for each } m \in \mathbb{N}_0.$$

Proof. We consider the case $m = 1$. For all $n \in \mathbb{N}_0$ such that $R_{n+1}(y) \neq 0$, it obviously follows

$$\frac{H(n+1,y)}{R^2_{n+1}(y)\, h(n+1)} = \frac{H(n,y)}{R^2_{n+1}(y)\, h(n+1)} + 1,$$

and we obtain that

$$\lim_{n\to\infty} \frac{R^2_{n+1}(y)\, h(n+1)}{H(n,y)} = 0.$$

Iterating this consideration we get the assertion for each $m \in \mathbb{N}_0$. ◇

Theorem 4.14.33 *Let $(R_n(y))_{n\in\mathbb{N}_0}$ generate a polynomial hypergroup and let $y \in S$. Assume that property (H_y) is satisfied. Then $(F^y_{2n}(x))_{n\in\mathbb{N}_0}$ is a selective approximate identity with respect to y.*

Proof. Since $F^y_{2n}(x) \geq 0$ for all $x \in S$, $n \in \mathbb{N}_0$, we have only to prove that $\lim\limits_{n\to\infty} a^y_{2n,k} = R_k(y)$ for all $k \in \mathbb{N}_0$, see Corollary 4.14.29. To show this we proceed similar to Proposition 4.14.16.

Let $n \geq k$. Since

$$\sum_{l=|k-j|}^{k+j} g(j,k;l)\, R_l(y) \;=\; R_k(y)\, R_j(y),$$

we obtain from

$$a^y_{2n,k} \;=\; \frac{\mu^y_{2n,k}}{\mu^y_{2n,0}} \;=\; \frac{1}{H(n,y)} \left(\sum_{j=0}^{n} \sum_{l=|k-j|}^{k+j} g(j,k;l)\, R_l(y)\, R_j(y)\, h(j) \right.$$

$$\left. -\; \sum_{j=n+1-k}^{n} \sum_{l=n+1}^{k+j} g(j,k;l)\, R_l(y)\, R_j(y)\, h(j) \right)$$

$$=\; R_k(y) \;-\; \frac{1}{H(n,y)} \sum_{j=n+1-k}^{n} \sum_{l=n+1}^{k+j} g(j,k;l)\, R_l(y)\, R_j(y)\, h(j)$$

Proposition 4.14.32, and $g(j,k;l)h(j) = g(l,k;j)h(l)$ imply that $\lim\limits_{n\to\infty} a^y_{2n,k} = R_k(y)$. ◇

Theorem 4.14.34 *Let $(R_n(y))_{n\in\mathbb{N}_0}$ generate a polynomial hypergroup. Assume that property (H_y) is satisfied for all $y \in S$, and that the convergence*

$$\lim_{n\to\infty} \frac{R_n^2(y)\, h(n)}{H(n,y)} \;=\; 0$$

is uniform with respect to $y \in S$. Define

$$F_{2n}\varphi(y) \;=\; \int_S F^y_{2n}(x)\, \varphi(x)\, d\pi(x).$$

Then $F_{2n}\varphi \in C(S)$ and $\lim\limits_{n\to\infty} \|F_{2n}\varphi - \varphi\|_\infty = 0$ for all $\varphi \in C(S)$.

Proof. At first notify that $y \mapsto a^y_{2n,k}$ is a continuous function on S. Since

$$\lim_{n\to\infty} \frac{R_n^2(y)\,h(n)}{H(n,y)} = 0$$

is uniform, the construction of $a^y_{2n,k}$ shows that $\lim\limits_{n\to\infty} a^y_{2n,k} = R_k(y)$ is uniform, too. Since $F^y_{2n}(x) \geq 0$, we obtain $\|F_{2n}\varphi\|_\infty \leq \|\varphi\|_\infty$, and by Theorem 4.14.31 the proof is complete. ◇

In order to apply Theorem 4.14.33 and Theorem 4.14.34 we have to examine property (H_y). A result of Nevai et al. [511] yields a sufficient condition such that (H_y) is uniformly true for all $y \in S$. We will use the notation of section 4.3.

Theorem 4.14.35 (Nevai, Totik, Zhang)
Suppose $\mathbb{N}_0$ is a polynomial hypergroup belonging to the Nevai class with R_n-parameters a, c. If $a > 0$, $c > 0$, then

$$\lim_{n\to\infty} \max_{y\in S} \frac{R_n^2(y)\,h(n)}{H(n,y)} = 0. \qquad (*)$$

Proof. Using the terminology of Nevai we have $\pi \in M(ba_0 + b_0, 2\sqrt{caa_0})$, where $a = \lim\limits_{n\to\infty} a_n$, $b = \lim\limits_{n\to\infty} b_n$, $c = \lim\limits_{n\to\infty} c_n$, $a_0 + b_0 = 1$, a_0 the starting parameter. By Theorem 2.1(2) of [511] we obtain exactly statement $(*)$. Note that $2\sqrt{caa_0} > 0$. ◇

Combining Theorem 4.14.34 and Theorem 4.14.35 we obtain

Corollary 4.14.36 *Suppose $\mathbb{N}_0$ is a polynomial hypergroup belonging to the Nevai class with R_n-parameters $a > 0$ and $c > 0$. Then*

$$\lim_{n\to\infty} \|F_{2n}\varphi - \varphi\|_\infty = 0$$

for each $\varphi \in C(S)$. In particular $(F^y_{2n})_{n\in\mathbb{N}_0}$ is a selective approximate identity with respect to every $y \in S$.

Remarks:

(1) Corollary 4.14.36 is applicable for many polynomial hypergroups on $\mathbb{N}_0$, see subsections 4.5.2, 4.5.3, 4.5.4, 4.5.5, 4.5.7, 4.5.9, 4.5.10 and 4.5.11.
(2) The operators F_{2n} are strongly related to Nevai's G-operator, see [511,Corollary 4.3.1].

Finally we consider the little q-Legendre polynomials $R_n(x;q)$ with parameter $0 < q < 1$, see subsection 4.5.12. Considering $y = 1 \in S$, we know that $h(n)$ grows exponentially. Hence property (H_1) cannot hold. Considering $y \in S, y \neq 1$, we notify that the polynomial hypergroup $\mathbb{N}_0$ generated by $(R_n(x;q))_{n\in\mathbb{N}_0}$ belongs to the Nevai class with $a = c = 0$ and $b = 1$, i.e. $\pi \in M(1,0)$. Hence by [509,Ch.4.1,Theorem 7(ii)] it follows that property (H_y) holds for each $y \in S\backslash\{1\}$.

15 Dual convolution and some representation results

15.1 *Examples with dual convolution*

In this section we collect examples of polynomial hypergroups with the property that the dual space S satisfies (F). We will also present additional properties of the dual convolution on S.

(1) **Jacobi polynomials $R_n^{(\alpha,\beta)}(x)$**
We have already noted that $(R_n^{(\alpha,\beta)}(x))_{n\in\mathbb{N}_0}$ generates a polynomial hypergroup on $\mathbb{N}_0$ if $(\alpha,\beta) \in J = \{(\alpha,\beta) : \alpha \geq \beta > -1,\ \alpha+\beta+1 \geq 0\}$. This is a result of George Gasper, see [252,Theorem]. George Gasper also showed in [254,Theorem 1] that for $x, y \in]-1,1[$ there exist $\omega(x,y) \in M(S)$ such that

$$R_n^{(\alpha,\beta)}(x)\, R_n^{(\alpha,\beta)}(y) \;=\; \int_S R_n(z)\, d\omega(x,y)(z) \qquad \text{for all } n \in \mathbb{N}_0$$

and

$$\|\omega(x,y)\| \;\leq\; M \qquad \text{(independent of } x, y)$$

if and only if $(\alpha,\beta) \in J$, and $\omega(x,y) \in M^+(S)$ if and only if $(\alpha,\beta) \in J_s$, where

$$J_s \;=\; \{(\alpha,\beta) \in J : \beta \geq -\frac{1}{2} \text{ or } \alpha+\beta \geq 0\}.$$

Note that $[-1,1] = S = D_s = D$. In particular, $\omega(x,y) \in M^1(S)$ if $(\alpha,\beta) \in J_s$.
For $\alpha = \beta \geq -\frac{1}{2}$ the explicit form of the measures $\omega(x,y)$ is given already at the end of Chapter 1. We used Theorem 1.2.3 to derive that $[-1,1]$ is a commutative hypergroup with $x = 1$ as unit element, the identity map as involution and $\omega(x,y)$ as convolution.
The case $\alpha > \beta$ and $x = -1$ was not mentioned in [254]. In [398,(2.5)] the appropriate measures $\omega(-1,y)$ are determined. To show that for $\alpha > \beta$, $(\alpha,\beta) \in J_s$ the quadruple $([-1,1], \omega, \mathrm{id}, 1)$ is a hypergroup, we

have to check only the axioms (H3) and (H6). From [254] and [398] it follows that

$$\omega(x,y) =$$

$$\begin{cases} \Big[xy - \sqrt{(1-x^2)(1-y^2)}, xy \\ \quad +\sqrt{(1-x^2)(1-y^2)}\Big] & \text{if } x, y \in]-1,1[\,,\, x+y \geq 0 \\ \Big[-1, xy + \sqrt{(1-x^2)(1-y^2)}\Big] & \text{if } x, y \in]-1,1[\,,\, x+y < 0 \\ [-1,-y] & \text{if } x = -1. \end{cases}$$

Obviously for $x = 1$ we set $\omega(1,y) = \epsilon_y$.
Since $xy + \sqrt{(1-x^2)(1-y^2)} = 1$ is equivalent to $x = y$, we see that (H6) is satisfied. Using that $xy - \sqrt{(1-x^2)(1-y^2)} = -1$ is equivalent to $x + y = 0$, one can show easily that (H3) is valid, too.
Collecting the results from above (mainly due to Gasper) we have:
The dual space $S = [-1,1]$ satisfies property (F) whenever $(\alpha,\beta) \in J$, and $S = [-1,1]$ is a hypergroup (under pointwise multiplication) whenever $(\alpha,\beta) \in J_s$.
By Proposition 3.4.13 it follows that the Pontryagin duality is valid whenever $(\alpha,\beta) \in J_s$.
Moreover, the property that the dual space $S = [-1,1]$ is a hypergroup for $(\alpha,\beta) \in J_s$ characterizes this polynomial hypergroup on $\mathbb{N}_0$. In fact, in [96,Corollary 3.6.3] it is proved that the only polynomial hypergroups on $\mathbb{N}_0$, such that the dual space $S = [-1,1]$ is a hypergroup (under pointwise multiplication) are generated by Jacobi polynomials $(R_n^{(\alpha,\beta)}(x))_{n\in\mathbb{N}_0}$ with parameters $(\alpha,\beta) \in J_s$. This result was announced in [156].
Remark: Results of Chapter 3 with the assumption that $\hat{K}$ is a hypergroup (e.g. Theorem 3.5.16, Theorem 3.8.6 or Theorem 3.9.11) applied to polynomial hypergroups on $K = \mathbb{N}_0$ are exactly results for Jacobi polynomials with $(\alpha,\beta) \in J_s$.

(2) **Generalized Chebyshev polynomials $T_n^{(\alpha,\beta)}(x)$**
We know that $(T_n^{(\alpha,\beta)}(x))_{n\in\mathbb{N}_0}$ generates a polynomial hypergroup on $\mathbb{N}_0$, if $\alpha \geq \beta > -1$ and $\alpha+\beta+1 \geq 0$, see subsection 4.5.3. T.P.Laine has shown that for $\alpha \geq \beta > -1$ and $\alpha+\beta+1 > 0$ or $\alpha = \beta = -\frac{1}{2}$ the dual space $S = [-1,1]$ satisfies property (F), see [398,Theorem 1(iii)]. Note that for $\beta = -\frac{1}{2}$ the polynomials $T_n^{(\alpha,-\frac{1}{2})}(x)$ are exactly the ultraspherical polynomials $R_n^{(\alpha,\alpha)}(x)$ which are studied above in (1).

Moreover, it is shown by Laine that the measures $\omega(x,y)$ are positive for $x,y \in [-1,1]$ if $\alpha > \beta \geq -\frac{1}{2}$, see [398,Theorem 1(iv)]. However, for $1/\sqrt{2} < x < 1$ we can derive that

$$\text{supp } \omega(x,-x) \;=\; [-1, 1-2x^2] \cup [2x^2-1, 1].$$

Thus the hypergroup axiom (H6) does not hold. In the same way one obtains that $(x,y) \mapsto \text{supp } \omega(x,y)$ is not continuous in $(x,1)$ for $x \in]-1,1[,\ x \neq 0$.

(3) **Cosh-polynomials $R_n(x)$**
Fix a parameter $a \geq 0$. Cosh-polynomials $R_n(x)$ are determined by the recurrence coefficients

$$c_n \;=\; \frac{\cosh(a(n-1))}{2\cosh(an)\cosh(a)} \quad \text{and } a_n \;=\; \frac{\cosh(a(n+1))}{2\cosh(an)\cosh(a)}$$

for $n \in \mathbb{N}$, see subsection 4.5.7.3. The polynomials $R_n(x)$ generate a polynomial hypergroup on $\mathbb{N}_0$. We know that (up to a multiplicative constant) $d\pi(x) = f(x)dx$ for $x \in [-2\sqrt{\gamma}, 2\sqrt{\gamma}]$, where $\frac{1}{2\sqrt{\gamma}} = \cosh(a)$ and $f(x) = \frac{1}{\sqrt{4\gamma - x^2}}$, see section 4.5.7.
Hence $S = [-2\sqrt{\gamma}, 2\sqrt{\gamma}]$. For $\gamma = \frac{1}{4}$, i.e. $a = 0$, we have $R_n(x) = T_n(x)$. Applying Corollary 3.11.15 we have already shown that S satisfies property (F) for every $a \geq 0$. Note that $1 \notin S$ if $a > 0$, and $D_s = [-1,1]$, $D = \{z \in \mathbb{C} : |z - 2\sqrt{\gamma}| + |z + 2\sqrt{\gamma}| \leq 2\}$.

15.2 *Approximation kernels and representation results*

It is in general not easy to derive property (F) for concrete examples. Following the construction of Gasper [254] for Jacobi polynomials one has to investigate kernels

$$K(x,y,z) \;=\; \sum_{k=0}^{\infty} \lambda_k \, R_k(x)\, R_k(y)\, R_k(z),$$

and finally to estimate these trilinear kernels. We also refer to the paper of Osilenker [527], where such kernels for general orthonormal polynomials are studied in detail. We shall show how approximation kernels may be used for representation results related to property (F) or an extension of the property $\overset{c}{\geq}$. Moreover we derive supplemens to the Schoenberg results for polynomial hypergroups, see Theorem 3.9.2.

Theorem 4.14.7 and Proposition 4.14.8 describe the relation of triple sums $A_n(x,y,z)$ and the representation property (F) on S. Combining Theorem 4.14.7 and Proposition 4.14.8 one has the following equivalence.

Corollary 4.15.1 *Assume that $(A_n)_{n\in\mathbb{N}_0}$ is an approximate identity with respect to $C(S)$, where A_n are defined by the triangular scheme $a_{n,k}$. Then S satisfies property (F) if and only if*

$$\sup_{n\in\mathbb{N}_0}\ \sup_{x,y\in S}\ \|A_n(x,y,\cdot)\|_1 \ < \ \infty.$$

In order to derive characterizations of inverse Fourier-Stieltjes transforms, we have only to replace the set of sequences $(R_n(x)\,R_n(y))_{n\in\mathbb{N}_0}$ by one bounded sequence $\kappa = (\kappa(n))_{n\in\mathbb{N}_0}$ and then to proceed as in Theorem 4.14.7 and Proposition 4.14.8. Define

$$A_n(\kappa, y) := \sum_{k=0}^{n} a_{n,k}\ \kappa(k)\ R_k(y)\ h(k) \qquad \text{for } y \in S.$$

Then we get

Theorem 4.15.2 *Let $\kappa = (\kappa(n))_{n\in\mathbb{N}_0}$ be a bounded sequence of complex numbers. If $\lim_{n\to\infty} a_{n,k} = 1$ for all $k \in \mathbb{N}_0$ and if*

$$\sup_{n\in\mathbb{N}_0}\ \|A_n(\kappa,\cdot)\|_1 \ = \ M \ < \ \infty,$$

then there exists a regular complex Borel measure $\mu_\kappa \in M(S)$ such that

$$\kappa(n) \ = \ \int_S R_n(x)\ d\mu_\kappa(x) \ = \ (\mu_\kappa)^\vee(n) \qquad \textit{for all } n \in \mathbb{N}_0.$$

Proposition 4.15.3 *Let $\kappa = (\kappa(n))_{n\in\mathbb{N}_0}$ be a bounded sequence of complex numbers. If*

$$\sup_{n\in\mathbb{N}_0}\ \sup_{u\in S}\ \|A_n(u,\cdot)\|_1 \ < \ \infty$$

and if

$$\kappa(n) \ = \ \int R_n(u)\ d\mu_\kappa(u) \qquad \textit{for all } n \in \mathbb{N}_0$$

for some $\mu_\kappa \in M(S)$, then

$$\sup_{n\in\mathbb{N}_0}\ \|A_n(\kappa,\cdot)\|_1 \ < \ \infty$$

.

Theorem 4.15.2 combined with Proposition 4.15.3 yields

Corollary 4.15.4 *Let $\kappa = (\kappa(n))_{n\in\mathbb{N}_0}$ be a bounded sequence of complex numbers. Assume that $(A_n)_{n\in\mathbb{N}_0}$ is an approximate identity with respect to $C(S)$, where A_n is defined by the triangular scheme $a_{n,k}$. Then $\kappa = (\mu_\kappa)^\vee$ for some $\mu_\kappa \in M(S)$ if and only if*

$$\sup_{n\in\mathbb{N}_0}\ \|A_n(\kappa,\cdot)\|_1 \ < \ \infty.$$

Counterexample.
We present now a concrete example, which is not an inverse Fourier-Stieltjes transform. Consider the polynomial hypergroup generated by the Chebyshev polynomials $T_n(x)$. We know that $a_{n,k} = 1 - \frac{k}{n+1}$, $k = 0, ..., n$, determines an approximate identity with respect to $C([-1,1])$, see Remark (4) in front of Theorem 4.14.18. Let $\kappa(n) = \sin(n\frac{\pi}{2})$, $n \in \mathbb{N}_0$. Then

$$A_n(\kappa, y) = \sum_{k=1}^{n} \left(1 - \frac{k}{n+1}\right) \sin\left(k\frac{\pi}{2}\right) T_k(y)\, h(k)$$

and

$$\|A_n(\kappa,\cdot)\|_1 = \int_{-1}^{1} |A_n(\kappa,y)|\, d\pi(y) \geq \int_0^1 |A_n(\kappa,y)|\, d\pi(y)$$

$$\geq \left|\int_0^1 A_n(\kappa,y)\, d\pi(y)\right| = \frac{2}{\pi}\left|\int_0^{\pi/2} \sum_{k=1}^{n} \left(1 - \frac{k}{n+1}\right) \sin\left(k\frac{\pi}{2}\right) \cos(kt)\, dt\right|$$

$$= \frac{2}{\pi}\left|\sum_{k=1}^{n} \left(1 - \frac{k}{n+1}\right) \sin\left(k\frac{\pi}{2}\right) \int_0^{\pi/2} \cos(kt)\, dt\right|$$

$$= \frac{2}{\pi}\sum_{k=1}^{n} \left(1 - \frac{k}{n+1}\right) \frac{\sin^2\left(k\frac{\pi}{2}\right)}{k}.$$

Hence

$$\|A_{2m}(\kappa,\cdot)\|_1 \geq \frac{2}{\pi}\left(\sum_{k=1}^{m} \frac{1}{2k-1} - \frac{m}{2m+1}\right) \longrightarrow \infty \text{ as } m \to \infty.$$

By Corollary 4.15.4 there does not exist a measure $\mu \in M([-1,1])$ such that $\kappa = \check{\mu}$. Obviously for each $\lambda = \check{\nu}$, $\nu \in M([-1,1])$, $\kappa + \lambda$ has not an inverse Fourier-Stieltjes transform, too.

Considering the hypergroup generated by an ultraspherical polynomial sequence $(R_n^{(\alpha,\alpha)}(x))_{n\in\mathbb{N}_0}$, $\alpha > \frac{1}{2}$, by Theorem 4.6.2 it follows that there does not exist a measure $\mu \in M([-1,1])$ such that $\kappa(n) = \int_{-1}^{1} R_n^{(\alpha,\alpha)}(x)\, d\mu(x)$.

Finally we consider an extension of property $\overset{c}{\geq}$, see section 4.6. We have to replace the single sequence $\kappa = (\kappa(n))_{n\in\mathbb{N}_0}$ by a set Φ of sequences $\kappa_x = (\kappa_x(n))_{n\in\mathbb{N}_0}$, $x \in S$, where $\kappa_x(n) = P_n(x)$ and $(P_n(x))_{n\in\mathbb{N}_0}$ is a sequence of polynomials $P_n(x)$ with degree $(P_n) = n$ and $P_0(x) = 1$. So

$$\Phi = \{\kappa_x = (\kappa_x(n))_{n\in\mathbb{N}_0} : \kappa_x(n) = P_n(x),\ x \in S\}.$$

Define

$$A_n(\kappa_x, y) := \sum_{k=0}^{n} a_{n,k}\, P_k(x)\, R_k(y)\, h(k)$$

for $x, y \in S$.

Theorem 4.15.5 *Let $(P_n(x))_{n\in\mathbb{N}_0}$, $x \in S$, be an arbitrary sequence of polynomials $P_n(x)$ with degree $(P_n) = n$ and $P_0(x) = 1$. If $\lim_{n\to\infty} a_{n,k} = 1$ for all $k \in \mathbb{N}_0$, and*

$$\sup_{n\in\mathbb{N}_0} \sup_{x\in S} \|A_n(\kappa_x, \cdot)\|_1 \ = \ M \ < \ \infty,$$

then there exists for each $x \in S$ a regular complex Borel measure $\mu_x \in M(S)$ such that

$$P_n(x) \ = \ \int_S R_n(y)\, d\mu_x(y) \ = \ (\mu_x)^\vee(n) \qquad \textit{for all } n \in \mathbb{N}_0,$$

and

$$\|\mu_x\| \ \leq \ M \quad \textit{for all } x \in S.$$

Proposition 4.15.6 *Let $(P_n(x))_{n\in\mathbb{N}_0}$, $x \in S$, be an arbitrary sequence of polynomials $P_n(x)$ with degree $(P_n) = n$ and $P_0(x) = 1$. If*

$$\sup_{n\in\mathbb{N}_)} \sup_{u\in S} \|A_n(u, \cdot)\|_1 \ = \ M_1 \ < \ \infty,$$

and if for each $x \in S$ there exists some $\mu_x \in M(S)$ such that

$$P_n(x) \ = \ \int_S R_n(y)\, d\mu_x(y) \ = \ (\mu_x)^\vee(n) \qquad \textit{for all } n \in \mathbb{N}_0$$

and if

$$\|\mu_x\| \ \leq \ M \qquad \textit{for all } x \in S,$$

then

$$\sup_{n\in\mathbb{N}_0} \sup_{x\in S} \|A_n(\kappa_x, \cdot)\|_1 \ < \ \infty.$$

Corollary 4.15.7 *Assume that $(A_n)_{n\in\mathbb{N}_0}$ is an approximate identity with respect to $C(S)$, where A_n are defined by the triangular scheme $a_{n,k}$. Further let $(P_n(x))_{n\in\mathbb{N}_0}$, $x \in S$, be a sequence of polynomials $P_n(x)$ with degree $P_n = n$ and $P_0(x) = 1$. Then there exists for each $x \in S$ a measure $\mu_x \in M(S)$ such that*

$$P_n(x) \ = \ \int_S R_n(y)\, d\mu_x(y) \ = \ (\mu_x)^\vee(n) \qquad \textit{for all } n \in \mathbb{N}_0$$

and $\sup_{x\in S} \|\mu_x\| < \infty$ if and only if

$$\sup_{n\in\mathbb{N}_0} \sup_{x\in S} \|A_n(\kappa_x, \cdot)\|_1 \ < \ \infty.$$

Example.
Consider $(R_n(x;a))_{n\in\mathbb{N}_0}$ the Grinspun polynomials with $a > 2$, see subsection 4.5.10(a). One might expect that $T_n \overset{c}{\geq} R_n(\cdot\,;a)$ is valid for each $a > 2$, or the weaker property that for each $x \in [-1,1]$ there exists a measure $\mu_x \in M([-1,1])$ such that

$$R_n(x;a) \;=\; \int_{-1}^{1} T_n(y)\,d\mu_x(y).$$

In fact, for $x = 0$ an easy calculation shows that

$$R_n(0;a) \;=\; \int_{-1}^{1} T_n(y)\,d\mu_0(y) \;=\; (\mu_0)^\vee(n) \qquad \text{for every } n \in \mathbb{N}_0,$$

where

$$\mu_0 \;=\; \frac{a-2}{a-1}\,\pi_T \;+\; \frac{1}{a-1}\,\epsilon_0 \;\in\; M^1([-1,1])$$

with π_T the normalized orthogonalization measure of $(T_n(x))_{n\in\mathbb{N}_0}$ and ϵ_0 the point measure of $0 \in S = [-1,1]$.

For $x = \sin(\frac{\pi}{4}) = \frac{1}{\sqrt{2}} = \cos(\frac{\pi}{4})$ we will use Corollary 4.15.4 to derive that there does not exist any measure $\mu \in M([-1,1])$ such that

$$R_n\left(\frac{1}{\sqrt{2}}\,;a\right) \;=\; \int_{-1}^{1} T_n(y)\,d\mu(y) \;=\; \check{\mu}(n) \qquad \text{for each } n \in \mathbb{N}_0,$$

whenever $a > 2$.

With the triangular scheme $a_{n,k} = 1 - \frac{k}{n-1}$ and $\kappa(n) = R_n(\frac{1}{\sqrt{2}};a)$ we have to investigate

$$A_n(\kappa,y) := \sum_{k=0}^{n}\left(1 - \frac{k}{n+1}\right) R_k\left(\frac{1}{\sqrt{2}};a\right) T_k(y)\,h(k)$$

$$= 1 + \left(1 - \frac{1}{n+1}\right) \cos\left(\frac{\pi}{4}\right) \cos(t)\,2$$

$$+ \sum_{k=2}^{n}\left(1 - \frac{k}{n+1}\right)\left[\frac{a}{2(a-1)}\cos\left(k\,\frac{\pi}{4}\right) + \frac{a-2}{2(a-1)}\cos\left((k-2)\,\frac{\pi}{4}\right)\right]$$

$$\times \cos(kt)\,2,$$

where $y = \cos(t)$. Obviously,

$$\frac{a}{2(a-1)}\cos\left(k\,\frac{\pi}{4}\right) + \frac{a-2}{2(a-1)}\cos\left((k-2)\,\frac{\pi}{4}\right)$$

$$= \frac{a-2}{2(a-1)}\left(\cos\left(k\,\frac{\pi}{4}\right) + \sin\left(k\,\frac{\pi}{4}\right)\right) + \frac{1}{a-1}\cos\left(k\,\frac{\pi}{4}\right),$$

and we obtain by an elementary calculation

$$\int_{1/\sqrt{2}}^{1} A_n(\kappa, y)\, d\pi(y) = \frac{1}{\pi}\left[\frac{\pi}{4} + \frac{a-2}{a-1}\left(1 - \frac{1}{n+1}\right)\right.$$

$$\left. + \frac{a-2}{a-1}\sum_{k=2}^{n}\left(1 - \frac{k}{n+1}\right)\frac{1}{k}\,(\varrho_k + \sigma_k) + \frac{2}{a-1}\sum_{k=1}^{n}\left(1 - \frac{k}{n+1}\right)\sigma_k\right]$$

$$= \frac{1}{\pi}\left[\frac{\pi}{4} + \frac{a-2}{a-1}\left(1 - \frac{1}{n+1}\right)\right] + I_1(n) + I_2(n),$$

where $\varrho_k = \sin^2(k\frac{\pi}{4})$, $\sigma_k = \cos(k\frac{\pi}{4})\,\sin(k\frac{\pi}{4}) = \frac{1}{2}\sin(k\frac{\pi}{2})$.

Since $\sigma_k = 0$ for $k \in 2\mathbb{N}$, $\sigma_k = \frac{1}{2}$ for $k \in 4\mathbb{N}+1$ and $\sigma_k = -\frac{1}{2}$ for $k \in 4\mathbb{N}_0 - 1$, the sequence $(I_2(n))_{n\in\mathbb{N}}$ is positive and convergent. Moreover, $\varrho_k + \sigma_k = 0$ for $k = 4l-1$ and $k = 4l$, $l \in \mathbb{N}$, and $\varrho_k + \sigma_k = 1$ for $k = 2$, $k = 4l-3$ and $k = 4l-2$, $l = 2, 3, \ldots$. It follows that $(I_1(n))_{n\in\mathbb{N}}$ is positive and unbounded. Since

$$\|A_n(\kappa, \cdot)\|_1 \geq \int_{1/\sqrt{2}}^{1} A_n(\kappa, y)\, d\pi(y)$$

we conclude by Corollary 4.15.4 that there does not exist a measure $\mu \in M([-1,1])$ with $R_n(\frac{1}{\sqrt{2}}; a) = \int_{-1}^{1} T_n(y)\, d\mu(y)$ for all $n \in \mathbb{N}_0$, provided $a > 2$. We formulate this result explicitly.

Proposition 4.15.8 *Let $(R_n(x;a))_{n\in\mathbb{N}_0}$ be the Grinspun polynomials with $a > 2$. Then $T_n \overset{c}{\geq} R_n(\cdot\,; a)$ is* **not** *valid. Even more, for $x = \frac{1}{\sqrt{2}}$ there does* **not** *exist a measure $\mu \in M([-1,1])$ such that*

$$R_n\left(\frac{1}{\sqrt{2}}; a\right) = \int_{-1}^{1} T_n(y)\, d\mu(y) \qquad \textit{for each } n \in \mathbb{N}_0.$$

For ultraspherical polynomials $R_n^{(\alpha,\alpha)}(x)$, $\alpha > -\frac{1}{2}$, sequences $(\kappa(n))_{n\in\mathbb{N}_0}$ with $\kappa(n) \sim o(\frac{1}{n})$ are presented in Proposition 4.6.3 and Proposition 4.6.4. They are inverse Fourier transforms of very special functions depending on the parameter α.

For the case $\alpha = -\frac{1}{2}$, i.e. the Chebyshev polynomials $T_n(x)$ of the first kind, we derive that a convexity condition of $\kappa = (\kappa(n))_{n\in\mathbb{N}_0}$ already implies that κ is an inverse Fourier transform. We will mainly use the Polya criterion, see Theorem 4.1 in [364].

Theorem 4.15.9 *Assume that $\kappa = (\kappa(n))_{n\in\mathbb{N}_0}$ is a sequence of nonnegative numbers tending to zero such that $\kappa(n-1)+\kappa(n+1)-2\kappa(n) \geq 0$ for $n \in \mathbb{N}$. Then there exists a nonnegative function $\varphi \in L^1([-1,1],\pi_T)$ such that*

$$\kappa(n) = \int_{-1}^{1} T_n(y)\,\varphi(y)\,d\pi_T(y) = \check{\varphi}(n) \qquad \textit{for all } n \in \mathbb{N}_0.$$

Proof. To apply Theorem 4.1 of [364] we put $\beta = (\beta(n))_{n\in\mathbb{Z}}$ with $\beta(n) = \beta(-n) = \kappa(n)$ for $n \in \mathbb{N}_0$. The limit function $f \in L^1(\mathbb{T})$ constructed in the proof of Theorem 4.1 satisfies in addition $f(e^{it}) = f(e^{-it})$ for almost all $t \in [-\pi,\pi]$. Now let $x \in [-1,1]$ and denote $\varphi(x) = f(e^{it})$ where $t \in [0,\pi]$ is defined by $x = \cos t$. Then Theorem 4.1 gives

$$\int_{-1}^{1} T_n(y)\,\varphi(y)\,d\pi_T(y) = \frac{1}{2\pi}\int_{-\pi}^{\pi} f(e^{it})\,e^{-int}\,dt = \beta(n) = \kappa(n).$$

◇

Example.

Let $1 < \alpha < 2$. Consider

$$\kappa(n) = \frac{2(\alpha-1)}{\alpha+(2-\alpha)\,n}, \qquad n \in \mathbb{N}_0.$$

Obviously $\kappa(n) > 0$ for each $n \in \mathbb{N}_0$ and $\lim_{n\to\infty} \kappa(n) = 0$. An elementary computation shows

$$\kappa(n-1)+\kappa(n+1)-2\kappa(n) = \frac{4\,(\alpha-1)\,(2-\alpha)^2}{[\alpha+(2-\alpha)(n-1)]\,[\alpha+(2-\alpha)(n+1)]\,[\alpha+(2-\alpha)n]} > 0$$

for every $n \in \mathbb{N}$. By Theorem 4.15.9 there exists a nonnegative function $\varphi \in L^1([-1,1],\pi_T)$ such that

$$\kappa(n) = \check{\varphi}(n) = \int_{-1}^{1} T_n(y)\,\varphi(y)\,d\pi_T(y) \qquad \text{for every } n \in \mathbb{N}_0.$$

We use this relation to derive the following result for the Geronimus polynomials $R_n(x) = R_n(x;\frac{1}{4},\alpha)$, $1 < \alpha < 2$, see subsection 4.5.7.1.

Proposition 4.15.10 *Let $R_n(x;\frac{1}{4},\alpha)$ be the Geronimus polynomials and $1 < \alpha < 2$. For every $x \in [-1,1]$ there exists a measure $\mu_x \in M([-1,1])$ such that*

$$R_n\left(x;\frac{1}{4},\alpha\right) = \int_{-1}^{1} T_n(y)\,d\mu_x(y) \quad \textit{and} \quad \sup_{x\in[-1,1]} \|\mu_x\| \leq M_\alpha.$$

Proof. We know already that

$$R_n\left(x;\frac{1}{4},\alpha\right) = \frac{2(\alpha-1)}{\alpha+(2-\alpha)n}\,T_n(x) + \left(1-\frac{2(\alpha-1)}{\alpha+(2-\alpha)n}\right)R_n^{(\frac{1}{2},\frac{1}{2})}(x),$$

see subsection 4.5.7.1. Furthermore $T_n \overset{c}{\geq} R_n^{(\frac{1}{2},\frac{1}{2})}$, see [22]. Let $\nu_x \in M^1([-1,1])$ be the probability measure satisfying $R_n^{(\frac{1}{2},\frac{1}{2})}(x) = (\nu_x)^\vee(n)$, $n \in \mathbb{N}_0$, and let $\varphi \in L^1([-1,1],\pi_T)$ as above. Define $\mu_x \in M([-1,1])$ by

$$\mu_x = \varphi * \epsilon_x + \nu_x - \varphi * \nu_x,$$

where $*$ is the convolution on $[-1,1]$ with respect to $T_n(x)$. Then

$$R_n\left(x;\frac{1}{4},\alpha\right) = (\varphi*\epsilon_x)^\vee(n) + (\nu_x)^\vee(n) + (\varphi*\nu_x)^\vee(n) = (\mu_x)^\vee(n)$$

for each $n \in \mathbb{N}_0$. Further,

$$\|\mu_x\| \leq 2\,\|\varphi\|_1 + 1 = 2\kappa(0) + 1 = \frac{5\alpha-4}{\alpha}.$$

◇

16 Absolute convergence and uniform convergence of orthogonal series

Assuming that S satisfies property (F) we have studied homogeneous Banach spaces on S in section 4.13 and approximate identities with respect to homogeneous Banach spaces in subsection 4.14.1. Homogeneous Banach spaces are Banach algebras, where the multiplication is the convolution induced by property (F).

Now we will study

$$A(S) = \{\hat{d}|S : d \in l^1(h)\} = \{\varphi \in C(S) : \check{\varphi} \in l^1(h)\}$$

and

$$U(S) = \{\varphi \in C(S) : \|\mathcal{D}_n\varphi - \varphi\|_\infty \to 0\}$$

for general polynomial hypergroups.

Obviously $A(S)$ with norm $\|\hat{d}|S\|_A = \|d\|_1$ is a Banach algebra with respect to pointwise multiplication. Since $\|R_k|S\|_\infty \leq 1$, we have $\|\varphi\|_\infty \leq \|\varphi\|_A$ for each $\varphi \in A(S)$. Moreover, we know that $\|\varphi\|_1 \leq \|\varphi\|_A$, see the proof of Theorem 4.13.16.

The space $U(S)$ with norm $\|\varphi\|_U := \sup_{n\in\mathbb{N}_0} \|\mathcal{D}_n\varphi\|_\infty$ is a Banach space, see the proof of Theorem 4.13.23. Moreover, we have $\|\varphi\|_\infty \leq \|\varphi\|_U$ and $\|\varphi\|_1 \leq \|\varphi\|_U$ for each $\varphi \in U(S)$.

16.1 *Absolute convergence*

Properties of $A(S)$ are already investigated in section 4.1, see Theorem 4.1.5 and Theorem 4.1.6 (with $D = S$). In addition we mention a result which follows immediately from Theorem 3.11.19: If $(R_n(x))_{n\in\mathbb{N}_0}$ generates a polynomial hypergroup and $S = D_s$, then $A(S) \subsetneq C(S)$.

We derive now, that we can cancel the assumption $S = D_s$, to have $A(S) \subsetneq C(S)$. We will use functional-analytic properties of $C(S)$.

Proposition 4.16.1 *Assume that $(R_n(x))_{n\in\mathbb{N}_0}$ generates a polynomial hypergroup. If $C(S)$ is not topological isomorphic to l^1, then $A(S) \subsetneq C(S)$.*

Proof. Assume as a relation of sets, that $A(S) = C(S)$. Then the identity mapping id: $A(S) \to C(S)$ is continuous. By the open mapping theorem it follows that id: $A(S) \to C(S)$ is a topological isomorphism. Since $A(S)$ is isometric isomorphic to $l^1(h)$, and since $l^1(h)$ is isometric isomorphic to l^1, we have a contradiction. ◇

Theorem 4.16.2 *Let $(R_n(x))_{n\in\mathbb{N}_0}$ generate a polynomial hypergroup. Then $A(S)$ is a proper subspace of $C(S)$.*

Proof. The support of the Plancherel measure π is a compact and infinite subset of $\mathbb{R}$. We use Miljutin's Theorem [3,Theorem 4.4.8] and the fact that $C([0,1])$ does not have an unconditional basis [3,Proposition 3.5.4(ii)] and obtain that $C(S)$ does not have an unconditional basis. Applying Proposition 4.16.1 we have to derive that $C(S)$ cannot be topological isomorphic to l^1.

Suppose that there exists a topological isomorphism T from l^1 onto $C(S)$. The standard vector basis $(\epsilon_n)_{n\in\mathbb{N}_0}$ of l^1 is an unconditional basis for l^1. By [481,Proposition 4.2.14] we get that $\{T(\epsilon_n)\}_{n\in\mathbb{N}_0}$ is an unconditional basis for $C(S)$. ◇

Remark: In subsection 4.16.2 we will extend the above result by comparing $A(S)$ with $U(S)$.

Now we continue to derive a very special feature of $A(S)$, the Wiener-Tauberian property. The proof is based on Theorem 4.1.6, a Wiener theorem for polynomial hypergroups. Recall that for $\varphi = \mathcal{F}d$, $d \in l^1(h)$ and $k \in \mathbb{N}_0$, we have $R_k \cdot \varphi = \mathcal{F}(L_k d)$, where $L_k d$ is the translate of $d \in l^1(h)$.

Theorem 4.16.3 *Assume that $(R_n(x))_{n\in\mathbb{N}}$ generates a polynomial hypergroup on $\mathbb{N}_0$. Suppose that $S = D_s = D$. Let $\varphi \in A(S)$, $\varphi \neq 0$. Then for* *every $\psi \in A(S)$ and $\varepsilon > 0$ there exist $n \in \mathbb{N}$ and $\gamma_0, ..., \gamma_n \in \mathbb{C}$ such that*

$$\left\| \psi - \sum_{k=0}^{n} \gamma_k \, (R_k \cdot \varphi) \, h(k) \right\|_A < \varepsilon$$

if and only if $\varphi(x) \neq 0$ for every $x \in S$.

Proof. Suppose that $\varphi(x) \neq 0$ for each $x \in S = D$. By Theorem 4.1.6 the inverse $1/\varphi$ is an element of $A(S)$. Given $\psi \in A(S)$ consider $\psi/\varphi \in A(S)$. Let $\varepsilon > 0$. Since $(\psi/\varphi)^\vee \in l^1(h)$ there exists $\gamma \in l^1(h)$ and $n \in \mathbb{N}$ such that $\gamma(k) = 0$ for $k > n$, and $\|(\psi/\varphi)^\vee - \gamma\|_1 < \varepsilon/\|\varphi\|_A$. Then

$$\|\check{\psi} - \gamma\check{\varphi}\|_1 \; \leq \; \|\check{\varphi}\|_1 \; \|(\psi/\varphi)^\vee - \gamma\|_1 \; < \; \varepsilon,$$

i.e. $\|\psi - \hat{\gamma}\varphi\|_A < \varepsilon$. Since

$$\hat{\gamma}\varphi \; = \; \left(\sum_{k=0}^{n} \gamma(k) \; R_k \; h(k)\right) \varphi \; = \; \sum_{k=0}^{n} \gamma(k) \; (R_k \cdot \varphi) \; h(k)$$

we have shown

$$\left\| \psi - \sum_{k=0}^{n} \gamma(k) \; (R_k \cdot \varphi) \; h(k) \right\|_A < \varepsilon.$$

Now suppose that $\varphi(x_0) = 0$ for some $x_0 \in S$. Then $(R_k \cdot \varphi)(x_0) = 0$ for each $k \in \mathbb{N}$. If the linear span of $\{R_k \cdot \varphi : k \in \mathbb{N}_0\}$ would be dense in $A(S)$, then the linear span of $\{L_k d : k \in \mathbb{N}_0\}$ were dense in $l^1(h)$, where $\varphi = \hat{d}$, and we would conclude that $\hat{\epsilon}_0 = 0$, where ϵ_0 is the point measure of $0 \in \mathbb{N}_0$, a contradiction to $\hat{\epsilon}_0 = 1$. ◇

Remark: $S = D_s = D$ is satisfied, whenever the polynomial hypergroup $\mathbb{N}_0$ is of subexponential growth, see Theorem 4.2.4.

Now we collect the local information to make the global statement that $\varphi \in A(S)$. We shall assume that $l^1(h)$ is D-regular, see section 3.5, and apply a striking result on regular commutative Banach algebras.

Theorem 4.16.4 *Assume that $(R_n(x))_{n \in \mathbb{N}_0}$ generates a polynomial hypergroup on $\mathbb{N}_0$. Suppose that $l^1(h)$ is D-regular. Let $\varphi \in C(S)$. If for every $x \in S$ there exists a function $\psi_x \in A(S)$ such that $\psi_x = \varphi$ on a neighbourhood of x, then $\varphi \in A(S)$.*

Proof. Given $x \in S$. Let $\psi_x \in A(S)$ and V_x an open subset of S such that $\varphi|V_x = \psi_x|V_x$. S is compact, and so there exists a finite set $\{x_1, ..., x_n\} \subseteq S$ such that $S = \bigcup_{j=1}^{n} V_{x_j}$. $l^1(h)$ is D-regular by assumption.

Therefore we can use Corollary 4.2.12 of [356], which says that there exist $d_1, ..., d_n \in l^1(h)$ such that

$$(\mathcal{F}d_1 + \cdots + \mathcal{F}d_n)|S = 1 \quad \text{and} \quad \mathcal{F}d_j|S/V_{x_j} = 0 \quad \text{for each } j = 1, ..., n.$$

Therefore we obtain on S

$$\varphi = \sum_{j=1}^{n} \varphi\, \mathcal{F}d_j|S = \sum_{j=1}^{n} \psi_{x_j}\, \mathcal{F}d_j|S \in A(S).$$

◇

In section 3.5 we have shown that for commutative hypergroups K the $\frac{1}{2}$-subexponential growth implies the $\mathcal{X}^b(K)$-regularity of $L^1(K)$, see Theorem 3.5.13. A direct calculation shows that for polynomial hypergroups on $\mathbb{N}_0$ the $\frac{1}{2}$-subexponential growth is equivalent to the existence of $0 \leq \beta < \frac{1}{2}$ such that $h(n) = O(\exp(n^\beta))$, compare with Proposition 4.2.3. Hence we have the following result.

Corollary 4.16.5 *Assume that $(R_n(x))_{n\in\mathbb{N}_0}$ generates a polynomial hypergroup on $\mathbb{N}_0$ such that the Haar weights $h(n)$ satisfy $h(n) = O(\exp(n^\beta))$ for some $\beta \in [0, \frac{1}{2}[$. Let $\varphi \in C(S)$. If for every $x \in S$ there exists a function $\psi_x \in A(S)$ such that $\psi_x = \varphi$ on a neighbourhood of x, then $\varphi \in A(S)$.*

Riesz' Theorem 4.1.5 yields a characterization of $A(S)$. The next result gives for a subclass of polynomial hypergroups a sufficient condition for $\varphi \in C(S)$ to be a member of $A(S)$. We will apply the selective modified Fejér kernel with respect to $y = 1$ as specified in Proposition 4.14.30.

Proposition 4.16.6 *Assume that*

$$\sum_{k=1}^{\infty} \frac{1}{h(k)\, a_k} = \infty.$$

Let $\varphi \in C(S)$ with $\check{\varphi}(n) \geq 0$ for every $n \in \mathbb{N}_0$. Then $\varphi \in A(S)$.

Proof. By Proposition 4.14.30 we have $1 \in S$ and

$$\lim_{n\to\infty} \mathcal{F}_n\varphi(1) = \lim_{n\to\infty} \sum_{k=0}^{n} \left(1 - \frac{b_{k-1,0}}{b_{n,0}}\right) \check{\varphi}(k)\, h(k) = \varphi(1),$$

where the coefficients $b_{k,0}$ are defined in Proposition 4.14.20. The kernel functions satisfy

$$\mathcal{F}_n(x) = \frac{1}{b_{n,0}} \left(\mathcal{D}_0(x) + \frac{1}{h(1)\, a_1}\, \mathcal{D}_1(x) + \cdots + \frac{1}{h(n)\, a_n}\, \mathcal{D}_n(x)\right),$$

where

$$b_{n,0} = 1 + \sum_{k=1}^{\infty} \frac{1}{h(k)\, a_k}.$$

Since $\check{\varphi}(k) \geq 0$, the partial sums

$$\mathcal{D}_n\varphi(1) = \sum_{k=0}^{n} \check{\varphi}(k)\, h(k)$$

will converge, provided $(\mathcal{D}_n\varphi(1))_{n\in\mathbb{N}_0}$ is bounded.
Assume that $(\mathcal{D}_n\varphi(1))_{n\in\mathbb{N}_0}$ is unbounded. Then for each $M > 0$ there exists $n_0 \in \mathbb{N}$ such that $\mathcal{D}_n\varphi(1) \geq M$ for all $n \geq n_0$. Then for each $n \geq n_0$ we have

$$\mathcal{F}_n\varphi(1) = \frac{1}{b_{n,0}} \left(\mathcal{D}_0\varphi(1) + \sum_{k=1}^{n_0} \frac{1}{h(k)\, a_n}\, \mathcal{D}_k\varphi(1) \right)$$

$$+ \frac{1}{b_{n,0}} \sum_{k=n_0+1}^{n} \frac{1}{h(k)\, a_k}\, \mathcal{D}_k\varphi(1).$$

Since $\sum\limits_{k=1}^{\infty} \frac{1}{h(k)a_k} = \infty$, it follows that $\limsup\limits_{n\to\infty} \mathcal{F}_n\varphi(1) \geq M$. This contradicts the convergence of $\mathcal{F}_n\varphi(1)$ towards $\varphi(1)$. Therefore we have proved $\varphi \in A(S)$. ◇

Remark: We want to point out that property (F) for S was not used. The main restriction is $\sum\limits_{k=1}^{\infty} \frac{1}{h(k)a_k} = \infty$. Hence for Jacobi polynomials $R_n^{(\alpha,\beta)}(x)$ we have to suppose that $-\frac{1}{2} \leq \alpha \leq 0$. Obviously, if $\{h(n) : n \in \mathbb{N}_0\}$ is bounded, then $\sum\limits_{k=1}^{\infty} \frac{1}{h(k)\, a_n} = \infty$.

Finally we refer to Theorem 4.6.1, which implies for ultraspherical polynomials $R_n^{(\alpha,\alpha)}(x)$, $\alpha \geq -\frac{1}{2}$, the following result,

$$A_\gamma([-1,1]) \subseteq A_\alpha([-1,1]),$$

where $A_\alpha([-1,1])$ denotes the space of all $\varphi \in C([-1,1])$ with absolutely convergent expansions with respect to $R_n^{(\alpha,\alpha)}(x)$, and $\gamma \geq \alpha \geq -\frac{1}{2}$. Note that $R_n^{(\alpha,\alpha)} \overset{d}{\geq} R_n^{(\gamma,\gamma)}$.

16.2 Uniform convergence

If $\varphi \in C(S)$ and $\varphi = \hat{d}|S$ for $d \in l^1(h)$, then $\check{\varphi}(k) = d(k)$ for all $k \in \mathbb{N}_0$, and hence

$$\mathcal{D}_n\varphi(x) = \sum_{k=0}^{n} d(k)\, R_k(x)\, h(k).$$

Thus $\mathcal{D}_n\varphi(x)$ converges uniformly on S towards $\sum_{k=0}^{\infty} d(k)R_k(x)h(k) = \hat{d}(x) = \varphi(x)$. Moreover,

$$|\mathcal{D}_n\varphi(x)| \le \sum_{k=0}^{n} |d(k)|\, h(k) \le \|\varphi\|_A,$$

which gives $\|\varphi\|_U \le \|\varphi\|_A$ and $A(S) \subseteq U(S) \subseteq C(S)$.

We know already that $A(S)$ is a proper subset of $C(S)$. Now we even will derive that $A(S) \subsetneq U(S)$. The paper [517] of J. Obermaier was very relevant for the proofs.

Proposition 4.16.7 *Let $(R_n(x))_{n\in\mathbb{N}_0}$ generate a polynomial hypergroup. For each $N \in \mathbb{N}_0$ and all $C > 0$ there exists $M \in \mathbb{N}_0$, $M > N$, and $\beta_{N+1}, ..., \beta_M \in \mathbb{C}$ such that*

$$\sum_{k=N+1}^{M} |\beta_k|\, h(k) > C \cdot \sup_{x\in S} \left| \sum_{k=N+1}^{M} \beta_k\, R_k(x)\, h(k) \right|.$$

Proof. Suppose that there exist $N \in \mathbb{N}_0$ and $C > 0$ such that

$$\sum_{k=N+1}^{M} |\beta_k|\, h(k) \le C \cdot \sup_{x\in S} \left| \sum_{k=N+1}^{M} \beta_k\, R_k(x)\, h(k) \right|$$

for each $M > N$ and all $\beta_{N+1}, ..., \beta_M \in \mathbb{C}$. Fix such an $N \in \mathbb{N}_0$ and let $M \in \mathbb{N}_0$ arbitrary. Since norms on finite dimensional spaces are equivalent, there exists some $D > 0$ such that

$$\sum_{k=0}^{N} |\beta_k|\, h(k) \le D \sup_{x\in S} \left| \sum_{k=0}^{N} \beta_k\, R_k(x)\, h(k) \right|$$

for all $\beta_1, ..., \beta_N \in \mathbb{C}$.

Setting $E = \max\{C, D\}$ we obtain for every $M \in \mathbb{N}_0$ and $\beta_0, ..., \beta_M \in \mathbb{C}$,

$$\sum_{k=0}^{M} |\beta_k|\, h(k)$$

$$\le E \cdot \left(\sup_{x\in S} \left| \sum_{k=0}^{\min\{M,N\}} \beta_k\, R_k(x)\, h(k) \right| + \sup_{x\in S} \left| \sum_{k=N+1}^{M} \beta_k\, R_k(x)\, h(k) \right| \right).$$

Obviously, for $\varphi(x) = \sum_{k=0}^{M} \beta_k R_k(x) h(k)$ we have $\mathcal{D}_N\varphi(x) = \sum_{k=0}^{\min\{M,N\}} \beta_k R_k(x) h(k)$ and $(\mathrm{id} - \mathcal{D}_N)\varphi(x) = \sum_{k=N+1}^{M} \beta_k R_k(x) h(k)$, where id: $C(S) \to C(S)$ is the identity mapping, and it follows that there exists a constant $F > 0$ such that

$$\sum_{k=0}^{M} |\beta_k|\ h(k) \ \le\ F \cdot \sup_{x\in S} \left| \sum_{k=0}^{M} \beta_k\ R_k(x)\ h(k) \right|$$

for each $M \in \mathbb{N}_0$ and $\beta_0, ..., \beta_M \in \mathbb{C}$.

Now by [481,Theorem 4.3.6] we obtain that $(R_n(x))_{n\in\mathbb{N}_0}$ is a basic sequence in $C(S)$, which is equivalent to the standard unit vector basis $(\epsilon_n)_{n\in\mathbb{N}_0}$ for l^1. (Have in mind that $\sup_{x\in S} |R_n(x)| \le 1$ and set $\alpha_k = \beta_k h(k)$ in the notation of [481,Theorem 4.3.6].)

By the theorem of Stone-Weierstraß– see Theorem 3.2.1 – we know that the linear span of $\{R_n|S : n \in \mathbb{N}_0\}$ is dense in $C(S)$. Hence by [481,Proposition 4.3.2] the Banach spaces $C(S)$ and l^1 are topological isomorphic, which is impossible as we have already shown in the proof of Theorem 4.16.2. $\diamond$

Below all polynomials P are restricted to S, e.g. $\|P\|_\infty = \sup_{x\in S} |P(x)|$.

Theorem 4.16.8 (Obermaier) *Let $(R_n(x))_{n\in\mathbb{N}_0}$ generate a polynomial hypergroup. Then $A(S)$ is a proper subspace of $U(S)$.*

Proof. Suppose that $A(S) = U(S)$. Since $\|\varphi\|_U \le \|\varphi\|_A$ for all $\varphi \in A(S)$. the identity mapping id: $A(S) \to U(S)$ is continuous. By the open mapping theorem the norms $\|\cdot\|_A$ and $\|\cdot\|_U$ are equivalent. By Proposition 4.16.7 it follows that for each $N \in \mathbb{N}_0$ and all $C > 0$ there exist $M > N$ and $\beta_{N+1}, ..., \beta_M \in \mathbb{C}$ such that

$$\left\| \sum_{k=N+1}^{M} \frac{\beta_k}{\sum_{k=N+1}^{M} |\beta(k)|\, h(k)}\ R_k\ h(k) \right\|_\infty \ <\ \frac{1}{C}.$$

This shows that we can construct polynomials

$$Q_n(x) \ =\ \sum_{k=l_n}^{u_n} \gamma_k\ R_k(x)\ h(k)$$

with $u_n < l_{n+1}$ and

$$\sum_{k=l_n}^{u_n} |\gamma_k|\ h(k) \ =\ 1 \qquad \text{for all } n \in \mathbb{N}$$

with the property $\lim_{n\to\infty} \|Q_n\|_\infty = 0$. Obviously, $\|\frac{1}{N}\sum_{n=1}^{N} Q_n\|_A = 1$ is true for each $N \in \mathbb{N}$. Using the special form of the polynomials Q_n we get

$$\begin{aligned}
&\left\|\sum_{n=1}^{N} Q_n\right\|_U \\
&= \sup\left\{\left\|\sum_{n=1}^{m} Q_n + \sum_{k=l_{m+1}}^{r} \gamma_k\, R_k\, h(k)\right\|_\infty : 0 \le m \le N-1,\ l_{m+1} \le r \le u_{m+1}\right\} \\
&\qquad \le \sup\left\{\sum_{n=1}^{m} \|Q_m\|_\infty + 1 :\ 0 \le m \le N-1\right\} \le 1 + \sum_{n=1}^{N} \|Q_n\|_\infty .
\end{aligned}$$

Applying the property that the Césaro means of sequences converging to zero also converges to zero, we get

$$\lim_{N\to\infty} \left\|\frac{1}{N}\sum_{n=1}^{N} Q_n\right\|_U = 0.$$

This contradicts the equivalence of the norms $\|\cdot\|_A$ and $\|\cdot\|_U$. ◇

Example. Concrete examples of $\varphi \in U(S)\backslash A(S)$ for Jacobi polynomials $R_n^{(\alpha,\beta)}(x)$,
$(\alpha,\beta) \in J\backslash\{(-\frac{1}{2},-\frac{1}{2})\}$ can be derived from Theorem 3.4 and equation (3.11) in [517]. Denoting the Haar weights of $(R_n^{(\alpha,\beta)}(x))_{n\in\mathbb{N}_0}$ by $h^{(\alpha,\beta)}(n)$, see subsection 4.5.2, and setting

$$d(n) = \frac{(-1)^n}{h^{(\alpha,\beta)}(n)}\left(\frac{1}{n} + \frac{1}{n+1}\right), \qquad n \in \mathbb{N},$$

then $\sum_{n=1}^{\infty} d(n)\, R_n^{(\alpha,\beta)} h^{(\alpha,\beta)}(n)$ converges uniformly on $S = [-1,1]$, but not absolutely.

Investigating the inclusion $U(S) \subseteq C(S)$, properties of the Dirichlet kernel

$$\mathcal{D}_n(x) = \sum_{k=0}^{n} R_k(x)\, h(k)$$

can be used, see subsection 4.14.2.3. The operator norm of $\mathcal{D}_n : C(S) \to C(S)$ is denoted by $\|\mathcal{D}_n\|^{C(S)}$, and

$$\mathcal{D}_n(x,y) = \sum_{k=0}^{n} R_k(x)\, R_k(y)\, h(k).$$

Applying the Banach-Steinhaus theorem we obtain

Theorem 4.16.9 *Let $(R_n(x))_{n\in\mathbb{N}_0}$ generate a polynomial hypergroup. Equivalent are:*

(1) $U(S) = C(S)$.
(2) $\|\mathcal{D}_n\varphi\|_\infty \leq M \, \|\varphi\|_\infty$ *for all* $\varphi \in C(S)$, $n \in \mathbb{N}_0$.
(3) $\|\cdot\|_U$ *and* $\|\cdot\|_\infty$ *are equivalent.*
(4) $\sup_{n\in\mathbb{N}_0} \|\mathcal{D}_n\|^{C(S)} = \sup_{n\in\mathbb{N}_0} \sup_{x\in S} \|\mathcal{D}_n(x,\cdot)\|_1 < \infty.$

Proof. (1) $\Leftrightarrow$ (2) $\Leftrightarrow$ (3) are direct consequences of the Banach-Steinhaus theorem. If $\varphi = \lim_{n\to\infty} \mathcal{D}_n\varphi$ for all $\varphi \in C(S)$, it follows

$$\|\varphi\|_\infty = \lim_{n\to\infty} \|\mathcal{D}_n\varphi\|_\infty \leq \sup_{n\in\mathbb{N}_0} \|\mathcal{D}_n\|^{C(S)} \, \|\varphi\|_\infty \quad \text{for all } \varphi \in C(S).$$

Hence $\sup_{n\in\mathbb{N}_0} \|\mathcal{D}_n\|^{C(S)} < \infty$. Proposition 4.14.2 says that

$$\sup_{n\in\mathbb{N}_0} \|\mathcal{D}_n\|^{C(S)} = \sup_{n\in\mathbb{N}_0} \sup_{x\in S} \|\mathcal{D}_n(x,\cdot)\|_1.$$

Hence (1) $\Rightarrow$ (4) is shown. Obviously (4) $\Rightarrow$ (2) is true. $\diamond$

Corollary 4.16.10 *Assume that $(R_n(x))_{n\in\mathbb{N}_0}$ generates a polynomial hypergroup and suppose $S = [-1,1]$ and $\pi' > 0$ almost everywhere. If $\{h(n) : n \in \mathbb{N}_0\}$ is unbounded, then $U(S)$ is a proper subspace of $C(S)$.*

Proof. Combine the statements of Theorem 4.14.27 and Theorem 4.16.9. $\diamond$

Corollary 4.16.10 can be considerably improved by using a result on approximation by a special type of bases of $C(S)$.

A basis $(P_n)_{n\in\mathbb{N}_0}$ of $C(S)$ is called a **Faber basis** if P_n is a real polynomial with degree equal to n. A result (from 1914) due to Georg Faber, see [211], says that in case of $S = [a,b]$ there does not exist a Faber basis of $C(S)$. Therefore the following extension of Corollary 4.16.10 is valid.

Theorem 4.16.11 (Faber) *Assume that $(R_n(x))_{n\in\mathbb{N}_0}$ generates a polynomial hypergroup. If $S = [a,b]$ then $U(S)$ is a proper subspace of $C(S)$.*

Proof. Considering an approximation in $C(S)$ with $P_n(x) = R_n(x)h(n)$, the coefficients of $\varphi \in C(S)$ are uniquely determined by $\check{\varphi}(n)$. Hence $U(S) = C(S)$ yields that $(P_n)_{n\in\mathbb{N}_0}$ is a basis of $C(S)$. In the case $S = [a,b]$ we get a contradiction to Faber's result in [211]. $\diamond$

Considering the little q-Legendre polynomials $(R_n(x;q))_{n\in\mathbb{N}_0}$, $0 < q < 1$, of subsection 4.5.12, we know that $S = \{1\} \cup \{1-q^k : k \in \mathbb{N}_0\}$. Applying Theorem 3 of [515] we get examples for $U(S) = C(S)$.

Theorem 4.16.12 (Obermaier) *For the little q-Legendre polynomials* $R_n(x;q)$,
$0 < q < 1$, *holds* $U(S) = C(S)$.

Proof. To use Theorem 3 of [515] we have to have in mind that the polynomials $R_n(x;q)$ are exactly the polynomials $R_n(1-x)$ studied in section 3 of [515]. ◇

Now we show that $U(S)$ is character-invariant. This property could be the starting point for investigating a normed space lying between $A(S)$ and $U(S)$. Note that for the trigonometric approximation on the torus $\mathbb{T} = \{e^{it} : t \in [0, 2\pi[\}$ the Banach space $U(\mathbb{T})$ is not a Banach algebra with respect to pointwise multiplication, see [352,Ch.I.6]. For polynomial hypergroups $U(S)$ can be a Banach algebra with respect to pointwise multiplication, see Theorem 4.16.12.

Lemma 4.16.13 *Suppose that* $(R_n(x))_{n\in\mathbb{N}_0}$ *generates a polynomial hypergroup on* $\mathbb{N}_0$. *Let* $\varphi \in L^1(S,\pi)$. *Then holds*

$$\mathcal{D}_n(R_1 \cdot \varphi) = R_1 \cdot \mathcal{D}_n\varphi + a_n h(n) \left(\check{\varphi}(n+1)\, R_n - \check{\varphi}(n)\, R_{n+1}\right)$$

on S.

Proof. We have

$$R_1 \cdot (\mathcal{D}_n\varphi) = \check{\varphi}(0)\, R_1 + \sum_{k=1}^{n} \check{\varphi}(k)\, h(k)\, (a_k R_{k+1} + b_k R_k + c_k R_{k-1})$$

and

$$\mathcal{D}_n(R_1\cdot\varphi) = \check{\varphi}(1)\, R_0 + \sum_{k=1}^{n} (a_k\check{\varphi}(k+1) + b_k\check{\varphi}(k) + c_k\check{\varphi}(k-1))\, R_k\, h(k).$$

Since $h(k)c_k = h(k-1)a_{k-1}$, it follows

$$R_1 \cdot (\mathcal{D}_n\varphi) - \mathcal{D}_n(R_1 \cdot \varphi) = h(n)\, a_n\, \check{\varphi}(n)\, R_{n+1} - h(n)\, a_n\, \check{\varphi}(n+1)\, R_n.$$

◇

Remark.
Lemma 4.16.13 also follows by applying the Christoffel-Darboux formula.

Proposition 4.16.14 *Assume that* $(R_n(x))_{n\in\mathbb{N}_0}$ *generates a polynomial hypergroup and* $1 \in S$. *Let* $\varphi \in U(S)$. *Then* $R_n \cdot \varphi \in U(S)$ *for all* $n \in \mathbb{N}_0$.

Proof. Using Lemma 4.16.13 we obtain

$$\|\mathcal{D}_n(R_1\cdot\varphi)-R_1\cdot\varphi\|_\infty \le \|\mathcal{D}_n(R_1\cdot\varphi)-R_1\cdot\mathcal{D}_n\varphi\|_\infty + \|R_1\cdot\mathcal{D}_n\varphi-R_1\cdot\varphi\|_\infty$$

$$\le h(n)\, a_n\, \|\check{\varphi}(n+1)\, R_n \; - \; \check{\varphi}(n)\, R_{n+1}\|_\infty \; + \; \|\mathcal{D}_n\varphi - \varphi\|_\infty$$

$$\le h(n)\, a_n\, (|\check{\varphi}(n+1)| \; + \; |\check{\varphi}(n)|) \; + \; \|\mathcal{D}_n\varphi - \varphi\|_\infty.$$

Since $1 \in S$ we have

$$\sum_{k=0}^{n} \check{\varphi}(k)\, h(k) \; = \; \mathcal{D}_n\varphi(1) \; \longrightarrow \; \varphi(1) \qquad \text{when } n \to \infty.$$

Necessarily it follows $h(n)\,|\check{\varphi}(n)| \to 0$ and therefore

$$a_n h(n)\, |\check{\varphi}(n)| \; \to \; 0 \quad \text{ and } a_n h(n)\, |\check{\varphi}(n{+}1)| \; = \; c_{n+1} h(n{+}1)\, |\check{\varphi}(n{+}1)| \; \to \; 0.$$

Thus

$$\lim_{n\to\infty} \|\mathcal{D}_n(R_1 \cdot \varphi) - R_1 \cdot \varphi\|_\infty \; = \; 0 \quad \text{ and } R_1 \cdot \varphi \in U(S).$$

Since

$$R_{k+1} \; = \; \frac{1}{a_k}\, R_1 R_k \; - \; \frac{b_k}{a_k}\, R_k \; - \; \frac{c_k}{a_k}\, R_{k-1},$$

we obtain by induction that $R_k \cdot \varphi \in U(S)$. ◇

We define

$$RU(S) := \; \{\varphi \in U(S) : \; \sup_{m\in\mathbb{N}_0} \; \|R_m \cdot \varphi\|_U < \infty\}.$$

If $\varphi \in A(S)$, then

$$|\mathcal{D}_n(R_m \cdot \varphi)(x)| \; = \; \left|\sum_{k=0}^{n} L_m\check{\varphi}(k)\; R_k(x)\; h(k)\right| \; \le \; \sum_{k=0}^{\infty} |L_m\check{\varphi}(k)|\; h(k)$$

$$\le \; \sum_{k=0}^{\infty} |\check{\varphi}(k)|\; h(k) \; < \; \infty$$

for each $x \in S$; $n, m \in \mathbb{N}_0$. Hence $A(S) \subseteq RU(S) \subseteq U(S)$. Moreover, $\|\varphi\|_{RU} := \sup_{m\in\mathbb{N}_0} \|R_m \cdot \varphi\|_U$ is a norm on the vector space $RU(S)$, satisfying $\|\varphi\|_U \le \|\varphi\|_{RU}$ for $\varphi \in RU(S)$ and $\|\varphi\|_{RU} \le \|\varphi\|_A$ for $\varphi \in A(S)$.

Theorem 4.16.15 *Assume that $(R_n(x))_{n\in\mathbb{N}_0}$ generates a polynomial hypergroup and $1 \in S$. If $U(S) = RU(S)$, then $\|\cdot\|_{RU}$ and $\|\cdot\|_U$ are equivalent norms on $U(S)$. In particular, $(U(S), \|\cdot\|_{RU})$ is a Banach space, provided $U(S) = RU(S)$.*

Proof. Write $\mathcal{M}_m \in \mathcal{B}(U(S))$ for the multiplication operator $\mathcal{M}_m(\varphi) = R_m \cdot \varphi$, $\varphi \in U(S)$. Using the Banach-Steinhaus theorem, $M := \sup_{m\in\mathbb{N}_0} \|\mathcal{M}_m\|^{U(S)}$ is finite. It follows

$$\|\varphi\|_U \leq \|\varphi\|_{RU} = \sup_{m\in\mathbb{N}_0} \|R_m \cdot \varphi\|_U \leq M\,\|\varphi\|_U.$$

$\diamond$

Certainly it is an interesting topic to give a detailed account on the relations between $A(S)$, $RU(S)$ and $U(S)$.

17 Almost-convergent sequences with respect to polynomial hypergroups

Almost convergence was formulated by Lorentz [457] for divergent bounded sequences. Lorentz calls a bounded sequence $\alpha = (\alpha(n))_{n\in\mathbb{N}_0}$ of complex numbers almost convergent to $d(\alpha)$, if $\mu(\alpha)$ equals $d(\alpha)$, where μ runs through the set of all Banach limits, that is the set of all invariant means μ of the semigroup $\mathbb{N}_0$. The concept of almost convergence is also studied in the context of amenable semigroups and groups.

In the following let $K = \mathbb{N}_0$ be a polynomial hypergroup generated by $(R_n(x))_{n\in\mathbb{N}_0}$. Invariant means or strongly invariant means play a decisive role for characterizing $\mathcal{L}$-almost convergence of a sequence $\alpha \in l^\infty$. Invariant means or strongly invariant means are already studied in section 3.12 for general commutative hypergroups, and for polynomial hypergroups in section 4.8. A mean on l^∞ is an element $\mu \in (l^\infty)^*$ such that $\mu(1) = 1 = \|\mu\|$. An invariant mean is a mean μ, which satisfies

$$\mu(L_n\beta) = \mu(\beta) \qquad \text{for all } n \in \mathbb{N}_0 \text{ and } \beta \in l^\infty.$$

A strongly invariant mean is a mean μ, which satisfies

$$\mu((L_n\beta)\cdot\gamma) = \mu(\beta\cdot(L_n\gamma)) \qquad \text{for all } n \in \mathbb{N}_0 \text{ and } \beta,\gamma \in l^\infty.$$

The set of all invariant means is denoted by $\mathcal{L}(\mathbb{N}_0)$ and the set of all strongly invariant means is denoted by $\mathcal{SL}(\mathbb{N}_0)$. Since polynomial hypergroups are commutative, $\mathcal{L}(\mathbb{N}_0)$ is nonempty, see Theorem 3.12.1. Obviously $\mathcal{SL}(\mathbb{N}_0) \subseteq \mathcal{L}(\mathbb{N}_0)$. $\mathcal{SL}(\mathbb{N}_0)$ is nonempty if and only if $1 \in S = \operatorname{supp} \pi$, see Corollary 4.8.11. $\mathcal{L}(\mathbb{N}_0)$ and $\mathcal{SL}(\mathbb{N}_0)$ are weak $*$-compact convex subsets of $(l^\infty)^*$.

17.1 *Følner condition for* $(S_n)_{n\in\mathbb{N}_0}$ *and almost convergence*

Extending the result of Lorentz to polynomial hypergroups our main concern is the investigation of the convergence

$$\lim_{n\to\infty} \frac{1}{H(n)} \sum_{k=0}^{n} L_m\alpha(k)\, h(k) \;=\; d(\alpha) \;\text{ uniformly in } m \in \mathbb{N}_0, \qquad (Lo1)$$

where $\alpha \in l^\infty$ and $d(\alpha) \in \mathbb{C}$. We call (Lo1) a **Lorentz condition.**

Denote $S_n = \{0, 1, ..., n\}$ and

$$f_n := \frac{1}{H(n)}\, \chi_{S_n} \in P^1(h) \;=\; \{f \in l^1(h) : f \geq 0,\; \|f\|_1 = 1\}.$$

Proposition 4.17.1 *Assume that the Lorentz condition (Lo1) holds for some* $\alpha \in l^\infty$ *with constant* $d(\alpha) \in \mathbb{C}$*. Then* $\mu(\alpha) = d(\alpha)$ *for all* $\mu \in \mathcal{L}(\mathbb{N}_0)$*.*

Proof. Since

$$\frac{1}{H(n)} \sum_{k=0}^{n} L_m\alpha(k)\, h(k) \;=\; f_n * \alpha(m) \quad \text{for each } m \in \mathbb{N}_0,$$

it follows that (Lo1) is equivalent to $\lim_{n\to\infty} \|f_n * \alpha - d(\alpha)\, 1\|_\infty = 0$. Let $\mu \in \mathcal{L}(\mathbb{N}_0)$. We have (since $\mu(L_m\alpha) = \mu(\alpha)$)

$$\mu(f_n * \alpha) \;=\; \frac{1}{H(n)} \sum_{k=0}^{n} \mu(\alpha)\, h(k) \;=\; \mu(\alpha)$$

and

$$|\mu(f_n * \alpha - d(\alpha)\, 1| \;\leq\; \|f_n * \alpha - d(\alpha)\, 1\|_\infty \;\longrightarrow 0 \quad \text{as } n \to \infty.$$

Therefore $\mu(\alpha) \;=\; \lim_{n\to\infty} \mu(f_n * \alpha) = d(\alpha)$. ◇

This result suggests the following definition.

Definition 1. A sequence $\alpha \in l^\infty$ is called **$\mathcal{L}$-almost convergent** to a constant $d(\alpha) \in \mathbb{C}$, if $\mu(\alpha) = d(\alpha)$ for all $\mu \in \mathcal{L}(\mathbb{N}_0)$. The set of all $\mathcal{L}$-almost convergent sequences is denoted by F. Furthermore, $F_0 = \{\alpha \in F : d(\alpha) = 0\}$.

We will apply that $l^1(h)$ can be isometrically embedded into $(l^\infty)^*$ via the mapping $f \mapsto j(f)$, where

$$j(f)(\varphi) \;=\; \sum_{k=0}^{\infty} \varphi(k)\, f(k)\, h(k),$$

see section 4.8. The set $j(P^1(h))$ is weak $*$-dense in the set of all means on l^∞.

It is easy to see that F is a $\|\cdot\|_\infty$-closed subspace of l^∞, invariant under the translation operator L_m.

Before proving the main theorem we consider characters of the polynomial hypergroup generated by $(R_n(x))_{n\in\mathbb{N}_0}$.

Write $\alpha_z(n) := R_n(z)$ for fixed $z \in D$. It is easy to derive that α_z are elements of F. In fact, since $L_m\alpha_z(n) = \alpha_z(m)\alpha_z(n)$ we get for each $\mu \in \mathcal{L}(\mathbb{N}_0)$

$$\mu(\alpha_z) = \mu(L_m\alpha_z) = R_m(z)\,\mu(\alpha_z) \tag{1}$$

for each $m \in \mathbb{N}_0$ and $z \in D$. If $z \neq 1$ there exists some $m \in \mathbb{N}_0$ such that $R_m(z) \neq 1$ (use the recursion formula of $R_n(x)$). Hence (1) is only possible for $z \in D\backslash\{1\}$ if $\mu(\alpha_z) = 0$ for all $\mu \in \mathcal{L}(\mathbb{N}_0)$. Thus we infer that $\alpha_z \in F$ with constant $d(\alpha_z) = 0$, if $z \in D\backslash\{1\}$ and $\alpha_1 \in F$ with $d(\alpha_1) = 1$.

We will use that the canonical sequence $(S_n)_{n\in\mathbb{N}_0}$ satisfies Følner condition (F_1), see Theorem 4.7.9, whenever $\lim\limits_{n\to\infty} \frac{a_n h(n)}{H(n)} = 0$ holds. If we confine ourselves to real-valued sequences, we indicate this by subscript $\mathbb{R}$.

Proposition 4.17.2 *Assume that $(R_n(x))_{n\in\mathbb{N}_0}$ generates a polynomial hypergroup and satisfies $\frac{a_n h(n)}{H(n)} \to 0$. For each $\alpha \in F$ and $m \in \mathbb{N}_0$ the limit*

$$\lim_{n\to\infty} \frac{1}{H(n)} \sum_{k=0}^{n} L_m\alpha(k)\, h(k)$$

exists and is equal to $d(\alpha)$ independent of $m \in \mathbb{N}_0$.

Proof. For real-valued $\beta \in l^\infty_\mathbb{R}$ define

$$\mathcal{P}(\beta) := \limsup_{n\to\infty} \sum_{k=0}^{n} f_n(k)\,\beta(k)\,h(k).$$

$\mathcal{P}$ satisfies $\mathcal{P}(\beta_1+\beta_2) \leq \mathcal{P}(\beta_1) + \mathcal{P}(\beta_2)$ and $\mathcal{P}(\lambda\beta) = \lambda\mathcal{P}(\beta)$ for $\lambda \geq 0$. Hence $p(\beta) := \mathcal{P}(-\beta) + \mathcal{P}(\beta)$ is a convex function on $l^\infty_\mathbb{R}$. By the Hahn-Banach extension theorem ($\mathbb{R}$-algebraic version) there exists a linear functional μ_0 on $l^\infty_\mathbb{R}$ such that $\mu_0(\beta) \leq p(\beta)$ for all $\beta \in l^\infty_\mathbb{R}$. That means $-\mathcal{P}(-\beta) \leq \mu_0(\beta) \leq \mathcal{P}(\beta)$ for all $\beta \in l^\infty_\mathbb{R}$.

Since $f_n \in P^1(h)$ it follows that $\mathcal{P} \geq 0$, $\mathcal{P}(1) = 1$. In particular, μ_0 is a mean on $l^\infty_\mathbb{R}$. Fix $m \in \mathbb{N}_0$ and $\beta \in l^\infty_\mathbb{R}$. Then

$$\begin{aligned}\mathcal{P}(L_m\beta - \beta) &= \lim_{l\to\infty} \sum_{k=0}^{n_l} f_{n_l}(k)\,(L_m\beta(k) - \beta(k))\; h(k)\\ &= \lim_{l\to\infty} \sum_{k=0}^{n_l} (L_m f_{n_l}(k) - f_{n_l}(k))\;\beta(k)\,h(k).\end{aligned}$$

Since $\frac{a_n h(n)}{H(n)} \to 0$, Theorem 4.7.9 says $\lim_{n\to\infty} \|L_m f_n - f_n\|_1 = 0$, since the Følner condition (F_1) is satisfied by the canonical sequence $(S_n)_{n\in\mathbb{N}_0}$. It follows $\mathcal{P}(L_m\beta - \beta) = 0 = -\mathcal{P}(-L_m\beta + \beta)$, and we conclude that μ_0 is an invariant mean on $l^\infty_{\mathbb{R}}$.

Now assume that there is a sequence $\beta \in F_{\mathbb{R}}$ such that

$$-\mathcal{P}(-\beta) = \liminf_{n\to\infty} \sum_{k=0}^{n} f_n(k)\,\beta(k)\,h(k) < \limsup_{n\to\infty} \sum_{k=0}^{n} f_n(k)\,\beta(k)\,h(k) = \mathcal{P}(\beta).$$

Then the Hahn-Banach extension theorem yields two invariant means μ_0 and μ_1 on $l^\infty_{\mathbb{R}}$ with $\mu_0(\beta) \neq \mu_1(\beta)$, which is impossible since $\beta \in F_{\mathbb{R}} \subseteq F$. So far we have shown that the limit

$$\lim_{n\to\infty} \sum_{k=0}^{n} f_n(k)\,\alpha(k)\,h(k)$$

exists for all $\alpha \in F$, and this limit is equal to the unique value $d(\alpha)$. Finally

$$\lim_{n\to\infty} \sum_{k=0}^{n} f_n(k)\,\alpha(k)\,h(k) \;=\; \lim_{n\to\infty} \sum_{k=0}^{n} f_n(k)\,L_1\alpha(k)\,h(k),$$

since

$$\sum_{k=0}^{n} (L_1\alpha(k) - \alpha(k))\,h(k) \;=\; a_n h(n)\,(\alpha(n+1) - \alpha(n)).$$

Using the recursion formula for the operators L_m we obtain by induction

$$\lim_{n\to\infty} \sum_{k=0}^{n} f_n(k)\,\alpha(k)\,h(k) \;=\; \lim_{n\to\infty} \sum_{k=0}^{n} f_n(k)\,L_m\alpha(k)\,h(k)$$

for all $m \in \mathbb{N}_0$. ◇

Remark. Note that $\frac{a_n h(n)}{H(n)}$ tend to zero, whenever condition (H) holds or a_n tend to zero.

To derive our main result it remains to show that the convergence in Proposition 4.17.2 is uniform with respect to $m \in \mathbb{N}_0$.

Lemma 4.17.3 *Let $\alpha \in F_{\mathbb{R}}$ and $\varphi \in P^1(h)$. Then*

$$\inf_{m\in\mathbb{N}_0} \sum_{k=0}^{\infty} \varphi(k)\,L_m\alpha(k)\,h(k) \;\le\; d(\alpha) \;\le\; \sup_{m\in\mathbb{N}_0} \sum_{k=0}^{\infty} \varphi(k)\,L_m\alpha(k)\,h(k).$$

Proof. Since $\varphi * \alpha(m) = \sum_{k=0}^{\infty} L_m\alpha(k)\,\varphi(k)\,h(k)$ we have for each $\mu \in \mathcal{L}(\mathbb{N}_0)$,

$$\mu(\varphi * \alpha) \;=\; \sum_{k=0}^{\infty} \mu(L_m\alpha)(k)\,\varphi(k)\,h(k) \;=\; d(\alpha)\sum_{k=0}^{\infty} \varphi(k)\,h(k) \;=\; d(\alpha).$$

Hence $\varphi * \alpha \in F_{\mathbb{R}}$ and $d(\varphi * \alpha) = d(\alpha)$. Since for each mean ν and $\beta \in l_\nu^\infty$ holds

$$\inf_{m\in\mathbb{N}_0} \beta(m) \;\leq\; \nu(\beta) \;\leq\; \sup_{m\in\mathbb{N}_0} \beta(m),$$

we obtain

$$\inf_{m\in\mathbb{N}_0} \varphi * \alpha(m) \;\leq\; d(\alpha) \;\leq\; \sup_{m\in\mathbb{N}_0} \varphi * \alpha(m).$$

◇

Now define for $\alpha \in F_{\mathbb{R}}$

$$\overline{M}(\alpha) := \inf_{\varphi\in P^1(h)} \sup_{m\in\mathbb{N}_0} \sum_{k=0}^{\infty} \varphi(k)\, L_m\alpha(k)\, h(k)$$

$$\underline{M}(\alpha) := \sup_{\varphi\in P^1(h)} \inf_{m\in\mathbb{N}_0} \sum_{k=0}^{\infty} \varphi(k)\, L_m\alpha(k)\, h(k).$$

By Lemma 4.17.3 follows for $\alpha \in F_{\mathbb{R}}$

$$\underline{M}(\alpha) \;\leq\; d(\alpha) \;\leq\; \overline{M}(\alpha). \tag{2}$$

As already used, the Følner condition (F_1) yields

$$\lim_{n\to\infty} \|L_m f_n - f_n\|_1 \;=\; 0 \qquad \text{for each } m \in \mathbb{N}_0.$$

It follows immediately that for each $\varphi \in P^1(h)$

$$\lim_{n\to\infty} \|\varphi * f_n - f_n\|_1 \;=\; 0, \tag{3}$$

provided $\frac{a_n h(n)}{H(n)} \to 0$.

Proposition 4.17.4 *Assume that $(R_n(x))_{n\in\mathbb{N}_0}$ generates a polynomial hypergroup and $\frac{a_n h(n)}{H(n)} \to 0$. For $\alpha \in F_{\mathbb{R}}$, $\alpha \neq 0$ and $\varepsilon > 0$ there exists some $n_0 \in \mathbb{N}_0$ such that*

$$\sup_{m\in\mathbb{N}_0} \sum_{k=0}^{\infty} f_n(k)\, L_m\alpha(k)\, h(k) \;<\; \overline{M}(\alpha) + \varepsilon$$

and

$$\inf_{m\in\mathbb{N}_0} \sum_{k=0}^{\infty} f_n(k)\, L_m\alpha(k)\, h(k) \;>\; \underline{M}(\alpha) - \varepsilon$$

for all $n \geq n_0$.

Proof. Let $\varepsilon > 0$. Choose $\varphi \in P^1(h)$ such that

$$\sup_{m\in\mathbb{N}_0} \sum_{k=0}^{\infty} \varphi(k)\, L_m\alpha(k)\, h(k) \;<\; \overline{M}(\alpha) + \frac{\varepsilon}{2}.$$

By (3) there exists $n_0 \in \mathbb{N}_0$ such that

$$\|\varphi * f_n - f_n\|_1 \;<\; \frac{\varepsilon}{2\,\|\alpha\|_\infty} \qquad \text{for } n \geq n_0.$$

We have

$$\sum_{k=0}^{\infty} \varphi * f_n(k)\, L_m\alpha(k)\, h(k) \;=\; \sum_{k=0}^{\infty}\sum_{j=0}^{\infty} L_k\varphi(j)\, f_n(j)\, h(j)\, L_m(\alpha)(k)\, h(k)$$

$$= \sum_{j=0}^{\infty} f_n(j) \sum_{k=0}^{\infty} L_j\varphi(k)\, L_m\alpha(k)\, h(k)\, h(j)$$

$$= \sum_{j=0}^{\infty} f_n(j) \sum_{k=0}^{\infty} \varphi(k)\, L_j \circ L_m\alpha(k)\, h(k)\, h(j).$$

Since

$$L_j \circ L_m \;=\; \sum_{i=|j-m|}^{j+m} g(j,m;i)\, L_i$$

with nonnegative coefficients $g(j,m;i)$ such that $\sum_{i=|j-m|}^{j+m} g(j,m;i) = 1$, see Proposition 1.1.13, we obtain

$$\sum_{k=0}^{\infty} \varphi * f_n(k)\, L_m\alpha(k)\, h(k)\, h(k) \;<\; \overline{M}(\alpha) + \frac{\varepsilon}{2}.$$

Consequently,

$$\sup_{m\in\mathbb{N}_0} \sum_{k=0}^{\infty} f_n(k)\, L_m\alpha(k)\, h(k) < \overline{M}(\alpha) + \frac{\varepsilon}{2} + \|f_n - \varphi * f_n\|_1\, \|\alpha\|_\infty < \overline{M}(\alpha) + \varepsilon$$

for $n \geq n_0$. In a similar way it is proved that

$$\inf_{m\in\mathbb{N}_0} \sum_{k=0}^{\infty} f_n(k)\, L_m\alpha(k)\, h(k) \;>\; \underline{M}(\alpha) - \varepsilon$$

for n large enough. ◇

Theorem 4.17.5 *Assume that $(R_n(x))_{n\in\mathbb{N}_0}$ generates a polynomial hypergroup and suppose that $\frac{a_n h(n)}{H(n)} \to 0$. Then $\alpha \in l^\infty$ is $\mathcal{L}$-almost convergent to the constant $d(\alpha)$ if and only if*

$$\lim_{n\to\infty} \frac{1}{H(n)} \sum_{k=0}^{n} L_m\alpha(k)\, h(k) \;=\; d(\alpha)$$

uniformly in $m \in \mathbb{N}_0$.

Proof. Assume $\alpha \in F$. It is sufficient to consider real-valued α's. Let $\varepsilon > 0$. By Proposition 4.17.4 it follows

$$\overline{M}(\alpha) \;\le\; \sup_{m\in\mathbb{N}_0} \sum_{k=0}^{\infty} f_n(k)\, L_m\alpha(k)\, h(k) \;<\; \overline{M}(\alpha) + \varepsilon \tag{4}$$

for n large enough. By Proposition 4.17.2 we know $\sum_{k=0}^{\infty} f_n(k)L_m\alpha(k)h(k)$ converges towards $d(\alpha)$, and (4) implies that the convergence is uniform in $m \in \mathbb{N}_0$.

The converse implication is already shown in Proposition 4.17.1. ◇

Using the equality

$$f_n * \alpha(m) \;=\; \frac{1}{H(n)} \sum_{k=0}^{n} L_m\alpha(k)\, h(k)$$

we can formulate the result of Theorem 4.17.5 in the following way.

Corollary 4.17.6 *Assume that $(R_n(x))_{n\in\mathbb{N}_0}$ generates a polynomial hypergroup and suppose that $\frac{a_n h(n)}{H(n)} \to 0$. Equivalent are*

(i) $\alpha \in F$.

(ii) $\lim_{n\to\infty} \frac{1}{H(n)} \sum_{k=0}^{n} L_m\alpha(k)\, h(k) \;=\; d(\alpha)$ *uniformly in $m \in \mathbb{N}_0$, i.e. (Lo1) is valid.*

(iii) $\lim_{n\to\infty} \|f_n * \alpha - d(\alpha)\,1\|_\infty \;=\; 0.$

17.2 *Properties of almost convergence*

Almost convegence is based on properties of sequences $f_n = \frac{1}{H(n)}\,\chi_{S_n} \in l^\infty$, which can be represented by the infinite matrix $D = (d_{n,k})_{n,k\in\mathbb{N}_0}$ with $d_{n,k} = \frac{1}{H(n)}$ for $k = 0, ..., n$ and $d_{n,k} = 0$ for $k = n+1, n+2, ...$.

We call an infinite matrix $B = (b_{n,k})_{n,k\in\mathbb{N}_0}$ **regular with respect to** $(R_n(x))_{n\in\mathbb{N}_0}$, if

$$b_{n,k} \to 0 \qquad \text{and} \quad \sum_{k=0}^{\infty} b_{n,k}h(k) \longrightarrow 1 \qquad \text{as } n \to \infty$$

and

$$\sum_{k=0}^{\infty} |b_{n,k}|\; h(k) \;\le\; M \;<\; \infty \qquad \text{for all } n \in \mathbb{N}_0.$$

Obviously D is regular with respect to $(R_n(x))_{n\in\mathbb{N}_0}$. For a regular matrix we denote by B_n the n-th row, i.e.

$$B_n \;=\; (b_{n,k})_{k\in\mathbb{N}_0} \;\in l^1(h).$$

Furthermore we write $\alpha \in F_B$ whenever $B_n * \alpha \in l^\infty$ converges to $d_B(\alpha)\,1$ in l^∞, where $d_B(\alpha) \in \mathbb{C}$ is some constant.

Theorem 4.17.7 *Assume that $(R_n(x))_{n\in\mathbb{N}_0}$ generates a polynomial hypergroup and suppose that B is a regular matrix with respect to $(R_n(x))_{n\in\mathbb{N}_0}$. Then $F_B \subseteq F$.*

Proof. If $B_n * \alpha$ converges in l^∞ towards $d_B(x)\,1$ we have for each $\mu \in \mathcal{L}(\mathbb{N}_0)$

$$\mu(B_n * \alpha) \;\longrightarrow\; d_B(\alpha)\;\mu(1) \;=\; d_B(\alpha).$$

Since $L_k\alpha(m) = L_m\alpha(k)$ we get

$$B_n * \alpha \;=\; \sum_{k=0}^{\infty} b_{n,k}\; L_k\alpha\; h(k),$$

where the series converges in l^∞ with respect to the sup-norm. Hence

$$\mu(B_n * \alpha) \;=\; \sum_{k=0}^{\infty} b_{n,k}\;\mu(L_k\alpha)\;h(k) \;\longrightarrow\; \mu(\alpha)$$

as $n \to \infty$. It follows $\mu(\alpha) = d_B(\alpha)$ for all $\mu \in \mathcal{L}(\mathbb{N}_0)$. ◇

Corollary 4.17.8 *Let $(R_n(x))_{n\in\mathbb{N}_0}$ generate a polynomial hypergroup satisfying $\frac{a_n h(n)}{H(n)} \to 0$. If B is a regular matrix with respect to $(R_n(x))_{n\in\mathbb{N}_0}$ and $\alpha \in F_B$, then*

$$\lim_{n\to\infty} \frac{1}{H(n)} \sum_{k=0}^{n} L_m\alpha(k)\; h(k) \;=\; d_B(\alpha)$$

uniformly in $m \in \mathbb{N}_0$, i.e. (Lo1) is valid.

Remark. Our notation of regular matrices with respect to $(R_n(x))_{n\in\mathbb{N}_0}$ is the appropriate generalization of that given in [457,§2]. There are many further summation methods, which may be studied in the context of almost convergence with respect to polynomial hypergroups.

Following the methods used for semigroups (see[283]) and locally compact groups (see [753]) we prove $F = \mathbb{C}\,1 \bigoplus F_0$, where $F_0 = \{\alpha \in F : d(\alpha) = 0\}$ and F_0 is equal to the closed linear span of $\{\beta - L_n\beta : \beta \in l^\infty,\ n \in \mathbb{N}_0\}$. This characterization is valid for general polynomial hypergroups. We will use the Arens product in $(l^\infty)^* = (l^1(h))^{**}$ induced by the convolution of $l^1(h)$ on l^∞. If $\beta \in l^1(h)$, $\alpha \in l^\infty \cong (l^1(h))^*$ and $\varphi \in (l^\infty)^*$ define $\alpha \cdot \varphi(\beta) := \varphi(\beta * \alpha)$. Obviously, $\alpha \cdot \varphi \in (l^1(h))^* \cong l^\infty$ and $\|\alpha \cdot \varphi\|_\infty \le \|\varphi\|\ \|\alpha\|_\infty$. For $\varphi, \psi \in (l^\infty)^*$ and $\alpha \in l^\infty$ define $\psi * \varphi(\alpha) := \psi(\alpha \cdot \varphi)$. Then $\psi * \varphi \in (l^\infty)^*$ and $\|\psi * \varphi\| \le \|\psi\|\ \|\varphi\|$.

Proposition 4.17.9 *Let $\psi, \varphi \in (l^\infty)^*$. Then*

(i) For $\alpha \in l^\infty$, $\beta \in l^1(h)$, $m \in \mathbb{N}_0$ we have
$$(L_m\alpha) \cdot \varphi(\beta) = \alpha \cdot \varphi(L_m\beta) = L_m(\alpha \cdot \varphi)(\beta).$$
(ii) If ψ is invariant, then $\psi * \varphi$ is invariant.
(iii) If ψ is a mean and φ is invariant, then $\psi * \varphi = \varphi$.

Proof.

(i) Using $(L_m\beta) * \alpha = \beta * (L_m\alpha)$ for $\beta \in l^1(h)$, $\alpha \in l^\infty$, it follows
$$(L_m\alpha) \cdot \varphi(\beta) = \varphi(\beta * (L_m\alpha)) = \varphi((L_m\beta) * \alpha) = \alpha \cdot \varphi(L_m\beta).$$
Moreover,
$$\sum_{k=0}^{\infty} (\alpha \cdot \varphi)(k)\, L_m\beta(k)\, h(k) = \sum_{k=0}^{\infty} L_m(\alpha \cdot \varphi)(k)\, \beta(k)\, h(k),$$
which is exactly the second equation of (i).

(ii) Using (i) we get for each $\alpha \in l^\infty$
$$\psi * \varphi(L_m\alpha) = \psi((L_m\alpha) \cdot \varphi) = \psi(L_m(\alpha \cdot \varphi)) = \psi(\alpha \cdot \varphi) = \psi * \varphi(\alpha).$$

(iii) Recall that $P^1(h) \subseteq (l^\infty)^*$ is weak $*$-dense in the set of all means on l^∞. Given a mean ψ on l^∞ let $(\beta_i)_{i\in I}$ be some net in $P_1(h)$ converging towards ψ in the weak $*$-topology. If φ is invariant we get
$$\psi * \varphi(\alpha) = \lim_i \beta_i(\alpha \cdot \varphi) = \lim_i \varphi(\beta_i * \alpha) = \varphi(\alpha),$$
where we used for the last equality that
$$\beta_i * \alpha = \sum_{k=0}^{\infty} L_k\alpha\ \beta_i(k)\, h(k),$$
which implies $\varphi(\beta_i * \alpha) = \varphi(\alpha)$.

◇

Theorem 4.17.10 *For every polynomial hypergroup the following is valid:*

(i) $F_0 := \{\alpha \in F : d(\alpha) = 0\}$ *= the closed linear span of* $\{\beta - L_N\beta : \beta \in l^\infty,\ n \in \mathbb{N}_0\}$.
(ii) $F = \mathbb{C}1 \bigoplus F_0$.

Proof.

(i) Denote by L the linear span of all sequences of the form $\beta - L_n\beta$, $\beta \in l^\infty$, $n \in \mathbb{N}_0$. If $\alpha \in \overline{L}$ it follows that $\mu(\alpha) = 0$ for all $\mu \in \mathcal{L}(\mathbb{N}_0)$, i.e. $\alpha \in F_0$. Conversely suppose that $\alpha \in F_0$. We have to show that $\alpha \in \overline{L}$. Otherwise there would exist $\varphi \in (l^\infty)^*$ such that $\varphi(\overline{L}) = 0$ and $\varphi(\alpha) \neq 0$. Then $\varphi(L_n\beta) = \varphi(\beta)$ for all $\beta \in l^\infty$, $n \in \mathbb{N}_0$. Since l^∞ is a unital C^*-algebra, φ is a linear combination of four means φ_i, $i = 1, ..., 4$, see e.g. [170,Corollary 3.2.17]. Now let $\psi \in \mathcal{L}(\mathbb{N}_0)$. By Proposition 4.17.9 (ii) we have $\psi * \varphi_i(L_n\beta) = \psi * \varphi_i(\beta)$ for all $\beta \in l^\infty$, $n \in \mathbb{N}$, $i = 1, ..., 4$. Hence $\psi * \varphi_i$ are elements of $\mathcal{L}(\mathbb{N}_0)$, which implies $\varphi_i * \psi(\alpha) = 0$. By Proposition 4.17.9 (iii) we obtain $\varphi(\alpha) = \psi * \varphi(\alpha) = 0$, a contradiction.

(ii) If $\alpha \in F$ then $\alpha - d(\alpha)1 \in F_0$. To show that the sum $\mathbb{C}1 + AC_0$ is direct, suppose that $d1 \in F_0$. Then $d = \mu(d1)$ for each $\mu \in \mathcal{L}(\mathbb{N}_0)$, which proves (ii).

◇

Hence the following equivalence is valid:

Corollary 4.17.11 *Assume that* $(R_n(x))_{n\in\mathbb{N}_0}$ *generates a polynomial hypergroup satisfying* $\frac{a_n h(n)}{H(n)} \to 0$. *Let* $\alpha \in l^\infty$. *Then* $\lim_{n\to\infty} \|f_n * \alpha\|_\infty = 0$ *if and only if* α *is an element of the* $\|\cdot\|_\infty$*-closure of the linear span of* $\{\beta - L_n\beta : \beta \in l^\infty,\ n \in \mathbb{N}_0\}$.

Considering the character sequences $(\alpha_z(n))_{n\in\mathbb{N}_0}$ we have the following results:
If $\frac{a_n h(n)}{H(n)} \to 0$, then Theorem 4.17.5 implies that $\frac{1}{H(n)} \sum_{k=0}^{n} R_k(z)h(k)$ tends to zero for each $z \in D\backslash\{1\}$. (This follows also from the Christoffel-Darboux formula, see Proposition 4.10.3.)

If $\frac{a_n h(n)}{H(n)} \to 0$ is not valid, the situation is much more complicated. For illustration we give an example. The polynomials connected with homogeneous trees $R_n(x; q)$ with $q \geq 1$ are determined by $a_n = \frac{q}{q+1}$, $b_n = 0$, $c_n = \frac{1}{q+1}$, $n \in \mathbb{N}$, and $a_0 = 1$, see subsection 4.5.7.2. The Haar weights are

$h(0) = 1,\ h(n) = q^{n-1}(q+1)$. For $q > 1$ we have $H(n) = \frac{(q+1)q^n-2}{q-1}$ and then $\lim\limits_{n\to\infty} \frac{a_n h(n)}{H(n)} = \frac{q-1}{q+1} > 0$. The character $\alpha_{-1}(n) = R_n(-1;q) = (-1)^n$ is an element of F with $d(\alpha_{-1}) = 0$. But

$$\frac{1}{H(n)} \sum_{k=0}^{n} R_k(-1;q)\, h(k) \;=\; \frac{q}{q+1}\, \frac{h(n)}{H(n)}\, (-1)^n$$

even does not converge.

The following property of invariant means $\mu \in \mathcal{L}(\mathbb{N}_0)$ is very useful.

Proposition 4.17.12 *Let A be a finite subset of $\mathbb{N}_0$. Then $\mu(\chi_A) = 0$ for each $\mu \in \mathcal{L}(\mathbb{N}_0)$.*

Proof. Consider $\epsilon_0(n) = \delta_{0,n}$. Then $L_m\epsilon_0(n) = \delta_{m,n}\,\frac{1}{h(m)}$. For $\mu \in \mathcal{L}(\mathbb{N}_0)$ we have $\mu(L_m\epsilon_0) = \mu(\epsilon_0)$ and hence $\mu(\chi_{S_m}) = \mu(\epsilon_0)\, H(n) \geq n\mu(\epsilon_0)$. Since $\mu(\chi_{S_n}) \leq 1$, it follows $\mu(\epsilon_0) = 0$ and thus $\mu(\chi_A) = 0$. ◇

We introduce a stronger form of almost convergence, see [459] and [370].

Definition 2. A sequence $\alpha \in l^\infty$ is almost convergent to $d_S(\alpha) \in \mathbb{C}$ in the strong sense, if $|\alpha - d_S(\alpha)\,1| \in F_0$. Then we write $\alpha \in FS$.

Proposition 4.17.13 *Let $\alpha \in l^\infty$ and $c \in \mathbb{C}$. Then α is almost convergent to $d_S(\alpha) = c$ in the strong sense if and only if for each $\varepsilon > 0$ there exists a set $A \subseteq \mathbb{N}$ such that $\mu(\chi_A) = 0$ for all $\mu \in \mathcal{L}(\mathbb{N}_0)$ and $|\alpha(n) - c| < \varepsilon$ for all $n \in \mathbb{N}\backslash A$.*

Proof. Recall that $\mu(\epsilon_0) = 0$ for all $\mu \in \mathcal{L}(\mathbb{N}_0)$. We begin assuming that for each $\varepsilon > 0$ there exists a set $A \subseteq \mathbb{N}$ such that $\mu(\chi_A) = 0$ for all $\mu \in \mathcal{L}(\mathbb{N}_0)$ and $|\alpha(n) - c| < \varepsilon$ for all $n \in \mathbb{N}\backslash A$. Then we get

$$\mu(|\alpha - c\,1|) \;\leq\; \varepsilon\, \mu(\chi_{\mathbb{N}_0\backslash A}) \;=\; \varepsilon$$

for each $\mu \in \mathcal{L}(\mathbb{N}_0)$, and it follows $\mu(|\alpha - c\,1|) = 0$ for all $\mu \in \mathcal{L}(\mathbb{N}_0)$, i.e. α is almost convergent to $d_S(\alpha) = c$ in the strong sense.

Conversely assume that α is almost convergent to $d_S(\alpha) = c$ in the strong sense. We have only to consider the case $0 < \varepsilon \leq \|\alpha - d_S(\alpha)\,1\|_\infty$. Let $A = \{n \in \mathbb{N} : |\alpha(n) - c| \geq \varepsilon\}$. Then $\mu(|\alpha - c\,1|) \geq \varepsilon\mu(A)$ for every $\mu \in \mathcal{L}(\mathbb{N}_0)$. Hence $\mu(A)$ has to be zero, and $|\alpha(n) - c| < \varepsilon$ for $n \in \mathbb{N}\backslash A$ obviously is true. ◇

Remark: We have the following inclusions: Every convergent sequence α with limit α_∞ is almost convergent with $d_S(\alpha) = \alpha_\infty$ in the strong sense. A sequence $\alpha \in l^\infty$ which is almost convergent to $d_S(\alpha)$ in the strong sense is almost convergent to $d(\alpha)$. In fact, if $\varepsilon > 0$ and $m \in \mathbb{N}$

such that $|\alpha(n) - \alpha_\infty| < \varepsilon$ for $n > m$, Proposition 4.7.13 yields that α is almost convergent to $d_S(\alpha) = \alpha_\infty$ in the strong sense, since $\mu(\chi_{S_m}) = 0$ for each $\mu \in \mathcal{L}(\mathbb{N}_0)$. Moreover, if $\mu(|\alpha - d_S(\alpha)\,1|) = 0$ for all $\mu \in \mathcal{L}(\mathbb{N}_0)$, it follows that $|\mu(\alpha - d_S(\alpha)\,1)| = 0$ for each $\mu \in \mathcal{L}(\mathbb{N}_0)$, and hence $\mu(\alpha) = d_S(\alpha)$ for all $\mu \in \mathcal{L}(\mathbb{N}_0)$, and $d(\alpha) = d_S(\alpha)$.

In addition to Proposition 4.17.13 we have another equivalence result for almost convergence in the strong sense.

Proposition 4.17.14 *Let $\alpha \in l^\infty$ and $c \in \mathbb{C}$. The following statements are equivalent.*

(1) α is almost convergent to $d_S(\alpha) = c$ in the strong sense.
(2) $\mu(\alpha \cdot \beta) = c\,\mu(\beta)$ for all $\beta \in l^\infty$ and $\mu \in \mathcal{L}(\mathbb{N}_0)$.

Proof. Suppose that $\mu(|\alpha - c\,1|) = 0$ for all $\mu \in \mathcal{L}(\mathbb{N}_0)$, and let $\beta \in l^\infty$ arbitrary. Since $|(\alpha - c\,1)\beta| \leq \|\beta\|_\infty\,|\alpha - c\,1|$, it follows $\mu((\alpha - c\,1)\beta) = 0$. Hence $\mu(\alpha \cdot \beta) = c\mu(\beta)$ for all $\mu \in \mathcal{L}(\mathbb{N}_0)$.

Conversely assume $\mu(\alpha \cdot \beta) = c\mu(\beta)$ for all $\beta \in l^\infty$, $\mu \in \mathcal{L}(\mathbb{N}_0)$. Then $\mu((\alpha - c\,1)\beta) = 0$ for all $\beta \in l^\infty$, $\mu \in \mathcal{L}(\mathbb{N}_0)$. Choosing $\beta(n) = \frac{|\alpha(n)-c|}{\alpha(n)-c}$ whenever $\alpha(n) \neq c$, we get $\mu(|\alpha - c\,1|) = 0$. ◇

Examples.

We consider again the characters $\alpha_z(n) = R_n(z)$, $z \in D$, and assume that the polynomial hypergroup is **symmetric**. We know already that α_{-1} is an element of F with $d(\alpha_{-1}) = 0$. But α_{-1} is not almost convergent in the strong sense, i.e. $\alpha_{-1} \notin FS$. In fact, assuming $\alpha_{-1} \in FS$, the constant $d_S(\alpha_{-1})$ has to be $d(\alpha_{-1}) = 0$. Since $|\alpha_{-1}(n)| = 1$ for each $n \in \mathbb{N}_0$, this is not true.

For the polynomial hypergroup generated by the Jacobi polynomials $R_n^{(\alpha,\beta)}(x)$, where $(\alpha,\beta) \in J = \{(\alpha,\beta) : \alpha \geq \beta > -1,\ \alpha+\beta+1 \geq 0\}$ we know that $|R_n^{(\alpha,\beta)}(x)| = O(n^{-\alpha-\frac{1}{2}})$ for each $x \in]-1,1[$, see [641, (4.1.1) and (8.21.18)]. Hence for $(\alpha,\beta) \in J\backslash\{(-\frac{1}{2},-\frac{1}{2})\}$ and $x \in]-1,1[$ we obtain $\alpha_x \in FS$ with $d_S(\alpha_x) = 0$. Obviously for $x = 1$ we have $\alpha_1 \in FS$ with $d_S(\alpha_1) = 1$ for each $(\alpha,\beta) \in J$.

If $x = -1$ and $\alpha = \beta$ we know from above that $\alpha_{-1} \in F\backslash FS$. If $x = -1$ and $\alpha > \beta$ we have $\alpha_{-1}(n) \to 0$, see [641, (4.11) and (4.14)]. Therefore $\alpha_{-1} \in FS$ with $d_S(\alpha_{-1}) = 0$ whenever $\alpha > \beta$.

It remains to check the case $\alpha = -\frac{1}{2} = \beta$, i.e. $R_n^{(-\frac{1}{2},-\frac{1}{2})}(x) = T_n(x)$. From above we know that $\alpha_1 \in FS$ and $\alpha_{-1} \in F\backslash FS$.

Let $x \in]-1,1[$. Suppose that $\alpha_x \in FS$. Since $d_S(\alpha_x) = d(\alpha_x) = 0$, $\alpha_x \in FS$ implies $|\alpha_x| \in F_0$. Since for $(T_n(x))_{n\in\mathbb{N}_0}$ we have $\frac{a_n h(n)}{H(n)} \to 0$, by Proposition 4.17.2 it is necessary that

$$\lim_{n\to\infty} \frac{1}{2n+1} \sum_{k=0}^{n} |T_k(x)|\, h(k) \;=\; 0$$

holds. It follows that

$$\lim_{n\to\infty} \frac{1}{2n+1} \sum_{k=0}^{n} T_k^2(x)\, h(k) \;=\; 0.$$

However, the example below Corollary 4.10.6 shows that

$$\lim_{n\to\infty} \frac{1}{2n+1} \sum_{k=0}^{n} T_k^2(x)\, h(k) \;=\; \frac{1}{2}.$$

Therefore we have $\alpha_x \in F\backslash FS$ for each $x \in]-1,1[$.

We extend this result shown for $(T_n(x))_{n\in\mathbb{N}_0}$ to general polynomial hypergroups, where

$$\mathcal{T} \;=\; \{y \in D_S : \limsup_{n\to\infty} \frac{1}{H(n)} \sum_{k=0}^{n} R_k^2(y)\, h(k) > 0\}$$

plays a central role, see subsection 4.9.

Proposition 4.17.15 *Assume that $(R_n(x))_{n\in\mathbb{N}_0}$ generates a polynomial hypergroup with $\frac{a_n h(n)}{H(n)} \to 0$ and $\mathcal{T}\backslash\{1\} \neq \emptyset$. For each $x \in \mathcal{T}\backslash\{1\}$ holds $\alpha_x \in F\backslash FS$.*

Proof. For $x \in \mathcal{T}\backslash\{1\}$ we have

$$\limsup_{n\to\infty} \frac{1}{H(n)} \sum_{k=0}^{n} |R_k(x)|\, h(k) \;>\; 0,$$

and for $x \in FS$ it is necessary that

$$\lim_{n\to\infty} \frac{1}{H(n)} \sum_{k=0}^{n} |R_k(x)|\, h(k) \;=\; 0.$$

Hence $\alpha_x \notin FS$. ◇

Remark. In Corollary 4.9.8 or Corollary 4.11.18 examples with $]-1,1] \subseteq \mathcal{T}$ or $[-1,1] = \mathcal{T}$ are studied.

Proving results about multipliers for F in subsection 4.17.3 we will use properties of elements of FS, mainly Proposition 4.17.13.

17.3 Multipliers for almost convergent sequences

Before we characterize multipliers of F, we derive some complementary properties of invariant means. Define the translation on $(l^\infty)^*$ by $L_m\nu(\varphi) := \nu(L_m\varphi)$ for $\varphi \in l^\infty$, $m \in \mathbb{N}_0$. Obviously $L_m\nu \in (l^\infty)^*$. If $\lim\limits_{n\to\infty} \frac{a_n h(n)}{H(n)} = 0$, we have used that

$$\lim_{n\to\infty} \|L_m(\jmath(f_n)) - \jmath(f_n)\| = \lim_{n\to\infty} \|L_m f_n - f_n\|_1 = 0 \quad \text{for all } m \in \mathbb{N},$$

where $f_n = \frac{1}{H(n)} \chi_{S_n}$. In particular, the existence of some $\mu \in \mathcal{SL}(\mathbb{N}_0) \subseteq \mathcal{L}(\mathbb{N}_0)$ follows from this property, whenever $\frac{a_n h(n)}{H(n)} \to 0$. Generalizing this access we consider **finite means** $g \in P^1(h)$, i.e. $\{k \in \mathbb{N}_0 : |g(k)| > 0\}$ is a finite set. Obviously $(f_n)_{n\in\mathbb{N}_0}$ is a sequence of finite means.

Definition 3. We call a polynomial hypergroup $K = \mathbb{N}_0$ **strongly amenable**, if there exists a net $(g_i)_{i\in I}$ of finite means $g_i \in P^1(h)$ convergent in norm to invariance, that is

$$\lim_i \|L_m g_i - g_i\|_1 = 0 \quad \text{for all } m \in \mathbb{N}_0.$$

From subsection 4.7 we know that $K = \mathbb{N}_0$ is strongly amenable, whenever
$\lim\limits_{n\to\infty} \frac{a_n h(n)}{H(n)} = 0.$

Lemma 4.17.16 *Assume that the polynomial hypergroup is strongly amenable with a net $(g_i)_{i\in I}$ of finite means $g_i \in P^1(h)$ convergent in norm to invariance. Then there exists a sequence*

$$(g_{i_n})_{n\in\mathbb{N}} \subseteq (g_i)_{i\in I} \text{ such that } \lim_{n\to\infty} \|L_m g_{i_n} - g_{i_n}\|_1 = 0 \text{ for all } m \in \mathbb{N}.$$

Proof. There exists $i_1 \in I$ such that $\|L_1 g_{i_1} - g_{i_1}\|_1 < 1$. Then there also exist $i_2^1, i_2^2 \in I$ such that

$$\|L_1 g_i - g_i\|_1 < \frac{1}{2} \quad \text{if } i \geq i_2^1$$

and

$$\|L_2 g_i - g_i\|_1 < \frac{1}{2} \quad \text{if } i \geq i_2^2.$$

Let $i_2 \in I$ such that $i_2 \geq i_2^1$ and $i_2 \geq i_2^2$. Then

$$\|L_m g_{i_2} - g_{i_2}\|_1 < \frac{1}{2} \quad \text{for } m = 1, 2.$$

If $i_1, ..., i_{k-1} \in I$ are chosen such that

$$\|L_m g_{i_l} - g_{i_l}\|_1 < \frac{1}{l} \qquad \text{for } m = 1, 2, ..., l; \quad l = 1, 2, ..., k-1,$$

then choose $i_k \in I$ in the following way: There exist $i_k^1, ..., i_k^k$ such that $i \geq i_k^l$, $l = 1, 2, ..., k$ implies $\|L_m g_i - g_i\|_1 < \frac{1}{k}$ for $m = 1, 2, ..., k$. Let $i_k \geq i_k^1, ..., i_k^k$. Then
$\|L_m g_{i_k} - g_{i_k}\|_1 < \frac{1}{k}$ for $m = 1, ..., k$. Thus the sequence $(g_{i_n})_{n\in\mathbb{N}}$ satisfies the stated property. Given $m \in \mathbb{N}$ we have

$$\|L_m g_{i_n} - g_{i_n}\|_1 < \frac{1}{n} \quad \text{for all } n \geq m,$$

so that

$$\lim_{n\to\infty} \|L_m g_{i_n} - g_{i_n}\|_1 = 0.$$

◇

Proposition 4.17.17 *Assume that there exist a sequence $(g_n)_{n\in\mathbb{N}}$, $g_n \in l^1(h)$ such that $\lim_{n\to\infty} \|L_m g_n - g_n\|_1 = 0$ for each $m \in \mathbb{N}_0$. If μ is a weak $*$-cluster point of $\{j(g_n) : n \in \mathbb{N}\}$, then $L_m\mu = \mu$ for each $m \in \mathbb{N}_0$.*

Proof. Fix $m \in \mathbb{N}_0$ and let $\varphi \in l^\infty$, $\varepsilon > 0$. Then

$$\begin{aligned}|(L_m\mu - \mu)(\varphi)| &\leq |L_m(\mu - j(g_n))(\varphi)| \\ &\quad + |(L_m(j(g_n)) - j(g_n))(\varphi)| + |(j(g_n) - \mu)(\varphi)|.\end{aligned}$$

We have

$$|(L_m(j(g_n)) - j(g_n))(\varphi)| \leq \|L_m g_n - g_n\|_1 \|\varphi\|_\infty.$$

Hence there exists n_1 such that $|(L_m(j(g_n)) - j(g_n))(\varphi)| < \frac{\varepsilon}{3}$ for all $n \geq n_1$ and $n_2 \geq n_1$ such that

$$j(g_{n_2}) \in \{\nu \in (l^\infty)^* : |(\nu - \mu)(L_m\varphi)| < \frac{\varepsilon}{3} \text{ and } |(\nu - \mu)(\varphi)| < \frac{\varepsilon}{3}\}.$$

Hence $|(L_m\mu - \mu)(\varphi)| < \varepsilon$. Therefore $L_m\mu = \mu$. ◇

The subspace F of l^∞, in general, is not closed under multiplication. We study this aspect of F by investigating the multipliers of F.

Definition 4. An element $\beta \in l^\infty$ such that $\beta \cdot F \subseteq F$ is called **multiplier** of F. The set of multipliers of F is denoted by MF.

Proposition 4.17.18 *Let $K = \mathbb{N}_0$ be a strongly amenable polynomial hypergroup. Then $\beta \in l^\infty$ is a multiplier of F if and only if $\beta \in F$ and*

$$\mu(\beta \cdot \alpha) = \mu(\beta)\,\mu(\alpha) = d(\beta)\,d(\alpha)$$

for all $\alpha \in F$ and $\mu \in \mathcal{L}(\mathbb{N}_0)$.

Proof. Each $\beta \in F$, which fulfils $\mu(\beta \cdot \alpha) = \mu(\beta)\mu(\alpha) = d(\beta)d(\alpha)$ for all $\alpha \in F$ and $\mu \in \mathcal{L}(\mathbb{N}_0)$ is obviously a multiplier of F.

Conversely, Let $\beta \in MF$. Since $1 \in F$ we have $\beta = \beta \cdot 1 \in F$. We have to show that for each $\alpha \in F$ and $\mu \in \mathcal{L}(\mathbb{N}_0)$ it follows $\mu(\beta \cdot \alpha) = \mu(\beta)\mu(\alpha)$. By Lemma 4.17.16 there exists a sequence $(g_n)_{n\in\mathbb{N}}$ of finite means $g_n \in P^1(h)$ such that

$$\lim_{n\to\infty} \left(j(g_n)(\varphi) \; - \; j(g_n)\left(L_m\varphi\right)\right) \; = \; 0 \qquad\qquad (i)$$

for each $\varphi \in l^\infty$, $m \in \mathbb{N}_0$.

It follows that for every $\alpha \in F$

$$\lim_{n\to\infty} j(g_n)(\alpha) \; = \; d(\alpha) \qquad\qquad (ii)$$

is true.

To show this we can use Theorem 4.17.10. If $\alpha \in F_0 = \overline{L}$, ($L$ is the linear span of $\{\varphi - L_m\varphi \, : \, \varphi \in l^\infty, \; m \in \mathbb{N}_0\}$) note that by (i) we get $\lim\limits_{n\to\infty} j(g_n)(\alpha) = 0 = d(\alpha)$. For $\alpha = c\,1$, $c \in \mathbb{C}$, obviously $\lim\limits_{n\to\infty} j(g_n)(\alpha) = c = d(\alpha)$ is valid.

Next let $k \in \mathbb{N}$ be fixed and define

$$\psi_n := \; g_n \cdot \beta \; - \; L_k(g_n \cdot \beta) \; \in \; l^1(h).$$

We will show that $(\psi_n)_{n\in\mathbb{N}}$ is a weak Cauchy sequence in $l^1(h)$. (See [189,Definition II.3.25].) For $\varphi \in l^\infty$ we obtain

$$j(\psi_n)(\varphi) \; = \; j(g_n \cdot \beta \; - \; L_k(g_n \cdot \beta))(\varphi) \; = \; j(g_n \cdot \beta)(\varphi) \; - \; j(g_n \cdot \beta)(L_k\varphi)$$

$$= \; j(g_n \cdot \beta)(\varphi \; - L_k\varphi) \; = \; j(g_n)(\beta \cdot (\varphi \; - L_k\varphi)).$$

Since β is a multiplier and $\varphi - L_k\varphi \in F_0 \subseteq F$, we have $\beta \cdot (\varphi - L_k\varphi) \in F$. By equation (ii) we get

$$\lim_{n\to\infty} j(\psi_n)(\varphi) \; = \; \lim_{n\to\infty} j(g_n) \left(\beta \cdot (\varphi - L_k\varphi)\right) \; = \; d(\beta \cdot (\varphi - L_k\varphi)) \qquad (iii)$$

for all $\varphi \in l^\infty$. In particular, $(\psi_n)_{n\in\mathbb{N}}$ is a weak Cauchy sequence in $l^1(h)$. Since $l^1(h)$ is weakly complete, see [189,Theorem IV.8.6]), there exists $\psi \in l^1(h)$ such that $\psi = \lim\limits_{n\to\infty} \psi_n$ in the weak topology. Obviously, $\psi(m) = \lim\limits_{n\to\infty} \psi_n(m)$ for each $m \in \mathbb{N}_0$. On the other hand, $\lim\limits_{n\to\infty} g_n(m) = \lim\limits_{n\to\infty} j(g_n)(\chi_{\{m\}}) = 0$, since $\chi_{\{m\}} \in F_0$ by Proposition 4.17.12. Hence

$$\psi(m) = \lim_{n\to\infty} (g_n(m)\,\beta(m) - L_k(g_n\cdot\beta)(m)) = \lim_{n\to\infty} L_k(g_n\cdot\beta)(m)$$

$$= \lim_{n\to\infty} \sum_{l=|k-m|}^{k+m} g(k,m;l)\,g_n(l)\,\beta(l) = 0.$$

So $\psi \equiv 0$, and by (iii) it follows for each $\mu \in \mathcal{L}(\mathbb{N}_0)$ and $\varphi \in l^\infty$

$$\mu(\beta\cdot(\varphi - L_k\varphi)) = d(\beta\cdot(\varphi - L_k\varphi)) = \lim_{n\to\infty} j(\psi_n)(\varphi) = 0$$

$$= \mu(\beta)\,\mu(\varphi - L_k\varphi).$$

Since $F = F_0 \oplus \mathbb{C}\,1$, $F_0 = \overline{L}$ and $\mu(1) = 1$ for each $\mu \in \mathcal{L}(\mathbb{N}_0)$, it follows $\mu(\beta\cdot\alpha) = \mu(\beta)\mu(\alpha)$ for each $\mu \in \mathcal{L}(\mathbb{N}_0)$. ◇

In order to derive further equivalence conditions for $\varphi \in l^\infty$ being a multiplier of F, we use the Stone-Čech compactification $\beta(\mathbb{N}_0)$ of $K = \mathbb{N}_0$. Based on the Stone-Čech compactification one can associate with an invariant mean a unique positive Borel measure on $\beta(\mathbb{N}_0)$. A concise description of the 'measure-theoretic character' of invariant means is given in [184]. For convenience we collect the main steps:

The Stone-Čech compactification $\beta(\mathbb{N}_0)$ is a compact Hausdorff space such that

(1) $\mathbb{N}_0$ (with discrete topology) is homeomorphic to a dense subset of $\beta(\mathbb{N}_0)$ and
(2) each $\varphi \in l^\infty$ can uniquely be extended to a continuous function $\hat{\varphi} \in C(\beta(\mathbb{N}_0))$.

$\beta(\mathbb{N}_0)$ is (up to homeomorphism) determined uniquely by property (1) and (2). If π is the inclusion map of $\mathbb{N}_0$ into $\beta(\mathbb{N}_0)$, then $\overline{\pi(A)}$ denotes the closure of $\pi(A)$ in $\beta(\mathbb{N}_0)$. Sets of the form $\overline{\pi(A)}$ are closed-open and they form an open basis for $\beta(\mathbb{N}_0)$.

The Banach space $C(\beta(\mathbb{N}_0))$ of continuous functions on the compact Hausdorff space $\beta(\mathbb{N}_0)$ is isometrically isomorphic to l^∞. The mapping π_* : $C(\beta(\mathbb{N}_0)) \to l^\infty$, defined by $\pi_*(f)(n) = f(\pi(n))$ for $f \in C(\beta(\mathbb{N}_0))$, $n \in \mathbb{N}_0$, in an onto, isometrical isomorphism. Moreover, $(l^\infty)^*$ is isometrically isomorphic to $C(\beta(\mathbb{N}_0))^*$ via π^*, the adjoint map of π_*. (For $\mu \in (l^\infty)^*$ and $f \in C(\beta(\mathbb{N}_0))$, we have $\pi^* : (l^\infty)^* \to C(\beta(\mathbb{N}_0))^*$, $(\pi^*\mu)(f) = \mu(\pi_* f)$.) $C(\beta(\mathbb{N}_0))^*$ can be identified with $M(\beta(\mathbb{N}_0))$, the space of all complex regular Borel measures on $\beta(\mathbb{N}_0)$. Hence we have $\mu(\pi_* f) = \int_{\beta(\mathbb{N}_0)} f d\mu$ for each $\varphi \in C(\beta(\mathbb{N}_0))$. If $\mu \in (l^\infty)^*$ is a mean, then $\pi^*\mu$ is a probablity measure. supp $\mu \subseteq \beta(\mathbb{N}_0)$ denotes the support of the probability measure $\pi^*\mu$.

Proposition 4.17.19 *Let $K = \mathbb{N}_0$ be a strongly amenable polynomial hypergroup. Then $\beta \in l^\infty$ is a multiplier of F if and only if $\beta \in F^0 \oplus \mathbb{C}\,1$, where $F^0 := \{\alpha \in l^\infty : |\alpha| \in F_0\}$.*

Proof. Let $\beta \in F^0 \oplus \mathbb{C}\,1 \subseteq F$, i.e. there exists a constant $d(\beta) \in \mathbb{C}$ and a sequence $\beta_0 \in F_0$ such that $\beta = \beta_0 + d(\beta)\,1$. Given $\alpha \in F$ and $\mu \in \mathcal{L}(\mathbb{N}_0)$ we have

$$|\mu(\beta_0 \cdot \alpha)| \;\leq\; \mu(|\beta_0 \cdot \alpha|) \;\leq\; \|\alpha\|_\infty\, \mu(|\beta_0|) \;=\; 0.$$

Hence $\mu(\beta_0 \cdot \alpha) = 0$ and

$$\mu(\beta \cdot \alpha) \;=\; \mu((\beta_0 \,+\, d(\beta)\,1) \cdot \alpha) \;=\; d(\beta)\,\mu(\alpha) \;=\; d(\beta)\,d(\alpha).$$

By Proposition 4.17.18 holds $\beta\alpha \in F$.

Proving the converse implication we consider $\varphi \in l^\infty$ as a continuous function on the Stone-Čech compactification $\beta(\mathbb{N}_0)$, and a mean on l^∞ as a probability measure on $\beta(\mathbb{N}_0)$.

Let $\beta \in l^\infty$ such that $\beta \cdot F \subseteq F$. Obviously $\beta \in F$, since $1 \in F$. To show that β is an element of $F^0 \oplus \mathbb{C}\,1$, it suffices to derive that for each $\mu \in \mathcal{L}(\mathbb{N}_0)$ there exists a constant $d_\mu(\beta)$, such that

$$\beta \;\equiv\; d_\mu(\beta) \qquad \text{on supp } \mu. \qquad\qquad (*)$$

Indeed if $(*)$ is valid, then there exists an element $\alpha \in l^\infty$ such that $\beta = d_\mu(\beta)\,1 + \alpha$ and $\alpha \equiv 0$ on supp μ for each $\mu \in \mathcal{L}(\mathbb{N}_0)$. It follows

$$\mu(|\alpha|) \;=\; \mu(|\beta \cdot \chi_{\mathbb{N}_0 \setminus \text{supp } \mu}|) \;\leq\; \|\beta\|_\infty\, \mu(\chi_{\mathbb{N}_0 \setminus \text{supp } \mu}) \;=\; 0$$

for each $\mu \in \mathcal{L}(\mathbb{N}_0)$, i.e. $\alpha \in F^0$. It remains to show that $(*)$ is true. Fix $\mu \in \mathcal{L}(\mathbb{N}_0)$. By Proposition 4.17.18 we have $\mu(\beta \cdot \alpha) = \mu(\beta)\mu(\alpha)$ for all $\alpha \in F$. In particular, we have for each $c \in \mathbb{C}$

$$\mu((\beta \,-\, c\,1)^2) \;=\; (\mu(\beta) \,-\, c)^2.$$

Choosing the constant $d_\mu(\beta) = \mu(\beta)$ we get

$$\mu((\beta \,-\, d_\mu(\beta)\,1)^2) \;=\; 0.$$

In particular $\beta \equiv d_\mu(\beta)$ on supp μ. ◇

Remark: We refer to [142] where the basic steps for proving Proposition 4.17.18 and Proposition 4.17.19 were used in the context of semigroups.

Combining Proposition 4.17.13, Proposition 4.17.14, Proposition 4.17.18 and Proposition 4.17.19 we obtain the following characterization of multipliers of F.

Theorem 4.17.20 *Let $K = \mathbb{N}_0$ be a strongly amenable polynomial. For $\beta \in l^\infty$ the following statements are equivalent*

(1) β is a multiplier of F.
(2) $\beta \in F$ and $\mu(\beta \cdot \alpha) = \mu(\beta)\,\mu(\alpha) = d(\beta)\,\mu(\alpha)$ for all $\alpha \in l^\infty$, $\mu \in \mathcal{L}(\mathbb{N}_0)$.
(3) $\beta = d_\mu(\beta)$ on supp μ for all $\mu \in \mathcal{L}(\mathbb{N}_0)$. ($d_\mu(\beta)$ is defined by $(*)$.)
(4) There exists a constant $c \in \mathbb{C}$ such that for a given $\varepsilon > 0$ there exists a set $A \subseteq \mathbb{N}$ such that $\mu(\chi_A) = 0$ for every $\mu \in \mathcal{L}(\mathbb{N}_0)$ and $|\beta(n) - c| < \varepsilon$ for all $n \in \mathbb{N}\backslash A$.
(5) $\beta \in FS$, i.e. β is almost convergent to $d_{\mathcal{S}}(\beta) \in \mathbb{C}$ in the strong sense.
(6) $\beta \in F^0 \oplus \mathbb{C}1$.

Proof. The equivalence of (2), (4) and (5) follows by Proposition 4.17.13 and Proposition 4.17.14. The equivalence of (1), (2) and (6) is contained in the proof of Proposition 4.17.19, where Proposition 4.17.18 is used. Finally note that $FS = F^0 \oplus \mathbb{C}1$. ◇

Chapter 5

Weakly Stationary Random Fields on a Commutative Hypergroup

1 Definition and Representation

Let K be a commutative hypergroup with convolution ω, involution $\tilde{}$ and unit element e.

Definition *A family $(X_a)_{a \in K}$ of square integrable complex-valued random variables on a probability space $(\Omega, \mathcal{F}, P)$ is called a* **weakly stationary** *random field on K, if the* **covariance function** *given by $d(a,b) := E(X_a \overline{X_b})$ is a bounded and continuous function on $K \times K$ and fulfils*

$$d(a,b) = \int_K d(x,e)\, d\omega(a,\tilde{b})(x) \qquad \textit{for } a,b \in K.$$

$E(X)$ denotes the expected value of a random variable X.

We will write $d(a)$ instead of $d(a,e)$.

Note that $d(a,b)$ is the usual covariance only if $E(X_a) = 0$ for all $a \in K$.

Often we shall write $d(a,b) = < X_a, X_b >$ to emphasize the Hilbert space properties of $L^2(\Omega, P)$.

We begin by proving a representation theorem for the covariance function d.

Theorem 5.1.1 *Let $(X_a)_{a \in K}$ be a weakly stationary random field on a commutative hypergroup K. For the covariance function d there exists a unique bounded regular positive Borel measure μ on the dual space $\widehat{K}$ such that*

$$d(a) = \int_{\widehat{K}} \alpha(a)\, d\mu(\alpha) \qquad \textit{for every } a \in K.$$

Proof. The function $d : K \to \mathbb{C}$ is bounded and continuous by assumption. Given $c_1, \ldots, c_n \in \mathbb{C}$; $a_1, \ldots, a_n \in K$ we see immediately

$$\sum_{i=1}^{n}\sum_{j=1}^{n} c_i \overline{c_j}\, \omega(a_i, \tilde{a_j})(d) = \sum_{i=1}^{n}\sum_{j=1}^{n} c_i \overline{c_j}\, d(a_i, a_j) = E\left(\left|\sum_{i=1}^{n} c_i X_{a_i}\right|^2\right)$$

This means $d \in P^b(K)$, and by Theorem 3.7.1 the statement follows. ◇

The measure of the above theorem is called **spectral measure** of $(X_a)_{a\in K}$. Moreover we have

$$d(a, b) = \int_{\widehat{K}} \alpha(a)\overline{\alpha(b)}\, d\mu(\alpha).$$

In fact by Fubini's theorem we get

$$d(a, b) = \int_K d(x)\, d\omega(a, \tilde{b})(x) = \int_K \int_{\widehat{K}} \alpha(x)\, d\mu(\alpha)\, d\omega(a, \tilde{b})(x) = \int_{\widehat{K}} \alpha(a)\overline{\alpha(b)}\, d\mu(\alpha).$$

Consider the Hilbert space $L^2(\widehat{K}, \mu)$. The Fourier-Stieltjes transform $\widehat{\epsilon_a}(\alpha) = \overline{\alpha(a)}$ of ϵ_a is an element of $L^2(\widehat{K}, \mu)$ and $\{\widehat{\epsilon_a} : a \in K\}$ is a linear independent subset by the uniqueness theorem. Moreover the linear span of $\{\widehat{\epsilon_a} : a \in K\}$ is dense in $L^2(\widehat{K}, \mu)$. In fact, if for $f \in L^2(\widehat{K}, \mu)$ and every $a \in K$ we assume $0 = \int_{\widehat{K}} f(\alpha)\alpha(a)\, d\mu(\alpha) = (f\mu)\check{}(a)$, then $f\mu = 0$. (Note that $L^2(\widehat{K}, \mu) \subseteq L^1(\widehat{K}, \mu)$.) That means $f = 0$ in $L^2(\widehat{K}, \mu)$.

Now define a mapping on the linear span of $\{\widehat{\epsilon_a} : a \in K\}$

$$\Phi : span\{\widehat{\epsilon_a} : a \in K\} \to span\{X_a : a \in K\} \subseteq L^2(\Omega, P)$$

by

$$\Phi\left(\sum_{k=1}^{n} c_k \widehat{\epsilon_{a_k}}\right) := \sum_{k=1}^{n} c_k X_{a_k},$$

where $n \in \mathbb{N}, c_1, \ldots, c_n \in \mathbb{C}, a_1, \ldots, a_n \in K$.

This map is well-defined because of the linear independency of $\{\widehat{\epsilon_a} : a \in K\}$. Obviously it is linear and surjective. Denoting the scalar products on $L^2(\widehat{K}, \mu)$ and $L^2(\Omega, P)$ by $<.,.>_\mu$ and $<.,.>$ respectively, we have for $a, b \in K$

$$<\widehat{\epsilon_a}, \widehat{\epsilon_b}>_\mu = \int_{\widehat{K}} \alpha(a)\overline{\alpha(b)}\, d\mu(\alpha) = d(a, b) = <X_a, X_b>.$$

Hence Φ is isometric. Extending Φ continuously to $L^2(\widehat{K}, \mu)$ we get an isometric isomorphism from $L^2(\widehat{K}, \mu)$ onto $H(X_a, K)$, where $H(X_a, K)$ is the closure of $span\{X_a : a \in K\}$ in $L^2(\Omega, P)$.

Remark: A general result for such mappings between Hilbert spaces is already contained in [1].

We denote this extension again with Φ. By virtue of $\Phi : L^2(\widehat{K}, \mu) \to H(X_a, K)$ we can define an orthogonal stochastic measure Z on $\widehat{K}$. Put for each Borel set $A \subseteq \widehat{K}$

$$Z(A) := \Phi(\chi_A)$$

A reference for stochastic measures is for example [623]. For the corresponding stochastic integral we have

$$\int_{\widehat{K}} \psi(\alpha)\, dZ(\alpha) = \Phi(\psi)$$

for every $\psi \in L^2(\widehat{K}, \mu)$. Hence

$$\int_{\widehat{K}} \alpha(a)\, dZ(\alpha) = \int_{\widehat{K}} \widehat{\epsilon_{\tilde{a}}}(\alpha)\, dZ(\alpha) = \Phi(\epsilon_{\tilde{a}}) = X_a.$$

Theorem 5.1.2 *Let $(X_a)_{a\in K}$ be a weakly stationary random field on a commutative hypergroup K. Then there exists an orthogonal stochastic measure $Z : \mathcal{B}(\widehat{K}) \to L^2(\Omega, P)$, $\mathcal{B}(\widehat{K})$ being the Borel σ-Algebra of $\widehat{K}$, such that*

$$X_a = \int_{\widehat{K}} \alpha(a)\, dZ(\alpha). \tag{1}$$

Moreover $\|Z(A)\|_2^2 = \mu(A)$ for all $A \in \mathcal{B}(\widehat{K})$, where μ is the spectral measure of $(X_a)_{a\in K}$. The orthogonal stochastic measure Z is uniquely determined by (1).

Proof. It remains to derive the uniqueness statement. Let

$$\int_{\widehat{K}} \alpha(a)\, dZ_1(\alpha) = X_a = \int_{\widehat{K}} \alpha(a)\, dZ_2(\alpha)$$

and let $\mu_1(A) = \|Z_1(A)\|_2^2$ and $\mu_2(A) = \|Z_2(A)\|_2^2$ for $A \in \mathcal{B}(\widehat{K})$.
Since $span\{\widehat{\epsilon_a} : a \in K\}$ is dense in $L^2(\widehat{K}, \mu_1 + \mu_2)$, there we have for each $A \in \mathcal{B}(\widehat{K})$ a sequence $(\psi_n)_{n\in\mathbb{N}}$ in $span\{\widehat{\epsilon_a} : a \in K\}$ with $\psi_n \to \chi_A$ in $L^2(\widehat{K}, \mu_1)$ and in $L^2(\widehat{K}, \mu_2)$ simultaneously. By (1) we have for all $n \in \mathbb{N}$

$$\int_{\widehat{K}} \psi_n(\alpha)\, dZ_1(\alpha) = \int_{\widehat{K}} \psi_n(\alpha)\, dZ_2(\alpha)$$

and $Z_1(A) = Z_2(A)$ follows. ◇

Corollary 5.1.3 *Let $(X_a)_{a\in K} \subseteq L^2(\Omega, P)$ be a random field on a commutative hypergroup K. The following statements are equal:*

(i) $(X_a)_{a\in K}$ *is weakly stationary.*
(ii) $E(X_a\overline{X_b}) = \int\limits_{\widehat{K}} \alpha(a)\overline{\alpha(b)}\, d\mu(\alpha)$ *for every* $a, b \in K$*, where* μ *is a bounded positive Borel measure on* $\widehat{K}$*.*
(iii) $X_a = \int\limits_{\widehat{K}} \alpha(a)\, dZ(\alpha)$ *for every* $a \in K$*, where* Z *is an orthogonal stochastic measure on* $\widehat{K}$*.*

Proof. "(i) ⇒ (ii)" and "(ii) ⇒ (iii)" followed by the considerations just before. Assuming (iii) we get

$$E(X_a\overline{X_b}) = \int\limits_{\widehat{K}} \alpha(a)\overline{\alpha(b)}\, d\mu(\alpha)$$

where μ is defined by $\mu(A) := \|Z(A)\|_2^2$, $A \in \mathcal{B}(\widehat{K})$. Hence by Fubini's theorem

$$E(X_a\overline{X_b}) = \int\limits_{\widehat{K}} \int\limits_{K} \alpha(y)\, d\omega(a,\tilde{b})(y)\, d\mu(\alpha) = \int\limits_{K} E(X_y\overline{X_e})\, d\omega(a,\tilde{b})(y).$$

◇

Let μ and Z be the spectral measure and the stochastic measure of the random field $(X_a)_{a\in K}$. For $a, b \in K$ define

$$\epsilon_b * X_a := \int\limits_{\widehat{K}} \alpha(a)\overline{\alpha(b)}\, dZ(\alpha)$$

and the linear extension $T_b : span\{X_a : a \in K\} \to H(X_a, K)$

$$T_b\left(\sum_{k=1}^{n} c_k X_{a_k}\right) := \sum_{k=1}^{n} c_k\, \epsilon_b * X_{a_k}. \tag{2}$$

This mapping is well defined. In fact the injectivity of the map Φ yields $\sum_{k=1}^{n} c_k \widehat{\epsilon_{a_k}} = \sum_{l=1}^{m} d_l \widehat{\epsilon_{a_l}}$ whenever $\sum_{k=1}^{n} c_k X_{a_k} = \sum_{l=1}^{m} d_l X_{a_l}$, and hence for all $\alpha \in \widehat{K}$

$$\sum_{k=1}^{n} c_k \alpha(a_k)\overline{\alpha(b)} = \sum_{\ell=1}^{m} d_\ell \alpha(a_\ell)\overline{\alpha(b)}.$$

Next we observe that T_b is continuous and $\|T_b\| \leq 1$. To prove this and the following properties of T_b it is convenient to apply the isometric isomorphism $\Phi : L^2(\widehat{K}, \mu) \to H(X_a, K)$. The operator T_b corresponds uniquely to the multiplication operator $M_b : span\{\widehat{\epsilon_{\tilde{a}}} : a \in K\} \to L^2(\widehat{K}, \mu)$ with

$$M_b(\psi) := \widehat{\epsilon_b}\psi, \qquad \psi \in span\{\widehat{\epsilon_{\tilde{a}}} : a \in K\}.$$

Since $\|\widehat{\epsilon_b}\|_\infty = 1$ we have $\|M_b\| \leq 1$ and $\|T_b\| \leq 1$. The continuous extensions of T_b and M_b to $H(X_a, K)$ and $L^2(\widehat{K}, \mu)$ are also denoted by T_b and M_b respectively. So we have the commuting diagram

$$\begin{array}{ccc} H(X_a, K) & \xrightarrow{T_b} & H(X_a, K) \\ \uparrow \Phi & & \uparrow \Phi \\ L^2(\hat{K}, \mu) & \xrightarrow[M_b]{} & L^2(\hat{K}, \mu) \end{array}$$

stating for every $\psi \in L^2(\widehat{K}, \mu)$

$$T_b(\Phi(\psi)) = \int_{\widehat{K}} \psi(\alpha)\overline{\alpha(b)}\, dZ(\alpha) = \Phi \circ M_b(\psi).$$

Theorem 5.1.4 *Let $(X_a)_{a\in K}$ be a weakly stationary random field on a commutative hypergroup K. Let Φ be the corresponding isometric isomorphism from $L^2(\widehat{K}, \mu)$ onto $H(X_a, K)$ and Z the corresponding orthogonal stochastic measure. For $b \in K$ the translation operators $T_b : H(X_a, K) \to H(X_a, K)$ have the following properties*

(1) $T_b^* = T_{\tilde{b}}$ and $T_0 = id$.
(2) The mapping $b \mapsto T_b(\Phi(\psi))$ is continuous from K into $H(X_a, K)$ for every $\psi \in L^2(\widehat{K}, \mu)$.
(3) The operators T_b are normal and commuting. The equation

$$T_b \circ T_a(\Phi(\psi)) = \int_K T_c(\Phi(\psi))\, d\omega(a, b)(c)$$

holds for every $\psi \in L^2(\widehat{K}, \mu)$ and $a, b \in K$. In particular $T_bX_a = \int_K X_c\, d\omega(a, b)(c)$.

Proof.

(1) Obviously, $T_0 = id$. $T_b^* = T_{\tilde{b}}$ holds, since $< M_b(\psi), \psi >_\mu = < \widehat{\epsilon_b}\psi, \psi >_\mu$ $= < \psi, \widehat{\epsilon_{\tilde{b}}}\psi >_\mu = < \psi, M_{\tilde{b}}\psi >_\mu$ for all $\phi, \psi \in L^2(\widehat{K}, \mu)$.

(2) We show that for each $\psi \in L^2(\widehat{K}, \mu)$, $\psi \neq 0$, the mapping $b \mapsto M_b(\psi)$ is continuous from K into $L^2(\widehat{K}, \mu)$. Given $\varepsilon > 0$ choose $C \subseteq \widehat{K}$ compact such that

$$\int_{\widehat{K}\backslash C} |\psi(\alpha)|^2 \, d\mu(\alpha) < \varepsilon/(8\|\psi\|_{2,\mu}^2).$$

For $a, b \in K$ we have

$$\begin{aligned} \|M_b(\psi) - M_a(\psi)\|_2^2 &= \int_{\widehat{K}} |\alpha(b) - \alpha(a)|^2 |\psi(\alpha)|^2 \, d\mu(\alpha) \\ &\leq \frac{\varepsilon}{2} + \int_C |\alpha(b) - \alpha(a)|^2 |\psi(\alpha)|^2 \, d\mu(\alpha). \end{aligned}$$

Refering to Theorem 3.4.9 and using the compactness of C, we obtain a neighborhood U_a of $a \in K$ such that for all $\alpha \in C$

$$|\alpha(b) - \alpha(a)|^2 < \varepsilon/(2\|\psi\|_{2,\mu}^2) \qquad \text{for all } b \in U_a,$$

and we see that $a \mapsto M_a(\psi)$ is continuous.

(3) The integral of (3) is a vector-valued integral of a continuous function on K. For $\phi, \psi \in L^2(\widehat{K}, \mu)$ we have

$$< M_b(M_a(\psi)), \phi >_\mu = \int_{\widehat{K}} \overline{\alpha(b)}\,\overline{\alpha(a)}\psi(\alpha)\overline{\phi(\alpha)} \, d\mu(\alpha)$$

$$= \int_{\widehat{K}} \int_K \overline{\alpha(c)} \, d\omega(a,b)(c) \psi(\alpha)\overline{\phi(\alpha)} \, d\mu(\alpha) = < \int_K M_c(\psi) \, d\omega(a,b)(c), \phi >_\mu .$$

Hence $M_b(M_a(\psi)) = \int_K M_c(\psi) \, d\omega(a,b)(c)$. The normality and commutativity of T_b, $b \in K$, are now obvious.

Putting $\psi = 1$ we get the additional statement. Note that $\Phi(1) = X_e$.

◇

For $\rho \in L^\infty(\widehat{K}, \mu)$ define the multiplication operator M_ρ on $L^2(\widehat{K}, \mu)$ such that $M_\rho\psi := \rho\psi$, and let $\mathcal{M} := \{M_\rho : \rho \in L^\infty(\widehat{K}, \mu)\}$. Obviously $M_b = M_{\widehat{\epsilon_b}} \in \mathcal{M}$. The set $\mathcal{M}$ is a C^*-subalgebra of $\mathcal{B}(L^2(\widehat{K}, \mu))$, that is closed with respect to the weak operator topology, i.e. $\mathcal{M}$ is a W^*-algebra, see [185, Def.4.46]. By means of the isometric isomorphism Φ the operator $M_\rho \in \mathcal{B}(L^2(\widehat{K}, \mu))$ is equivalent to the operator $T_\rho = \Phi \circ M_\rho \circ \Phi^{-1} : H(X_a, K) \to H(X_a, K)$. Applying [185,4.20] we have the following facts:

(i) $\|T_\rho\| = \|M_\rho\| = \|\rho\|_\infty$.
(ii) $T_\rho^* = T_{\overline{\rho}}$.
(iii) $T_\rho \circ T_\phi = T_{\rho\phi}$.
(iv) T_ρ is a projection if and only if ρ is a characteristic function.
(v) For the spectrum $\sigma(T_\rho)$ holds $\sigma(T_\rho) = \mathcal{R}(\rho)$, where $\mathcal{R}(\rho)$ is the essential range of ρ (see [185, 4.24 and 2.61]). In particular $\sigma(T_\rho) = \overline{\rho(\operatorname{supp}\mu)}$ if $\rho \in C^b(\widehat{K})$.

If $\rho = \chi_A$ is the characteristic function of a Borel set $A \subseteq \widehat{K}$, we define

$$E(A) := \Phi \circ M_{\chi_A} \circ \Phi^{-1}.$$

A straightforward calculation shows that $E : \mathcal{B}(\widehat{K}) \to \mathcal{B}(H(X_a, K))$ is a resolution of the identity on $\widehat{K}$, see [609]. We have

$$E(A)(\Phi(\psi)) = \Phi(\chi_A\psi) \qquad \text{for every } \psi \in L^2(\widehat{K}, \mu),$$

and hence

$$\langle E(A)(\Phi(\psi)), \Phi(\phi)\rangle = \langle \Phi(M_{\chi_A}\psi), \Phi(\phi)\rangle = \langle \chi_A\psi, \phi\rangle_\mu \quad \text{for all } \psi, \phi \in L^2(\widehat{K}, \mu).$$

For $\rho \in L^\infty(\widehat{K}, \mu)$ we have therefore

$$\int_{\widehat{K}} \rho(\alpha)\, dE(\alpha)\Phi(\psi) = \Phi(\rho\psi)$$

and

$$\langle \int_{\widehat{K}} \rho(\alpha)\, dE(\alpha)\Phi(\psi), \Phi(\phi)\rangle = \langle \rho\psi, \phi\rangle_\mu.$$

In particular follows

$$\int_{\widehat{K}} \alpha(a)\, dE(\alpha) = T_a, \qquad \int_{\widehat{K}} \alpha(a)\, dE(\alpha)X_e = X_a$$

and

$$\langle \int_{\widehat{K}} \alpha(a)\, dE(\alpha)X_e, X_b\rangle = \langle X_a, X_b\rangle$$

Thus we can add to the statements before:

Theorem 5.1.5 *Let $(X_a)_{a\in K}$ be a weakly stationary random field on a commutative hypergroup K. Then there exists a resolution $E : \mathcal{B}(\widehat{K}) \to \mathcal{B}(H(X_a, K))$ of the identity on $\widehat{K}$ such that*

$$T_a = \int_{\widehat{K}} \alpha(a)\, dE(\alpha), \tag{3}$$

where T_a, $a \in K$, is the translation operator induced by (2).

Moreover we have for each $A \in \mathcal{B}(\widehat{K})$

$$E(A)X_e = Z(A)$$

and hence $\langle E(A)X_e, X_e\rangle = \mu(A)$. The resolution E of the identity is uniquely determined by (3).

Proof. Since $X_e = \Phi(1)$ we obtain

$$E(A)\Phi(1) = \Phi(\chi_A) = Z(A)$$

and $\mu(A) = \|Z(A)\|_2^2 = \langle E(A)X_e, E(A)X_e\rangle = \langle E(A)X_e, X_e\rangle$.

To prove the uniqueness assume that

$$\int\limits_{\widehat{K}} \alpha(a)\, dE_1(\alpha) = T_a = \int\limits_{\widehat{K}} \alpha(a)\, dE_2(\alpha)$$

Similar as in the proof of Theorem 5.1.4 approximate χ_A by $\psi_n \in span\{\widehat{\epsilon_a} : a \in K\}$ in $L^2(\widehat{K}, \mu_1)$ and $L^2(\widehat{K}, \mu_2)$ simultaneously. Then we get

$$\int\limits_{\widehat{K}} \psi_n(\alpha)\, dE_1(\alpha) = \int\limits_{\widehat{K}} \psi_n(\alpha)\, dE_2(\alpha),$$

and $E_1(A) = E_2(A)$ follows. ◇

Corollary 5.1.6 *Let $(X_a)_{a\in K} \subseteq L^2(\Omega, P)$ be a random field on a commutative hypergroup K. The following statement (iv) is equivalent to the three conditions of Corollary 5.1.3.*

(iv) $T_a = \int\limits_{\widehat{K}} \alpha(a)\, dE(\alpha)$ for every $a \in K$, where E is a resolution of the identity on $\widehat{K}$, and T_a is the translation operator induced by (2).

Proof. It remains to show that (iv) implies one of the conditions of Corollary 5.1.3. We have

$$\begin{aligned}\langle X_a, X_b\rangle &= \langle \int\limits_{\widehat{K}} \alpha(a)\, dE(\alpha)X_e, X_b\rangle = \int\limits_{\widehat{K}} \alpha(a)\Phi^{-1}X_e(\alpha)\overline{\Phi^{-1}X_b(\alpha)}\, d\mu(\alpha)\\ &= \int\limits_{\widehat{K}} \alpha(a)\overline{\alpha(b)}\, d\mu(\alpha).\end{aligned}$$

◇

Remark. We have seen that every weakly stationary random field $(X_a)_{a\in K}$ on a commutative hypergroup K defines a family $(T_a)_{a\in K}$ of operators on $H(X_a, K)$ with the properties (1), (2) and (3) of Theorem 5.1.4. We can find X_a from T_a by putting $X_a := T_a X_e$.
Conversely for a Hilbert space $H \subseteq L^2(\Omega, P)$ we have:
If $(T_a)_{a\in K}$ is a family of operators on H enjoying the properties

(1) $T_e = id$, $T_b^* = T_{\tilde{b}}$, $\|T_b\| \leq 1$.
(2) $b \mapsto T_b(X)$ is continuous for each $X \in H$.
(3) $T_b \circ T_a(X) = \int_K T_c(X)\, d\omega(a,b)(c)$ for every $X \in H$,

then each $X \in H$ defines a weakly stationary random field $(X_a)_{a\in K}$ on K by $X_a := T_a(X)$. In fact $d(a,b) = \langle T_a(X), T_b(X)\rangle$ is a bounded and continuous function on $K \times K$ and

$$d(a,b) = \langle T_a(X), T_b(X)\rangle = \langle T_{\tilde{b}} \circ T_a(X), T_e(X)\rangle = \int_K d(c,e)\, d\omega(a,\tilde{b})(c).$$

2 Occurrence of Random Fields on Hypergroups

We give only a short list of occasions where one meets weakly stationary random fields on a commutative hypergroup. Here we want to recall that every locally compact Abelian group is a commutative hypergroup.

2.1 *Real and imaginary parts*

Let $(Y_n)_{n\in\mathbb{Z}}$ be a weakly stationary complex-valued process on $\mathbb{Z}$ with symmetry, that is $Y_{-n} = \overline{Y}_n$. The random sequence $(U_n)_{n\in\mathbb{N}_0}$ of the real parts $U_n = \text{Re } Y_n = \frac{1}{2}(Y_n + Y_{-n})$, $n \in \mathbb{N}_0$ is no longer weakly stationary in the usual sense. However, one can easily check that

$$E(U_m U_n) = \frac{1}{2}\, E(U_{n+m} U_0) + \frac{1}{2}\, E(U_{|n-m|} U_0).$$

That means $(U_n)_{n\in\mathbb{N}_0}$ is a weakly stationary random field on the polynomial hypergroup $\mathbb{N}_0$ generated by the Chebyshev polynomials $T_n(x)$ of the first kind, see subsection 1.2.1.

Denote the imaginary part of Y_n by

$$V_n = \text{Im } Y_n = \frac{1}{2i}\,(Y_n - Y_{-n}).$$

Since $V_0 = 0$ we have

$$E(U_n U_0) + iE(V_n U_0) = E(Y_n \overline{Y}_0) = \int_{-\pi}^{\pi} \cos(nt)\, d\mu(t) + i\int_{-\pi}^{\pi} \sin(nt)\, d\mu(t),$$

where $\mu \in M^+(\,]-\pi, \pi])$ is the spectral measure of $(Y_n)_{n\in\mathbb{Z}}$.

The random sequence $(U_n)_{n\in\mathbb{N}_0}$ is weakly stationary with respect to $T_n(x) = \cos(nt)$, where $x = \cos t$ for $x \in [-1,1]$, $t \in [0,\pi]$. Hence we have a unique spectral representation

$$E(U_n U_0) = \int_0^\pi \cos(nt)\, d\nu(t),$$

$\nu \in M^+([0,\pi])$. Now it is clear that $\nu = \mu|[0,\pi] + \mu_1|\,]0,\pi[$ where μ_1 is the image measure of μ under the mapping $t \mapsto -t$. Further we observe that

$$E(U_m U_n) + E(V_m V_n) = \operatorname{Re} E(Y_m \overline{Y}_n)$$

$$= \int_{-\pi}^\pi \cos((m-n)t)\, d\mu(t) = \int_0^\pi \cos((m-n)t)\, d\nu(t)$$

$$= \int_0^\pi \cos(mt)\, \cos(nt)\, d\nu(t) + \int_0^\pi \sin(mt)\, \sin(nt)\, d\nu(t).$$

Since $E(U_m U_n) = \int\limits_0^\pi \cos(mt)\cos(nt)\, d\nu(t)$, it follows that

$$E(V_m V_n) = \int_0^\pi \sin(mt)\, \sin(nt)\, d\nu(t).$$

Define for $n \in \mathbb{N}_0$

$$X_n := \frac{1}{n+1}\, V_{n+1} = \frac{1}{n+1} \operatorname{Im} Y_{n+1}.$$

Then

$$E(X_m X_n) = \int_0^\pi \frac{\sin((m+1)t)}{(m+1)\sin t}\, \frac{\sin((n+1)t)}{(n+1)\sin t}\, (\sin t)^2\, d\nu(t).$$

By Corollary 5.1.4 we see that $(X_n)_{n\in\mathbb{N}_0}$ is a weakly stationary random field on the polynomial hypergroup $\mathbb{N}_0$ generated by the Chebyshev polynomials $R_n^{(\frac{1}{2},\frac{1}{2})}(x)$ of the second kind. Recall that

$$R_n^{(\frac{1}{2},\frac{1}{2})}(x) = \frac{\sin((n+1)t)}{(n+1)\sin t}, \qquad x = \cos t,$$

compare subsection 4.5.1.

The corresponding spectral measure is $(\sin t)^2 d\nu(t)$.

2.2 *Arithmetic mean estimates*

Consider a weakly stationary process $(X_n)_{n\in\mathbb{Z}}$ with constant mean $M = E(X_n)$, $n \in \mathbb{Z}$. We assume that $M = 0$. The usual estimates of the mean are given by

$$Y_n := \frac{1}{2n+1} \sum_{k=-n}^{n} X_k, \qquad n \in \mathbb{N}_0.$$

The stochastic process $(Y_n)_{n\in\mathbb{N}_0}$ is no longer weakly stationary in the usual sense related to the group $\mathbb{Z}$, but it is a weakly stationary random field on the polynomial hypergroup $\mathbb{N}_0$ generated by the Jacobi polynomials $R_n^{(\alpha,\beta)}(x)$ with parameters $\alpha = \frac{1}{2}$, $\beta = -\frac{1}{2}$. In fact, let ν be the spectral measure of the process $(X_n)_{n\in\mathbb{Z}}$. Then we have for the covariance of $(Y_n)_{n\in\mathbb{N}_0}$:

$$\begin{aligned}
d(n,m) &= E\left(\left(\frac{1}{2n+1}\sum_{k=-n}^{n} X_k\right)\left(\frac{1}{2m+1}\sum_{\ell=-m}^{m}\overline{X_\ell}\right)\right) \\
&= \int_0^{2\pi}\left(\frac{1}{2n+1}\sum_{k=-n}^{n} e^{ikt}\right)\left(\frac{1}{2m+1}\sum_{\ell=-m}^{m} e^{i\ell t}\right) d\nu(t) \\
&= \int_0^{2\pi} \frac{\sin\left((2n+1)\frac{t}{2}\right)}{(2n+1)\sin\left(\frac{t}{2}\right)}\, \frac{\sin\left((2m+1)\frac{t}{2}\right)}{(2m+1)\sin\left(\frac{t}{2}\right)}\, d\nu(t) \\
&= \int_0^{2\pi} R_n^{(\frac{1}{2},-\frac{1}{2})}(\cos t) R_m^{(\frac{1}{2},-\frac{1}{2})}(\cos t)\, d\nu(t) \\
&= \sum_{k=|n-m|}^{n+m} g(n,m;k) \int_0^{2\pi} R_k^{(\frac{1}{2},-\frac{1}{2})}(\cos t)\, d\nu(t) \\
&= \sum_{k=|n-m|}^{n+m} g(n,m;k)\, d(k,0),
\end{aligned}$$

where $g(n,m;k)$ are the linearization coefficients of the product $R_n^{(\frac{1}{2},-\frac{1}{2})} \cdot R_m^{(\frac{1}{2},-\frac{1}{2})}$.

2.3 *Other mean estimates*

In an obvious way we can generalize the construction of 5.2.2. Given again a weakly stationary process $(X_n)_{n\in\mathbb{Z}}$ on the additive group $\mathbb{Z}$, we get a weakly stationary random field $(Y_n)_{n\in\mathbb{N}_0}$ on the polynomial hypergroup generated by an orthogonal sequence $(R_k)_{k\in\mathbb{N}_0}$ by setting

$$Y_n := \sum_{k=-n}^{n} a_{n,k} X_k,$$

where $R_n(\cos t) = \sum_{k=-n}^{n} a_{n,k} e^{ikt}$ with $a_{n,k} = a_{n,-k}$ for $k = -n, \ldots, n$. Note that $1 = R_n(1) = \sum_{k=-n}^{n} a_{n,k}$.

As before follows

$$d(n,m) = E\left(\left(\sum_{k=-n}^{n} a_{n,k} X_k\right)\left(\sum_{k=-m}^{m} a_{m,k} \overline{X_k}\right)\right) = \sum_{k=|n-m|}^{n+m} g(n,m;k)\, d(k,0),$$

where the $g(n,m;k)$ are now the linearization coefficients of $R_n \cdot R_m$.

For example, formulas (2.7), (3.15) and Gegenbauer's formula (7.5) of [18] yield for the n-th Jacobi polynomial with $\lambda > -\frac{1}{2}$:

$$R_{2n}^{(\lambda-\frac{1}{2},\lambda-\frac{1}{2})}\left(\cos\frac{t}{2}\right) = \frac{(2n)!}{(2\lambda)_{2n}} \sum_{k=0}^{2n} \frac{(\lambda)_{2n-k}(\lambda)_k}{(2n-k)!\; k!} \cos\left((n-k)t\right)$$

$$= \frac{(2n)!}{(2\lambda)_{2n}} \sum_{k=-n}^{n} \frac{(\lambda)_{n+k}(\lambda)_{n-k}}{(n+k)!(n-k)!}\, e^{ikt},$$

where as usual $(a)_n = 1\cdot 2\cdot\cdots\cdot(a+n-1)$ denotes the Pochhammer symbol. Using (3.13) in [18] we have

$$R_n^{(\lambda-\frac{1}{2},\lambda-\frac{1}{2})}(\cos t) = \sum_{k=-n}^{n} a_{n,k} e^{ikt},$$

where $a_{n,k} = \frac{(\lambda)_{n+k}(\lambda)_{n-k}(2n)!}{(n+k)!(n-k)!(2\lambda)_{2n}}$.

For $\lambda = 1$ we get the arithmetic means of 5.2.2.

The center of the averages Y_n in 5.2.2 and 5.2.3 is the zero index. Finding another index $\ell \in \mathbb{Z}$ and defining $Y_n^\ell := \sum_{k=-n}^{n} a_{n,k} X_{k+\ell}$ we have $E(Y_n^\ell \cdot \overline{Y_m^\ell}) = E(Y_n \cdot \overline{Y_m})$. Hence the covariance of the averages is independent of the average-center.

2.4 *Coefficients of random orthogonal expansions for density estimation*

Suppose that the distribution of a random variable X is absolutely continuous with respect to a positive Borel measure π on the interval $[-1,1]$, i.e.

$$P(X \in A) = \int_A f(x)\, d\pi(x)$$

for all Borel sets $A \subseteq [-1,1]$, where $f \in L^1([-1,1],\pi)$, $f \geq 0$.

We assume that $|\text{supp}\,\pi| = \infty$. Consider the uniquely determined sequence $(p_n)_{n\in\mathbb{N}_0}$ of polynomials that are orthonormal with respect to π and let as before $R_n(x) = p_n(x)/p_n(1)$. We assume that $(R_n)_{n\in\mathbb{N}_0}$ generates a polynomial hypergroup on $\mathbb{N}_0$.

Given independent random variables $X_1, X_2, \ldots, X_N$ equally distributed as X the unknown density function $f(x)$ can be estimated by the random orthogonal expansion

$$f_N(\omega; x) := \sum_{k=0}^{q(N)} a_{N,k}\; c_{N,k}(\omega)\; p_k(x).$$

$q(N)$ is the truncation point, and $a_{N,k}$ are numerical coefficients to be chosen in an appropriate manner, see [176], [550] or [433]. The random coefficients are given by

$$c_{N,k}(\omega) := \frac{1}{N}\sum_{j=1}^{N} p_k(X_j(\omega)).$$

Define $C_k := c_{N,k}/p_k(1)$ for every $k \in \mathbb{N}_0$. The family $(C_k)_{k\in\mathbb{N}_0}$ is a weakly stationary random field on the polynomial hypergroup $\mathbb{N}_0$ generated by $(R_k)_{k\in\mathbb{N}_0}$. In fact we have $C_n = \frac{1}{N}\sum_{j=1}^{N} R_n(X_j)$ and therefore

$$\begin{aligned}
E(C_n \cdot C_m) &= \frac{1}{N^2}\sum_{i,j=1}^{N} E\left(R_n(X_i)\, R_m(X_j)\right) \\
&= \frac{1}{N^2}\sum_{i,j=1}^{N} \int_{-1}^{1} R_n(x)R_m(x)f(x)d\pi(x) \\
&= \int_{-1}^{1} R_n(x)R_m(x)f(x)\,d\pi(x) \\
&= \sum_{k=|n-m|}^{n+m} g(n,m;k)\int_{-1}^{1} R_k(x)R_0(x)f(x)\,d\pi(x) \\
&= \sum_{k=|n-m|}^{n+m} g(n,m;k)E(C_k \cdot C_0).
\end{aligned}$$

Estimating the density function f by random orthogonal expansion is hence strongly related to estimating the spectral measure $f\pi$ of the random field $(C_k)_{k\in\mathbb{N}_0}$ where $(C_k)_{k\in\mathbb{N}_0}$ is weakly stationary with respect to $R_k = p_k/p_k(1)$. It should be noted that the mean $E(C_k)$ is not constant.

2.5 *Stationary radial stochastic processes on homogeneous trees*

Polynomials connected with homogeneous trees we have already studied in subsection 4.5.7.2. We denote by $\mathcal{X}$ a homogeneous tree of degree $q \geq 1$ with metric d. Let G be the isometry group of $\mathcal{X}$, $t_0 \in \mathcal{X}$ a fixed point of $\mathcal{X}$ and let H be the stabiliser of t_0 in G. We identify $\mathcal{X}$ with the coset space G/H and call mappings on $\mathcal{X}$ **radial** if they depend only on $|t| := d(t, t_0)$. A square integrable complex-valued stochastic process $(X_t)_{t\in\mathcal{X}}$ is called **stationary**, if there exists a function $\phi : \mathbb{N}_0 \to \mathbb{R}$ such that

$$E(X_s\overline{X_t}) = \phi(d(s,t))$$

for all s and t in $\mathcal{X}$, see [11], [12]. It is known, see [12], that one has the spectral representation

$$E(X_s\overline{X_t}) = \int_{-1}^{1} R_{d(s,t)}(x)\, d\mu(x)$$

where $(R_n(x))_{n\in\mathbb{N}_0}$ are the orthogonal polynomials (Cartier-Dunau polynomials) corresponding to homogeneous trees with recurrence coefficients $a_n = \frac{q}{q+1}$ and $c_n = \frac{1}{q+1}$, and μ is a unique measure on $[-1,1]$.

Now we assume that the stationary stochastic process $(X_t)_{t\in\mathcal{X}}$ is radial, i.e. $X_t = X_s$ if $|t| = |s|$. Putting $X_n := X_t$ whenever $|t| = n$ we get a well-defined random field on $\mathbb{N}_0$. Since

$$\int_H R_{d(h(s),t)}(x)\, d\beta(h) = R_{|s|}(x)R_{|t|}(x),$$

where β is the Haar measure on the compact stabiliser H, and $h(s)$ is the action of $h \in H$ on $\mathcal{X} \cong G/H$, the above spectral representation gives

$$E(X_n\overline{X_m}) = \int_{-1}^{1} R_n(x)R_m(x)\, d\mu(x).$$

By Corollary 5.1.4 we obtain that $(X_n)_{n\in\mathbb{N}_0}$ is weakly stationary on the hypergroup $\mathbb{N}_0$.

If the stationary process $(X_t)_{t\in\mathcal{X}}$ on the homogeneous tree fails to be radial, a symmetrization procedure yields a radial one. In fact set

$$Y_t = \int_H X_{h(t)}\, d\beta(h).$$

Clearly Y_t is radial and with the spectral measure μ from above we have

$$\begin{aligned} E(Y_s\overline{Y_t}) &= \int\limits_H \int\limits_H E(X_{h(s)}\overline{X_{k(t)}})\, d\beta(h)\, d\beta(k) \\ &= \int\limits_H \int\limits_H \int\limits_{-1}^{1} R_{d(h(s),k(t))}(x)\, d\mu(x)\, d\beta(h)\, d\beta(k) \\ &= \int\limits_{-1}^{1} \int\limits_H \int\limits_H R_{d(k^{-1}h(s),t)}(x)\, d\beta(h)\, d\beta(k)\, d\mu(x) \\ &= \int\limits_{-1}^{1} R_{|s|}(x) R_{|t|}(x)\, d\mu(x). \end{aligned}$$

That means $(Y_n)_{n\in\mathbb{N}_0}$ is a weakly stationary random field on the hypergroup $\mathbb{N}_0$.

2.6 *Differences in sequences with stationary increments*

Random processes with stationary increments play an important role in applications, e. g. Brownian motion or Poisson process. A random sequence $(X_k)_{k\in\mathbb{Z}}$ is called a sequence with stationary increments if $E(X_{n+k} - X_k)$ depend only on $n \in \mathbb{Z}$ and $E\left((X_{n_1+k} - X_k)(\overline{X_{n_2+k}} - \overline{X_k})\right)$ depend only on $n_1, n_2 \in \mathbb{Z}$, see [756,section 23]. It is readily seen, see [765,(4.227)], that

$$X_{n+k} - X_k = \int\limits_{-\pi}^{\pi} e^{ikt}\, \frac{e^{int} - 1}{e^{it} - 1}\, dZ(t),$$

where Z is an orthogonal stochastic measure on $]-\pi, \pi]$. Replacing $k = -m$ and $n = 2m + 1$ for arbitrary $m \in \mathbb{N}_0$ we get

$$X_{m+1} - X_{-m} = \int\limits_0^{\pi} \frac{\sin\left((2m+1)\frac{t}{2}\right)}{\sin\left(\frac{t}{2}\right)}\, d\tilde{Z}(t),$$

where $\tilde{Z}$ is the orthogonal stochastic measure on $[0, \pi]$ defined by

$$\tilde{Z} = Z|\,[0,\pi] + Z_1, \quad Z_1(]a,b]) = Z([-b,-a[) \qquad \text{for }]a,b] \subseteq]0,\pi[, \quad Z_1(\{0\}) = 0.$$

Putting $Y_m := \frac{1}{2m+1}(X_{m+1} - X_{-m})$ we have a weakly stationary random field $(Y_m)_{m\in\mathbb{N}_0}$ on the polynomial hypergroup $\mathbb{N}_0$ generated by the Jacobi polynomials $R_m^{(\frac{1}{2},-\frac{1}{2})}(x)$.

2.7 Continuous arithmetic means

Of course there is a continuous version of the estimates of 5.2.2. Let $(X_t)_{t\in\mathbb{R}}$ be a continuous weakly stationary process with constant mean $M = E(X_t)$, $t \in \mathbb{R}$. The mean M is usually estimated by

$$Y_0 := X_0 \qquad \text{and} \qquad Y_r := \frac{1}{2r}\int_{-r}^{r} X_t\, dt, \quad r > 0.$$

If μ is the spectral measure of $(X_t)_{t\in\mathbb{R}}$, we obtain for $r, s > 0$

$$\begin{aligned} E(Y_r\overline{Y_s}) &= \int_{\mathbb{R}} \frac{1}{2r}\int_{-r}^{r} e^{iu\lambda}\, du \frac{1}{2s}\int_{-s}^{s} e^{-iv\lambda}\, dv\, d\mu(\lambda) \\ &= \int_{\mathbb{R}} \frac{\sin(\lambda r)}{\lambda r}\frac{\sin(\lambda s)}{\lambda s}\, d\mu(\lambda) = \int_{\mathbb{R}_0^+} \varphi_r^{(\frac{1}{2})}(\lambda)\varphi_s^{(\frac{1}{2})}(\lambda)\, d\tilde{\mu}(\lambda), \end{aligned}$$

where $\varphi_r^{(\frac{1}{2})}(\lambda) = \frac{\sin(\lambda r)}{\lambda r}$ are the characters of the Bessel-Kingman hypergroup with parameter $\alpha = 1/2$, see subsection 1.2.2 (or Theorem 8.2.1 in Ch.8), and $\tilde{\mu} = \mu_1 + \mu_2$ with $\mu_1 = \mu\,|\mathbb{R}_0^+$ and μ_2 is the image measure of $\mu\,|\,]-\infty, 0[$ under the mapping $t \mapsto -t$. Hence $(Y_t)_{t\in\mathbb{R}_0^+}$ is a weakly stationary random field on the Bessel-Kingman hypergroup $\mathbb{R}_0^+$ with parameter $\alpha = 1/2$ with spectral measure $\tilde{\mu}$.

2.8 Other continuous means

Generalizing the preceding simple mean estimates we suppose that a family $(F_r)_{r\geq 0}$ of continuous functions $F_r : \mathbb{R}_0^+ \to \mathbb{C}$ is given such that

(i) $(F_r)_{r\in\mathbb{R}_0^+}$ is exactly the character space of a commutative hypergroup on $\mathbb{R}_0^+$.

(ii) the F_r have a Laplace representation of the following kind

$$F_r(x) = \int_{-r}^{r} e^{ix\tau}\, da_r(\tau),$$

where a_r is a symmetric complex-valued Borel measure on $[-r, r]$ (symmetric means $a_r(A) = a_r(-A)$).

Since $F_r(0) = 1$, we necessarily have $a_r([-r, r]) = 1$. If $(X_t)_{t\in\mathbb{R}}$ is a continuous weakly stationary process with constant mean $M = E(X_t)$, $t \in \mathbb{R}$, the mean can be estimated by

$$Y_0 := X_0 \qquad \text{and} \qquad Y_r := \int_{-r}^{r} X_t \, da_r(t), \qquad r > 0.$$

As before we obtain

$$E(Y_r \overline{Y_s}) = \int_{\mathbb{R}_0^+} F_r(\lambda) F_s(\lambda) \, d\tilde{\mu}(\lambda).$$

Therefore $(Y_t)_{t \in \mathbb{R}_0^+}$ is a weakly stationary random field on a Sturm-Liouville hypergroup on $\mathbb{R}_0^+$, see section 8.2. Concerning the Laplace representation we refer to [96,Definition 3.5.56].

2.9 *Isotropic random fields*

A homogeneous random field $(X_a)_{a \in \mathbb{R}^d}$ on the additive group $\mathbb{R}^d$ is called **isotropic** if the covariance function depends only on the radius $r = \|a\|_2$ of the vector a, see [756]. Even if the random field depends only on the radius, we shall call $(X_a)_{a \in \mathbb{R}^d}$ **radial**. Hence putting $Y_r := X_a$ whenever $r = \|a\|_2$ we obtain a well-defined weakly stationary random field on the Bessel-Kingman hypergroup $\mathbb{R}_0^+$ with parameter $\alpha = \frac{d}{2} - 1$. To check this statement let β be the normalized Haar measure on the compact group $SO(d)$. The spectral measure $\mu \in M^+(\mathbb{R}^d)$ is invariant under the action of $SO(d)$ and so $\tilde{\mu}([0,t[) := \mu(U_t(0))$, where $t > 0$ and $U_t(0) := \{x \in \mathbb{R}^d : \|x\|_2 < t\}$, defines a bounded positive Borel measure on $\mathbb{R}_0^+$ that is the spectral measure of $(Y_t)_{t \in \mathbb{R}_0^+}$. In fact

$$\begin{aligned}
E(Y_r \overline{Y_s}) &= \int_{SO(d)} \int_{SO(d)} E(X_{h(a)} \overline{X_{k(b)}}) \, d\beta(h) \, d\beta(k) \\
&= \int_{SO(d)} \int_{SO(d)} \int_{\mathbb{R}^d} e^{i\langle h(a) - k(b), x \rangle} \, d\mu(x) \, d\beta(h) \, d\beta(k) \\
&= \int_{\mathbb{R}^d} \int_{SO(d)} \int_{SO(d)} e^{i\langle k^{-1}h(a) - b, k^{-1}x \rangle} \, d\beta(h) \, d\beta(k) \, d\mu(x) \\
&= \int_{\mathbb{R}_0^+} \varphi_r^{(\alpha)}(\lambda) \varphi_s^{(\alpha)}(\lambda) \, d\tilde{\mu}(\lambda),
\end{aligned}$$

see subsection 1.2.2 or example 1 of subsection 8.1.2, where in the above calculation $\|a\|_2 = r$, $\|b\|_2 = s$.

Remark:

(1) We point out that these important examples of 5.2.9 are a special case of a general theorem concerning weakly stationary random fields on orbit spaces of locally compact groups.

(2) In the next chapter we will restrict our investigations of weakly stationary random fields to the analysis of weakly stationary random sequences, where the parameter set $\mathbb{N}_0$ is equipped with a polynomial hypergroup structure. Such sequences can be used in the analysis of time series. Concerning further results for weakly stationary random fields we refer to [319], [443] and [444].

Chapter 6

Weakly stationary random sequences on a polynomial hypergroup

This chapter deals with problems of time series analysis such as estimation of mean, covariance or spectral function as well as filtering, prediction, extrapolation and interpolation in the context of weakly stationary random fields on a polynomial hypergroup on $\mathbb{N}_0$. It seems to be suitable to call a random field on $\mathbb{N}_0$ a random sequence. We also call a weakly stationary random sequence on the polynomial hypergroup $\mathbb{N}_0$ generated by $(R_n)_{n\in\mathbb{N}_0}$ briefly a **weakly $\mathbf{R_n}$-stationary random sequence**. We start by generalizing the concepts of moving average and autoregression.

1 Moving Averages and Autoregression

A sequence $(Z_n)_{n\in\mathbb{N}_0}$ of square integrable random variables is called white noise with respect to the polynomial hypergroup on $\mathbb{N}_0$ generated by $(R_n)_{n\in\mathbb{N}_0}$, briefly **$\mathbf{R_n}$-white noise**, if $E(Z_n) = 0$, $E(Z_n\overline{Z_m}) = 0$ for $n \neq m$ and $E(Z_n\overline{Z_n}) = \frac{1}{h(n)} = g(n,n;0)$. In this case the spectral measure is the orthogonalization measure π on D_s of the orthogonal polynomial sequence $(R_n)_{n\in\mathbb{N}_0}$. In fact,

$$E(Z_n\overline{Z_m}) = \int_S R_n(x)R_m(x)\,d\pi(x),$$

where $S = \operatorname{supp}\pi \subseteq D_s$.

Given a family $(X_k)_{k\in\mathbb{Z}}$ of orthogonal random variables with finite second moments μ_k^2 such that $\sum\limits_{k=-\infty}^{\infty} \mu_k^2 < \infty$, and given a sequence $(x_k)_{k\in\mathbb{Z}}$ of points $x_k \in D_s$ we can form a weakly R_n-stationary random sequence by putting for $n \in \mathbb{N}_0$

$$Y_n := \sum_{k=-\infty}^{\infty} R_n(x_k)X_k. \tag{1}$$

In fact Y_n is a well defined square integrable random variable and for the covariance we have

$$E(Y_n\overline{Y_m}) = \sum_{k=-\infty}^{\infty} \mu_k^2 R_n(x_k) R_m(x_k) = \sum_{j=|n-m|}^{n+m} g(n,m;j)\, E(Y_j\overline{Y_0}).$$

Therefore $(Y_n)_{n\in\mathbb{N}_0}$ is weakly stationary with respect to $(R_n)_{n\in\mathbb{N}_0}$. It is called $\mathbf{R_n}$**-oscillation**. It is readily shown that

$$d(n) = \sum_{k=-\infty}^{\infty} \mu_k^2 R_n(x_k) = \int_{D_s} R_n(x)\, d\mu(x),$$

where $\mu = \sum_{k=-\infty}^{\infty} \mu_k^2 \epsilon_{x_k}$ is the spectral measure.

We will take an interest mainly in moving averages and autoregression. They depend heavily on the translation operators T_m, which we have already introduced in section 5.1. Since in case of polynomial hypergroup the action of the operators T_m can be formulated more directly we repeat their definition. For any sequence $(Y_n)_{n\in\mathbb{N}_0}$ of square integrable random variables we write

$$\epsilon_n * Y_m \;=\; \sum_{k=|n-m|}^{n+m} g(n,m;k)\, Y_k$$

and for the linear extension

$$T_n\Big(\sum_{k=1}^{m} c_k Y_{m_k}\Big) \;=\; \sum_{k=1}^{m} c_k\ \epsilon_n * Y_{m_k}.$$

The continuous extension to $H(Y_n,\mathbb{N}_0)$ is denoted by T_n again. Of course all the statements of Theorem 5.1.4 hold.

Given some R_n-white noise $(Z_n)_{n\in\mathbb{N}_0}$, a sequence $(X_n)_{n\in\mathbb{N}_0}$ is called a moving average sequence on $\mathbb{N}_0$ generated by $(R_n)_{n\in\mathbb{N}_0}$, briefly $\mathbf{R_n}$**-moving average sequence** if there exists $(a_n)_{n\in\mathbb{N}_0} \in l^2(h)$ such that for every $n \in \mathbb{N}_0$

$$X_n := \sum_{k=0}^{\infty} a_k\ \epsilon_n * Z_k\ h(k). \tag{2}$$

Firstly we note that our definition of $(X_n)_{n\in\mathbb{N}_0}$ is correct. Since

$$\|\, X_0 \,\|_2^2 = \sum_{k=0}^{\infty} \|\, a_k Z_k h(k) \,\|_2^2 = \sum_{k=0}^{\infty} |a_k|^2 h(k) \ < \infty,$$

we have $X_0 \in L^2(\Omega, P)$, and thus $X_n = T_n(X_0) \in L^2(\Omega, P)$. Furthermore

$$E(X_n\overline{X}_m) = E(T_n(X_0)\overline{T_m(X_0)}) = E(T_m(T_n(X_0))\overline{X}_0)$$

$$= \sum_{k=|n-m|}^{n+m} g(n,m;k)\, E(T_k(X_0)\overline{X}_0)$$

i.e. $(X_n)_{n\in\mathbb{N}_0}$ is weakly R_n-stationary. Further note that $\epsilon_n * Z_k = \epsilon_k * Z_n$. Moreover $E(X_n) = 0$, since $E(Z_n) = 0$ for each $n \in \mathbb{N}_0$. Hence $(X_n)_{n\in\mathbb{N}_0}$ is a centered random sequence.

Lemma 6.1.1 *Let $(Z_n)_{n\in\mathbb{N}_0}$ be R_n-white noise and $(a_n)_{n\in\mathbb{N}_0} \in l^2(h)$. For the R_n-moving average sequence $(X_n)_{n\in\mathbb{N}_0}$ holds*

$$X_n = \sum_{k=0}^{\infty} a_k\, \epsilon_n * Z_k\, h(k) = \sum_{k=0}^{\infty} \epsilon_n * a(k)\, Z_k\, h(k).$$

Proof. For $l \in \mathbb{N}_0$ set $\delta_l(j) = \delta_{l,j}\,\frac{1}{h(l)}$, $j \in \mathbb{N}_0$. Then

$$\epsilon_n * \delta_l(k) = g(n,k;l)\,\frac{1}{h(l)} = \sum_{j=|n-k|}^{n+k} g(n,k;j) < Z_j, Z_l > = < \epsilon_n * Z_k, Z_l > .$$

Hence using Theorem 2.2.6 we obtain

$$< X_n, Z_l > = \sum_{k=0}^{\infty} a_k < \epsilon_n * Z_k, Z_l > h(k) = \sum_{k=0}^{\infty} a_k\, \epsilon_n * \delta_l(k)\, h(k)$$

$$= \sum_{k=0}^{\infty} \epsilon_n * a(k)\, \delta_l(k)\, h(k) = \sum_{k=0}^{\infty} \epsilon_n * a(k) < Z_k, Z_l > h(k)$$

$$= < \sum_{k=0}^{\infty} \epsilon_n * a(k)\, Z_k h(k), Z_l > .$$

Since the linear span of the Z_l's is dense in $H(Z_n, \mathbb{N}_0)$ the assertion follows. ◇

Now it is easy to determine the covariance-sequence and the spectral measure. For every $n \in \mathbb{N}_0$ we have by means of Lemma 6.1.1:

$$d(n) = E(X_n\overline{X}_0) = \sum_{j,k=0}^{\infty} \epsilon_n * a(k)\, \overline{a}_j\, E(Z_k\overline{Z}_j)\, h(k)\, h(j)$$

$$= \sum_{k=0}^{\infty} \epsilon_n * a(k)\, \overline{a}_k\, h(k) = a * a^{\star}(n),$$

where $a^\star(n) = \overline{a(n)}$ and $a * a^\star(n)$ is the convolution of two sequences in $l^2(h)$, compare Proposition 2.4.3.

Applying Proposition 3.3.10 we see that $|\mathcal{P}(a)|^2(x)\, d\pi(x)$ is the spectral measure, i.e.

$$d(n) = a * a^\star(n) = \int_S R_n(x)\, |\mathcal{P}(a)|^2(x)\, d\pi(x).$$

Note that $\mathcal{P}(a) = \lim\limits_{N\to\infty} \sum\limits_{k=0}^{N} a_k R_k h(k)$, where the limit has to be taken in $L^2(D_s, \pi)$.

Putting together we have the following results for R_n-moving average sequences.

Theorem 6.1.2 *Let $(Z_n)_{n\in\mathbb{N}_0}$ be R_n-white noise and $a = (a_n)_{n\in\mathbb{N}_0} \in l^2(h)$. For the R_n-moving average sequence $(X_n)_{n\in\mathbb{N}_0}$ defined by*

$$X_n = \sum_{k=0}^{\infty} a_k\, \epsilon_n * Z_k\, h(k)$$

hold:

(i) $X_n = \sum\limits_{k=0}^{\infty} \epsilon_n * a(k)\, Z_k\, h(k)$.

(ii) $d(n) = a * a^\star(n)$ *for every* $n \in \mathbb{N}_0$.

(iii) $d(n,m) = \int_{D_s} R_n(x)\, R_m(x)\, |\mathcal{P}(a)|^2(x)\, d\pi(x)$.

If the R_n-moving average sequence $(X_n)_{n\in\mathbb{N}_0}$ is of finite order, say $q \in \mathbb{N}$, that means $a_q \neq 0$, $a_{q+1} = 0$, $a_{q+2} = 0, \ldots$, the spectral density $f(x)$ with respect to π is a polynomial of degree $2q$,

$$f(x) = \left|\sum_{k=0}^{q} a_k R_k(x)\, h(k)\right|^2.$$

We start the investigation of autoregression with the simplest case, a first order R_n-autoregressive random sequence. Assume again that $(Z_n)_{n\in\mathbb{N}_0}$ is R_n-white noise. A random sequence $(X_n)_{n\in\mathbb{N}_0}$ is called a **first order $\mathbf{R_n}$-autoregressive random sequence** if it satisfies for any $n \in \mathbb{N}_0$

$$X_n = \alpha\, \epsilon_1 * X_n + Z_n, \tag{3}$$

where $\alpha \in \mathbb{C}$. We will prove that weakly stationary random sequences on the corresponding polynomial hypergroup $\mathbb{N}_0$ exist and are R_n-moving average sequences, provided $|\alpha| < 1$.

Lemma 6.1.3 *Let ϵ_i be the point measure in $i \in \mathbb{N}_0$, and define recursively $\epsilon_1^0 = \epsilon_0$, $\epsilon_1^k = \epsilon_1 * \epsilon_1^{k-1}$ for $k \in \mathbb{N}$, where $*$ denotes the convolution in the polynomial hypergroup $\mathbb{N}_0$ generated by $(R_n)_{n\in\mathbb{N}_0}$. Further let $\alpha \in \mathbb{C}$. Then for every $m \in \mathbb{N}$*

$$\sum_{k=0}^{m} \alpha^k \epsilon_1^k = \sum_{i=0}^{m} \beta_{m,i}\, \epsilon_i \tag{4}$$

where

$$\beta_{m,i} := \sum_{j=0}^{m} \alpha^j \, (R_1^j)^\vee(i)\, h(i) = \sum_{j=i}^{m} \alpha^j \, (R_1^j)^\vee(i)\, h(i).$$

Proof. The k-fold convolution product ϵ_1^k can be written

$$\epsilon_1^k = \sum_{j=0}^{k} a_{kj}\epsilon_j,$$

where $a_{kj} = (R_1^k)^\vee(j)\, h(j)$.

Hence we have by reordering the summation

$$\sum_{k=0}^{m} \alpha^k \, \epsilon_1^k = \sum_{k=0}^{m} \alpha^k \sum_{j=0}^{k} a_{kj}\epsilon_j = \sum_{i=0}^{m} \gamma_{mi}\epsilon_i$$

with

$$\gamma_{mi} = \sum_{j=i}^{m} \alpha^j \, a_{ji} = \sum_{j=i}^{m} \alpha^j \, (R_1^j)^\vee(i)\, h(i) = \beta_{mi}.$$

◇

Now let $(Z_n)_{n\in\mathbb{N}_0}$ be R_n-white noise. Given $\alpha \in \mathbb{C}$, $|\alpha| < 1$, consider $g_\alpha \in C(D_s) \subseteq L^2(D_s, \pi)$

$$g_\alpha(x) = \frac{1}{1-\alpha R_1(x)} = \sum_{k=0}^{\infty} R_1^k(x)\, \alpha^k, \qquad x \in D_s.$$

Then the sequence of Fourier coefficients $(\check{g}_\alpha(k))_{k\in\mathbb{N}_0}$ belongs to $l^2(h)$, and we can define

$$X_n = \sum_{k=0}^{\infty} \check{g}_\alpha(k)\, \epsilon_n * Z_k \, h(k) = \sum_{k=0}^{\infty} \check{g}_\alpha(k)\, \epsilon_k * Z_n \, h(k) \tag{5}$$

for each $n \in \mathbb{N}_0$.

We have to introduce auxiliary linear combinations of the Z_k's. Put $n, m \in \mathbb{N}_0$

$$Y_{n,m} := \sum_{k=0}^{m} \beta_{mk}\, \epsilon_k * Z_n, \tag{6}$$

where $\beta_{m,k}$ is defined as in Lemma 6.1.3.

Proposition 6.1.4 *If* $|\alpha| < 1$, *and* X_n *and* $Y_{n,m}$ *are defined as in (5) and (6), respectively, we have for every* $n \in \mathbb{N}_0$

$$X_n = \lim_{m\to\infty} Y_{n,m}.$$

Proof. Write for the partial sums of X_n

$$X_{n,m} := \sum_{k=0}^{m} \check{g}_\alpha(k)\, \epsilon_k * Z_n\, h(k).$$

Since $\check{g}_\alpha(k) = \sum_{j=k}^{\infty} \alpha^j\, (R_1^j)^\vee(k)$, we have for $k \leq m$

$$\check{g}_\alpha(k)\, h(k) - \beta_{m,k} = \sum_{j=m+1}^{\infty} \alpha^j\, (R_1^j)^\vee(k)\, h(k) = (g_{m+1,\alpha})^\vee(k)\, h(k),$$

where $g_{m+1,\alpha} \in C(D_s)$ is given by

$$g_{m+1,\alpha}(x) = \frac{(\alpha R_1(x))^{m+1}}{1 - \alpha R_1(x)}, \qquad x \in D_s.$$

Since $g_{m+1,\alpha} \in L^2(D_s, \pi)$ the Plancherel isomorphism yields

$$|(g_{m+1,\alpha})^\vee(k)|^2\, h(k) \leq \|\, g_{m+1,\alpha}\, \|_2^2 \leq C_1 \int_{D_s} |\alpha R_1(x)|^{2m+2}\, d\pi(x)$$
$$\leq C_2\, |\alpha|^{2m+2},$$

with C_1 and C_2 being appropriate constants independent of $k, m \in \mathbb{N}_0$.

Now $\|\, \epsilon_k * Z_n\, \|_2 = \|\, T_n(\epsilon_k * Z_0)\, \|_2 \leq \|\, \epsilon_k * Z_0\, \|_2 = \|\, Z_k\, \|_2 = 1/\sqrt{h(k)}$ and thus we obtain

$$\|\, X_{n,m} - Y_{n,m}\, \|_2^2 = \|\, \sum_{k=0}^{m} (\check{g}_\alpha(k)\, h(k) - \beta_{m,k})\, \epsilon_k * Z_n\, \|_2^2$$

$$\leq \sum_{k=0}^{m} |\check{g}_\alpha(k)\, h(k) - \beta_{m,k}|^2\, \frac{1}{h(k)} = \sum_{k=0}^{m} |(g_{m+1,\alpha})^\vee(k)|^2\, h(k)$$

$$\leq C_2 \sum_{k=0}^{m} |\alpha|^{2m+2} = C_2\,(m+1)\,|\alpha|^{2m+2}.$$

Hence $\lim_{m\to\infty} \|\, X_{n,m} - Y_{n,m}\, \|_2 = 0$. ◇

Now we are able to prove the extension of the well-known result on $AR(1)$-time series to the first order R_n-autoregressive random sequences.

Theorem 6.1.5 *Let $(Z_n)_{n\in\mathbb{N}_0}$ be R_n-white noise, and let $\alpha \in \mathbb{C}$, $|\alpha| < 1$. If $(X_n)_{n\in\mathbb{N}_0}$ is a weakly R_n-stationary random sequence that fulfils the first order autoregression*

$$X_n = \alpha\, \epsilon_1 * X_n + Z_n, \tag{3}$$

then X_n is a R_n-moving average sequence given by

$$X_n = \sum_{k=0}^{\infty} \check{g}_\alpha(k)\, \epsilon_n * Z_k\, h(k), \qquad n \in \mathbb{N}_0 \tag{5}$$

where $g_\alpha(x) = \dfrac{1}{1-\alpha R_1(x)}$, $x \in D_s$.

Conversely each process $(X_n)_{n\in\mathbb{N}_0}$ defined by (5) with $|\alpha| < 1$ satisfies (3).

Proof. Assume $(X_n)_{n\in\mathbb{N}_0}$ weakly R_n-stationary satisfying (3). Applying identity (4) of Lemma 6.1.3 the autoregression formula gives

$$X_n = \alpha^{m+1}\epsilon_1^{m+1} * X_n + \sum_{k=0}^{m} \alpha^k\, \epsilon_1^k * Z_n$$

$$= \alpha^{m+1}\epsilon_1^{m+1} * X_n + Y_{n,m},$$

with the notation of (6). Thus

$$\| X_n - Y_{n,m} \|_2 = |\alpha|^{m+1}\ \| \epsilon_1^{m+1} * X_n \|_2 .$$

Using the weakly R_n-stationarity of $(X_n)_{n\in\mathbb{N}_0}$ it follows that $\{\| \epsilon_1^{m+1} * X_n \|_2 \colon m, n \in \mathbb{N}_0\}$ is bounded. Hence $\lim\limits_{m\to\infty} \| X_n - Y_{n,m} \|_2 = 0$, and Proposition 6.1.4 implies that X_n can be written as in (5).

Conversely assume that X_n has a representation as in (5). By Lemma 6.1.3 it follows

$$Y_{n,m} - \alpha\, \epsilon_1 * Y_{n,m} = Z_n - \alpha^{m+1}\epsilon_1^{m+1} * Z_n,$$

and Proposition 6.1.4 implies

$$X_n = \alpha\, \epsilon_1 * X_n + Z_n.$$

◇

We want to point out that we did not use the fact that $1-\alpha R_1(x)$, $x \in D$, has an inverse element in the Banach algebra $A(D)$. In fact Wiener's theorem 4.1.6 implies that $\sum_{k=0}^{\infty} |\check{g}_\alpha(k)|\, h(k) < \infty$, which could be applied in proving Theorem 6.1.5.

The autoregression formula (3) can be written in matrix form as

$$\begin{pmatrix} X_0 \\ X_1 \\ X_2 \\ \vdots \\ \vdots \end{pmatrix} = \alpha \begin{pmatrix} 0 & 1 & 0 & 0 & 0 & \cdots & \cdots \\ c_1 & b_1 & a_1 & 0 & 0 & \cdots & \cdots \\ 0 & c_2 & b_2 & a_2 & 0 & \cdots & \cdots \\ \cdots & \cdots & \ddots & \ddots & \ddots & \cdots & \cdots \\ \cdots & \cdots & \cdots & \ddots & \ddots & \ddots & \cdots \end{pmatrix} \begin{pmatrix} X_0 \\ X_1 \\ X_2 \\ \vdots \\ \vdots \end{pmatrix} + \begin{pmatrix} Z_0 \\ Z_1 \\ Z_2 \\ \vdots \\ \vdots \end{pmatrix}$$

or in the notation of section 4.1

$$(\mathrm{Id} - \alpha L_1)\, X(\omega) \;=\; Z(\omega)$$

where $\omega \in \Omega$ and $X(\omega) = (X_0(\omega), X_1(\omega), ...)$, $Z(\omega) = (Z_0(\omega), Z_1(\omega), ...)$. As operator on $l^2(h)$ the translation L_1 has norm $\| L_1 \| \leq 1$. Hence $\mathrm{Id} - \alpha L_1$ is invertible in $\mathcal{B}(l^2(h))$ whenever $|\alpha| < 1$. Assuming $X(\omega), Z(\omega) \in l^2(h)$ we have

$$X(\omega) \;=\; (\mathrm{Id} - \alpha L_1)^{-1}\, Z(\omega).$$

The special form of the coefficients $\check{g}_\alpha(k)$ in Theorem 6.1.5 makes an easy computation possible by means of the associated orthogonal polynomial sequence. Since we apply throughout the normalization that fits to the hypergroup structure, our associated orthogonal polynomials differ slightly from those usually used, compare [138].

Lemma 6.1.6 *Let $(R_n)_{n\in\mathbb{N}_0}$ be an orthogonal polynomial sequence defined by (1) and (2) of subsection 1.2.1, orthogonal with respect to π. For $z \in \mathbb{C}$ and $k \in \mathbb{N}$ define*

$$Q_{k-1}(z) := \int_S \frac{R_k(x) - R_k(z)}{R_1(x) - R_1(z)}\, d\pi(x).$$

Then $(Q_k)_{k\in\mathbb{N}_0}$ is an orthogonal polynomial sequence fulfilling the recursion relation

$$Q_0(z) = 1, \qquad Q_1(z) \;=\; \frac{R_1(z) - b_1}{a_1}$$

and for $k \in \mathbb{N}$

$$Q_1(z)\, Q_k(z) \;=\; \frac{a_{k+1}}{a_1}\, Q_{k+1}(z) + \frac{b_{k+1} - b_1}{a_1}\, Q_k(z) + \frac{c_{k+1}}{a_1}\, Q_{k-1}(z). \quad (7)$$

(We call $(Q_k)_{k\in\mathbb{N}_0}$ the **associated** *polynomial sequence of $(R_n)_{n\in\mathbb{N}_0}$.)*

Proof. Applying the recursion formula of R_n we obtain

$$a_1 Q_1(z) = \int_S \frac{a_1 R_2(x) - a_1 R_2(z)}{R_1(x) - R_1(z)}\, d\pi(x)$$

$$= -b_1 + \int_S \frac{R_1^2(x) - R_1^2(z)}{R_1(x) - R_1(z)}\, d\pi(x) = -b_1 + R_1(z),$$

and for $k \geq 2$

$$R_1(z)\, Q_{k-1}(z) = \int_S \frac{R_1(z)\, R_k(x) - R_1(z)\, R_k(z)}{R_1(x) - R_1(z)}\, d\pi(x)$$

$$= \int_S \frac{R_1(x)\, R_k(x) - R_1(z)\, R_k(z)}{R_1(x) - R_1(z)}\, d\pi(x) = a_k Q_k(z) + b_k Q_{k-1}(z) + c_k Q_{k-2}(z)$$

and the recursion formula above follows. Favard's theorem implies that $Q_k(z)$ are orthogonal polynomials. ◇

Proposition 6.1.7 *Let $\check{g}_\alpha(k)$, $k \in \mathbb{N}_0$, be the coefficients given in the R_n-moving average expansion (5) of a first order R_n-autoregressive random sequence $(X_n)_{n\in\mathbb{N}_0}$. For $|\alpha| < 1$, $\alpha \neq 0$ and $R_1(x) = \frac{x}{a_0} - \frac{b_0}{a_0}$ set $x_\alpha := \frac{a_0}{\alpha} + b_0$. Then $\check{g}_\alpha(k)$ can be calculated by*

$$\check{g}_\alpha(k) = R_k(x_\alpha)\, \check{g}_\alpha(0) - \frac{1}{\alpha}\, Q_{k-1}(x_\alpha), \qquad k \in \mathbb{N}, \tag{8}$$

where the polynomials $Q_k(x)$ are defined by equation (7) above.

Proof.

$$\check{g}_\alpha(k) = \int_S \frac{R_k(x)}{1 - \alpha R_1(x)}\, d\pi(x) = \frac{-1}{\alpha} \int_S \frac{R_k(x)}{R_1(x) - R_1(x_\alpha)}\, d\pi(x)$$

$$= \frac{-1}{\alpha} \left(Q_{k-1}(x_\alpha) + R_k(x_\alpha) \int_S \frac{1}{R_1(x) - R_1(x_\alpha)}\, d\pi(x) \right)$$

$$= \frac{-1}{\alpha}\, Q_{k-1}(x_\alpha) + R_k(x_\alpha)\, \check{g}_\alpha(0).$$

◇

We close our study of first order autoregression by determining covariance and spectral measure.

Because of $(\check{g}_\alpha * (\check{g}_\alpha)^\star)^\wedge(x) = |g_\alpha(x)|^2$ we have for the spectral measure μ

$$d\mu(x) = \frac{d\pi(x)}{|1 - \alpha R_1(x)|^2}. \tag{9}$$

In order to calculate $d(n, m)$ we derive another series representation of the X_n. On that way we extend the recurrence relation (7).

Lemma 6.1.8 *Let $(R_n)_{n\in\mathbb{N}_0}$ be an orthogonal polynomial sequence generating a polynomial hypergroup on $\mathbb{N}_0$. Let $(Q_n)_{n\in\mathbb{N}_0}$ be the associated orthogonal polynomial sequence. For $z \in \mathbb{C}$ set $q_n := Q_{n-1}(z)$, $n \in \mathbb{N}$ and $q_0 = Q_{-1}(z) = 0$. Then for $q = (q_n)_{n\in\mathbb{N}_0}$ and $n \in \mathbb{N}$ we have*

$$\epsilon_n * q(k) = \begin{cases} R_k(z)\, Q_{n-1}(z) & \text{for } k = 0, 1, ..., n-1 \\ R_n(z)\, Q_{k-1}(z) & \text{for } k = n, n+1, ... \end{cases}$$

Proof. For $k \geq n$ follows

$$R_n(z)\, Q_{k-1}(z) = \int_{D_s} R_k(x) \frac{R_n(z) - R_n(x)}{R_1(x) - R_1(z)}\, d\pi(x)$$

$$+ \int_{D_s} \frac{R_n(x)\, R_k(x) - R_n(z)\, R_k(z)}{R_1(x) - R_1(z)}\, d\pi(x).$$

Since $\dfrac{R_n(z) - R_n(x)}{R_1(x) - R_1(z)}$ is a polynomial in x of degree at most $n-1$, the first summand is zero and we get using the linearization of the product $R_n R_k$:

$$R_n(z)\, Q_{k-1}(z) = \sum_{j=k-n}^{k+n} g(n, k; j)\, q_j = \epsilon_n * q(k).$$

For $0 \leq k < n$ we have

$$\epsilon_n * q(k) = \epsilon_k * q(n) = R_k(z)\, Q_{n-1}(z).$$

◇

Theorem 6.1.9 *Let $(X_n)_{n\in\mathbb{N}_0}$ be a first order R_n-autoregressive random sequence with $|\alpha| < 1$. Using the notation of Theorem 6.1.5 and Proposition 6.1.7 the following hold.*

(i)

$$\epsilon_n * \check{g}_\alpha(k) = \begin{cases} R_k(x_\alpha)\, \check{g}_\alpha(n) & \text{for } k = 0, 1, ..., n-1 \\ R_n(x_\alpha)\, \check{g}_\alpha(k) & \text{for } k = n, n+1, ... \end{cases}$$

(ii)

$$X_n = \check{g}_\alpha(n) \sum_{k=0}^{n-1} R_k(x_\alpha)\, Z_k\, h(k) + R_n(x_\alpha) \sum_{k=n}^{\infty} \check{g}_\alpha(k)\, Z_k\, h(k)$$

$$\text{and} \quad X_n - R_n(x_\alpha)\, X_0 = \sum_{k=0}^{n-1} (\check{g}_\alpha(n)\, R_k(x_\alpha) - R_n(x_\alpha)\, \check{g}_\alpha(k))\, Z_k\, h(k).$$

(iii)

$$E(X_n\overline{Z}_k) \; = \epsilon_n * \breve{g}_\alpha(k) \;\; \textit{for } k,n \in \mathbb{N}_0$$

$$\textit{and} \quad E\left((X_n - R_n(x_\alpha)X_0)\cdot \overline{Z}_k\right) \; = 0 \qquad \textit{for } k \geq n.$$

(iv)

$$d(n) \; = \; \left(\breve{g}_\alpha * (\breve{g}_\alpha)^\star\right)(n)$$

$$= \; \breve{g}_\alpha(n) \sum_{k=0}^{n-1} \overline{\breve{g}_\alpha(k)}\, R_k(x_\alpha)\, h(k) \; + R_n(x_\alpha) \sum_{k=n}^{\infty} |\breve{g}_\alpha(k)|^2 \, h(k).$$

(v)

$$d(n,m) \; = \; \breve{g}_\alpha(n)\, \overline{\breve{g}_\alpha(m)} \sum_{k=0}^{n-1} |R_k(x_\alpha)|^2 \, h(k)$$

$$+ \; R_n(x_\alpha)\, \overline{\breve{g}_\alpha(m)} \sum_{k=n}^{m-1} \breve{g}_\alpha(k)\, \overline{R_k(x_\alpha)}\, h(k)$$

$$+ \; R_n(x_\alpha)\, \overline{R_m(x_\alpha)} \sum_{k=m}^{\infty} |\breve{g}_\alpha(k)|^2 \, h(k) \qquad \textit{for } n \leq m.$$

(For $n > m$ $d(n,m) \; = \overline{d(m,n)}$ is valid.)

Proof. By Proposition 6.1.7 and Lemma 6.1.8 the equality of (i) follows immediately. The representation of X_n in (ii) is attained by Lemma 6.1.1 and (i).

The statement (iii) is obviously true. To derive (iv) and (v) observe that for $n \leq m$

$$E(X_n\overline{X}m) = \overline{\breve{g}_\alpha(m)} \sum_{k=0}^{m-1} \overline{R_k(x_\alpha)}\, E(X_n\overline{Z}_k)\, h(k)$$

$$+ \overline{R_m(x_\alpha)} \sum_{k=m}^{\infty} \overline{\breve{g}_\alpha(k)}\, E(X_n\overline{Z}_k)\, h(k)$$

$$= \overline{\breve{g}_\alpha(m)}\, \breve{g}_\alpha(n) \sum_{k=0}^{n-1} |R_k(x_\alpha)|^2 \, h(k) + \overline{\breve{g}_\alpha(m)}\, R_n(x_\alpha) \sum_{k=n}^{m-1} \breve{g}_\alpha(k)\, \overline{R_k(x_\alpha)}\, h(k)$$

$$+ \; \overline{R_m(x_\alpha)}\, R_n(x_\alpha) \sum_{k=m}^{\infty} |\breve{g}_\alpha(k)|^2 \, h(k).$$

◇

In case of $R_n(x) = R_n^{(-1/2,-1/2)}(x) = T_n(x)$ the Chebyshev polynomials of the first kind, the associated orthogonal polynomials $Q_k(x)$ are Chebyshev polynomials of the second kind

$$Q_k(x) = (k+1)\, R_k^{(1/2,1/2)}(x).$$

Hence the calculations of the quantities in Theorem 6.1.9 yield various trigonometric and hyperbolic formulas.

Now we investigate random sequences $(X_n)_{n\in\mathbb{N}_0}$ satisfying autoregressive equation of order $p \in \mathbb{N}$, i.e.

$$X_n - a_1\, \epsilon_1 * X_n - a_2\, \epsilon_2 * X_n - \ldots - a_p\, \epsilon_p * X_n = Z_n, \qquad (10)$$

for $n \in \mathbb{N}_0$, where $(Z_n)_{n\in\mathbb{N}_0}$ is R_n-white noise, $*$ is the convolution corresponding to $(R_n)_{n\in\mathbb{N}_0}$ and $a_1, \ldots, a_p \in \mathbb{C}$, $a_p \neq 0$. Then $(X_n)_{n\in\mathbb{N}_0}$ is called a **p-order $\mathbf{R_n}$-autoregressive random sequence.**

We begin with introducing very useful operators defined by functions $g \in A(D)$. Remind that $g \in A(D)$ means $g(x) = \sum_{k=0}^{\infty} \check{g}(k)\, R_k(x)\, h(k)$ for $x \in D$, such that $\sum_{k=0}^{\infty} |\check{g}(k)|\, h(k) < \infty$.

Proposition 6.1.10 *Let $(Y_n)_{n\in\mathbb{N}_0}$ be a family of square integrable random variables, such that $\sup_{n\in\mathbb{N}_0} \| Y_n \|_2^2 =: M < \infty$.*

Given a polynomial hypergroup on $\mathbb{N}_0$ generated by $(R_n)_{n\in\mathbb{N}_0}$, let $g \in A(D)$. Then the series

$$g(T)\, Y_n := \sum_{k=0}^{\infty} \check{g}(k)\, \epsilon_n * Y_k\, h(k)$$

converges to some random variable in $H(Y_n, \mathbb{N}_0) \subseteq L^2(\Omega, P)$.

Proof. If $l < m$ then

$$\| \sum_{k=l+1}^{m} \check{g}(k)\, \epsilon_n * Y_k\, h(k) \|_2^2 = \sum_{i,j=l+1}^{m} \check{g}(i)\, \overline{\check{g}}(j)\, < \epsilon_n * Y_i, \epsilon_n * Y_j >\, h(i)\, h(j)$$

$$\leq \sum_{i,j=l+1}^{m} \check{g}(i)\, \overline{\check{g}(j)}\, \| Y_i \|_2\, \| Y_j \|_2\, h(i)\, h(j) \leq M \left| \sum_{i=l+1}^{m} \check{g}(i)\, h(i) \right|^2.$$

Hence by $g \in A(D)$ we have $\| \sum_{k=l+1}^{m} \check{g}(k)\, \epsilon_n * Y_k\, h(k) \|_2^2 \to 0$ as $l, m \to \infty$. By the Cauchy criterion $g(T)\, Y_n$ converges in $H(Y_n, \mathbb{N}_0)$. ◇

A slight generalization of Lemma 6.1.1 says that

$$g(T)\, Y_n = \sum_{k=0}^{\infty} \check{g}(k)\, \epsilon_n * Y_k\, h(k) = \sum_{k=0}^{\infty} \epsilon_n * \check{g}(k)\, Y_k\, h(k).$$

If $f, g \in A(D)$, then $f \cdot g \in A(D)$ and $\check{f} * \check{g}(k) = (f \cdot g)^{\vee}(k)$ for each $k \in \mathbb{N}_0$.

Proposition 6.1.11 *Let $(Z_n)_{n\in\mathbb{N}_0}$ be R_n-white noise. If $f,g \in A(D)$ then $f(T)(g(T)\, Z_n)$ and $g(T)(f(T)\, Z_n)$ are well-defined in $H(Z_n, \mathbb{N}_0)$, and*

$$f(T)(g(T)\, Z_n) \;=\; fg(T)\, Z_n \;=\; g(T)(f(T)\, Z_n).$$

Proof. For $Y_n := g(T)\, Z_n$ we have $\sup\limits_{n\in\mathbb{N}_0} \| Y_n \|_2^2 \leq \| \check{g} \|_1^2$. By the Proposition above $f(T)\, Y_n \;=\; f(T)(g(T)\, Z_n)$ is a well-defined element of $H(Y_n, \mathbb{N}_0) \subseteq H(Z_n, \mathbb{N}_0)$. This is also true for $fg(T)\, Z_n$ and $g(T)(f(T)\, Z_n)$.

For every $l \in \mathbb{N}_0$ follows that

$$< f(T)\, Y_n, Z_l > = \sum_{k=0}^{\infty} \check{f}(k) \;< \epsilon_n * Y_k, Z_l >\; h(k)$$

$$= \sum_{k=0}^{\infty} \check{f}(k) \sum_{j=0}^{\infty} \check{g}(j) \;< \epsilon_k * \epsilon_n * Z_j, Z_l >\; h(j)\, h(k)$$

$$= \sum_{j=0}^{\infty} \sum_{k=0}^{\infty} \check{f}(k)\, \epsilon_k * \check{g}(j)\, h(k) \;< \epsilon_n * Z_j, Z_l >\; h(j)$$

$$= \sum_{j=0}^{\infty} \check{f} * \check{g}(j) \;< \epsilon_n * Z_j, Z_l >\; h(j) \;= < \sum_{j=0}^{\infty} (fg)^{\vee}(j)\, \epsilon_n * Z_j\, h(j), Z_l >,$$

where $\check{f} * \check{g}(j) \;=\; (fg)^{\vee}(j)$ is used. Therefore $f(T)(g(T)\, Z_n) \;=\; fg(T)\, Z_n$. $\diamond$

Now it is easy to derive the following result on R_n-autoregressive random sequences $(X_n)_{n\in\mathbb{N}_0}$.

Theorem 6.1.12 *Let $(Z_n)_{n\in\mathbb{N}_0}$ be R_n-white noise, and let $P(x) \;=\; 1 - a_1 R_1(x) - \ldots - a_p R_p(x)$ be a polynomial with $a_p \neq 0$ such that $P(x) \neq 0$ for every $x \in D$.*

If $(X_n)_{n\in\mathbb{N}_0}$ is a weakly R_n-stationary random sequence on $\mathbb{N}_0$ that fulfils for each $n \in \mathbb{N}_0$ the p-order autoregression

$$X_n \;- a_1\, \epsilon_1 * X_n \;- a_2\, \epsilon_2 * X_n \;- \ldots - a_p\, \epsilon_p * X_n \;=\; Z_n \qquad (10)$$

then $(X_n)_{n\in\mathbb{N}_0}$ is a R_n-moving average sequence given by

$$X_n \;=\; \sum_{k=0}^{\infty} \check{g}(k)\, \epsilon_n * Z_k\, h(k), \qquad n \in \mathbb{N}_0 \qquad (11)$$

where $g \in A(D)$ *is the unique function with* $g(x) = \frac{1}{P(x)}$ *for* $x \in D$. *(Compare Wiener's Theorem 4.1.6).*

Conversely, each process $(X_n)_{n \in \mathbb{N}_0}$ *defined by Eq. (11) with a function* $g \in A(D)$ *with* $g(x) = \frac{1}{P(x)}$ *for* $x \in D$, *satisfies the autoregressive equation (10).*

Proof. At first let $(X_n)_{n \in \mathbb{N}_0}$ fulfil Eq. (10). By

$$\check{P}(k) = -a_k/h(k)\, \chi_{\{0,\ldots,p\}}$$

(with $a_0 = -1$), we get

$$P(T)\, X_n = \epsilon_n * X_0 - a_1\, \epsilon_n * X_1 - \ldots - a_p\, \epsilon_n * X_p$$

$$= X_n - a_1\, \epsilon_1 * X_n - \ldots - a_p\, \epsilon_p * X_n = Z_n.$$

If $g \in A(D)$ fulfils $g(x)\, P(x) = 1$ for all $x \in D_s$ we have by Proposition 6.1.11

$$\sum_{k=0}^{\infty} \check{g}(k)\, \epsilon_n * Z_k\, h(k) = g(T)\, Z_n = g(T)\, (P(T)\, X_n)$$

$$= g\, P(T)\, X_n = X_n.$$

Conversely if $X_n = g(T) Z_n$, $g(x)P(x) = 1$ for $x \in D_s$, then

$$P(T)\, X_n = P(T)\, (g(T)\, Z_n) = Z_n.$$

But $P(T)X_n = X_n - a_1\epsilon_1 * X_n - a_2\epsilon_2 * X_n - \cdots - a_p\epsilon_p * X_n$, and the theorem is shown completely. ◇

Remark. Since $(\check{g} * (\check{g})^*)^\wedge (x) = |g(x)|^2$ we have for the spectral measure μ of $(X_n)_{n \in \mathbb{N}_0}$:

$$d\mu(x) = \frac{1}{|P(x)|^2}\, d\pi(x).$$

2 Mean Estimation

Throughout $(R_n)_{n \in \mathbb{N}_0}$ will be an orthogonal polynomial sequence that generates a polynomial hypergroup on $\mathbb{N}_0$. $(X_n)_{n \in \mathbb{N}_0}$ will be a weakly stationary random sequence on this polynomial hypergroup. In this section we will assume that the means of the X_n are constant, i.e. $EX_n = M$ for all $n \in \mathbb{N}_0$. We consider several estimators of M appropriate for this type of random sequence, and examine some of their properties.

The most general form of linear unbiased estimators for the mean of $(X_n)_{n\in\mathbb{N}_0}$ is given by

$$M_n := \sum_{k=0}^{n} b_{n,k} X_k \qquad \text{with } \sum_{k=0}^{n} b_{n,k} = 1, \tag{1}$$

where the b's are complex numbers.

A simple calculation shows that the mean squared error is given by

$$\begin{aligned} E(|M_n - M|^2) &= \sum_{i,j=0}^{n} b_{n,i}\overline{b_{n,j}} E\left((X_i - M)\,\overline{(X_j - M)}\right) \\ &= \int_{D_s} |B_n(x)|^2 \, d\mu(x), \end{aligned} \tag{2}$$

where $B_n(x) = \sum_{k=0}^{n} b_{n,k} R_k(x)$ and μ is the spectral measure of $(X_n - M)_{n\in\mathbb{N}_0}$. The careful reader will notice that now we use the spectral measure of $Y_n := X_n - M$. But the difference is minor, since $(X_n)_{n\in\mathbb{N}_0}$ and $(Y_n)_{n\in\mathbb{N}_0}$ are both weakly stationary with respect to $(R_n)_{n\in\mathbb{N}_0}$ and $E(Y_n \cdot \overline{Y}_0) = E(X_n \cdot \overline{X}_0) - |M|^2$.

We give a sufficient condition such that the estimators M_n are consistent in mean-square, i.e.

$$E(|M_n - M|^2) = \| M_n - M \|_2^2 \to 0 \qquad \text{as } n \to \infty.$$

For that we bring a well-known theorem of Toeplitz (see [374,p.75]) into use.

Theorem 6.2.1 *Let $(X_n)_{n\in\mathbb{N}_0}$ be a weakly stationary random sequence on a polynomial hypergroup $\mathbb{N}_0$ generated by $(R_n)_{n\in\mathbb{N}_0}$. Assume that $E(X_n) = M$ for every $n \in \mathbb{N}_0$. Define the estimators M_n for the mean M as in (1), and suppose that all $b_{n,k} \geq 0$. If $E\left((X_n - M)\,\overline{(X_0 - M)}\right) \to 0$, and if*

$$\sum_{k=0}^{n} b_{n,k}^2 / h(k) \to 0 \tag{3}$$

as $n \to \infty$, the estimators M_n are consistent, i.e. $E(|M_n - M|^2)$ tend to zero.

Proof. For brevity write $a_n = (a_{n,k})_{k\in\mathbb{N}_0}$ with $a_{n,k} := b_{n,k}/h(k)$ for $k \leq n$ and $a_{n,k} = 0$ for $k > n$, and for the covariance write $d(n) := E\left((X_n - M)\,\overline{(X_0 - M)}\right)$. We have

$$E(|M_n - M|^2) = \int_{D_s} |B_n(x)|^2\, d\mu(x), \qquad B_n(x) = \sum_{k=0}^{n} a_{n,k} R_k(x)\, h(k) = \widehat{a_n}(x).$$

Hence

$$E(|M_n - M|^2) = \int_{D_s} (a_n * a_n^\star)^\wedge(x)\, d\mu(x) = \sum_{k=0}^{2n} a_n * a_n^\star(k)\, h(k)\, d(k).$$

Since $d(n) \to 0$, Toeplitz's theorem implies the statement when we have $a_n * a_n^\star(k)\; h(k) \to 0$ for each $k \in \mathbb{N}_0$ and $\sum_{k=0}^{2n} a_n * a_n^\star(k)\; h(k) = 1$ as $n \to \infty$. Note that $a_n * a_n^\star(k) \geq 0$. Now we see that

$$\sum_{k=0}^{2n} a_n * a_n^\star(k)\, h(k) = (a_n * a_n^\star)^\wedge(1) = |B_n(1)|^2 = |\sum_{k=0}^{n} b_{n,k}|^2 = 1.$$

In order to show that $a_n * a_n^\star(k)\; h(k)$ converges to zero for any $k \in \mathbb{N}_0$ as n tends to infinity, we use that $a_n * a_n^\star$ is positive definite and bounded. Hence $|a_n * a_n^\star(k)| \leq a_n * a_n^\star(0)$ for all $k \in \mathbb{N}_0$. Since $a_n * a_n^\star(0) = \sum_{l=0}^{n} b_{n,l}^2/h(l)$, assumption (3) gives $a_n * a_n^\star(k)\; h(k) \to 0$ as $n \to \infty$. ◇

Setting, in particular, $b_{n,k} = 0$ for $k \neq n$ and $b_{n,n} = 1$, we obtain:

Corollary 6.2.2 *If* $E\left((X_n - M) \cdot \overline{(X_0 - M)}\right) \to 0$ *and* $h(n) \to \infty$ *as* $n \to \infty$, *we have*
$E(|X_n - M|^2) \to 0$ *as* $n \to \infty$.

If we use the arithmetic means of $(X_n)_{n\in\mathbb{N}_0}$, i.e. $b_{n,k} = \frac{1}{n+1}$ for $k = 0, ..., n$, Theorem 6.2.1 yields:

Corollary 6.2.3 *Let* $\tilde{X}_n := \frac{1}{n+1} \sum_{k=0}^{n} X_k$. *If* $E\left((X_n - M) \cdot \overline{(X_0 - M)}\right) \to 0$ *then* $E(|\tilde{X}_n - M|^2) \to 0$ *as* $n \to \infty$.

Remark: If the spectral measure μ of $(X_n - M)$ is absolutely continuous with respect to the orthogonalization measure π of $(R_n)_{n\in\mathbb{N}_0}$, i.e. $\mu = f\pi$, $f \in L^1(D_s, \pi)$, we know that $d(n) = \int_S R_n(x)\, f(x)\, d\pi(x) = \hat{f}(n) \to 0$ as $n \to \infty$, see Proposition 3.3.2 (ii).

It is natural to ask for the efficiency of the various unbiased and consistent estimators for the mean. The investigation of the asymptotics of the estimation errors leads us to introduce the following special weights $a_{n,k}$, where $H(n) = \sum_{j=0}^{n} h(j)$ and

$$a_{n,k} := \frac{h(k)}{H(n)}. \tag{4}$$

We have $\sum_{k=0}^{n} a_{n,k}^2/h(k) = \dfrac{1}{H(n)} \to 0$ with $n \to \infty$, i.e. the $a_{n,k}$ fulfil the crucial assumption (3) of the theorem above. It turns out that for weakly stationary random sequences $(X_n)_{n\in\mathbb{N}_0}$ (with constant mean) on a polynomial hypergroup, these $a_{n,k}$ are the appropriate substitute for the arithmetic mean used in the case of classical weakly stationary processes. Growth condition (H) for $h(n)$ is important for the subsequent investigations, see section 4.7. We recall that the orthogonal polynomial sequence $(R_n)_{n\in\mathbb{N}_0}$ satisfies condition (H), if

$$\lim_{n\to\infty} \frac{h(n)}{H(n)} = 0.$$

Property (H) allows to construct kernel functions on D_s that are a suitable substitute of the Fejér-kernel. For classical weakly stationary processes the Fejér-kernel is used to study the asymptotic behaviour of the estimation error.

We begin by recalling some consequences and properties of condition (H).

For

$$\chi_n(k) = \begin{cases} 1 & \text{for } k = 0, ..., n \\ 0 & \text{for } k = n+1, n+2, ... \end{cases}$$

holds

(a) $\displaystyle\lim_{n\to\infty} \frac{\chi_n * \chi_n(k)}{\chi_n * \chi_n(0)} = 1$ for each $k \in \mathbb{N}_0$,
whenever (H) is valid. (a) is shown in Proposition 4.14.17.

(b) If condition (H) holds, then $1 \in \operatorname{supp} \pi = S$, see Corollary 4.2.12.

(c) If
$$\frac{h(n+1)}{h(n)} = \frac{a_n}{c_{n+1}} \to 1 \quad \text{for } n \to \infty,$$
then (H) is fulfilled. If
$$\frac{h(n+1)}{h(n)} = \frac{a_n}{c_{n+1}} \to \gamma > 1 \quad \text{for } n \to \infty,$$
then (H) does not hold. See Theorem 4.7.1 and Corollary 4.7.2.

Theorem 6.2.4 *Let $(X_n)_{n\in\mathbb{N}_0}$ be a weakly stationary random sequence on a polynomial hypergroup $\mathbb{N}_0$ generated by $(R_n)_{n\in\mathbb{N}_0}$. Assume that $E(X_n) = M$ for every $n \in \mathbb{N}_0$. Define the estimators A_n for the mean M by the weights $a_{n,k}$ of (4), i.e.*

$$A_n = \sum_{k=0}^{n} a_{n,k} X_k, \qquad a_{n,k} = h(k)/H(n).$$

Then

$$H(n)\, E(|A_n - M|^2) = \sum_{k=0}^{2n} \frac{\chi_n * \chi_n(k)}{\chi_n * \chi_n(0)}\, \check{\mu}(k)\, h(k) = \frac{1}{H(n)} \int_{D_s} \mathcal{D}_n^2(x)\, d\mu(x),$$

where μ is the spectral measure of $(X_n - M)_{n\in\mathbb{N}_0}$ and $\mathcal{D}_n(x) = \sum_{k=0}^{n} R_k(x)\, h(k)$ is the Dirichlet kernel. $\dfrac{\chi_n * \chi_n(k)}{\chi_n * \chi_n(0)}$ are the Fejér-weights considered in (a) above.

Proof. Write $d(j,k) = E\left((X_j - M)\overline{(X_k - M)}\right) = \int_{D_s} R_j(x)\, R_k(x)\, d\mu(x)$.
Then

$$H(n)\, E(|A_n - M|^2) = \frac{1}{H(n)} \sum_{j,k=0}^{n} d(j,k)\, h(j)\, h(k)$$

$$= \frac{1}{H(n)} \sum_{j,k=0}^{n} \epsilon_j * \check{\mu}(k)\, h(j)\, h(k) = \frac{1}{H(n)} \check{\mu} * \chi_n * \chi_n(0)$$

$$= \sum_{k=0}^{2n} \frac{\chi_n * \chi_n(k)}{\chi_n * \chi_n(0)}\, \check{\mu}(k)\, h(k).$$

The second equality holds by $\mathcal{D}_n^2(x) = (\widehat{\chi_n})^2(x) = (\chi_n * \chi_n)^\wedge(x) = \sum_{k=0}^{2n} \chi_n * \chi_n(k)\, R_k(x)\, h(k)$. ◇

Remark: If in Theorem 6.2.4 additionally property (H) holds, we know that $\dfrac{\chi_n * \chi_n(k)}{\chi_n * \chi_n(0)} \to 1$ as $n \to \infty$. If, for example the covariance sequence $d = (d(n))_{n\in\mathbb{N}_0}, d_n = E\left((X_n - M)\, \overline{(X_0 - M)}\right)$ is an element of $l^1(h)$, one immediately obtains

$$\lim_{n\to\infty} H(n)\, E(|A_n - M|^2) = \sum_{k=0}^{\infty} d(k)\, h(k) = f(1),$$

where $f(x) = \sum_{k=0}^{\infty} d(k)\, R_k(x)\, h(k)$ is the spectral density, i.e. $d\mu(x) = f(x)\, d\pi(x)$. We can prove this limit relation without the assumption $d \in l^1(h)$.

Theorem 6.2.5 *Assume that in addition to the assumptions of Theorem 6.2.4 property (H) holds. Further let $d\mu(x) = f(x)\, d\pi(x)$, where f is a continuous spectral density function on D_s.*
Then

$$\lim_{n\to\infty} H(n)\, E(|A_n - M|^2) = f(1).$$

(Note that $1 \in S$.)

Proof. Denote $F_{2n}(x) := \sum_{k=0}^{2n} \frac{\chi_n * \chi_n(k)}{\chi_n * \chi_n(0)} R_k(x)\, h(k)$ the Fejér-type kernel with respect to $(R_n)_{n\in\mathbb{N}_0}$, see subsection 4.14.2. Define a sequence of linear functionals φ_n on $C(S)$ by

$$\varphi_n(g) = \int_S F_{2n}(x)\, g(x)\, d\pi(x), \qquad g \in C(S).$$

The φ_n are continuous, since $|\varphi_n(g)| \leq \| g \|_\infty \int_S |F_{2n}(x)|\, d\pi(x) = \| g \|_\infty$. (Note that $F_{2n}(x) \geq 0$.)

If $P(x)$ is some polynomial with degree less than $2l$, we get for $n \geq l$ by (a)

$$\varphi_n(P) = \sum_{k=0}^{2n} \frac{\chi_n * \chi_n(k)}{\chi_n * \chi_n(0)} \check{P}(k)\, h(k)$$

$$= \sum_{k=0}^{2l} \frac{\chi_n * \chi_n(k)}{\chi_n * \chi_n(0)} \check{P}(k)\, h(k) \longrightarrow \sum_{k=0}^{2l} \check{P}(k)\, h(k) = P(1)$$

as $n \to \infty$. Now by the theorem of Weierstraß polynomials are dense in $C(S)$. Hence for the continuous spectral density f and any $\varepsilon > 0$ there is a polynomial P such that $\| f - P \|_\infty < \varepsilon$. Then we have

$$\begin{aligned}|\varphi_n(f) - f(1)| &\leq |\varphi_n(f) - \varphi_n(P)| + |\varphi_n(P) - P(1)| + |P(1) - f(1)| \\ &\leq 2\varepsilon + |\varphi_n(P) - P(1)|\end{aligned}$$

By Theorem 6.2.4 we see that

$$\lim_{n\to\infty} H(n)\, E(|A_n - M|^2) \;=\; \lim_{n\to\infty} \varphi_n(f) \;= f(1).$$

$\diamond$

By the above theorem we get the exact rate of convergence of A_n to M provided $f(1) > 0$. For many examples $1/H(n)$ can be easily calculated. If $R_n(x) = R_n^{(\alpha,\beta)}(x)$ are the Jacobi polynomials, $\alpha \geq \beta > -1$, $\alpha+\beta+1 \geq 0$, we know, see subsection 4.5.2, that

$$h(n) \;=\; \frac{(2n+\alpha+\beta+1)\;(\alpha+\beta+1)_n(\alpha+1)_n}{(\alpha+\beta+1)\; n!(\beta+1)_n} \;\approx\; \frac{n^{2\alpha+1}}{2^{\alpha+\beta}(\Gamma(\alpha+1))^2}$$

from which we can deduce that

$$H(n) \;=\; \frac{(\alpha+\beta+2)_n(\alpha+2)_n}{(\beta+1)_n n!} \;\approx\; \frac{n^{2\alpha+2}}{2^{\alpha+\beta+1}(\alpha+1)\;(\Gamma(\alpha+1))^2}.$$

Therefore we know in that case $E(|A_n - M|^2) \;\sim\; \frac{1}{n^{2\alpha+2}}$, provided that $f(1) > 0$.

A straightforward calculation shows that linear unbiased and consistent estimators M_n will never exist if $\mu(\{1\}) > 0$, where μ is the spectral measure on D_s. In fact if $a = \mu(\{1\}) > 0$ we write $\mu = \tilde{\mu} + a\epsilon_1$, where ϵ_1 is the point measure of $x = 1 \in D_s$. Then for any linear unbiased estimator M_n we have

$$E(|M_n - M|^2) \;=\; \int_S |B_n(x)|^2 d\tilde{\mu}(x) \;+ a|B_n(1)|^2 \geq a > 0.$$

If condition (H) is fulfilled and $\mu(\{1\}) = 0$ we can derive that the special estimators A_n are always consistent. In fact from the Christoffel-Darboux formula (Proposition 4.10.3 (i)) we obtain for $x \neq 1$

$$\frac{1}{H(n)}\left|\sum_{k=0}^{n} R_k(x)\,h(k)\right| = a_0 a_n\,\frac{h(n)}{H(n)}\left|\frac{R_{n+1}(x)-R_n(x)}{x-1}\right| \leq \frac{h(n)}{H(n)}\,\frac{2a_0}{|x-1|}.$$

Hence by property (H) we have

$$\frac{1}{H(n)}\mathcal{D}_n(x) \;\longrightarrow\; \chi_{\{1\}}(x)$$

pointwise for each $x \in D_s$, where $\chi_{\{1\}}(x) \;= \delta_{x,1}$. Lebesgue's theorem of dominated convergence gives (compare Theorem 6.2.4)

$$E(|A_n - M|^2) \;=\; \frac{1}{(H(n))^2}\int_{D_s} \mathcal{D}_n^2(x)\; d\mu(x) \;\longrightarrow\; \mu(\{1\}).$$

In summary we have just shown the following result.

Theorem 6.2.6 *Let $(X_n)_{n\in\mathbb{N}_0}$ be a weakly stationary random sequence on a polynomial hypergroup $\mathbb{N}_0$ generated by $(R_n)_{n\in\mathbb{N}_0}$. Assume that $E(X_n) = M$ for every $n \in \mathbb{N}_0$. If the spectral measure μ of $(X_n - M)_{n\in\mathbb{N}_0}$ has positive measure in $x = 1$, i.e. $\mu(\{1\}) > 0$, then there do not exist consistent linear unbiased estimators of M. If $\mu(\{1\}) = 0$ the estimators A_n of Theorem 6.2.4 are always consistent.*

There is a description of the best linear unbiased estimators for M. We search for $b_{n,k}$ such that $\sum_{k=0}^{n} b_{n,k} = 1$ and

$$\int_{D_s} |B_n(x)|^2 \, d\mu(x)$$

is minimized, where $B_n(x) = \sum_{k=0}^{n} b_{n,k} R_k(x)$, compare (2). Obviously, this is equivalent to determining a polynomial B such that degree$(B) \leq n$, $B(1) = 1$ and $\int_{D_s} |B(x)|^2 d\mu(x)$ is minimal.

Theorem 6.2.7 *Let $(X_n)_{n\in\mathbb{N}_0}$ be a weakly stationary random sequence on the polynomial hypergroup $\mathbb{N}_0$ generated by $(R_n)_{n\in\mathbb{N}_0}$. Assume that $E(X_n) = M$ for all $n \in \mathbb{N}_0$, and let μ be the spectral measure of $(X_n - M)_{n\in\mathbb{N}_0}$ with $|supp\ \mu| = \infty$. Let $(q_k)_{k\in\mathbb{N}_0}$ be the orthonormal polynomial sequence with respect to μ. Denote*

$$B_n^\star(x) := \frac{1}{\sum_{j=0}^{n} |q_j(1)|^2} \sum_{j=0}^{n} \overline{q_j(1)}\, q_j(x).$$

Then $B_n^\star(x)$ is the minimizing polynomial in

$$\min\{\int_{D_s} |B(x)|^2 d\mu(x) : \ \textit{degree}\ (B) \leq n, \ B(1) = 1\}.$$

Hence the best linear unbiased estimators $M_n^\star$ for M have the form

$$M_n^\star = \int_{D_s} B_n^\star(x)\, dZ(x)$$

(where Z is the orthogonal stochastic measure corresponding to $(X_n - M)_{n\in\mathbb{N}_0}$), and for the estimation error holds

$$E(|M_n^\star - M|^2) = \frac{1}{\sum_{k=0}^{n} |q_k(1)|^2}.$$

Proof. Let $B(x) = \sum_{k=0}^{n} b_k q_k(x)$ such that $B(1) = \sum_{k=0}^{n} b_k q_k(1) = 1$. Applying Cauchy's inequality we get

$$1 \leq \sum_{k=0}^{n} |b_k|^2 \sum_{k=0}^{n} |q_k(1)|^2,$$

where equality holds exactly when

$$b_k = a\, \overline{q_k(1)} \qquad \text{with} \quad a = \frac{1}{\sum_{k=0}^{n} |q_k(1)|^2}.$$

Since $\int_{D_s} |B(x)|^2 d\mu(x) = \sum_{k=0}^{n} |b_k|^2$ we have

$$\int_{D_s} |B(x)|^2 d\mu(x) \geq \frac{1}{\sum_{k=0}^{n} |q_k(1)|^2} = a$$

with equality if and only if $B(x) = a \sum_{k=0}^{n} \overline{q_k(1)}\, q_k(x)$. ◇

With the help of the above theorem we have the following result on the existence of consistent mean estimators, compare also Theorem 6.2.6.

Corollary 6.2.8 *Let $(X_n)_{n\in\mathbb{N}_0}$ be a weakly stationary random sequence on a polynomial hypergroup $\mathbb{N}_0$ generated by $(R_n)_{n\in\mathbb{N}_0}$. Assume that $E(X_n) = M$ for every $n \in \mathbb{N}_0$, and let μ be the spectral measure of $(X_n - M)_{n\in\mathbb{N}_0}$. There exist consistent linear unbiased estimators of M if and only if $\mu(\{1\}) = 0$.*

Proof. It is more or less folklore in the theory of orthonormal polynomials $q_k(x)$ that $\sum_{k=0}^{\infty} |q_k(x_0)|^2 = 1/\mu(\{x_0\})$. For the sake of completeness (and the lack of an appropriate reference) we give here a proof for $x_0 = 1$. Let U_n be neighbourhoods of $1 \in D_s$ with $\bigcap_{n=1}^{\infty} U_n = \{1\}$, and let $f_n \in C(D_s)$ such that $f_n(1) = 1$ and $f_n|D_s\backslash U_n = 0$. Plancherel' theorem says

$$\int_{D_s} |f_n(x)|^2 d\mu(x) = \sum_{k=0}^{\infty} |\tilde{f}(k)|^2,$$

where $\tilde{f}(k) = \int_{D_s} f(x)\ q_k(x)\ d\mu(x)$. By Lebesgue's dominated convergence theorem we obtain

$$\mu(\{1\}) = \lim_{n\to\infty} \int_{D_s} |f_n(x)|^2 d\mu(x) = \lim_{n\to\infty} \sum_{k=0}^{\infty} |\tilde{f}_n(k)|^2 = \mu(\{1\})^2 \sum_{k=0}^{\infty} |q_k(1)|^2,$$

since $\lim_{n\to\infty} \tilde{f}_n(k) = \lim_{n\to\infty} \int_{U_n} f_n(x)\ q_k(x)\ d\mu(x) = q_k(1)\ \mu(\{1\})$. Hence $1/\sum_{k=0}^{\infty} |q_k(1)|^2 = \mu(\{1\})$, which is also true if $\mu(\{1\}) = 0$. Thus by Theorem 6.2.7 the best linear unbiased estimators $M_n^\star$ are consistent exactly when $\mu(\{1\}) = 0$. ◇

We note that for $\mu = \pi$ we have $q_k(x) = \sqrt{h(k)}\ R_k(x)$. Hence

$$\frac{1}{\sum_{k=0}^{n} |q_k(1)|^2} = \frac{1}{H(n)} = \frac{1}{\chi_n * \chi_n(0)}.$$

For the sake of brevity we introduce the following notation

$$\lambda_n(\pi) = \frac{1}{H(n)} \qquad \text{and} \qquad \lambda_n(\mu) = \frac{1}{\sum_{k=0}^{n} |q_k(1)|^2}.$$

For the efficiency of the estimators $A_n = \lambda_n(\pi) \sum_{k=0}^{n} X_k h(k)$ we can derive the following results.

Proposition 6.2.9 *Let $(R_n)_{n\in\mathbb{N}_0}$ generate a polynomial hypergroup on $\mathbb{N}_0$ such that condition (H) holds. Let μ be the measure on D_s defined by $d\mu(x) = f(x)\, d\pi(x)$, where f is a nonnegative continuous function on D_s with $f(1) > 0$. Then*

$$\limsup_{n\to\infty} \frac{\lambda_n(\mu)}{\lambda_n(\pi)} \le f(1).$$

If f has in addition at most finitely many zeros of finite order then

$$\liminf_{n\to\infty} \frac{\lambda_n(\mu)}{\lambda_n(\pi)} \ge f(1).$$

Proof. Since the polynomial $B_n(x) = \lambda_n(\pi) \sum_{k=0}^{n} R_k(x)\ h(k) = \lambda_n(\pi)\ \mathcal{D}_n(x)$ is of degree n and $B_n(1) = 1$, we have

$$\lambda_n(\mu) \le \lambda_n^2(\pi) \int_{D_s} \mathcal{D}_n^2(x)\ f(x)\ d\pi(x) = \lambda_n(\pi) \sum_{k=0}^{2n} \frac{\chi_n * \chi_n(k)}{\chi_n * \chi_n(0)}\ \check{f}(k)\ h(k).$$

As shown in Theorem 6.2.5 we have $\lim\limits_{n\to\infty} \sum\limits_{k=0}^{2n} \dfrac{\chi_n * \chi_n(k)}{\chi_n * \chi_n(0)} \check{f}(k)\, h(k) = f(1)$.

Hence $\limsup\limits_{n\to\infty} \dfrac{\lambda_n(\mu)}{\lambda_n(\pi)} \le f(1)$.

For the second assertion note that we can choose a polynomial $P(x)$ with $P(1) = 1$ such that $g(x) := f^{-1}(x)\ |P(x)|^2$ is a continuous function on D_s. Let degree $(P) = m$. For any polynomial B with degree $(B) \le n$ and $B(1) = 1$ we have

$$B(x)\ P(x) = \int_{D_s} \sum_{k=0}^{n+m} R_k(x)\ h(k)\ R_k(y)\ B(y)\ P(y)\ d\pi(y).$$

Setting $x = 1$, Cauchy-Schwarz's inequality implies

$$1 = |B(1)\ P(1)|^2$$

$$\le \int_{D_s} |B(y)|^2 f(y)\ d\pi(y) \int_{D_s} f^{-1}(y)\ |P(y)|^2 \left(\sum_{k=0}^{n+m} R_k(y)\ h(k)\right)^2 d\pi(y)$$

$$= \int_{D_s} |B(y)|^2 d\mu(y) \left(\sum_{k=0}^{2n+2m} \frac{\chi_{n+m} * \chi_{n+m}(k)}{\chi_{n+m} * \chi_{n+m}(0)}\ \check{g}(k)\ h(k)\right) \frac{1}{\lambda_{n+m}(\pi)}.$$

By condition (H) we have

$$\lim_{n\to\infty} \frac{\lambda_n(\pi)}{\lambda_{n+m}(\pi)} = 1$$

as well as $\lim\limits_{n\to\infty} \sum\limits_{k=0}^{2n+2m} \dfrac{\chi_{n+m} * \chi_{n+m}(k)}{\chi_{n+m} * \chi_{n+m}(0)}\ \check{g}(k)\ h(k) = \dfrac{1}{f(1)}$.

Therefore

$$1 \le \liminf_{n\to\infty} \frac{\lambda_n(\mu)}{\lambda_n(\pi)} \frac{1}{f(1)}.$$

◇

Theorem 6.2.10 *Let $(X_n)_{n\in\mathbb{N}_0}$ be a weakly stationary random sequence on a polynomial hypergroup $\mathbb{N}_0$ generated by $(R_n)_{n\in\mathbb{N}_0}$ with property (H). Assume that $E(X_n) = M$ for every $n \in \mathbb{N}_0$, and let $d\mu(x) = f(x)\ d\pi(x)$ be the spectral measure of $(X_n - M)_{n\in\mathbb{N}_0}$. Further suppose that f is continuous, has at most finitely many zeros of finite order and satisfies $f(1) > 0$. Then the estimators A_n of Theorem 6.2.4 and the best estimators $M_n^\star$ have equivalent asymptotic efficiency, i.e.*

$$\lim_{n\to\infty} H(n)\ E(|A_n - M|^2) = f(1) = \lim_{n\to\infty} H(n)\ E(|M_n^\star - M|^2).$$

Proof. The first equality is shown in Theorem 6.2.5, whereas Proposition 6.2.9 and Theorem 6.2.7 imply the second one. ◇

Remark: Concerning the estimation of the covariance sequence $(d(n))_{n\in\mathbb{N}_0}$ of a weakly R_n-stationary random sequence we refer to [332].

3 Prediction

As before fix an orthogonal polynomial sequence $(R_n)_{n\in\mathbb{N}_0}$ which determines a polynomial hypergroup on $\mathbb{N}_0$. Let $(X_n)_{n\in\mathbb{N}_0}$ be a weakly stationary random sequence on the polynomial hypergroup $\mathbb{N}_0$. For brevity we write $H = H(X_n, \mathbb{N}_0)$, compare section 6.1. The problem in one-step prediction can be formulated as follows. Given $n \in \mathbb{N}_0$ denote the linear space generated by $X_0, ..., X_n$ by $H_n = \text{span}\,\{X_0, ..., X_n\} \subseteq H$. We want to characterize that $\hat{X}_{n+1} \in H_n$ with the minimum property

$$\| \hat{X}_{n+1} - X_{n+1} \|_2 = \min\{\| Y - X_{n+1} \|_2\colon Y \in H_n\}. \tag{1}$$

It is well-known that $\hat{X}_{n+1} = P_{H_n} X_{n+1}$, where P_{H_n} is the orthogonal projection from H onto H_n.

The difficulty now is to determine the coefficients $b_{n,k}$, $k = 0, ..., n$ in the representation

$$\hat{X}_{n+1} = \sum_{k=0}^{n} b_{n,k} X_k, \tag{2}$$

and to decide whether the prediction error

$$\delta_n := \| \hat{X}_{n+1} - X_{n+1} \|_2 \tag{3}$$

converges to zero as n tends to infinity. We call $(X_n)_{n\in\mathbb{N}_0}$ **asymptotic R_n-deterministic** if $\delta_n \to 0$ as $n \to \infty$.

It is well known from Hilbert space theory that $\hat{X}_{n+1}$ can be characterized by the property

$$E\left((X_{n+1} - \hat{X}_{n+1}) \cdot \overline{Y}\right) = < X_{n+1} - \hat{X}_{n+1}, Y > = 0$$

for all $Y \in H_n$.

Evaluating the scalar product for $Y = X_0, ..., X_n$ we get a linear equation for
$b = (b_{n,0}, ..., b_{n,n})^T$ of the form

$$\Phi^T b = \varphi, \tag{4}$$

where $\varphi = \left(E(X_{n+1}\overline{X}_0), E(X_{n+1}\overline{X}_1), ..., E(X_{n+1}\overline{X}_n)\right)^T$ and Φ is the $(n+1) \times (n+1)$-matrix $\Phi = \left(E(X_i\overline{X}_j)\right)_{0\le i,j\le n}$.

Provided Φ is non-singular we can compute the coefficients $b_{n,k}$ by inverting Φ, i.e. $b = (\Phi^{-1})^T\varphi$. Matrices of the form as Φ are the generalizations of Toeplitz-matrices in connection with the orthogonal polynomial sequence $(R_n)_{n\in\mathbb{N}_0}$. The (i, j)-entry is of the form $T_i a(j)$, where $a = (a(n))_{n\in\mathbb{N}_0}$ is a sequence and $T_i a(j) = \sum_{k=|i-j|}^{i+j} g(i,j;k)\, a(k)$. In equation (4) the matrix is determined by the "covariance"-sequence $d = (d(n))_{n\in\mathbb{N}_0} = \left(E(X_n\overline{X}_0)\right)_{n\in\mathbb{N}_0}$.

Proposition 6.3.1 *Let $(X_n)_{n\in\mathbb{N}_0}$ be a weakly stationary random sequence on the polynomial hypergroup $\mathbb{N}_0$. The matrix $\Phi = \left(E(X_i\overline{X}_j)\right)_{0\le i,j\le n}$ is singular for some $n \in \mathbb{N}_0$ if and only if the spectral measure μ has finite support.*

Proof. If Φ is singular there exists a vector $c = (c_0, ..., c_n)^T$ such that $\| \sum_{k=0}^{n} c_k X_k \|^2 = c^\star \Phi^T c = 0$. Applying the isometric isomorphism between H and $L^2(D_s, \mu)$ we obtain

$$\int_{D_s} \left| \sum_{k=0}^{n} c_k R_k(x) \right|^2 d\mu(x) = 0.$$

This is only possible if supp $\mu \subseteq \{x \in D_s : \sum_{k=0}^{n} c_k R_k(x) = 0\}$. Hence supp μ is a finite set.

Conversely if supp μ is a finite set, say of n points in D_s, it is possible to find $c_k \in \mathbb{C}$ such that supp μ coincides with the zero-set of $\sum_{k=0}^{n} c_k R_k(x)$. The reverse argument of before yields $c^\star \Phi^T c = 0$, i.e. Φ is singular. ◇

The isometric isomorphism of H onto $L^2(D_s, \mu)$ allows to transfer the prediction problem completely to a problem of orthogonal polynomials. Denote

$$\mathcal{P}_{n+1}^1 := \{Q : Q(x) = R_{n+1}(x) - \sum_{k=0}^{n} \beta_{n,k} R_k(x),\ \beta_{n,k} \in \mathbb{C}\}.$$

We determine the polynomial $Q^\star \in \mathcal{P}_{n+1}^1$ such that

$$\| Q^\star \|_2 = \min\{\| Q \|_2 : Q \in \mathcal{P}_{n+1}^1\}. \tag{5}$$

The coefficients $b_{n,k}$ in the representation $Q^\star(x) = R_{n+1}(x) - \sum_{k=0}^{n} b_{n,k} R_k(x)$ are exactly the coefficients of the representation of $\hat{X}_{n+1}$ in (2). The prediction error δ_n of (3) is equal to $\| Q^\star \|_2$.

Theorem 6.3.2 *Let $(X_n)_{n\in\mathbb{N}_0}$ be a weakly stationary random sequence on the polynomial hypergroup $\mathbb{N}_0$ generated by $(R_n)_{n\in\mathbb{N}_0}$. Assume that the spectral measure μ has an infinite support, and let $(q_n)_{n\in\mathbb{N}_0}$ be the orthonormal* *polynomial sequence with respect to* μ, *and denote* $\varrho_n = \varrho_n(\mu)$ *the positive leading coefficient of* $q_n(x)$, *i.e.* $q_n(x) = \varrho_n x^n + \dots$. *Further denote* $\sigma_n = \sigma_n(\pi)$ *the leading coefficient of* $R_n(x)$, *i.e.* $R_n(x) = \sigma_n x^n + \dots$. *Then the minimizing polynomial* $Q^\star \in \mathcal{P}^1_{n+1}$ *in (5) is given by*

$$Q^\star(x) = \frac{\sigma_{n+1}}{\varrho_{n+1}} q_{n+1}(x).$$

Hence

$$\hat{X}_{n+1} = \int_{D_s} \left(R_{n+1}(x) - \frac{\sigma_{n+1}}{\varrho_{n+1}} q_{n+1}(x) \right) dZ(x) \tag{6}$$

and for the prediction error holds

$$\delta_n = \frac{\sigma_{n+1}}{\varrho_{n+1}}. \tag{7}$$

Proof. All polynomials $Q \in \mathcal{P}^1_{n+1}$ can be written as

$$Q(x) = \frac{\sigma_{n+1}}{\varrho_{n+1}} q_{n+1}(x) + \sum_{k=0}^{n} \beta_k q_k(x).$$

Hence $\| Q \|_2^2 = \int_{D_s} |Q(x)|^2 d\mu(x) = |\beta_0|^2 + \dots + |\beta_n|^2 + \left| \frac{\sigma_{n+1}}{\varrho_{n+1}} \right|^2$ and $\| Q \|_2$ will be minimal for $\beta_0 = \dots = \beta_n = 0$. ◇

We note that by Theorem 6.3.2

$$X_n - \hat{X}_n = \frac{\sigma_n}{\varrho_n} \int_{D_s} q_n(x)\, dZ(x) \tag{8}$$

for $n \in \mathbb{N}$. Defining $\hat{X}_0 := 0$, we call $(X_n - \hat{X}_n)_{n \in \mathbb{N}_0}$ innovation, compare [122,§5.2].

Corollary 6.3.3 *Under the conditions of Theorem 6.3.2 the one-step predictor* $\hat{X}_{n+1}$ *has the innovation representation*

$$\hat{X}_{n+1} = \sum_{k=0}^{n} < R_{n+1}, q_k > \frac{\varrho_k}{\sigma_k} (X_k - \hat{X}_k). \tag{9}$$

Proof. Writing $\left(R_{n+1}(x) - \frac{\sigma_{n+1}}{\varrho_{n+1}} q_{n+1}(x) \right)$ as a Fourier expansion with respect to $(q_n)_{n \in \mathbb{N}_0}$, we obtain from (8)

$$\begin{aligned} \hat{X}_{n+1} &= \sum_{k=0}^{n} < R_{n+1}, q_k > \int_{D_s} q_k(x)\, dZ(x) \\ &= \sum_{k=0}^{n} < R_{n+1}, q_k > \frac{\varrho_k}{\sigma_k} (X_k - \hat{X}_k). \end{aligned}$$

◇

Solving the prediction problem we get involved with two orthogonal polynomial sequences: The orthogonal polynomial sequence $(R_n)_{n\in\mathbb{N}_0}$ with orthogonality measure π and the orthonormal polynomial sequence $(q_n)_{n\in\mathbb{N}_0}$ with the spectral measure μ as orthogonality measure. In particular for the asymptotic behaviour of the prediction error δ_n we have to deduce the asymptotics of σ_n (the leading coefficient of $R_n(x)$) and of ϱ_n (the leading coefficient of $q_n(x)$). We assume throughout that supp μ is an infinite set.

If the three term recurrence relations are given by

$$R_1(x)\, R_n(x) \;=\; a_n(\pi)\, R_{n+1}(x) \;+\; b_n(\pi)\, R_n(x) \;+\; c_n(\pi)\, R_{n-1}(x)$$

for $n = 1, 2, ...$ starting with

$$R_0(x) \;= 1, \qquad R_1(x) \;= \frac{1}{a_0(\pi)}\,(x - b_0(\pi))\,,$$

and

$$xq_n(x) \;=\; A_{n+1}(\mu)\, q_{n+1}(x) \;+\; B_n(\mu)\, q_n(x) \;+\; A_n(\mu)\, q_{n-1}(x)$$

for $n = 0, 1, 2, ...$ starting with

$$q_{-1}(x) \;= 0, \qquad q_0(x) \;= 1/\sqrt{\mu(D_s)},$$

we obtain from (7) for $n = 1, 2, 3, ...$

$$\delta_n \;=\; \delta_{n-1}\,\frac{A_{n+1}(\mu)}{a_n(\pi)\, a_0(\pi)} \tag{10}$$

starting with $\delta_0 = \dfrac{A_1(\mu)}{\sqrt{\mu(D_s)}\, a_0(\pi)}$.

Hence we have

$$\delta_n \;=\; \frac{1}{\sqrt{\mu(D_s)}}\,\frac{1}{(a_0(\pi))^n}\prod_{k=0}^{n}\frac{A_{k+1}(\mu)}{a_k(\pi)}. \tag{11}$$

A standard procedure in the Hilbert space $L^2(D_s, \mu)$ also yields

$$\delta_n^2 \;=\; \Delta_{n+1}/\Delta_n, \tag{12}$$

where Δ_n is the Gramian determinant $\det\left(E(X_i\overline{X}_j)\right)_{0\le i,j\le n}$. Note that $\Delta_n > 0$ for each n if and only if supp μ is infinite, see Proposition 6.3.1.

In case that the Haar weights $h(n)$ tend to infinity it is easy to derive sufficient conditions for $\delta_n \to 0$.

Lemma 6.3.4 *Let $(R_n)_{n\in\mathbb{N}_0}$ determine a polynomial hypergroup on $\mathbb{N}_0$, and assume that $g(n,n;0) = h(n)^{-1} \to 0$. Then for every $k,l \in \mathbb{N}_0$ holds $g(n,n+l;k) \to 0$ as $n \to \infty$.*

Proof. For $n \geq k$ follows by the Cauchy-Schwarz inequality

$$g(n,n+l;k) = h(k) \int_{D_s} R_n(x)\, R_{n+l}(x)\, R_k(x)\, d\pi(x)$$

$$\leq h(k)\, (g(n,n;0))^{1/2} \left(\int_{D_s} (R_{n+l}(x)\, R_k(x))^2\, d\pi(x) \right)^{1/2}.$$

Since $|R_{n+l}(x)\, R_k(x)| \leq 1$ for $x \in D_s$, we see that $g(n,n+l;k) \to 0$ as $n \to \infty$. ◇

Theorem 6.3.5 *Let $(X_n)_{n\in\mathbb{N}_0}$ be a weakly stationary random sequence on the polynomial hypergroup $\mathbb{N}_0$ generated by $(R_n)_{n\in\mathbb{N}_0}$. Assume that $h(n) \to \infty$ as $n \to \infty$. If $d(n) = E(X_n\overline{X}_0)$ tends to zero, the random sequence $(X_n)_{n\in\mathbb{N}_0}$ is asymptotic R_n-deterministic.*

Proof. We have $E(X_n\overline{X}_n) = \sum_{k=0}^{2n} g(n,n;k)\, d(k)$ and $\sum_{k=0}^{2n} g(n,n;k) = 1$. Toeplitz's theorem and Lemma 6.3.4 yield that $E(X_n\overline{X}_n) \to 0$ as $n \to \infty$. The optimality condition of the one-step predictor gives

$$\delta_n^2 \leq \int_{D_s} R_{n+1}^2(x)\, d\mu(x) = E(X_{n+1}\overline{X}_{n+1}).$$

◇

Corollary 6.3.6 *Let $(X_n)_{n\in\mathbb{N}_0}$ be a weakly stationary random sequence on the polynomial hypergroup $\mathbb{N}_0$ generated by $(R_n)_{n\in\mathbb{N}_0}$, and let $h(n) \to \infty$ as $n \to \infty$. If the spectral measure μ is absolutely continuous with respect to π, i.e. $\mu = f\pi$, then $(X_n)_{n\in\mathbb{N}_0}$ is asymptotic R_n-deterministic.*

Proof. Apply Proposition 3.3.2 (ii). ◇

Proving Theorem 6.3.5 we have shown by the way a result worthwhile to be noted explicitly.

Proposition 6.3.7 *Let $(R_n)_{n\in\mathbb{N}_0}$ be an orthogonal polynomial sequence generating a polynomial hypergroup on $\mathbb{N}_0$. Assume that $h(n) \to \infty$ as $n \to \infty$. Given any bounded positive definite sequence $(a(n))_{n\in\mathbb{N}_0} = a$ the following conditions are equivalent:*

(i) $a(n) \to 0$ *as* $n \to \infty$.
(ii) $L_n a(n+l) \to 0$ *as* $n \to \infty$ *for every* $l \in \mathbb{N}_0$.

Proof. Let $a(n) = \int_{D_s} R_n(x)\, d\mu(x), \;\; \mu \in M^+(D_s)$. If (ii) holds we have (with $l = 0$):

$$|a(n)| = \left| \int_{D_s} R_n(x)\, d\mu(x) \right| \leq \mu(D_s)^{1/2} \left(\int_{D_s} R_n^2(x)\, d\mu(x) \right)^{1/2}$$
$$= \mu(D_s)^{1/2} L_n a(n) \longrightarrow 0 \quad \text{as } n \to \infty.$$

If conversely (i) holds, we obtain

$$L_n a(n+l) \;=\; \sum_{k=l}^{2n+l} g(n, n+l; k)\, a(k).$$

Now apply Lemma 6.3.4 and Toeplitz's theorem to derive $L_n a(n+l) \to 0$ as $n \to \infty$. ◇

We can improve the inequality $\delta_n^2 \leq E(X_{n+1}\overline{X}_{n+1})$, which was used in Theorem 6.3.5. Denote the Gramian determinant

$$\Delta(X_n, ..., X_{n+m}) \;=\; \det \begin{pmatrix} < X_n, X_n > \cdots & < X_n, X_{n+m} > \\ \vdots & \vdots \\ < X_{n+m} > \;\cdots & < X_{n+m}, X_{n+m} > \end{pmatrix}.$$

By (12) and [486,p.46] we have a sequence of upper bounds for δ_n :

$$\delta_n^2 = \frac{\Delta(X_0, ..., X_{n+1})}{\Delta(X_0, ..., X_n)} \leq \frac{\Delta(X_1, ..., X_{n+1})}{\Delta(X_1, ..., X_n)}$$
$$\leq ... \leq \frac{\Delta(X_n, X_{n+1})}{\Delta(X_n)} \leq \Delta(X_{n+1}). \tag{13}$$

By means of (13) we can at least partly deal with spectral measures that contain discrete or singular parts.

Theorem 6.3.8 *Let $(X_n)_{n\in\mathbb{N}_0}$ be a weakly stationary random sequence on the polynomial hypergroup $\mathbb{N}_0$ generated by $(R_n)_{n\in\mathbb{N}_0}$, where $h(n) \to \infty$ as $n \to \infty$. Assume that the spectral measure has the form*

$$\mu \;=\; f\pi \;+\; \mu_0 \;+\; \mu_1,$$

where $f \in L^1(D_s, \pi)$, supp $\mu_0 \subseteq D_{s,0} := \{x \in D_s : \;|R_n(x)| \to 0 \;\text{ for }\; n \to \infty\}$ and supp $\mu_1 \subseteq D_s \backslash D_{s,0}$ is a finite set (or empty). If $m = |$supp $\mu_1|$ furthermore suppose that there is some $n_0 \in \mathbb{N}$ such that

$$\inf_{n \geq n_0} \det \begin{pmatrix} < R_n, R_n >_{\mu_1} & \cdots & < R_n, R_{n+m-1} >_{\mu_1} \\ \vdots & & \vdots \\ < R_{n+m-1}, R_n >_{\mu_1} & \cdots & < R_{n+m-1}, R_{n+m-1} >_{\mu_1} \end{pmatrix} > 0.$$

Then the random sequence $(X_n)_{n\in\mathbb{N}_0}$ is asymptotic R_n-deterministic.

Proof. If $\mu_1 = 0$ we have $d(n,n) = \int_{D_s} R_n^2(x)\ f(x)\ d\pi(x) + \int_{D_s} R_n^2(x)\ d\mu_0(x)$. By the theorem of dominated convergence and Corollary 6.3.6 we have $d(n,n) \to 0$ and the statement follows.
If $m = |\text{supp}\ \mu_1|$ we consider the Gramian determinants $\Delta(X_n, ..., X_{n+m-1})$ and $\Delta(X_n, ..., X_{n+m})$. In general expanding the determinants we obtain

$$\Delta(X_n, ..., X_{n+k})$$
$$= \det \begin{pmatrix} < R_n, R_n >_{\mu_1} & \cdots & < R_n, R_{n+k} >_{\mu_1} \\ \vdots & & \vdots \\ < R_{n+k}, R_n >_{\mu_1} & \cdots & < R_{n+k}, R_{n+k} >_{\mu_1} \end{pmatrix} + \sum_{\sigma \in I_k} \varphi_\sigma,$$

where $|I_k|$ is finite and independent of n, and where the φ_σ's are products of $< R_{n+i_1}, R_{n+j_1} >_{f\pi+\mu_0}$ and $< R_{n+i_2}, R_{n+j_2} >_{\mu_1}$ containing at least one factor $< R_{n+i}, R_{n+j} >_{f\pi+\mu_0}$. Since all $< R_{n+i}, R_{n+j} >_{f\pi+\mu_0}$ tend to zero as $n \to \infty$, we conclude from the assumption, that $1/\Delta(X_n, ..., X_{n+m-1})$ is bounded for $n \geq n_0$. Since $|\text{supp}\ \mu_1| = m$ the $(m+1) \times (m+1)$-determinant

$$\det \begin{pmatrix} < R_n, R_n >_{\mu_1} & \cdots & < R_n, R_{n+m} >_{\mu_1} \\ \vdots & & \vdots \\ < R_{n+m}, R_n >_{\mu_1} & \cdots & < R_{n+m}, R_{n+m} >_{\mu_1} \end{pmatrix}$$

has to be zero. Hence $\Delta(X_n, ..., X_{n+m})$ tends to zero with $n \to \infty$, and by (13) we get $\delta_n \to 0$. $\diamond$

A typical situation where Theorem 6.3.8 can be used, is when $D_s = [-1, 1]$ and
$|R_n(x)| \to 0$ when $n \to \infty$ for every $x \in]-1, 1[$. If π is an even measure we have $R_n(-1) = (-1)^n$ and of course $R_n(1) = 1$. For the discrete measure $\mu_1 = \alpha\epsilon_{-1} + \beta\epsilon_1;\ \ \alpha, \beta > 0$ we have

$$\det \begin{pmatrix} < R_n, R_n >_{\mu_1} & < R_n, R_{n+1} >_{\mu_1} \\ < R_{n+1}, R_n >_{\mu_1} & < R_{n+1}, R_{n+1} >_{\mu_1} \end{pmatrix} = 4\alpha\beta.$$

Therefore we have asymptotic R_n-determinacy independent of the spectral measure μ. A sufficient condition for $|R_n(x)| \to 0$ for every $x \in]-1, 1[$ can be derived by bringing the Turan determinant into use. In subsection 4.11 we have already studied the Turan determinants in order to derive conditions for the boundedness of the Haar weights. Now we will investigate the impact of $h(n) \to \infty$ on the convergence properties of $(R_n(x))_{n\in\mathbb{N}_0}$, $x \in S$. It is appropriate to modify the Turan determinants a little bit and put

$$\Theta_n(x) := h(n) \left(R_n^2(x) - \frac{a_n}{a_{n-1}} R_{n-1}(x)\, R_{n+1}(x) \right) \tag{14}$$

for $n \in \mathbb{N}$.

We restrict to the symmetric case, i.e. $b_n = 0$, and assume further that $a_n \to a$ as $n \to \infty$, and of course that $(R_n)_{n\in\mathbb{N}_0}$ generates a polynomial hypergroup on $\mathbb{N}_0$. In Theorem 4.3.3 using theorems of Blumenthal and Poincaré we have proven that the only possible values of a are $\frac{1}{2} \le a < 1$ and

$$\text{supp}\, \pi = [-\gamma, \gamma] \qquad \text{with } \gamma := 2\sqrt{a(1-a)}$$

and $D_s = [-1,1]$. Note that $\gamma = 1$ exactly when $a = \frac{1}{2}$. Concerning $\Theta_n(x)$ we can prove in that case:

Proposition 6.3.9 *Assume that $(R_n)_{n\in\mathbb{N}_0}$ is even and generates a polynomial hypergroup on $\mathbb{N}_0$. Further let $a_n \to a$ as $n \to \infty$. The following inequalities for $\Theta_n(x)$ are valid:*

(i) $\Theta_n(x)/h(n) \le C_1 \left(R_{n-1}^2(x) + R_n^2(x) + R_{n+1}^2(x)\right)$ *for all $x \in \mathbb{R}$, where $C_1 > 0$ is a constant independent of x and n.*

(ii) $|\Theta_n(x) - \Theta_{n-1}(x)|/h(n) \le C_2\, |a_{n-1}c_n - a_{n-2}c_{n-1}| \left(R_{n-1}^2(x) + R_n^2(x)\right)$ *for all $x \in [-1,1]$, where $C_2 > 0$ is a constant independent of x and n.*

(iii) Given some $\delta \in\,]0,\gamma[$ there is $N \in \mathbb{N}$ such that $\Theta_n(x)/h(n) \ge C_3 \left(R_{n-1}^2(x) + R_n^2(x)\right)$ *for all $x \in [-\gamma+\delta, \gamma-\delta]$ and $n \ge N$, where $C_3 > 0$ is a constant independent of x and n.*

Proof. (Compare with the calculations of Proposition 4.11.7)

(i) Since $\lim_{n\to\infty} \frac{a_n}{a_{n-1}} = 1$ and $2|R_{n-1}(x)\, R_{n+1}(x)| \le R_{n-1}^2(x) + R_{n+1}^2(x)$, the inequality of (i) follows immediately.

(ii) By using the recurrence relation we get

$$\Theta_n(x) = h(n)\, R_n^2(x) + h(n-1)\, R_{n-1}^2(x) - \frac{x}{a_{n-1}} h(n)\, R_{n-1}(x)\, R_n(x) \tag{15}$$

and

$$\Theta_n(x) = h(n)\, R_n^2(x) + h(n+1) \frac{a_n c_{n+1}}{a_{n-1} c_n} R_{n+1}^2(x) - h(n) \frac{a_n x}{a_{n-1} c_n} R_n(x)\, R_{n+1}(x). \tag{16}$$

Applying (16) to Θ_n gives

$$\Theta_n(x) - \Theta_{n-1}(x) = h(n)\left(1 - \frac{a_{n-1}c_n}{a_{n-2}c_{n-1}}\right) R_n^2(x)$$

$$+ \left(h(n-1)\frac{a_{n-1}}{a_{n-2}c_{n-1}} - h(n)\frac{1}{a_{n-1}}\right) xR_{n-1}(x)\, R_n(x)$$

$$= h(n)\left(\frac{a_{n-2}c_{n-1} - a_{n-1}c_n}{a_{n-2}c_{n-1}} R_n^2(x) + \frac{a_{n-1}c_n - c_{n-1}a_{n-2}}{c_{n-1}a_{n-2}a_{n-1}} xR_{n-1}(x)\, R_n(x)\right).$$

Since $a_n \to a,\; c_n \to 1-a > 0$ and by $|x| \le 1$ we obtain

$$|\Theta_n(x) - \Theta_{n-1}(x)|/h(n) \le C_2\, |a_{n-1}c_n - a_{n-2}c_{n-1}|\, \left(R_{n-1}^2(x) + R_n^2(x)\right).$$

(iii) Applying (15) we have

$$\Theta_n(x) = h(n)\left(R_n(x) - \frac{x}{2a_{n-1}} R_{n-1}(x)\right)^2 + \left(1 - \frac{x^2}{4c_n a_{n-1}}\right) h(n-1)\, R_{n-1}^2(x)$$

and

$$\Theta_n(x) = h(n-1)\left(R_{n-1}(x) - \frac{x}{2c_n} R_n(x)\right)^2 + \left(1 - \frac{x^2}{4c_n a_{n-1}}\right) h(n)\, R_n^2(x).$$

In particular it follows that

$$\Theta_n(x) \ge \left(1 - \frac{x^2}{4c_n a_{n-1}}\right) h(n-1)\, R_{n-1}^2(x)$$

and

$$\Theta_n(x) \ge \left(1 - \frac{x^2}{4c_n a_{n-1}}\right) h(n)\, R_n^2(x).$$

Fix $\delta \in\,]0,\gamma[$ and let $x \in [-\gamma+\delta, \gamma-\delta]$. Since $c_n a_{n-1} \to a(1-a) = \frac{\gamma^2}{4}$, there is a constant C_3 and $N \in \mathbb{N}$ such that

$$\Theta_n(x)/h(n) \ge C_3\left(R_{n-1}^2(x) + R_n^2(x)\right)$$

for all $n \ge N$.

◇

In case of $a_n \to a,\; \frac{1}{2} < a < 1,\;$ supp $\pi = [-\gamma,\gamma]$ is a proper subset of $D_s = [-1,1]$. Dealing with $x \in D_s \backslash \text{supp}\, \pi$ we prove:

Proposition 6.3.10 *Assume that $(R_n)_{n\in\mathbb{N}_0}$ is even and generates a polynomial hypergroup on $\mathbb{N}_0$. Further let $a_n \to a$ such that $\frac{1}{2} < a < 1$. Then $R_n(x) \to 0$ as $n \to \infty$ for each $x \in\,]-1,-\gamma] \cup [\gamma, 1[$.*

Proof. It is sufficient to consider $x \in [\gamma, 1[$. Fix $\alpha_0 \in]\gamma, 1[$. By the separating property of zeros of $R_n(x)$, see [138], we get $R_n(\alpha_0) > 0$. Now $Q_n(x) = R_n(\alpha_0 x)/R_n(\alpha_0)$ defines a polynomial hypergroup on $\mathbb{N}_0$. The recurrence relation of the $Q_n(x)$ is given by

$$xQ_n(x) = \tilde{a}_n Q_{n+1}(x) + \tilde{c}_n Q_{n-1}(x),$$

where $\tilde{a}_n = a_n \dfrac{R_{n+1}(\alpha_0)}{R_n(\alpha_0)\,\alpha_0}$, $\tilde{c}_n = c_n \dfrac{R_{n-1}(\alpha_0)}{R_n(\alpha_0)\,\alpha_0}$.

By Poincaré's theorem, compare the proof of Theorem 4.3.3, we have

$$\lim_{n\to\infty} \frac{R_n(\alpha_0)}{R_{n-1}(\alpha_0)} < 1.$$

Since $\tilde{a}_n$ is convergent, the dual space of the polynomial hypergroup generated by $(Q_n)_{n\in\mathbb{N}_0}$ is $[-1, 1]$. In particular

$$|R_n(\alpha_0 x)/R_n(\alpha_0)| \le 1$$

for all $x \in [-1, 1]$, $n \in \mathbb{N}_0$. That means $\lim\limits_{n\to\infty} |R_n(\alpha_0 x)| \le \lim\limits_{n\to\infty} R_n(\alpha_0) = 0$ for all $x \in [-1, 1]$. Hence $\lim\limits_{n\to\infty} R_n(x) = 0$ for all $x \in [\gamma, 1[$. ◇

Theorem 6.3.11 *Assume that $(R_n)_{n\in\mathbb{N}_0}$ is even and generates a polynomial hypergroup on $\mathbb{N}_0$. Let $a_n \to a$, where $\frac{1}{2} \le a < 1$ and suppose that $h(n) \to \infty$ and*

$$\sum_{n=1}^{\infty} |a_n c_{n+1} - a_{n-1} c_n| < \infty. \tag{17}$$

Then every weakly stationary random sequence $(X_n)_{n\in\mathbb{N}_0}$ on the polynomial hypergroup $\mathbb{N}_0$ generated by $(R_n)_{n\in\mathbb{N}_0}$ is asymptotic R_n-deterministic.

Proof. At first we deal with $a = \frac{1}{2}$, i.e. $\operatorname{supp} \pi = [-1, 1] = D_s$. Fix some $x \in]-1, 1[$. By Proposition 6.3.9 (ii) and (iii) we have

$$|\Theta_n(x) - \Theta_{n-1}(x)| \le h(n)\, C_2\, |a_{n-1}c_n - a_{n-2}c_{n-1}| \left(R_{n-1}^2(x) + R_n^2(x)\right)$$

$$\le \frac{C_2}{C_3} |a_{n-1}c_n - a_{n-2}c_{n-1}|\, \Theta_n(x) = \varepsilon_n \Theta_n(x)$$

for all $n \ge N$, where $\varepsilon_n := \dfrac{C_2}{C_3} |a_{n-1}c_n - a_{n-2}c_{n-1}|$. Thus

$$\frac{1}{1+\varepsilon_n}\, \Theta_{n-1}(x) \le \Theta_n(x) \le \frac{1}{1-\varepsilon_n}\, \Theta_{n-1}(x)$$

for all $n \geq N$ (we can assume $\varepsilon_n < 1$). Since $\sum_{n=2}^{\infty} \varepsilon_n$ is convergent, $\Theta_n(x)$ is convergent. Applying once more Proposition 6.3.9 (iii) and $h(n) \to \infty$ we get $R_n(x) \to 0$. The statements directly after Theorem 6.3.8 yield that every weakly stationary $(X_n)_{n\in\mathbb{N}_0}$ is asymptotic R_n-deterministic.

For $\frac{1}{2} < a < 1$ we use also Proposition 6.3.9 (ii) and (iii) to show that $R_n(x) \to 0$ for every $x \in \,]-\gamma, \gamma[$. By Proposition 6.3.10 we get $R_n(x) \to 0$ for all $x \in \,]-1, -\gamma] \cup [\gamma, 1[$, and hence $(X_n)_{n\in\mathbb{N}_0}$ is asymptotic R_n-deterministic also in case of $\frac{1}{2} < a < 1$. ◇

Remarks:

(i) Orthogonal polynomials with property (17) are called of bounded variation.

(ii) When only $a_n \to a$, $\frac{1}{2} < a < 1$, holds, we obtain from Proposition 6.3.10 that every weakly stationary random sequence $(X_n)_{n\in\mathbb{N}_0}$ with spectral measure μ, $\operatorname{supp} \mu \subseteq [-1, -\gamma] \cup [\gamma, 1]$, is asymptotic R_n-deterministic.

Examples:

(1) **Ultraspherical polynomials:** Let $R_n^{(\alpha,\alpha)}(x)$ be the ultraspherical polynomials with $\alpha > -\frac{1}{2}$. They are subclass of the Jacobi polynomials $R_n^{(\alpha,\beta)}(x)$, see example (2) below, with even orthogonalization measure. Nevertheless the asymptotic behaviour of $R_n^{(\alpha,\beta)}(x)$ is well-known, we want to bring Theorem 6.3.11 into use. Since
$a_n = \dfrac{n+2\alpha+1}{2n+2\alpha+1}$, $c_n = \dfrac{n}{2n+2\alpha+1}$ we can directly compute that

$$a_n c_{n+1} - a_{n-1} c_n = \frac{4\alpha^2 - 1}{(2n+2\alpha-1)\,(2n+2\alpha+1)\,(2n+2\alpha+3)}.$$

Hence $\sum_{n=1}^{\infty} |a_n c_{n+1} - a_{n-1} c_n| < \infty$.

Moreover $a_n \to \frac{1}{2}$ and $h(n) = \dfrac{(2n+2\alpha+1)\,(2\alpha+1)_n}{(2\alpha+1)\,n!} \to \infty$.

Therefore every weakly stationary random sequence $(X_n)_{n\in\mathbb{N}_0}$ (weakly stationary with respect to $(R_n^{(\alpha,\alpha)})_{n\in\mathbb{N}_0}$, $\alpha > -\frac{1}{2}$) is asymptotic $R_n^{(\alpha,\alpha)}$-deterministic.

(2) **Jacobi polynomials:** Let $R_n^{(\alpha,\beta)}(x)$ be the Jacobi polynomials with $\alpha \geq \beta > -1$, $\alpha + \beta + 1 \geq 0$. By [641] formula (4.1.1) and (8.21.8) we have

$$|R_n^{(\alpha,\beta)}(x)| = O(n^{-\alpha-\frac{1}{2}})$$

for all $x \in]-1,1[$. Thus $R_n^{(\alpha,\beta)}(x) \to 0$ for $x \in]-1,1[$ provided $\alpha > -\frac{1}{2}$. Also when $\alpha > \beta$ we have by [641] formula (4.1.1) and (4.1.14)

$$R_n^{(\alpha,\beta)}(-1) = (-1)^n \binom{n+\beta}{n} / \binom{n+\alpha}{n}.$$

Hence $R_n^{(\alpha,\beta)}(-1)$ tends to zero if $\alpha > \beta$. The consideration of Theorem 6.3.8 implies that every $(X_n)_{n\in\mathbb{N}_0}$ is asymptotic $R_n^{(\alpha,\beta)}$-deterministic. So we see that — apart from the case $\alpha = \beta = -\frac{1}{2}$ — without any assumption on the spectral measure we have asymptotic $R_n^{(\alpha,\beta)}$ — deterministic behaviour of $(X_n)_{n\in\mathbb{N}_0}$. This result can also be obtained by Theorem 6.3.13 below, using that the orthogonalization measure $d\pi(x) = c_{\alpha,\beta}(1-x)^\alpha(1+x)^\beta \chi_{[-1,1]} dx$ satisfies the Szegö-condition. (See [335,Theorem 4.2].)

(3) **Polynomials connected with homogeneous trees:** Let $R_n(x;a)$, $a \geq 2$, be the orthogonal polynomials connected with homogeneous trees, see subsection 4.5.7.2. They are defined by the recurrence coefficients $a_n = \dfrac{a-1}{a}$, $b_n = 0$, $c_n = \dfrac{1}{a}$ for $n \in \mathbb{N}$, and $a_0 = 1$, $b_0 = 0$. Since $a = 2$ gives the Chebyshev polynomials of first kind, we consider $a > 2$. We have $h(n) = a(a-1)^{n-1} \to \infty$ and $a_n c_{n+1} - a_{n-1}c_n = 0$ for $n = 2,3,\ldots$. Hence Theorem 6.3.11 can be applied, i.e. every weakly stationary random sequence $(X_n)_{n\in\mathbb{N}_0}$ (weakly stationary with respect to $R_n(x;a)$, $a > 2$) is asymptotic in $R_n(\ ;a)$-deterministic.

Another access using the Kolmogorov-Szegö-property on $\mathbb{T}$ is investigated now. To each orthonormal polynomial sequence $(p_n)_{n\in\mathbb{N}_0}$, orthonormal with respect to ν, such that supp $\nu \subseteq [-1,1]$ and $|\text{supp } \nu| = \infty$, one can associate a unique polynomial sequence $(\psi_n)_{n\in\mathbb{N}_0}$ on $\mathbb{T} := \{z \in \mathbb{C} : |z| = 1\}$ orthonormal with respect to a measure α given by

$$d\alpha(t) := |\sin t|\, d\nu(\cos t), \qquad t \in]-\pi,\pi] \tag{18}$$

where we identify $]-\pi,\pi]$ with $\mathbb{T}$ by the mapping $t \mapsto e^{it}$. Denote the positive leading coefficients of $\psi_n(t)$ by $\kappa_n(\alpha)$.

We cite now some important results on orthogonal polynomials on $\mathbb{T}$, that we will use for orthogonal polynomials on $[-1,1]$. The original references are [271] and [641], see also [415,Ch.14].

If the Radon Nikodym derivative α' of α fulfills the Kolmogorov-Szegö-property, i.e. $\ln(\alpha') \in L^1(\mathbb{T})$ or equivalently

$$\int_{-\pi}^{\pi} \ln(\alpha'(t))\, dt > -\infty,$$

then the leading coefficients $\kappa_n(\alpha)$ converge monotone increasing towards

$$\kappa(\alpha) := \exp\left(-\frac{1}{4\pi}\int_{-\pi}^{\pi} \ln(\alpha'(t))\, dt\right).$$

If the Kolmogorov-Szegö-property does not hold we have $\lim_{n\to\infty} \kappa_n(\alpha) = \infty$.

This result can be transferred to $[-1, 1]$. In fact we have, see [271,Theorem 9.2]:

Proposition 6.3.12 *Let $(p_n)_{n\in\mathbb{N}_0}$ be an orthonormal polynomial sequence with respect to a measure ν with supp $\nu \subseteq [-1,1]$, $|\text{supp } \nu| = \infty$. Let ν' be the Radon Nikodym derivative and $\kappa_n(\nu)$ the leading coefficients of p_n. Then alternatively we have*

(i) $\displaystyle\int_{-1}^{1} \frac{\ln(\nu'(x))}{\sqrt{1-x^2}}\, dx > -\infty$, *and there exist positive constants C_1, C_2 such that*

$$C_1 \le \frac{\kappa_n(\nu)}{2^n} \le C_2 \qquad \text{for all } n \in \mathbb{N}_0$$

or

(ii) $\displaystyle\int_{-1}^{1} \frac{\ln(\nu'(x))}{\sqrt{1-x^2}}\, dx = -\infty, \qquad$ *and* $\displaystyle\lim_{n\to\infty} \frac{\kappa_n(\nu)}{2^n} = \infty.$

Proof. Consider the orthonormal polynomial sequence $(\psi_n)_{n\in\mathbb{N}_0}$ on $\mathbb{T}$ with leading coefficients $\kappa_n(\alpha)$, where α is the measure as in (18) above. The coefficients $\kappa_n(\nu)$ and $\kappa_n(\alpha)$ are connected by the inequalities, see [271,Eq.(9.9)]:

$$\frac{\kappa_{2n-1}(\alpha)}{2\sqrt{\pi}} \le \frac{\kappa_n(\nu)}{2^n} \le \frac{\kappa_{2n}(\alpha)}{\sqrt{\pi}}, \qquad n \in \mathbb{N}. \tag{19}$$

Moreover we have

$$\int_{-\pi}^{\pi} \ln(\alpha'(t))\, dt = 2\int_{-1}^{1} \frac{\ln(\nu'(x))}{\sqrt{1-x^2}} dx,$$

and the assertions (i) and (ii) follow by the convergence of $\kappa_n(\alpha)$ towards $\kappa(\alpha)$ or towards infinity, respectively. $\diamond$

For the orthonormal version $p_n(x)$ of $R_n(x)$ we have $p_n(x) = \sqrt{h(n)}\, R_n(x)$. Hence for the leading coefficients $\sigma_n(\pi)$ of $R_n(x)$ and the positive leading coefficients $\varrho_n(\pi)$ of $p_n(x)$ it follows $\sigma_n(\pi) = \frac{\varrho_n(\pi)}{\sqrt{h(n)}}$. We obtain the following result.

Theorem 6.3.13 *Let $(X_n)_{n\in\mathbb{N}_0}$ be a weakly stationary random sequence on the polynomial hypergroup $\mathbb{N}_0$ generated by $(R_n)_{n\in\mathbb{N}_0}$. Assume that $\operatorname{supp}\pi \subseteq [-1,1]$ and that π fulfils the Kolmogorov-Szegö property on $[-1,1]$, i.e.*

$$\int_{-1}^{1} \frac{\ln(\pi'(x))}{\sqrt{1-x^2}}\, dx \;>\; -\infty.$$

Then $(X_n)_{n\in\mathbb{N}_0}$ is asymptotically R_n-deterministic if and only if

$$\lim_{n\to\infty} \frac{\sqrt{h(n)}\,\varrho_n(\mu)}{2^n} \;=\; \infty,$$

where $\varrho_n(\mu)$ is the positive leading coefficient of $q_n(x)$, $(q_n(x))_{n\in\mathbb{N}_0}$ being the orthonormal polynomials with respect to the spectral measure μ of $(X_n)_{n\in\mathbb{N}_0}$.

Proof. We know by (7) that

$$\delta_n \;=\; \frac{\sigma_{n+1}(\pi)}{\varrho_{n+1}(\mu)} \;=\; \frac{\varrho_{n+1}(\pi)}{\sqrt{h(n+1)}\,\varrho_{n+1}(\mu)}.$$

Since π fulfils the Kolmogorov-Szegö property, by Proposition 6.3.12 there exist positive constants C_1, C_2 such that

$$0 \;<\; C_1 \;\le\; \frac{\varrho_{n+1}(\pi)}{2^{n+1}} \;\le\; C_2 \qquad \text{for all } n\in\mathbb{N}_0.$$

Hence

$$0 \;<\; C_1 \;\le\; \sqrt{h(n+1)}\,\frac{\varrho_{n+1}(\mu)}{2^{n+1}}\,\delta_n \;\le\; C_2 \qquad \text{for all } n\in\mathbb{N}_0,$$

and the statement of equivalence follows. ◇

Corollary 6.3.14 *Under the conditions of Theorem 6.3.13, the following statements are true:*

(i) If $\lim_{n\to\infty} h(n) = \infty$, then $(X_n)_{n\in\mathbb{N}_0}$ is asymptotically R_n-deterministic without any assumption on the spectral measure μ. For the prediction error we have at least $\delta_n = O(\frac{1}{\sqrt{h(n+1)}})$ as $n\to\infty$.

(ii) If $\{h(n) : n\in\mathbb{N}_0\}$ is bounded, then $(X_n)_{n\in\mathbb{N}_0}$ is asymptotically R_n-deterministic if and only if μ does not fulfil the Kolmogorov-Szegö property on $[-1,1]$.

(iii) If $\{h(n) : n\in\mathbb{N}_0\}$ is unbounded, then $(X_n)_{n\in\mathbb{N}_0}$ is asymptotically R_n-deterministic provided μ does not fulfil the Kolmogorov-Szegö property on $[-1,1]$.

Proof. The three statements follow immediately from

$$0 < C_1 \leq \sqrt{h(n+1)}\, \frac{\varrho_{n+1}(\mu)}{2^n}\, \delta_n \leq C_2,$$

shown in the proof of the above theorem and by Proposition 6.3.12 applied to the spectral measure μ. ◇

Finally we show if $(h(n))_{n\in\mathbb{N}_0}$ does not converge to infinity, then π fulfils the Kolmogorov-Szegö property. This result may be seen as a supplement of subsection 4.11, where the boundedness of $\{h(n) : n \in \mathbb{N}_0\}$ is studied.

Proposition 6.3.15 *Let $(R_n)_{n\in\mathbb{N}_0}$ generate a polynomial hypergroup on $\mathbb{N}_0$, and suppose that $D_s = [-1,1]$. If π does not fulfil the Kolmogorov-Szegö property, then $\lim_{n\to\infty} h(n) = \infty$.*

Proof. Write

$$R_n(x) = \sum_{k=0}^{n} a_{n,k}\, T_k(x),$$

where $a_{n,k}$ are the connection coefficients of the representation of $R_n(x)$ by the Chebyshev polynomials $T_k(x)$ of the first kind. Note that $T_k(x) = 2^{k-1}x^k + \cdots$. Since $|R_n(x)| \leq 1$ for every $x \in D_s$, $n \in \mathbb{N}_0$, we obtain applying the orthogonalization measure π^T of $(T_k(x))_{k\in\mathbb{N}_0}$,

$$1 \geq \int_{-1}^{1} R_n^2(x)\, d\pi^T(x) \geq a_{n,n}^2 \int_{-1}^{1} T_n^2(x)\, d\pi^T(x) = \frac{a_{n,n}^2}{2}.$$

Comparing the leading coefficients of $R_n(x)$ and $T_n(x)$ we obtain

$$a_{n,n} = \frac{\sigma_n(\pi)}{2^{n-1}} = \frac{\varrho_n(\pi)}{\sqrt{h(n)}\, 2^{n-1}}.$$

Hence $2h(n) \geq (\varrho_n(\pi)/2^{n-1})^2$. By Proposition 6.3.12 we have $\lim_{n\to\infty} \varrho_n(\pi)/2^n = \infty$ and $\lim_{n\to\infty} h(n) = \infty$ follows. ◇

4 Translation-invariant linear filtering

If $X = (X_n)_{n\in\mathbb{Z}}$ is a stochastic sequence on the group $\mathbb{Z}$, a sequence $Y = (Y_n)_{n\in\mathbb{Z}}$ is obtained by linear filtering, where a matrix $C = (c_{n,m})_{n,m\in\mathbb{Z}}$ is acting on X, i.e. $Y = CX$. If there exists a sequence $A = (a_k)_{k\in\mathbb{Z}}$ such that $c_{n,m} = a_{n-m}$, A is called a time-invariant linear filter, see [122,§4.10].

Subsequently we investigate an appropriate generalization of time-invariant filtering of weakly R_n-stationary random sequences. Based on filtering we can describe various relations between stochastic sequences. The following result generalizes Theorem 6.1.2, where R_n-moving average sequences are studied.

Proposition 6.4.1 *Let $(X_n)_{n\in\mathbb{N}_0}$ be a weakly R_n-stationary random sequence and $T_n : H \to H$ the corresponding translation operators on $H = H(X_n, \mathbb{N}_0)$. Assume that the spectral measure μ of $(X_n)_{n\in\mathbb{N}_0}$ is of the form $\mu = f\pi$, $f \in L^\infty(D_s, \pi)$, and $a = (a_k)_{k\in\mathbb{N}_0} \in l^2(h)$. Define*

$$Y_n = \sum_{k=0}^{\infty} a_k \, T_k X_n \, h(k) \qquad \text{for } n \in \mathbb{N}_0.$$

Then $(Y_n)_{n\in\mathbb{N}_0}$ is a well-defined random sequence in $H \subseteq L^2(\Omega, P)$ with the property $\|Y_n\|_2 \le \|f\|_\infty^{\frac{1}{2}} \, \|a\|_2$ for each $n \in \mathbb{N}_0$.

For the cross-correlation $\langle Y_n, X_m\rangle$ we have

$$\langle Y_n, X_m\rangle = \int_{D_s} R_n(x)\, R_m(x)\, \mathcal{P}(a)(x)\, d\mu(x),$$

i.e. the cross-correlation $\langle Y_n, X_m\rangle$ is R_n-stationary.

Proof. From section 5.1 we know that $\|T_n\| \le 1$. Since $T_k X_n = T_n X_k$ and $a \in l^2(h)$, we get that

$$Y_n = \sum_{k=0}^{\infty} a_k \, T_k X_n \, h(k) = T_n \left(\sum_{k=0}^{\infty} a_k \, X_k \, h(k) \right)$$

is a well-defined element of $H \subset L^2(\Omega, P)$. Moreover, using the isometric isomorphism Φ from $L^2(D_s, \mu)$ onto H we obtain

$$\|Y_n\|_2^2 = \int_{D_s} \left| \sum_{k=0}^{\infty} a_k \, R_k(x)\, R_n(x)\, h(k) \right|^2 f(x)\, d\pi(x)$$

$$\le \|f\|_\infty \int_{D_s} |\mathcal{P}(a)(x)|^2 \, d\pi(x) = \|f\|_\infty \, \|a\|_2^2$$

for each $n \in \mathbb{N}_0$. Since

$$\langle T_n X_k, X_m\rangle = \sum_{j=|n-k|}^{n+k} g(n,k;j)\, \langle X_j, X_m\rangle$$

$$= \sum_{j=|n-k|}^{n+k} g(n,k;j) \int_{D_s} R_j(x)\, R_m(x)\, d\mu(x) = \int_{D_s} R_n(x)\, R_k(x)\, R_m(x)\, d\mu(x),$$

we have

$$\langle Y_n, X_m\rangle = \sum_{k=0}^{\infty} a_k \, \langle T_n X_k, X_m\rangle \, h(k)$$

$$= \int_{D_s} R_n(x)\, R_m(x) \sum_{k=0}^{\infty} a_k \, R_k(x)\, h(k)\, d\mu(x) = \int_{D_s} R_n(x)\, R_m(x)\, \mathcal{P}(a)(x)\, d\mu(x).$$

$\diamond$

Theorem 6.4.2 *Let $(X_n)_{n\in\mathbb{N}_0}$ be a weakly R_n-stationary random sequence with spectral measure μ. Define $Y_n = \sum_{k=0}^{\infty} a_k\ T_k X_n\ h(k)$ as in Proposition 6.4.1.*

(1) Assume that $\mu = f\pi$, $f \in L^\infty(D_s, \pi)$ and $a = (a_k)_{k\in\mathbb{N}_0} \in l^2(h)$. Then $(Y_n)_{n\in\mathbb{N}_0}$ is a weakly R_n-stationary random sequence with spectral measure

$$d\nu(x) = |\mathcal{P}(a)(x)|^2\ d\mu(x).$$

(2) Now let μ be an arbitrary spectral measure on D_s and assume that $a = (a_k)_{k\in\mathbb{N}_0} \in l^1(h)$. Then $(Y_n)_{n\in\mathbb{N}_0}$ is a weakly R_n-stationary random sequence with spectral measure

$$d\nu(x) = |\hat{a}(x)|^2\ d\mu(x).$$

Proof.

(1) Using Proposition 6.4.1 we obtain

$$\begin{aligned}\langle Y_n, Y_m\rangle &= \langle Y_n, \sum_{k=0}^{\infty} a_k\ T_m X_k\ h(k)\rangle \\ &= \sum_{k=0}^{\infty} \overline{a_k} \sum_{j=|m-k|}^{m+k} g(m,k;j)\ \langle Y_n, X_j\rangle\ h(k) \\ &= \sum_{k=0}^{\infty} \overline{a_k} \sum_{j=|m-k|}^{m+k} g(m,k;j) \int_{D_s} R_n(x)\ R_j(x)\ \mathcal{P}(a)(x)\ d\mu(x)\ h(k) \\ &= \int_{D_s} R_n(x)\ R_m(x)\ \mathcal{P}(a)(x) \sum_{k=0}^{\infty} \overline{a_k}\ R_k(x)\ h(k)\ d\mu(x) \\ &= \int_{D_s} R_n(x)\ R_m(x)\ |\mathcal{P}(a)(x)|^2\ d\mu(x).\end{aligned}$$

Hence $(Y_n)_{n\in\mathbb{N}_0}$ is R_n-stationary with spectral measure $|\mathcal{P}(a)(x)|^2 d\mu(x)$.

(2) Since

$$\langle X_n, X_n\rangle = d(n,n) \le d(0) = \langle X_0, X_0\rangle$$

(see Theorem 5.1.1 and the inequalities for positive definite functions in section 2.4), we get

$$\|Y_n\|_2 \le \sum_{k=0}^{\infty} |a_k|\ \|T_k X_n\|_2\ h(k) \le \|X_0\|_2\ \|a\|_1 < \infty.$$

As $l^1(h) \subseteq l^2(h)$, we can now proceed exactly as in the proof of (1).

◇

Remark:

(1) In the case $X_n = Z_n$ (Z_n the R_n-white noise), $(Y_n)_{n\in\mathbb{N}_0}$ is the R_n-moving average sequence defined in Theorem 6.1.2.
(2) In the case $a = (a_k)_{k\in\mathbb{N}_0} \in l^1(h)$, the covariance sequence d_Y of $(Y_n)_{n\in\mathbb{N}_0}$ is given by $d_Y(n) = a * a^* * d(n)$, where d is the covariance sequence of $(X_n)_{n\in\mathbb{N}_0}$.

We shall use the notation

$$a * X_n := Y_n = \sum_{k=0}^{\infty} a_k \, T_k X_n \, h(n)$$

for the sequence $(Y_n)_{n\in\mathbb{N}_0}$ studied in Theorem 6.4.2. The transition from $(X_n)_{n\in\mathbb{N}_0}$ to $(a * X_n)_{n\in\mathbb{N}_0}$ is called **filtering**. Concerning the composition of filtering we have the following result.

Proposition 6.4.3 *Let $(X_n)_{n\in\mathbb{N}_0}$ be a weakly R_n-stationary random sequence with spectral measure μ of the form $\mu = f\pi$, $f \in L^\infty(D_s, \pi)$. If $a = (a_k)_{k\in\mathbb{N}_0} \in l^1(h)$, $b = (b_k)_{k\in\mathbb{N}_0} \in l^2(h)$, then for $Y_n = (a * b) * X_n$ is valid:*
$(Y_n)_{n\in\mathbb{N}_0}$ is a weakly R_n-stationary random sequence with spectral measure $d\nu(x) = |\hat{a}(x)\mathcal{P}(b)(x)|^2 d\mu(x)$. Furthermore we have

$$(a * b) * X_n = a * (b * X_n) = b * (a * X_n).$$

Proof. By Theorem 2.2.8 we know that $a * b \in l^2(h)$ and $\|a * b\|_2 \leq \|a\|_1 \|b\|_2$, and by Theorem 6.4.2 it follows that $(Y_n)_{n\in\mathbb{N}_0}$ is a weakly R_n-stationary random sequence with spectral measure $d\nu(x) = |\mathcal{P}(a*b)(x)|^2 d\mu(x)$. Since $\mathcal{P}(a)\cdot\mathcal{P}(b) = \hat{a}\cdot\mathcal{P}(b)$ it follows $\hat{a}\cdot\mathcal{P}(b) \in L^1(D_s,\pi)\cap L^2(D_s,\pi)$. By Proposition 3.3.10 we know that $a * b = (\mathcal{P}(a)\cdot\mathcal{P}(b))^\vee = (\hat{a}\cdot\mathcal{P}(b))^\vee$. Since $\check{\varphi} = \mathcal{P}^{-1}(\varphi)$ for each $\varphi \in L^1(D_s,\pi)\cap L^2(D_s,\pi)$, see the proof of Theorem 3.3.9, we obtain $\mathcal{P}(a * b) = \mathcal{P}(\mathcal{P}^{-1}(\hat{a}\cdot\mathcal{P}(b))) = \hat{a}\cdot\mathcal{P}(b)$. Thus we have shown that $d\nu(x) = |\hat{a}(x)\cdot\mathcal{P}(b)(x)|^2 d\mu(x)$.

To show the last statement we use the orthogonal stochastic measure $Z : \mathcal{B}(D_s) \to L^2(\Omega, P)$ of $(X_n)_{n\in\mathbb{N}_0}$. By the considerations before Corollary 5.1.3 we obtain

$$(a * b) * X_n = \int_{D_s} R_n(x)\, \hat{a}(x)\, \mathcal{P}(b)(x)\, dZ(x)$$

$$= a * \left(\int_{D_s} R_n(x)\, \mathcal{P}(b)(x)\, dZ(x)\right) = a * (b * X_n).$$

The equality $(a * b) * X_n = b * (a * X_n)$ follows in the same way. ◇

Many properties of the filtering of X_n to $a * X_n$ are determined by the continuous function $\hat{a} \in C(D_s)$ for $a \in l^1(h)$, respectively by $\mathcal{P}(a) \in L^2(D_s, \pi)$ for $a \in l^2(h)$. $\hat{a}$ and $\mathcal{P}(a)$ are called **transfer functions**.

The results of Proposition 4.6.1, Theorem 4.6.2 and Proposition 4.6.3 are the basis for investigating R_n-ARMA sequences (section 6.5) and decomposition, extrapolation and interpolation (section 6.6).

Finally we investigate the effect on random sequences, when we change the polynomial hypergroup structure via connection coefficients. Consider two polynomial hypergroups on $\mathbb{N}_0$ generated by $(R_n(x))_{n\in\mathbb{N}_0}$, respectively by $(P_n(x))_{n\in\mathbb{N}_0}$. There exist $c_{R,P}(n,k) \in \mathbb{R}$, the connection coefficients of the representation

$$R_n(x) = \sum_{k=0}^{n} c_{R,P}(n,k)\, P_k(x).$$

Since $R_n(1) = 1 = P_n(1)$ we have $\sum\limits_{k=0}^{n} c_{R,P}(n,k) = 1$. Denote by D_s^R and D_s^P the dual spaces $\hat{\mathbb{N}}_0$ with respect to $(R_n(x))_{n\in\mathbb{N}_0}$, respectively to $(P_n(x))_{n\in\mathbb{N}_0}$.

Proposition 6.4.4 *Let $(X_n^P)_{n\in\mathbb{N}_0}$ be a weakly P_n-stationary random sequence with spectral measure μ satisfying supp $\mu \subseteq D_s^R$. Define*

$$X_n^R := \sum_{k=0}^{n} c_{R,P}(n,k)\, X_k^P.$$

Then $(X_n^R)_{n\in\mathbb{N}_0}$ is a weakly R_n-stationary random sequence with the same spectral measure μ. Moreover, if $(X_n^P)_{n\in\mathbb{N}_0}$ has constant mean M, then $(X_n^R)_{n\in\mathbb{N}_0}$ has constant mean M, too.

Proof. Since $M = E(X_k^P)$ for each $k \in \mathbb{N}_0$, and $\sum\limits_{k=0}^{n} c_{R,P}(n,k) = 1$, it follows $E(X_n^R) = M$. The isometric isomorphism $\Phi : L^2(D_s^P, \mu) \to H = H(X_n^P, \mathbb{N}_0)$ yields

$$d^R(n,m) := E(X_n^R \cdot X_m^R) = E\left(\left(\sum_{k=0}^{n} c_{R,P}(n,k)\, X_k^P\right) \cdot \left(\sum_{k=0}^{m} c_{R,P}(n,k)\, X_k^P\right)\right)$$

$$= \int_{D_s^P} \left(\sum_{k=0}^{n} c_{R,P}(n,k)\, P_k(x)\right) \cdot \left(\sum_{k=0}^{m} c_{R,P}(n,k)\, P_k(x)\right)\, d\mu(x)$$

$$= \int_{D_s^R} R_n(x)\, R_m(x)\, d\mu(x).$$

Linearization of

$$R_n(x)\, R_m(x) \;=\; \sum_{k=|n-m|}^{n+m} g(n,m;k)\, R_k(x)$$

shows, that

$$d^R(n,m) \;=\; \sum_{k=|n-m|}^{n+m} g(n,m;k)\, d^R(k,0).$$

Since supp $\mu \subseteq D_s^R$ we get a spectral representation

$$d^R(n) \;=\; \int_{D_s^R} R_n(x)\, d\mu(x),$$

and since $|d^R(n)| \leq \mu(D_s^R)$, d^R is a bounded function on $\mathbb{N}_0 \times \mathbb{N}_0$. ◇

5 $\mathbf{R_n}$-ARMA sequences

In section 6.1 we already studied p-order R_n-autoregressive random sequences. Supposing that the autoregressive polynomial $P(x) = 1 - a_1R_1(x) - \ldots - a_pR_p(x)$ is not zero for all $x \in D$, Theorem 6.1.12 gives a complete characterization of such random sequences as R_n-moving average sequences. The assumption $P(x) \neq 0$ for all $x \in D$ yields that $g(x) = \frac{1}{P(x)}$ is an element of $A(D)$, i.e. $\check{g} \in l^1(h)$, see Theorem 4.1.6. Using the results of section 6.4 about filtering we can weaken the assumption.

Theorem 6.5.1 *Let $(Z_n)_{n\in\mathbb{N}_0}$ be R_n-white noise, and let $b_0, \ldots, b_p \in \mathbb{C}$ such that $B(x) = \sum_{k=0}^{p} b_kR_k(x)$ does not have zeros on $S =$ supp π. Then the p-order autoregression equation*

$$b_oX_n \,+\, b_1T_1X_n \,+\, \ldots \,+\, b_pT_pX_n \;=\; Z_n \qquad (*)$$

is solved by exactly one weakly R_n-stationary random sequence $(X_n)_{n\in\mathbb{N}_0}$, where $(X_n)_{n\in\mathbb{N}_0}$ is a R_n-moving average sequence of the form

$$X_n \;=\; \sum_{k=0}^{\infty} a_k\, T_kZ_n\, h(k),$$

where $a = (a_k)_{k\in\mathbb{N}_0} \in l^2(h)$ satisfying

$$\frac{1}{B(x)} \;=\; \sum_{k=0}^{\infty} a_k\, R_k(x)\, h(x) \qquad in\ L^2(D_s,\pi)$$

(we set $\frac{1}{B(x)} = 0$ for $x \in D_s \backslash S$). The spectral measure of $(X_n)_{n\in\mathbb{N}_0}$ is

$$d\mu(x) \;=\; \frac{1}{|B(x)|^2}\, d\pi(x).$$

Proof. Since $B(x) \neq 0$ for each $x \in \operatorname{supp} \pi$, we have $\frac{1}{B(x)} \in C(S) \subseteq L^2(D_s, \pi)$. In particular, there exists a sequence $a = (a_k)_{k\in\mathbb{N}_0} \in l^2(h)$ such that

$$\frac{1}{B(x)} \;=\; \sum_{k=0}^{\infty} a_k\, R_k(x)\, h(k)$$

in $L^2(D_s, \pi)$. Applying Theorem 6.4.2(1) we know that $(X_n)_{n\in\mathbb{N}_0}$ is a weakly R_n-stationary random sequence with spectral measure

$$d\mu(x) \;=\; |\mathcal{P}(a)(x)|^2\, d\pi(x) \;=\; \frac{1}{|B(x)|^2}\, d\pi(x).$$

Let $b_{p+k} = 0$ for $k = 1, 2, \ldots$. Then Proposition 6.4.3 implies $b_n * X_n = b*(a*Z_n) = (b*a)*Z_n = Z_n$, i.e. $(X_n)_{n\in\mathbb{N}_0}$ solves the p-order autoregression equation $(*)$.

It remains to show that $(X_n)_{n\in\mathbb{N}_0}$ is the unique weakly R_n-stationary random sequence solving $(*)$. Since $Z_n \in H(X_n, \mathbb{N}_0)$, the translation operators T_k determined by $(Z_n)_{n\in\mathbb{N}_0}$, respectively by $(X_n)_{n\in\mathbb{N}_0}$, coincide. Let $(Y_n)_{n\in\mathbb{N}_0}$ be a second weakly R_n-stationary random sequence satisfying $(*)$. By Proposition 6.4.3 we obtain

$$Y_n \;=\; (a*b)*Y_n \;=\; a*(b*Y_n) \;=\; a*Z_n \;=\; a*(b*X_n) \;=\; (a*b)*X_n \;=\; X_n.$$

◇

Definition. Let $(Z_n)_{n\in\mathbb{N}_0}$ be a R_n-white noise, and let $a_0, \ldots, a_q \in \mathbb{C}$, $a_q \neq 0$, and $b_0, \ldots, b_p \in \mathbb{C}$, $b_p \neq 0$. A weakly R_n-stationary random sequence $(X_n)_{n\in\mathbb{N}_0}$ is called **R_n-ARMA sequence**, if the following equations are satisfied,

$$\begin{aligned} & b_0 X_n + b_1 T_1 X_n + \ldots + b_p T_p X_n \\ &= a_0 Z_n + a_1 T_1 Z_n h(1) + \ldots + a_q T_q Z_n h(q). \qquad (**) \end{aligned}$$

Sometimes $(X_n)_{n\in\mathbb{N}_0}$ is called R_n-ARMA$[p, q]$-sequence.

The result of Theorem 6.5.1 can be extended to R_n-ARMA sequences.

Corollary 6.5.2 *Let $(Z_n)_{n\in\mathbb{N}_0}$ be a R_n-white noise, and let $b_0, \ldots, b_p \in \mathbb{C}$ such that $B(x) = \sum_{k=0}^{p} b_k R_k(x)$ does not have zeros on $S = \operatorname{supp} \pi$. Furthermore let $a_0, \ldots, a_q \in \mathbb{C}$. Then the R_n-ARMA regression equation $(**)$ is solved by exactly one weakly R_n-stationary sequence with spectral measure*

$$d\mu(x) \;=\; \left| \frac{\hat{a}(x)}{B(x)} \right|^2 d\pi(x),$$

where

$$\hat{a}(x) = \sum_{k=0}^{q} a_k \, R_k(x) \, h(k) \qquad \textit{and}$$

$$\frac{1}{B(x)} = \sum_{k=0}^{\infty} c_k \, R_k(x) \, h(k) \in L^2(D_s, \pi)$$

(with $\frac{1}{B(x)} = 0$ *for* $x \in D_s \backslash S$*).*

Proof. Define

$$Y_n = \sum_{k=0}^{q} a_k \, T_k Z_n \, h(k) = a * Z_n,$$

where $a = (a_0, ..., a_q, 0, ...) \in l^1(h)$. The spectral measure of $(Y_n)_{n\in\mathbb{N}_0}$ is $d\nu(x) = |\hat{a}(x)|^2 d\pi(x)$. Replacing Z_n by $Y_n = a * Z_n$ we can argue as in the proof of Theorem 6.5.1. Note that the assumptions of Theorem 6.4.2(1) and Theorem 6.4.3 are satisfied. ◇

Next we show that a special property of the spectral measure of a weakly R_n-stationary random sequence $(X_n)_{n\in\mathbb{N}_0}$ yields a R_n-white noise $(Z_n)_{n\in\mathbb{N}_0}$ such that $(X_n)_{n\in\mathbb{N}_0}$ has a series representation based on $(Z_n)_{n\in\mathbb{N}_0}$.

Theorem 6.5.3 *Let* $(X_n)_{n\in\mathbb{N}_0}$ *be a weakly* R_n*-stationary random sequence with spectral measure* $d\mu(x) = f(x)d\pi(x)$, $f \in L^1(D_s, \pi)$ *and* $f(x) > 0$ *for* π*-almost all* $x \in D_s$*. Then there exists a* R_n*-white noise sequence* $(Z_n)_{n\in\mathbb{N}_0}$ *and a sequence* $a = (a_k)_{k\in\mathbb{N}_0} \in l^2(h)$ *such that*

$$X_n = \sum_{k=0}^{\infty} a_k \, T_k Z_n \, h(k),$$

where the series converges in $L^2(\Omega, P)$*.*

Proof. Obviously there exists a function $\varphi \in L^2(D_s, \pi)$, $\varphi > 0$ π-almost everywhere, such that $f(x) = \varphi(x)^2$. The element $\varphi \in L^2(D_s, \pi)$ has a series representation in $L^2(D_s, \pi)$

$$\varphi(x) = \sum_{k=0}^{\infty} a_k \, R_k(x) \, h(k),$$

where $a_k = \mathcal{P}^{-1}(\varphi)(k)$.

Let $Z : \mathcal{B}(D_s) \to L^2(\Omega, P)$ be the orthogonal stochastic measure determined by $(X_n)_{n\in\mathbb{N}_0}$, see Theorem 5.1.2. Since $\frac{1}{\varphi} \in L^2(D_s, f\pi) = L^2(D_s, \mu)$,

$$Z_n := \int_{D_s} \frac{R_n(x)}{\varphi(x)} \, dZ(x)$$

are well-defined elements of $L^2(\Omega, P)$.

Now define

$$\tilde{Z}(A) := \int_{D_s} \frac{\chi_A(x)}{\varphi(x)}\, dZ(x) \qquad \text{for each Borel set } A \subseteq D_s.$$

It is routine to show that $\tilde{Z} : \mathcal{B}(D_s) \to L^2(\Omega, P)$ is an orthogonal stochastic measure determined by $(Z_n)_{n\in\mathbb{N}_0}$. By Corollary 5.1.3 we obtain that the spectral measure of $(Z_n)_{n\in\mathbb{N}_0}$ is a weakly R_n-stationary random sequence with spectral measure $\frac{1}{\varphi^2}\mu = \pi$. Therefore $(Z_n)_{n\in\mathbb{N}_0}$ is a R_n-white noise. Using the basic properties of stochastic integrals with respect to the orthogonal stochastic measure $\tilde{Z}$, see [623,Ch.VI,2], we obtain

$$\sum_{k=0}^{\infty} a_k\, T_k Z_n\, h(k) = \sum_{k=0}^{\infty} a_k\, h(k) \int_{D_s} R_k(x)\, R_n(x)\, d\tilde{Z}(x)$$

$$= \int_{D_s} R_n(x)\, \varphi(x)\, d\tilde{Z}(x) = \int_{D_s} R_n(x)\, dZ(x) = X_n.$$

◇

Corollary 6.5.4 *Let $A(x)$, $B(x)$ be polynomials of degree q, respectively of degree p, and suppose that $B(x)$ has no zeros in $S = supp\,\pi$. Assume that $(X_n)_{n\in\mathbb{N}_0}$ is a weakly R_n-stationary random sequence with spectral measure $d\mu(x) = |\frac{A(x)}{B(x)}|^2 d\pi(x)$. Then there exists a R_n-white noise $(Z_n)_{n\in\mathbb{N}_0}$ such that $(X_n)_{n\in\mathbb{N}_0}$ is a R_n-ARMA[p,q]-sequence based on $(Z_n)_{n\in\mathbb{N}_0}$.*

Proof. Theorem 6.5.3 yields that $(X_n)_{n\in\mathbb{N}_0}$ has a series representation based on $(Z_n)_{n\in\mathbb{N}_0}$:

$$X_n = \sum_{k=0}^{\infty} c_k\, T_k Z_n\, h(k),$$

where $c = (c_k)_{k\in\mathbb{N}_0} \in l^2(h)$ and

$$\frac{A(x)}{B(x)} = \sum_{k=0}^{\infty} c_k\, R_k(x)\, h(k) \in L^2(D_s, \pi).$$

Moreover we have $c = a * b^{-1}$, where $a = \check{A}^{\cdot}$, $b^{-1} = \mathcal{P}^{-1}(\frac{1}{B(\cdot)})$ and $c = \mathcal{P}^{-1}(\frac{A(\cdot)}{B(\cdot)})$. By Proposition 6.4.3 it follows $b{*}X_n = b{*}((a{*}b^{-1}){*}Z_n) = a{*}Z_n$, i.e. the R_n-ARMA regression equation $(**)$ is fulfilled. ◇

Theorem 6.5.3 may be used to get more information about the relation between random sequences $(X_n^P)_{n\in\mathbb{N}_0}$ and $(X_n^R)_{n\in\mathbb{N}_0}$ determined by the connection coefficients $c_{R,P}(n,k)$, see Proposition 6.4.4.

Corollary 6.5.5 *Let $(R_n(x))_{n\in\mathbb{N}_0}$ and $(P_n(x))_{n\in\mathbb{N}_0}$ be orthogonal polynomials generating two polynomial hypergroups on $\mathbb{N}_0$. Denote the corresponding orthogonalization measures by π^R respectively π^P. Assume $\pi^P = f\pi^R$, $f \in L^1(D_s, \pi^R)$, $f > 0$ π^R-almost everywhere. If $(Z_n^P)_{n\in\mathbb{N}_0}$ is a P_n-white noise and $X_n^R = \sum_{k=0}^{n} c_{R,P}(n,k) Z_k^P$, then there exists a R_n-white noise $(Z_n^R)_{n\in\mathbb{N}_0}$ and $a \in l^2(h^R)$ such that*

$$X_n^R = \sum_{k=0}^{\infty} a_k \, T_k Z_n^R \, h^R(k).$$

($h^R(k)$ are the Haar weights with respect to $(R_n(x))_{n\in\mathbb{N}_0}$.)

Proof. By Proposition 6.4.4 it follows that $(X_n^R)_{n\in\mathbb{N}_0}$ is a weakly R_n-stationary random sequence with spectral measure $\mu = \pi^P = f\pi^R$. Theorem 6.5.3 yields the statement. ◇

Remark.

(1) Corollary 6.5.5 can be applied for many pairs of polynomial hypergroups on $\mathbb{N}_0$. If $R_n(x) = T_n(x)$ are the Chebyshev polynomials of the first kind, the assumptions in Corollary 6.5.5 are satisfied for $P_n(x) = R_n^{(\alpha,\beta)}(x)$, the Jacobi polynomials with $(\alpha,\beta) \in J = \{(\alpha,\beta) : \alpha \geq \beta > -1,\ \alpha+\beta+1 \geq 0\}$. This is also true for the case that $P_n(x)$ belong to the class of generalized Chebyshev polynomials $T_n^{(\alpha,\beta)}(x)$, $(\alpha,\beta) \in J$. Further examples for suitable $(P_n(x))$ are Geronimus polynomials, Grinspun polynomials and others.
Obviously, considering general $R_n(x)$, there are many possibilities to find appropriate $P_n(x)$, such that Corollary 6.5.5 can be applied. For example, if $R_n(x) = R_n^{(\alpha,\alpha)}(x)$ are ultraspherical polynomials with parameter $\alpha > -\frac{1}{2}$, choose $P_n(x) = R_n^{(\beta,\beta)}(x)$, $\beta > \alpha$.

(2) The statement of Corollary 6.5.5 can be extended by replacing $(Z_n^P)_{n\in\mathbb{N}_0}$ (and therefore π^P replacing by a spectral measure ν^P) by a weakly P_n-stationary random sequence $(X_n^P)_{n\in\mathbb{N}_0}$.

6 Decomposition

We start with a decomposition of the 'time' domain $\mathbb{N}_0$ of a weakly R_n-stationary random sequence, which is very useful in the context of prediction, extrapolation and interpolation. For the time decomposition of weakly stationary sequences on the group $\mathbb{Z}$ we refer to Wold's expansion, see [623,Ch.VI,5].

Let $(X_n)_{n\in\mathbb{N}_0}$ be a weakly R_n-stationary random sequence. We denote by H_n the closure of span $\{X_k : k \geq n\}$ in $L^2(\Omega, P)$ and $S(X) := \bigcap_{n\in\mathbb{N}_0} H_n$. Recall that for $n = 0$ we have $H_0 = H = H(X_n, \mathbb{N}_0)$.

Definition. A weakly R_n-stationary random sequence is called **deterministic** (or **singular**), if $S(X) = H$, **purely nondeterministic** (or **regular**) if $S(X) = \{0\}$, and **nondeterministic** if $S(X) \neq H$.

Let T_n be the translation operator acting on H. For each $m, n \in \mathbb{N}_0$, $m \leq n$, we have $T_m(H_n) \subseteq H_{n-m}$. In particular, it follows that $T_m(S(X)) \subseteq S(X)$. We denote the orthogonal projection from H onto H_n by P_{H_n}. Note that for each $Y_0 \in H$ we have

$$\|P_{H_n} Y_0 - Y_0\|_2 = \inf \{ \|Y - Y_0\|_2 : Y \in H_n\}.$$

Lemma 6.6.1 *Let $(X_n)_{n\in\mathbb{N}_0}$ be a weakly R_n-stationary random sequence. For each $n \in \mathbb{N}$ is true*

$$\|P_{H_n} X_{n-1} - X_{n-1}\|_2 \leq \frac{1}{c_n} \|P_{H_{n+1}} X_n - X_n\|_2,$$

where c_n are the recurrence coefficients of $(R_n(x))_{n\in\mathbb{N}_0}$, see (1) in subsection 1.2.1.

Proof. We have

$$T_1 X_n = a_n X_{n+1} + b_n X_n + c_n X_{n-1} \qquad \text{for } n \in \mathbb{N}.$$

Hence

$$X_{n-1} = \frac{1}{c_n} T_1 X_n - \frac{a_n}{c_n} X_{n+1} - \frac{b_n}{c_n} X_n,$$

and it follows

$$\|P_{H_n} X_{n-1} - X_{n-1}\|_2 = \inf \{ \|Y - X_{n-1}\|_2 : Y \in H_n\}$$

$$= \inf \{ \|Y - \frac{1}{c_n} T_1 X_n\|_2 : Y \in H_n\},$$

since $\frac{a_n}{c_n} X_{n+1} + \frac{b_n}{c_n} X_n \in H_n$.

Moreover, since

$$\|T_1 Y - \frac{1}{c_n} T_1 X_n\|_2 \leq \|Y - \frac{1}{c_n} X_n\|_2,$$

we get

$$\frac{1}{c_n} \|P_{H_{n+1}} X_n - X_n\|_2 = \frac{1}{c_n} \inf \{ \|Y - X_n\|_2 : Y \in H_{n+1}\}$$

$$= \inf \{ \|Y - \frac{1}{c_n} X_n\|_2 : Y \in H_{n+1}\}.$$

Furthermore, $T_1(H_{n+1}) \subseteq H_n$ yields

$$\inf\{\,\|Y - \frac{1}{c_1} T_1 X_n\|_2 : Y \in H_n\} \le \inf\{\,\|T_1 Y - \frac{1}{c_n} T_1 X_n\|_2 : Y \in H_{n+1}\}.$$

Combining the inequalities and equalities above, we obtain

$$\begin{aligned}
\|P_{H_n} X_{n-1} - X_{n-1}\|_2 &= \inf\{\,\|Y - \frac{1}{c_n} T_1 X_n\|_2 : Y \in H_n\} \\
&\le \inf\{\,\|T_1 Y - \frac{1}{c_n} T_1 X_n\|_2 : Y \in H_{n+1}\} \\
&\le \inf\{\,\|Y - \frac{1}{c_n} X_n\|_2 : Y \in H_{n+1}\} \\
&= \frac{1}{c_n} \inf\{\,\|Y - X_n\|_2 : Y \in H_{n+1}\} = \frac{1}{c_n} \|P_{H_{n+1}} X_n - X_n\|_2.
\end{aligned}$$

◇

The following conditions, which are easy to check, describe deterministic (respectively nondeterministic) weakly R_n-stationary random sequences.

Proposition 6.6.2 *Let $(X_n)_{n\in\mathbb{N}_0}$ be a weakly R_n-stationary random sequence. Then*

(1) If $(X_n)_{n\in\mathbb{N}_0}$ is nondeterministic, then there exists $m \in \mathbb{N}_0$ such that
$$H_0 = H_1 = \cdots = H_m \supsetneq H_{m+1} \supsetneq H_{m+2} \supsetneq \cdots .$$
(2) $(X_n)_{n\in\mathbb{N}_0}$ is deterministic if and only if $X_0 \in H_n$ for all $n \in \mathbb{N}_0$.

Proof.

(1) Since $S(X) \neq H$ there exists a minimal $m \in \mathbb{N}_0$ such that $H_m \supsetneq H_{m+1}$. Suppose that $H_{m+1} = H_{m+2}$. Then $\|P_{H_{m+2}} X_{m+1} - X_{m+1}\|_2 = 0$. Lemma 6.6.1 implies that $\|P_{H_{m+1}} X_m - X_m\|_2 = 0$, too. Hence $X_m \in H_{m+1}$, and thus $H_m = H_{m+1}$, a contradiction. By a routine induction argument, statement (1) follows.

(2) Let $S(X) = H$. Using $T_n(H_{n+l}) \subseteq H_l$ we get for every $k, l, n \in \mathbb{N}_0$

$$\begin{aligned}
\|P_{H_l} T_n X_k - T_n X_k\|_2 &= \inf\{\,\|Y - T_n X_k\|_2 : Y \in H_l\} \\
\le \inf\{\,\|T_n Y - T_n X_k\|_2 : Y \in H_{n+l}\} &\le \inf\{\,\|Y - X_k\|_2 : Y \in H_{n+l}\} \\
&= \|P_{H_{n+l}} X_k - X_k\|_2.
\end{aligned}$$

Put $k = 0$. We obtain

$$\|P_{H_l} X_n - X_n\|_2 \le \|P_{H_{n+l}} X_0 - X_0\|_2. \qquad (*)$$

Now suppose that $X_0 \in H_m$ for all $m \in \mathbb{N}_0$. By $(*)$ it follows that $X_n \in H_l$ for all $n, l \in \mathbb{N}_0$, i.e. $H = H_0 = S(X)$.

Conversely, if $H_0 = S(X)$, we have $X_0 \in H_m$ for each $m \in \mathbb{N}_0$.

◇

Theorem 6.6.3 *Let $(X_n)_{n\in\mathbb{N}_0}$ be a weakly R_n-stationary random sequence. Then there exists a decomposition*

$$X_n = X_n^r + X_n^s, \qquad n \in \mathbb{N}_0, \qquad (*)$$

where $X_n^r := T_n P_{S(X)^\perp} X_0$ and $X_n^s := T_n P_{S(X)} X_0$, where the random sequences $(X_n^r)_{n\in\mathbb{N}_0}$ and $(X_n^s)_{n\in\mathbb{N}_0}$ fulfil the following properties,

(1) $(X_n^r)_{n\in\mathbb{N}_0}$ is weakly R_n-stationary and purely nondeterministic (regular).
(2) $(X_n^s)_{n\in\mathbb{N}_0}$ is weakly R_n-stationary and deterministic (singular).
(3) $(X_n^r)_{n\in\mathbb{N}_0}$ and $(X_n^s)_{n\in\mathbb{N}_0}$ are orthogonal.
(4) Denoting by H_n^r the closure of span$\{X_k^r : k \geq n\}$ and by H_n^s the closure of span$\{X_k^s : k \geq n\}$, we have $H_n^r \subseteq H_n$ and $H_n^s \subseteq H_n$ for all $n \in \mathbb{N}_0$.

The decomposition $(*)$ is uniquely determind by the properties (1), (2), (3) and (4).

Proof. Since the operators $T_n \in \mathcal{B}(H)$ are self-adjoint and $T_n(S(X)) \subseteq S(X)$, we get immediately that $T_n(S(X)^\perp) \subseteq S(X)^\perp$.

Hence $X_n^r \in S(X)^\perp \subseteq H$ and $X_n^s \in S(X) \subseteq H$ are well-defined elements of H and $\langle X_n^r, X_m^s \rangle = 0$ for $m, n \in \mathbb{N}_0$. Moreover, since

$$T_n \circ T_m = \sum_{k=|n-m|}^{n+m} g(n, m; k)\, T_k$$

it follows that $(X_n^r)_{n\in\mathbb{N}_0}$ and $(X_n^s)_{n\in\mathbb{N}_0}$ are weakly R_n-stationary random sequences. We also have for each $n \in \mathbb{N}_0$,

$$X_n = T_n\left(P_{S(X)^\perp} X_0 + P_{S(X)} X_0\right) = X_n^r + X_n^s.$$

Since $X_k^s \in S(X)$ for each $k \in \mathbb{N}_0$, we have $H_n^s \subseteq S(X) \subseteq H_n$ for every $n \in \mathbb{N}_0$. This also implies $X_k^r = X_k - X_k^s$ for each $k \in \mathbb{N}_0$. Thus $H_n^r \subseteq H_n$ for every $n \in \mathbb{N}_0$, which yields $S(X^r) = \{0\}$.

So far, (1), (3) and (4) are shown. To derive that $(X_n^s)_{n\in\mathbb{N}_0}$ is deterministic, note that $H_0^s \subseteq S(X)$ on the one hand. By $S(X) \subseteq H_n^s \oplus H_n^r$ and $H_n^r \subseteq S(X)^\perp$ we get $S(X) \subseteq \bigcap_{n\in\mathbb{N}_0} H_n^s$ on the other hand. It follows

$$\bigcap_{n\in\mathbb{N}_0} H_n^s \subseteq H_0^s \subseteq S(X) \subseteq \bigcap_{n\in\mathbb{N}_0} H_n^s.$$

Hence we have shown that $H^s = H_0^s = S(X^s)$, and (2) is proved.

To derive the uniqueness of the decomposition, let

$$X_n = A_n + B_n,$$

$(A_n)_{n\in\mathbb{N}_0}$ purely nondeterministic, $(B_n)_{n\in\mathbb{N}_0}$ deterministic, such that (1),...,(4) are valid.

Denote by H_n^A the closure of span $\{A_k : k \geq n\}$ and H_n^B analogous. We have

$$H_n^B \subseteq H_n \subseteq H_n^A \oplus H_n^B \subseteq H_n^A \oplus \bigcap_{k\in\mathbb{N}_0} H_k^B.$$

Since $\bigcap_{n\in\mathbb{N}_0} H_n^A = \{0\}$, intersecting these inclusions leads to

$$\bigcap_{n\in\mathbb{N}_0} H_n^B \subseteq S(X) \subseteq \{0\} \oplus \bigcap_{k\in\mathbb{N}_0} H_k^B, \quad \text{i.e.} \quad \bigcap_{n\in\mathbb{N}_0} H_n^B = S(X).$$

Thus $H_0^B = S(X)$ and then $A_n \in S(X)^\perp$ for each $n \in \mathbb{N}_0$. Consequently,

$$B_n = P_{S(X)}(A_n + B_n) = P_{S(X)}(X_n^r + X_n^s) = X_n^s \quad \text{for all } n \in \mathbb{N}_0.$$

By $A_n = X_n - B_n = X_n - X_n^s = X_n^r$ the uniqueness statement is proven. ◇

Using Proposition 6.6.2 and Theorem 6.6.3 we immediately obtain:

Corollary 6.6.4 *Let $(X_n)_{n\in\mathbb{N}_0}$ be a weakly R_n-stationary random sequence. Then the following equivalences are true,*

(1) $(X_n)_{n\in\mathbb{N}_0}$ is deterministic (singular) $\Leftrightarrow X_0 \in S(X)$ $\Leftrightarrow P_{S(X)}X_0 = X_0 \Leftrightarrow X_n = X_n^s$.

(2) $(X_n)_{n\in\mathbb{N}_0}$ is purely nondeterministic (regular) $\Leftrightarrow S(X) = \{0\}$ $\Leftrightarrow P_{S(X)^\perp}X_0 = X_0 \Leftrightarrow X_n = X_n^r$.

It is time to present examples:

(a) R_n-white noise $(Z_n)_{n\in\mathbb{N}_0}$ is purely nondeterministic. In fact, $(R_n)_{n\in\mathbb{N}_0}$ is an orthogonal basis of $L^2(D_s,\pi)$. Hence $R_0,...,R_{n-1}$ cannot be approximated by R_k, $k \geq n$. Now apply the isometric isomorphism from $L^2(D_s,\pi)$ onto H.

(b) Let $Y_n = R_n(x)X$ be an R_n-oscillation determined by a single point $x \in D_s$ and a random variable $X \in L^2(\Omega,P)$, $X \neq 0$. Then $(Y_n)_{n\in\mathbb{N}_0}$ is deterministic. The spectral measure is the point measure ϵ_x (multiplied by some constant).

The examples (a) and (b) are very special. The R_n-white noise $(Z_n)_{n\in\mathbb{N}_0}$ is the most basic weakly R_n-stationary random sequence expanded on the spectral domain D_s with respect to π. R_n-moving average sequences are series expansions based on an R_n-white noise. We will investigate weakly R_n-stationary random sequences in order to derive which properties of the spectral measure imply singularity or regularity. We concentrate on the case where the spectral measure μ has the form $\mu = f\pi,\ f \in L^1(D_s,\pi)$. f is called **spectral density**.

Given a weakly R_n-stationary random sequence $(X_n)_{n\in\mathbb{N}_0}$ with spectral measure μ, we use the following notation: For any subset $A \subseteq \mathbb{N}_0$ we denote by $H_A(R)$ the closure in $L^2(D_s,\mu)$ of span $\{R_n : n \in A\}$, and by $H_A(X)$ the closure in H of span $\{X_n : n \in A\}$.

Theorem 6.6.5 *Let $(X_n)_{n\in\mathbb{N}_0}$ be a weakly R_n-stationary random sequence with spectral measure $\mu = f\pi,\ f \in L^1(D_s,\pi)$ such that $\frac{1}{f} \in L^1(D_s,\pi)$. Then $(X_n)_{n\in\mathbb{N}_0}$ is purely nondeterministic (regular).*

Proof. Since

$$\int_{D_s} \left(\frac{1}{f(x)}\right)^2 d\mu(x) = \int_{D_s} \frac{1}{f(x)}\, d\pi(x),$$

we have $\frac{1}{f} \in L^2(D_s,\mu)$. The functions $\varphi_n := R_n/f$ are also in $L^2(D_s,\mu)$ and $\varphi_n \neq 0$ because of

$$\|\varphi_n\|_{2,\mu}^2 = \int_{D_s} \frac{R_n^2(x)}{f(x)}\, d\pi(x) > 0.$$

(Note that $R_n(x)$ has only n zeros.)

Now fix $n \in \mathbb{N}_0$. Then

$$\langle \varphi_n, R_m\rangle_\mu = \langle R_n, R_m\rangle_\pi = 0 \qquad \text{for } m \neq n.$$

Hence $\varphi_n \in H_{A_n}(R)^\perp$, where $A_n = \mathbb{N}_0\backslash\{n\}$, and therefore $H_{A_n}(R) \subsetneq L^2(D_s,\mu)$. Consider the orthonormal polynomial sequence $(q_n)_{n\in\mathbb{N}_0}$ with respect to the spectral measure μ. Then $q_n \notin H_{A_n}(R)$. To prove this, note that we can find coefficients $a_{n,k}$ with $a_{n,n} \neq 0$ such that $R_n = \sum_{k=0}^{n-1} a_{n,k}R_k + a_{n,n}q_n$. Thus assuming $q_n \in H_{A_n}(R)$ implies $R_n \in H_{A_n}(R)$, a contradiction. Hence $q_n \notin H_{B_{n+1}}(R)$, where $B_{n+1} = \{n+1, n+2, ...\}$. This is true for each $n \in \mathbb{N}_0$, which implies

$$\bigcap_{n\in\mathbb{N}_0} H_{B_{n+1}}(R) = \{0\}, \quad \text{and then} \quad \bigcap_{n\in\mathbb{N}_0} H_{B_n} = \{0\}.$$

Applying the isometric isomorphism Φ we get $S(X) = \{0\}$. $\diamond$

We apply Theorem 6.6.5 considering the essential infimum of the spectral density f, where

$$\text{ess }\inf_{\pi}(f) = \sup\{c \in [0,\infty] : \pi(\{x \in D_s : f(x) < c\}) = 0\}.$$

Corollary 6.6.6 *Let $(X_n)_{n\in\mathbb{N}_0}$ be a weakly R_n-stationary random sequence with spectral measure $\mu = f\pi$, $f \in L^1(D_s,\pi)$ such that ess $\inf_\pi(f) > 0$. Then $(X_n)_{n\in\mathbb{N}_0}$ is purely nondeterministic (regular).*

Proof. If ess $\inf_\pi(f) > 0$, then $\frac{1}{f} \leq c$ π-almost everywhere for some $c > 0$, and thus $\frac{1}{f} \in L^1(D_s,\pi)$. ◇

Theorem 6.6.7 *Let $(X_n)_{n\in\mathbb{N}_0}$ be a weakly R_n-stationary random sequence with spectral measure $\mu = f\pi$, $f \in L^1(D_s,\pi)$. Equivalent are*

(1) $(X_n)_{n\in\mathbb{N}_0}$ is deterministic (singular).
(2) For all polynomials $p \neq 0$ holds $\frac{p}{f} \notin L^1(D_s,\pi)$.

Proof. The case $X_n = 0$ for all $n \in \mathbb{N}_0$ is trivial. To prove (1) ⇒ (2) let $p \neq 0$ be a polynomial such that $\frac{p}{f} \in L^1(D_s,\mu)$. Assume that the degree of p is n. Then

$$p = \sum_{k=0}^{n} d_k\, R_k\, h(k)$$

with $d_n = 1$ and $d_0,\ldots,d_{n-1} \in \mathbb{C}$. Let $A_n := \{k \in \mathbb{N}_0 : k \geq n\}$. We decompose $R_n = S + Q$, where $S \in H_{A_{n+1}}(R)$ and $Q \in (H_{A_{n+1}}(R))^{\perp}$. We show

$$Q \neq 0 \qquad \text{in } L^2(D_s,\mu).$$

Since p is bounded on D_s, it follows $\frac{|p|^2}{f} \in L^1(D_s,\pi)$ and $\int\limits_{D_s} \frac{|p|^2}{f}\, d\pi > 0$. (Note that $\operatorname{supp}\pi$ is not finite.) Put

$$C := \frac{1}{\int_{D_s} \frac{|p|^2}{f}\, d\pi} \qquad \text{and } g := C\,\frac{p}{f}.$$

It follows

$$\|g\|_{2,\mu}^2 = C^2 \int_{D_s} \frac{|p|^2}{f^2}\, d\mu = C^2 \int_{D_s} \frac{|p|^2}{f}\, d\pi = C$$

and

$$C \int_{D_s} \frac{p}{f}\, R_n\, d\mu = C \int_{D_s} p\, R_n\, d\pi = C,$$

since $d_n = 1$. Therefore,

$$0 < \|g\|_{2,\mu}^2 = C = C \int_{D_s} \frac{p}{f} R_n \, d\mu = \langle g, R_n \rangle_\mu. \qquad (i)$$

Moreover, for $k \geq n+1$ we have $\langle \frac{p}{f}, R_k \rangle_\mu = \langle p, R_k \rangle_\pi = 0$. It follows $g \in (H_{A_{n+1}}(R))^\perp$.

Since Q is the orthogonal projection of R_n onto $(H_{A_{n+1}}(R))^\perp$, we obtain

$$\|R_n - Q\|_{2,\mu}^2 \leq \|R_n - g\|_{2,\mu}^2. \qquad (ii)$$

Combining (i) and (ii) it follows

$$\begin{aligned} 0 < \|g\|_{2,\mu}^2 &= \|R_n\|_{2,\mu}^2 - \left(\|R_n\|_{2,\mu}^2 + \|g\|_{2,\mu}^2 - \langle R_n, g \rangle_\mu - \langle g, R_n \rangle_\mu \right) \\ &= \|R_n\|_{2,\mu}^2 - \|R_n - g\|_{2,\mu}^2 \leq \|R_n\|_{2,\mu}^2 - \|R_n - Q\|_{2,\mu}^2 \\ &= \|R_n\|_{2,\mu}^2 - \|S\|_{2,\mu}^2 = \|Q\|_{2,\mu}^2. \end{aligned}$$

In particular, we have shown that $Q \neq 0$ in $L^2(D_s, \mu)$. However, this implies that $H_{A_n}(R) \supsetneq H_{A_{n+1}}(R)$. The isometric isomorphism Φ yields $H_n \supsetneq H_{n+1}$, a contradiction to the assumption that $(X_n)_{n \in \mathbb{N}_0}$ is deterministic.

To prove (2) $\Rightarrow$ (1) we assume that $(X_n)_{n \in \mathbb{N}_0}$ is not deterministic. By Theorem 6.6.3 there exists $n \in \mathbb{N}_0$ such that $X_0 \notin H_n$. Applying the isometric isomorphism Φ this means $1 = R_0 \notin H_{A_n}(R)$. We decompose $R_0 = S + Q$ in $L^2(D_s, \mu)$, where $S \in H_{A_n}(R)$ and $Q \in (H_{A_n}(R))^\perp$. Since $R_0 \notin H_{A_n}(R)$ it follows $Q \neq 0$ in $L^2(D_s, \mu)$ and

$$0 = \langle Q, R_k \rangle_\mu = \int_{D_s} Q \, R_k \, f \, d\pi \qquad \text{for all } k \geq n.$$

By $Q \in L^2(D_s, \mu) \subseteq L^1(D_s, \mu)$ we get $Qf \in L^1(D_s, \pi)$ and the inversion formula of Theorem 3.3.8 yields

$$Qf = \sum_{k=0}^{n-1} d_k \, R_k \, h(k) \qquad \text{in } L^1(D_s, \pi),$$

where $d_k = \int_{D_s} Qf \, R_k d\pi$ for $k \leq n-1$. Moreover,

$$d_0 = \int_{D_s} Q R_0 \, d\mu = \int_{D_s} Q \, \overline{(S+Q)} \, d\mu = \|Q\|_{2,\mu}^2 \neq 0.$$

In particular, Qf is a nontrivial polynomial. Finally, the polynomial $p = |Qf|^2$ fulfils

$$\int_{D_s} \frac{p}{f} \, d\pi = \int_{D_s} |Q|^2 \, d\mu = \|Q\|_{2,\mu}^2 < \infty,$$

which contradicts assumption (2). $\diamond$

Corollary 6.6.8 *Let $(X_n)_{n\in\mathbb{N}_0}$ be a weakly R_n-stationary random sequence with spectral measure $\mu = f\pi$, $f \in L^1(D_s, \pi)$. Assume that π is a continuous measure (i.e. $\pi(\{x\}) = 0$ for each $x \in D_s$). If $(X_n)_{n\in\mathbb{N}_0}$ is purely nondeterministic (regular), then $(X_n)_{n\in\mathbb{N}_0}$ is an R_n-moving average process.*

Proof. If $X_n = 0$ for all $n \in \mathbb{N}_0$ the statement is obviously true. By Theorem 6.6.7 there exists a polynomial $p \neq 0$ such that $\frac{p}{f} \in L^1(D_s, \pi)$. Since π is a continuous measure it follows that $f > 0$ π-almost everywhere. By Theorem 6.5.3 the corollary is proved. ◇

Chapter 7

Difference equations and stationary sequences on polynomial hypergroups

This chapter deals with problems of difference equations modelling dynamical processes of a special form. Elementary introductions to difference equations can be found in [205] or [368]. A central topic will be a stationarity condition of sequences in Hilbert spaces H. In case $H = L^2(\Omega, P)$ there will be several connections to problems of stochastic modelling in Chapters 5 and 6.

1 Difference equations induced by polynomial hypergroups

Throughout $(R_n)_{n\in\mathbb{N}_0}$ will be an orthogonal polynomial sequence with recurrence coefficients a_n, b_n, c_n generating a polynomial hypergroup. Further let $A : B \to B$ be a linear operator defined on a vector space B. We consider difference equations and define $X(n) \in B$ by

$$a_n X(n+1) \; + \; b_n X(n) \; + \; c_n X(n-1) \; = \; A(X(n)) \qquad (I)$$

for $n \in \mathbb{N}$, where

$$X(0) \in B \quad \text{and} \quad X(1) \; = \; A(X(0)). \qquad (II)$$

The **inherent structure** of (I) is determined by a_n, b_n, c_n and the **dynamics** by $A : B \to B$. Concerning the triple (a_n, b_n, c_n) see subsection 1.2.1.

Each polynomial P of degree n has a unique representation

$$P(z) \; = \; \sum_{k=0}^{n} d_k \, (R_1(z))^k \quad \text{with } d_0, ..., d_n \in \mathbb{C}.$$

We define $P(A) : B \to B$ by

$$P(A) := \sum_{k=0}^{n} d_k \, A^k.$$

Obviously $P(A)$ is a linear operator from B to B, and $R_0(A) = \text{id}$, $R_1(A) = A$. If $P(z)$ and $Q(z)$ are two polynomials, then an elementary calculation shows

$$P(A) \circ Q(A) \;=\; (P \cdot Q)(A) \;=\; Q(A) \circ P(A). \qquad (III)$$

Theorem 7.1.1 *Given the inherent structure by a_n, b_n, c_n and the dynamics by $A : B \to B$. Fix $X(0) \in B$ and set $X(1) = A(X(0))$. Then there is a unique solution $(X(n))_{n\in\mathbb{N}_0}$, $X(n) \in B$, of the difference equations (I), (II). The solution is given by*

$$X(n) \;=\; R_n(A)(X(0)). \qquad (IV)$$

Proof. The existence and uniqueness statement follows directly by the difference equations (I), (II). We have to show that (IV) is true. Clearly $R_0(A)(X(0)) = X(0)$ and $R_1(A)(X(0)) = A(X(0)) = X(1)$. Now we use induction and assume that $X(k) = R_k(A)(X(0))$ is already shown for $k = 0, ..., n$. We have

$$R_{n+1}(z) \;=\; \frac{1}{a_n}\, R_1(z)\, R_n(z) \;-\; \frac{b_n}{a_n}\, R_n(z) \;-\; \frac{c_n}{a_n}\, R_{n-1}(z),$$

and hence

$$R_{n+1}(A) \;=\; \frac{1}{a_n}\, (R_1 \cdot R_n)(A) \;-\; \frac{b_n}{a_n}\, R_n(A) \;-\; \frac{c_n}{a_n}\, R_{n-1}(A).$$

By (III) we obtain

$$R_{n+1}(A) \;=\; \frac{1}{a_n}\, R_1(A) \circ R_n(A) \;-\; \frac{b_n}{a_n}\, R_n(A) \;-\; \frac{c_n}{a_n}\, R_{n-1}(A).$$

By the assumption it follows

$$\begin{aligned} R_{n+1}(A)\,(X(0)) \;&=\; \frac{1}{a_n}\, R_1(A)\,(X(n)) \;-\; \frac{b_n}{a_n}\, X(n) \;-\; \frac{c_n}{a_n}\, X(n-1) \\ &=\; \frac{1}{a_n}\, A(X(n)) \;-\; \frac{b_n}{a_n}\, X(n) \;-\; \frac{c_n}{a_n}\, X(n-1). \end{aligned}$$

By (I) we get $R_{n+1}(A)(X(0)) = X_{n+1}$. ◇

The solutions of the difference equations follow a discrete flow of special type.

We call $\varphi : \mathbb{N}_0 \times B \to B$ a R_n**-flow**, if

$$\varphi(0, X) \;=\; X \quad \text{and} \quad \varphi(m, \varphi(n, X)) \;=\; \sum_{k=|n-m|}^{n+m} g(m, n; k)\, \varphi(k, X) \qquad (V)$$

for all $X \in B$.

(Notify that by defining $\varphi(\omega(m, n), X) := \sum_{k=|n-m|}^{n+m} g(m, n; k)\varphi(k, X)$ we have $\varphi(m, \varphi(n, X)) = \varphi(\omega(m, n), X)$.)

Theorem 7.1.2 *Let $A : B \to B$ be a linear operator, and let $X \in B$.*

(1) By $\varphi(n, X) := R_n(A)(X)$ a R_n-flow is defined.
(2) If $\varphi : \mathbb{N}_0 \times B \to B$ is a R_n-flow, and if $\varphi(1, \cdot)$ is a linear operator, then $\varphi(n, X) = R_n(A)$, where $A(X) = \varphi(1, X)$.

Proof.

(1) We have $\varphi(0, X) = R_0(A)(X) = X$ and by (III)

$$\begin{aligned}\varphi(m, \varphi(n, X)) &= \varphi(m, R_n(A)(X)) \\ &= R_m(R_n(A)(X)) = (R_M \cdot R_n)(A)(X)\end{aligned}$$

$$= \sum_{k=|n-m|}^{n+m} g(m, n; k)\, R_k(A)(X) \;=\; \sum_{k=|n-m|}^{n+m} g(m, n; k)\, \varphi(k, X).$$

(2) We use induction. Obviously $\varphi(0, X) = X = R_0(A)(X)$ and $\varphi(1, X) = A(X) = R_1(A)(X)$. Assume that $\varphi(k, X) = R_k(A)(X)$ for $k = n-1, n$. Since $\varphi(\omega(1, n), X) = \varphi(1, \varphi(n, X))$, we obtain

$$\begin{aligned}\varphi(n+1, X) \;&=\; \frac{1}{a_n}\, \varphi(1, \varphi(n, X)) \;-\; \frac{b_n}{a_n}\, \varphi(n, X) \;-\; \frac{c_n}{a_n}\, \varphi(n-1, X) \\ &= \frac{1}{a_n}\, \varphi(1, R_n(A)(X)) \;-\; \frac{b_n}{a_n}\, R_n(A)(X) \;-\; \frac{c_n}{a_n}\, R_{n-1}(A)(X) \\ &= \frac{1}{a_n}\, R_1(R_n(A)(X)) \;-\; \frac{b_n}{a_n}\, R_n(A)(X) \;-\; \frac{c_n}{a_n}\, R_{n-1}(A)(X) \\ &= \frac{1}{a_n}\, (R_1 \cdot R_n)(A)(X) \;-\; \frac{b_n}{a_n}\, R_n(A)(X) \;-\; \frac{c_n}{a_n}\, R_{n-1}(A)(X) \\ &= R_{n+1}(A)(X).\end{aligned}$$

◇

Let $C : B \to B$ be an invertible linear operator. Consider two sequences $(Y(n))_{n\in\mathbb{N}_0}$ and $(Z(n))_{n\in\mathbb{N}_0}$ in B, defined by an initial value $Y(0) = Z(0) \in B$ and $Y(n+1) = C(Y(n))$, $Z(n+1) = C^{-1}(Z(n))$ for each $n \in \mathbb{N}_0$. Furthermore, define $A = \frac{1}{2}(C + C^{-1})$.

We present two examples that show how to determine the solution of a difference equation (I), (II), with special inherent structure and dynamics A.

1. Set $X(n) := \frac{1}{2}(Y(n) + Z(n))$ for every $n \in \mathbb{N}_0$, the arithmetic mean of $Y(n)$ and $Z(n)$. It is easily checked that $X(1) = A(X(0))$ and $\frac{1}{2}X(n+1) + \frac{1}{2}X(n-1) = A(X(n))$ for all $n \in \mathbb{N}$.

The Chebyshev polynomials $T_n(z)$ of the first kind satisfy

$$\frac{1}{2}\,T_{n+1}(z) \,+\, \frac{1}{2}\,T_{n-1}(z) \;=\; T_1(z)\,T_n(z)$$

for $n \in \mathbb{N}$, and $T_1(z) = z$, $T_0(z) = 1$. Hence it follows that $(X(n))_{n\in\mathbb{N}_0}$, $X(n) = T_n(A)(X(0))$ is the unique solution of (I), (II), where $a_0 = 1$, $a_n = \frac{1}{2} = c_n$ for $n \in \mathbb{N}$, and $b_n = 0$ for $n \in \mathbb{N}_0$ is the inherent structure, and $A = \frac{1}{2}(C + C^{-1})$ is the dynamics.

2. Define $X(0) = \frac{1}{2}(C - C^{-1})(Y(0))$ and $X(n) = \frac{1}{2n+2}(Y(n+1) - Z(n+1))$ for $n \in \mathbb{N}$. Since $Y(0) = Z(0)$, it follows $X(1) = A(X(0))$ and for $n \in \mathbb{N}$

$$\begin{aligned}
&\frac{n+2}{2n+2}\,X(n+1) \,+\, \frac{n}{2n+2}\,X(n-1)\\
&= \frac{1}{2(2n+2)}\,(Y(n+2) - Z(n+2) + Y(n) - Z(n))\\
&= \frac{1}{2(2n+2)}\,(C(Y(n+1)) \,-\, C^{-1}(Z(n+1)) \,+\, C^{-1}(Y(n+1))\\
&\qquad - C(Z(n+1)))\\
&= \frac{1}{2}\,C(X(n)) \,+\, \frac{1}{2}\,C^{-1}(X(n)) \;=\; A(X(n)).
\end{aligned}$$

The Chebyshev polynomials $U_n(z) = R_n^{(\frac{1}{2},\frac{1}{2})}(z)$ of the second kind satisfy

$$\frac{n+2}{2n+2}\,U_{n+1}(z) \,+\, \frac{n}{2n+2}\,U_{n-1}(z) \;=\; U_1(z)\,U_n(z)$$

for $n \in \mathbb{N}$ and $U_1(z) = z$, $U_0(z) = 1$. Thus we have shown that $(X(n))_{n\in\mathbb{N}_0}$, $X(n) = U_n(A)(X(0))$ is the unique solution of (I), (II), where $a_0 = 1$, $a_n = \frac{n+2}{2n+2}$, $c_n = \frac{n}{2n+2}$ for $n \in \mathbb{N}$, and $b_n = 0$ for $n \in \mathbb{N}_0$ is the inherent structure, and $A = \frac{1}{2}(C + C^{-1})$ is the dynamics.

For a third example we consider $(Y(n))_{n\in\mathbb{N}_0}$, $(Z(n))_{n\in\mathbb{N}_0}$, defined as above. The operator $A : B \to B$ is now given by $A = \frac{1}{3}(C + \mathrm{id} + C^{-1})$.

3. Set $X(0) := Y(0)$ and

$$X(n) := \frac{1}{2n+1}\left(Y(0) \,+\, \sum_{k=1}^{n}(Y(k) \,+\, Z(k))\right).$$

Obviously $X(1) = A(X(0))$. Having in mind the Jacobi polynomials $R_n^{(\frac{1}{2},-\frac{1}{2})}(z)$, see subsection 4.5.2, we consider

$$\begin{aligned}
&\frac{2n+3}{3(2n+1)}\,X(n+1) \,+\, \frac{1}{3}\,X(n) \,+\, \frac{2n-1}{3(2n+1)}\,X(n-1)\\
&= \frac{1}{3(2n+1)}\left(Y(0) + \sum_{k=1}^{n+1}(Y(k) \,+\, Z(k))\right) \,+\, \frac{1}{3}\,X(n)
\end{aligned}$$

$$+ \frac{1}{3(2n+1)} \left(Y(0) + \sum_{k=1}^{n-1} (Y(k) + Z(k)) \right)$$

$$= \frac{1}{3(2n+1)} \left(C^{-1}(Y(1)) + \sum_{k=0}^{n} (C(Y(k)) + C^{-1}(Z(k))) \right) + \frac{1}{3} X(n)$$

$$+ \frac{1}{3(2n+1)} \left(C(Z(1)) + \sum_{k=2}^{n} (C^{-1}(Y(k)) + C(Z(k))) \right)$$

$$= \frac{1}{3} C(X(n)) + \frac{1}{3} X(n) + \frac{1}{3} C^{-1}(X(n)) = A(X(n)).$$

The Jacobi polynomials $R_n(z) = R_n^{(\frac{1}{2},-\frac{1}{2})}(z)$ satisfy

$$\frac{2n+3}{3(2n+1)} R_{n+1}(z) + \frac{1}{3} R_n(z) + \frac{2n-1}{3(2n-1)} R_{n-1}(z) = R_1(z) R_n(z)$$

for $n \in \mathbb{N}$ and $R_1(z) = \frac{2}{3}z + \frac{1}{3}$, $R_0(z) = 1$. Thus $(X(n))_{n\in\mathbb{N}_0}$, $X(n) = R_n^{(\frac{1}{2},-\frac{1}{2})}(A)(X(0))$ is the unique solution of (I), (II), where $a_0 = \frac{3}{2}$, $a_n = \frac{2n+3}{3(2n+1)}$, $b_n = \frac{1}{3}$, $c_n = \frac{2n-1}{3(2n+1)}$ for $n \in \mathbb{N}$, and $b_0 = -\frac{1}{2}$ and $b_n = \frac{1}{3}$ for $n \in \mathbb{N}$ is the inherent structure, and $A = \frac{1}{3}(C + \mathrm{id} + C^{-1})$ is the dynamics.

Remark. Note that the property $g(m,n;k) \geq 0$ was not used when proving Theorem 7.1.1 and Theorem 7.1.2.

For the next result the assumption that $(R_n(z))_{n\in\mathbb{N}_0}$ generates a polynomial hypergroup is essential.

Theorem 7.1.3 *Assume that $B = \mathbb{C}$ and $A : B \to B$ is given by $Ax = ax$ for some $a \in \mathbb{C}$. Let $(X(n))_{n\in\mathbb{N}_0}$ be the solution of (I), (II), with initial value $X(0) \in \mathbb{C}\backslash\{0\}$. Then the following dichotomy is valid, depending on either $a \in D$ or $a \notin D$, where D is the dual space of the polynomial hypergroup generated by $(R_n(z))_{n\in\mathbb{N}_0}$, i.e.*

$$D = \{z \in \mathbb{C} : |R_n(z)| \leq 1 \text{ for all } n \in \mathbb{N}_0\} :$$

(i) $(X(n))_{n\in\mathbb{N}_0}$ is a bounded sequence in $\mathbb{C}$ with bound $|X(0)|$ exactly if $a \in D$, and

(ii) $(X(n))_{n\in\mathbb{N}_0}$ is unbounded in $\mathbb{C}$, exactly if $a \in \mathbb{C}\backslash D$.

Proof. Combine Theorem 7.1.1 and Theorem 4.1.1 (i). $\diamond$

Further results on difference equations with inherent structure based on polynomial hypergroups can be found in Chapter 10 of [647].

2 $\mathbf{R_n}$-stationary sequences and boundedness

If B is a Hilbert space H, the sequences $(X(n))_{n\in\mathbb{N}_0}$ defined by (I), (II) often satisfy an interesting property, which we know already from Chapter 5 and 6, where $H = L^2(\Omega, P)$. In the sequel H is an arbitrary Hilbert space.

Let $(X(n))_{n\in\mathbb{N}_0}$ be a sequence in H. We write

$$T_n X(m) := \sum_{k=|n-m|}^{n+m} g(m,n;k)\, X(k) \qquad (VI)$$

for $m, n \in \mathbb{N}_0$. Obviously $T_n X(m) = T_m X(n)$.

Definition 1 A sequence $(X(n))_{n\in\mathbb{N}_0}$ in H is called $\mathbf{R_n}$**-stationary**, if

$$\langle X(m), X(n)\rangle = \langle T_n X(m), X(0)\rangle$$

for all $m, n \in \mathbb{N}_0$.

Using the recurrence relation of $(R_n(z))_{n\in\mathbb{N}_0}$ we get

$$T_{n+1} = \frac{1}{a_n}\, T_1 \circ T_n - \frac{b_n}{a_n}\, T_n - \frac{c_n}{a_n}\, T_{n-1} \quad \text{for } n \in \mathbb{N}, \qquad (VII)$$

where $T_0 = \mathrm{id}$.

Theorem 7.2.1 *Let $(X(n))_{n\in\mathbb{N}_0}$ be a sequence in H. The following conditions are equivalent:*

(1) $(X(n))_{n\in\mathbb{N}_0}$ is R_n-stationary.
(2) $\langle T_1 X(m), X(n)\rangle = \langle X(m), T_1 X(n)\rangle$ for all $m, n \in \mathbb{N}_0$.
(3) $\langle T_k X(m), X(n)\rangle = \langle X(m), T_k X(n)\rangle$ for all $k, m, n \in \mathbb{N}_0$.

Proof. According to the definition of the translation operators T_k and the associativity of the convolution on the polynomial hypergroup we get, assuming the R_n-stationarity:

$$\langle T_1 X(m), X(n)\rangle = \langle a_m X(m+1) + b_m X(m) + c_m X(m-1), X(n)\rangle$$

$$= \langle T_n(T_1 X(m)), X(0)\rangle = \langle X(m), T_1 X(n)\rangle \quad \text{for } m, n \in \mathbb{N}_0.$$

Thus (1) $\Rightarrow$ (2) is shown.

(2) $\Rightarrow$ (1): We show the R_n-stationarity of $(X(n))_{n\in\mathbb{N}_0}$ by induction with respect to n. For $n = 0$ we have $\langle X(m), X(0)\rangle = \langle T_0 X(m), X(0)\rangle$ for $m \in \mathbb{N}_0$, and for $n = 1$ we get by (2)

$$\langle X(m), X(1)\rangle = \langle X(m), T_1 X(0)\rangle = \langle T_1 X(m), X(0)\rangle$$

for $m \in \mathbb{N}_0$. Assume $\langle X(m), X(k)\rangle = \langle T_k X(m), X(0)\rangle$ for $k = n-1, n$ and every $m \in \mathbb{N}_0$. It follows by (2) and (VII)

$$\begin{aligned}\langle X(m), X(n+1)\rangle &= \frac{1}{a_n}\langle X(m), T_1 X(n)\rangle - \frac{b_n}{a_n}\langle X(m), X(n)\rangle \\ &\quad - \frac{c_n}{a_n}\langle X(m), X(n-1)\rangle \\ = \frac{1}{a_n}\langle T_1 X(m), X(n)\rangle &- \frac{b_n}{a_n}\langle X(m), X(n)\rangle - \frac{c_n}{a_n}\langle X(m), X(n-1)\rangle \\ = \frac{1}{a_n}\langle T_n(T_1 X(m)), X(0)\rangle &- \frac{b_n}{a_n}\langle T_n X(m), X(0)\rangle - \frac{c_n}{a_n}\langle T_{n-1} X(m), X(0)\rangle \\ &= \langle T_{n+1} X(m), X(0)\rangle.\end{aligned}$$

It remains to prove (2) $\Rightarrow$ (3). Now apply induction with respect to k. Assume (3) is true for $k-1$ and k. Using (VII) for k we get

$$\begin{aligned}\langle T_{k+1} X(m), X(n)\rangle &= \frac{1}{a_k}\langle T_1(T_k X(m)), X(n)\rangle - \frac{b_k}{a_k}\langle T_k X(m), X(n)\rangle \\ &\quad - \frac{c_k}{a_k}\langle T_{k-1} X(m), X(n)\rangle = \langle X(m), T_{k+1} X(n)\rangle.\end{aligned}$$

Thus (3) is shown. ◇

Considering a solution $(X(n))_{n\in\mathbb{N}_0}$ of the difference equation (I), (II) with dynamics $A : H \to H$, Theorem 7.2.1 yields the following equivalence statement.

Corollary 7.2.2 *Let $(X(n))_{n\in\mathbb{N}_0}$ be a solution of (I), (II) with dynamics $A : H \to H$. Then $(X(n))_{n\in\mathbb{N}_0}$ is R_n-stationary if and only if A is symmetric on span$\{X(n) : n \in \mathbb{N}_0\}$.*

Proof. Solutions of (I), (II) are determined by $T_1 X(n) = AX(n)$. Thus Theorem 7.2.1 says that $(X(n))_{n\in\mathbb{N}_0}$ is R_n-stationary if and only if $\langle AX(m), X(n)\rangle = \langle X(m), AX(n)\rangle$ for each $n, m \in \mathbb{N}_0$, which is equivalent to $\langle AX, Y\rangle = \langle X, AY\rangle$ for all $X, Y \in \text{span}\{X(n) : n \in \mathbb{N}_0\}$. ◇

Noting that $\{X(n) : n \in \mathbb{N}_0\}$ may be linearly dependent, we show the following result.

Lemma 7.2.3 *Let $(X(n))_{n\in\mathbb{N}_0}$ be a R_n-stationary sequence in H. Then for each $m \in \mathbb{N}_0$ the mapping T_m defined on $\{X(n) : n \in \mathbb{N}_0\}$ can be linearly extended to span$\{X(n) : n \in \mathbb{N}_0\}$ by setting*

$$T_m\left(\sum_{k=0}^{N} d_k\, X(k)\right) := \sum_{k=0}^{N} d_k\, T_m X(k).$$

Proof. We just have to show that T_m is well-defined. If $Y = \sum_{k=0}^{N} d_k X(k) = 0$, by Theorem 7.2.1(3) it follows that

$$\langle T_m Y, X(n)\rangle = \sum_{k=0}^{N} d_k \langle T_m X(k), X(n)\rangle = \sum_{k=0}^{N} d_k \langle X(k), T_m X(n)\rangle$$

$$= \langle Y, T_m X(n)\rangle = 0 \qquad \text{for all } X(n),\ n \in \mathbb{N}_0.$$

Hence $T_m Y = 0$. ◇

Denote $L(X) = \text{span}\{X(n) : n \in \mathbb{N}_0\}$. We will apply Bochner's representation theorem, see Theorem 3.7.1, to extend the linear operators T_m defined on $L(X)$ to $\overline{L(X)} \subseteq H$.

Theorem 7.2.4 *Let $(X(n))_{n\in\mathbb{N}_0}$ be a bounded R_n-stationary sequence in H. Then there exists a unique bounded regular positive Borel measure μ on D_s such that*

$$\langle X(n), X(m)\rangle = \int_{D_s} R_n(x)\, R_m(x)\, d\mu(x)$$

for all $n, m \in \mathbb{N}_0$.

Conversely, let μ be a bounded regular positive Borel measure on D_s. Then there exists a bounded R_n-stationary sequence $(X(n))_{n\in\mathbb{N}_0}$ in $L^2(D_s, \mu)$.

Proof. In the same way as in the proof of Theorem 5.1.1 we obtain that $d = (d(n))_{n\in\mathbb{N}_0}$, $d(n) = \langle X(n), X(0)\rangle$, is a bounded positive definite sequence with respect to $(R_n(x))_{n\in\mathbb{N}_0}$. By Theorem 3.7.1 there exists a unique Borel measure $\mu \in M^+(D_s)$ such that

$$d(n) = \int_{D_s} R_n(x)\, d\mu(x) = \check{\mu}(n).$$

Since $(X(n))_{n\in\mathbb{N}_0}$ is R_n-stationary, it follows

$$\langle X(n), X(m)\rangle = \langle T_n X(m), X(0)\rangle = \sum_{k=|n-m|}^{n+m} g(n,m;k)\, \langle X(k), X(0)\rangle$$

$$= \sum_{k=|n-m|}^{n+m} g(n,m;k)\, d(k) = \int_{D_s} R_n(x)\, R_m(x)\, d\mu(x).$$

Now we prove the second statement. Let $d(n) = \int\limits_{D_s} R_n(x) d\mu(x) = \check{\mu}(n)$. Then $d = (d(n))_{n\in\mathbb{N}_0}$ is a bounded positive definite sequence with respect to $(R_n(x))_{n\in\mathbb{N}_0}$, and

$$L_n d(m) \;=\; \int_{D_s} R_m(x)\, R_n(x)\, d\mu(x).$$

Consider the Hilbert space $L^2(D_s, \mu)$ and $X(n) = R_n|D_s \in L^2(D_s, \mu)$. It follows (with the scalar product in $L^2(D_s, \mu)$)

$$\langle X(m), X(n)\rangle \;=\; \int_{D_s} R_m(x)\, R_n(x)\, d\mu(x) \;=\; L_n d(m)$$

and

$$\langle T_n X(m), X(0)\rangle \;=\; \int_{D_s} \left(\sum_{k=|m-n|}^{m+n} g(m,n;k)\, R_k(x) \right) R_0(x)\, d\mu(x)$$

$$=\; \int_{D_s} R_m(x)\, R_n(x)\, d\mu(x) \;=\; \langle X(m), X(n)\rangle.$$

Thus $(X(n))_{n\in\mathbb{N}_0}$ is a bounded R_n-stationary sequence in $L^2(D_s, \mu)$ (and μ is the spectral measure of $(X(n))_{n\in\mathbb{N}_0}$, and $H(X) = L^2(D_s, \mu)$.) $\diamond$

$\{R_n : n \in \mathbb{N}_0\}$ is a linear independent subset and span $\{R_n : n \in \mathbb{N}_0\}$ is dense in $L^2(D_s, \mu)$. Now define a mapping

$$\Phi:\ \text{span}\,\{R_n : n \in \mathbb{N}_0\} \longrightarrow L(X) \subseteq H$$

by

$$\Phi\left(\sum_{k=1}^{N} c_k\, R_{n_k}\right) := \sum_{k=1}^{N} c_k\, X(n_k), \qquad (VIII)$$

where $N \in \mathbb{N}_0$; $c_1, ..., c_N \in \mathbb{C}$; $n_1, ..., n_N \in \mathbb{N}_0$.

This map is well-defined, linear and surjective. Denoting the scalar product on $L^2(D_s, \mu)$ by $\langle\,\cdot\,,\,\cdot\,\rangle_\mu$, it follows

$$\langle R_n, R_m\rangle_\mu \;=\; \langle X(n), X(m)\rangle.$$

Hence Φ is isometric and can be continuously extended to $L^2(D_s, \mu)$. Thus we get an isometric isomorphism – which again is denoted by Φ – from $L^2(D_s, \mu)$ onto $H(X) := \overline{L(X)} \subseteq H$. We call μ the **spectral measure** of $(X(n))_{n\in\mathbb{N}_0}$.

Furthermore, consider the linear mapping $T_m : L(X) \to L(X)$ defined by Lemma 7.2.3. T_m satisfies $\langle T_m Y, Z\rangle = \langle Y, T_m Z\rangle$ for $Y, Z \in L(X)$.

Assuming boundedness of $(X(n))_{n\in\mathbb{N}_0}$ we can apply $\Phi : L^2(D_s,\mu) \to H(X)$ in order to extend T_m to a self-adjoint contraction on $H(X)$. In fact, we have

$$\Phi^{-1}\left(T_m\left(\sum_{k=0}^{N} d_k\, X(k)\right)\right) = R_m \cdot \sum_{k=0}^{N} d_k\, R_k,$$

and multiplication by $R_m(z)$ is a symmetric contraction. Hence $T_m : L(X) \to L(X)$ can be extended to a self-adjoint contraction on $H(X)$, which also is denoted by T_m.

We have shown the following properties of Φ and T_m.

Theorem 7.2.5 *Let $(X(n))_{n\in\mathbb{N}_0}$ be a bounded R_n-stationary sequence in H. Then*

(i) The extension of the mapping Φ in (VIII) is an isometric isomorphism of $L^2(D_s,\mu)$ onto $H(X)$.
(ii) For each $m \in \mathbb{N}_0$ the extension of $T_m : L(X) \to L(X)$ is a self-adjoint operator on $H(X)$ with $\|T_m\| \le 1$.
(iii) $X(n) = T_n(X(0))$ for all $n \in \mathbb{N}_0$.
(iv) For the spectrum $\sigma(T_m)$ of $T_m \in \mathcal{B}(H(X))$ is valid

$$\sigma(T_m) = R_m(\operatorname{supp}\mu).$$

Proof. (i) and (ii) are shown. To prove (iii) note that $X(1) = T_1(X(0))$, and using (VII) the statement of (iii) follows by induction on n. In order to prove (iv) consider the multiplication operators $M_m : L^2(D_s,\mu) \to L^2(D_s,\mu)$,

$$M_m(\psi) := R_m \cdot \psi \qquad \text{for } \psi \in L^2(D_s,\mu).$$

We have $T_m = \Phi \circ M_m \circ \Phi^{-1}$ on $H(X)$. By [185,Corollary 4.24] it follows

$$\sigma(T_m) = \sigma(M_m) = R_m(\operatorname{supp}\mu).$$

◇

Corollary 7.2.6 *Let $(X(n))_{n\in\mathbb{N}_0}$ be a bounded R_n-stationary sequence in H. Then*

$$\|X(n)\| \le \|X(0)\|$$

for each $n \in \mathbb{N}_0$.

Now we can add the following results concerning boundedness of solutions of difference equations with inherent structure determined by $(R_n(z))_{n\in\mathbb{N}_0}$, and dynamics $A : H \to H$. Note that the properties of these solutions depend on the choice of the starting vector $X(0)$.

Theorem 7.2.7 *Let $(X(n))_{n\in\mathbb{N}_0}$ be a solution of a difference equation (I), (II) in H, where the dynamics $A : H \to H$ is an arbitrary linear operator.*

(1) If $A : H \to H$ is closed and if $(X(n))_{n\in\mathbb{N}_0}$ is a bounded R_n-stationary sequence, then $A|H(X)$ is a bounded self-adjoint linear operator on $H(X)$, and the spectrum $\sigma(A|H(X))$ of $A|H(X)$ satisfies

$$a_0\, \sigma(A|H(X)) \,+\, b_0 \,\subseteq\, D_s.$$

Moreover, we have

$$X(n) \;=\; R_n(A)(X(0)) \;=\; T_n(X(0)).$$

(2) If $A|H(X)$ is a bounded self-adjoint operator on $H(X)$, and the spectrum $\sigma(A(H(X))$ satisfies $a_0\sigma(A|H(X)) + b_0 \subseteq D_s$, then $(X(n))_{n\in\mathbb{N}_0}$ with $X(n) = R_n(A|H(X))(X(0))$ is a bounded R_n-stationary sequence.

Proof.

(1) Since $(X(n))_{n\in\mathbb{N}_0}$ is R_n-stationary, it follows that $\langle AX, Y\rangle = \langle X, AY\rangle$ for all $X, Y \in L(X)$, see Corollary 7.2.2. By (VIII) we have

$$\begin{aligned}\Phi(&a_n R_{n+1} + b_n R_n + c_n R_{n-1})\\ &= a_n X(n+1) \,+\, b_n X(n) \,+\, c_n X(n-1) \;=\; AX(n).\end{aligned}$$

Hence, if $Y = \sum_{k=0}^{N} d_k X(k) \in L(X)$, then

$$AY \;=\; \Phi\left(R_1 \cdot \sum_{k=0}^{N} d_k\, R_k\right) \;=\; T_1 Y,$$

i.e. A and T_1 coincide on $L(X)$. Now use that $A : H \to H$ is closed. That means, if $x_n \in H$ and $x_n \to x$ and $Ax_n \to y$ as $n \to \infty$, then $Ax = y$. This implies that A and T_1 coincide on $H(X)$, too.
In particular, $A|H(X)$ is a bounded self-adjoint linear operator on $H(X)$, see Theorem 7.2.5(ii). Furthermore we get

$$\sigma(A|H(X)) \;=\; R_1(\operatorname{supp}\mu),$$

see Theorem 7.2.5(iv). Hence

$$\operatorname{supp}\mu \;=\; a_0\,\sigma(A|H(X)) \,+\, b_0 \,\subseteq D_s.$$

Combining Theorem 7.1.1 and Theorem 7.2.5(iii) it follows $X(n) = R_n(A)(X(0)) = T_n(X(0))$.

(2) Now we use the spectral theorem for bounded self-adjoint operators, see [609,12.24], and that $|R_n(x)| \leq 1$ for every $x \in D_s$. Since $a_0\sigma(A|H(X)) + b_0 \subseteq D_s$, we obtain for the operators $R_n(A|H(X))$ that

$$\|R_n(A|H(X))\| \leq \sup\{\,|R_n(x)| : x \in a_0\,\sigma(A|H(X)) + b_0\} \leq 1.$$

By Theorem 7.1.1 it follows $(X(n))_{n\in\mathbb{N}_0}$ is a bounded R_n-stationary sequence in H.

◇

Considering ergodic properties of bounded R_n-stationary sequences in a Hilbert space H, we can use the results of section 4.10.

For any sequence $(X(n))_{n\in\mathbb{N}_0}$ in H and $N \in \mathbb{N}_0$ denote by $C_N \in H$

$$C_N := \frac{1}{H(N)} \sum_{k=0}^{N} X(k)\,h(k).$$

Corollary 7.2.8 *Assume that* $\frac{a_0 h(n)}{H(n)}$ *tends to zero, or that* $R_n(x) \to 0$ *for all* $x \in D_s\backslash\{1\}$. *If* $(X(n))_{n\in\mathbb{N}_0}$ *is a bounded* R_n*-stationary sequence in* H, *the limit* $C_\infty := \lim_{N\to\infty} C_N$ *exists in* $H(X)$ *(in the sense of norm-convergence).*

Proof. Applying the isometric isomorphism $\Phi : L^2(D_s,\mu) \to H(X)$, the statement follows directly from Theorem 4.10.4. ◇

In the case $\frac{a_n h(n)}{H(n)} \to 0$ we can give more information on the limit C_∞.

Theorem 7.2.9 *Suppose that* $\frac{a_n h(n)}{H(n)} \to 0$. *Let* $(X(n))_{n\in\mathbb{N}_0}$ *be a bounded* R_n*-stationary sequence in* H. *Then*

$$C_\infty = P_F(X(0)),$$

where P_F *denotes the orthogonal projection in* $H(X)$ *onto the space* $F := \{Y \in H(X) : T_1 Y = Y\}$.

Proof. Denote $E := \{T_1 Y - Y : Y \in H(X)\}$. Since $T_1 \in B(H(X))$ is self-adjoint, it follows that $F = E^\perp$. Let $Z_0 = T_1 Y - Y$ for some $Y \in H(X)$. Using $a_k + b_k + c_k = 1$ and $h(k+1) = h(k)\,\frac{a_k}{c_{k+1}}$, a straightforward calculation yields

$$\sum_{k=0}^{N} T_k Z_0\,h(k) = \sum_{k=0}^{N} (T_k(T_1 Y) - T_k Y)\,h(k)$$

$$= a_N T_{N+1} Y\,h(N) - c_{N+1} T_N Y\,h(N+1) = a_N h(N)\,(T_{N+1} Y - T_N Y).$$

The assumptions imply

$$\frac{1}{H(N)} \sum_{k=0}^{N} T_k Z_0 \, h(k) \to 0 \qquad \text{as } N \to \infty. \qquad (*)$$

Now let $Z_0 \in \overline{E}$. Choose $Y_n \in H(X)$ such that $Z_0 = \lim_{n\to\infty} T_1 Y_n - Y_n$. Since $\|T_k\| \leq 1$, $(*)$ also holds for $Z_0 \in \overline{E}$. By $\overline{E} = (E^\perp)^\perp = F^\perp$ we have $X(0) = Z_0 + P_F(X(0))$, where $Z_0 \in \overline{E}$. Since $T_k Y = Y$ for every $Y \in F$, we obtain

$$C_N = \frac{1}{H(N)} \sum_{k=0}^{N} T_k(X(0)) \, h(k)$$

$$= \frac{1}{H(N)} \sum_{k=0}^{N} T_k Z_0 \, h(k) + P_F(X(0)) \to P_F(X(0))$$

as $N \to \infty$. ◇

Remarks:

(1) The proof of Theorem 7.2.9 is based on the proof of [545,Theorem 1.2].
(2) Referring to $\Phi : L^2(D_s, \mu) \to H(X)$ and Theorem 4.10.4, we see that $\Phi(\chi_{\{1\}}) = P_F(X(0))$.

We call $P_F(X(0))$ the **mean element** of the bounded R_n-stationary sequence $(X(n))_{n\in\mathbb{N}_0}$ and write $\mathcal{M}(X) := P_F(X(0))$.

Lemma 7.2.10 *Let $(X(n))_{n\in\mathbb{N}_0}$ be a bounded R_n-stationary sequence in H. Denote (as in the proof of Theorem 7.2.9)*

$$E := \{T_1 Y - Y : Y \in H(X)\}.$$

Then we have $T_n Y - Y \in E$ for each $n \in \mathbb{N}$ and $Y \in H(X)$.

Proof. We use induction on n. Applying (VII), we get

$$T_{n+1}Y - Y = \frac{1}{a_n} (T_n(T_1 Y) - T_1 Y) - \frac{b_n}{a_n} (T_n Y - Y)$$

$$- \frac{c_n}{a_n} (T_{n-1} Y - Y) + \frac{1}{a_n} (T_1 Y - Y),$$

and the assertion is proven. ◇

Proposition 7.2.11 *Suppose that $\frac{a_n h(n)}{H(n)} \to 0$. Let $(X(n))_{n\in\mathbb{N}_0}$ be a bounded R_n-stationary sequence in H. Then*

$$\langle \mathcal{M}(X), X(m) \rangle = \langle \mathcal{M}(X), X(0) \rangle = \|\mathcal{M}(X)\|_2^2$$

for every $m \in \mathbb{N}_0$.

Proof. We have

$$\begin{aligned}
&\langle \frac{1}{H(N)} \sum_{k=0}^{N} X(k)\, h(k), X(m) - X(0)\rangle \\
&= \langle \frac{1}{H(N)} \sum_{k=0}^{N} (T_m X(k) \,-\, X(k))\, h(k), X(0)\rangle \\
&= \langle \frac{1}{H(N)} \sum_{k=0}^{N} T_k Z_0\, h(k), X(0)\rangle,
\end{aligned}$$

where $Z_0 = T_m X(0) - X(0)$. By Lemma 7.2.10 we know $Z_0 \in E$. Since

$$\frac{1}{H(N)} \sum_{k=0}^{N} T_k Z_0\, h(k) \;\to\; 0 \qquad \text{as } N \to \infty$$

(see $(*)$ in the proof of Theorem 7.2.9), it follows

$$\langle \mathcal{M}(X), X(0)\rangle \;=\; \langle \mathcal{M}(X), X(m)\rangle \qquad \text{for all } m \in \mathbb{N}_0.$$

Therefore $\langle \mathcal{M}(X), X(0)\rangle \;=\; \langle \mathcal{M}(X), \mathcal{M}(X)\rangle$. ◇

Remark: By Corollary 4.10.5 it follows

$$\|\mathcal{M}(X)\|_2^2 \;=\; \|\chi_{\{1\}}\|_{2,\mu}^2 \;=\; \mu(\{1\}),$$

and by Corollary 4.10.6 we have $\|\mathcal{M}(X)\|_2^2$ is equal to

$$\inf\left\{ \sum_{i,j=1}^{n} c_i \bar{c}_j \langle X(m_i), X(m_j)\rangle : n \in \mathbb{N}; c_1, ..., c_n \in \mathbb{C}, c_1 + \cdots + c_n = 1; \right.$$
$$\left. m_1, ..., m_n \in \mathbb{N}_0 \right\}.$$

The investigations of bounded R_n-stationary sequences $(X(n))_{n\in\mathbb{N}_0}$ in H are based on the fact that $(d(n))_{n\in\mathbb{N}_0}$, $d(n) = \langle X(n), X(0)\rangle$ is a bounded positive definite sequence, and so we can apply Bochner's theorem in order to obtain

$$\begin{aligned}
\langle X(m), X(n)\rangle &= \int_{D_s} R_m(x)\, R_n(x)\, d\mu(x) \;=\; \sum_{k=|n-m|}^{n+m} g(m,n;k)\, d(k) \\
&= L_n d(m),
\end{aligned}$$

see Theorem 7.2.4.

This suggests to consider **double** sequences $(a(m,n))_{m,n\in\mathbb{N}_0}$, $a(m,n) \in \mathbb{C}$, which satisfy

$$\sum_{m,n=0}^{N} c_m\, \bar{c}_n\, a(m,n) \;\geq\; 0$$

for any finite set of complex numbers $c_0, ..., c_N$. Then we call the $(a(m,n))_{m,n\in\mathbb{N}_0}$ a **positive definite double sequence.** This inequality already implies that $a(m,n) \geq 0$ and $a(m,n) = \overline{a(n,m)}$. In fact, we get

$$a(m,m) + |\lambda|^2 a(n,n) + \lambda\, a(n,m) + \bar{\lambda}\, a(m,n) \geq 0 \qquad \text{for any } \lambda \in \mathbb{C},$$

and $a(m,n) = \overline{a(n,m)}$ follows. Now we can refer to [212,Appendix] for an elementary proof of the following result:

Theorem 7.2.12 *Let $(a(m,n))_{m,n\in\mathbb{N}_0}$ be a double sequence of complex numbers. There exists a sequence $(X(n))_{n\in\mathbb{N}_0}$ of elements in a Hilbert space H such that*

$$a(m,n) = \langle X(m), X(n)\rangle \qquad \text{for all } m,n \in \mathbb{N}_0$$

if and only if $(a(m,n))_{m,n\in\mathbb{N}_0}$ is positive definite.

This equivalence relation is first of all independent of any algebraic structure on $\mathbb{N}_0$. When $a(m,n) = \varphi(m+n)$, where $(\varphi(n))_{n\in\mathbb{N}_0}$ is a positive definite sequence with respect to the semigroup $(\mathbb{N}_0, +)$, see [71], the algebraic structures of semigroups play an important role. Replacing $\mathbb{N}_0$ by $\mathbb{Z}$ and putting $a(m,n) = \varphi(m-n)$, $(\varphi(n))_{n\in\mathbb{Z}}$ a positive definite sequence with respect to the group $(\mathbb{Z}, +)$, properties of discrete commutative groups are relevant.

Our interest is concentrated on polynomial hypergroups. We emphasize that our results are essential for investigating weakly stationary random sequences on polynomial hypergroups, or difference equations induced by polynomial hypergroups.

In general the product of two positive definite functions on a hypergroup K is not positive definite. For positive definite double sequences on $\mathbb{N}_0$ we can show the following property see [212,p.594].

Proposition 7.2.13 *Let $(a(m,n))_{m,n\in\mathbb{N}_0}$, $(b(m,n))_{m,n\in\mathbb{N}_0}$ be two positive definite double sequences. Then $(a(m,n)\cdot b(m,n))_{m,n\in\mathbb{N}_0}$ is a positive definite double sequence, too.*

Proof. By Theorem 7.2.12 there exist $X(m)$ and $X(n)$ in a Hilbert space H, such that $a(m,n) = \langle X(m), X(n)\rangle$. Hence we can write

$$a(m,n) = \langle X(m), X(n)\rangle = \sum_{i,j=0}^{\infty} x_{m,j}\, \overline{x_{n,j}},$$

where $\sum_{j=0}^{\infty} |x_{n,j}|^2 < \infty$ for each $n \in \mathbb{N}_0$.

Then it follows for each $N \in \mathbb{N}_0$ and any $\varrho_0, ..., \varrho_N \in \mathbb{C}$

$$\sum_{m,n=0}^{N} a(m,n) \cdot b(m,n)\, \varrho_m\, \bar{\varrho}_n = \sum_{j=0}^{\infty} \left(\sum_{m=0}^{N} \sum_{n=0}^{N} b(m,n)\, (x_{m,j}\varrho_m)\, \overline{(x_{n,j}\varrho_n)} \right) \geq 0,$$

i.e. $(a(m,n) \cdot b(m,n))_{m,n\in\mathbb{N}_0}$ is a positive definite double sequence. ◇

Remark. Let $(\varphi(n))_{n\in\mathbb{N}_0}$ and $(\psi(n))_{n\in\mathbb{N}_0}$ be positive definite sequences with respect to $(R_n(x))_{n\in\mathbb{N}_0}$. By Proposition 7.2.13 $(a(m,n))_{m,n\in\mathbb{N}_0}$, where $a(m,n) := L_n\varphi(m) \cdot L_n\psi(m)$, is a positive definite double sequence. By Theorem 7.2.12 we know that $a(m,n) = \langle X(m), X(n)\rangle$ for a sequence of vectors X_n in a Hilbert space. In general $(X(n))_{n\in\mathbb{N}_0}$ is not R_n-stationary. In fact, $\langle X(m), X(0)\rangle = a(m,0) = \varphi(m)\psi(m)$. Assuming R_n-stationarity of $(X(n))_{n\in\mathbb{N}_0}$, it follows

$$L_n\varphi(m) \cdot L_n\psi(m) = \langle X(m), X(n)\rangle = \langle T_nX(m), X(0)\rangle$$

$$= \sum_{k=|n-m|}^{n+m} g(m,n;k)\, \langle X(k), X(0)\rangle = \sum_{k=|n-m|}^{n+m} g(m,n;k)\, \varphi(k) \cdot \psi(k)$$

$$= L_n(\varphi \cdot \psi)(m)$$

for all $m, n \in \mathbb{N}_0$, which is only true for very special φ or ψ.

3 Examples and autoregressive positive definite sequences

We present examples of bounded positive definite sequences $(d(n))_{n\in\mathbb{N}_0}$ with respect to $(R_n(x))_{n\in\mathbb{N}_0}$ and corresponding bounded R_n-stationary sequences $(X(n))_{n\in\mathbb{N}_0}$ in appropriate Hilbert spaces. We begin with a simple example, which is included to clarify the relations between $(d(n))_{n\in\mathbb{N}_0}$ and $(X(n))_{n\in\mathbb{N}_0}$.

Example 1. Let $d(n) = \delta_{0,n} = \epsilon_0(n)$. Then $d = (d(n))_{n\in\mathbb{N}_0} = (1,0,0,...)$ is a bounded positive definite sequence with respect to $(R_n(x))_{n\in\mathbb{N}_0}$. Furthermore we get $L_nd(m) = \frac{1}{h(n)}\, \delta_{m,n} = \frac{1}{h(n)}\, \epsilon_n(m)$. Considering the Hilbert space $H = L^2(D_s, \pi)$ and $(X(n))_{n\in\mathbb{N}_0}$, where $X(n) = R_n|D_s \in H$, we have (using VI of subsection 7.2)

$$\langle X(m), X(n)\rangle = \int_{D_s} R_m(x)\, R_n(x)\, d\pi(x) = \langle T_nX(m), X(0)\rangle = L_nd(m).$$

π is the spectral measure of $(X(n))_{n\in\mathbb{N}_0}$, and $H(X) = L^2(D_s, \pi)$.

Another possibility is to consider the 'dual' Hilbert space $H = l^2(h)$ and $(X(n))_{n\in\mathbb{N}_0}$, where $X(n) = \epsilon_n$, $\epsilon_n(m) = \frac{1}{h(n)}\,\delta_{m,n}$. It is straightforward to show that

$$\langle X(m), X(n)\rangle \;=\; \frac{1}{h(n)}\,\delta_{m,n} \;=\; \langle T_n X(m), X(0)\rangle \;=\; L_n d(m).$$

Note that $\mathcal{P}(\epsilon_n) = R_n|D_s$, where $\mathcal{P} : l^2(h) \to L^2(D_s, \pi)$.

Example 2. Choose a sequence $(x_k)_{k\in\mathbb{N}}$ of points $x_k \in D_s$ and a sequence $(\lambda_k)_{k\in\mathbb{N}}$ of complex numbers $\lambda_k \in \mathbb{C}$ such that $\sum_{k=1}^{\infty} |\lambda_k|^2 < \infty$. Let

$$d(n) \;=\; \sum_{k=1}^{\infty} |\lambda_k|^2\, R_n(x_k) \qquad \text{for each } n \in \mathbb{N}_0.$$

Then $d = (d(n))_{n\in\mathbb{N}_0}$ is a bounded positive definite sequence with respect to $(R_n(x))_{n\in\mathbb{N}_0}$, and

$$L_n d(m) \;=\; \sum_{k=1}^{\infty} |\lambda_k|^2\, R_m(x_k)\, R_n(x_k).$$

Denote by $\mu_d = \sum_{k=1}^{\infty} |\lambda_k|^2 \epsilon_{x_k}$ the discrete measure on D_s (ϵ_{x_k} is the point measure at $x_k \in D_s$). In the Hilbert space $H = L^2(D_s, \mu_d)$ consider $X(n) = R_n|D_s \in H$. With the scalar product $\langle\ ,\ \rangle$ in H we have

$$\langle X_m, X_n\rangle \;=\; \sum_{k=1}^{\infty} |\lambda_k|^2\, R_m(x_k)\, R_n(x_k) \;=\; \langle T_n X(m), X(0)\rangle \;=\; L_n d(m).$$

One can construct a bounded R_n-stationary sequence $(X(n))_{n\in\mathbb{N}_0}$ in every infinite dimensional Hilbert space H with $\langle X(m), X(n)\rangle = L_n d(m)$. In fact, let $(Z(k))_{k\in\mathbb{N}}$ be an orthonormal sequence in H, and put

$$X(n) \;=\; \sum_{k=1}^{\infty} \lambda_k\, R_n(x_k)\, Z(k) \;\in H \qquad \text{for } n \in \mathbb{N}_0.$$

Obviously,

$$\langle X(m), X(n)\rangle = \sum_{k=1}^{\infty} |\lambda_k|^2\, R_m(x_k)\, R_n(x_k) = \langle T_n X(m), X(0)\rangle = L_n d(m).$$

Example 3. Let $\alpha = (\alpha(n))_{n\in\mathbb{N}_0} \in l^2(h)$. Put

$$d(n) \;=\; \alpha * \alpha^*(n) \;=\; \sum_{k=0}^{\infty} L_n\alpha(k)\,\overline{\alpha(k)}\,h(k) \qquad \text{for } n \in \mathbb{N}_0.$$

Then $d = (d(n))_{n\in\mathbb{N}_0}$ is a bounded positive definite sequence with respect to $(R_n(x))_{n\in\mathbb{N}_0}$. Note that by Proposition 3.3.10 we know that

$$\alpha * \alpha^* = (\mathcal{P}(\alpha)\,\overline{\mathcal{P}(\alpha)})^\vee = (|\mathcal{P}(\alpha)|^2)^\vee.$$

Hence, even $\lim_{n\to\infty} d(n) = 0$ is true by Proposition 3.3.2(ii). Defining $\mu := |\mathcal{P}(\alpha)|^2\pi$ we get a bounded regular positive Borel measure on D_s. Considering $H = L^2(D_s,\mu)$ and $X(n) = R_n|D_s \in H$ we get by Theorem 7.2.4 that $(X(n))_{n\in\mathbb{N}_0}$ is a bounded R_n-stationary sequence in H.

Now let H be an arbitrary infinite-dimensional Hilbert space H. In section 6.1 we have studied R_n-moving sequences $(X(n))_{n\in\mathbb{N}_0}$ in $L^2(\Omega, P)$, see Theorem 6.1.2. Following the construction of R_n-moving average sequences $(X(n))_{n\in\mathbb{N}_0}$ of section 6.1 in any infinite-dimensional Hilbert space H leads to

$$X(n) := \sum_{k=0}^{\infty} \alpha(k)\, T_n Z(k)\, h(k),$$

where $\langle Z(m), Z(n)\rangle = \frac{1}{h(n)}\,\delta_{m,n}$. The proof of Theorem 6.1.2 shows that $(X(n))_{n\in\mathbb{N}_0}$ is a bounded R_n-stationary sequence in H.

Before we consider R_n-stationary sequences $(X(n))_{n\in\mathbb{N}_0}$ satisfying an autoregressive equation, we have to analyse corresponding positive definite sequences. The Christoffel-Darboux formula, see Proposition 4.10.3 with $y = 1$, says

$$\frac{1}{a_n h(n)} \sum_{k=0}^{n} R_k(x)\, h(k) = \frac{R_{n+1}(x) - R_n(x)}{R_1(x) - 1}, \qquad x \neq 1.$$

The summation of this equality from $n = 1$ to N yields (see also Proposition 4.14.20)

$$\frac{R_{N+1}(x) - 1}{R_1(x) - 1} = \sum_{k=0}^{N} d_{N,k}\, R_k(x)\, h(k), \qquad x \neq 1,$$

where

$$d_{N,k} = \sum_{j=k}^{N} \frac{1}{a_j h(j)} \qquad \text{for } k = 1, ..., N \tag{1}$$

and

$$d_{N,0} = 1 + \sum_{j=1}^{N} \frac{1}{a_j h(j)}. \tag{2}$$

Using $d_{N,k}$ we can create another class of bounded R_N-stationary sequences.

Example 4. Fix $N \in \mathbb{N}_0$ and set

$$f_N(x) := \frac{R_{N+1}(x) - 1}{R_1(x) - 1}, \qquad x \in D_s.$$

Since f_N is nonnegative on D_s, the sequence $d_N = (d_N(k))_{k \in \mathbb{N}_0}$ given by $d_N(k) = d_{N,k}$ for $k = 0, ..., N$ and $d_N(k) = 0$ for $k = N+1, ...,$ is positive definite with respect to $(R_n(x))_{n \in \mathbb{N}_0}$, and obviously bounded. Considering $H = L^2(D_s, f_N \pi)$ and $X(n) = R_n | D_s \in H$, Theorem 7.2.4 says that $(X(n))_{n \in \mathbb{N}_0}$ is a bounded R_n-stationary sequence in H.

For many examples of polynomial hypergroups the sequence $(d_{N,0})_{N \in \mathbb{N}}$ converges. If $h(n) = O(n^a)$, $a > 1$ and $a_j > \eta > 0$ for each $j \in \mathbb{N}$, then the sequence $(d_{N,0})_{n \in \mathbb{N}}$ is convergent. For instance, for the polynomial hypergroup generated by the ultraspherical polynomials $R_n^{(\alpha,\alpha)}(x)$ the sequence $(d_{N,0})_{N \in \mathbb{N}}$ converges if $\alpha > 0$, see subsection 4.5.1. Applying Lemma 4.2.7 and Corollary 4.2.12 we get that $(d_{N,0})_{N \in \mathbb{N}}$ converges whenever $1 \notin \operatorname{supp} \pi$.

Now we assume that $(d_{N,0})_{N \in \mathbb{N}}$ is convergent and write

$$d_\infty(0) := \lim_{N \to \infty} d_{N,0} = 1 + \sum_{j=1}^{\infty} \frac{1}{a_j h(j)}. \tag{3}$$

Then

$$d_\infty(k) := \lim_{N \to \infty} d_{N,k} = \sum_{j=k}^{\infty} \frac{1}{a_j h(j)}$$

also converges for each $k \in \mathbb{N}$.

For each $k < N$ we have

$$|d_\infty(0) - d_N(0)| = |d_\infty(1) - d_N(1)| = \cdots = |d_\infty(k) - d_N(k)|.$$

Hence the convergence of $d_{N,0} = d_N(0)$ towards $d_\infty(0)$ implies convergence of $d_{N,k} = d_N(k)$ towards $d_\infty(k)$, which is uniform on finite subsets of $\mathbb{N}_0$. Hence d_∞ is a bounded and positive definite sequence with respect to $(R_n(x))_{n \in \mathbb{N}_0}$, see Proposition 2.4.9. Furthermore, $\lim_{m \to \infty} d_\infty(m) = 0$. In fact, for each $m \in \mathbb{N}$ we have $d_\infty(m) = d_\infty(0) - d_{m-1,0}$, and hence

$$\lim_{m \to \infty} d_\infty(m) = d_\infty(0) - d_\infty(0) = 0.$$

Bochner's theorem yields a positive regular Borel measure $\mu_\infty \in M(D_s)$, such that

$$d_\infty(n) = \int_{D_s} R_n(x)\, d\mu_\infty(x) = (\mu_\infty)^\vee(n). \tag{4}$$

In order to establish a characterizing autoregressive equation for d_∞ we have to show an auxiliary result, which is a small extension of Lemma 6.3.4.

Lemma 7.3.1 *Assume that $h(N) \to \infty$ as $N \to \infty$. Then for each $m, n \in \mathbb{N}_0$ there holds $g(N, m; n) \to 0$ as $N \to \infty$.*

Proof. The Cauchy-Schwarz inequality implies

$$g(N, m; n) = h(n) \int_{D_s} R_n(x)\, R_m(x)\, R_N(x)\, d\pi(x)$$

$$\leq h(n)\, g(N, N; 0)^{\frac{1}{2}} \left(\int_{D_s} (R_n(x)\, R_m(x))^2\, d\pi(x) \right)^{\frac{1}{2}}.$$

Since $g(N, N; 0) = \frac{1}{h(N)}$ and $|R_n(x) R_m(x)| \leq 1$ for all $x \in D_s$, it follows $g(N, m; n) \to 0$ as $N \to \infty$. ◇

Definition. We say that $(R_n(x))_{n\in\mathbb{N}_0}$ satisfies condition (C) if

$$\sum_{k=1}^{\infty} \frac{1}{a_k h(k)} < \infty.$$

Since $0 < a_k \leq 1$, condition (C) implies $h(k) \to \infty$ as $k \to \infty$.

Theorem 7.3.2 *(1) Suppose that $(R_n(x))_{n\in\mathbb{N}_0}$ satisfies condition (C). Then the sequence $d_\infty = (d_\infty(n))_{n\in\mathbb{N}_0}$ is bounded and positive definite with respect to $(R_n(x))_{n\in\mathbb{N}_0}$ and satisfies*

$$L_n d_\infty(m) - L_1(L_n d_\infty)(m) = \frac{1}{h(n)}\, \delta_{n,m}$$

for all $m, n \in \mathbb{N}_0$.

(2) Conversely, if $d = (d(n))_{n\in\mathbb{N}_0}$ is a sequence with $\lim_{m\to\infty} d(m) = 0$ satisfying

$$d(m) - L_1 d(m) = \delta_{0,m},$$

then $d = d_\infty$. In particular, $(R_n(x))_{n\in\mathbb{N}_0}$ satisfies condition (C).

Proof.

(1) Using $f_N(x) = \frac{R_{N+1}(x)-1}{R_1(x)-1}$ of Example 4, we have for $m, n \in \mathbb{N}_0$

$$\frac{1}{h(n)}\, (\delta_{n,m} - g(N{+}1, m; n)) = \int_{D_s} R_m(x)\, R_n(x)\, (1 - R_{N+1}(x))\, d\pi(x)$$

$$= \int_{D_s} R_n(x)\, R_m(x)\, (1 - R_1(x))\, f_N(x)\, d\pi(x)$$

$$= L_n d_N(m) - L_1(L_n d_\infty)(m).$$

Since $h(N) \to \infty$ as $N \to \infty$, Lemma 7.3.1 yields that $g(N{+}1, m; n) \to 0$, and hence $\frac{1}{h(n)} \delta_{n,m} = L_n d_\infty(m) - L_1(L_n d_\infty)(m)$.

(2) Applying induction we show that for $n \in \mathbb{N}$

$$d(n+1) = d(n) - \frac{1}{a_n h(n)}.$$

At first notice that $d(1) = d(0) - 1$. For $n = 1$ we have $a_1 d(2) + b_1 d(1) + c_1 d(0) = L_1 d(1) = d(1)$. Hence

$$d(2) = \frac{1}{a_1}\,(d(1) - b_1 d(1) - c_1 d(1) - c_1) = d(1) - \frac{c_1}{a_1} = d(1) - \frac{1}{a_1 h(1)}.$$

Suppose that the assumption is true for $n-1$ and n. Then

$$a_n d(n+1) + b_n d(n) + c_n d(n-1) = L_1 d(n) = d(n),$$

and therefore

$$\begin{aligned} d(n+1) &= \frac{1}{a_n}\left(d(n) - b_n d(n) - c_n d(n) - c_n \frac{1}{a_{n-1} h(n-1)}\right) \\ &= d(n) - \frac{1}{a_n h(n)}. \end{aligned}$$

Now it is obvious that $d(n) = d(0) - 1 - \sum_{j=1}^{n-1} \frac{1}{a_j h(j)}$. Since $\lim_{n\to\infty} d(n) = 0$, we have $d(0) = 1 + \sum_{j=1}^{\infty} \frac{1}{a_j h(j)}$ and $d(n) = d_\infty(n)$ follow immediately, and (C) is shown as well.

◇

Now we can construct bounded R_n-stationary sequences in Hilbert spaces related to d_∞.

Example 5. Using the positive Borel measure $\mu_\infty \in M(D_s)$ such that $(\mu_\infty)^\vee = d_\infty$ (see (4)), and setting $X(n) = R_n|D_s$, we get a bounded R_n-stationary sequence in the Hilbert space $H = L^2(D_s, \mu_\infty)$. μ_∞ is the spectral measure of $(X(n))_{n\in\mathbb{N}_0}$ and $H(X) = L^2(D_s, \mu_\infty)$.

If H is any infinite dimensional Hilbert space, we can construct an R_n-invariant sequence related to d_∞ in the following way.

Theorem 7.3.3 *Suppose that condition (C) is fulfilled. Let H be an infinite dimensional Hilbert space and $(Z(n))_{n\in\mathbb{N}_0}$ an orthonormal sequence in H Define recursively*

$$X(0) := Z(0) + \sum_{j=1}^{\infty} \frac{1}{(a_j h(j))^{\frac{1}{2}}}\, Z(j), \qquad X(1) := X(0) - Z(0)$$

and

$$X(n+1) := X(n) - \frac{1}{(a_n h(n))^{\frac{1}{2}}}\, Z(n). \tag{5}$$

Then $(X(n))_{n\in\mathbb{N}_0}$ is a bounded R_n-stationary sequence in H, such that

$$\langle X(m), X(n)\rangle = L_n d_\infty(m).$$

Proof. Note that $X(0) \in H$ and

$$\|X(0)\|^2 = 1 + \sum_{j=1}^{\infty} \frac{1}{a_j h(j)} = d_\infty(0).$$

Moreover, $\langle X(1), X(0)\rangle = d_\infty(0) - 1 = d_\infty(1)$, and

$$\langle X(n+1), X(0)\rangle = \langle X(n), X(0)\rangle - \frac{1}{a_n h(n)}.$$

Hence $\langle X(n), X(0)\rangle = d_\infty(n)$ for each $n \in \mathbb{N}_0$. From (5) follows

$$X(n) = \sum_{j=n}^{\infty} \frac{1}{(a_j h(j))^{\frac{1}{2}}} Z(j) \qquad \text{for } n \in \mathbb{N}. \tag{6}$$

It remains to show that $(X(n))_{n\in\mathbb{N}_0}$ is R_n-stationary in H. By Theorem 7.2.1 we have to show $\langle T_1 X(m), X(n)\rangle = \langle X(m), T_1 X(n)\rangle$ for $m, n \in \mathbb{N}_0$. It is sufficient to consider $m < n$. Using (5) and (6) we obtain

$$T_1 X(n) = \sum_{j=n}^{\infty} \frac{1}{(a_j h(j))^{\frac{1}{2}}} Z(j) - \frac{a_n}{(a_n h(n))^{\frac{1}{2}}} Z(n) + \frac{c_n}{(a_{n-1} h(n-1))^{\frac{1}{2}}} Z(n-1).$$

It follows

$$\langle T_1 X(n), X(m)\rangle = \sum_{j=n}^{\infty} \frac{1}{a_j h(j)} - \frac{1}{h(n)} + \frac{c_n}{a_{n-1} h(n-1)}$$

$$= \sum_{j=n}^{\infty} \frac{1}{a_j h(j)} = d_\infty(n),$$

whereas

$$\langle X(n), T_1 X(m)\rangle = \sum_{j=n}^{\infty} \frac{1}{a_j h(j)} = d_\infty(n).$$

Thus it is shown that the sequence $(X(n))_{n\in\mathbb{N}_0}$ defined by (5) and (6) is a bounded R_n-stationary sequence, which satisfies

$$\langle X(m), X(n)\rangle = L_n d_\infty(m) = \int_{D_s} R_n(x)\, R_m(x)\, d\mu_\infty(x).$$

◇

Finally we present one more property which implies that condition (C) is satisfied. Define

$$q(x) = \frac{1}{1 - R_1(x)} \qquad \text{for } x \neq 1.$$

Proposition 7.3.4 *If $q \in L^1(D_s, \pi)$, then*

$$\check{q}(n) \;=\; \sum_{j=n}^{\infty} \frac{1}{a_j h(j)} \qquad \textit{for each } n \in \mathbb{N}_0.$$

In particular, condition (C) is satisfied and the spectral measure μ_∞ is identical to $q \cdot \pi$.

Proof. From Proposition 3.3.2(ii) we know that for each $m \in \mathbb{N}_0$

$$(R_m \cdot q)^{\vee}(k) \;=\; \int_{D_s} R_k(x)\, R_m(x)\, q(x)\, d\pi(x)$$

tends to zero as $k \to \infty$. Hence for $N \in \mathbb{N}_0$ and $m = 0, ..., N$ it follows

$$\begin{aligned} d_{N,m} &= \int_{D_s} \left(\sum_{k=0}^{N} d_{N,k}\, R_k(x)\, h(k) \right) R_m(x)\, d\pi(x) \\ &= \int_{D_s} \frac{1 - R_{N+1}(x)}{1 - R_1(x)}\, R_m(x)\, d\pi(x) \\ &= \int_{D_s} R_m(x)\, q(x)\, d\pi(x) \;-\; \int_{D_s} R_{N+1}(x)\, R_m(x)\, q(x)\, d\pi(x). \end{aligned}$$

With $N \to \infty$ we obtain

$$d_\infty(m) \;=\; \lim_{N\to\infty} d_{N,m} \;=\; \int_{D_s} R_m(x)\, q(x)\, d\pi(x) \;=\; \check{q}(m).$$

◇

Remark: Theorem 7.3.3 can directly be used for the description of first-order R_n-autoregressive random sequences, see section 6.1.

4 Multipliers of bounded R_n-stationary sequences

Throughout this section we assume that $(X(n))_{n\in\mathbb{N}_0}$ is a bounded R_n-stationary sequence in a Hilbert space H such that $\operatorname{span}\{X(n) : n \in \mathbb{N}_0\}$ is dense in H. As before, $T_n \in \mathcal{B}(H)$ are the operators on H with $X(n) = T_n(X(0))$ and $\Phi : L^2(D_s, \mu) \to H$ is the isometric isomorphism satisfying $\Phi(R_n) = X(n)$, see Theorem 7.2.5.

Definition. $A \in \mathcal{B}(H)$ is called a **multiplier of** $(X(n))_{n\in\mathbb{N}_0}$ if

$$A \circ T_n \;=\; T_n \circ A \qquad \text{for all } n \in \mathbb{N}_0.$$

Obviously, the set of all multipliers of $(X(n))_{n\in\mathbb{N}_0}$ is a closed $*$-subalgebra of $\mathcal{B}(H)$.

Proposition 7.4.1 *Let $A \in \mathcal{B}(H)$ be a multiplier of $(X(n))_{n\in\mathbb{N}_0}$. Define*

$$Y(n) := \; A\, X(n) \qquad \textit{for } n \in \mathbb{N}_0.$$

Then $(Y(n))_{n\in\mathbb{N}_0}$ is a bounded R_n-stationary sequence in H.

Proof. Note that

$$Y(n) = A\,X(n) = A\,T_n(X(0)) = T_n(A\,X(0)) = T_n(Y(0))$$

for all $n \in \mathbb{N}_0$. Since $T_1 \in \mathcal{B}(H)$ is self-adjoint, we have $\langle T_1(Y(m)), Y(n)\rangle = \langle Y(m), T_1(Y(n))\rangle$. We have to show that the recursion formula of the $X(n)$ is transmitted to the $Y(n)$. In fact,

$$\begin{aligned} T_1(Y(m)) &= T_1(A\,X(m)) \\ &= A\,T_1(X(m)) = a_m Y(m+1) + b_m Y(m) + c_m Y(m-1) \end{aligned}$$

for all $m \in \mathbb{N}$. Therefore we obtain

$$\begin{aligned} &\langle a_m Y(m+1) + b_m Y(m) + c_m Y(m-1), Y(n)\rangle \\ &= \langle T_1(Y(m)), Y(n)\rangle = \langle Y(m), T_1(Y(n))\rangle \\ &= \langle Y(m), a_n Y(n+1) + b_n Y(n) + c_n Y(n-1)\rangle, \end{aligned}$$

and the R_n-stationarity of $(Y(n))_{n\in\mathbb{N}_0}$ is shown. Boundedness holds obviously, $\|Y(n)\| = \|A\,X(n)\| \le \|A\|$. ◇

Each $g \in L^\infty(D_s,\mu)$ determines uniquely a bounded linear operator $A_g \in \mathcal{B}(H)$ by setting

$$A_g(Z) = \Phi(g \cdot \Phi^{-1}(Z)) \tag{1}$$

for $Z \in H$. Since for $\psi \in L^2(D_s,\mu)$ we have

$$\begin{aligned} &\langle \Phi(R_n \cdot \psi), X(m)\rangle \\ &= \int_{D_s} R_n \cdot \psi(x)\, R_m(x)\, d\mu(x) = \int_{D_s} \psi(x)\, R_n(x)\, R_m(x)\, d\mu(x) \\ &= \langle \Phi(\psi), \Phi(R_n \cdot R_m)\rangle = \langle \Phi(\psi), T_n X(m)\rangle = \langle T_n \Phi(\psi), X(m)\rangle \end{aligned}$$

for all $m \in \mathbb{N}_0$, it follows $\Phi(R_n \cdot \psi) = T_n \Phi(\psi)$ for all $\psi \in L^2(D_s,\mu)$ and $n \in \mathbb{N}_0$. Hence

$$\begin{aligned} A_g \circ T_n(Z) &= \Phi(g \cdot \Phi^{-1}(T_n Z)) = \Phi(g \cdot R_n \cdot \Phi^{-1}(Z)) \\ &= \Phi(R_n \cdot g \cdot \Phi^{-1}(Z)) = T_n \circ A_g(Z) \end{aligned}$$

for all $Z \in H$. Thus we have shown

Proposition 7.4.2 *Let $g \in L^\infty(D_s,\mu)$ and define $A_g \in \mathcal{B}(H)$ as in (1). Then A_g is a multiplier of $(X(n))_{n\in\mathbb{N}_0}$.*

We will show that each multiplier A of $(X(n))_{n\in\mathbb{N}_0}$ is of the form A_g, $g \in L^\infty(D_s,\mu)$.

Proposition 7.4.3 *Let $A \in \mathcal{B}(H)$ be a multiplier of $(X(n))_{n\in\mathbb{N}_0}$. Then*

$$A \circ A_g \ = \ A_g \circ A$$

for all $g \in L^\infty(D_s, \mu)$.

Proof. Consider the multiplication operator $M_n(\psi) = R_n \cdot \psi$ for $\psi \in L^2(D_s, \mu)$. We have $T_n = \Phi \circ M_n \circ \Phi^{-1}$ on $H = H(X)$. Therefore

$$A_{R_n}(Z) \ = \ \Phi(R_n \cdot \Phi^{-1}(Z)) \ = \ \Phi \circ M_n \circ \Phi^{-1}(Z) \ = \ T_n(z) \qquad \text{for all } Z \in H.$$

It follows $A \circ A_P = A_P \circ A$ for all polynomials P, and then $A \circ A_g = A_g \circ A$ for all continuous functions $g \in C(D_s)$.

Let $\varphi, \psi \in L^2(D_s, \mu)$ and consider

$$\kappa := \ \Phi^{-1}(A^* \Phi(\psi)) \in \ L^2(D_s, \mu)$$

and

$$\lambda := \ \Phi^{-1}(A\, \Phi(\varphi)) \in \ L^2(D_s, \mu).$$

For every $g \in C(D_s)$ we obtain

$$\int_{D_s} g(x)\, \varphi(x)\, \overline{\kappa(x)}\, d\mu(x) \ = \ \langle A_g(\Phi(\varphi)), A^*(\Phi(\psi))\rangle$$

$$= \ \langle A \circ A_g(\Phi(\varphi), \Phi(\psi)) \rangle \ = \ \langle A_g \circ A(\Phi(\varphi)), \Phi(\psi)\rangle$$

$$= \ \int_{D_s} g(x)\, \lambda(x)\, \overline{\psi(x)}\, d\mu(x).$$

The functions $\varphi\bar{\kappa}$ and $\lambda\bar{\psi}$ are elements of $L^1(D_s, \mu)$, and the regular Borel measures $(\varphi\bar{\kappa})\mu \in M(D_s)$ and $(\lambda\bar{\psi})\mu \in M(D_s)$ coincide. Therefore we have

$$\langle A \circ A_g(\Phi(\varphi)), \Phi(\psi)\rangle \ = \ \langle A_g \circ A(\Phi(\varphi)), \Phi(\psi)\rangle$$

for $g \in L^\infty(D_s, \mu)$, too. It follows

$$A \circ A_g \ = \ A_g \circ A \qquad \text{for every } g \in L^\infty(D_s, \mu).$$

◇

Theorem 7.4.4 *Let $(X(n))_{n\in\mathbb{N}_0}$ be a bounded R_n-stationary sequence in a Hilbert space $H = H(X)$. Let $A \in \mathcal{B}(H)$. Equivalent are*

(i) A is a multiplier of $(X(n))_{n\in\mathbb{N}_0}$.
(ii) There exists a unique $g \in L^\infty(D_s, \mu)$ such that $A = A_g$.

Proof. By Proposition 7.4.2 the implication $(ii) \Rightarrow (i)$ is shown. Conversely, assume that $A \in \mathcal{B}(H)$ is a multiplier of the sequence $(X(n))_{n\in\mathbb{N}_0}$. Set

$$g := \Phi^{-1}(A(X(0)) \in L^2(D_s, \mu).$$

We prove that $g \in L^\infty(D_s, \mu)$. Firstly, note that for every $f \in L^\infty(D_s, \mu) \subseteq L^2(D_s, \mu)$ it follows

$$A_f(X(0)) = \Phi(f \cdot \Phi^{-1}(X(0))) = \Phi(f \cdot R_0) = \Phi(f),$$

and by Proposition 7.4.3

$$\begin{aligned} A(\Phi(f)) &= A \circ A_f(X(0)) = A_f \circ A(X(0)) \\ &= A_f(\Phi(g)) = \Phi(f \cdot g). \qquad (*) \end{aligned}$$

Consider $f_n = \chi_{E_n}$, where $E_n = \{x \in D_s : |g(x)| \geq \|A\| + \frac{1}{n}\}$. By $(*)$ we get

$$\begin{aligned} \|A\|\,\|f_n\|_2 &\geq \|A(\Phi(f_n)\|_2 = \|\Phi(f_n \cdot g)\|_2 = \left(\int_{D_s} |g\,\chi_{E_n}|^2\,d\mu\right)^{\frac{1}{2}} \\ &\geq \left(\|A\| + \frac{1}{n}\right)\left(\int_{D_s} |\chi_{E_n}|^2\,d\mu\right)^{\frac{1}{2}} = \left(\|A\| + \frac{1}{n}\right)\|f_n\|_2. \end{aligned}$$

Therefore $\|f_n\|_2 = 0$, and we get

$$\mu\left(\{x \in D_s : |g(x)| > \|A\|\}\right) = 0$$

and we infer that $g \in L^\infty(D_s, \mu)$. Equation $(*)$ says

$$A(\Phi(f)) = A_g(\Phi(f)) \qquad \text{for } f \in L^\infty(D_s, \mu).$$

Since $C(D_s)$ is dense in $L^2(D_s, \mu)$ it follows that $A = A_g$. The uniqueness statement follows from the fact, that $A_{g_1} = A_{g_2}$ implies $g_1 = g_2$ μ-almost everywhere. ◇

Remark. The correspondence $g \mapsto A_g$ between $L^\infty(D_s, \mu)$ and the space of all multipliers of $(X(n))_{n\in\mathbb{N}_0}$ is a Banach $*$-algebra isomorphism. In particular, we have $A_{g_1 g_2} = A_{g_1} \circ A_{g_2}$ and $A_{\bar{g}} = A_g^*$. Moreover, the Banach $*$-subalgebra of multipliers of $(X(n))_{n\in\mathbb{N}_0}$ (contained in $\mathcal{B}(H)$) is commutative. The proof of Theorem 7.4.4 shows that $\mu(\{x \in D_s : |g(x)| > \|A_g\|\}) = 0$. Hence $\|A_g\| \geq \|g\|_\infty$. Since$\|A_g\| \leq \|g\|_\infty$ is obviously true, the mapping $g \mapsto A_g$ is isometric.

We can use Theorem 7.4.4 to derive a characterization of the closed linear subspaces E of $H = H(X)$ which are invariant under all T_n, $n \in \mathbb{N}_0$.

Corollary 7.4.5 *Let $(X(n))_{n\in\mathbb{N}_0}$ be a bounded R_n-stationary sequence in a Hilbert space $H = H(X)$. Equivalent are*

(i) $E \subseteq H$ *is a closed linear subspace of H such that $Y \in E$ implies $T_n(Y) \in E$ for each $n \in \mathbb{N}_0$.*
(ii) There exists a Borel set $M \subseteq D_s$ such that the orthogonal projection P_E onto E is equal to A_{χ_M}, i.e.

$$E = \{Y \in H : \Phi^{-1}(Y) = 0 \ \mu\text{-almost everywhere on } D_s \backslash M\}.$$

Proof. Suppose (ii) is valid. It is easy to check that E is a closed linear subspace of H, which is invariant under T_n. Supposing (i) the orthogonal projection P_E is a multiplier of the sequence $(X(n))_{n\in\mathbb{N}_0}$, which is easily checked. By Theorem 7.4.4 there exists $g \in L^\infty(D_s, \mu)$ such that $P_E = A_g$. Since P_E is a projection, we have $A_g = A_{g^2}$. It follows $g = g^2$ μ-almost everywhere. Setting $M = g^{-1}(\{1\})$, we see that $g = \chi_M$ μ-almost everywhere, and (ii) is shown. $\diamond$

5 The imaginary part of T_n-stationary sequences

We consider now the special case $R_n^{(-\frac{1}{2},-\frac{1}{2})}(x) = T_n(x) = \cos(nt)$, $x = \cos t$, the Chebyshev polynomials of the first kind, i.e. the real part of e^{int}. It is quite natural to analyze which T_n-stationary sequences $(X(n))_{n\in\mathbb{N}_0}$ can be supplemented to a stationary sequence with respect to the group $\mathbb{Z}$. We will search for U_n-stationary sequences $(Y(n))_{n\in\mathbb{N}_0}$ such that for $n \in \mathbb{Z}$ we get a stationary sequence $(Z(n))_{n\in\mathbb{Z}}$,

$$Z(n) := X(|n|) + in\, Y(|n| - 1), \tag{1}$$

with respect to the group $\mathbb{Z}$, i.e.

$$\langle Z(m), Z(n)\rangle = \langle Z(m-n), Z(0)\rangle \tag{2}$$

for $m, n \in \mathbb{Z}$.

$U_n(x)$ are the Chebyshev polynomials of the second kind,

$$U_n(x) = R_n^{(\frac{1}{2},\frac{1}{2})}(x) = \frac{\sin((n+1)t)}{(n+1)\sin t},$$

see subsection 4.5.1.

Definition. Let $(X(n))_{n\in\mathbb{N}_0}$ be a T_n-stationary sequence in a Hilbert space H. We say that $(X(n))_{n\in\mathbb{N}_0}$ allows an imaginary part, if there exists a U_n-stationary sequence $(Y(n))_{n\in\mathbb{N}_0}$ in H such that $(Z(n))_{n\in\mathbb{Z}}$ defined by (1) is a stationary sequence in H, i.e. (2) holds.

We need some formulas involving Chebyshev polynomials. For recursion formulas and linearization formulas we refer to subsection 4.5.1. We will use the following identities, which are easily obtained via the trigonometric addition formulas:

$$T_n(x) \;=\; \frac{n+1}{2}\, U_n(x) \;-\; \frac{n-1}{2}\, U_{n-2}(x) \qquad \text{for } n \geq 2, \tag{3}$$

$$n\, U_{n-1}(x)\,(1-x^2) \;=\; x\, T_n(x) \;-\; T_{n+1}(x) \qquad \text{for } n \geq 1. \tag{4}$$

Let $(X(n))_{n\in\mathbb{N}_0}$ be a bounded T_n-stationary sequence in a Hilbert space $H = H(X)$. We will use the results of section 7.1 and section 7.2. Defining $A_m \in \mathcal{B}(H)$ as the extension of

$$A_m\left(\sum_{k=0}^{N} d_k\, X(k)\right) \;=\; \Phi\left(T_m \cdot \sum_{k=0}^{N} d_k\, T_k\right)$$

on $L(X)$ to $H(X)$ yields a sequence $(A_m)_{m\in\mathbb{N}_0}$ of self-adjoint contractions on $H(X)$. (In order to avoid confusion we use the notation A_m for the translation operators of T_n-stationary sequences.)

Given $A = A_1$, we know that $\mathrm{id} - A^2$ is a positive operator on $H = H(X)$. Hence $B = \sqrt{\mathrm{id} - A^2}$ is a well-defined self-adjoint operator on H satisfying $A^2 + B^2 = \mathrm{id}$, $AB = BA$, see [609,Theorem 12.33].

Proposition 7.5.1 *Let $(X(n))_{n\in\mathbb{N}_0}$ be a bounded T_n-stationary sequence in a Hilbert space $H = H(X)$, and $A_n \in \mathcal{B}(H)$, $n \in \mathbb{N}_0$, the corresponding translation operators. Setting $A = A_1$ and $B = \sqrt{id - A^2}$ define*

$$B_n \;=\; U_n(A)\, B. \tag{5}$$

Then by $W_0 = id$, $W_n = A_n + inB_{n-1}$, $W_{-n} = A_n - inB_{n-1}$ for $n \in \mathbb{N}$, a group of unitary operators on H is given. This implies

$$W_n \;=\; W_1^n \qquad \text{for all } n \in \mathbb{Z}.$$

Proof. Since

$$W_1\, W_{-1} \;=\; W_{-1}\, W_1 \;=\; (A_1 + iB_0)\,(A_1 - iB_0)$$

$$=\; (A + iB)\,(A - iB) \;=\; A^2 \,+\, B^2 \;=\; \mathrm{id},$$

it is sufficient to prove that for each $n \in \mathbb{Z}$ holds $W_1 W_n = W_{n+1}$.

We begin with $n \in \mathbb{N}$. For $n = 1$ we have

$$W_1\, W_1 \;=\; (A_1{+}iB_0)\,(A_1{+}iB_0) \;=\; A^2 - B^2 + 2iAB \;=\; 2A^2 - \mathrm{id}{+}2iAB$$

$$=\; A_2 \,+\, 2iAB \;=\; A_2 \,+\, 2iB_1 \;=\; W_2,$$

where we used $B^2 = \text{id} - A^2$ and $B_1 = U_1(A)B = AB$. Applying (4) we obtain

$$W_1\,W_n \;=\; A\,A_n \;+\; inA\,U_{n-1}(A)\,B \;+\; iB\,A_n \;-\; n\,U_{n-1}(A)\,B^2$$

$$= \; A\,A_n \;+\; inA\,U_{n-1}(A)\,B \;+\; iB\,A_n \;-\; nU_{n-1}(A) \;+\; nU_{n-1}(A)\,A^2$$

$$= \; A\,A_n \;+\; inA\,U_{n-1}(A)\,B \;+\; iA_n\,B \;-\; A\,A_n \;+\; A_{n+1}$$

$$= \; A_{n+1} \;+\; inA\,U_{n-1}(A)\,B \;+\; iA_n\,B.$$

By means of equation (3), which is valid for $n \geq 2$, follows

$$W_1\,W_n = A_{n+1} \;+\; inA\,U_{n-1}(A)\,B \\ +i\left(\frac{n+1}{2}\,U_n(A)\,B \;-\; \frac{n-1}{2}\,U_{n-2}(A)\,B\right).$$

Finally the recursion formula of $(U_n(x))_{n\in\mathbb{N}_0}$ can be written for $n \geq 2$ as

$$x\,U_{n-1}(x) \;=\; \frac{n+1}{2n}\,U_n(x) \;+\; \frac{n-1}{2n}\,U_{n-2}(x),$$

and we get

$$W_1\,W_n \;=\; A_{n+1} \;+\; i(n+1)\,U_n(A)\,B \;=\; A_{n+1} \;+\; i(n+1)\,B_n \;=\; W_{n+1}.$$

For negative integers it follows

$$W_{-n} \;=\; W_n^* \;=\; (W_1\,W_{n-1})^* \;=\; W_{n-1}^*\,W_1^* \;=\; W_{-1}\,W_{-n+1}$$

and hence $W_1 W_{-n} = W_{-n+1}$ for each $n \in \mathbb{N}$. This completes the proof. ◇

Theorem 7.5.2 *Let $(X(n))_{n\in\mathbb{N}_0}$ be a bounded T_n-stationary sequence in a Hilbert space $H = H(X)$, and $A_n \in \mathcal{B}(H)$, $n \in \mathbb{N}_0$, the corresponding translation operators. Then $(X(n))_{n\in\mathbb{N}_0}$ allows an imaginary part $(Y(n))_{n\in\mathbb{N}_0}$, which is given by*

$$Y(n) \;=\; U_n(A)\;B(X(0)) \qquad \textit{for all } n \in \mathbb{N}_0,$$

where $A = A_1$, $B = \sqrt{id - A^2}$.

Proof. Consider $W_n = A_{|n|} + inU_{|n|-1}(A)B$ for $n \in \mathbb{Z}$. By Proposition 7.5.1 we know that $(W_n)_{n\in\mathbb{Z}}$ is a group of unitary operators on H. In particular,

$$\langle W_m(X(0)), W_n(X(0))\rangle \;=\; \langle W_1^m(X(0)), W_1^n(X(0))\rangle$$

$$= \; \langle W_1^{m-n}(X(0)), X(0)\rangle \;=\; \langle W_{m-n}(X(0)), X(0)\rangle$$

for all $m, n \in \mathbb{Z}$. Since

$$\begin{aligned} W_n(X(0)) &= A_{|n|}(X(0)) + inU_{|n|-1}(A)\,B(X(0)) \\ &= X(|n|) + iY(|n|-1) = Z(n), \end{aligned}$$

it follows that $(Z(n))_{n\in\mathbb{Z}}$ is a stationary sequence with respect to the group $\mathbb{Z}$.

It remains to check that $(Y(n))_{n\in\mathbb{N}_0}$ is a U_n-stationary sequence in H. It holds

$$\begin{aligned} \langle Y(m), Y(n)\rangle &= \langle U_m(A)\,B(X(0)), U_n(A)\,B(X(0))\rangle \\ &= \langle U_n(A)\,U_m(A)\,Y(0), Y(0)\rangle. \end{aligned}$$

Since

$$U_n(A) \circ U_m(A) = U_n \cdot U_m(A) = \sum_{k=|n-m|}^{n+m} \tilde{g}(n,m;k)\,U_k(A),$$

where $\tilde{g}(n,m;k)$ are the linearization coefficients of $U_n(x))_{n\in\mathbb{N}_0}$, the U_n-stationarity of $(Y(n))_{n\in\mathbb{N}_0}$ follows. ◇

We can use the spectral measure $\mu \in M(D_s)$ of $(X(n))_{n\in\mathbb{N}_0}$ and the isometric isomorphism Φ of $L^2(D_s,\mu)$ onto $H(X)$ to give an alternative proof of the existence of an imaginary part of $(X(n))_{n\in\mathbb{N}_0}$.

Proposition 7.5.3 *Let $(X(n))_{n\in\mathbb{N}_0}$ be a bounded T_n-stationary sequence in $H = H(X)$ with spectral measure $\mu \in M(D_s)$. Denote*

$$f_n : D_s \to \mathbb{R}, \quad f_n(x) = U_n(x)\,\sqrt{1-x^2} \qquad \textit{for } n \in \mathbb{N}_0.$$

Then $(Y(n))_{n\in\mathbb{N}_0}$ with $Y(n) := \Phi(f_n)$ is an imaginary part of $(X(n))_{n\in\mathbb{N}_0}$.

Proof. Obviously $f_n \in L^2(D_s,\mu)$ and $Y(n) = \Phi(f_n) \in H(X)$.

$$\begin{aligned} \langle Y(m), Y(n)\rangle = \int_{D_s} f_m(x)\,f_n(x)\,d\mu(x) &= \int_{D_s} U_m(x)\,U_n(x)\,(1-x^2)\,d\mu(x) \\ &= \sum_{k=|m-n|}^{m+n} \tilde{g}(m,n;k) \int_{D_s} U_k(x)\,U_0(x)\,(1-x^2)\,d\mu(x) \\ &= \sum_{k=|m-n|}^{m+n} \tilde{g}(m,n;k) \int_{D_s} f_k(x)\,f_0(x)\,d\mu(x) \\ &= \sum_{k=|m-n|}^{m+n} \tilde{g}(m,n;k)\,\langle Y(k), Y(0)\rangle, \end{aligned}$$

where $\tilde{g}(m,n;k)$ are the linearization coefficients of $(U_n(x))_{n\in\mathbb{N}_0}$. Hence $(Y(n))_{n\in\mathbb{N}_0}$ is a U_n-stationary sequence in H. Now define

$$Z(n) \;=\; X(|n|) \;+\; inY(|n|-1) \qquad \text{for } n \in \mathbb{Z}.$$

With $x = \cos t,\ t \in [0,\pi]$,

$$\begin{aligned} T_{|n|}(x) \;+\; inU_{|n|-1}(x)\,\sqrt{1-x^2} \;&=\; T_{|n|}(\cos t) \;+\; inU_{|n|-1}(\cos t)\,\sin t \\ &=\; \cos(|n|t) \;+\; i\,\mathrm{sign}\,(n)\,\sin(|n|t) \;=\; e^{int} \end{aligned}$$

is valid, and so $Z(n) = \Phi(e^{in\arccos(x)}) \in H(X)$. This yields

$$\langle Z(m), Z(n)\rangle \;=\; \int_{D_s} e^{im\arccos(x)}\, e^{-in\arccos(x)}\, d\mu(x) \;=\; \langle Z(m-n), Z(0)\rangle.$$

◇

Chapter 8

Further hypergroup examples

1 Hypergroups based on group structures

Many examples of hypergroups can be derived from locally compact groups. By a locally compact group we shall mean a topological group with a locally compact topology which is also Hausdorff, see [232,Corollary 2.3]. The construction of these hypergroups is based on continuous actions of a compact group H on a locally compact group G.

1.1 *Continuous actions*

A **continuous action** of a compact group H on a locally compact Hausdorff space X is a continuous mapping $(b, x) \mapsto b(x)$, $H \times X \to X$, such that

$$e(x) = x, \qquad c(b(x)) = (bc)(x) \qquad \text{for all } x \in X \text{ and } b, c \in H.$$

Let $[x]_H$ denote the H-orbit of x in X, i.e. $[x]_H = \{b(x) : b \in H\}$. Obviously each H-orbit is an element of $\mathcal{C}(X)$ (see (H2) of the Definition of hypergroups). We write $X_H = \{[x]_H : x \in X\}$. On X_H there are two topologies: the quotient topology induced by the topology on X on the one hand and the relative topology induced by the Michael topology on $\mathcal{C}(X)$ on the other hand. For the proof of the following topological results we refer to [350,Theorem 8.1A].

Proposition 8.1.1 *Let $(b, x) \mapsto b(x)$, $H \times X \to X$ be a continuous action of a compact group H on a locally compact Hausdorff space X. Then X_H is a closed subset of $\mathcal{C}(X)$. The quotient topology on X_H and the relative topology on X_H coincide. X_H is a locally compact Hausdorff space, and the mapping $x \mapsto [x]_H$, $X \to X_H$ is continuous and open from X onto X_H. This mapping is also proper.*

In order to study the relations between $M(X_H)$ and $M(X)$ we use the basics of measures and integration and refer to section 1.1. Let σ be the normalized Haar measure on H.

Proposition 8.1.2 *Let $(b,x) \mapsto b(x)$, $H \times X \to X$ be a continuous action of a compact group H on a locally compact Hausdorff space X. For each $x \in X$ define $\varphi_x \in M_c^1(X)$ by*

$$\varphi_x(f) = \int_H f(b(x))\, d\sigma(b) \qquad \text{for each } f \in C^b(X).$$

Then

(1) The mapping $[x]_H \mapsto \varphi_x$, $X_H \to M_c^1(X)$ is well-defined and continuous with respect to the Bernoulli topology.
(2) The mapping $[x]_H \to \varphi_x$ can be uniquely extended to a linear mapping $\mu \mapsto \varphi(\mu)$, $M(X_H) \to M(X)$, which is positive continuous, i.e. $\varphi(\mu) \in M^+(X)$ when $\mu \in M^+(X_H)$ and the restricted mapping from $M^+(X_H)$ to $M^+(X)$ is continuous with respect to the Bernoulli topologies. (One readily sees that $\{\varphi_\mu : \mu \in M(X_H)\} = \{\nu \in M(X) : \nu = \nu \circ b \text{ for all } b \in H\}$.)

Proof.

(1) Obviously $\varphi_x \in M^1(X)$ and supp $\varphi_x \subseteq [x]_H$. Hence $\varphi_x \in M_c^1(X)$. By the invariance of the Haar measure $d\sigma$ the mapping $[x]_H \mapsto \varphi_x$, $X_H \to M_c^1(X)$ is well-defined. Let $x_0 \in X$, $f \in C^b(X)$, $\varepsilon > 0$. There exists a neighbourhood U of x_0 such that $|f(b(x)) - f(b(x_0))| < \varepsilon$ for $x \in U$ and every $b \in H$. It follows $|\varphi_x(f) - \varphi_{x_0}(f)| \leq \varepsilon$ for all $x \in U$, i.e. $x \mapsto \varphi_x(f)$, $X \to \mathbb{C}$ is continuous. Since X_H has the quotient topology, we obtain that $[x]_H \mapsto \varphi_x(f)$, $X_H \to \mathbb{C}$ is continuous, see e.g. [367, Ch.3, Theorem 9]. So we have shown that $[x]_H \mapsto \varphi_x$, $X_H \to M_c^1(X)$ is continuous with respect to the Bernoulli topology.
(2) An extension of $[x]_H \mapsto \varphi_x$ can be defined explicitly by setting for $\mu \in M(X_H)$

$$\varphi(\mu)(f) = \int_{X_H} \varphi_x(f)\, d\mu([x]_H)$$

for each $f \in C^b(X)$. Applying Riesz' theorem we get the desired linear extension. Proceeding as in the proof of Theorem 1.1.4 we see that this linear extension is unique and positive continuous.

◇

Remark: The measures $\varphi_x \in M_c^1(X)$ can be used to define an operator on the Banach space $C^b(X)$. In fact, $f \mapsto f^H$, $C^b(X) \to C^b(X)$, where

$$f^H(x) = \varphi_x(f) = \int_H f(b(x))\, d\sigma(b), \qquad x \in X,$$

is a well-defined bounded linear operator on $C^b(X)$ into $C^b(X)$, and f^H is an H-invariant function on X.

1.2 *Hypergroups induced by $[FIA]_B^-$-groups*

Let G be a locally compact group. Denote by $\mathrm{Aut}(G)$ the group of topological automorphisms of G. The appropriate topology making $\mathrm{Aut}(G)$ a topological group, is the so-called Birkhoff topology. For $C \subseteq G$ compact and an open neighbourhood U of $e \in G$, define

$$M(C,U) := \{a \in \mathrm{Aut}(G) : a(x)\, x^{-1} \in U \text{ and } a^{-1}(x)\, x^{-1} \in U\} \quad \text{for } x \in C.$$

The family of all such $M(C,U)$'s constitutes a fundamental system of neighbourhoods of $\mathrm{id} \in \mathrm{Aut}(G)$. (We want to point to the fact that using the compact-open topology on $\mathrm{Aut}(G)$ the inversion in $\mathrm{Aut}(G)$ may be not continuous.) Obviously the Birkhoff topology is finer than the compact-open topology.

If $\mathrm{Aut}(G)$ is equipped with the compact-open topology the evaluation map $(a,x) \mapsto a(x)$, $\mathrm{Aut}(G) \times G \to G$ is continuous. Hence this map is also continuous in case $\mathrm{Aut}(G)$ bears the Birkhoff topology. We will consider subgroups B of $\mathrm{Aut}(G)$. The subgroup of all inner automorphisms $y \mapsto xyx^{-1}$, $G \to G$, $x \in G$ is denoted by $I(G)$. Throughout we will assume that the subgroups $B \subseteq \mathrm{Aut}(G)$ carry the Birkhoff topology.

Definition. A locally compact group G is called

(1) $[FIA]_B^-$ group if B is relatively compact in the Birkhoff topology.
(2) $[FC]_B^-$ group if the B-orbits $\{a(x) : a \in B\}$ are relatively compact in G.
(3) $[SIN]_B$ group if $e \in G$ has a neighbourhood basis consisting of B-invariant sets.

If $B = I(G)$ we usually omit the B.

There is an important result of Grosser and Moskowitz characterizing $[FIA]_B^-$ groups. In [286,Theorem 4.1] it is shown that $G \in [FIA]_B^-$ if and only if $G \in [FC]_B^- \cap [SIN]_B$.

Let $G \in [FIA]_{\overline{B}}^{-}$. We will apply the results of subsection 8.1.1 to the compact group $H = \overline{B} \subseteq \mathrm{Aut}(G)$, and $X = G$. We define for $x, y \in G$ the $M(G_H)$-valued integral

$$\omega(\,[x]_H, [y]_H) := \int_H \epsilon_{[a(x)y]_H}\, d\sigma(a). \tag{1}$$

The operation given in (1) is a well-known mapping $([x]_H, [y]_H) \mapsto \omega([x]_H, [y]_H)$, $G_H \times G_H \to M(G_H)$. In fact, since $a(x)b(y) = b(b^{-1}(a(x)y))$ for all $x, y \in G$ and $a, b \in H$, we obtain for any $b \in H$ by the invariance property of σ, that

$$\int_H \epsilon_{[a(x)b(y)]_H}\, d\sigma(a) \;=\; \int_H \epsilon_{[b^{-1}(a(x))y]_H}\, d\sigma(a) \;=\; \int_H \epsilon_{[a(x)y]_H}\, d\sigma(a).$$

Hence we also obtain for $f \in C^b(G)$,

$$\varphi(\omega(\,[x]_H, [y]_H))(f) \;=\; \int_{G_H} \varphi_z(f)\, d\omega(\,[x]_H, [y]_H)(\,[z]_H)$$

$$= \int_{G_H} f^H(z)\, d\omega(\,[x]_H, [y]_H)(\,[z]_H) \;=\; f^H(xy) \;=\; \epsilon_x * \epsilon_y(f^H) \tag{2}$$

where $\epsilon_x * \epsilon_y$ is the convolution of the point measures ϵ_x and ϵ_y in the group algebra $M(G)$.

Theorem 8.1.3 *Let $G \in [FIA]_{\overline{B}}^{-}$, $H = \overline{B} \subseteq Aut(G)$. Define on the orbit space G_H the convolution $\omega : G_H \times G_H \to M_c^1(G_H)$ as in (1), the involution $[x]_H^{\sim} := [x^{-1}]_H$ and the unit element $[e]_H := \{e\}$, then G_H is a hypergroup.*

Proof. Obviously $\omega([x]_H, [y]_H)$ is an element of $M_c^1(G_B)$. That the continuity properties of (H, ω) are satisfied, follows by (2) and since G_H bears the quotient topology. The associativity also follows directly from (2). Moreover, the properties of (H, $^{\sim}$) and (H,e) follow by (2) and by $a(x^{-1}) = (a(x))^{-1}$ for $x \in G$ and $a \in H$. ◇

Denote by $C_0(G, B)$ (respectively, $C^b(G, B)$) the Banach space of all B-invariant functions $f \in C_0(G)$ (respectively $C^b(G)$). Obviously, $f \in C_0(G)$ is an element of $C_0(G, B)$ if and only if f is constant on the H-orbits. It follows that $C_0(G, B)$ is isometrically isomorphic to the Banach space $C_0(G_H)$, where G_H is given the quotient topology. The same result is true for the pair $C^b(G, B)$ and $C^b(G_H)$. Hence the dual space

$$(C_0(G, B))^* \;\cong\; Z_B(M(G)) := \;\{\mu \in M(G) : \;\mu \circ b = \mu \text{ for all } b \in B\}$$

is isometrically isomorphic to $C_0(G_H)^* \;\cong\; M(G_H)$, see also Proposition 8.1.2.

For $G \in [FIA]_{\overline{B}}^-$ the Haar measure λ on G is $\overline{B}$-invariant, see [287,Proposition 2.4]. Hence λ may be regarded as a left-invariant Haar measure on G_H.

From now on we shall in general assume that $I(G)$ (the inner automorphism of G) is contained in B. In that case G_H is a commutative hypergroup, since $[a(x)y]_H = [y(a(x)y)y^{-1}]_H = [ya(x)]_H$ for all $x, y \in G$ and $a \in H$.

Theorem 8.1.4 *Let $G \in [FIA]_{\overline{B}}^-$, $H = \overline{B}$. Then the three dual spaces $\mathcal{S}(G_H)$, $\widehat{G_H}$ and $\mathcal{X}^b(G_H)$ coincide.*

Proof. Denote the mapping $x \mapsto [x]_H$ by κ. Let $C \subseteq G_H$ be a compact subset. Then $\kappa^{-1}(C) \subseteq G$ is compact, since κ is a proper mapping. By $I(G) \subseteq B$ it follows $G \in [FC]^-$. Hulanicki [339] observes that $G \in [FC]^-$ is of subexponential growth, see also [491,Theorem 4.12]. Since $\lambda(C) = \lambda(\kappa^{-1}(C))$ it follows that G_H is a commutative hypergroup of subexponential growth. By Theorem 3.4.6 we obtain $\mathcal{S}(G_H) = \widehat{G_H} = \mathcal{X}^b(G_H)$. ◇

Our next goal is to derive that $\widehat{G_H}$ is a hypergroup under pointwise multiplication, see subsection 3.4. Proving this result we have to derive a series of special properties of $[FIA]_{\overline{B}}^-$ groups. The basic tool is the identification of $\widehat{G_H}$ with $E(G, B)$. Using the notation of subsection 2.4, denote by $P^0(G, B)$ the set of all B-invariant continuous positive definite functions on G such that $\varphi(e) \leq 1$, endowed with the weak $*$-topology. $P^0(G, B)$ is compact and convex. The set of all non-zero extremal points of $P^0(G, B)$ is denoted by $E(G, B)$. Note that for $\psi \in E(G, B)$ we have $\psi(e) = 1$ and $\text{ex}P^0(G, B) = E(G, B) \cup \{0\}$, and on $E(G, B)$ the weak $*$-topology is equal to the topology of uniform convergence on compact subsets of G, see Theorem 2.4.12 or [180,13.5.2]. The functions $\psi \in E(G, B)$ are called B-characters in [491] and [453]. Proposition 1.1 of [453] contains the result that $\psi \in C^b(G, B)$, $\psi \neq 0$, is an element of $E(G, B)$ if and only if ψ satisfies the product formula

$$\psi(x)\,\psi(y) = \int_H \psi(a(x)y)\,d\sigma(a) \qquad \text{for all } x, y \in G.$$

(The proof of this result uses also the subexponential growth of $G \in [FC]^-$.)

Each $\psi \in E(G, B)$ may be regarded as a continuous function ψ^B on G_H. It follows that the identification $\psi \mapsto \psi^B$ is a homeomorphism from $E(G, B)$ onto $\mathcal{S}(G_H) = \widehat{G_H} = \mathcal{X}^b(G_H)$.

In a first step we will derive a product formula (P) for all $\alpha, \beta \in \widehat{G_H}$, see subsection 3.4.

Proposition 8.1.5 *Let $G \in [FIA]_{\overline{B}}^-$, $I(G) \subseteq B$, $H = \overline{B}$. For all $\alpha, \beta \in \widehat{G_H}$ there exists a probability measure $\omega(\alpha, \beta) \in M^1(\widehat{G_H})$ such that*

$$\alpha(\,[x]_H)\, \beta(\,[x]_H) \;=\; \int_{\widehat{G_H}} \tau(\,[x]_H)\, d\omega(\alpha, \beta)(\tau) \qquad (P)$$

for all $[x]_H \in G_H$.

Proof. Considering α and β as elements of $E(G, B)$ we obtain that $\gamma := \alpha \cdot \beta$ is a B-invariant continuous positive definite function with $\gamma(e) = 1$. Hence γ may be regarded as a continuous function on G_H with $\gamma([e]_H) = 1$. We shall show that $\gamma \in P^1(G_H)$.

Let $\lambda_1, ..., \lambda_n \in \mathbb{C}$ and $x_1, ..., x_n \in G$. Then

$$\sum_{i,j=1}^{n} \lambda_i \overline{\lambda_j}\, \omega(\,[x_i]_H, [x_j]\tilde{}_H\,)(\gamma) \;=\; \sum_{i,j=1}^{n} \lambda_i \overline{\lambda_j}\, \omega(\,[x_i], [b(x_j)^{-1}]_H)(\gamma)$$

$$= \int_B \sum_{i,j=1}^{n} \lambda_i \overline{\lambda_j}\, \gamma(a(x_i) \cdot b(x_j)^{-1})\, d\sigma(a) \qquad \text{for every } b \in H.$$

We can select $b = a$, and it follows

$$\sum_{i,j=1}^{n} \lambda_i \overline{\lambda_j}\, \omega(\,[x_i]_H, [x_j]\tilde{}_H\,)(\gamma) \;=\; \int_B \sum_{i,j=1}^{n} \lambda_i \overline{\lambda_j}\, \gamma(a(x_i) \cdot a(x_j^{-1}))\, d\sigma(a)$$

$$= \sum_{i,j=1}^{n} \lambda_i \overline{\lambda_j}\, \gamma(x_i x_j^{-1}) \;\geq\; 0.$$

Hence $\gamma \in P^1(G_H)$. By Bochner's theorem (Theorem 3.7.1) there exists a unique positive measure $a \in M^+(\widehat{G_H})$ such that

$$\alpha(\,[x]_H) \cdot \beta(\,[x]_H) \;=\; \gamma(\,[x]_H) \;=\; \int_{\widehat{G_H}} \tau(\,[x]_H)\, da(\tau) \qquad \text{for all } [x]_H \in G_H.$$

Since $\gamma([e]_H) = 1$ it follows that $a \in M^1(\widehat{G_H})$ and we set $\omega(\alpha, \beta) = a$. ◇

In order to show that $\omega(\alpha, \beta) \in M^1(\widehat{G_H})$ has compact support, we have to consider an arbitrary group D of automorphisms on G with $D \subseteq B$, where $G \in [FIA]_{\overline{B}}^-$ with $I(G) \subseteq B$. Of course we have $G \in [FIA]_{\overline{D}}^-$. We will investigate the relations between the dual spaces of $G_{\overline{D}}$ and G_H. Note that $G_{\overline{D}}$ is a hypergroup, but in general not a commutative one. The mapping $s : P^0(G,D) \to P^0(G,B)$, $s(\varphi) = \varphi^H$ is continuous and affine. Applying the Krein-Milman theorem, we obtain that for each $\varphi \in \overline{P^0(G,B)}$ there exists a probability measure μ with support contained in $\overline{\operatorname{ex} P^0(G,D)}$ such that

$$\langle f, \varphi \rangle = \int_{\operatorname{ex} P^0(G,D)} \langle f, \tau \rangle \, d\mu(\tau) \qquad \text{for all } f \in L^1(G),$$

where $\langle f, \varphi \rangle = \int\limits_G f(x) \, \overline{\varphi(x)} \, d\lambda(x)$.

Lemma 8.1.6 *Let $G \in [FIA]_B^-$, and let D be a subgroup of B. Let $\alpha \in E(G,B)$. Then there exists a probability measure μ with compact support in $E(G,D)$ such that*

$$\alpha(x) = \int_{E(G,D)} \kappa(x) \, d\mu(\kappa) \qquad \textit{for all } x \in G.$$

Proof. Since $\alpha \in E(G,B) \subseteq \overline{P^0(G,D)}$ there exists a probability measure μ such that $\operatorname{supp} \mu \subseteq \overline{\operatorname{ex} P^0(G,D)}$ and

$$\langle f, \alpha \rangle = \int_{\operatorname{ex} P^0(G,D)} \langle f, \kappa \rangle \, d\mu(\kappa) \qquad \text{for all } f \in L^1(G).$$

The image measure $s(\mu)$ satisfies $\operatorname{supp} s(\mu) = s(\operatorname{supp} \mu) \subseteq \overline{\operatorname{ex}(P^0(G,B))}$ and

$$\langle f, \alpha \rangle = \int_{\operatorname{ex} P^0(G,B)} \langle f, \tau \rangle \, ds(\mu)(\tau) \qquad \text{for all } f \in L^1(G).$$

Since $\overline{\operatorname{ex} P^0(G,B)} = E(G,B) \cup \{0\}$ and $\alpha \in E(G,B)$ it follows that $s(\mu) = \epsilon_\alpha$, the point measure at α. In particular, $s(\kappa) = \kappa^H = \alpha$ for each $\kappa \in \operatorname{supp} \mu$. Therefore $0 \notin \operatorname{supp} \mu$, and since $\overline{\operatorname{ex} P^0(G,D)} = E(G,D) \cup \{0\}$, it follows that $\operatorname{supp} \mu$ is a compact subset in $E(G,D)$. (Recall that on $E(G,D)$ the weak $*$-topology and the topology of uniform convergence on compact subsets of G coincide.) Finally we infer that $\alpha(x) = \int\limits_{E(G,D)} \kappa(x) \, d\mu(\kappa)$ for all $x \in G$. ◇

In the case that $I(G) \subseteq D \subseteq B$ it follows immediately

Corollary 8.1.7 *Let $G \in [FIA]_B^-$, D a subgroup of B such that $I(G) \subseteq D \subseteq B$, $H = \bar{B}$.*

(1) For each $\alpha \in \widehat{G_H}$ there exists a unique probability measure μ_α on $\widehat{G_{\bar{D}}}$ with compact support such that

$$\alpha([x]_H) = \int_{\widehat{G_{\bar{D}}}} \tau([x]_{\bar{D}}) \, d\mu_\alpha(\tau) \qquad \textit{for all } x \in G.$$

(2) Let $\varphi \in C^b(G_H)$ such that

$$\varphi([x]_H) = \int_{\widehat{G_H}} \alpha([x]_H)\, db(\alpha) \qquad \textit{for all } x \in G,$$

where $b \in M(\widehat{G_H})$. Then there exists some $a \in M(\widehat{G_{\bar{D}}})$ such that

$$\varphi([x]_{\bar{D}}) = \int_{\widehat{G_{\bar{D}}}} \tau([x]_{\bar{D}})\, da(\tau) \qquad \textit{for all } x \in G.$$

Proof.

(1) Is a direct consequence of Lemma 8.1.6.
(2) We will apply Theorem 3.9.2. Let $g \in C_c(G_{\bar{D}})$. Then

$$\left|\int_{G_{\bar{D}}} \varphi([x]_{\bar{D}})\, g([x]_{\bar{D}})\, d\lambda(x)\right| = \left|\int_G \varphi(x)\, g(x)\, d\lambda(x)\right|$$

$$= \left|\int_G \int_{\widehat{G_H}} \alpha([x]_H)\, db(\alpha)\, g(x)\, d\lambda(x)\right| \leq \|b\| \sup_{\alpha \in \widehat{G_H}} \left|\int_G \alpha(x)\, g(x)\, d\lambda(x)\right|$$

$$= \|b\| \sup_{\alpha \in \widehat{G_H}} \left|\int_G \int_{\widehat{G_{\bar{D}}}} \tau([x]_{\bar{D}})\, d\mu_\alpha(\tau)\, g(x)\, d\lambda(x)\right|$$

$$= \|b\| \sup_{\alpha \in \widehat{G_H}} \left|\int_{\widehat{G_{\bar{D}}}} \int_{G_{\bar{D}}} \tau([x]_{\bar{D}})\, g([x]_{\bar{D}})\, d\lambda(x)\, d\mu_\alpha(\tau)\right| \leq \|b\|\, \|\hat{g}\|_\infty,$$

where $\hat{g}(\tau)$ is the Fourier transform of g defined on the commutative hypergroup $G_{\bar{D}}$. Now Theorem 3.9.2 yields a measure $a \in M(\widehat{G_{\bar{D}}})$ such that

$$\varphi([x]_{\bar{D}}) = \int_{\widehat{G_{\bar{D}}}} \tau([x]_{\bar{D}})\, da(\tau) \qquad \text{for all } x \in G.$$

◇

Remark. Corollary 8.1.7(2) is a 'connection theorem' for $[FIA]_B$-groups. Compare with the results for polynomial hypergroups, section 4.6, e.g. Theorem 4.6.2.

For the proof of the next Lemma we use a result concerning the restriction of $\alpha \in E(G,B)$ to a closed B-invariant subgroup, see [298,Lemma 1.3] or [491,Proposition 2.9].

Lemma 8.1.8 *Let $G \in [FIA]_B^-$, $I(G) \subseteq B$, and let N be a closed subgroup of G. Suppose D is a subgroup of B containing the inner automorphisms of G induced by elements of N, and that N is D-invariant. Denote $D' = D|N$. Then for each $\alpha \in E(G,B)$ the measure $\mu_{\alpha|N}$ on $E(N,D')$ representing $\alpha|N$ has compact support.*

Proof. By Lemma 8.1.6 there exists a probability measure μ with compact support in $E(G,D)$ such that

$$\alpha(x) = \int_{E(G,B)} \kappa(x)\, d\mu(\kappa) \qquad \text{for all } x \in G.$$

Applying Lemma 1.3 of [298] we get $\varphi|N \in E(N,D')$ for each $\varphi \in E(G,D)$. Since the restriction map $r_N : P^1(G) \to P^1(N)$ is continuous, the image measure $r_N(\mu)$ has compact support in $E(N,D')$ and

$$\alpha(y) = \int_{E(N,D')} \beta(y)\, dr_N(\mu)(\beta) \qquad \text{for all } y \in N.$$

By the uniqueness of the representing measure we have $\mu_{\alpha|N} = r_N(\mu)$. $\diamond$

Now we can get more information about the probability measure $\omega(\alpha,\beta) \in M^1(\widehat{G_H})$ of Proposition 8.1.5.

Proposition 8.1.9 *Let $G \in [FIA]_B^-$, $I(G) \subseteq B$ and $\alpha,\beta \in \widehat{G_H}$. Then there exists a unique probability measure $\omega(\alpha,\beta) \in M_c^1(\widehat{G_H})$ with compact support, such that*

$$\alpha([x]_H)\, \beta([x]_H) = \int_{\widehat{G_H}} \tau([x]_H)\, d\omega(\alpha,\beta)(\tau) \qquad (P)$$

for all $[x]_H \in G_H$.

Proof. The existence of a measure $\omega(\alpha,\beta) \in M^1(\widehat{G_H})$ satisfying (P) is already shown. The uniqueness of $\omega(\alpha,\beta)$ with this property follows by Theorem 3.3.3. It remains to show that $\omega(\alpha,\beta)$ has compact support. Let

$$N = \{(x,x) : x \in G\} \subseteq G \times G$$

be the diagonal group in $G \times G$ and

$$D = \{(\sigma,\sigma) : \sigma \in B\} \subseteq B \times B.$$

Then D is a subgroup of $B \times B$, contains the inner automorphisms induced by N, and N is D-invariant. Identifying N with G we get $D|G = B$. Obviously $G \times G \in [FIA]_{B\times B}^-$ and $\alpha \otimes \beta \in E(G \times G, B \times B)$, where

$$\alpha \otimes \beta(x,y) = \alpha(x) \cdot \beta(y).$$

We have $\alpha \otimes \beta|G = \alpha \cdot \beta$ and by Lemma 8.1.8 it follows that the representing measure $\mu_{\alpha\otimes\beta}|G =: \mu_{\alpha\cdot\beta}$ has compact support, i.e.

$$\alpha(x) \cdot \beta(x) = \int_{E(G,B)} \tau(x)\, d\mu_{\alpha\cdot\beta}(\tau)$$

and $\mu_{\alpha\cdot\beta} \in M_c^1(E(G,B))$. Finally, $\omega(\alpha,\beta)$ is the image measure of $\mu_{\alpha\cdot\beta}$ under the homeomorphism $\psi \mapsto \psi^B$, $E(G,B) \to \widehat{G_H}$. In particular, $\omega(\alpha,\beta)$ has compact support in $\widehat{G_H}$. ◇

Applying Proposition 3.4.10 it remains to show that the axioms (H3) and (H6) are satisfied, to conclude that $\widehat{G_H}$ is a hypergroup under pointwise multiplication.

$$\Lambda_H : \widehat{G_H} \times \widehat{G_H} \to \mathcal{C}(\widehat{G_H}), \quad \Lambda_H((\alpha,\beta)) = \text{supp}\ \omega(\alpha,\beta)$$

is continuous, where $\mathcal{C}(\widehat{G_H})$ is equipped with the Michael topology, if and only if

$$\Lambda_B : E(G,B) \times E(G,B) \to \mathcal{C}(E(G,B)), \quad \Lambda_B((\alpha,\beta)) = \text{supp}\,\mu_{\alpha\cdot\beta}$$

is continuous. The proof of the continuity of Λ_B is based on a characterization of compactly generated $[FC]_B^-$-groups and is rather involved. We omit the detailed proof and refer to Lemma 2, Lemma 3 and Corollary 2 of [293].

Proposition 8.1.10 *Let $G \in [FIA]_B^-$, $I(G) \subseteq B$. Then the mapping*

$$\Lambda_H : \widehat{G_H} \times \widehat{G_H} \to \mathcal{C}(\widehat{G_H}), \quad \Lambda_H((\alpha,\beta)) = \text{supp}\,\omega(\alpha,\beta)$$

is continuous.

It remains to prove '$1 \in \text{supp}\,\omega(\alpha,\bar{\beta})$ if and only if $\alpha = \beta$'. Again we have to use structure theory for groups satisfying compactness conditions. We omit a detailed proof, but describe the main tools used to show the statements.

Proposition 8.1.11 *Let $G \in [FIA]_B^-$, $I(G) \subseteq B$. Assume that G contains a compact B-invariant subgroup N such that G/N is B-fixed (i.e. $\sigma(x)N \subseteq xN$ for all $x \in G$, $\sigma \in H = \bar{B}$). Then $1 \in \text{supp}\,\omega(\alpha,\bar{\beta})$ exactly when $\alpha = \beta$ for all $\alpha,\beta \in \widehat{G_B} \cong E(G,B)$.*

For the proof we refer to [293] and note that Proposition 2.3 of [453] is applied. We also refer to [359,Lemma 4].

Proposition 8.1.12 *Let $G \in [FIA]_B^-$, $I(G) \subseteq B$. Then $1 \in \text{supp}\,\omega(\alpha,\bar{\beta})$ exactly when $\alpha = \beta$ for all $\alpha,\beta \in \widehat{G_B} \cong E(G,B)$.*

Proof. At first we consider $B = I(G)$ and open, compactly generated normal subgroups N of G. Denote by $r_N : P^1(G) \to P^1(N)$ the restriction mapping from G to N. It is straightforward to show that

$$\text{supp}\,\omega(\alpha,\bar{\beta}) = \bigcap_N r_N^{-1}(\text{supp}\ \omega(\alpha|N, \bar{\beta}|Na)).$$

Applying the structure theorem of [287,Theorem 3.20] we can use Proposition 8.1.11 to derive that $1 \in \operatorname{supp} \omega(\alpha, \bar{\beta})$ if and only if $\alpha = \beta$. Using the mapping

$$s: E(G) \to E(G,B), \qquad s(\varphi) = \varphi^H,$$

Lemma 3 of [293] yields the statement. $\diamond$

Collecting all the results given above we can add to Theorem 8.1.3,

Theorem 8.1.13 *Let $G \in [FIA]_{\bar{B}}^{-}$, $B \supseteq I(G)$. Then $\widehat{G_H}$ is a hypergroup under pointwise multiplication.*

We choose a Haar measure m on G_H. Recall that any Haar measure on G is $\bar{B}$-invariant, see [287,Proposition 2.4]. We know that the Plancherel measure π on $\mathcal{S}(G_H) = \widehat{G_H} = \mathcal{X}^b(G_H)$ is uniquely defined by the equation

$$\int_{G_H} |f(x)|^2 \, dm(x) = \int_{\widehat{G_H}} |\hat{f}(\alpha)|^2 \, d\pi(\alpha) \qquad (*)$$

for all $f \in L^1(G_H) \cap L^2(G_H)$, see Theorem 3.2.6. Moreover, π is a Haar measure on the hypergroup $\widehat{G_H}$. We choose the Haar measure $\pi \in \widehat{G_H}$ such that $(*)$ is satisfied.

As already mentioned in section 3.5, a general result for commutative hypergroups yields the following property of $L^1(G_H) = L^1(G_H, m)$.

Corollary 8.1.14 *Let $G \in [FIA]_{\bar{B}}^{-}$, $B \supseteq I(G)$. Then $L^1(G_H)$ is $\mathcal{X}^b(G_H)$-regular.*

Proof. Follows directly by Theorem 8.1.4 and Theorem 3.5.16. $\diamond$

In section 3.4 we introduced the mapping $i: K \to \hat{\hat{K}}$, where $i(x)(\alpha) := \overline{\alpha(x)}$ for $\alpha \in \hat{K}$, assuming that $\hat{K}$ is a hypergroup under pointwise multiplication, see Theorem 3.4.12. We know that $i: k \to \hat{\hat{K}}$ is a homeomorphism from K onto $i(K) \subseteq \hat{\hat{K}}$. Identifying $\hat{\hat{K}}$ with $\Delta_s(L^1(\hat{K}))$, see Theorem 3.1.1, we have $i(x) = h_x$, where

$$h_x(\varphi) = \int_{\hat{K}} \varphi(\alpha) \, \alpha(x) \, d\pi(\alpha) \qquad \text{for } \varphi \in L^1(\hat{K}).$$

By Theorem 2 of [360] we have the following result in case that $B = I(G)$:

Identifying $E(G)$ with $\widehat{G_H}$, the map $G_H \to \Delta_s(L^1(\widehat{G_H}))$, $[x]_H \mapsto h_{[x]_H}$, where

$$h_{[x]_H}(\varphi) = \int_{\widehat{G_H}} \varphi(\alpha) \, \alpha(x) \, d\pi(\alpha) \qquad \text{for } \varphi \in L^1(\widehat{G_H}),$$

is a homeomorphism from G_H onto $\Delta_s(L^1(\widehat{G_H}))$.

Proposition 8.1.15 *Let $G \in [FIA]^-$. Then the Pontryagin duality is valid for G_H, $H = \overline{I(G)}$.*

Applying again the mapping $s : E(G) \to E(G, B)$, $s(\varphi) = \varphi^H$, Proposition 8.1.15 yields, see [292,Proposition 2]:

Theorem 8.1.16 *Let $G \in [FIA]_B^-$, $B \supseteq I(G)$. Then the Pontryagin duality is valid for G_H, $H = \bar{B}$.*

Examples.

(1) Let $G = \mathbb{R}^n$ and $B = SO(n)$, $n \geq 2$, the special orthogonal group on $\mathbb{R}^n$. $SO(n)$ is a compact linear group, see [304, Ch.II, (4.25)] and G is a $[FIA]_B^-$ group. For each $n \geq 2$ considering the hypergroups G_B we get involved with normalized Bessel functions,

$$j_\alpha(z) \;=\; \sum_{k=0}^{\infty} \frac{(-1)^k\,\Gamma(\alpha+1)}{2^{2k}k!\,\Gamma(k+\alpha+1)}\, z^{2k}, \qquad z \in \mathbb{C},$$

where $\alpha = \frac{n}{2} - 1$, see subsection 1.2.2. In fact, G_B and $\widehat{G_B}$ can be identified with $\mathbb{R}_0^+ = [0, \infty[$ via integration with respect to the normalized Haar measure on $SO(n)$. We have to combine results of Theorem 1.2.3, where

$$\varphi_x(s) \;=\; j_\alpha(xs) \qquad \text{for } x, s \in \mathbb{R}_0^+ \text{ and } \alpha = \frac{n}{2} - 1,$$

with the construction of the hypergroups $(\mathbb{R}^n)_{SO(n)}$. We conclude that $(\mathbb{R}^n)_{SO(n)}$ can be identified with the Bessel-Kingman hypergroup on $\mathbb{R}_0^+$ of order $\alpha = \frac{n}{2} - 1$. Moreover, for these hypergroups the Pontryagin duality is valid, where each non-constant character $\varphi \in ((\mathbb{R}^n)_{SO(n)})^\wedge$ is given by $\varphi = \varphi_s$, where $s \in]0, \infty[$ and $\varphi_s(x) = j_\alpha(xs)$, $x \in \mathbb{R}_0^+$. (For arbitrary $\alpha > -\frac{1}{2}$ the Bessel-Kingman hypergroup is discussed in subsection 8.2.)

(2) The following example of a $[FIA]_B^-$ group is due to Dunkl-Ramirez, see [200,Ch.I,3]. It is based on the method of symmetrization on hypergroups, see [199].
Fix a prime p. Consider the set $\mathbb{Q}_p$ of p-**adic numbers**. $\mathbb{Q}_p$ is the completion of $\mathbb{Q}$ with respect to the p-**adic norm**. The p-adic norm of $r \in \mathbb{Q}\backslash\{0\}$ is given by $|r|_p = p^{-m}$, where $r = p^m q$ with $m \in \mathbb{Z}$, and q is a rational number whose numerator and denominator are not divisible by p. Set $|0|_p = 0$. Since addition and multiplication of $\mathbb{Q}$ are continuous with respect to $|\cdot|_p$, these arithmetic operations extend to

the completion of $\mathbb{Q}$ with respect to $|\cdot|_p$. This completion is a field denoted by $\mathbb{Q}_p$, th field of p-adic numbers. We refer to [304,§10] or [232,p.34-36]. The closed ball

$$\bar{B}(1,0) = \{x \in \mathbb{Q}_p : |x|_p \leq 1\}$$

is called the ring of **p-adic integers** and is denoted by Δ_p. Δ_p is a compact ring and each $x \in \Delta_p$ can be written as

$$x = x_0 + x_1 p + \cdots + x_n p^n + \cdots, \qquad \text{with } x_j \in \{0, 1, ..., p-1\}.$$

Furthermore, let $B = \{y \in \mathbb{Q}_p : |x|_p = 1\}$. Now write $G = \Delta_p$. Then G is a compact additive group and B is a compact multiplicative group acting on G by multiplication. Hence $G \in [FIA]_B^-$.
Moreover, if $x, y \in \Delta_p$ we have $x = uy$ for some $u \in B$ exactly when $|x|_p = |y|_p$. Hence Δ_p is the union of countably many B-orbits $\{\xi_j : j = 0, 1, ..., \infty\}$, where $\xi_j = \{x \in \Delta_p : |x|_p = p^{-j}\}$ for $j = 0, 1, \ldots$. We conclude that G_B is a compact commutative hypergroup homeomorphic to $\mathbb{N}_0^* = \mathbb{N}_0 \cup \{\infty\}$, the one-point compactification of $\mathbb{N}_0$.
We collect some results concerning $G_B = \mathbb{N}_0 \cup \{\infty\}$ shown in [200].
The unit element e of G_B is ∞ (corresponding to $0 \in \Delta_p$) and the involution $\tilde{}$ is the identity mapping. The convolution ω is given by $\omega(m, \infty) = \delta_m$ for each $m \in \mathbb{N}_0 \cup \{\infty\}$. For each $m, n \in \mathbb{N}_0$ we have $\omega(m, n) = \epsilon_{\min(m,n)}$ if $m \neq n$, and

$$\omega(n,n)(l) = \begin{cases} 0 & \text{for } l < n \\ \frac{p-2}{p-1} & \text{for } l = n \\ \frac{1}{p^k} & \text{for } l = n + k > n. \end{cases}$$

The normalized Haar measure on $G_B = \mathbb{N}_0 \cup \{\infty\}$ is given by

$$m(\{\infty\}) = 0 \quad \text{and} \quad m(\{k\}) = \frac{1}{p^k}\left(1 - \frac{1}{p}\right) \quad \text{for } k \in \mathbb{N}_0.$$

The dual space $\widehat{G_B} = (\mathbb{N}_0 \cup \{\infty\})^\wedge$ can be identified with $\mathbb{N}_0$. In fact, $\{\alpha_n : n \in \mathbb{N}_0\}$ is the set of all characters on G_B, where

$$\alpha_n(m) = \begin{cases} 1 & \text{if } m \geq n \text{ or } m = \infty \\ -\frac{1}{p-1} & \text{if } 0 \leq m = n-1 \\ 0 & \text{if } 0 \leq m \leq n-2. \end{cases}$$

Applying Theorem 8.1.13 we obtain that $\widehat{G_B}$, which can be identified with $\mathbb{N}_0$, is a hypergroup under pointwise multiplication. By Theorem 8.1.16 we even get that $\widehat{\widehat{G_B}} \cong \widehat{\mathbb{N}_0}$ can be identified with $G_B = \mathbb{N}_0 \cup \{\infty\}$, the one-point compactification of $\mathbb{N}_0$.

Remark. For the special case that $B = I(G)$ many locally compact groups belong to the class of $[FIA]^-$ groups. Obviously commutative locally compact groups or compact groups belong to $[FIA]^-$.

(1) If G is a Z-**group**, i.e. a locally compact group such that G/Z is compact, where $Z = Z(G)$ denotes the center of G,

$$Z(G) = \{z \in G : zx = xz \text{ for all } x \in G\}.$$

Z is a closed normal subgroup of G, and $I(G)$ is compact. Thus $G_{I(G)}$ is a commutative hypergroup such that the Pontryagin duality is valid.

(2) $[FD]^-$ is the class of locally compact groups such that $\overline{G'}$ is compact, where G' denotes the commutator subgroup of G. It is easy to show that $[FD]^- \subseteq [FC]^-$. Thus if

$$G \in [FD]^- \cap [SIN] \subseteq [FC]^- \cap [SIN] = [FIA]^-,$$

we have that $G_{I(G)}$ is a commutative hypergroup satisfying the Pontryagin duality.

.

1.3 *Hypergroups induced by spherical projectors*

In subsection 8.1.2 we used a mapping from a $[FIA]_B^-$-group G onto a set of H-orbits in G to construct a special class of hypergroups. In [500], Muruganandam used a projection from $C_c(G)$ into $C_c(G)$ to construct a large class of hypergroups. We now recall from [500] the properties of this projection on $C_c(G)$.

A function $f \in C_c(G)$ is called positive, if f is nonnegative and is not identically zero. A linear map $A : C_c(G) \to C_c(G)$ is called **positivity preserving**, if $A(f)$ is positive whenever f is positive.

Definition 1. Let G be a locally compact group with left Haar measure λ. A linear map $P : C_c(G) \to C_c(G)$ is called a **spherical projector**, if it satisfies the following conditions, valid for all $f, g \in C_c(G)$.

(SH_1) (1) $P^2 = P$ and P is positivity preserving,

(2) $P(P(f) \cdot g) = P(f)\, P(g)$,

(3) $$\int_G P(f)(x)\, g(x)\, d\lambda(x) = \int_G f(x)\, P(g)(x)\, d\lambda(x)$$

(4) $$\int_G P(f)(x)\, d\lambda(x) = \int_G f(x)\, d\lambda(x)$$

(SH_2) $P(P(f) * P(g)) = P(f) * P(g)$.

(SH_3) Let $P^* : M(G) \to M(G)$ denote the adjoint map of the extension of P to $C_0(G)$ (see Proposition 8.1.17 below), and let $O_x =$ supp $P^*(\epsilon_x)$, where ϵ_x is the point measure of $x \in G$. Then for all $x, y \in G$,

(1) either $O_x \cap O_y = \emptyset$ or $O_x = O_y$,
(2) if $y \in O_x$, then $y^{-1} \in O_{x^{-1}}$ and if $O_{xy} = O_e$ then $O_y = O_{x^{-1}}$,
(3) the map $x \mapsto O_x$ is a continuous function from G into $\mathcal{C}(G)$, the set of all nonempty compact subsets of G equipped with the Michael topology.

Proposition 8.1.17 *Let $P : C_c(G) \to C_c(G)$ be a spherical projector. Then P can be extended to norm decreasing linear operators on the Banach spaces $L^1(G, \lambda)$ and $C_0(G)$.*

Proof. Since $P : C_c(G) \to C_c(G)$ is positivity preserving, it follows that $|P(f)| \leq P(|f|)$, and hence $(SH_1)(4)$ yields $\|P(f)\|_1 \leq \|f\|_1$ for each $f \in C_c(G)$. Therefore P can be extended to a norm decreasing linear operator $P : L^1(G, \lambda) \to L^1(G, \lambda)$. Obviously, by $|P(f)| \leq P(|f|)$ for all $f \in C_c(G)$, we get for the $\| \cdot \|_\infty$-completion $C_0(G)$ of $C_c(G)$ a norm decreasing operator $P : C_0(G) \to C_0(G)$. ◇

Considering the adjoint operators of $P \in \mathcal{B}(L^1(G, \lambda))$ or $P \in \mathcal{B}(C_0(G))$ we get norm decreasing linear operators $P^* \in \mathcal{B}(L^\infty(G, \lambda))$ respectively $P^* \in \mathcal{B}(M(G))$.

Definition 2. Let $P : C_c(G) \to C_c(G)$ be a spherical projector. A function $f \in C_c(G)$ is called P-**radial** if $P(f) = f$. A measure $\mu \in M(G)$ is called P-**radial** if $P^*(\mu) = \mu$. The spherical projector P is called **ultra-spherical** if the modular function Δ of G is P-radial.

Remarks: (1) Using $(SH)_2$ we observe that the convolution $f * g$ of P-radial functions $f, g \in C_c(G)$ is again P-radial. $(SH_3)(2)$ implies that$f \in C_c(G)$ is P-radial exactly when $\check{f}$ is P-radial, where $\check{f}(x) = f(x^{-1})$.

We have the following supplement to property (SH_2).

Lemma 8.1.18 *Let $P : C_c(G) \to C_c(G)$ be a spherical projector. Then*

$$P(f * P(g)) = P(f) * P(g) = P(P(f) * P(g)) \qquad \text{for } f, g \in C_c(G).$$

Proof. We have only to show that the first equality is true. Let $f \in C_c(G)$. If $g, h \in C_c(G)$ are P-radial, then $f * g$ is P-radial, and therefore we have (using $(SH_1)(3)$),

$$\int_G P(f * g)(x)\, h(x)\, d\lambda(x) = \int_G f * g(x)\, h(x)\, d\lambda(x)$$

$$= \int_G f(x)\,(g * \check{h})^\vee(x)\,d\lambda(x)$$

$$= \int_G P(f)(x)\,(g * \check{h})^\vee(x)\,d\lambda(x) \;=\; \int_G (P(f) * g)(x)\,h(x)\,d\lambda(x).$$

(For the second equality we refer to Proposition 2.2.15.)

For arbitrary $g, h \in C_c(G)$ it follows

$$\int_G P(f * P(g))(x)\,h(x)\,d\lambda(x) \;=\; \int_G P(f * P(g))(x)\,P(h)(x)\,d\lambda(x)$$

$$= \int_G P(f) * P(g)(x)\,P(h)(x)\,d\lambda(x) \;=\; \int_G P(P(f) * P(g))(x)\,h(x)\,d\lambda(x)$$

for all $h \in C_c(G)$. Thus $P(f * P(g)) = P(P(f) * P(g))$. ◇

We have to derive some properties of the measures $P^*(\mu) \in M(G)$, mainly of $P^*(\epsilon_x)$. Basic facts concerning measures and their support can be found in the introductory section 1.1.

From $(SH_1)(3)$ it follows that the adjoint operator $P^* \in \mathcal{B}(L^\infty(G,\lambda))$ of $P \in \mathcal{B}(L^1(G,\lambda))$ is an extension of $P : C_c(G) \to C_c(G)$. We shall denote this adjoint operator restricted to $C^b(G)$ also by P. Applying $(SH_1)(2)$ we get $P(1) = 1$.

Given $x \in G$, let $\mu_x := P^*(\epsilon_x)$. By $(SH_1)(1)$ μ_x is a probability measure on G. We will write $O_x := \operatorname{supp}\mu_x$ for the supports of μ_x. Note that for these support sets we have by $(SH_3)(1)$ either $O_x \cap O_y = \emptyset$ or $O_x = O_y$.

Lemma 8.1.19 *Let $P : C_c(G) \to C_c(G)$ be a spherical projector. Let $\mu \in M(G)$ and let $g \in C_c(G)$ be P-radial. Then*

$$P(\mu * g) \;=\; P^*(\mu) * g.$$

Proof. Let $h \in C_c(G)$. Since $\check{g}$ is P-radial, we get

$$\int_G h(x)\,P(\mu * g)(x)\,d\lambda(x) \;=\; \int_G P(h)(x)\,\mu * g(x)\,d\lambda(x)$$

$$= \int_G \int_G P(h)(x)\,\check{g}(x^{-1}y)\,d\lambda(x)\,d\mu(y) \;=\; \int_G (P(h) * \check{g})(y)\,d\mu(y)$$

$$= \int_G (h * \check{g})(y)\,dP^*(\mu)(y) \;=\; \int_G h(x)\,(P^*(\mu) * g)(x)\,d\lambda(x).$$

(The fourth equality holds by Lemma 8.1.18.) Thus $P(\mu * g) = P^*(\mu) * g$ is shown. ◇

Let $(k_i)_{i\in I}$ be a net of funcions of $C_c(G)$ with $k_i \geq 0$, $\int\limits_G k_i(x)\,d\lambda(x) = 1$ and $\operatorname{supp} k_i \to \{e\}$, see [232,Proposition 2.42] or Proposition 2.1.6. The net $(P(k_i))_{i\in I}$ satisfies $P(k_i) \geq 0$, $\int\limits_G P(k_i)(x)\,d\lambda(x) = 1$ and using $(SH_1)(2)$ and $P(k_i) \in C_c(G)$ it also follows that $\operatorname{supp} P(k_i) \to \{e\}$.

Proposition 8.1.20 *Let $P : C_c(G) \to C_c(G)$ be a spherical projector. If $y \notin O_x$ then $O_y \cap O_x = \emptyset$. In particular it follows $x \in O_x$.*

Proof. Let $y \notin O_x$. Applying Proposition 2.42 of [232] and using that $\mu_x \in M^1(G)$ we obtain for nets $(k_i)_{i\in I}$ as above

$$\int_G k_i * \mu_x(z)\, d\mu_y(z) = \int_G P(k_i * \mu_x)(z)\, d\epsilon_y(z) = \int_G P(k_i) * \mu_x(z)\, d\epsilon_y(z)$$
$$= P(k_i) * \mu_x(y) \to \mu_x(y) = 0.$$

Since $k_i \geq 0$ it follows that $\operatorname{supp} \mu_x \cap \operatorname{supp} \mu_y = \emptyset$. ◇

Proposition 8.1.21 *Let $P : C_c(G) \to C_c(G)$ be a spherical projector. If $O_x = O_y$ then $\mu_x = \mu_y$.*

Proof. Let $O_x = O_y$ and assume that $\mu_x \neq \mu_y$. Then there exists a positive function $f \in C^b(G)$ such that

$$P(f)(x) = \int_{O_x} f(z)\, d\mu_x(z) \neq \int_{O_x} f(z)\, d\mu_y(z) = P(f)(y).$$

This means that $P(f)$ is not constant on O_x. Let y_0 be one point of O_x where $P(f)$ attains its maximum on O_x. Since $P^2 = P$ and $O_{y_0} = O_x$ by Proposition 8.1.20, we have

$$P(f)(y_0) = \int_{O_x} P(f)(z)\, d\mu_{y_0}(z).$$

This is not possible, since $P(f)$ is not constant on O_x. ◇

We collect conclusions of the results from above, which are relevant to define hypergroup structures induced by a spherical projector.

Theorem 8.1.22 *Let $P : C_c(G) \to C_c(G)$ be a spherical projector. Then the following statements hold.*

(1) $y \in O_x$ is equivalent to $O_y = O_x$, and $y \notin O_x$ is equivalent to $O_x \cap O_y = \emptyset$.

(2) $O_x = O_y$ is equivalent to $\mu_x = \mu_y$.

Proof.

(1) Recall axiom $(SH_3)(1)$. Hence Propositin 8.1.20 yields

$$y \in O_x \iff O_x = O_y \quad \text{and} \quad y \notin O_x \iff O_x \cap O_y = \emptyset.$$

(2) By Proposition 8.1.21 we get $O_x = O_y \iff \mu_x = \mu_y$.

◇

Corollary 8.1.23 *Let $P : C_c(G) \to C_c(G)$ be a spherical projector. Let $f \in C_c(G)$. Then f is P-radial if and only if f is constant on each support set O_x.*

Proof. Let $f \in C_c(G)$ be P-radial, and let $x \in O_y$. Theorem 8.1.22 (1) and (2) imply $\mu_x = \mu_y$, and hence

$$f(x) = P(f)(x) = \int_G f(z)\,d\mu_x(z) = \int_G f(z)\,d\mu_y(z) = P(f)(y) = f(y).$$

Conversely, suppose that f is constant on each support set O_x. Since μ_x is a probability measure, we get

$$f(x) \;=\; \int_{O_x} f(z)\,d\mu_x(z) \;=\; P(f)(x).$$

◇

Now let $H = \{O_x : x \in G\}$ be the set consisting of all supports O_x, $x \in G$, equipped with the quotient topology under the map $x \mapsto O_x$, $G \to H$. We call this quotient map q, i.e. $q(x) \equiv O_x$. Note that by $(SH_3)(3)$ the support sets O_x are compact.

Similar as in Proposition 8.1.1 we have the following properties.

The quotient topology on H and the relative topology induced by the Michael topology on $\mathcal{C}(G)$ coincide. Furthermore H is a locally compact Hausdorff space, and the mapping $q : G \to H$, $q(x) = O_x$ is continuous and open. We identify the Banach space $M(H)$ with the closed subspace $\{\mu \in M(G) : \mu \; P\text{-radial}\}$ of $M(G)$ and $C_c(H)$ is identified with $\{f \in C_c(G) : f \; P\text{-radial}\}$. Subsequently we will use these identifications without annotation. Writing $\dot{x}$ instead of O_x by this identification, the point measure $\epsilon_{\dot{x}}$ corresponds to $\mu_x = P^*(\epsilon_x)$. The convolution of $\dot{x} \in H$ and $\dot{y} \in H$ is defined by

$$\omega(\dot{x},\dot{y}) := \; P^*(\mu_x * \mu_y). \qquad (*)$$

In order to prove that ω is a hypergroup convolution on H, we have to analyse the effect of translations of $f \in C_c(G)$ and the consequences for the convolution ω.

Proposition 8.1.24 *Let $P : C_c(G) \to C_c(G)$ be a spherical projector.*

(1) Let $f \in C_c(G)$ be P-radial and let $y \in G$. Then the function $x \mapsto P(L_x f)(y)$ is P-radial.
(2) Let $f \in C_c(H)$. Then

$$\int_H f(\dot{z})\,d\omega(\dot{x},\dot{y})(\dot{z}) \;=\; P(L_x f)(y).$$

Proof.

(1) By Lemma 8.1.19 we obtain

$$P(L_x f)(y) = P(\epsilon_{x^{-1}} * f)(y) = P^*(\epsilon_{x^{-1}}) * f(y) = \int_G f(z^{-1}y)\, d\mu_{x^{-1}}(z).$$

By $(SH_3)(2)$ we have $z \in O_{x^{-1}}$ if and only if $z^{-1} \in O_x$. Furthermore applying Theorem 8.1.22 we get $z \in O_{x^{-1}}$ is equivalent to $\mu_z = \mu_{x^{-1}}$. Hence it follows that $x \mapsto P(L_x f)(y)$ is P-radial.

(2) For $f \in C_c(H)$ we obtain by (1)

$$\int_H f(\dot{z})\, d\omega(\dot{x}, \dot{y})(\dot{z}) = \int_G f(z)\, d(\mu_x * \mu_y)(z) = \int_G f(uv)\, d\mu_x(u)\, d\mu_y(v)$$

$$= \int_G \int_G P(L_u f)(v)\, d\epsilon_y(v)\, d\mu_x(u) = \int_G P(L_u f)(y)\, d\mu_x(u)$$

$$= P(L_x f)(y).$$

◇

We will use Proposition 8.1.24 to derive axiom $(H, \tilde{})$. To check axiom (H, ω) note that Proposition 8.1.24(2) and $P(1) = 1$ yield that ω is a probability measure on H. Next we show that $\omega(\dot{x}, \dot{y})$ also has compact support. To prove this we derive a property of the quotient map $q : G \to H$ concerning the support of measures $\mu \in M(G)$.

Proposition 8.1.25 *Let $P : C_c(G) \to C_c(G)$ be a spherical projector and let $\mu \in M(G)$. Then $\operatorname{supp} P^*(\mu) = q(\operatorname{supp} \mu)$.*

Proof. First we prove

$$q(\operatorname{supp} \mu) \subseteq \operatorname{supp} P^*(\mu).$$

Let $z \in \operatorname{supp} \mu$. We have to show that $\dot{z} = q(z) \in \operatorname{supp} P^*(\mu)$. Let W be a neighbourhood of $q(z)$ and $f \in C_c(H)$ such that $\operatorname{supp} f \subseteq W$. Since $z \in q^{-1}(W)$ and $\operatorname{supp} f \subseteq q^{-1}(W)$ it follows $\int_G f(x) d\mu(x) \neq 0$. Therefore we have

$$\int_H f(\dot{x})\, dP^*(\mu)(\dot{x}) = \int_G f(x)\, d\mu(x) \neq 0.$$

This is valid for each neighbourhood W of $q(z)$, and it follows $q(z) \in \operatorname{supp} P^*(\mu)$.

To show the reverse inclusion we assume that $\dot{z} = q(z) \notin q(\operatorname{supp}\mu)$. Let W be a neighbourhood of $\dot{z}$ such that $W \cap q(\operatorname{supp}\mu) = \emptyset$. Let $f \in C_c(H)$ such that $\operatorname{supp} f \subseteq W$. As a function defined on G, f is P-radial. Hence $\operatorname{supp}\mu \cap \operatorname{supp} f = \emptyset$ and so $\int\limits_G f(x)d\mu(x) = 0$. It follows

$$\int_H f(\dot{x})\, dP^*(\mu)(\dot{x}) \;=\; \int_G f(x)\, d\mu(x) \;= 0,$$

and $\dot{z} \notin \operatorname{supp} P^*(\mu)$ is shown. ◇

Considering $\mu = \mu_x * \mu_y$ we know that $\operatorname{supp}\mu_x * \mu_y = \operatorname{supp}\mu_x \cdot \operatorname{supp}\mu_y = O_x \cdot O_y$. Since $O_x \cdot O_y$ is compact, we obtain by Proposition 8.1.25 that $\operatorname{supp}\omega(\dot{x},\dot{y}) = \operatorname{supp} P^*(\mu_x * \mu_y) = q(O_x \cdot O_y)$ is compact. The associativity law of $\omega(\dot{x},\dot{y})$ follows from the associativity of the convolution of $\mu_x * \mu_y$.

By $(SH_3)(3)$ the map $(\dot{x},\dot{y}) \mapsto (O_x, O_y)$ from $H \times H$ into $\mathcal{C}(G) \times \mathcal{C}(G)$ is continuous. The product map on the group is a continuous surjection. It follows that the mapping $(K_1, K_2) \mapsto K_1 \cdot K_2$, $\mathcal{C}(G) \times \mathcal{C}(G) \to \mathcal{C}(G)$ is continuous with respect to the Michael topology. In particular $(O_x, O_y) \mapsto O_x \cdot O_y$ is continuous. As composition of continuous maps, the mapping $(\dot{x},\dot{y}) \mapsto O_x \cdot O_y$, $H \times H \to \mathcal{C}(G)$ is continuous, and also $(\dot{x},\dot{y}) \mapsto q(O_x \cdot O_y) = \operatorname{supp}\omega(\dot{x},\dot{y})$, $H \times H \to \mathcal{C}(H)$ is continuous. Hence the axiom (H,ω) holds.

The involution $\tilde{}$ on H is defined by

$$(\dot{x})^{\sim} := (x^{-1})^{\cdot}. \qquad (**)$$

Obviously $(\dot{x})^{\sim\sim} = \dot{x}$ and $\dot{x} \mapsto (\dot{x})^{\sim}$ is a homeomorphism on H. Let $\dot{x}, \dot{y} \in H$. For each $f \in C_c(H)$ we have by Proposition 8.1.24(2)

$$\int_H f(\dot{z})\, d\omega(\dot{x},\dot{y}))^{\sim}(\dot{z}) \;=\; \int_H f((z^{-1})^{\cdot})\, d\omega(\dot{x},\dot{y})(\dot{z}) \;=\; P(L_x \check{f})(y)$$

$$= P(L_{y^{-1}} f)(x^{-1}) \;=\; \int_H f(\dot{z})\, d\omega((\dot{y})^{\sim}, (\dot{x})^{\sim})(\dot{z}).$$

Hence $(\omega(\dot{x},\dot{y}))^{\sim} = \omega((\dot{y})^{\sim}, (\dot{x})^{\sim})$, and the axiom $(H, \tilde{})$ holds.

Finally we show that (H, e) is valid. We prove that $\dot{e}$ is the unit element of H. Since $\dot{e} = O_e$ we have $\omega(\dot{e},\dot{x}) = \epsilon_{\dot{x}} = \omega(\dot{x},\dot{e})$. By Proposition 8.1.25 follows $\operatorname{supp}\omega(\dot{x},(x^{-1})^{\cdot}) = q(O_x \cdot O_{x^{-1}})$. Since $x \in O_x$, $x^{-1} \in O_{x^{-1}}$ it follows $\dot{e} \in \operatorname{supp}\omega(\dot{x},(\dot{x})^{\sim})$. Conversely let $\dot{e} \in \operatorname{supp}\omega(\dot{x},\dot{y})$. Again by Proposition 8.1.25 we know that there exist $x_1 \in O_x$ and $y_1 \in O_y$ such that $q(x_1 \cdot y_1) = \dot{e}$, i.e. $O_{x_1 \cdot y_1} = O_e$. By axiom $(SH_3)(2)$ it follows $O_{y_1} = O_{x_1^{-1}}$, which means $\dot{y} = (\dot{x})^{\sim}$. This means that axiom (H, e) is valid.

Therefore we have proved the following result.

Theorem 8.1.26 *Let $P : C_c(G) \to C_c(G)$ be a spherical projector. Then the space H, consisting of the support sets $\dot{x} = O_x$, $x \in G$, is a hypergroup, where the* **convolution** ω *is given by*

$$\omega(\dot{x}, \dot{y}) \;=\; P^*(\mu_x * \mu_y),$$

the **involution** $\tilde{}$ *is given by*

$$(\dot{x})^{\sim} \;=\; (x^{-1})^{\cdot}$$

and the **unit element** *is* $\dot{e}$.

We will call the hypergroup H obtained by Theorem 8.1.26 a **hypergroup induced by a spherical projector**.

Remark. The support set O_e has a special property: O_e is a subgroup of G. In fact, $O_e = q^{-1}(\{\dot{e}\})$ and $\{\dot{e}\}$ is a subgroup of the hypergroup H. Therefore O_e is a subgroup of G.

Proposition 8.1.27 *Let H be a hypergroup induced by a spherical projector $P : C_c(G) \to C_c(G)$. The positive regular Borel measure $m \in M^\infty(H)$ defined by*

$$\int_H f(\dot{x})\, dm(\dot{x}) \;=\; \int_G f(x)\, d\lambda(x) \qquad \text{for all } f \in C_c(H)$$

is a (left-invariant) Haar measure on H.

Proof. By Proposition 8.1.24 and $(SH_1)(4)$ we get for $f \in C_c(H)$

$$\int_H L_{\dot{x}} f(\dot{y})\, dm(\dot{y}) \;=\; \int_H P(L_x f)(y)\, dm(\dot{y})$$

$$=\; \int_G L_x f(y)\, d\lambda(y) \;=\; \int_G f(y)\, d\lambda(y) \;=\; \int_H f(\dot{y})\, dm(\dot{y}).$$

◇

Let Δ be the modular function of G. Set

$$\Delta_P := \; (P(\Delta^{-1}))^\vee.$$

Concerning properties of the modular function on hypergroups we refer to Theorem 2.1.10 and the considerations before Theorem 2.2.16. For modular functions on groups we refer to [232,Ch.2.4]. Note that $P(\Delta^{-1})$ is well-defined since Δ is $]0,\infty[$-valued.

Proposition 8.1.28 *Let H be a hypergroup induced by a spherical projector $P : C_c(G) \to C_c(G)$. Then Δ_P is the modular function on H.*

Proof. At first note that by $(SH1)(2)$ we have for $f \in C_c(H)$

$$P(\check{f}) \cdot P(\Delta^{-1}) = P(\check{f} \cdot \Delta^{-1}).$$

Hence for the Haar measure m defined in Proposition 8.1.27 we get

$$\int_H f((\dot{x})^\vee) \cdot (\Delta_P)^\vee(\dot{x})\, dm(\dot{x}) = \int_H \check{f}(\dot{x}) \cdot \Delta_P((\dot{x})^\vee)\, dm(\dot{x})$$

$$= \int_H P(\check{f} \cdot \Delta^{-1})(\dot{x})\, dm(\dot{x}) = \int_G f(x^{-1})\, \Delta(x^{-1})\, d\lambda(x)$$

$$= \int_G f(x)\, d\lambda(x) = \int_H f(\dot{x})\, dm(\dot{x}).$$

◇

According to Proposition 2.2.16 the involution of every $f \in L^1(H) = L^1(H,m)$ is defined by

$$f^*(\dot{x}) = \Delta_P((\dot{x})^\vee)\, \overline{f((\dot{x})^\vee)}.$$

Note that the involution of G and H coincide for functions of $L^1(H)$.

Example 1(a): Set of double cosets

Let K be a compact subgroup (with normalized Haar measure) of a locally compact group G. Consider the map $P : C_c(G) \to C_c(G)$ defined by

$$P(f)(y) = \int_K \int_K f(uyv)\, du\, dv \qquad \text{for } f \in C_c(G),\ x \in G.$$

We have $O_y = KyK$ and $y \mapsto KyK$ is a continuous map from G into $\mathcal{C}(G)$. That the axioms (SH_1), (SH_2) and (SH_3) hold, is verified without difficulties. Moreover, the modular function Δ is P-radial. Thus $P : C_c(G) \to C_c(G)$ is an ultraspherical projector that induces a hypergroup structure on the set

$$G//K = \{O_y = KyK : y \in G\}$$

of double cosets. That $H = G//K$ is a hypergroup is shown already in [350,Theorem 8.2B], see also [96,Theorem 1.1.9].

Example 1(b): Set of orbits

Particularly interesting examples of double coset hypergroups are the orbit hypergroups which are defined as follows.
Let G be an arbitrary locally compact group and B a compact subgroup of all topological automorphisms of G, see [304,Section26]. The orbit hypergroup G_B is the space of all orbits,

$$B(x) = \{\beta(x) : \beta \in B\}, \quad x \in G,$$

endowed with the quotient topology, see [96,Theorem 1.1.11]. (The special case $G \in [FIA]_B^-$ is studied in subsection 8.1.2.) Now consider the semidirect product $G \rtimes B$, the group defined by the product rule

$$(x, \alpha) \cdot (y, \beta) \;=\; (x\,\alpha(y), \alpha\beta) \qquad \text{for } x, y \in G,\ \alpha, \beta \in B.$$

Then the map $B(x) \mapsto B(x, \beta)B$ is a topological isomorphism between G_B and the double coset hypergroup $(G \rtimes B)//B$, see [599].

So we see that the class of hypergroups induced by ultraspherical projectors contains all double coset hypergroups and orbit hypergroups. In [500] and [172] some topics of harmonic analysis for hypergroups induced by spherical or ultraspherical projectors are investigated.

Commutativity is an important property of hypergroups. The commutativity of double coset hypergroups is equivalent to the fact that the pair (G, K) is a **Gelfand pair**, see [96,Example 4.3.23]. This suggests to study commutativity of hypergroups induced by spherical projectors. Below we extend the analysis of Gelfand pairs (see e.g. [178,Ch.6]) to hypergroups induced by spherical projectors.

Proposition 8.1.29 (C.Berg) *Let $P : C_c(G) \to C_c(G)$ be an ultraspherical projector, and let H be the induced hypergroup. If H is commutative, then G is unimodular.*

Proof. We know that

$$\int_G f(x)\, d\lambda(x) \;=\; \int_G f(x^{-1})\, \Delta(x^{-1})\, d\lambda(x) \qquad \text{for all } f \in C_c(G),$$

and we have to show

$$\int_G f(x)\, d\lambda(x) \;=\; \int_G f(x^{-1})\, d\lambda(x)$$

whenever H is commutative. Since

$$\int_G f(x)\, d\lambda(x) \;=\; \int_G P(f)(x)\, d\lambda(x)$$

by $(SH_1)(4)$, and

$$\int_G f(x^{-1})\, \Delta(x^{-1})\, d\lambda(x) \;=\; \int_G P(\check{f} \cdot \Delta^{-1})(x)\, d\lambda(x)$$

$$= \int_G P(\check{f})(x)\, P(\Delta^{-1})(x)\, d\lambda(x) \;=\; \int_G P(\check{f})(x)\, \Delta(x^{-1})\, d\lambda(x)$$

by $(SH_1)(4)$, $(SH_1)(2)$, we can assume that f is P-radial. Choose a P-radial function g in $C_c(G)$ such that $g(x) = 1$ for each $x \in \operatorname{supp} f \cup (\operatorname{supp} f)^{-1}$. Since H is commutative, we get

$$\int_G f(x)\, d\lambda(x) \;=\; f * g(e) \;=\; g * f(e) \;=\; \int_G f(x^{-1})\, d\lambda(x).$$

◇

A sufficient condition for the commutativity of H is given next.

Theorem 8.1.30 *Let $P : C_c(G) \to C_c(G)$ be a spherical projector. Assume that there exists a continuous automorphism a on G, such that $a \circ a = id$ and $a(x) \in O_{x^{-1}}$ for each $x \in G$. Then the hypergroup H induced by P is commutative.*

Proof. For $f \in C_c(G)$ set $f^a(x) := f(a(x))$ for $x \in G$. The extension of the mapping $f \mapsto \int\limits_G f^a(y) d\lambda(y)$, $C_c(G) \to \mathbb{C}$ yields a positive Radon measure λ^a on G. Since

$$\int_G (L_x f)^a(y)\, d\lambda(y) \;=\; \int_G f(x\, a(y))\, d\lambda(y) \;=\; \int_G f^a(y)\, d\lambda(y),$$

the measure λ^a is left-invariant. Hence there exists $c > 0$ such that

$$\int_G f^a(y)\, d\lambda(y) \;=\; c \int_G f(\lambda)\, d\lambda(y).$$

Since $a \circ a = \text{id}$ we get $c = 1$. Hence the Haar measure λ and λ^a coincide, and it follows that $g^a * f^a = (g * f)^a$ for $f, g \in C_c(G)$. On the other hand we have $\check{g} * \check{f} = (f * g)^\vee$. Now let $f, g \in C_c(G)$ be P-radial. Hence f and g are constant on each support set O_x, see Corollary 8.1.23. Since $a(x) \in O_{x^{-1}}$, it follows $\check{f} = f^a$, $\check{g} = g^a$ and $(f * g)^\vee = (f * g)^a$. Therefore,

$$(f * g)^a \;=\; (f * g)^\vee \;=\; \check{g} * \check{f} \;=\; g^a * f^a \;=\; (g * f)^a.$$

Since a is an automorphism, $f * g = g * f$ follows, and thus H is commutative.

◇

Three dual spaces of commutative hypergroups are studied in section 3.1. For commutative hypergroups induced by a spherical projector P we have additional information depending on P.

Definition 3. Let $P : C_c(G) \to C_c(G)$ be a spherical projector. A P-radial function $\varphi \in C^b(G)$, $\varphi \neq 0$, is called P**-spherical function**, if

$$h_\varphi(f) := \int_G f(x)\, \overline{\varphi(x)}\, d\lambda(x), \qquad f \in C_c(G),$$

satisfies

$$h_\varphi(f * g) \;=\; h_\varphi(f)\, h_\varphi(g) \qquad \text{for all } P\text{-radial } f, g \in C_c(G).$$

Proposition 8.1.31 *Let $P : C_c(G) \to C_c(G)$ be a spherical projector and let $\varphi \in C^b(G)$ be a P-radial function, $\varphi \neq 0$. Then φ is a P-spherical function if and only if*

$$P(L_y\varphi)(x) \;=\; \varphi(x)\,\varphi(y)$$

for every $x, y \in G$. In particular, $\varphi(e) = 1$.

Proof. Let $f, g \in C_c(G)$. Then

$$\begin{aligned} h_\varphi(P(f) * P(g)) &= \int_G P(f)(y) \int_G P(g)(x)\,\varphi(y\,x)\,d\lambda(x)\,d\lambda(y) \\ &= \int_G g(x) \int_G P(f)(y)\,P(L_y\varphi)(x)\,d\lambda(y)\,d\lambda(x) \\ &= \int_G \int_G P(L_y\varphi)(x)\,f(y)\,g(x)\,d\lambda(y)\,d\lambda(x), \end{aligned}$$

where the last equation follows by Proposition 8.1.24(1). It follows

$$h_\varphi(P(f) * P(g)) \;-\; h_\varphi(P(f)) \cdot h_\varphi(P(g))$$

$$= \int_G \int_G (P(L_y\varphi)(x) - \varphi(x)\,\varphi(y))\,f(y)\,g(x)\,d\lambda(y)\,d\lambda(x).$$

This is valid for all $f, g \in C_c(G)$. Hence the statement is shown, and $\varphi(e) = 1$ follows, since $\varphi(x)\varphi(e) = P(\varphi)(x) = \varphi(x)$ for all $x \in G$. ◇

Remark. *Ab initio* we supposed that a P-spherical function φ is a bounded function, so that $P(\varphi)$ is well-defined. In addition we get that the sup-norm $\|\varphi\|_\infty$ is equal to 1, see section 3.1.

Theorem 8.1.32 *Let $P : C_c(G) \to C_c(G)$ be a spherical projector. Assume that the hypergroup H induced by P is commutative. Then each P-spherical function φ is an element $\mathcal{X}^b(H)$ and each $\alpha \in \mathcal{X}^b(H)$ is a P-spherical function.*

Proof. If φ is a P-spherical function, we extend the functional h_φ from $C_c(G)$ to $L^1(G, \lambda)$. Furthermore we get by Proposition 8.1.24(2) that

$$\int_H \varphi(\dot{z})\,d\omega(\dot{x}, \dot{y})(z) \;=\; P(L_x\varphi)(y)$$

(extending the equality to $C^b(H)$). Proposition 8.1.31 yields $\omega(\dot{x}, \dot{y})(\varphi) = \varphi(\dot{x})\varphi(\dot{y})$ for all $\dot{x}, \dot{y} \in H$, i.e. $\varphi \in \mathcal{X}^b(H)$. Conversely, if $\alpha \in \mathcal{X}^b(H)$ we get (after identification of α with a P-radial continuous function on G) that $P(L_x\alpha)(y) = \alpha(x)\alpha(y)$, i.e. α is a P-spherical function. ◇

Remark. If the P-spherical function φ satisfies $\varphi(x) = \overline{\varphi(x^{-1})}$, then $\varphi \in \hat{H}$.

2 Hypergroups and special functions

We will present examples of hypergroups which are associated with special functions. Sometimes these hypergroups are induced by groups in addition. Moreover, the special functions often are eigenfunctions of a differential-operator.

Example 1: Bessel-Kingman hypergroups

This class of hypergroups is already introduced in subsection 1.2.2. In Example 1 of subsection 8.1.2 we considered the orbit space G_B, where $G = \mathbb{R}^n$ and $B = SO(n)$, and we showed that G_B are Bessel-Kingman hypergroups of order $\alpha = \frac{n}{2} - 1$.

We start with collecting some informations about Bessel functions of the first kind, see for example [207]. The Bessel functions $J_\alpha(z)$ of the first kind are defined by the convergent series

$$J_\alpha(z) \; = \; \left(\frac{z}{2}\right)^\alpha \sum_{k=0}^{\infty} (-1)^k \, \frac{1}{k!\,\Gamma(k+\alpha+1)} \, \left(\frac{z}{2}\right)^{2k},$$

where $\alpha, z \in \mathbb{C}$. $J_\alpha(z)$ satisfies the differential equation

$$z^2 w''(z) \; + \; z w'(z) \; + \; (z^2 - \alpha^2)\, w(z) \; = \; 0.$$

Moreover, $z^{-\alpha} J_\alpha(z)$ is an entire function.

We will use the **normalized** version of $J_\alpha(z)$ (and assume that $\alpha > -\frac{1}{2}$), i.e.

$$j_\alpha(z) \; = \; \Gamma(\alpha+1)\, 2^\alpha z^{-\alpha} \, J_\alpha(z) \; = \; \sum_{k=0}^{\infty} \frac{(-1)^k\, \Gamma(\alpha+1)}{2^{2k} k!\, \Gamma(k+\alpha+1)} \, z^{2k},$$

with $z \in \mathbb{C}$. Fix $\alpha > -\frac{1}{2}$ and consider for $x, s \in \mathbb{R}_0^+ = [0, \infty[$

$$\varphi_x^\alpha(s) := \; j_\alpha(xs).$$

The definition of the hypergroup-convolution is based on a product formula of the function φ_x^α. In fact, by Sonine's product formula (see subsection 1.2.2) we obtain for all $s \in \mathbb{R}_0^+$

$$\varphi_x^\alpha(s)\, \varphi_y^\alpha(s) \; = \; \int_{\mathbb{R}_0^+} \varphi_z(s)\, d\omega_\alpha(x,y)(z), \qquad (P)$$

where the probability measure $\omega_\alpha(x, y)$ is explicitly given in the following theorem.

Theorem 8.2.1 *Let* $\alpha > -\frac{1}{2}$. *Then*

$$H_\alpha := \; ([0, \infty[\, , \omega_\alpha, \tilde{}\,, 0)$$

with 0 as unit element, the identity map as involution and

$$d\omega_\alpha(x,y)(z) = c_\alpha 2z\,\frac{[(x+y+z)(x+y-z)(x-y+z)(-x+y+z)]^{\alpha-\frac{1}{2}}}{(2xy)^{2\alpha}} \times \chi_{[|x-y|,x+y]}(z)$$

as convolution (with $c_\alpha = \frac{\Gamma(\alpha+1)}{\Gamma(\alpha+\frac{1}{2})\Gamma(\frac{1}{2})}$*), is a commutative hypergroup.*

Proof. We have to check that (F1),...,(F7) of Theorem 1.2.3 are satisfied. Obviously (F1), (F2) and (F3) are satisfied. (F4) is exactly the statement (d) of Lemma 1.1 in [617]. (F5) is the product formula

$$\varphi_x^\alpha(s)\,\varphi_y^\alpha(s) \;=\; \int_0^\infty \varphi_z(s)\,d\omega_\alpha(x,y)(z).$$

Note that $\omega_\alpha(x,y)$ is a probability measure. (F6) is true, since $0 \in \operatorname{supp}\omega_\alpha(x,y) = [|x-y|, x+y]$ exactly when $x=y$, and (F7) holds evidently by $\operatorname{supp}\omega_\alpha(x,y) = [|x-y|, x+y]$. ◇

H_α is called **Bessel-Kingman hypergroup of order** α.

Theorem 8.2.2 *Let* $\alpha > -\frac{1}{2}$. *For the Bessel-Kingman hypergroup* H_α *holds*

(1) A Haar measure on $H_\alpha = [0,\infty[$ *is given by*

$$dm_\alpha(x) \;=\; x^{2\alpha+1}\,dx.$$

(2) The dual space $\hat{H}_\alpha$ *is equal to* $\{\varphi_x^\alpha : x \in [0,\infty[\}$ *and is homeomorphic to* $H_\alpha = [0,\infty[$ *.*

(3) $\mathcal{S}(H_\alpha) = \hat{H}_\alpha = \mathcal{X}^b(H_\alpha)$, *and the Pontryagin duality is valid for* H_α.

Proof.

(1) One has to show that for $f \in C_c([0,\infty[)$ holds

$$\int_0^\infty f(y)\,y^{2\alpha+1}\,dy = \int_0^\infty L_x f(y)\,y^{2\alpha+1}\,dy = \int_0^\infty \omega_\alpha(x,y)(f)\,y^{2\alpha+1}\,dy.$$

Using the explicit form of ω_α, a direct calculation shows that this equality is true.

(2) Since $\varphi_x^\alpha(s) = j_\alpha(xs)$, a dual version of the product formula (P) is valid:

$$\varphi_x^\alpha(s)\,\varphi_x^\alpha(t) \;=\; \int_0^\infty \varphi_x^\alpha(u)\,d\omega_\alpha(s,t)(u) \qquad \text{for all } x \in \mathbb{R}_0^+.$$

Since φ_x^α are real-valued and bounded functions, it follows that $\{\varphi_x^\alpha : x \in [0,\infty[\,\} \subseteq \hat{H}_\alpha$. Theorem 4.4 of [617] yields the following fact. If

ψ is a non-trivial algebra homomorphism of $M(H_\alpha)$ onto the complex numbers, then there exists an element $y \in [0, \infty[$ such that

$$\psi(\mu) = \int_0^\infty \varphi_y^\alpha(x)\, d\mu(x).$$

By Theorem 6.3F of [350] we get that $\Delta(M(H_\alpha)) = \mathcal{X}^b(H_\alpha) \supseteq \hat{H}_\alpha$, and therefore $\{\varphi_x^\alpha : x \in [0, \infty[\,\} \supseteq \hat{H}_\alpha$. Finally, equipped with the topology of uniform convergence on compact subsets the set $\{\varphi_x^\alpha : x \in [0, \infty[\,\}$ is homeomorphic to $[0, \infty[$.

(3) Applying Theorem 3.4.6 we get by (1) that $\mathcal{S}(H_\alpha) = \hat{H}_\alpha = \mathcal{X}^b(H_\alpha)$. By (2) it follows that the Pontryagin duality is valid for H_α.

◇

Remark. Considering $\alpha \to -\frac{1}{2}$ leads to the **Chebyshev hypergroup on $\mathbb{R}_0^+$**. The special functions are $\varphi_x^{-\frac{1}{2}}(s) = \cos(xs)$ and the convolution $\omega_{-\frac{1}{2}}$ is given by

$$\omega_{-\frac{1}{2}}(x, y) = \frac{1}{2}\, \epsilon_{|x-y|} + \frac{1}{2}\, \epsilon_{x+y},$$

see subsection 1.2.2. The statements of Theorem 8.2.2 are valid for $\alpha = -\frac{1}{2}$, too.

Example 2: Naimark hypergroup

In [501, Ch.IV, §20] there is studied $L^1([0, \infty[\,, \sinh(2x)dx)$, furnished with a multiplication such that $L^1([0, \infty[\,, \sinh(2x)dx)$ is a commutative Banach algebra. (Note that Naimark adjoins an identity to this L^1-algebra.) Jewett showed in [350,9.5] that this multiplication is based on a hypergroup convolution, also see [96,Example 2.2.49].

Let $x, y \in \mathbb{R}_0^+ = [0, \infty[$ and let $s \in \mathbb{C},\ s \neq 0$. The product formula

$$2\, \frac{\sin(sx)}{s} \cdot \frac{\sin(sy)}{s} = \int_{|x-y|}^{x+y} \frac{\sin(su)}{s}\, du \tag{1}$$

follows by an elementary calculation, see [501,p.280].

When $s = i$ we get

$$2\, (\sinh(x))\, (\sinh(y)) = \int_{|x-y|}^{x+y} \sinh(u)\, du. \tag{2}$$

Define for $s \in \mathbb{C}$ and $x \in \mathbb{R}_0^+$

$$\varphi_x(s) = \left(\sum_{k=0}^{\infty} s^k\, \frac{(-1)^k}{(2k+1)!}\, x^{2k} \right) \Big/ \left(\sum_{k=0}^{\infty} \frac{1}{(2k+1)!}\, x^{2k} \right),$$

and let $t \in \mathbb{C}$ such that $s = t^2$. If $x > 0$ and $s \neq 0$ then

$$\varphi_x(s) \;=\; \frac{\sin(tx)}{t \cdot \sinh(x)}.$$

Moreover, $\varphi_x(-1) = 1$ for each $x \in \mathbb{R}_0^+$, and $\varphi_x(0) = \frac{x}{\sinh(x)}$ for $x > 0$. For each $s \in \mathbb{C}$ the function $x \mapsto \varphi_x(s)$, $\mathbb{R}_0^+ \to \mathbb{C}$ is continuous, and $\varphi_0(s) = 1$ for all $s \in \mathbb{C}$.

Formulae (1) and (2) yield for $x, y > 0$ the following product formula (P)

$$\varphi_x(s)\, \varphi_y(s) \;=\; \int_{\mathbb{R}_0^+} \varphi_z(s)\, d\omega(x,y)(z),$$

where the probability measure $\omega(x, y)$ is explicitly given in the theorem below.

Theorem 8.2.3 $K = ([0,\infty[\,,\omega,\tilde{\ },0)$ *is a commutative hypergroup with 0 as unit element, the identity map as involution and*

$$\omega(x,y) \;=\; \frac{1}{2\,(\sinh(x))(\sinh(y))} \int_{|x-y|}^{x+y} (\sinh(u))\, \epsilon_u \, du \qquad (\textit{ for } x, y > 0)$$

as convolution.

Proof. In order to apply Theorem 1.2.3 we have to restrict ourselves to those $s \in \mathbb{C}$ such that $x \mapsto \varphi_x(s)$ is a bounded function on $[0,\infty[$. By Naimark's proof this is exactly the case when $|\operatorname{Im} t| \leq 1$, where $t^2 = s$. Furthermore, the description of the positive functionals on the L^1-algebra of Naimark (see [501,p.281]) yields that (F4) of Theorem 1.2.3 is satisfied. Now it is straightforward to show that conditions (F1),...,(F7) of Theorem 1.2.3 are satisfied. (Note that $\omega(x, y)$ is defined by integrating a measure-valued function.) $\diamond$

The hypergroup K of Theorem 8.2.3 is called **Naimark hypergroup**. The three dual spaces of K are different. We write $\alpha_s(x) := \varphi_x(s)$. Since $x \mapsto \varphi_x(s)$, $\mathbb{R}_0^+ \to \mathbb{C}$ is bounded if and only if $|\operatorname{Im} t| \leq 1$ where $t^2 = s$, it follows immediately that

$$\mathcal{X}^b(K) \;=\; \{\alpha_s : s = c + id,\ c \geq \frac{d^2}{4} - 1\},$$

$$\hat{K} \;=\; \{\alpha_s : s \in [-1,\infty[\,\}.$$

For the detailed proof of the following statements we refer to [350,9.5]. We will identify $\hat{K}$ with $[-1,\infty[$ and $\mathcal{X}^b(K)$ with $\{s \in \mathbb{C} : s = c + id,\ c \geq \frac{d^2}{4} - 1\}$.

Theorem 8.2.4 *For the Naimark hypergroup K holds*

(1) A Haar measure on $K = [0, \infty[$ is given by $dm(x) = (\sinh(x))^2 dx$.
(2) The Plancherel measure π_K on $[-1, \infty[$ associated with m is given by

$$\int_{-1}^{\infty} h(u)\, d\pi_K(u) \;=\; \frac{1}{\pi} \int_0^{\infty} h(u)\, \sqrt{u}\, du$$

for all $h \in C_c([-1, \infty[)$. In particular, $\mathcal{S}(K)$ is homeomorphic to $[0, \infty[$ and

$$\mathcal{S}(K) \subsetneq \hat{K} \subsetneq \mathcal{X}^b(K).$$

Remark:

(1) The Naimark hypergroup is isomorphic to the double coset hypergroup $SL(2, \mathbb{C})//SU(2)$, where $SL(2, \mathbb{C})$ is the group of all 2×2-matrices with complex entries and determinant equal to 1, and $SU(2)$ is the subgroup of unitary matrices in $SL(2, \mathbb{C})$, see [350,15.2].
(2) With a different parametrization of the dual spaces of K one gets another representation of K (as a member of the class of Sturm-Liouville hypergroups), see [96, 3.5.66].

Example 3: Dual Jacobi hypergroups

The hypergroups of Example 1 and 2 are defined on $[0, \infty[$. Examples defined on the compact set $K = [-1, 1]$ are already presented in subsection 1.2.2, the dual ultraspherical hypergroups. In subsection 4.15.1 this result is extended for dual Jacobi hypergroups. Based on [254,Theorem 1] a dual hypergroup convolution can be derived on $K = [-1, 1]$ for Jacobi polynomials $R_n^{(\alpha,\beta)}(x)$, provided $(\alpha, \beta) \in J$ and $\beta \geq -\frac{1}{2}$ or $\alpha+\beta \geq 0$. The explicit form of $\omega(x, y)$ is given in subsection 4.15.1.

Note that for these dual Jacobi hypergroups $H_{(\alpha,\beta)}$ we have $(H_{(\alpha,\beta)})^\wedge = \mathbb{N}_0$, and the Pontryagin duality is valid for $H_{(\alpha,\beta)}$.

The dual ultraspherical hypergroups $H_{(\alpha,\alpha)}$ are isomorphic to the double coset hypergroups $SO(d)//SO(d-1)$ for $d \geq 3$ and $\alpha = (d-3)/2$. Here we identify $SO(d-1)$ with the subgroup of all elements of $SO(d)$ which fix the vector $(1, 0, ..., 0)^t \in \mathbb{R}^d$. The bijection between $[-1, 1]$ and $SO(d)//SO(d-1)$ is given by

$$[-1, 1] \to SO(d)//SO(d-1), \quad x \mapsto SO(d-1)\, a(x)\, SO(d-1)$$

where

$$a(x) \;=\; \begin{pmatrix} \begin{pmatrix} x & -\sqrt{1-x^2} \\ \sqrt{1-x^2} & x \end{pmatrix} & 0_{2,d-2} \\ 0_{d-2,2} & 1_{d-2,d-2} \end{pmatrix}$$

and $0_{n,m}$ and $1_{n,m}$ denote, respectively, the zero and identity matrices of size $n \times m$, see e.g. [172]. In particular, the Pontryagin duality is valid for $SO(d)//SO(d-1)$.

Examples 1, 2 and 3 are one-dimensional hypergroups. In [760] elementary facts about one-dimensional hypergroups are proved. For general results on polynomial hypergroups in several variables we refer to [765] and [96,Ch.3.1].

Now we discuss an example of a two-dimensional hypergroup based on orthogonal polynomials defined on the unit disc. These polynomials are based on Jacobi polynomials $R_n^{(\alpha,\beta)}(x)$ and are very important in applications, i.e. in geometrical optics see [754].

Example 4. Disc polynomial hypergroups

For $\alpha > -1$ and $m, n \in \mathbb{N}_0$ define the (m,n)-th disc polynomial for $z = x + iy \in \mathbb{C}$

$$R_{m,n}^{\alpha}(z,\bar{z}) := \begin{cases} R_n^{(\alpha,m-n)}(2z\bar{z}-1)\; z^{m-n} & \text{for } m \geq n \\ R_m^{(\alpha,n-m)}(2z\bar{z}-1)\; \bar{z}^{n-m} & \text{for } m < n. \end{cases}$$

The following properties hold obviously,

$$\overline{R_{m,n}^{\alpha}(z,\bar{z})} \;=\; R_{m,n}^{\alpha}(\bar{z},z) \;=\; R_{n,m}^{\alpha}(z,\bar{z}) \qquad (i)$$

$$R_{m,n}^{\alpha}(re^{i\varphi}, re^{-i\varphi}) \;=\; e^{i(m-n)\varphi}\, R_{m,n}^{\alpha}(r,r) \qquad (ii)$$

$$R_{m,n}^{\alpha}(0,0) \;=\; \frac{(-1)^m m!\alpha!}{(m+\alpha)!}\,\delta_{m,n} \quad \text{and } R_{m,n}^{\alpha}(1,1) \;=\; 1. \qquad (iii)$$

The first equality of (iii) follows directly from (4.1.4) in [641]. By orthogonalization of the sequence $1, z, \bar{z}, z^2, z\bar{z}, \bar{z}^2, \ldots$ on the unit disc $D := \{z \in \mathbb{C} : \; |z| \leq 1\}$ with respect to the measure

$$d\pi_\alpha(z) := \frac{\alpha+1}{\pi}\,(1-z\bar{z})^\alpha\, dz \qquad \text{for } z \in D$$

we get the family $(R_{m,n}^{\alpha}(z,\bar{z}))_{m,n \in \mathbb{N}_0^2}$.

Till now we have used the notation $R_{m,n}^{\alpha}(z,\bar{z})$ to emphasize that polynomials of the kind

$$\sum_{m,n} b_{m,n}\, z^m\, \bar{z}^n$$

are studied. To simplify notation we will write $Q_{m,n}^{\alpha}(z) := R_{m,n}^{\alpha}(z,\bar{z})$. The measure π_α is a probability measure on D and can be written as

$$d\lambda_\alpha(x,y) \;=\; \frac{\alpha+1}{\pi}\,(1-x^2-y^2)^\alpha\, dx\, dy,$$

where $z = x + iy \in D$. The orthogonality relations are given by

$$\frac{\alpha+1}{\pi} \int_D Q^\alpha_{m,n}(z)\, Q^\alpha_{k,l}(\bar{z})\, (1 - z\bar{z})^\alpha\, dz \;=\; \frac{1}{h^\alpha_{m,n}}\, \delta_{m,k}\, \delta_{n,l}$$

with

$$h^\alpha_{m,n} \;=\; \frac{(m+n+\alpha+1)(\alpha+1)_m(\alpha+1)_n}{(\alpha+1)\; m!\; n!}\,. \qquad (iv)$$

The polynomial of degree zero is $Q^\alpha_{0,0}(z) = 1$. The polynomials of degree one are $Q^\alpha_{1,0}(z) = z$ and $Q^\alpha_{0,1}(z) = \bar{z}$.

The basic recurrence relations are

$$z\, Q^\alpha_{m,n}(z) \;=\; \frac{\alpha+m+1}{\alpha+m+n+1}\, Q^\alpha_{m+1,n}(z) \;+\; \frac{n}{\alpha+m+n+1}\, Q^\alpha_{m,n-1}(z)$$

for $m \in \mathbb{N}_0,\; n \in \mathbb{N}$, and

$$\bar{z}\, Q^\alpha_{m,n}(z) \;=\; \frac{\alpha+n+1}{\alpha+m+n+1}\, Q^\alpha_{m,n+1}(z) \;+\; \frac{m}{\alpha+m+n+1}\, Q^\alpha_{m-1,n}(z)$$

for $m \in \mathbb{N},\; n \in \mathbb{N}_0$.

Now let $\alpha \geq 0$. Our goal is to show that $(Q_{m,n}(z))_{m,n\in\mathbb{N}_0^2}$ generates a polynomial hypergroup on $K_\alpha = \mathbb{N}_0 \times \mathbb{N}_0$. The main tool is a linearization formula shown by Koornwinder. In fact, using Corollary 5.2 of [381] and $\overline{Q^\alpha_{m,n}} = Q^\alpha_{n,m}$ yields

$$\begin{aligned} &Q^\alpha_{m_1,n_1}(z)\, Q^\alpha_{m_2,n_2}(z) \\ &\quad = \sum_{(m_3,n_3)} g((m_1,n_1),(m_2,n_2);(m_3,n_3))\, Q_{n_3,m_3}(z) \qquad (A1) \end{aligned}$$

with $g((m_1,n_1),(m_2,n_2);(m_3,n_3)) \geq 0$. The pairs (m_3,n_3) in the sum satisfy

$$m_1 + m_2 + m_3 \;=\; n_1 + n_2 + n_3 \qquad \text{and}$$

$$|m_1 + n_1 - m_2 - n_2| \;\leq\; m_3 + n_3 \;\leq\; m_1 + n_1 + m_2 + n_2.$$

Hence we define

$$\begin{aligned} &\omega_\alpha((m_1,n_1),(m_2,n_2)) \\ &\quad := \sum_{(m_3,n_3)} g((m_1,n_1),(m_2,n_2);(m_3,n_3))\, \epsilon_{(n_3,m_3)}. \qquad (v) \end{aligned}$$

Theorem 8.2.5 *Let $\alpha \geq 0$. $K_\alpha = (\mathbb{N}_0 \times \mathbb{N}_0, \omega_\alpha, \tilde{\ }, (0,0))$ is a commutative hypergroup with $(0,0)$ as unit element, $(m,n)^\sim = (n,m)$ as involution, and $\omega_\alpha((m_1,n_1),(m_2,n_2))$ as convolution.*

Proof. The linearization formula (A1) (and $Q^\alpha_{n,m}(1) = 1$) yields $\omega_\alpha((m_1,n_1),(m_2,n_2)) \in M^1_c(\mathbb{N}_0 \times \mathbb{N}_0)$. Moreover, $(0,0) \in \operatorname{supp}\omega_\alpha((m_1,n_1),(m_2,n_2))$ if and only if $(m_1,n_1) = (n_2,m_2)$. The associativity law is shown as in Theorem 1.2.2. Therefore (H,ω_α) is satisfied. The properties of $(H,\tilde{\ })$ and $(H,(0,0))$ hold obviously. ◇

K_α is called **disc polynomial hypergroup of order** α. Given $z \in \mathbb{C}$ denote

$$\varphi^\alpha_z : \mathbb{N}_0 \times \mathbb{N}_0 \to \mathbb{C}, \qquad \varphi^\alpha_z((m,n)) := Q^\alpha_{m,n}(z).$$

Theorem 8.2.6 *Let $\alpha \geq 0$. For the disc polynomial hypergroup K_α hold*

(1) A Haar measure on K_α is given by the weights $h^\alpha_{m,n}$ of (iv).
(2) The dual space $\hat{K}_\alpha$ is equal to $\{\varphi^\alpha_z : z \in D\}$ and is homeomorphic to D.

Proof.

(1) By Theorem 2.1.1 we know that a Haar measure h^α on $\mathbb{N}_0 \times \mathbb{N}_0$ with $h^\alpha((0,0)) = 1$ has the weight $\omega_\alpha((m,n),(n,m);(0,0))^{-1}$ on the point (m,n). Since

$$\omega_\alpha((m,n),(n,m);(0,0))^{-1} = \int_D Q^\alpha_{m,n}(z)\, Q^\alpha_{m,n}(\bar{z})\, d\pi_\alpha(z)$$

the weight $h^\alpha((m,n))$ is exactly $h^\alpha_{m,n}$ of equality (iv).

(2) We have only to show that $|z| > 1$ implies $|\varphi^\alpha_z((m,n))| > 1$ for some $(m,n) \in \mathbb{N}_0 \times \mathbb{N}_0$. For $r = |z| > 1$ we get by (ii)

$$\varphi^\alpha_z((1,1)) = Q^\alpha_{1,1}(z) = R^\alpha_{1,1}(re^{i\varphi}, re^{-i\varphi}) = R^\alpha_{1,1}(r,r) = R^{(\alpha,0)}_1(r).$$

Since the Jacobi polynomial $R^{(\alpha,0)}_1(x)$ is strictly greater that 1 for each $x \in]1,\infty[$ the statement follows.

◇

Identifying $\hat{K}_\alpha$ with $D = \{z \in \mathbb{C} : |z| \leq 1\}$ equipped with the orthogonalization measure π_α one can employ the following product formula.

Let $\alpha > 0$ and $v, w \in D$. Then

$$Q^\alpha_{m,n}(v)\, Q^\alpha_{m,n}(w) = \int_D Q^\alpha_{m,n}(z)\, \kappa^\alpha_{v,w}(z)\, d\pi_\alpha(z) \qquad (A2)$$

where

$$\kappa^\alpha_{v,w}(z) = \begin{cases} \dfrac{\alpha}{\alpha+1} \dfrac{(1-|v|^2-|w|^2-|z|^2+2\,\mathrm{Re}\,(vw\bar{z}))^{\alpha-1}}{(1-|v|^2)^\alpha\,(1-|w|^2)^\alpha\,(1-|z|^2)^\alpha} \\ 0 \end{cases}$$

with the first value assigned exactly when z is in the disc with center vw and radius $\sqrt{(1-|v|^2)(1-|w|^2)}$, see [361,p.107(6)] or [160] or [94]. The product formula (A2) follows from the representation

$$Q_{m,n}^{\alpha}(re^{i\varphi}) \;=\; r^{|m-n|}\, e^{i(m-n)\varphi}\, R_{m\wedge n}^{(\alpha,|m-n|)}(2r^2-1),$$

where $m \wedge n = \min\{m,n\}$ and the **dual** product formula of the Jacobi polynomials $R_k^{(\alpha,l)}(x)$, $l \in \mathbb{N}$, $x \in [-1,1]$. We refer to subsection 4.15.1.

Theorem 8.2.7 *Let $\alpha > 0$.*

(1) The dual space $\hat{K}_\alpha = D$ of the disc polynomial hypergroup K_α is a commutative hypergroup. In fact, $D_\alpha := (D, \hat{\omega}_\alpha, \tilde{}, 1)$ is a commutative hypergroup with $1 \in D$ as unit element, $\tilde{z} = \bar{z}$ and

$$d\hat{\omega}_\alpha(v,w)(z) \;=\; \kappa_{v,w}^{\alpha}(z)\, d\pi_\alpha(z) \qquad \text{for } |v| < 1 \text{ and } |w| < 1$$

and

$$\hat{\omega}_\alpha(v,w) \;=\; \epsilon_{vw} \qquad \text{for } |v| = 1 \text{ or } |w| = 1.$$

(2) A Haar measure on D_α is π_α.
(3) The Pontryagin duality is valid for K_α.

Proof.

(1) This statement is formulated in several publications, e.g. [160] or [161] or [96,p.142] or [99]. We can use Theorem 1.2.3 to show that D_α is a hypergroup. Consider

$$\varphi_z^{\alpha} : \mathbb{N}_0 \times \mathbb{N}_0 \to \mathbb{C}, \quad \varphi_z^{\alpha}((m,n)) \;=\; Q_{m,n}^{\alpha}(z).$$

Condition (F1) of Theorem 1.2.3 is fulfilled with $e = 1 \in D$ and $s_0 = (0,0) \in \mathbb{N}_0 \times \mathbb{N}_0$. Obviously (F2) and (F3) are satisfied. Applying Theorem 3.3.3 for the hypergroup K_α we get that (F4) is true. (F5) is the product formula (A2).
In order to check (F6), note that in case of $|v| < 1$ and $|w| < 1$ the special form of $\kappa_{v,w}^{\alpha}$ yields that $1 \in \operatorname{supp} \hat{\omega}_\alpha(v,\bar{w})$ if and only if $v = w$. If $|v| = 1$ or $|w| = 1$, condition (F6) obviously is satisfied. In order to check (F7), we can use the remark in section 1.1 concerning the Michael topology in case of K being a metric space.

(2) The orthogonalization measure π_α is the Plancherel measure corresponding to K_α, compare with Theorem 4.1.3. Theorem 3.4.11 yields that π_α is a Haar measure on D_α.

(3) The Pontryagin duality of K_α is valid by Proposition 3.4.13.

◇

Finally we give a short survey of one-dimensional hypergroups defined on $[0,\infty[$, which are generated by eigenfunctions of a Sturm-Liouville problem. A general reference for Sturm-Liouville hypergroups is [96,Ch.3.5] containing many examples. We also refer to [760], [766], [235] and [647, Ch.4].

A **Sturm-Liouville function** is a mapping $A : [0,\infty[\to \mathbb{R}$ such that $A \in C([0,\infty[)$, $A(x) > 0$ for all $x > 0$ and $A \in C^1(]0,\infty[)$. (i.e. A is a continuously differentiable function on $]0,\infty[$.) Given a Sturm-Liouville function A the **Sturm-Liouville differential operator** L_A is defined by

$$L_A f := -\frac{1}{A}\,(A \cdot f')' = -f'' - \frac{A'}{A}\,f' \qquad \text{for } f \in C^2(\,]0,\infty[\,).$$

Considering one-dimensional hypergroups K defined on $[0,\infty[$ we know from [760, Proposition 2.3 and Corollary 2.4]

(i) the unit element e is equal to 0.
(ii) the involution $\tilde{}$ is equal to the identity mapping.
(iii) $\omega(x,y) = \omega(y,x)$ for all $x, y \in K$, i.e. K is commutative.

Moreover, by [760, Theorem 3.7] it is no restriction to suppose that K is **normalized**, which means that $\min\operatorname{supp}\omega(x,y) = |x-y|$ and $\max\operatorname{supp}\omega(x,y) = |x+y|$ for all $x, y \in K$.

Definition. A normalized hypergroup on $K = [0,\infty[$ is called a **Sturm-Liouville hypergroup** if there exists a Sturm-Liouville function $A : [0,\infty[\to \mathbb{R}$ such that given any real-valued C^∞-function f on $[0,\infty[$ the function $u_f \in C([0,\infty[\times [0,\infty[)$ defined by

$$u_f(x,y) = \int_0^\infty f(z)\, d\omega(x,y)(z)$$

is for all $x, y \in [0,\infty[$ twice continuously differentiable, and u_f is a solution of the Cauchy problem,

$$\partial_1^2 u(x,y) + \frac{A'(x)}{A(x)}\,\partial_1 u(x,y) = \partial_2^2 u(x,y) + \frac{A'(y)}{A(y)}\,\partial_2 u(x,y)$$

$$\text{and} \qquad \partial_2 u(x,0) = 0$$

for all $x, y > 0$.

A Sturm-Liouville hypergroup with defining Sturm-Liouville function A is denoted by $([0,\infty[\,,\omega(A))$.

For a Sturm-Liouville hypergroup the conditions (i), (ii) and (iii) yield that $u_f(x,0) = f(x) = u_f(0,x)$ and $\partial_1 u_f(0,x) = 0$ for all $x > 0$.

If the following two assumptions (SL1) **and** (SL2) for A are satisfied, then a Sturm-Liouville hypergroup $([0,\infty[\,,\omega(A))$ exists. (SL1) and (SL2) are given by

(SL1) **Either** (SL1a) **or** (SL1b) are satisfied, where

(SL1a) (Singularity at 0) In a neighbourhood U of 0 we have

$$\frac{A'(x)}{A(x)} = \frac{\alpha_0}{x} + \alpha_1(x) \qquad \text{for all } x \in U,\ x > 0,$$

where $\alpha_0 > 0$ and $\alpha_1 \in C^\infty(\mathbb{R})$ with $\alpha_1(-x) = -\alpha_1(x)$ for all $x \in \mathbb{R}$ (which implies that $A(0) = 0$).

(SL1b) (Regularity at 0) We have $\alpha_0 = 0$ and $\alpha_1 \in C^1([0,\infty[\,)$.

(SL2) There exists $\beta \in C^1([0,\infty[\,)$ such that $\beta(0) \geq 0$, $\frac{A'}{A} - \beta$ is nonnegative and decreasing on $]0,\infty[$ and

$$q := \frac{1}{2}\,\beta' - \frac{1}{4}\,\beta^2 + \frac{A'}{2A}\,\beta$$

is decreasing on $]0,\infty[\,$.

For the proof of the following theorem we refer to [766, Theorem 3.11] or [96, Theorem 3.5.45].

Theorem 8.2.8 *Let A be a Sturm-Liouville function satisfying (SL1) and (SL2). Then*

(1) $K = ([0,\infty[\,,\omega(A))$ is a Sturm-Liouville hypergroup,
(2) A Haar measure m on $([0,\infty[\,,\omega(A))$ is given by

$$dm(x) = A(x)\,dx.$$

In order to determine the dual spaces $\mathcal{X}^b(K)$ and $\hat{K}$ of $K = ([0,\infty[\,,\omega(A))$ we have to characterize solutions (eigenfunctions) of the following Sturm-Liouville boundary problem:

$$L_A\varphi(x) = -\lambda\,\varphi(x), \qquad \varphi(0) = 1, \quad \varphi'(0) = 0 \qquad (*)$$

for $x > 0$ with $\lambda \in \mathbb{C}$. A straightforward calculation yields

Proposition 8.2.9 *Let $K = ([0,\infty[\,,\omega(A))$ be a Sturm-Liouville hypergroup. Then a continuous function $\varphi : [0,\infty[\to \mathbb{C}$ satisfies*

$$\int_0^\infty \varphi(z)\,d\omega(x,y)(z) = \varphi(x)\,\varphi(y)$$

for all $x,y \in [0,\infty[$ if and only if $\varphi :]0,\infty[\to \mathbb{C}$ is a C^2-function and is a solution of $(*)$.

We refer to [96, 3.5.23] or [647, Theorem 4.2].

Corollary 8.2.10 *Let* $K = ([0,\infty[\,,\omega(A))$ *be a Sturm-Liouville hypergroup. Then* $\mathcal{X}^b(K)$ *is exactly the set of all* **bounded** *continuous complex-valued functions* $\varphi : [0,\infty[\,\to \mathbb{C}$ *such that* $\varphi|\,]0,\infty[\, \in C^2(\,]0,\infty[\,)$ *satisfying* $(*)$ *with* $\lambda \in \mathbb{C}$. *Moreover,* $\varphi \in \hat{K}$ *if and only if* $\varphi \in \mathcal{X}^b(K)$ *is real-valued, which is exactly the case if* $\lambda \in \mathbb{R}$.

For a subclass of the Sturm-Liouville hypergroups we get more information concerning the dual spaces $\mathcal{S}(K)$, $\hat{K}$ and $\mathcal{X}^b(K)$. Let A be a Sturm-Liouville function such that (SL1a) is satisfied and $\frac{A'}{A} \geq 0$ is decreasing and A is increasing with $\lim\limits_{x\to\infty} A(x) = \infty$. Then A is called a **Chebli-Trimèche function**. Note that in this case condition (SL2) is fulfilled with $\beta = 0$. By Theorem 8.2.8 there exists a Sturm-Liouville hypergroup $K = ([0,\infty[\,,\omega(A))$ with defining Chebli-Trimèche function A. These hypergroups are called **Chebli-Trimèche hypergroups**. By [96, Theorem 3.5.54] we can derive the following properties of the dual spaces.

Theorem 8.2.11 *Let* $K = ([0,\infty[\,,\omega(A))$ *be a Chebli-Trimèche hypergroup with* $A \in C^2([0,\infty[\,)$. *Then*

(1) $\mathcal{S}(K)$ *is homeomorphic to* $[0,\infty[$,
(2) K *is of subexponential growth if and only if* $\lim\limits_{x\to\infty} \frac{A'(x)}{A(x)} = 0$. *Therefore,* $\lim\limits_{x\to\infty} \frac{A'(x)}{A(x)} = 0$ *implies that* $\mathcal{S}(K) = \hat{K} = \mathcal{X}^b(K)$ *is homeomorphic to* $[0,\infty[$.

Finally we mention that the Bessel-Kingman hypergroups H_α, $\alpha > -\frac{1}{2}$, belong to the class of Sturm-Liouville hypergroups.

Bibliography

[1] N.I.Achieser and I.M.Glasmann: *Theorie der linearen Operatoren im Hilbert-Raum.* Akademie-Verlag, Berlin, 1975.

[2] A.Achour and K.Triméche: Opérateurs de translation généralisée associés à un opérateur différentiel singulier sur un intervalle borné. C.R. Acad. Sci. A288, 399-402 (1979)

[3] F.Albiac and N.J.Kalton: *Topics in Banach Space Theory.* Springer Graduate Texts in Math. 233, 2016.

[4] H.A.Ali and F.M.Kandil: Convergence of exponential convex functions on hypergroups. Far East J. Math. Sci. 43, 225-234 (2010)

[5] M.Amini: Fourier transform of unbounded measures on hypergroups. Boll. Unione Mat. 10, 819-828 (2007)

[6] M.Amini: Harmonic functions on [IN] and central hypergroups. Monatsh. Math. 169, 267-284 (2013)

[7] M.Amini and C.H.Chu: Harmonic functions on hypergroups. J. Funct. Analysis 261, 1835-1864 (2011)

[8] M.Amini and A.Medghalchi: Fourier algebras on tensor hypergroups. Contemporary Mathematics 363, 1-14 (2004)

[9] M.Amini and A.R.Medghalchi: Amenability of compact hypergroup algebras. Math. Nachr. 287, 1609-1617 (2014)

[10] H.Annabi and K.Trimèche: Convolution généralisée sur le disque unité. C. R. Acad. Sci. A278, 21-24 (1974)

[11] J.P.Arnaud: Fonctions spheriques et fonctions definies positive sur l"arbre homogene. C. R. Acad. Sci. A290, 99-101 (1980)

[12] J.P.Arnaud: Stationary processes indexed by a homogeneous tree. Ann. of Probability 22, 195-218 (1994)

[13] J.P.Arnaud, J.L.Dunau and G.Letac: Arbres homogènes et couple de Gelfand. Publication 02, Laboratory of Statistics and Probability, Univ. Paul Sabatier, Toulouse, 1983.

[14] R.Askey: Orthogonal expansions with positive coefficients. Proc. Amer. Math. Soc. 16, 1191-1194 (1965)

[15] R.Askey: Orthogonal polynomals and positivity. In: *Studies in Applied Mathematics 6, Special Functions and Wave Propagation.* Ed. D.Ludwig et

al., SIAM, Philadelphia, 64-85, 1970.

[16] R.Askey: Linearization of the product of orthogonal polynomials. In: *Problems in Analysis.* Ed. R.Gunning, Princeton University Press, Princeton, 131-138 (1970)

[17] R.Askey: Orthogonal expansions with positive coefficients II. SIAM J. Math. Anal. 2, 340-346 (1971)

[18] R.Askey: *Orthogonal Polynomials and Special Functions.* SIAM, Philadelphia, 1975.

[19] R.Askey: An integral of products of Legendre functions and a Clebsch-Gordan sum. Letters in Math. Physics 6, 299-302 (1982)

[20] R.Askey: Orthogonal polynomials old and new, and some combinatorial connections. In: *Enumeration and Design.* Ed. D.M.Jackson and S.A.Vanstone, Academic Press, New York, 67-84, 1984.

[21] R.Askey and N.H.Bingham: Gaussian processes on compact symmetric spaces. Z. Wahrscheinlichkeitstheorie verw. Gebiete 37, 127-143 (1976)

[22] R.Askey and J.Fitch: Integral representations for Jacobi polynomials and some applications. J. Math. Anal. Appl. 26, 411-437 (1969)

[23] R.Askey and G.Gasper: Linearization of the product of Jacobi polynomials, III. Can. J. Math. 23, 332-338 (1971)

[24] R.Askey and G.Gasper: Jacobi polynomial expansions of Jacobi polynomials with non-negative coefficients. Proc. Camb. Phil. Soc. 70, 243-255 (1971)

[25] R.Askey and G.Gasper: Positive Jacobi polynomial sums, II. American Journal of Math. 98, 709-737 (1976)

[26] R.Askey and G.Gasper: Convolution structure for Laguerre polynomials. J. D'Analyse Math. 31, 48-68 (1977)

[27] R.Askey, G.Gasper and M.E.H.Ismail: A positive sum from summability theory. Journal of Appr. Theory 13, 413-420 (1975)

[28] R.Askey and M.E.H.Ismail: The Rogers q-ultraspherical polynomials. In: *Approximation Theory III.* Ed. E.Cheney, Academic Press, New York, 175-182, 1983.

[29] R.Askey and M.E.H.Ismail: A generalization of ultraspherical polynomials. In: *Studies in Pure Mathematics.* Ed. P.Erdös, Birkhäuser, Basel, 55-78, 1983.

[30] R.Askey and M.E.H.Ismail: Recurrence relations, continued fractions and orthogonal polynomials. Mem. Amer. Math. Soc. vol. 300, 1984.

[31] R.Askey and H.Pollard: Some absolutely monotonic and completely monotonic functions. SIAM J. Math. Anal. 5, 58-63 (1974)

[32] R.Askey and St.Wainger: On the behaviour of special classes of ultraspherical expansions I. Journal d'Analyse Math. 15, 193-220 (1965)

[33] R.Askey and St.Wainger: A transplantation theorem for ultraspherical coefficients. Pacific J. of Math. 16, 393-405 (1966)

[34] R.Askey and St.Wainger: A transplantation theorem between ultraspherical series. Illinois J. Math. 10, 322-344 (1966)

[35] R.Askey and St.Wainger: A dual convolution structure for Jacobi polynomials. In: *Orthogonal Expansions and their Continuous Analogues.* South-

ern Illinois University Press, Edwardsville, 25-36, 1968.

[36] R.Askey and St.Wainger: A convolution structure for Jacobi series. Amer. J. Math. 91, 463-483 (1969)

[37] R.Askey and J.Wilson: Some basic hypergeometric orthogonal polynomials that generalize Jacobi polynomials. Mem. Amer. Math. Soc. 319 (1985)

[38] R.A.Askey and J.Wimp: Associated Laguerre and Hermite polynomials. Proc. Roy. Soc. Edinburgh 96A, 15-37 (1984)

[39] M.Assal: Generalized wave equations in the setting of Bessel-Kingman hypergroups. Fractional Calculus & Applied Analysis 11, 249-257 (2008)

[40] M.Assal: Pseudo-differential operators associated with Laguerre hypergroups. Journal of Comp. Appl. Math. 233, 617-620 (2009)

[41] M.Assal and H.Ben Abdallah: Generalized weighted Besov spaces on the Bessel hypergroup. Journal of Function Spaces and Appl. 4, 91-111 (2006)

[42] M.Assal and M.M.Nessibi: Sobolev type spaces on the dual of the Laguerre hypergroup. Potential Analysis 20, 85-103 (2004)

[43] A.Azimifard: On the α-amenability of hypergroups. Mh. Math. 155, 1-13 (2008)

[44] A.Azimifard: On multipliers for the Hilbert space of a hypergroup. C. R. Math. Acad. Sci. Soc. R. Can. 30, 84-88 (2008)

[45] A.Azimifard: α-Amenable hypergroups. Math. Z. 265, 971-982 (2010)

[46] A.Azimifard, E.Samei and N.Spronk: Amenability properties of the centres of group algebras. J. Funct. Analysis 256, 1544-1564 (2009)

[47] A.Bakali and M.Akkhouchi: Une généralisation de paires de Guelfand. Bollettino U.M.I. 7-B, 795-822 (1992)

[48] D.Bakry and N.Huet: The hypergroup property and representation of Markov kernels. Lecture Notes in Math. Vol. 1934, Springer, Berlin, 295-347, 2008.

[49] D.Bakry and O.Mazet: Characterization of Markov semigroups on $\mathbb{R}$ associated to some families of orthogonal polynomials. In: *Séminaire de Probabilities XXXVII*, Lecture Notes in Math. Vol. 1832, Springer, Berlin, 60-80, 2003.

[50] M.L.Bami, M.Pourgholamhossein and H.Samea: Fourier algebras on locally compact hypergroups. Math. Nachr. 282, 16-25 (2009)

[51] M.L.Bami and H.Samea: Amenability and essential amenability of certain convolution Banach algebras on compact hypergroups. Bull. Belg. Math. Soc. 16, 145-152 (2009)

[52] E.Bannai and T.Ito: *Algebraic Combinatorics I – Association Schemes.* Benjamin and Cummings, Menlo Park, 1984.

[53] R.B.Bapat and V.S.Sunder: On hypergroups of matrices. Linear and Multilinear Algebra 29, 125-140 (1991)

[54] P.Barrucand and P.Dickinson: On the associated Legendre polynomials. In: *Orthogonal Expansions and their Continuous Analogues.* Southern Illinois University Press, Edwardsville, 43-50, 1968.

[55] H.Bavinck: Convolution operators for Fourier-Jacobi expansions. In: *Linear Operators and Approximation*, ISNM, Vol. 20, Birkhäuser, Basel, 371-380 (1972)

[56] H.Bavinck: On positive convolution operators for Jacobi series. Tôhoku Math. J. 24, 55-69 (1972)

[57] Y.Ben Natan, Y.Benyamini, H.Hedenmalm and Y.Weit: Wiener's tauberian theorem in $L^1(G//K)$ and harmonic functions in the unit disk. Bull. Amer. Math. Soc. 32, 43-49 (1995)

[58] Y.Ben Natan, Y.Benyamini, H.Hedenmalm and Y.Weit: Wiener's tauberian theorem for spherical functions on the automorphism group of the unit disk. Ark. Mat. 34, 199-224 (1996)

[59] N.Ben Salem: Convolution semigroups and central limit theorem associated with a dual convolution structure. Journal of Theoretical Probability 7, 417-436 (1994)

[60] N.Ben Salem and M.N.Lazhari: Limit theorems for some hypergroup structures on $\mathbb{R}^n \times [0, \infty[$. Contemporary Mathematics 183, 1-13 (1995)

[61] G.Benke: On the hypergroup structure of central $\Lambda(p)$ sets. Pacific J. of Math. 50, 19-27 (1974)

[62] C.Bennett and R.Sharpley: *Interpolation of Operators.* Academic Press, Orlando, 1988.

[63] Y.Benyamini and Y.Weit: Harmonic analysis on spherical functions on $SU(1,1)$. Ann. Inst. Fourier 42, 671-694 (1992)

[64] Yu.M.Berezansky: Nuclear spaces of test functions connected with hypercomplex systems and representations of such systems. Contemporary Mathematics 183, 15-20 (1995)

[65] Yu.M.Berezansky: Some generalizations of the classical moment problem. Integr. Equ. and Oper. Theory 44, 255-289 (2002)

[66] Yu.M.Berezansky and A.A.Kalyuzhny: Hypercomplex systems with locally compact bases. Sel. Math. Sov. 4, 151-200 (1985)

[67] Yu.M.Berezansky and A.A.Kalyuzhny: Hypercomplex systems and hypergroups: connections and distinctions. Contemporary Mathematics 183, 21-44 (1995)

[68] Yu.M.Berezansky and A.A.Kalyuzhny: *Harmonic Analysis in Hypercomplex Systems.* Kluwer Academic Press, Dordrecht, 1998.

[69] C.Berg: Studies definies negatives et espaces Dirichlet sur la sphere. Sém. Brelot-Choquet-Deny, *Theorie du Potential,* 13e annee, 1969/1970.

[70] C.Berg: Dirichlet forms on symmetric spaces. Ann. Inst. Fourier 23, 135-156 (1973)

[71] C.Berg, J.P.Christensen and P.Ressel: *Harmonic Analysis on Semigroups. Theory of Positive Definite and Related Functions.* Springer, Berlin, 1984.

[72] C.Berg and G.Forst: *Potential Theory on Locally Compact Abelian Groups,* Springer, Berlin, 1975.

[73] C.Berg and R.Szwarc: Bounds on Turán determinants. J. of Appr. Theory 161, 127-141 (2009)

[74] G.Berschneider: Decomposition of conditionally positive definite functions on commutative hypergroups. Monatsh. Math. 166, 329-340 (2012)

[75] J.J.Betancor: Transference results for multipliers, maximal multipliers and transplantation operators associated with Fourier-Bessel expansions and Hankel transform. Revista de la Union Mat. Argentina 45, 89-102 (2004)

[76] J.J.Betancor, J.D.Betancor and J.M.Méndez: Distributional Fourier transform and convolution associated to Chébli-Trimèche hypergroups. Mh. Math. 134, 265-286 (2002)

[77] J.J.Betancor, J.D.Betancor and J.M.R.Méndez: Chébli-Trimèche hypergroups and W-type spaces. Journal of Math. Anal. and Appl. 271, 359-373 (2002)

[78] J.J.Betancor, J.D.Betancor and J.M.R.Méndez: Hypercyclic and chaotic convolution operators on Chébli-Trimèche hypergroups. Rocky Mountain J. of Math. 34, 1207-1237 (2004)

[79] J.D.Betancor, J.J.Betancor and J.M.R.Méndez: Convolution operators on Schwartz spaces for Chébli-Trimèche hypergroups. Rocky Mountain J. of Math. 37, 723-761 (2007)

[80] J.J.Betancor and L.Rodriguez-Mesa: Weighted inequalities for Hankel convolution operators. Illinois Journal of Math. 44, 230-245 (2000)

[81] J.J.Betancor and L.Rodriguez-Mesa: Density properties of Hankel translations of positive definite functions. Arch. Math. 75, 456-463 (2000)

[82] J.J.Betancor and K.Stempak: Relating multipliers and transplantation for Fourier-Bessel expansions and Hankel transform. Tôhoku Math. J. 53, 109-129 (2001)

[83] N.H.Bingham: Factorization theory and domains of attraction for generalized convolution algebras. Proc. London Math. Soc. 23, 16-30 (1971)

[84] N.H.Bingham: Random walks on spheres. Z. Wahrscheinlichkeitstheorie und verw. Gebiete 22, 169-192 (1972)

[85] N.H.Bingham: Positive definite functions on spheres. Math. Proc. Camb. Phil. Soc. 73, 145-156 (1973)

[86] N.H.Bingham: Integrability theorems for convolutions. J. London Math. Soc. 18, 502-510 (1978)

[87] B.Di Blasio: Hypergroups associated to harmonic NA groups. J. Austral. Math. Soc. 72, 209-216 (2002)

[88] W.R.Bloom: Infinitely divisible measures on hypergroups. *Probability Measures on Groups VI*, Lecture Notes in Math. Vol. 928, Springer, Berlin, 1-15, 1982.

[89] W.R.Bloom: Idempotent measures on commutative hypergroups. *Probability Measures on Groups VIII*, Lecture Notes in Math. Vol. 1210, Springer, Berlin, 13-23, 1986.

[90] W.R.Bloom and H.Heyer: The Fourier transform for probability measures on hypergroups. Rendiconti di Matematica 2, Serie VII, 315-334 (1982)

[91] W.R.Bloom and H.Heyer: Convergence of convolution products of probability measures on hypergroups. Rendiconti di Matematica 2, Serie VII, 547-563 (1982)

[92] W.R.Bloom and H.Heyer: Convolution semigroups and resolvent families of measures on hypergroups. Math. Z. 188, 449-474 (1985)

[93] W.R.Bloom and H.Heyer: Non-symmetric translation invariant Dirichlet forms on hypergroups. Bull. Austral. Math. Soc. 36, 61-72 (1987)

[94] W.R.Bloom and H.Heyer: Continuity of convolution semigroups on hypergroups. Journal of Theoretical Probability 1, 271-286 (1988)

[95] W.R.Bloom and H.Heyer: Characterisation of potential kernels of transient convolution semigroups on a commutative hypergroup. *Probability Measures on Groups IX*, Lecture Notes in Math. Vol. 1379, Springer, Berlin, 21-35, 1989.

[96] W.R.Bloom and H.Heyer: *Harmonic Analysis of Probability Measures on Hypergroups.* de Gruyter, Berlin, 1995.

[97] W.R.Bloom and H.Heyer: Negative definite functions on a commutative hypergroup. Prob. and Math. Stat. 16, 157-177 (1995)

[98] W.R.Bloom and H.Heyer: An application of Sobolev spaces for commutative hypergroups. In: *Trends in Contemporary Infinite Dimensional Analysis and Quantum Probability*, Instituto Italiano di Cultura, Kyoto, 81-100, 2000.

[99] W.R.Bloom and H.Heyer: Polynomial hypergroup structures and applications to probability theory. Publ. Math. Debrecen 72, 199-225 (2008)

[100] W.R.Bloom, H.Heyer and Y.Wang: Positive definite functions and the Lévy continuity theorem for commutative hypergroups. *Probability Measures on Groups X*, Plenum Press, New York, 19-38, 1991.

[101] W.R.Bloom and P.Ressel: Positive definite and related functions on hypergroups. Can. J. Math. 43, 242-254 (1991)

[102] W.R.Bloom and P.Ressel: Exponentially bounded positive-definite functions on commutative hypergroups. J. Aust. Math. Soc. 61, 238-248 (1996)

[103] W.R.Bloom and P.Ressel: Representations of negative definite functions on polynomial hypergroups. Arch. Math. 78, 318-328 (2002)

[104] W.R.Bloom and P.Ressel: Negative definite and Schoenberg functions on commutative hypergroups. J. Aust. Math. Soc. 79, 25-37 (2005)

[105] W.R.Bloom and S.Selvanathan: Hypergroup structures on the set of natural numbers. Bull. Austral. Math. Soc. 33, 89-102 (1986)

[106] W.R.Bloom and M.E.Walter: Isomorphisms of hypergroups. J. Austral. Math. Soc. (Series A) 52, 383-400 (1992)

[107] W.R.Bloom and Zengfu Xu: The Hardy-Littlewood maximal function for Chébli-Trimèche hypergroups. Contemporary Mathematics 183, 45-70 (1995)

[108] W.R.Bloom and Zengfu Xu: Hardy spaces on Chébli-Trimèche hypergroups. Methods Funct. Anal. Topology 3, 1-26 (1997)

[109] W.R.Bloom and Zengfu Xu: Fourier transforms of Schwartz functions on Chébli-Trimèche hypergroups. Mh. Math. 125, 89-109 (1998)

[110] W.R.Bloom and Zengfu Xu: Fourier multipliers for local Hardy spaces on Chébli-Trimèche hypergroups. Can. J. Math. 50, 897-928 (1998)

[111] W.R.Bloom and Zengfu Xu: Fourier multipliers for L^p on Chébli-Trimèche hypergroups. Proc. London. Math. Soc. 80, 643-664 (2000)

[112] W.R.Bloom and Zengfu Xu: Maximal functions on Chébli-Trimèche hypergroups: Infin. Dimens. Anal. Quantum Probab. Relat. Top. 3, 403-434 (2000)

[113] G.Blower: Stationary processes for translation operators. Proc. London Math. Soc. 72, 697-720 (1996)

[114] S.Bochner: Positive zonal functions on spheres. Proc. Nat. Acad. Sci. 40,

1141-1147 (1954)

[115] S.Bochner: Sturm-Liouville and heat equations whose eigenfunctions are ultraspherical polynomials or associated Bessel functions. In: *Proc. of the Conference on Differential Equations.* University of Maryland Book Store, College Park, Maryland, 23-48, 1956.

[116] M.Bouhaik and L.Gallardo: Une loi des grandes nombres et un théorème limite central pour les chaines de Markov sur $\mathbb{N}_0^2$ associées aux polynômes discaux. C. R. Acad. Sci. A310, 739-744 (1990)

[117] M.Bouhaik and L.Gallardo: Un théorème limite central dans un hypergroupe bidimensionnel. Ann. Inst. Henri Poincaré 28, 47-61 (1992)

[118] B.L.J.Braaksma and H.S.V.de Snoo: Generalized translation operators associated with a singular differential operator. In: *Ordinary and Partial Differential Equations.* Lecture Notes in Math. Vol. 415, Springer, Berlin, 62-77, 1974.

[119] J.Braconnier: Sur les groupes topologiques localement compact. J. Math. Pures Appl. 27, 1-85 (1948)

[120] D.M.Bressoud: Linearization and related formulas for q-ultraspherical polynomials. SIAM J. Math. Anal. 12, 161-168 (1981)

[121] D.M.Bressoud: On partitions, orthogonal polynomials and the expansion of certain infinite products. Proc. London Math. Soc. 42, 478-500 (1981)

[122] P.J.Brockwell and R.A.Davis: *Time Series: Theory and Methods.* Springer Series in Statistics, New York, 1991.

[123] R.Bürger: Contributions to duality theory on groups and hypergroups. In: *Topics in Modern Harmonic Analysis* Vols I/II, 1st Naz. Alta Mat. Francesco Severi, Rome, 1055-1070, 1983.

[124] J.Bustoz and M.E.H.Ismail: The associated ultraspherical polynomials and their q-analogues. Can. J. Math. 34, 718-736 (1982)

[125] P.L.Butzer and R.J.Nessel: *Fourier-Analysis and Approximation.* Birkhäuser Verlag, Basel, 1971.

[126] L.Carlitz: The product of two ultraspherical polynomials. Proc. Glasgow Math. Assoc. 5, 76-79 (1961)

[127] P.Cartier: Harmonic analysis on trees. Proc. Sympos. Pure Math. 26, Amer. Math. Soc., Providence, 419-424, 1974.

[128] D.I.Cartwright, G.Kuhn and P.M.Soardi: A product formula for spherical representations of a group of automorphisms of a homogeneous tree, I. Trans. Amer. Math. Soc. 353, 349-364 (2000)

[129] D.I.Cartwright and W.Mlotkowski: Harmonic analysis for groups acting on triangle buildings. J. Austral. Math. Soc. 56, 345-383 (1994)

[130] W. zu Castell and F.Filbir: De la Vallée Poussin Means for the Hankel transform. *Advanced Problems in Constructive Approximation.* Ed. M.D.Buhmann and D.H.Mache. Intern. Series of Num. Appr. Vol. 142, Birkhäuser, 27-38 (2002)

[131] W. zu Castell, F.Filbir and Y. Xu: Cesàro means of Jacobi expansions on the parabolic biangle. J. of Appr. Theory 159, 167-179 (2009)

[132] F.Cazzaniga and C.Meaney: A local property of absolutely convergent Jacobi polynomial series. Tôhoku Math. Journ. 34, 389-406 (1982)

[133] Yu.A.Chapovsky and L.I.Vainerman: Compact quantum hypergroups. J. Operator Theory 41, 261-289 (1999)

[134] D.K.Chang and M.M.Rao: Bimeasures and nonstationary processes. In: *Real and Stochastic Analysis.* Ed. M.M.Rao, Wiley, New York, 7-118, 1986.

[135] H.Chébli: Sur la positivité des operateurs de translation généralisée associes à un opérateur de Sturm-Liouville sur $]0,\infty[$. C. R. Acad. Sci. A 275, 601-604 (1972)

[136] H.Chébli: Opérateurs de translation généralisée et semi-groupes de convolution. Lecture Notes in Math. Vol. 404, Springer, Berlin, 35-59 (1974)

[137] H.Chébli: Sturm-Liouville hypergroups. Contemporary Mathematics 183, 71-88 (1995)

[138] T.S.Chihara: *An Introduction to Orthogonal Polynomials.* Gordon and Breach, New York, 1978.

[139] A.K.Chilana and A.Kumar: Spectral synthesis in Segal algebras on hypergroups. Pacific J. Math. 80, 59-76 (1979)

[140] A.K.Chilana and A.Kumar: Ultra-strong Ditkin sets in hypergroups. Proc. Amer. Math. Soc. 77, 353-358 (1979)

[141] A.K.Chilana and K.Ross: Spectral synthesis in hypergroups. Pacific J. Math. 76, 313-328 (1978)

[142] C.Chou and J.P.Duran: Multipliers for the space of almost-convergent functions on a semigroup. Proc. Amer. Math. Soc. 39, 125-128 (1973)

[143] J.M.Cohen: Radial functions on free products. J. Funct. Analysis 59, 167-174 (1984)

[144] J.M.Cohen and A.R.Trenholme: Orthogonal polynomials with a constant recursion formula and an application to harmonic analysis. J. Funct. Analysis 59, 175-184 (1984)

[145] I.Colojoară: An imprimitivity theorem for hypergroups. Math. Z. 197, 395-402 (1988)

[146] W.C.Connett, C.Markett and A.L.Schwartz: Jacobi polynomials and related hypergroup structures. In: *Probability Measures on Groups X*, Ed. H.Heyer, Plenum Press, New York, 45-81, 1991.

[147] W.C.Connett, C.Markett and A.L.Schwartz: Convolution and hypergroup structures associated with a class of Sturm-Liouville systems. Trans. Amer. Math. Soc. 332, 365-390 (1993)

[148] W.C.Connett, C.Markett and A.L.Schwartz: Product formulas and convolution for angular and radial spheroidal wave functions. Trans. Amer. Math. Soc. 338, 695-710 (1993)

[149] W.C.Connett, C. Markett and A.L.Schwartz: Product formulas and convolutions for the radial oblate spheroidal wave functions. Meth. and Appl. of Analysis 6, 87-102 (1999)

[150] W.C.Connett and A.L.Schwartz: A multiplier theorem for ultraspherical series. Studia Math. 51, 51-70 (1974)

[151] W.C.Connett and A.L.Schwartz: A multiplier theorem for Jacobi expansions. Studia Math. 52, 243-261 (1975)

[152] W.C.Connett and A.L.Schwartz: A correction to the paper "A multiplier theorem for Jacobi expansions". Studia Math. 54, 107 (1975)

[153] W.C.Connett and A.L.Schwartz: The theory of ultraspherical multipliers. Mem. Amer. Math. Soc. 183, 1-92 (1977)
[154] W.C.Connett and A.L.Schwartz: The Littlewood-Paley theory for Jacobi expansions. Trans. Amer. Math. Soc. 251, 219-234 (1979)
[155] W.C.Connett and A.L.Schwartz: A Hardy-Littlewood maximal inequality for Jacobi type hypergroups. Proc. Amer. Math. Soc. 107, 137-143 (1989)
[156] W.Connett and A.L.Schwartz: Product formulas, hypergroups, and the Jacobi polynomials. Bull. Amer. Math. Soc. 22, 91-96 (1990)
[157] W.Connett and A.L.Schwartz: Analysis of a class of probability preserving measure algebras on compact intervals. Trans. Amer. Math. Soc. 320, 371-393 (1990)
[158] W.Connett and A.L.Schwartz: Positive product formulas and hypergroups associated with singular Sturm-Liouville problems on a compact interval. Colloq. Math. 60/61, 525-535 (1990)
[159] W.Connett and A.L.Schwartz: Interpolation of Banach algebras. In: *Proc. Int. Workshop on Interpolation Spaces and Related Topics*, Ed. M.Cwickel, Israel Math. Conference Proceedings 5, 41-55, 1992.
[160] W.Connett and A.L.Schwartz: Fourier analysis off groups. Contemporary Mathematics 137, 169-176 (1992)
[161] W.Connett and A.L.Schwartz: Continuous 2-variable polynomial hypergroups. Contemporary Mathematics 183, 89-109 (1995)
[162] W.Connett and A.L.Schwartz: Subsets of $\mathbb{R}$ which support hypergroups with polynomial characters. Journal of Comp. Appl. Math. 65, 73-84 (1995)
[163] W.Connett and A.L.Schwartz: Hypergroups and differential equations. In: *Lie Groups and Lie Algebras*, Vol. 433 of Math. Appl., Kluwer, Dordrecht, 109-115, 1998.
[164] W.Connett and A.L.Schwartz: Measure algebras associated with orthogonal polynomials. Contemporary Mathematics 254, 127-140 (2000)
[165] W.Connett and A.L.Schwartz: Partial differential equations satisfied by polynomials which have a product formula. Rocky Mountain Journal of Math. 32, 607-637 (2002)
[166] M.Cowling, St. Meda and A.G.Setti: An overview of harmonic analysis on the group of isometries of a homogeneous tree. Expo. Math. 16, 385-424 (1998)
[167] W.Czaja and G.Gigante: Continuous Gabor transform for strong hypergroups. J. of Fourier Analysis and Appl. 9, 321-339 (2003)
[168] A.Dachraoui: Weyl-Bessel transforms. Journal of Comp. Appl. Math. 133, 263-276 (2001)
[169] R.Daher and T.Kawazoe: An uncertainty principle on Sturm-Liouville hypergroups. Proc. Japan. Acad. 83, 167-169 (2007)
[170] H.G.Dales: *Banach Algebras and Automatic Continuity*. Clarendon Press, Oxford, 2000
[171] S.Degenfeld-Schonburg: On the Hausdorff-Young theorem for commutative hypergroups. Colloq. Math. 131, 219-231 (2013)
[172] S.Degenfeld-Schonburg, E.Kaniuth and R.Lasser: Spectral synthesis in Fourier algebras of ultraspherical hypergroups. J. Fourier Anal. Appl. 20,

258-281 (2014)

[173] L.Deleaval: Two results on the Dunkl maximal operator. Studia Math. 203, 47-68 (2011)

[174] J.Delsarte: Sur une extension de la formule de Taylor. J. Math. Pures et Appl. 17, 213-231 (1938)

[175] P.Delsarte: *An algebraic approach to the association schemes of coding theory.* Philips Res. Reports Suppl. 10, 1973.

[176] L.Devroye and L.Györfi: *Nonparametric Density Estimation: The L_1-View.* Wiley, New York, 1985.

[177] J.Dieudonné: Gelfand pairs and spherical functions. Int. J. Math. & Math. Sci. 2, 153-162 (1979)

[178] G.van Dijk: *Introduction to Harmonic Analysis and Generalized Gelfand Pairs.* Walter de Gruyter, New York, 2009.

[179] J.Dixmier: Opérateurs de rang fini dans les représentations unitaires. Publ. Math. Inst. Hautes Étud. Sci. 6, 305-317 (1960)

[180] J.Dixmier: *Les C^*-Algébras et leurs Représentations.* Gauthier-Villars, Paris, 1969.

[181] Y.Domar: Harmonic analysis based on certain commutative Banach algebras. Acta Math. 96, 1-66 (1956)

[182] J.Dougall: A theorem of Sonine in Bessel functions with two extensions to spherical harmonics. Proc. Edinburgh Math. Soc. 37, 33-47 (1919)

[183] J.Dougall: The product of two Legendre polynomials. Proc. Glasgow Math. Assoc. 1, 121-125 (1953)

[184] R.G.Douglas: On the measure-theoretic character of an invariant mean. Proc. Amer. Math. Soc. 16, 30-36 (1965)

[185] R.G.Douglas: *Banach Algebra Techniques in Operator Theory.* Academic Press, New York, 1972.

[186] B.Dreseler and W.Schempp: Approximation on double coset spaces. In: *Approximation Theory*, Banach Center Publications, Vol.4, Polish Scientific Publishers, Warsaw, 69-81, 1979.

[187] B.Dreseler and P.M.Soardi: A Cohen type inequality for ultraspherical series. Arch. Math. 38, 243-247 (1982)

[188] B.Dreseler and P.M.Soardi: A Cohen-type inequality for Jacobi expansions and convergence of Fourier series on compact symmetric spaces. J. Approx. Theory 35, 214-221 (1982)

[189] N.Dunford and J.T.Schwartz: *Linear Operators. I: General Theory.* Interscience, New York, 1958

[190] C.F.Dunkl: Operators and harmonic analysis on the sphere. Trans. Amer. Math. Soc. 125, 250-263 (1966)

[191] C.F.Dunkl: Existence and nonuniqueness of invariant means on $L^\infty(\hat{G})$. Proc. Amer. Math. Soc. 32, 525-530 (1972)

[192] C.F.Dunkl: The measure algebra of a locally compact hypergroup. Trans. Amer. Math. Soc. 179, 331-348 (1973)

[193] C.F.Dunkl: Structure hypergroups for measure algebras. Pacific J. of Math. 47, 413-425 (1973)

[194] C.F.Dunkl: On expansion in ultraspherical polynomials with nonnegative

coefficients. SIAM J. Math. Anal. 5, 51-52 (1974)
[195] C.F.Dunkl: A Krawtchouk polynomial addition theorem and wreath products of symmetric groups. Indiana University Math. Journal 25, 335-358 (1976)
[196] C.F.Dunkl: Orthogonal functions and some permutation groups. Proc. of Symposia in Pure Mathematics 34, 129-147 (1979)
[197] C.F.Dunkl: Cube group invariant spherical harmonics and Krawtchouk polynomials. Math. Z. 177, 561-577 (1981)
[198] C.F.Dunkl and D.Ramirez: Weakly almost periodic functionals carried by hypercosets. Trans. Amer. Math. Soc. 164, 427-434 (1972)
[199] C.F.Dunkl and D.Ramirez: Krawtchouk polynomials and the symmetrization of hypergroups. SIAM J. Math. Anal. 5, 351-366 (1974)
[200] C.F.Dunkl and D.Ramirez: A family of countably compact P_*-hypergroups. Trans. Amer. Math. Soc. 202, 339-356 (1975)
[201] G.K.Eagleson: A characterization theorem for positive definite sequences on the Krawtchouk polynomials. Austral. J. Statist. 11, 29-38 (1969)
[202] W.F.Eberlein: Characterizations of Fourier-Stieltjes transforms. Duke Math. J. 22, 465-468 (1955)
[203] M.Ehring: Large deviations on one dimensional hypergroups. In: *Probability Measures on Groups and Related Structures XI*, World Scientific, Singapore, 91-101, 1995.
[204] M.Ehring: A large deviation principle for polynomial hypergroups. J. London Math. Soc. 53, 197-208 (1996)
[205] S.N.Elaydi: *An Introduction to Difference Equations.* Springer, New York, 1999.
[206] H.Emamirad and G.S.Heshmati: Pseudomeasure character of the ultraspherical semigroups. Semigroup Forum 65, 336-347 (2002)
[207] A.Erdelyi, W.Magnus, F.Oberhettinger and F.G.Tricomi: *Higher Transcendental Functions*, Vol.1. Mac Graw-Hill, New York, 1953.
[208] A.Erdelyi, W.Magnus, F.Oberhettinger and F.G.Tricomi: *Higher Transcendental Functions*, Vol.2. Mac Graw-Hill, New York, 1953.
[209] K.Ey and R.Lasser: Facing linear difference equations through hypergroup methods. Journal of Difference Equations 13, 953-965 (2007)
[210] P.Eymard: A survey of Fourier algebras. Contemporary Mathematics 183, 111-128 (1995)
[211] G.Faber: Über die interpolatorische Darstellung stetiger Funktionen. Jahresbericht der DMV 23, 192-210 (1914)
[212] K.Fan: On positive definite sequences. Annals of Math. 47, 593-607 (1946)
[213] P.Feinsilver and R.Schott: Krawtchouk polynomials and finite probability theory. *Probability Measures on Groups X*, Plenum Press, New York, 129-136, 1991.
[214] J.Feng and H.Heyer: Large deviation principles for Markov chains on some compact hypergroups. In: *Probability Measures on Groups and Related Structures XI*, World Scientific, Singapore, 126-140, 1995.
[215] F.Filbir, R.Girgensohn, A.Saxena, A.I.Singh and R.Szwarc: Simultaneous preservation of orthogonality of polynomials by linear operators arising

from dilation of orthogonal polynomial systems. J. of Comp. Analysis and Appl. 2, 177-213 (2000)

[216] F.Filbir and R.Lasser: Reiter's condition P_2 and the Plancherel measure for hypergroups. Illinois J. of Math. 44, 20-32 (2000)

[217] F.Filbir, R.Lasser and J.Obermaier: Summation kernels for orthogonal polynomials. Ch. 15 *Handbook on Analytic Computational Methods in Applied Math.*, Birkhäuser, 709-749 (2000)

[218] F.Filbir, R.Lasser and R.Szwarc: Reiter's condition P_1 and approximate identities for polynomial hypergroups. Monatsh. Math. 143, 189-203 (2004)

[219] F.Filbir, R.Lasser and Szwarc: Hypergroups of compact type. Journal of Comp. and Applied Math. 178, 205-214 (2005)

[220] F.Filbir, H.N.Mhaskar and J.Prestin: On a filter for exponentially localized kernels based on Jacobi polynomials. Journal of Appr. Theory 160, 256-280 (2009)

[221] F.Filbir and W.Themistoclakis: On the construction of de la Vallée Poussin means for orthogonal polynomials using convolution structures. J. Comp. Anal. Appl. 6, 297-312 (2004)

[222] F.Filbir and W.Themistoclakis: Generalized de la Vallée Poussin operators for Jacobi weights. In: *Proceedings of the Intern. Conf. on Numerical Analysis and Appr. Theory*, 25-34, 2006.

[223] C.Finet: Ensembles de Riesz pour les groupes et les hypergroupes compact. C. R. Acad. Sci. Paris 310, 509-512 (1990)

[224] C.Finet: Lacunary sets for groups and hypergroups. J. Austral. Math. Soc. 54, 39-60 (1993)

[225] G.Fischer and R.Lasser: Homogeneous Banach spaces with respect to Jacobi polynomials. Rendiconti del Circolo Matematico di Palermo Serie II, Suppl. 76, 331-353 (2005)

[226] B.Fishel: Generalized translations associated with an unbounded selfadjoint operator. Math. Proc. Camb. Phil. Soc. 99, 519-528 (1986)

[227] B.Fishel: The Fourier-Plancherel transform as a spectral representation of a rigged Hilbert space. Math. Proc. Camb. Phil. Soc. 101, 567-572 (1987)

[228] M.Flensted-Jensen and T.Koornwinder: The convolution structure for Jacobi function expansions. Ark. Mat. 11, 245-262 (1973)

[229] M.Flensted-Jensen and T.Koornwinder: Jacobi functions: The addition formula and the positivity of dual convolution structure. Ark. Mat. 17, 139-151 (1979)

[230] P.Floris: A noncommutative discrete hypergroup associated with q-disk polynomials. Journal of Comp. Appl. Math. 68, 69-78 (1996)

[231] P.Floris and H.T.Koelink: A commuting q-analogue of the addition formula for disk polynomials. Constr. Approx. 13, 511-535 (1997)

[232] G.B.Folland: A Course in Abstract Harmonic Analysis. CRC Press, Boca Raton, 1995.

[233] J.Fournier and K.A.Ross: Random Fourier series on compact abelian hypergroups. J. Austral. Math. Soc. 37, 45-81 (1984)

[234] G.Frobenius: Über Gruppencharaktere. In: *Gesammelte Abhandlungen III*, Springer, Berlin, 1-37, 1968.

[235] F.Früchtl: Sturm-Liouville hypergroups and asymptotics. Mh. Math. 186, 11-36 (2018)

[236] Y.Funakoshi and S.Kawakami: Entropy of probability measures on finite commutative hypergroups. Bull. Nara Univ. Educ. 57, 17-20 (2008)

[237] L.Gallardo: Comportement asymptotique des marches aléatoires associées aux polynômes de Gegenbauer et applications. C.R. Acad. Sc. Paris 296, 887-890 (1983)

[238] L.Gallardo: Comportement asymptotique des marches aléatoires associées aux polynômes de Gegenbauer et applications. Adv. Appl. Prob. 16, 293-323 (1984)

[239] L.Gallardo: Exemples d'hypergroupes transients. Probability Measures on Groups VIII, Lecture Notes in Math. Vol. 1210, Springer, Berlin, 68-72, 1986

[240] L.Gallardo: The rate of escape of a polynomial random walk on $\mathbb{N}^2$. In: *Harmonic Analysis and Discrete Potential Theory.* Ed. M.A.Picardello. Plenum Press, New York, 233-247, 1992.

[241] L.Gallardo: Asymptotic behaviour of the paths of random walks on some commutative hypergroups. Contemporary Mathematics 183, 135-169 (1995)

[242] L.Gallardo: Chaines de Markov à derive stable et lois des grands nombres sur les hypergroupes. Ann. Inst. Henri Poincaré 32, 701-723 (1996)

[243] L.Gallardo: Asymptotic drift of the convolution and moment functions on hypergroups. Math. Z. 224, 427-444 (1997)

[244] L.Gallardo: Some methods to find moment functions on hypergroups. In: *Harmonic Analysis and Hypergroups.* Ed. K.A.Ross et al. Birkhäuser, Boston, 13-31, 1998.

[245] L.Gallardo: A central limit theorem for Markov chains and applications to hypergroups. Proc. Amer. Math. Soc. 127, 1837-1845 (1999)

[246] L.Gallardo and O.Gebuhrer: Lois de probabilité infiniment divisibles sur les hypergroupes commutatifs, discrets, denombrables. Probability Measures on Groups VII, Lecture Notes in Math. Vol. 1064, Springer, Berlin, 116-130, 1984.

[247] L.Gallardo and O.Gebuhrer: Marches aleatoires sur les hypergroupes localement compacts et analyse harmonique commutative. Publications de l'IRMA, Université Strasbourg, 1-93, 1985.

[248] L.Gallardo and O.Gebuhrer: Marches aléatoires et hypergroupes. Expo. Math. 5, 41-73 (1987)

[249] L.Gallardo and K.Trimèche: Lie theorems for one dimensional hypergroups. Integral Transforms and Special Functions 13, 71-92 (2002)

[250] L.Gallardo and K.Trimèche: One dimensional diffusive hypergroups with asymptotic drift. Integral Transforms and Special Functions 13, 101-108 (2002)

[251] R.O.Gandulfo and G.Gigante: Some multiplier theorems on the sphere. Collect. Math. 51, 157-203 (2000)

[252] G.Gasper: Linearization of the product of Jacobi polynomials I. Can. J. Math. 22, 171-175 (1970)

[253] G.Gasper: Linearization of the product of Jacobi polynomials II. Can. J. Math. 22, 582-593 (1970)

[254] G.Gasper: Positivity and the convolution structure for Jacobi series. Ann. of Math. 93, 112-118 (1971)

[255] G.Gasper: Banach algebras for Jacobi series and positivity of a kernel. Ann. of Math. 95, 261-280 (1972)

[256] G.Gasper: *Positivity and Special Functions. Theory and Application of Special Functions.* Math. Research Center Univ. Wisconsin Publ. 35, Academic Press, New York, 375-434, 1975.

[257] G.Gasper: Positive sums of the classical orthogonal polynomials. SIAM J. Math. Anal. 8, 423-447 (1977)

[258] G.Gasper: A convolution structure and positivity of a generalized translation operator for the continuous q-Jacobi polynomials. Conference on Harmonic Analysis, Wadsworth Intern. Group, Belmont, 44-59, 1983.

[259] G.Gasper: Roger's linearization formula for the continuous q-ultraspherical polynomials and quadratic transformation formulas. SIAM J. Math. Anal. 16, 1061-1071 (1985)

[260] G.Gasper and M.Rahman: Nonnegative kernels in product formulas for q-Racah polynomials I. J. Math. Anal. Appl. 95, 304-318 (1983)

[261] G.Gasper and M.Rahman: *Basic Hypergeometric Series.* Cambridge University Press, Cambridge, 1990.

[262] G.Gasper and W.Trebels: Multiplier criteria of Marcinkiewicz type for Jacobi expansions. Trans. Amer. Math. Soc. 231, 117-132 (1977)

[263] G.Gasper and W.Trebels: Multiplier criteria of Hörmander type for Jacobi expansions. Studia Math. 68, 187-197 (1980)

[264] G.Gasper and W.Trebels: A Riemann-Lebesgue lemma for Jacobi expansions. Contemporary Mathematics 190, 117-125 (1995)

[265] O.Gebuhrer: Bounded measure algebras: A fixed point approach. Contemporary Mathematics 183, 171-190 (1995)

[266] O.Gebuhrer and A.Kumar: The Wiener property for a class of discrete hypergroups. Math. Z. 202, 271-274 (1989)

[267] O.Gebuhrer and A.L.Schwartz. Sidon sets and Riesz sets for some measure algebras on the disk. Colloq. Math. 72, 269-279 (1997)

[268] O.Gebuhrer and A.L.Schwartz: Harmonic analysis on compact commutative hypergroups: The role of the maximum subgroup. Journal d'Analyse Math. 82, 175-206 (2000)

[269] O.Gebuhrer and R.Szwarc: On symmetry of discrete polynomial hypergroups. Proc. Amer. Math. Soc. 127, 1705-1709 (1999)

[270] J.Geronimus: On a set of polynomials. Ann. of Math. 3, 681-686 (1930)

[271] Y.L.Geronimus: *Polynomials Orthogonal on a Circle and Interval.* Pergamon, Oxford, 1960.

[272] A.Ghaffari: Weakly almost periodic functions on hypergroup algebras. Far East J. Math. Sci. 5, 277-287 (2002)

[273] A.Ghaffari: Convolution operators on the dual of hypergroup algebras. Comment. Math. Univ. Carolinae 44, 669-679 (2003)

[274] F.Ghahramani and A.R.Medgalchi: Compact multipliers on weighted hy-

pergroup algebras. Math. Proc. Camb. Phil. Soc. 98, 493-500 (1985)

[275] F.Ghahramani and A.R.Medgalchi: Compact multipliers on weighted hypergroup algebras II. Math. Proc. Camb. Phil. Soc. 100, 145-149 (1986)

[276] G.Gigante: Transference for hypergroups. Collect. Math. 52, 127-155 (2001)

[277] J.Gilewski and K.Urbanik: Generalized convolutions and generating functions. Bull. de L'Academie Polonaise des Sciences 56, 481-487 (1968)

[278] H.J.Glaeske and T.Runst: The discrete Jacobi transform of generalized functions. Math. Nachr. 132, 239-251 (1987)

[279] E.Görlich and C.Markett: A convolution structure for Laguerre series. Nederl. Akad. Wetensch. Indag. Math. 44, 161-171 (1982)

[280] P.Graczyk and C.R.E.Raja: Classical theorems of probability on Gelfand pairs – Khinchin and Cramer theorems. Israel J. of Math. 132, 61-107 (2002)

[281] C.C.Graham: The Fourier transform is onto only when the group is finite. Proc. Amer. Math. Soc. 38, 365-366 (1973)

[282] C.C.Graham and O.C.McGehee: *Essays in Commutative Harmonic Analysis.* Springer, New York, 1979.

[283] E.E.Granirer: Functional analytic properties of extremely amenable semigroups. Trans. Amer. Math. Soc. 137, 53-76 (1969)

[284] R.M.Green: On planar algebras arising from hypergroups. Journal of Algebra 263, 126-150 (2003)

[285] F.R.Greenleaf: Ergodic theorems and the construction of summing sequences in amenable locally compact groups. Comm. Pure Appl. Math. 26, 29-46 (1973)

[286] S.Grosser and M.Moskowitz: On central topological groups. Trans. Amer. Math. Soc. 127, 317-340 (1967)

[287] S.Grosser and M.Moskowitz: Compactness conditions in topological groups. J. Reine Angew. Math. 246, 1-40 (1971)

[288] N.Guillotin-Plantard: On the local time of random walks associated with Gegenbauer polynomials. J. Theor. Probab. 24, 1157-1169 (2011)

[289] E.R.Hansen: *A Table of Series and Products.* Prentice-Hall, Englewood Cliffs, 1975.

[290] G.H.Hardy, J.E.Littlewood and G.Pólya: *Inequalities.* Cambridge University Press, Cambridge, 1978.

[291] K.Hare: Sidonicity in compact, abelian hypergroups. Colloq. Math. 76, 171-180 (1998)

[292] K.Hartmann: $[FIA]_B^-$ Gruppen und hypergruppen. Mh. Math. 89, 9-17 (1980)

[293] K.Hartmann, R.W.Henrichs and R.Lasser: Duals of orbit spaces in groups with relatively compact inner automorphism groups are hypergroups. Mh. Math. 88, 229-238 (1979)

[294] C.Hassenforder: A study of some stationary Gaussian processes indexed by the homogeneous tree. *Probability Measures on Groups X*, Plenum Press, New York, 177-188, 1991.

[295] W.Hauenschild, E. Kaniuth and A.Kumar: Harmonic analysis on central hypergroups and induced representations. Pacific J. Math. 110, 83-112 (1984)

[296] W.Hazod: Probability on matrix-cone hypergroups: Limit theorems and structural properties. Journal of Applied Analysis 15, 205-245 (2009)

[297] S.Helgason: Lacunary Fourier series on noncommutative groups. Proc. Amer. Math. Soc. 9, 782-790 (1958)

[298] R.W.Henrichs: Über Fortsetzung positiv definiter Funktionen. Math. Ann. 232, 131-1250 (1978)

[299] P.Hermann: Induced representations of hypergroups. Math. Z. 211, 687-699 (1992)

[300] P.Hermann: Induced representations and hypergroup homomorphisms. Mh. Math. 116, 245-262 (1993)

[301] P.Hermann: Representations of double coset hypergroups and induced representations. Manuscripta Math. 88, 1-24 (1995)

[302] P.Hermann and M.Voit: Induced representations and duality results for commutative hypergroups. Forum Math. 7, 543-558 (1995)

[303] E.Hewitt: Fourier transform of the class L_p. Arkiv för Matematik 2, 571-574 (1954)

[304] E.Hewitt and K.A.Ross: *Abstract Harmonic Analysis I.* Springer, Berlin, 1963.

[305] E.Hewitt and K.A.Ross: *Abstract Harmonic Analysis II.* Springer, Berlin, 1970.

[306] H.Heyer: Convolution semigroups of probability measures on Gelfand pairs. Expo. Math. 1, 3-45 (1983)

[307] H.Heyer: Probability theory on hypergroups: a survey. In: *Probability Measures on Groups VII.* Ed. H.Heyer, Lecture Notes in Math. Vol. 1064, Springer, Berlin, 481-550, 1984.

[308] H.Heyer: Convolution semigroups of local type on a commutative hypergroup. Hokkaido Math. Journal 18, 321-337 (1989)

[309] H.Heyer: Infinitely divisible probability measures on a discrete Gelfand pair. Nagoya Math. Journal 116, 43-62 (1989)

[310] H.Heyer: Characterization of potential kernels of transient semigroups on a commutative hypergroup. In: *Probability Measures on Groups IX.* Ed. H.Heyer, Lecture Notes in Math. Vol. 1379, Springer, Berlin, 21-35, 1989.

[311] H.Heyer: Infinitely divisible probability measures on a discrete Gelfand pair. Nagoya Math. J. 116, 43-62 (1989)

[312] H.Heyer: Convolution semigroups and potential kernels on a commutative hypergroup. In: *The Analytical and Topological Theory of Semigroups.* Ed. K.H.Hofmann et al. De Gruyter, Berlin, 279-312, 1990.

[313] H.Heyer: Stationary random fields over hypergroups. In: *Gaussian Random Fields.* World Scientific, Singapore, 197-213, 1991.

[314] H.Heyer: Functional limit theorems for random walks on one-dimensional hypergroups. In: *Proceedings "Stability Problems for Stochastic Models".* Eds: Kalashnikov et al., Lecture Notes in Math. Vol. 1546, Springer, Berlin, 45-57, 1993.

[315] H.Heyer: Progress in the theory of probability on hypergroups. Contemporary Mathematics 183, 191-212 (1995)

[316] H.Heyer: The covariance distribution of a generalized random field over a

commutative hypergroup. Contemporary Mathematics 261, 73-82 (2000)

[317] H.Heyer: Stationary random fields over graphs and related structures. In: *Operator Theory: Advances and Applications*, Vol. 167, Birkhäuser-Verlag, Basel, 157-171, 2006.

[318] H.Heyer: Positive and negative definite functions on a hypergroup and its dual. *Conference Proceedings: Infinite Dimensional Harmonic Analysis 4*, Tokyo 2007, Eds. Hilgert et al., 63-96, World Scientific, 2008.

[319] H.Heyer: *Random Fields and Hypergroups.* Ch.2 of *Real and Stochastic Analysis. Current Trends.* Edited by M.M.Rao, pp. 85-182, World Scientific, 2014.

[320] H.Heyer, T.Jimbo, S.Kawakami and K.Kawasaki: Finite commutative hypergroups associated with actions of finite Abelian groups. Bull. Nara Univ. Educ. 54, 23-29 (2005)

[321] H.Heyer, Y.Katayama, S.Kawakami and K.Kawasaki; Extensions of finite commutative hypergroups. Scientiae Math. Japonicae 65, 373-385 (2007)

[322] H.Heyer and S.Kawakami: Paul Levy's continuity theorem: Some history and recent progress. Bull. Nara Univ. Educ. 54, 11-21 (2005)

[323] H.Heyer and S.Kawakami: Extensions of Pontryagin hypergroups. Prob. and Math. Stat. 26, 245-260 (2006)

[324] H.Heyer and S.Kawakami: A cohomology approach to the extension problem for commutative hypergroups. Semigroup Forum 83, 371-394 (2011)

[325] H.Heyer and S.Koshi: *Harmonic Analysis in the Disk Hypergroup.* Mathematical Seminar Notes, Tokyo Metropolitan University, 1993.

[326] H.Heyer and G.Pap: Martingale characterization of increment processses in a commutative hypergroup. Adv. in Pure and Applied Math. 1, 117-140 (2010)

[327] H.Heyer and Y.Wang: Measures of finite energy on a commutative hypergroup. Rend. Sem. Mat. Fis. Milano 59, 245-267 (1991)

[328] F.Heymann and R.Lasser: Convolution structure of the Bessel transform and the generalization of a theorem of Watson. Integral Transforms and Special Functions 21, 319-326 (2010)

[329] I.I.Hirschman Jr.: Harmonic Analysis and Ultraspherical Polynomials. In: *Symposium on Harmonic Analysis and Related Integral Transforms*, Cornell, 1-24, 1956.

[330] I.I.Hirschman Jr.: Variation diminishing transformations and ultraspherical polynomials. Journal d'Analyse Math. 8, 337-360 (1960)

[331] I.I.Hirschman Jr.: Integral equations on certain compact homogeneous spaces. SIAM J. Math. Anal. 3, 314-343 (1972)

[332] V.Hösel: On the estimation of covariance functions on P_n-weakly stationary processes. Stochastic Anal. Appl. 16, 607-629 (1998)

[333] V.Hösel and R.Lasser: One-step prediction for P_n-weakly stationary processes. Mh. Math. 113, 199-212 (1992)

[334] V.Hösel and R.Lasser: A Wiener theorem for orthogonal polynomials. J. Funct. Analysis 133, 395-401 (1995)

[335] V.Hösel and R.Lasser: Prediction of weakly stationary sequences on polynomial hypergroups. Ann. of Probability 31, 93-114 (2003)

[336] V.Hösel and R.Lasser: Approximation with Bernstein-Szegö polynomials. Numerical Functional Analysis and Optimization 27, 377-389 (2006)

[337] V.Hösel, M.Hofmann and R.Lasser: Means and Folner condition on polynomial hypergroups. Mediterr. J. Math. 7, 75-88 (2010)

[338] J.Huang and H.Liu: An analogue of Beurling's theorem for the Laguerre hypergroup. J. Math. Anal. Appl. 336, 1406-1413 (2007)

[339] A.Hulanicki: On positive functions on a group algebra multiplicative on a subalgebra. Studia Math. 37, 163-171 (1971)

[340] V.Hutson and J.S.Pym: Generalized translations associated with an operator. Math. Ann. 187, 241-258 (1970)

[341] V.Hutson and J.S.Pym: Generalized translations associated with a differential operator. Proc. London Math. Soc. 24, 548-576 (1972)

[342] V.Hutson and J.S.Pym: Measure algebra associated with a second-order differential operator. J. Funct. Anal. 12, 68-96 (1973)

[343] E.Hylleraas: Linearization of products of Jacobi polynomials. Math. Scand. 10, 189-200 (1962)

[344] R.Ichihara and S.Kawakami: Strong hypergroups of order four arising from extensions. Scientiae Mathematicae Japonicae 68, 371-381 (2008)

[345] R.Ichihara and S.Kawakami: Estimation of the orders of hypergroup extensions. Scientiae Mathematicae Japonicae 72, 307-317 (2010)

[346] R.Ichihara, S.Kawakami and M.Sakao: Hypergroup extensions of finite abelian groups by hypergroups of order two. Nihonkai Math. J. 21, 47-71 (2010)

[347] S.Igari and Y.Uno: Banach algebra related to the Jacobi polynomials. Tôhoku Math. J. 21, 668-673 (1969)

[348] R.Iltis: Some algebraic structure in the dual of a compact group. Can. J. Math. 20, 1499-1510 (1968)

[349] M.E.H.Ismail: *Classical and Quantum Orthogonal Polynomials in One Variable.* Cambridge University Press, Cambridge, 2005.

[350] R.I.Jewett: Spaces with an abstract convolution of measures. Adv. in Math. 18, 1-101 (1975)

[351] P.E.T.Jorgensen: Semigroups of measures in non-commutative harmonic analysis. Semigroup Forum 43, 263-290 (1991)

[352] J.P.Kahane: *Séries de Fourier absolument convergentes.* Springer, Berlin, 1970

[353] V.Kaimanovich and W.Woess: Construction of discrete, non-unimodular hypergroups. In: *Probability Measures on Groups and Related Structures XI*, World Scientific, Singapore, 196-209, 1995

[354] A.A.Kalyuzhnyi, G.B.Podkolzin and Y.A.Chapovsky: Harmonic analysis on a locally compact hypergroup. Methods Funct. Anal. Topology 16, 304-332 (2010)

[355] A.A.Kalyuzhnyi, G.B.Podkolzin and Y.A.Chapovsky: On infinitesimal structure of a hypergroup that originates from a Lie group. Methods Funct. Anal. Topology 17, 319-329 (2011)

[356] E.Kaniuth: *A Course in Commutative Banach Algebras.* Graduate Texts in Mathematics, Springer, 2009.

[357] E.Kaniuth, A.T.Lau and J.Pym: On φ-amenability of Banach algebras. Math. Proc. Camb. Phil. Soc. 144, 85-95 (2008)

[358] E.Kaniuth, A.T.Lau and A.Ülger: Homomorphisms of commutative Banach algebras and extensions to multiplier algebras with applications to Fourier algebras. Studia Math. 183, 35-62 (2007)

[359] E.Kaniuth and G.Schlichting: Zur harmonischen Analyse klassenkompakter Gruppen II. Inventiones Math. 10, 332-345 (1970)

[360] E.Kaniuth and D.Steiner: On complete regularity of group algebras. Math. Ann. 204, 305-329 (1973)

[361] Y.Kanjin: A convolution measure algebra on the unit disc. Tôhoku Math. Journ. 28, 105-115 (1976)

[362] Y.Kanjin: Banach algebra related to disk polynomials. Tôhoku Math. J. 37, 395-404 (1985)

[363] S.Karlin and J.McGregor: Random walks. Illinois J. Math. 3, 66-81 (1959)

[364] Y.Katznelson: *An Introduction to Harmonic Analysis.* Dover Publications, New York, 1976.

[365] S.Kawakami and W.Ito: Crossed products of commutative finite hypergroups. Bull. Nara Univ. Educ. 48, 1-6 (1999)

[366] S.Kawakami, I.Mikami, T.Tsurii and S.Yamanaka: Actions of finite hypergroups and applications to extension problem. Bull. Nara Univ. Educ. 60, 19-28 (2011)

[367] J.L.Kelley: *General Topology.* Springer, New York, 1975.

[368] W.G.Kelley and A.C.Peterson: *Difference Equations. An Introduction with Applications.* Academic Press, New York, 1991.

[369] M.Kennedy: A stochastic process associated with ultraspherical polynomials. Proc. Royal Irish Acad. 61, 89-100 (1961)

[370] L.Kerchy: Operators with regular norm sequences. Acta Sci. Math. (Szeged) 63, 571-605 (1997)

[371] J.E.Kingman: Random walks with spherical symmetry. Acta Math. 109, 11-53 (1963)

[372] I.A.Kipriyanov and A.A.Kulikov: The Paley-Wiener-Schwartz theorem for the Fourier-Bessel transform. Soviet Math. Dokl. 37, 13-17 (1988)

[373] H.Kleindienst and A. Lüchow: Multiplication theorem for orthogonal polynomials. Intern. J. of Quantum Chemistry 48, 239-247 (1993)

[374] K.Knopp: *Theorie und Anwendung der unendlichen Reihen.* Springer-Verlag, Berlin, 1964.

[375] T.H.Koornwinder: The addition formula for Jacobi polynomials I. Summary of results. Nederl. Akad. Wetensch. Proc. Ser. A75, Indag. Math. 34, 188-191 (1972)

[376] T.H.Koornwinder: Orthogonal polynomials in two variables which are eigenfunctions of two algebraically independent partial differential operators I. Indag. Math. (Proceedings) 77, 48-58 (1974)

[377] T.H.Koornwinder: Orthogonal polynomials in two variables which are eigenfunctions of two algebraically independent partial differential operators II. Indag. Math. (Proceedings) 77, 59-66 (1974)

[378] T.H.Koornwinder: Jacobi polynomials, II. An analytic proof of the product

formula. SIAM J. Math. Anal. 5, 125-137 (1974)

[379] T.H.Koornwinder: A new proof of a Paley-Wiener type theorem for Jacobi transform. Ark. Math. 13, 145-159 (1975)

[380] T.H.Koornwinder: The addition formula for Laguerre polynomials. SIAM J. Math. Anal. 8, 535-540 (1977)

[381] T.H.Koornwinder: Positivity proofs for linearization and connection coefficients of orthogonal polynomials satisfying an addition formula. J. London Math. Soc. 18, 101-114 (1978)

[382] T.H.Koornwinder: Krawtchouk polynomials, a unification of two different group theoretic interpretations. SIAM J. Math. Anal. 13, 1011-1023 (1982)

[383] T.H.Koornwinder: Group theoretic interpretations of Askey's scheme of hypergeometric orthogonal polynomials. *Probability Measures on Groups IX*, Lecture Notes in Math. Vol. 1379, Springer, Berlin, 46-72, 1989.

[384] T.H.Koornwinder: Orthogonal polynomials in connection with quantum groups. In: *Orthogonal Polynomial: Theory and Practice*, NATO ASI Series, 294, Ed. P. Nevai, Kluwer, Dordrecht, 257-292, 1990.

[385] T.H.Koornwinder: The addition formula for little q-Legendre polynomials and the SU(2) quantum group. SIAM J. Math. Anal. 22, 295-301 (1991)

[386] T.H.Koornwinder: Positive convolution structures associated with quantum groups. *Probability Measures on Groups* 10, ed. H.Heyer, Plenum Press, New York, 249-268, 1991.

[387] T.H.Koornwinder: Discrete hypergroups associated with compact quantum Gelfand pairs. Contemporary Mathematics 183, 213-235 (1995)

[388] T.H.Koornwinder and A.L.Schwartz: Product formulas and associated hypergroups for orthogonal polynomials on the simplex and on a parabolic biangle. Constr. Approx. 13, 537-567 (1997)

[389] H.Kortas and M.Sifi: Lévy-Khintchine formula and dual convolution semigroups associated with Laguerre and Bessel functions. Potential Analysis 15, 43-58 (2001)

[390] C.S.Kubrusly: *Spectral Theory of Operators on Hilbert Spaces.* Birkhäuser, Springer, New York, 2012.

[391] A.Kumar: A qualitative uncertainty principle for hypergroups. *Functional Analysis and Operator Theory*, Lecture Notes in Math. Vol. 1511, Springer, Berlin, 1-9, 1992.

[392] A.Kumar: A qualitative uncertainty principle for certain hypergroups. Glasnik Mat. 36, 33-38 (2001)

[393] A.Kumar and A.I.Singh: Spectral synthesis in products and quotients of hypergroups. Pacific J. of Math. 94, 177-192 (1981)

[394] A.Kumar and A.I.Singh: Counter examples in spectral synthesis on hypergroups. Rend. Mat. 8, 329-338 (1988)

[395] A.Kumar and A.I.Singh: A dichotomy theorem for random walks on hypergroups. In: *Probability Measures on Groups IX.* Ed. H.Heyer, Springer Lecture Notes 1379, 179-184 (1989)

[396] H.B.Kushner: The linearization of the product of two zonal polynomials. SIAM J. Math. Anal. 19, 687-717 (1988)

[397] J.Laali and J.Pym: Concepts of Arens regularity for general measure alge-

bras. Quart. J. Math. Oxford 47, 187-198 (1996)
[398] T.Laine: The product formula and convolution structure for the generalized Chebyshev polynomials. SIAM J. Math. Anal. 11, 133-146 (1980)
[399] W.C.Lang: The structure of hypergroup measure algebras. J. Austral. Math. Soc. 46, 319-342 (1989)
[400] R.Larsen: *An Introduction to the Theory of Multipliers.* Springer, Berlin, 1971.
[401] R.Larsen, T.S.Liu and J.K.Wang: On functions with Fourier transforms in L_p. Michigan Math. J. 11, 369-378 (1964)
[402] R.Lasser: Almost periodic functions on hypergroups. Math Ann. 252, 183-196 (1980)
[403] R.Lasser: Fourier-Stieltjes transforms on hypergroups. Analysis 2, 281-303 (1982)
[404] R.Lasser: Bochner theorems for hypergroups and their applications to orthogonal polynomial expansions. J. Approx. Theory 37, 311-325 (1983)
[405] R.Lasser: Orthogonal polynomials and hypergroups. Rend. Mat. 3, 185-209 (1983)
[406] R.Lasser: Linearization of the product of associated Legendre polynomials. SIAM J. Math. Anal. 14, 403-408 (1983)
[407] R.Lasser: Lacunarity with respect to orthogonal polynomial sequences. Acta Sci. Math. 47, 391-403 (1983)
[408] R.Lasser: On the Levy-Hincin formula on commutative hypergroups. *Probability Measures on Groups VII.* Ed. H.Heyer, Springer Lecture Notes 1064, 298-308 (1984)
[409] R.Lasser: On the problem of modified moments. Proc. Amer. Math. Soc. 90, 360-362 (1984)
[410] R.Lasser: Convolution semigroups on hypergroups. Pacific J. Math. 127, 353-371 (1987)
[411] R.Lasser: Applications of the theory of hypergroups. Math. Comput. Modelling 11, 210-211 (1988)
[412] R.Lasser: A modification of stationarity for stochastic processes induced by orthogonal polynomials. *Probability Measures on Groups IX.* Ed. H.Heyer, Springer Lecture Notes 1379, 185-191 (1989)
[413] R.Lasser: Orthogonal polynomials and hypergroups II - the symmetric case. Trans. Amer. Math. Soc. 341, 749-770 (1994)
[414] R.Lasser: On the modified moment problem with solution carried by $[-1, 1]$. *Probability Measures on Groups and Related Structures XI*, World Scientific Publishing. Ed. H.Heyer (Proceedings Volume of Oberwolfach Conference), 225-231 (1995)
[415] R.Lasser: *Introduction to Fourier Series.* Marcel Dekker, New York, 1996.
[416] R.Lasser: On the character space of commutative hypergroups. Jahresbericht der DMV 104, 3-16 (2002)
[417] R.Lasser: Almost periodic sequences with respect to orthogonal polynomials. Contemporary Math. 363, 201-212 (2004)
[418] R.Lasser: Discrete commutative hypergroups. In: *Advances in the Theory of Special Functions and Orthogonal Polynomials*, Eds. W. zu Castell,

F.Filbir, B.Forster, Nova Science Publishers, 55-102 (2005)

[419] R.Lasser: Amenability and weak amenability of L1-algebras of polynomial hypergroups. Studia Mathematica 182, 183-196 (2007)

[420] R.Lasser: On alpha-amenability of commutative hypergroups. *Conference Proceedings: Infinite Dimensional Harmonic Analysis 4*, Tokyo 2007, Eds. J.Hilgert et al, 184-195, World Scientific (2008)

[421] R.Lasser: Point derivations on the $L1$-algebra of polynomial hypergroups. Colloq. Math. 116, 15-30 (2009)

[422] R.Lasser: Various amenability properties of the $L1$-algebra of polynomial hypergroups and applications. Journal of Comp. Appl. Math. 233, 786-792 (2009)

[423] R.Lasser: On positive definite and stationary sequences with respect to polynomial hypergroups. J. of Appl. Analysis 17, 207-230 (2011)

[424] R.Lasser: Almost-convergent sequences with respect to polynomial hypergroups. Acta Math. Hungar. 138, 127-139 (2013)

[425] R.Lasser and M.Leitner: Stochastic processes indexed by hypergroups I. J. Theoret. Probab. 2, 301-311 (1989)

[426] R.Lasser and M.Leitner: On the estimation of the mean of weakly stationary and polynomial weakly stationary sequences. J. Multivariate Anal. 35, 31-47 (1990)

[427] R.Lasser, D.H.Mache and J.Obermaier: On approximation methods by using orthogonal polynomial expansions. *Advanced Problems in Constructive Approximation.* Ed. M.D.Buhmann and D.H.Mache. Intern. Series of Num. Appr. Vol. 142, Birkhäuser, 95-107 (2002)

[428] R.Lasser and J.Obermaier: On Fejér means with respect to orthogonal polynomials: A hypergroup theoretic approach. J. Approx. Theory, Special Volume: *Progress in Approx. Theory*, 551-565 (1991)

[429] R.Lasser and J.Obermaier: On the convergence of weighted Fourier expansions with respect to orthogonal polynomials. Acta Sci. Math. 61, 345-355 (1995)

[430] R.Lasser and J.Obermaier: Orthogonal expansions for L^p- and C-spaces. IWSF-Proceedings '99, Hong Kong, 194-206 (2000)

[431] R.Lasser and J.Obermaier: Strongly invariant means on commutative hypergroups. Colloq. Math. 129, 119-131 (2012)

[432] R.Lasser, J.Obermaier and H.Rauhut: Generalized hypergroups and orthogonal polynomials. Journal of the Australian Mathematical Society 82, 369-393 (2007)

[433] R.Lasser, J.Obermaier and W.Strasser: On the consistency of weighted orthogonal series density estimators with respect to L^1-norm. Nonparametric Statist. 3, 71-80 (1993)

[434] R.Lasser, J.Obermaier and J.Wagner: On the spectrum of tridiagonal operators and the support of orthogonalization measures. Arch. Math. 100, 289-299 (2013)

[435] R.Lasser and E.Perreiter: Homomorphisms of L1-algebras on signed polynomial hypergroups. Banach Journal of Mathematical Analysis 4(2), 1-10 (2010)

[436] R.Lasser and M.Rösler: A note on property (T) of orthogonal polynomials. Arch. Math. 60, 459-463 (1993)

[437] R.Lasser and M.Skantharajah: Reiter's condition for amenable hypergroups. Monatsh. Math. 163, 327-338 (2011)

[438] R.Lasser and J.Obermaier: Weighted shift operators, orthogonal polynomials and chain sequences. Acta Sci. Math. (Szeged) 86, 331-342 (2020)

[439] A.T.Lau: Analysis on a class of Banach algebras with applications to harmonic analysis on locally compact groups and semigroups. Fund. Math. 118, 161-175 (1983)

[440] M.N.Lazhari and K.Trimèche: Convolution algebras and factorization of measures on Chébli-Trimèche hypergroups. C. R. Math. Acad. Sci. Soc. R. Can. 17, 165-169 (1995)

[441] N.Leblanc: Classification des algebres de Banach associees aux operateurs differentiels de Sturm-Liouville. J. Funct. Analysis 2, 52-72 (1968)

[442] N.Leblanc: Algèbres de Banach associées à un opérateur différentiel de Sturm-Liouville. In: L'analyse harmonique dans le domaine complexe. Lecture Notes in Math. Vol. 336, 40-50 (1973)

[443] M.Leitner: Stochastic processes indexed by hypergroups II. J. Theoret. Probab. 4, 321-332 (1991)

[444] M.Leitner, Hyper-weakly harmonizable processes and operator families. Stochastic Anal. Appl. 13, 471-485 (1995)

[445] G.Letac: Problèmes classiques de probabilité sur un couple de Gelfand. Lecture Notes in Math. Vol. 861, Springer, Berlin, 93-120 (1981)

[446] G.Letac: Les fonctions spheriques d'es couple de Gelfand symetrique et les chaînes de Markov. Adv. Appl. Prob. 14, 272-294 (1982)

[447] G.Letac: Dual random walks and special functions on homogeneous trees. Institut Elie Cartan (Nancy) 7, 96-142 (1983)

[448] B.M.Levitan: *Generalized Translation Operators and Some of Their Applications.* Israel Program for Scientific Translations, Jerusalem, 1962.

[449] B.M.Levitan: Transmutation operators and the inverse spectral problem. Contemporary Mathematics 183, 237-244 (1995)

[450] Z.Li and L.Peng: Some representations of translations of the product of two functions for Hankel transforms and Jacobi transforms. Constr. Approx. 26, 115-125 (2007)

[451] M.Lindlbauer: On the rate of convergence of the laws of Markov chains associated with orthogonal polynomials. Journal Comp. Appl. Math. 99, 287-297 (1998)

[452] M.Lindlbauer and M.Voit: Limit theorems for isotropic random walks on triangle buildings. J. Aust. Math. Soc. 73, 301-333 (2002)

[453] J.Liukkonen and R.Mosak: Harmonic analysis and centers of group algebras. Trans. Amer. Math. Soc. 195, 147-163 (1974)

[454] G.L.Litvinov: Hypergroups and hypergroup algebras. J. Soviet Math. 38, 1734-1761 (1987)

[455] J.Löfström and J.Peetre: Approximation theorems connected with generalized translations. Math. Ann. 181, 255-268 (1969)

[456] L.Lorch: The Lebesgue constants for Jacobi series I. Proc. Amer. Math.

Soc. 10, 756-761 (1959)
[457] G.G.Lorentz: A contribution to the theory of divergent sequences. Acta Math. 80, 167-190 (1948)
[458] R.Ma: Heisenberg uncertainty principle on Chébli-Trimèche hypergroups. Pacific J. Math. 235, 289-296 (2008)
[459] I.J.Maddox: A new type of convergence. Math. Proc. Camb. Phil. Soc. 83, 61-64 (1978)
[460] M.Maggioni: Wavelet frames on groups and hypergroups via discretization of Calderón formulas. Monatsh. Math. 143, 299-331 (2004)
[461] N.H.Mahmoud: Partial differential equations with matricial coefficients and generalized translation operators. Trans. Amer. Math. Soc. 352, 3687-3706 (2000)
[462] C.Markett: Cohen type inequalities for Jacobi, Laguerre and Hermite expansions. SIAM J. Math. Anal. 14, 819-833 (1983)
[463] C.Markett: Product formulas for Bessel, Whittaker and Jacobi functions via the solutions of an associated Cauchy problem. Internat. Series Numer. Math. 65, 449-462 (1984)
[464] C.Markett: Norm estimates for generalized transform operators associated with a singular differential operator. Proc. Kon. Ned. Acad. Wet. Ser. A 87, 299-313 (1984)
[465] C.Markett: A new proof of Watson's product formula for Laguerre polynomials via a Cauchy problem associated with a singular differential operator. SIAM J. Math. Anal. 17, 1010-1032 (1986)
[466] C.Markett: Product formulas for eigenfunctions of singular Sturm-Liouville equations. In *Approximation Theory V*. Ed. C.K.Chui et al., Academic Press, New York, 467-470, 1986.
[467] C.Markett: Product formulas and convolution structure for Fourier-Bessel series. Constr. Approx. 5, 383-404 (1989)
[468] C.Markett: The product formula and convolution structure associated with the generalized Hermite polynomials. J. Approx. Theory 73, 199-217 (1993)
[469] C.Markett: Linearization of the product of symmetric orthogonal polynomials. Constr. Approx. 10, 317-338 (1994)
[470] C.Markett: New representation and factorizations of higher-order ultraspherical-type differential equations. J. Math. Appl. 421, 244-259 (2015)
[471] T.Martinez: Multipliers of Laplace transform type for ultraspherical expansions. Math. Nachr. 281, 978-988 (2008)
[472] A.Maté, P.Nevai and V.Totik: Necessary conditions for weighted mean convergence of Fourier series in orthogonal polynomials. Journal of Appr. Theory 46, 314-322 (1986)
[473] C.Meaney: Spherical functions and spectral synthesis. Compositio Mathematica 54, 311-329 (1985)
[474] A.R.Medghalchi: The second dual algebra of a hypergroup. Math. Z. 210, 615-624 (1992)
[475] A.R.Medghalchi: Isometric isomorphisms on the dual and second dual of a hypergroup. Acta Math. Hungar. 74, 167-175 (1997)
[476] A.R.Medghalchi: $M(X)^{**}$ determines X. Far East J. Math. Sci. 6, 109-114

(1998)
[477] A.R.Medghalchi and M.S.Modarres: Amenability of the second dual of hypergroup algebras. Acta Math. Hungar. 86, 335-342 (2000)
[478] A.R.Medghalchi and S.M.Tabatabaie: An extension of the spectral mapping theorem. Intern. J. of Math. and Math. Sciences, Volume 2008, Article ID531424, 8 pages.
[479] A.R.Medghalchi and S.M.Tabatabaie: Contraction semigroups on hypergroup algebras. Scientiae Math. Japonicae Online, 2008-50, 571-584.
[480] A.R.Medghalchi and S.M.Tabatabaie: Spectral subspaces on hypergroup algebras. Publ. Math. Debrecen 74, 307-320 (2009)
[481] R.E.Megginson: *An Introduction to Banach Space Theory*. Springer, New York, 1998
[482] S.Menges: Functional limit theorems for probability measures on hypergroups. Probab. Math. Statist. 25, 155-171 (2005)
[483] H.N.Mhaskar and S.Tikhonov: Wiener type theorems for Jacobi series with nonnegative coefficients. Proc. Amer. Math. Soc. 140, 977-986 (2012)
[484] C.A.Micchelli: A charcterization of M.W.Wilson criterion for non-negative expansions of orthogonal polynomials. Proc. Amer. Math. Soc. 71, 69-72 (1978)
[485] E.Michael: Topologies on spaces of subsets. Trans. Amer. Math. Soc. 71, 152-182 (1951)
[486] D.S.Mitrinovic: *Analytic Inequalities*. Springer, Berlin, 1970.
[487] W.Mlotkowski: Some class of polynomial hypergroups. *Quantum Probability*, Banach Center Publications, Vol. 73, 357-362 (2006)
[488] W.Mlotkowski: Nonnegative linearization for orthogonal polynomials with eventually constant Jacobi parameters. *Noncommutative Harmonic Analysis with Appl. to Probability II*, Banach Center Publications, Vol. 89, 223-230 (2010)
[489] W.Mlotkowski and R.Szwarc: Nonnegative linearization for polynomials orthogonal with respect to discrete measures. Constr. Approx. 17, 413-429 (2001)
[490] R.D.Mosak: Central functions in group algebras. Proc. Amer. Math. Soc. 29, 613-616 (1971)
[491] R.D.Mosak: The L^1- and C^*-algebras of $[\mathrm{FIA}]_B^-$ groups, and their representations. Trans. Amer. Math. Soc. 163, 277-310 (1972)
[492] R.D.Mosak: Ditkin's condition and primary ideals in central Beurling algebras. Monatsh. Math. 85, 115-124 (1978)
[493] M.A.Mourou and K.Trimèche: Inversion of the Weyl integral transform and the Radon transform on $\mathbb{R}^n$ using generalized wavelets. Mh. Math. 126, 73-83 (1998)
[494] B.Muckenhoupt: Mean convergence of Jacobi series. Proc. Amer. Math. Soc. 23, 306-310 (1969)
[495] J.R.Mc Mullen: An algebraic theory of hypergroups. Bull. Austral. Math. Soc. 20, 35-55 (1979)
[496] J.R.Mc Mullen: On the dual object of a compact connected group. Math. Z. 185, 539-552 (1984)

[497] J.R. Mc Mullen and J.F.Price: Reversible hypergroups. Rend. Sem. Mat. Fis. Milano 47, 67-85 (1977)
[498] G.J.Murphy: *C^*-algebras and Operator Theory.* Academic Press, Boston, 1990.
[499] V.Muruganandam: Fourier algebra of a hypergroup – I. J. Austral. Math. Soc. 82, 59-83 (2007)
[500] V.Muruganandam: Fourier algebra of a hypergroup – II. Spherical hypergroups. Math. Nachr. 281, 1590-1603 (2008)
[501] M.A.Naimark: *Normed Algebras.* Wolters-Noordhoff Publishing, Groningen, 1972.
[502] R.Nasr-Isfahani: Representations and positive definite functions on hypergroups. Serdica Math. J. 25, 283-296 (1999)
[503] R.Nasr-Isfahani: On exponentially bounded positive-definite functions on hypergroups. Arch. Math. 76, 455-457 (2001)
[504] R.Nasr-Isfahani: Integral representation for exponentially bounded negative-definite functions on hypergroups. Math. Nachr. 256, 82-87 (2003)
[505] C.Nebbia: Classification of all irreducible unitary representations of the stabilizer of the horicycles of a tree. Israel J. Math. 70, 343-351 (1990)
[506] M.M.Nessibi and B.Selmi: A Wiener-Tauberian and a Pompeiu type theorems on the Laguerre hypergroup. J. Math. Anal. Appl. 351, 232-243 (2009)
[507] M.M.Nessibi and K.Trimèche: Inversion of the Radon transform on the Laguerre hypergroup by using generalized wavelets. J. Math. Anal. Appl. 208, 337-368 (1997)
[508] M.M.Nessibi, L.T.Rachdi and K.Trimèche: The local central limit theorem on the product of the Chébli-Trimèche hypergroup and the Euklidean hypergroup $\mathbb{R}^n$. Jour. of Math. Sciences 9, 109-123 (1998)
[509] P.Nevai: Orthogonal Polynomials. Memoirs Amer. Math. 231 (1979)
[510] P.Nevai: Géza Freud, Orthogonal polynomials and Christoffel functions. A case study. Journal of Appr. Theory 48, 3-167 (1986)
[511] P.Nevai, V.Totik and J.Zhang: Orthogonal polynomials: Their growth relative to their sums. Journal of Appr. Theory 67, 215-234 (1991)
[512] N.Obata: Isometric operators on L^1-algebras of hypergroups. *Probability Measures on Groups X*, Plenum Press, New York, 315-328, 1991.
[513] N.Obata and N.J.Wildberger: Generalized hypergroups and orthogonal polynomials. Nagoya Math. J. 142, 67-93 (1996)
[514] J.Obermaier: The de la Vallée Poussin kernel for orthogonal polynomial systems. Analysis 21, 277-288 (2001)
[515] J.Obermaier: A continuous function space with a Faber basis. J. Appr. Theory 125, 303-312 (2003)
[516] J.Obermaier: A modified Fejér and Jackson summability method with respect to orthogonal polynomials. J. Approx. Theory 163, 554-567 (2011)
[517] J.Obermaier: On convergence and absolute convergence of Fourier series with respect to orthogonal polynomials. Bull. of Math. Analysis and Appl. 7, 1-10 (2015)
[518] J.Obermaier and R.Szwarc: Nonnegative linearization for little q-Laguerre polynomials and Faber basis, Journal of Comp. Appl. Math. 199, 89-94

(2007)
[519] J.Obermaier and R.Szwarc: Orthogonal polynomials of discrete variable and boundedness of the Dirichlet kernel. Constr. Approx. 27, 1-13 (2008)
[520] P.Oleszcuk: Laguerre entire functions and Sturm-Liouville hypergroups. Methods Funct. Anal. Topology 7, 67-79 (2001)
[521] A.Orosz: Difference equations on discrete polynomial hypergroups. Advances in Difference Equations, Volume 2006, Article ID51427, Pages 1-10.
[522] A.Orosz: Sine and cosine equation on discrete polynomial hypergroups. Aequationes Math. 72, 225-233 (2006)
[523] A.Orosz and L.Szekelyhidi: Moment function on polynomial hypergroups in several variables. Publ. Math. Debrecen 65, 429-438 (2004)
[524] A.Orosz and L.Szekelyhidi: Moment functions on polynomial hypergroups. Arch. Math. 85, 141-150 (2005)
[525] A.Orosz and L.Székelyhidi: Moment functions on Sturm-Liouville hypergroups. Ann. Univ. Sci. Budapest Sect. Comput. 29, 141-156 (2008)
[526] B.P.Osilenker: The generalized shift operator and a convolution structure for orthogonal polynomials. Soviet Math. Dokl. 37, 217-221 (1988)
[527] B.P.Osilenker: The representation of the trilinear kernel in general orthogonal polynomials and some applications. J. Appr. Theory 67, 93-114 (1991)
[528] B.P.Osilenker: The generalized Λ-translation in a multiple orthogonal polynomial system. Israel Math. Conference Proceedings, Vol. 5, Ed. M.Cwikel et al., AMS, 165-185, 1992.
[529] B.P.Osilenker: Generalized product formula for orthogonal polynomials. Contemporary Mathematics 183, 269-285 (1995)
[530] G.Pap and M.Voit: Edgeworth expansion on n-spheres and Jacobi hypergroups. Bull. Austral. Math. Soc. 58, 393-401 (1998)
[531] A.W.Parr: Compactly bounded convolution of measures. Proc. Amer. Math. Soc. 130, 2661-2667 (2002)
[532] A.L.Paterson: *Amenability.* American Math. Soc., Providence, 1988.
[533] L.Pavel: On quasi-positive definite functions and representations of hypergroups in QP_n spaces. Rocky Mountain Journal of Math. 27, 889-902 (1997)
[534] L.Pavel: Stationary hypergroups and amenability. Rendiconti del Circolo Matematico di Palermo Serie II, Suppl. 52, 695-705 (1998)
[535] L.Pavel: On hypergroups with Kazhdan's property (T), Math. Reports 2 (52), 345-350 (2000)
[536] L.Pavel: Induced representations of hypergroups and positive definite measures. C. R. Acad. Sci. Paris, t.331, Série I, 685-690 (2000)
[537] L.Pavel: Ergodic sequences of probability measures on commutative hypergroups. Int. J. Math. Sci. 7, 335-343 (2004)
[538] L.Pavel: Reiter's condition (P_2) and hypergroup representations. C.R. Acad. Sci. Ser. I 341, 475-480 (2005)
[539] L.Pavel: Weak containment and hypergroup algebras. Rev. Roumaine Math. Pures Appl. 52, 87-93 (2007)
[540] L.Pavel: Multipliers for the L_p-spaces of a hypergroup. Rocky Mountain Journal of Math. 37, 987-1000 (2007)

[541] L.Pavel: An ergodic property of amenable hypergroups. Bol. Soc. Mat. Mexicana 13, 123-129 (2007)

[542] L.Pavel: Compact hypergroup characters and finite universal Korovkin sets in L_1-hypergroup algebra. Mediterranean Journal of Math. 5, 341-356 (2008)

[543] E.Perreiter: L^1-algebras on commutative hypergroups: Structure and properties arising from harmonic analysis. Dissertation. TU-Muenchen, 1-57, 2011

[544] E.Perreiter: Spectra of L^1-convolution operators acting on L^p-spaces of commutative hypergroups. Math. Proc. Camb. Phil. Soc. 151, 503-519 (2011)

[545] K.Petersen: *Ergodic Theory.* Cambridge University Press, Cambridge, 1983.

[546] J.-P.Pier: *Amenable Locally Compact Groups.* Wiley, New York, 1984.

[547] G.B.Podkolzin: An infinitesimal algebra of the hypergroup generated by double cosets and nonlinear differential equations. Contemporary Mathematics 183, 287-297 (1995)

[548] F.Pollaczek: Sur une famille de polynômes orthogonaux à quatre parameters. C.R. Acad. Sci. Paris 230, 2254-2256 (1950)

[549] H.Pollard: The mean convergence of orthogonal series III. Duke Math. J. 16, 189-191 (1949)

[550] B.L.S.Prakasa Rao: *Nonparametric Functional Estimation.* Academic Press, Orlando, 1983.

[551] J.Pym: The convolution of functionals on spaces of bounded functions. Proc. London Math. Soc. 15, 84-104 (1965)

[552] J.Pym: Weakly separately continuous measure algebras. Math. Ann. 175, 207-219 (1968)

[553] J.Pym: Dual structures for measure algebras. Proc. London Math. Soc. 19, 625-660 (1969)

[554] M.Rahman: A positive kernel for Hahn-Eberlein polynomials. SIAM J. Math. Anal. 9, 891-905 (1978)

[555] M.Rahman: A generalization of Gasper's kernel for Hahn polynomials: application to Pollaczek polynomials. Canad. J. Math. 30, 133-146 (1978)

[556] M.Rahman: The linearization of the product of continuous q-Jacobi polynomials. Canad. J. Math. 33, 961-987 (1981)

[557] M.Rahman: A product formula for the continuous q-Jacobi polynomials. J. Math. Anal. Appl. 118, 309-322 (1986)

[558] M.Rahman and Q.M.Tariq: Addition formulas for q-Legendre-type functions. Methods and Appl. of Analysis 6, 3-20 (1999)

[559] M.Rahman and A.Verma: Product and addition formulas for the continuous q-ultraspherical polynomials. SIAM J. Math. Anal. 17, 1461-1474 (1986)

[560] D.L.Ragozin: Uniform convergence of spherical harmonic expansions. Math. Ann. 195, 87-94 (1972)

[561] D.L.Ragozin: Central measures on compact simple Lie groups. J. Funct. Analysis 10, 212-229 (1972)

[562] D.L.Ragozin: Rotation invariant measure algebras on Euclidean space. Indiana Univ. Math. J. 23, 1139-1154 (1974)

[563] D.L.Ragozin: Zonal measure algebras on isotropy irreducible homogeneous spaces. J. Funct. Analysis 17, 355-376 (1974)

[564] C.R.E.Raja: Normed convergence property for hypergroups admitting an invariant measure. Southeast Asian Bull. of Math. 26, 479-481 (2002)

[565] C.R.E.Raja: Krengel-Lin decomposition for probability measures on hypergroups. Bull. Sci. Math. 127, 283-291 (2003)

[566] M.M.Rao: Harmonizable processes: structure theory. Enseign. Math., 11. Series 28, 295-351 (1982)

[567] M.M.Rao: Bimeasures and harmonizable processes. *Probability Measures on Groups IX.* Lecture Notes in Math. Vol. 1379, Springer, Berlin, 254-298, 1989.

[568] H.Rau: Über die Lebesgueschen Konstanten der Reihenentwicklung nach Jacobischen Polynomen. J. Reine Angew. Math. 161, 237-254 (1929)

[569] H.Rauhut and M.Rösler: Radial multiresolution in dimension three. Constr. Approx. 22, 167-188 (2005)

[570] H.Reiter and J.D.Stegeman: *Classical Harmonic Anlysis and Locally Compact Groups.* Clarendon Press, Oxford, 2000.

[571] C.Rentzsch: A Lévi Khintchine type representation of convolution semigroups on commutative hypergroups. Probab. Math. Statist. 18, 185-198 (1998)

[572] C.Rentzsch: Lévy-Khintchine representations on local Sturm-Liouville hypergroups. Infinite Dimensional Analysis, Quantum Probability and Related Topics 2, 79-104 (1999)

[573] C.Rentzsch and M.Voit: Lévy processes on commutative hypergroups. Contemporary Mathematics 261, 83-105 (2000)

[574] C.R.Rickart: *General Theory of Banach Algebras.* van Nastrand, Princeton, 1960.

[575] D.Rider: Gap series on groups and spheres. Canad. J. Math. 18, 389-398 (1966)

[576] D.Rider: Translation-invariant Dirichlet algebras on compact groups. Proc. Amer. Math. Soc. 17, 977-983 (1966)

[577] D.Rider: Central Lacunary sets. Mh. Math. 76, 328-338 (1972)

[578] D.Rider: Central idempotent measures on compact groups. Trans. Amer. Math. Soc. 186, 459-479 (1973)

[579] L.J.Rogers: Second memoir on the expansion of certain infinite products. Proc. London Math. Soc. 25, 318-343 (1894)

[580] L.J.Rogers: Third memoir on the expansion of certain infinite products. Proc. London Math. Soc. 26, 15-32 (1895)

[581] R.Ronveaux: Orthogonal polynomials: connection and linearization coefficients. In: *Proc. of the International Workshop on Orthogonal Polynomials in Mathematical Physics.* Eds. M.Alfaro et al, 131-142, 1996.

[582] M.Rösler: Bessel-type signed hypergroups on $\mathbb{R}$. *Probability Measures on Groups XI.* World Scientific, Singapore, 292-304, 1994.

[583] M.Rösler: Convolution algebras which are not necessarily positivity-

preserving. Contemporary Mathematics 183, 299-318 (1995)
[584] M.Rösler: On the dual of a commutative signed hypergroup. Manuscripta Math. 88, 147-163 (1995)
[585] M.Rösler: Trigonometric convolution structures on $\mathbb{Z}$ derived from Jacobi polynomials. J. Comput. Appl. Math. 65, 357-368 (1995)
[586] M.Rösler: Generalized Hermite polynomials and the heat equation for Dunkl operators. Commun. Math. Phys. 192, 519-542 (1998)
[587] M.Rösler: Positivity of Dunkl's intertwining operator. Duke Math. J. 98, 445-463 (1999)
[588] M.Rösler: A positive radial product formula for the Dunkl kernel. Trans. Amer. Math. Soc. 355, 2413-2438 (2003)
[589] M.Rösler: Bessel convolutions on matrix cones. Composito Math. 143, 749-779 (2007)
[590] M.Rösler: Positive convolution structure for a class of Heckman-Opdam hypergeometric functions of type BC. J. Funct. Analysis 258, 2779-2800 (2010)
[591] M.Rösler and M.Voit: An uncertainty principle for ultraspherical expansions. J. Math. Anal. Appl. 209, 624-634 (1997)
[592] M.Rösler and M.Voit: Biorthogonal polynomials associated with reflection groups and a formula of Macdonald. J. Comp. Appl. Math. 99, 337-351 (1998)
[593] M.Rösler and M.Voit: Markov processes related with Dunkl operators. Advances in Applied Math. 21, 575-643 (1998)
[594] M.Rösler and M.Voit: Partial characters and signed quotient hypergroups. Canad. J. Math. 51, 96-116 (1999)
[595] M.Rösler and M.Voit: An uncertainty principle for Hankel transforms. Proc. Amer. Math. Soc. 127, 183-194 (1999)
[596] M.Rösler and M.Voit: Positivity of Dunkl's intertwining operator via the trigonometric setting. Int. Math. Research Notices 63, 3379-3389 (2004)
[597] M.Rösler and M.Voit: Deformation of convolution semigroups on commutative hypergroups. *Infinite Dimensional Harmonic Analysis* 3, Eds. H.Heyer et al, 249-264, World Scientific, Tübingen 2003 (2005)
[598] M.Rösler and M.Voit: $SU(d)$-biinvariant random walks on $SL(d,\mathbb{C})$ and their Euclidean counterparts. Acta Appl. Math. 90, 179-195 (2006)
[599] K.A.Ross: Hypergroups and centers of measure algebras. Symposia Math. Vol. XXII. Academic Press, London, 189-203, 1977.
[600] K.A.Ross: Centers of hypergroups. Trans. Amer. Math. Soc. 243, 251-269 (1978)
[601] K.A.Ross: Signed hypergroups – A survey. Contemporary Mathematics 183, 319-329 (1995)
[602] K.A.Ross: Hypergroups and signed hypergroups. In: *Harmonic Analysis and Hypergroups*. Ed. K.A.Ross et al. Birkhäuser, Boston, 77-91, 1998.
[603] K.A.Ross: LCA hypergroups. Topology Proceedings 24, 533-546 (1999)
[604] K.A.Ross and D.Xu: Norm convergence of random walks on compact hypergroups. Math. Z. 214, 415-423 (1993)
[605] K.A.Ross and D.Xu: Hypergroup deformations and Markov chains. J. The-

oret. Probab. 7, 813-830 (1994)
[606] K.A.Ross and D.Xu: Some Metropolis Markov chains are random walks on hypergroups. J. Math. Sciences 28, 194-234 (1994)
[607] A.Roukbi and D.Zeglami: d'Alembert's functional equations on hypergroups. Adv. Pure Appl. Math. 2, 147-166 (2011)
[608] W.Rudin: *Fourier Analysis on Groups.* Interscience Publishers, New York, 1962.
[609] W.Rudin: *Functional Analysis.* McGraw-Hill Book Company, New York, 1973.
[610] W.Rudin: *Real and Complex Analysis.* McGraw-Hill Book Company, New York, 1987.
[611] D.A.Salamon: *Measure and Integration.* EMS Textbooks in Mathematics, 2016.
[612] H.Samea: Weak amenability of convolution Banach algebras on compact hypergroups. Bull. Korean Math. Soc. 47, 307-317 (2010)
[613] H.Samea: Derivations on matrix algebras with applications to harmonic analysis. Taiwanese Journal of Math. 15, 2667-2687 (2011)
[614] Z.Sasvári: *Positive Definite and Definitizable Functions.* Akademie-Verlag, Berlin, 1994.
[615] S.Sawyer: Isotropic random walks in a tree. Z. Wahrscheinlichkeitsth. verw. Geb. 42, 279-292 (1978)
[616] H.P.Scheffler and Hm. Zeuner: Domains of attraction on Sturm-Liouville hypergroups of polynomial growth. Journal of Applied Analysis 5, 153-170 (1999)
[617] A.Schwartz: The structure of the algebra of Hankel transfroms and the algebra of Hankel-Stieltjes transforms. Can. J. Math. 23, 236-246 (1971)
[618] A.Schwartz: Generalized convolutions and positive definite functions associated with general orthogonal series. Pacific J. of Math. 55, 565-582 (1974)
[619] A.Schwartz: l^1-convolution algebras: representation and factorization. Zeitschr. f. Wahrscheinlichkeitstheorie 41, 161-176 (1977)
[620] A.Schwartz: Classification of one-dimensional hypergroups. Proc. Amer. Math. Soc. 103, 1073-1081 (1988)
[621] A.Schwartz: Three lectures on hypergroups Dehli, December 1995. In *Harmonic Analysis and Hypergroups.* Ed. K.A.Ross et al, Birkhäuser, Boston, 93-129, 1998.
[622] A.Schwartz: Partial differential equations and bivariate orthogonal polynomials. J. Symbolic Computation 28, 827-845 (1999)
[623] A.N.Shiryayev: *Probability.* Springer, New York, 1984.
[624] A.I.Singh: Modern Wiener-Tauberian theorems and applications in spectral synthesis. In: *Proc. Int. Symp. on Modern Analysis and Applications.* Ed. H.L.Manocha. New Delhi, 207-227, 1986.
[625] A.I.Singh: Completely positive hypergroup actions. Memoirs Amer. Math. Soc. Vol. 124, No. 593, xii+68 pages (1996)
[626] M.Skantharajah: Amenable hypergroups. Illinois J. of Math. 36, 15-46 (1992)
[627] P.M.Soardi: Limit theorems for random walks on discrete semigroups re-

lated to nonhomogeneous trees and Chebyshev polynomials. Math. Z. 200, 313-325 (1989)

[628] P.M.Soardi: Bernstein polynomials and random walks on hypergroups. In: *Probability Measures on Groups X*, Plenum Press, New York, 387-393, 1991.

[629] P.M.Soardi and W.Woess: Amenability, unimodularity and the spectral radius of random walks on infinite graphs. Math. Z. 205, 471-486 (1990)

[630] R.Spector: Apercu de la theorie des hypergroupes. *Analyse Harmonique sur les Groupes de Lie*, Seminaire Nancy-Strasbourg 1973-1975, Lecture Notes in Math. Vol. 497, Springer, Berlin, 643-673, 1975.

[631] R.Spector: Theorie axiomatique des hypergroupes. C. R. Acad. Sci. Paris 280, 1743-1744 (1975)

[632] R.Spector: Une classe d'hypergroupes dénombrables. C. R. Acad. Sci. A281, 105-106 (1975)

[633] R.Spector: Mesures invariantes sur les hypergroupes. Trans. Amer. Math. Soc. 239, 147-165 (1978)

[634] R.Srinivasan, V.S.Sunder and N.J.Wildberger: Discrete series of fusion algebras. J. Austral. Math. Soc. 72, 419-425 (2002)

[635] U.Stegmeir: Centers of group algebras. Math. Ann. 243, 11-16 (1979)

[636] K.Stempak: An algebra associated with the generalized sublaplacian. Studia Mathematica 88, 245-256 (1988)

[637] K.Stempak: Mean summability methods for Laguerre series. Trans. Amer. Math. Soc. 322, 671-690 (1990)

[638] V.S.Sunder: II_1 factors, their bimodules and hypergroups. Trans. Amer. Math. Soc. 330, 227-256 (1992)

[639] V.S.Sunder: On the relation between subfactors and hypergroups. Contemporary Mathematics 183, 331-339 (1995)

[640] V.S.Sunder and N.J.Wildberger: Actions of finite hypergroups. J. of Algebraic Combinatorics 18, 135-151 (2003)

[641] G.Szegö: *Orthogonal Polynomials.* Amer. Math. Soc., New York, 1959.

[642] L.Székelyhidi: Functional equations on hypergroups. In: *Functional Equations, Inequalities and Applications.* Ed. Th.M.Rassias, Kluwer, The Netherlands, 167-181 (2003)

[643] L.Székelyhidi: Spectral analysis and spectral synthesis on polynomial hypergroups. Mh. Math. 141, 33-43 (2004)

[644] L.Székelyhidi: Functional equations on Sturm-Liouville hypergroups. Math. Pannonica 17, 169-182 (2006)

[645] L.Székelyhidi: Superstability of moment functions on hypergroups. Nonlinear Funct. Anal. Appl. 11, 815-821 (2006)

[646] L.Székelyhidi: Spectral synthesis on multivariate polynomial hypergroups. Mh. Math. 153, 145-152 (2008)

[647] L.Székelyhidi: *Functional Equations on Hypergroups.* World Scientific, Singapore, 2013.

[648] R.Szwarc: Orthogonal polynomials and a discrete boundary value problem I. SIAM J. Math. Anal. 23, 959-964 (1992)

[649] R.Szwarc: Orthogonal polynomials and a discrete boundary value problem II. SIAM J. Math. Anal. 23, 965-969 (1992)

[650] R.Szwarc: Connection coefficients of orthogonal polynomials. Canad. Math. Bull. 35, 1-9 (1992)
[651] R.Szwarc: Convolution structures associated with orthogonal polynomials. J. Math. Anal. Appl. 170, 158-170 (1992)
[652] R.Szwarc: Linearization and connection coefficients of orthogonal polynomials. Mh. Math. 113, 319-329 (1992)
[653] R.Szwarc: Convolution structures and Haar matrices. J. Funct. Anal. 113, 19-35 (1993)
[654] R.Szwarc: Chain sequences and compact perturbations of orthogonal polynomials. Math. Z. 217, 57-71 (1994)
[655] R.Szwarc: Connection coefficients of orthogonal polynomials with applications to classical orthogonal polynomials. Contemporary Mathematics 183, 341-346 (1995)
[656] R.Szwarc: Nonnegative linearization of orthogonal polynomials. Colloq. Math. 69, 309-316 (1995)
[657] R.Szwarc: A lower bound for orthogonal polynomials with an application to polynomial hypergroups. J. Approx. Theory 81, 145-150 (1995)
[658] R.Szwarc: Uniform subexponential growth of orthogonal polynomials. J. Approx. Theory 81, 296-302 (1995)
[659] R.Szwarc: A counterexample to subexponential growth of orthogonal polynomials. Constr. Approx. 11, 381-389 (1995)
[660] R.Szwarc: Nonnegative linearization of the associated q-ultraspherical polynomials. Methods Appl. Anal. 2, 399-407 (1995)
[661] R.Szwarc: Nonnegative linearization and quadratic transformation of Askey-Wilson polynomials. Canad. Math. Bull. 39, 241-249 (1996)
[662] R.Szwarc: Chain sequences, orthogonal polynomials, and Jacobi matrices. J. Approx. Theory 92, 59-73 (1998)
[663] R.Szwarc: Positivity of Turán determinants for orthogonal polynomials. In: *Harmonic Analysis and Hypergroups*. K.A.Ross et al. eds., Birkhäuser, Boston, 165-182, 1998.
[664] R.Szwarc: Sharp estimates for Jacobi matrices and chain sequences. J. Approx. Theory 118, 94-105 (2002)
[665] R.Szwarc: Sharp estimates for Jacobi matrices and chain sequences II. J. Approx. Theory 125, 295-302 (2003)
[666] R.Szwarc: A necessary and sufficient condition for nonnegative product linearization of orthogonal polynomials. Constr. Appr. 19, 565-573 (2003)
[667] R.Szwarc: Strong nonnegative linearization of orthogonal polynomials. In: *Theory and Applications of Special Functions*. A Volume dedicated to Mizan Rahman. Developments in Math. 13, 461-477 (2005)
[668] R.Szwarc: Orthogonal polynomials and Banach algebras. In: *Advances in the Theory of Special Functions and Orthogonal Polynomials*, Eds. W.zu Castell, F.Filbir and B.Forster, Nova Science Publisher, 103-139 (2005)
[669] S.Thangavelu and Y.Xu: Convolution operator and maximal function for Dunkl transform. J. d'Analyse Mathematique 97, 25-56 (2005)
[670] N.V.Thu: Generalized independent increments processes. Nagoya Math. J. 133, 155-175 (1994)

[671] N.V.Thu: Generalized translation operators and Markov processes. Demonstratio Mathematica 34, 295-304 (2001)

[672] N.V.Thu: Hypergroups of orthogonal polynomials. Acta Math. Vietnamica 28, 11-15 (2003)

[673] H.Triebel: *Höhere Analysis.* VEB Deutscher Verlag der Wissenschaften, Berlin, 1972.

[674] K.Trimèche: The Radon transform and its dual associated with partial differential operators and applications to polynomials on the unit disk. Journal of Comp. Appl. Math. 49, 271-280 (1993)

[675] K.Trimèche: Generalized transmutation and translation operators associated with partial differential operators. Contemporary Mathematics 183, 347-372 (1995)

[676] K.Trimèche: Inversion of the Lions transmutation operators using generalized wavelets. Applied and Comp. Harmonic Analysis 4, 97-112 (1997)

[677] K.Trimèche: *Generalized Wavelets and Hypergroups.* Gordon and Breach, Amsterdam, 1997.

[678] K.Trimèche: Wavelets on hypergroups. In: *Harmonic Analysis and Hypergroups.* Ed. K.A.Ross et al., Birkhäuser, Boston, 183-213, 1998.

[679] K.Trimèche: Hypoelliptic distributions on Chébli-Trimèche hypergroups. Global Journal of Pure and Applied Math. 1, 251-271 (2005)

[680] K.Urbanik: Generalized convolutions. Studia Math. 23, 217-245 (1964)

[681] K.Urbanik: Generalized convolutions II. Studia Math. 45, 57-70 (1973)

[682] K.Urbanik: Generalized convolutions III. Studia Math. 80, 167-189 (1984)

[683] L.I.Vainerman: Duality for algebras with involution and generalized shift operators. Journal of Soviet Math. 42, 52-59 (1988)

[684] L.I.Vainerman: Hypercomplex systems and finite-difference operators. Selecta Math. 12, 49-56 (1993)

[685] L.I.Vainerman: Gelfand pairs of quantum groups, hypergroups and q-special functions. Contemporary Mathematics 183, 373-394 (1995)

[686] L.Vajday: Exponential monomials on Sturm-Liouville hypergroups. Banach J. Math. Anal. 4, 139-146 (2010)

[687] L.Vajday: Moment property of exponential monomials on Sturm-Liouville hypergroups. Ann. Funct. Anal. 1, 57-63 (2010)

[688] W.Van Assche: Asymptotics for orthogonal polynomials and three-term recurrences. In: *Orthogonal Polynomials: Theory and Practice.* Ed. P.Nevai, NATO ASI Series C, vol. 294, 435-462, 1990.

[689] M.Vogel: Spectral synthesis on algebras of orthogonal polynomial series. Math. Z. 194, 99-116 (1987)

[690] M.Vogel: Harmonic Analysis and spectral synthesis in central hypergroups. Math. Ann. 281, 369-385 (1988)

[691] M.Voit: Positive Characters on commutative hypergroups and some applications. Math. Z. 198, 405-421 (1988)

[692] M.Voit: Positive and negative definite functions on the dual space of a commutative hypergroup. Analysis 9, 371-387 (1989)

[693] M.Voit: Negative definite functions on commutative hypergroups. *Probability Measures on Groups IX.* Lecture Notes in Math. Vol. 1379, Springer,

Berlin, 376-398, 1989.
[694] M.Voit: Central limit theorems for a class of polynomial hypergroups. Adv. Appl. Prob. 22, 68-87 (1990)
[695] M.Voit: A law of the iterated logarithm for a class of polynomial hypergroups. Mh. Math. 109, 311-326 (1990)
[696] M.Voit: Laws of large numbers for polynomial hypergroups and some applications. J. Theor. Prob. 3, 245-266 (1990)
[697] M.Voit: Central limit theorems for random walks on $\mathbb{N}_0$ that are associated with orthogonal polynomials. J. Multivar. Anal. 34, 290-322 (1990)
[698] M.Voit: A generalization of orbital morphisms of hypergroups. *Probability Measures on Groups* 10, ed. H.Heyer, Plenum Press, New York, 425-433 (1991)
[699] M.Voit: Pseudoisotropic random walks on free groups and semigroups. J. Multivar. Anal. 38, 275-293 (1991)
[700] M.Voit: A law of the iterated logarithm for martingales. Bull. Austral. Math. Soc. 43, 181-185 (1991)
[701] M.Voit: On the Fourier transformation of positive, positive definite measures on commutative hypergroups, and dual convolution structures. Manuscripta Math. 72, 141-153 (1991)
[702] M.Voit: On the dual space of a commutative hypergroup. Arch. Math. 56, 380-385 (1991)
[703] M.Voit: Factorization of probability measures on symmetric hypergroups. J. Austral. Math. Soc. (Series A) 50, 417-467 (1991)
[704] M.Voit: Duals of subhypergroups and quotients of commutative hypergroups. Math. Z. 210, 289-304 (1992)
[705] M.Voit: Strong laws of large numbers for random walks associated with a class of one-dimensional convolution structures. Mh. Math. 113, 59-74 (1992)
[706] M.Voit: A positivity result and normalization of positive convolution structures. Math. Ann. 297, 677-692 (1993)
[707] M.Voit: A law for the iterated logarithm for Markov chains on $\mathbb{N}_0$ associated with orthogonal polynomials. J. Theor. Prob. 6, 653-669 (1993)
[708] M.Voit: An uncertainty principle for commutative hypergroups and Gelfand pairs. Math. Nachr. 164, 187-195 (1993)
[709] M.Voit: A formula of Hilb's type for orthogonal polynomials. J. Comp. Appl. Math. 49, 339-348 (1993)
[710] M.Voit: Projective and inductive limits of hypergroups. Proc. London Math. Soc. 67, 617-648 (1993)
[711] M.Voit: Central limit theorems for Markov processes associated with Laguerre polynomials. J. Math. Anal. Appl. 182, 731-741 (1994)
[712] M.Voit: Substitution of open subhypergroups. Hokkaido Math. J. 23, 143-183 (1994)
[713] M.Voit: Central limit theorems for Jacobi hypergroups. Contemporary Mathematics 183, 395-411 (1995)
[714] M.Voit: A central limit theorem for isotropic random walks on n-spheres for $n \to \infty$. J. Math. Anal. Appl. 189, 215-224 (1995)

[715] M.Voit: Limit theorems for random walks on the double coset spaces $U(n)//U(n-1)$ for $n \to \infty$. J. Comp. Appl. Math. 65, 449-459 (1995)
[716] M.Voit: Discrete hypergroups constructed by successive substitutions. *Probability Measures on Groups and Related Structures XI.* Singapore, World Scientific Publ. 392-405, 1995.
[717] M.Voit: A limit theorem for isotropic random walks on R^d as $d \to \infty$. Russ. J. Math. Phys. 3, 535-539 (1996)
[718] M.Voit: Asymptotic behavior of heat kernels on spheres of large dimensions. J. Multivariate Anal. 59, 230-248 (1996)
[719] M.Voit: Asymptotic distributions for the Ehrenfest Urn and related random walks. J. Appl. Probab. 33, 340-356 (1996)
[720] M.Voit: Limit theorems for compact two-point homogeneous spaces of large dimensions. J. Theor. Prob. 9, 353-370 (1996)
[721] M.Voit: Compact almost discrete hypergroups. Can. J. Math. 48, 210-224 (1996)
[722] M.Voit: Compact groups having almost discrete orbit hypergroups. Mh. Math. 122, 239-250 (1996)
[723] M.Voit: Properties of subhypergroups. Semigroup Forum 56, 373-391 (1998)
[724] M.Voit: Rates of convergence to Gaussian measures on n-spheres and Jacobi hypergroups. Ann. Probability 25, 457-477 (1997)
[725] M.Voit: A Lévy-type characterization of one-dimensional diffusions. Arch. Math. 70, 235-238 (1998)
[726] M.Voit: Hypergroups with invariant metric. Proc. Amer. Math. Soc. 126, 2635-2640 (1998)
[727] M.Voit: Asymptotics of heat kernels on projective spaces of large dimensions and on disk hypergroups. Math. Nachr. 194, 225-238 (1998)
[728] M.Voit: Berry-Esséen type inequalities for ultraspherical expansions. Publ. Math. Debrecen 54, 103-129 (1999)
[729] M.Voit: A Girsanov-type formula for Levy processes on commutative hypergroups. Conference Proceedings: Infinite Dimensional Harmonic Analysis. Eds. H.Heyer et al, 346-359, Graebner (2000)
[730] M.Voit: A product formula for orthogonal polynomials associated with infinite distance-transitive graphs. J. Approx. Theory 120, 337-354 (2003)
[731] M.Voit: Hypergroups on two tori. Semigroup Forum 76, 192-203 (2008)
[732] R.C.Vrem: Lacunarity on compact hypergroups. Math. Z. 164, 93-104 (1978)
[733] R.C.Vrem: Harmonic analysis on compact hypergroups. Pacific J. Math. 85, 239-251 (1979)
[734] R.C.Vrem: Continuous measures and lacunarity on hypergroups. Trans. Amer. Math. Soc. 269, 549-556 (1982)
[735] R.C.Vrem: Hypergroup joins and their dual objects. Pacific J. Math. 111, 483-495 (1984)
[736] R.C.Vrem: Connectivity and supernormality results for hypergroups. Math. Z. 195, 419-428 (1987)
[737] R.C.Vrem: L^p-improving measures on hypergroups. *Probability Measures*

on Groups IX, Lecture Notes in Mathematiccs Vol. 1379, 389-397 (1989)
[738] R.C.Vrem: Torsion and torsion-free hypergroups. Contemp. Math. 91, Amer. Math. Soc., Providence, 281-289 (1989)
[739] R.C.Vrem: Independent sets and lacunarity for hypergroups. J. Austral. Math. Soc. 50, 171-188 (1991)
[740] S.Waldron: Orthogonal polynomials on the disc. J. of Appr. Theory 150, 117-131 (2008)
[741] H.S.Wall: Hypergroups. Amer. J. of Math. 59, 77-98 (1937)
[742] H.S.Wall: *Analytic Theory of Continued Fractions.* van Nastrand, New York, 1948
[743] G.N.Watson: *A Treatise on the Theory of Bessel Functions.* Cambridge University Press, Cambridge, 1966
[744] A.Weinmann and R.Lasser: Lipschitz spaces with respect to Jacobi translation. Math. Nachr. 284, 2312-2326 (2011)
[745] N.J.Wildberger: Hypergroups and Harmonic Analysis. Centre Math. Anal. (ANU) 29, 238-253 (1992)
[746] N.J.Wildberger: Finite commutative hypergroups and applications from group theory to conformal field theory. Contemporary Mathematics 183, 413-434 (1995)
[747] N.J.Wildberger: Hypergroups, symmetric spaces, and wrapping maps. In: *Probability Measures on Groups and Related Structures XI*, World Scientific, Singapore, 406-425, 1995.
[748] N.J.Wildberger: Lagrange's theorem and integrality for finite hypergroups with applications for strongly regular graphs. Journal of Algebra 182, 1-37 (1996)
[749] N.J.Wildberger: Strong hypergroups of order three. Journal of Pure and Applied Algebra 174, 95-115 (2002)
[750] M.W.Wilson: Nonnegative expansions of polynomials. Proc. Amer. Math. Soc. 24, 100-102 (1970)
[751] S.Wolfenstetter: Weakly almost periodic functions on hypergroups. Mh. Math. 96, 67-79 (1983)
[752] S.Wolfenstetter: Spectral synthesis on algebras of Jacobi polynomial series. Arch. Math. 43, 364-369 (1984)
[753] J.C.S.Wong: Topologically stationary locally compact groups and amenability. Trans. Amer. Math. Soc. 144, 351-363 (1969)
[754] A.Wünsche: Generalized Zernike or disc polynomials. J. of Comp. and Appl. Math. 174, 135-163 (2005)
[755] Y.Xu: Generalized translation operator and approximation in several variables. J. Comp. Appl. Math. 178, 489-512 (2005)
[756] A.M.Yaglom: *Correlation Theory of Stationary and Related Random Functions 1.* Springer, New York, 1987.
[757] K.Yosida: *Functional Analysis.* Springer, Berlin, 1974.
[758] N.Youmbi: Semigroup of operators commuting with translations on compact commutative hypergroups. Int. Journal of Math. Analysis 3, 801-813 (2009)
[759] N.Youmbi: Some multipliers results on compact hypergroups. International

Math. Forum 5, 2569-2580 (2010)

[760] Hm.Zeuner: One-dimensional hypergroups. Advances in Math. 76, 1-18 (1989)

[761] Hm.Zeuner: Laws of large numbers of hypergroups on $\mathbb{R}_+$. Math. Ann. 283, 657-678 (1989)

[762] Hm.Zeuner: The central limit theorem for Chébli-Trimèche hypergroups. J. Theoret. Probab. 2, 51-63 (1989)

[763] Hm.Zeuner: Properties of the *cosh* hypergroup. *Probability Measures on Groups IX*. Lecture Notes in Math. Vol. 1379, Springer, Berlin, 425-434, 1989.

[764] Hm.Zeuner: Duality of commutative hypergroups. *Probability Measures on Groups X*. Plenum, New York, 467-488, 1991.

[765] Hm.Zeuner: Polynomial hypergroups in several variables. Arch. Math. 58, 425-434 (1992)

[766] Hm.Zeuner: Moment functions and laws of large numbers on hypergroups. Math. Z. 211, 369-407 (1992)

[767] Hm.Zeuner: Invariance principles for random walks on hypergroups on $\mathbb{R}_+$ and $\mathbb{N}$. J. Theoret. Probab. 7, 225-245 (1994)

[768] Hm.Zeuner: Domains of attraction with inner norming on Sturm-Liouville hypergroups. Journal of Applied Analysis 1, 213-221 (1995)

[769] Hm. Zeuner: Kolmogorov's three series theorem on one-dimensional hypergroups. Contemporary Mathematics 183, 435-441 (1995)

Index

CPSIA information can be obtained
at www.ICGtesting.com
Printed in the USA
BVHW060215291222
654929BV00002B/11